DIFFERENTIAL EQUATIONS

THEORY AND APPLICATIONS

DIFFERENTIAL EQUATIONS

THEORY AND APPLICATIONS

by

Ray Redheffer

University of California at Los Angeles

with

Dan Port

JONES AND BARTLETT PUBLISHERS

BOSTON

Permissions

We are indebted to Charles and Ray Eames for the X ray photo of a nautilus shell that is correlated with a logarithmic spiral in Chapter 3 and forms part of our logo. The satellite orbit of Chapter 1, Figure 1 was computed by M.C. Davidson. With permission from the McGraw-Hill Book Company, we have used a few problems and excerpts from the Sokolnikoff-Redheffer text *Mathematics of Physics and Modern Engineering*. The quotation on page 14 from Libby's article on radiocarbon dating is reprinted with permission from *The Encyclopædia Britannica*, 14th edition, copyright 1965 by Encyclopædia Britannica, Inc. Vol. 18, pages 904–905.

Editorial, Sales, and Customer Service Offices

Jones and Bartlett Publishers
20 Park Plaza
Boston, MA 02116

Printed in the United States of America
10 9 8 7 6 5 4 3 2 1

Library of Congress Cataloging-in-Publication Data

Redheffer, Raymond M.
    Differential equations : theory and applications / by Ray Redheffer.
       p.   cm.
    Includes index.
    ISBN 0-86720-200-9
    1. Differential equations.   I. Port, Dan.   II. Title.
QA371.R343 1991
515' .35--dc20   90-20767
                                                    CIP

ISBN: 0-86720-200-9

Cover Design: Hannus Design Associates
Cover Illustration: X ray photo of a nautilus shell, courtesy of Charles and Ray Eames.

To Heddy

TABLE OF CONTENTS

E= elementary   I= intermediate   A= advanced

To the instructor

THIS BOOK contains elementary, intermediate and advanced chapters. The classification is given in the table of contents and should be taken seriously. A major goal is to bridge the gap: easy to difficult, naive to sophisticated, elementary to advanced.

The elementary chapters and parts of the intermediate chapters are for average sophomores who have had one year of calculus but no differential equations, no advanced analysis, no linear algebra, and little or no physics. The intermediate and advanced chapters are for the same students, or some of them, at a different stage of their mathematical development. By the time they reach these chapters they have learned something about differential equations. They may also be involved in concurrent courses in linear algebra, advanced calculus, or one of the mathematical sciences. They are the same students, yet not the same.

Course outlines and suggestions for use accompany desk copies and examination copies. Suffice it to say here that the following twelve-lecture program gives access to the thirteen chapters listed below:

Lecture: 1.2, 2.1, 2.2, 2.3, 2.4, 4.1, 4.2, 6.1, 6.2, 6.3, 6.4, 6.6

Chapter: 7, 8, 9, 10, 11, 12, 14, 15, 16, 19, 21, 24, A.

Chapters 12 and 24 require knowledge of power series, Chapters 11, 15, and 16 presuppose some linear algebra, and Chapters 14, 24, and A require a certain degree of mathematical maturity. If students have this background, mastery of the preparatory material is sufficient for the thirteen chapters listed. A few sections in the first row are unnecessary for a few chapters in the second, and two lectures, for instance 2.1–2.2, 4.1–4.2, or 6.3–6.4, can sometimes be combined into one. Aside from this, the preparatory program is close to minimal and major reduction of it is not recommended.

The above remarks are merely guidelines, to show how the book is organized. Only you can determine what pace is right for your class, and whether the early part of the course should be enriched (but also slowed down) by the inclusion of additional topics from Chapters 1–6.

There are as many ways to teach as there are people teaching, but when writing a text one has to make definite choices. My choices will not appeal to everyone, but that is all the more reason for describing them here.

This book is user-friendly. Nearly every problem is provided with an answer and many have hints. Some are accompanied by the letter P for "propædeutic." This is a word of Greek origin suggesting a combination of "pedagogical", "preview" and "preparatory". The P problems sometimes enable students to do the homework even if the relevant material was not fully covered in class. They are designed to ensure that students' first experience of a new topic will be positive and reinforcing.

Not only P problems, but quite a few others, are carefully composed with a definite pedagogical end in view. For examples, see Problems 6 to 11 on pages 91–92, Problems 4 and 6 on page 193, and Problem 1 on page 592. Computer software tailored to the text is not yet available but is in preparation. It will start with direction fields for the equations of Chapter 2 and will end with the algorithms of Chapter 24.

The following discussion contains some expressions of opinion. There is no intent to impose these views on anyone. They are mentioned only because they influence the tone of the book.

Manipulation in algebra and also in calculus can be done by a computer; but a computer cannot impart the kind of understanding that is the goal of this text. Simple but powerful techniques are preferred to routine methods involving a lot of algebra; in other words, the book is oriented more toward ideas than manipulation. For illustrations see the introductory treatment of systems in Chapter 9, the coefficient tables for power series in Chapter 12, and the computation of Laplace transforms by differentiation in Chapter 22.

Applications are presented with a minimum of distracting detail and complications associated with diverse measuring units or with real-life numerical values are not emphasized. Of course such things are important, but they are best left to those who actually teach the disciplines in question. The special contribution of the mathematics course is to complement, not duplicate, what students see in courses on engineering and mathematical science.

Proofs are addressed to the student. They are meant to be read, and they need not be repeated in class. Most proofs occupy only a few lines and even so, the end is marked by an interline skip:

It has been my experience that students want to understand why the results they use are true. Usually the best way to explain why is to give a proof. The success of short proofs as a teaching device has been confirmed by anonymous student evaluations of courses using this text.

Long proofs, by contrast, are at the end of sections or chapters. For average students such proofs need be taken up neither in the lectures nor in the assigned reading. Their presence conveys the message that proofs are important, but the main reason for including them is that every large class is likely to have some students with both interest and talent in mathematics. It is a disservice to them to omit difficult proofs and merely give a reference. By the time they go to the library, find the reference, get used to the new author's notation, and study the proof, they could have read the proof in this book several times over.

To the student

UNDER the headings: calculation, understanding, and self-study, here are three of the principles that have guided the writing of this book.

Calculation

Many differential equations can be solved either by long calculations or by powerful methods that use ideas rather than manipulation. In a typical problem involving resonance, for example, a popular method might require that you find the eleventh derivative of

$$t^5(A\cos t + B\sin t)$$

where A and B are constant. Here such calculations are avoided by the exponential shift theorem and by the use of complex numbers; see Chapter 8, Section 2, Problem 4. It will take effort on your part to master the underlying ideas, but it also takes effort to find a high-order derivative as described above. Mastering the ideas seems to me a better use of your time.

As another illustration, a number of specific methods are introduced in rapid succession in Chapter 2. Classroom experience has shown that students learn these methods with ease, because only enough complication has been introduced to illustrate the principles. If the homework problems were so constructed as to lead to difficult integrals, the pace would have to be slower and the learning process would be impeded. Teaching the use of integral tables is a worthwhile goal, but it is not a goal of this text.

Difficult calculations can hardly be avoided in generalized power series (Chapter 13), matrix differential equations and eigenvectors (Chapters 15 to 17), Fourier series and boundary-value problems (Chapter 18), nonlinear analysis (Chapter 20), certain uses of the Laplace transform (Chapters 21 to 23), and numerical methods (Chapter 24). Even in such cases, however, the level of technical difficulty is kept as low as the subject allows.

Understanding

Balancing this diminished emphasis on manipulation is a strong emphasis on understanding. Mathematics is fun when you understand it, no fun when you don't. Since understanding is a primary goal, this book lacks graphical devices (equations in color, formulas in boxes, and so on) that encourage you to skip the text and rush to the homework. Such devices may seem to be user-friendly but they are not.

Again with understanding as the goal, nearly everything is proved. Seeing a lot of short proofs will perhaps be a novel experience for you, but they are essential. The purpose of a proof is both to show that a result is true and to show why it is true. A proof is not just a proof but is also an explanation.

New concepts are sometimes introduced briefly, without emphasis, so they will be more easily understood when encountered later. For example Chapter 1 introduces the following terms in a low-key way: separation of variables, substitution, integrating factor, complete solution, general solution, particular solution, initial condition, variation of parameters. You are not expected to learn all these terms then and there. In time they will become familiar.

You may be surprised to find an occasional problem in which you have nothing to do but read it, think about it, and compare your answer with mine. The object is communication. Offbeat problems like these—there are only a few—stem from my educational philosophy, which is directed more toward teaching than testing.

Self-study

Prerequisites for the more advanced parts of this book are developed independently in Chapters B–F. Each of these chapters is self-contained. It starts at the beginning and is intended to be a minicourse rather than a review. Many undergraduate schedules are too crowded to include these chapters officially, as part of the course outline, but you may find it helpful to read some of them on your own. Chapter B on power series gives a solid introduction to Chapter 12 and Chapters CDE give enough linear algebra for Chapters 11, 15, 16, and 17. Even if you know these subjects already, the development given here may help you to organize your knowledge and correlate it with the rest of the text.

Except for courses with such titles as "Mathematical Models" most class schedules will have time for only a few of the applications listed at the front of this book. You might enjoy reading about some of these on your own. Many of the items refer to problems rather than to the text, but a carefully composed problem is an effective means of communication even if you don't solve it. Reading for pleasure is not the same as studying for an examination.

Finally, the book contains a number of advanced topics that are of direct interest to research scientists, mathematicians, and engineers. It is unlikely that many of these topics will be in a standard undergraduate course. If you are heading for a career in mathematical science, however, you may want to keep the book after graduation and study the advanced parts as the need for them arises in your professional work.

Acknowledgement

On his own time, as a friend, my colleague Basil Gordon read the whole book, including the problems, and provided a great many corrections, criticisms and improvements. I doubt that a board of editors could have done as well. Nearly every one of his suggestions was accepted and the book is the better for it. Saying "thank you" is not enough, but I do say it.

It is a pleasure also to acknowledge the cooperation of our publisher, Jones and Bartlett. Faced with an extremely tight production schedule (two months) they took a deep interest in the project, from president Paul Prindle and vice president Carl Hesler right on down through the organization, yet never used this personal involvement as a means of overriding our wishes. They are the kind of publisher that authors look for but seldom find.

Dave Mallis introduced a number of subtle but important improvements in typography. He is the founder and president of *Publishing Experts* and, judging from our experience, his company is very appropriately named.

We are indebted to Monica McArthur for valuable assistance when needed. Young Lee, just out of high school, taught himself linear algebra by working through Chapters CDE and checked the text, examples, and problems of Chapters 15, 16, and 17. Fifteen years after his last math course, Gerald Jaimovich similarly checked Chapters 18 and 20.

Finally, let me describe my three-year collaboration with Dan Port. He was a junior at UCLA when we started our project and is now a graduate student at MIT. Since about age 12 he has been interested in computerized typesetting, computer graphics, numerical algorithms, and the design and maintenance of computers. He began by setting up a computer-monitor on which I could type directly, whereupon he reduced my plain typing to TEX, the layout having been decided by mutual consultation. Over a period of months he taught me, against resistance, until I could program TEX too; so about half the book is typeset by him, half by me. Teaching me was probably a greater achievement than it would have been to do it all himself.

Dan Port generated about 50 three-by-three matrices with simple eigenvalues and eigenvectors for Chapters 16 and 17, he checked the series solutions in Chapters 12 and 13, and he provided the computations needed in Chapter 24. Nearly every one of the 250 figures in the text is due to him.

For Chapter 24 he gave me a long, detailed outline of his views on numerical analysis, with excellent examples, and provided references to the literature exactly when needed. His outline showed an unerring instinct for what is truly important, and Chapter 24 is a collaborative enterprise. Except for a couple of sentences in that chapter, however, Dan Port did not write the book. His name on the title page is justified by the contributions mentioned above and by the fact that, without him, the book would not have come into existence.

Raymond Redheffer

The voice of the intellect is a soft one, but it does not rest until it has gained a hearing.

—Sigmund Freud

Thoughts that come with doves' footsteps guide the world.

—Friedrich Nietzsche

Chapter 1

INTRODUCTION

THE DERIVATIVE of a function represents a rate of change, and an equation involving derivatives is called a differential equation. Many natural laws give a relation between the rates of change of various quantities, rather than between the quantities themselves. It should be no surprise, then, that differential equations have had a major impact on the history of science and represent one of the main avenues by which mathematical methods are applied to the real world. This chapter describes some of the uses of differential equations, first in general terms, then in the context of a specific equation.

1. A survey of applications

The subject of differential equations has its primary historical origin in Newtonian mechanics, and we begin by saying something about Newton and his achievements. Isaac Newton was born on Christmas Day, 1642. It was a premature birth and, according to tradition, he was "small enough to fit into a quart mug." From this modest beginning, he became one of the chief architects of modern science. Newton's importance for this book is that much of his scientific work was based on differential equations, which he applied with astonishing skill to the study of Nature. For example, Newton's second law of motion

$$(\text{force}) = (\text{mass})(\text{acceleration})$$

is a differential equation. When combined with his law of gravitation, Newton's laws of motion enabled him to compute the orbits of the planets, the moon, and comets. As part of this same development he went on to estimate the weight and density of the sun and moon, to calculate the tides, and to explain the precession of the equinoxes.

The orbit of the moon is perturbed by the pull of the sun. Taking account of this perturbation, Newton explained two ancient observations that had gone unexplained since the time of Ptolemy. He also began the theory of planetary perturbations. An interesting sidelight is that the planet Uranus was discovered about fifty years after Newton's death, and about seventy years after that, perturbations in the orbit of Uranus led to the discovery of Neptune. Today the theory of perturbations is a major subject in differential equations and is applied to such diverse topics as weather prediction, electronic orbits, and ecological systems.

When only the sun and a single planet are considered, the differential equations obtained by Newton can be solved in a way that would now be considered elementary, though at that time the solution was a spectacular achievement. As a result, Newton could give a rather brief derivation of the three laws of Kepler, which Kepler had deduced by a lifetime spent in the study of astronomical observations. The problem of determining the motion of two gravitating masses, such as the sun and a planet, is called the *two-body problem*. The *three-body problem* is to determine the motion of three gravitating masses, such as earth-moon-sun. This problem is not in any sense elementary. It has occupied the attention of many outstanding mathematicians since Newton and still has its puzzles today. It was the only problem, Newton said, that made his head ache.

As you would surmise, the corresponding many-body problem for the solar system as a whole is even harder. For example, nobody has been able to prove that the solar system is stable, though the differential equations can be formulated with great precision. (Stability in this context means that things will stay much as they are now, without anything catastrophic such as a collision of two planets to upset the picture.) The differential equations that Newton formulated for the study of astronomy are the same as those that are now used to predict the orbits of man-

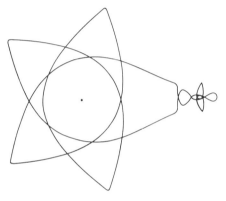

Figure 1

made satellites like the Moon Lander. Here too we are dealing with the many-body problem, and one reason why space travel is technically feasible is that today's computers have enough power to solve the relevant differential equations numerically.

Figure 1 shows a periodic earth-moon orbit of a satellite. This orbit lies in a plane. In general, orbits in the three-body problem do not lie in a plane, a fact that leads to even further complication.

Since Newton hated controversy, he hesitated to publish his results, and leadership in the field of differential equations quickly passed to the European continent, where it remained for over one hundred years. Some of the leading names are Leibniz, Johann and Daniel Bernoulli, Euler, Laplace, Lagrange, Fourier, Poincaré, Picard, Liapunov, and Volterra. It is chiefly through their work that the subject of elementary differential equations attained the form it has today.

These later researchers showed that differential equations apply not only to Newtonian mechanics, but to a truly astonishing variety of scientific fields. The main requirement is that the phenomena being described should vary continuously, rather than in discrete jumps. For example, differential equations apply to fluid flow, propagation of sound waves, and the flow of electricity in a cable. Fourier's book of 1822 entitled *The Analytic Theory of Heat* has been called the Bible of mathematical physics. It is devoted, in large part, to a study of various differential equations. By means of differential equations, Maxwell predicted the existence of electromagnetic waves (radio waves) before they were observed experimentally by Hertz. The theory of geometric optics depends on differential equations, and so does the theory of deformation of elastic structures (beams and the like). Chemical reactions ... radioactive decay ... propagation of neutrons in a reactor ... the list goes on and on.

The examples mentioned above are confined to such fields as physics, chemistry and engineering, which are sometimes referred to as the "exact sciences." What is meant by this phrase, for purposes of the present text, is that one can formulate the relevant differential equations with a high degree of certainty. Though the equation may be hard to solve, as is the case for the many-body problem, there is no serious doubt as to what the equation should be. To be sure, in setting up the equations of motion for a satellite the effects of relativity are usually neglected. But these effects are so minute that only an experiment of extreme sensitivity can detect the difference between the mechanics of Newton and that of Einstein. Similar remarks apply to other topics, such as the theory of heat flow. Although the molecules of a substance do not form a continuum, and hence do not correspond precisely to the mathematical model, the error is negligible in all ordinary cases. The equations for heat flow are not subject to controversy.

There are other fields, no less important, in which differential equations are still useful, but are not known with the same certainty as in the fields mentioned above. Among these are topics connected with economics, ecology, biology, and medicine. Often in such fields it is by no means obvious what the differential equation should be, and in not a few cases the formulation of the equation is more difficult than its solution.

For example, the successful program of eradicating smallpox received an early boost from Daniel Bernoulli, who set up a differential equation for the progress of an infectious disease. This equation enabled him to predict, in 1760, that a program of inoculation would increase life expectancy by about three years. (He also made some contributions to the theory of probability in the course of his calculation.) Even now, nobody would claim to know an exact differential equation for the problem he considered, in the sense that the equations of mathematical physics are exact. Nevertheless his work was of great importance, since it gave a scientific justification for the risk of vaccination.

Another example of this kind is the application of differential equations to the use of pesticides. It may happen that the target population—insects, for example—is actually increased by the pesticide, instead of being decreased. The trouble is that other insects that prey on the target population are killed off too. The reduction of predators may help the target population more than the pesticide harms it. Differential equations have given much insight into problems such as this, though their correct formulation is not as clear-cut as in problems of physical science.

Similar remarks apply when differential equations are used to study the fluctuations of animal populations, such as the Arctic fox, or the harvesting of renewable resources such as fish in the Adriatic. (Both these examples have played a role in the development of mathematical ecology.) Further examples are given by the mathematics of heart physiology, the transmission equations for nerve impulses, the growth laws for tumors—again the list could go on and on.

This section was written, in part, to alert you to the fact that when differential equations are applied to scientific problems, they may have varying degrees of reliability. Though strongly committed to a logical development of the subject of differential equations, we feel no obligation to examine the logical structure of the different fields to which these equations are applied. With that in mind, the derivations are presented as briefly as clarity allows, and it is left for you to decide whether you think the underlying assumptions are justified.

To conclude this general discussion, we remark that it would be a mistake to think that differential equations are important only in connection with other fields of science. The *Mathematical Reviews* publishes abstracts of about 75 new papers on differential equations per month, and the number has been increasing from year to year. Differential equations occupy a central place in modern mathematics, and they are interesting in their own right apart from applications.

2. Exponential growth

To bring the foregoing remarks into sharper focus, we consider a specific differential equation in detail. It is an equation with which you are probably familiar and it has been chosen, in part, for that reason. The equation is

$$\frac{dy}{dx} = ky, \qquad k = \text{const.}$$

This is called the equation of exponential growth (or decay) because its solutions are exponential functions.

On any interval on which $y \neq 0$ the equation can be written

$$\frac{dy}{y} = k\, dx.$$

If $y > 0$ on the interval, integration gives

$$\ln y = kx + C$$

where C is constant and where, as elsewhere in this book, ln denotes the natural logarithm; that is, the logarithm to base e. Taking exponentials we get

$$y = e^{kx+C} = e^{kx}e^{C} = e^{kx}c.$$

Here $c = e^C$ is another constant, necessarily positive because the exponential function takes only positive values. However, when we check the solution, we find that the sign of c does not matter. Namely,

$$y = ce^{kx} \quad \text{gives} \quad \frac{dy}{dx} = cke^{kx} = k(ce^{kx}) = ky.$$

In writing the equation as $dy/y = k\,dx$ we put everything involving y on one side of the equation and everything involving x on the other. It is said, then, that the equation is separated, and the procedure illustrates a general method known as the method of separation of variables.

In checking the equation we compute dy/dx from the given formula and verify that the equation holds at all values of x. This illustrates what is meant by a solution of a differential equation. A formal definition is given in Chapter 2.

We now give another method. When $y > 0$ it is possible to write $y = e^u$ where u, like y, is a differentiable function. By the chain rule

$$\frac{d}{dx}e^u = e^u \frac{du}{dx}.$$

Substituting $y = e^u$ into the differential equation $dy/dx = ky$ we get

$$e^u \frac{du}{dx} = ke^u.$$

Since an exponential never vanishes, the factor e^u can be divided out to give

$$\frac{du}{dx} = k.$$

Integration gives $u = kx + C$, where C is constant, and $y = e^u$ leads again to the solution found above.

The above calculation illustrates the method of substitution. As in this case, quite often a differential equation can be simplified by an appropriate change of variable. Although at first glance the correct choice of variable may seem mysterious, it can be suggested by the form of the equation or by the form of the integrals encountered in a more direct approach. When applied problems are formulated as differential equations, the question whether the solution is simple or complicated depends strongly on the choice of variables.

The foregoing methods can be modified so as to work when $y < 0$ throughout the interval, instead of $y > 0$; but they both break down if y changes sign. We now present another method that is free of this defect. Write the differential equation in the form

$$\frac{dy}{dx} - ky = 0$$

and multiply by e^{-kx}. The left side turns out to be the derivative of $e^{-kx}y$, so that we get

$$0 = e^{-kx}\left(\frac{dy}{dx} - ky\right) = e^{-kx}\frac{dy}{dx} + y(-ke^{-kx}) = \frac{d}{dx}(e^{-kx}y).$$

It is a familiar fact of calculus that if the derivative of a function is 0 on an interval then the function is constant on that interval. Hence

$$e^{-kx}y = c, \qquad c = \text{const.,}$$

and multiplying by e^{kx}, we get the same solution as before. This procedure does not assume $c > 0$, it does not assume $y \neq 0$, and it shows that every solution of the differential equation is given by the formula for some value of c. We summarize by writing

(1) $$\frac{dy}{dx} = ky \iff y = ce^{kx}$$

where k and c are constant. The symbol $\iff$ means "if and only if." It gives a brief way of saying that two statements are logically equivalent in the sense that either can be deduced from the other. The solutions for $k = 1$ and $k = -1$ are shown graphically in Figure 2.

From the standpoint of general procedure, the function e^{-kx} is called an integrating factor, because multiplying by this factor gives a function that is readily integrated. The method of integrating factors applies to broad classes of differential equations.

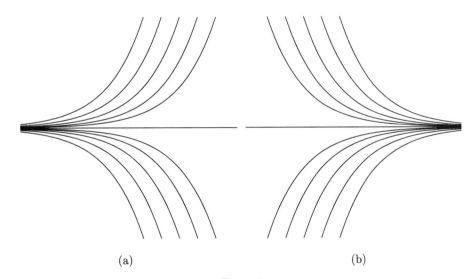

(a)   (b)

Figure 2

Since the solution (1) contains the arbitrary constant c, the function y is not uniquely determined by the differential equation and some other condition is needed. The most common condition is to require that the solution curve pass through some specified point (x_0, y_0) of the (x, y) plane, so that $y = y_0$ when $x = x_0$. This gives, in succession,

$$ce^{kx_0} = y_0, \qquad c = y_0 e^{-kx_0}.$$

Substitution into $y = ce^{kx}$ gives

$$y = y_0 e^{-kx_0} e^{kx} = y_0 e^{k(x-x_0)}.$$

Anticipating terminology that will be discussed again later, we remark that the family ce^{kx} is called the *complete solution* of $dy/dx = ky$ because every member of the family satisfies the differential equation, and every solution of the differential equation is a member of the family. The family ce^{kx} is called a *general solution* of the differential equation because the constant c in it can be chosen so as to satisfy the arbitrarily specified condition $y = y_0$ at the arbitrarily specified point $x = x_0$. (Hence, in this case the complete solution and the general solution coincide.) The statement that $y = y_0$ when $x = x_0$ is called an initial condition. It is often written in the form

$$y(x_0) = y_0.$$

Generally speaking, an initial condition holds at a single place or time, while the differential equation holds at all places or times. Both a differential equation and an initial condition are needed, as a rule, to determine a particular solution, that is, a solution of the form $y = \phi(x)$ not containing any arbitrary constants. Depending on context, the word "solution" might mean a general solution such as $y = ce^x$ or a particular solution such as $y = 2e^x$.

PROBLEMS

Problems marked P, for "preview", are intended to make later problems, or later parts of the text, easier for you to understand. Please read these problems whether you do them or not. Knowledge of them is assumed from that point on.

1P. In this book, derivatives are often denoted by primes, thus:

$$y' = \frac{dy}{dx}, \qquad y'' = \frac{d^2 y}{dx^2}, \qquad y''' = \frac{d^3 y}{dx^3}, \qquad y^{iv} = \frac{d^4 y}{dx^4}$$

When the independent variable is the time t, derivatives are often indicated by dots placed above the letters, thus:

$$\dot{u} = \frac{du}{dt}, \qquad \ddot{u} = \frac{d^2 u}{dt^2}$$

This notation goes back to Newton, who called u the *fluent* and $\dot{u}$ the *fluxion*. It is seldom used in this book, but may be used by your instructor and you are likely to meet it in other courses. For higher derivatives $y^{(j)}$ is used, with the convention that $y^{(0)} = y$. Thus,

$$y^{(j)} = \frac{d^j y}{dx^j}, \qquad j = 0, 1, 2, 3, 4, \cdots$$

Check that the functions e^{x^2}, $\sin x$, x^3, x^5, te^t satisfy the following differential equations, respectively:

$$y' = 2xy, \qquad y'' + y = 0 \qquad y^{(4)} = 0 \qquad y^{(5)} = 120 \qquad \ddot{u} + u = 2\dot{u}$$

2. If $y' = ky$ on an interval, where k is constant, show that y cannot change sign on the interval. That is, $y > 0$ throughout the interval, or $y = 0$ throughout the interval, or $y < 0$ throughout the interval.

Hint: Use (1). It will be found that the three possible behaviors correspond to three possibilities for the constant c.

3. A family of negative solutions for $y' = 25y$ is obtained by setting $y = -e^u$, where u is a differentiable function of x. Show that a graph of u versus x gives a family of parallel straight lines. What is their common slope?

4. Find a family of positive solutions of $y' = 6y$

(a) by setting $y = e^u$

(b) by use of a suitable integrating factor

(c) by separation of variables.

5. Find a family of negative solutions of $y' + 7y = 0$

(a) by setting $y = -e^u$

(b) by use of a suitable integrating factor

(c) by separation of variables.

Hint for (c): The integral of dy/y is $\ln|y| + c$.

6. Solve the problem $y' + 4y = 0$, $y(0) = 3$ by setting $y = 3e^u$ and also by separation of variables.

7. Solve the problem $y' - 4y = 0$, $y(0) = -3$ by setting $y = -3e^u$ and also by separation of variables.

8P. According to the text, e^{2x} is an integrating factor for $y' + 2y = 0$. Using this integrating factor, solve $y' + 2y = 6$.

Hint: Multiplication by e^{2x} gives $(e^{2x}y)' = 6e^{2x}$.

9P. Find a constant a such that $y = a$ satisfies $y' + 2y = 6$. Add to this the complete solution of $y' + 2y = 0$ and thus obtain the family of solutions found in Problem 8. The procedure illustrated here applies to a broad class of equations known as linear differential equations with constant coefficients.

10. By the procedure suggested in Problem 8 solve $y' + 3y = 8e^x$.

11. By trying $y = ae^x$ where a is constant, obtain a solution of $y' + 3y = 8e^x$. Add to this the complete solution of $y' + 3y = 0$ and thus obtain the result of Problem 10.

12. Solve $y' + 1 = y + e^{2x}$, $\qquad y(0) = 3$, and check.

13P. The left-hand curves in Figure 2 have the equations

$$y = e^{x-c}, \qquad y = 0, \qquad y = -e^{x-c}$$

so that changing c merely translates the curve $y = e^x$ or $y = -e^x$ horizontally. Similarly interpret the right-hand curves. What choice of c gives $y = 0$?

14P. Variation of parameters. According to the text, the family $y = ce^{kx}$ represents the complete solution of $y' = ky$. We are going to solve $y' = ky + f$ by the substitution $y = ue^{kx}$. Since the constant parameter c is now replaced by a variable u, this is called the method of *variation of parameters*.

Show that the suggested substitution leads to $u' = e^{-kx}f(x)$. If f is continuous an integration yields u and then $y = ue^{kx}$ is the solution sought.

15. (abcd) Solve Problems 8, 10, and 12 by variation of parameters.

Hint for 8: Check that the substitution $y = ue^{-2x}$ yields $u' = 6e^{2x}$.

─────────────────ANSWERS─────────────────

2. The three cases correspond to $c > 0$, $c = 0$, $c < 0$ respectively.

3. Since $u' = 25$, the lines are parallel and the common slope is 25.

4. $y = e^{6x+C}$, $y = ce^{6x}$ with $c > 0$, $y = e^{6x+C}$

5. $y = -e^{-7x+C}$, $y = ce^{-7x}$ with $c < 0$, $y = -e^{-7x+C}$

6. $y = 3e^{-4x}$ 7. $y = -3e^{4x}$ 8. $y = 3 + ce^{-2x}$ 9. $a = 3$

10. $y = 2e^x + ce^{-3x}$ 11. $a = 2$ 12. $y = 1 + e^x + e^{2x}$

13. Partial answer: None.

3. Four examples

In numerous applications the independent variable is the time t rather than a space variable x. When this is the case we write

$$y' = \frac{dy}{dt}, \qquad y'' = \frac{d^2y}{dt^2}$$

and so on. The main result (1) of the last section gives

(1) $$y' = ky \iff y = y_0 e^{kt},$$

where k is constant and where $y_0 = y(0)$ is the initial value. When convenient we shall use this result instead of repeating the derivation.

(a) Growth

Many organisms grow at a rate proportional to their size. If S denotes the size at time t, this means $dS/dt = kS$, where k is constant. It is a remarkable fact that this law of growth is wholly determined if we know the size at two different times. To see why, let the size be S_0 at time $t = 0$ and S_1 at time $t = t_1$. By change of notation in (1),

(2) $$S = S_0 e^{kt}.$$

Since the size is S_1 when $t = t_1$ we have

(3) $$S_1 = S_0 e^{kt_1}.$$

We could find k by taking logarithms in (3), but it is more efficient to find e^k instead. This is done as follows. By (3),

$$e^{kt_1} = \frac{S_1}{S_0}, \quad \text{hence} \quad e^k = (e^{kt_1})^{1/t_1} = \left(\frac{S_1}{S_0}\right)^{1/t_1}.$$

Using this value for e^k in (2) we get the desired result,

$$S = S_0(e^k)^t = S_0\left(\frac{S_1}{S_0}\right)^{t/t_1}.$$

As a check and aid to memory, note that $S(0) = S_0$, $S(t_1) = S_1$ by inspection, and the variable t occurs as an exponent.

(b) Bacteria

In a colony of bacteria each bacterium divides into two in a time interval of average length T. To find T it would be hopelessly inefficient, and also inaccurate, just to watch a bacterium through a microscope until you see it divide. What you really want is an average over many bacteria, not just a single observation.

This problem is solved as follows. Using a microscope slide ruled in squares, one can count the bacteria present at two different times and use a formula to compute T. The basis for the computation is a differential equation, but there is a logical difficulty, because the function N, being integer-valued, is not differentiable unless it is constant. To deal with this difficulty we replace N by a differentiable approximation, still called N, and we suppose that the hypothesis about doubling means

$$\frac{dN}{dt} = kN, \qquad k = \text{const.}$$

After these preliminaries let there be N_0 bacteria at time $t = 0$ and N_1 at time $t = t_1$. Since the initial population N_0 doubles in time T

$$0 \leq t \leq T \quad \text{corresponds to} \quad N_0 \leq N \leq 2N_0$$

and, by hypothesis,

$$0 \leq t \leq t_1 \quad \text{corresponds to} \quad N_0 \leq N \leq N_1.$$

Let us write the differential equation in the separated form

$$k\,dt = \frac{dN}{N}$$

and integrate between corresponding limits. The result is

$$\int_0^T k\,dt = \int_{N_0}^{2N_0} \frac{dN}{N}, \qquad \int_0^{t_1} k\,dt = \int_{N_0}^{N_1} \frac{dN}{N}$$

or, equivalently,

$$kT = \ln 2N_0 - \ln N_0, \qquad kt_1 = \ln N_1 - \ln N_0.$$

Since $\ln 2N_0 - \ln N_0 = \ln 2$, dividing the first equation by the second gives

$$\frac{T}{t_1} = \frac{\ln 2}{\ln N_1 - \ln N_0}.$$

Clearly, T can be found from this. The calculation illustrates the important method of *integrating between limits*, which is justified in Problem 12.

Many investigations involve integer-valued variables like the number N of bacteria in this example. In the following discussion of radioactivity, the primary variable is the number of atoms in a sample of material. Ecology is concerned with the number of individuals in a population, and further examples are easily given. In such cases the differential equation applies not to the original variable, but to a differentiable approximation thereto. Since this matter has now been brought to your attention, it will not be mentioned again.

(c) Radiocarbon dating

In a radioactive material each atom has a certain probability of disintegrating into a lower state. For example, radium goes through a number of states the last of which is lead. If you have twice as much material, twice as many atoms will disintegrate, on the average, in a given time interval. If you have three times as much material, three times as many will disintegrate, and so on. These considerations lead to the differential equation

$$\frac{dA}{dt} = kA, \qquad k = \text{const.}$$

where A is the amount of material at time t. The constant k is negative because the amount A decreases with time—the equation describes decay not growth. A measure of the rate of decay is given by the *half-life*, H. This is the time needed for A to decay to $A/2$, in other words, for half the material to change into the new form. As an illustration, the half-lives of some materials are:

radium	carbon 14	uranium
1620 years	5568 years	4,510,000,000 years

Obviously one cannot wait until half the material is gone to get the half-life. Instead, one measures the amount present at two different times. If the amount A_0 is present at time $t = 0$ and A_1 at time $t = t_1$, then

$$0 \leq t \leq H \quad \text{corresponds to} \quad A_0 \geq A \geq \frac{1}{2}A_0,$$

$$0 \leq t \leq t_1 \quad \text{corresponds to} \quad A_0 \geq A \geq A_1.$$

Since $\ln(1/2) = -\ln 2$, a development almost identical to that given for bacteria above leads to the formula

$$\frac{H}{t_1} = \frac{\ln 2}{\ln A_0 - \ln A_1}.$$

This is the way the half-life H is determined in practice.

If a radioactive material is combined with another material that is not radioactive, the ratio of radioactive material to the other satisfies the same differential equation as does the amount of radioactive material. More important, the rate of decomposition of the mixture is the same as that for the radioactive material alone. Suppose, for example, that a sample of carbon consists of A grams of carbon 14, which is radioactive, combined with C grams of ordinary carbon, which is not radioactive. The above remarks mean

$$\frac{dA}{dt} = kA \iff \frac{d}{dt}\left(\frac{A}{C}\right) = k\frac{A}{C}, \qquad \frac{dA}{dt} = kA \iff \frac{d}{dt}(A + C) = kA.$$

The equations follow from the fact that C is constant, independent of time. The easy verification is left for you.

These remarks give the mathematical basis for a method of dating historical events, which was discovered by the Nobel prize winner Willard Libby and is known as the method of *radiocarbon dating*. The best way to explain the procedure is to present it in the words of the discoverer himself. The following is quoted from an article that Libby wrote for the Encyclopædia Britannica:

"Radiocarbon age dating, developed in the late 1940s at The University of Chicago, is an example of the application of one of the newest sciences (atomic energy) to one of the oldest (archeology). The technique involves measuring the relative activities of radioactive carbon (C^{14}) in (1) present-day living organic matter and (2) the sample under investigation, and multiplying the logarithm of this ratio by the rate at which the activity of C^{14} decays with time. Careful measurements have shown that the activity of any given preparation of carbon-14 is reduced by exactly one-half during each interval of $5,568 \pm 30$ years. This value is called the half-life of C^{14}.

"Radiocarbon is produced in nature by an indirect process involving the interaction of cosmic rays from outer space with the nitrogen in the earth's atmosphere. The competing processes of formation and of decay of C^{14} have been going on for so long that the equilibrium has been established, and the world inventory of C^{14} is estimated at about 70 metric tons. Radiocarbon therefore has been introduced into the biosphere, and all living matter contains a small quantity of radiocarbon which averages 15.3 ± 0.1 disintegrations per minute per gram of contained carbon. This activity remains constant throughout the life of the organic matter because of the above-mentioned equilibrium processes.

"However, at death the introduction of radiocarbon into the specimen ceases, while the normal decay of the contained radiocarbon continues according to the half-life mentioned above. Therefore an archeological specimen (for example, a mummy or a tree) which yields 7.65 disintegrations per minute per gram of carbon instead of 15.3 is judged to be $5,568 \pm 30$ years old. If the material shows only one-fourth the radiocarbon content of living matter, the age of the specimen is $11,136 \pm 60$ years, etc.

"Age dating has revealed that there lived in North America over 9,000 years ago people who were capable of performing the finest basketry work; that the continents of North America and Europe were both glaciated in their northern latitudes 11,000 years ago; and that the calendars of the ancient Babylonians and the Mayas can be correlated with that of the Christians."

(d) Drifting of a canoe

This problem pertains to a canoe in still water on a windless day. For ease of exposition, you are nominated as the paddler. At time $t = 0$ you stop paddling just as you reach a line of flags stretched over the water to mark the finish line of a race. By time $t = t_0$ the canoe has drifted a distance s_0 beyond the line, and by time $t = 2t_0$ it has drifted a distance s_1 beyond the line. Both s_0 and s_1 are known to you. The problem is to show that the maximum distance the canoe can drift after you stop paddling is

$$s_\infty = \frac{s_0^2}{2s_0 - s_1}.$$

That is, if you wait long enough, the distance will be arbitrarily close to that value, but you will never quite reach $s = s_\infty$. The surprise of the result is that it does not require any knowledge of t_0, or of the velocity v_0 that the canoe had when you stopped paddling, or of the magnitude of the resistance that the water offers to the motion, or of the mass of the loaded canoe!

To see where the formula comes from, let us accept the fact that for low-speed motion the force of resistance due to the water is proportional to the velocity v. Since the acceleration is dv/dt, Newton's law gives

$$m\frac{dv}{dt} = -k_1 v, \qquad k_1 = \text{const.},$$

where m is the mass of the canoe. The constant $-k_1$ is negative because the force tends to retard the motion, hence is in a direction opposite to that associated with the velocity v. Resistance of this type is called *viscous damping*. Setting $k_1 = km$ and dividing by m we get

(4) $$\frac{dv}{dt} = -kv, \quad \text{or} \quad v = v_0 e^{-kt}.$$

It is important that k is a *positive constant*. Aside from this, the value of k will not enter into the final result, and that is what we meant when we said the result is independent of the resistance and of the mass.

Since the displacement s satisfies $ds/dt = v$ by the definition of velocity, the second formula (4) gives

(5)
$$s = \int_0^t v\, dt = v_0 \frac{e^{-kt}}{-k}\Big|_0^t = \frac{v_0}{k}(1 - e^{-kt}).$$

The formula shows that $s < v_0/k$ and that

$$s_\infty = \lim_{t\to\infty} s(t) = \frac{v_0}{k}.$$

Thus, our problem is to determine v_0/k given

(6)
$$s_0 = \frac{v_0}{k}(1 - e^{-kt_0}), \quad s_1 = \frac{v_0}{k}(1 - e^{-2kt_0}).$$

These formulas follow from (5) together with the hypothesis involving s_0 and s_1. If the second equation (6) is divided by the first the result is

$$\frac{s_1}{s_0} = \frac{1 - e^{-2kt_0}}{1 - e^{-kt_0}} = 1 + e^{-kt_0}$$

which gives

$$1 - e^{-kt_0} = 2 - (1 + e^{-kt_0}) = 2 - \frac{s_1}{s_0}.$$

Hence, by the first equation (6),

$$\frac{v_0}{k} = \frac{s_0}{1 - e^{-kt_0}} = \frac{s_0}{2 - (s_1/s_0)} = \frac{s_0^2}{2s_0 - s_1}.$$

Since $s_\infty = v_0/k$, this is the desired result.

PROBLEMS

1. The size of a growing organism, such as a clam, can be measured by its length L, its area A, or its volume V. If the shape remains unchanged, so that the clam remains similar to itself as it grows,

$$A = aL^2, \qquad V = bL^3 \qquad a, b \text{ const.}$$

As an illustration, for a cube of side L we have $a = 6$ and $b = 1$. For a sphere of diameter L the values are $a = \pi$, $b = \pi/6$. Show that if the first of the following laws of growth holds, where $k = k(t)$ is any function of t, then the second and third hold automatically:

$$\frac{dL}{dt} = kL \qquad \frac{dA}{dt} = 2kA \qquad \frac{dV}{dt} = 3kV$$

The point is that the result is independent of the shape.

Hint: $dA/dt = (d/dt)(aL^2) = 2aL\, dL/dt$.

2. If the equation for radioactive decay is written $dA/dt = -KA$ and the equation for bacterial growth is written $dN/dt = KN$, where K is a positive constant, show that the half-life H and the doubling time T satisfy

$$KH = KT = \ln 2$$

3. The rate of decomposition of a substance is proportional to the amount y at time t, so that $dy/dt = -ky$ where k is a positive constant. Find k if y changes from 1000 grams to 500 grams in half an hour, t being measured in hours.

4. Lambert's law of absorption. The percentage of incident light absorbed in passing through a layer of material is proportional to the thickness of the material. If one inch of material reduces the light to $1/3$ its original intensity, how much additional material is needed to reduce the intensity to $1/27$ of its initial value? Obtain the answer by inspection, and check by use of an appropriate differential equation.

5. If 16 grams of a radioactive substance were present at time $t = 1$ year, and 2 grams at time $t = 4$ years, how much was present initially, and what is the half-life? Solve by inspection, and check by use of a differential equation.

6. A geometric sequence is a sequence a, ab, $ab^2, ..., ab^n, ...$ such that the ratio of two consecutive terms is constant, and an arithmetic sequence is a sequence A, $A + B$, $A + 2B, ..., A + nB, ...$ such that the difference of two consecutive terms is constant. If y is a solution of the growth equation $dy/dt = ky$ with k constant, show that the function $y(t)$ transforms any arithmetic sequence into a geometric sequence. That is,

$$y(A + Bn) = ab^n, \qquad n = 0, 1, 2, \cdots$$

If $y(0) = 1$ express a and b in terms of k, A, B. This problem sheds light on Problems 4 and 5 above.

7. In this problem we assume no wind, still water, and viscous damping. The time t is in seconds, the displacement s in feet. A motorboat is going 30 miles per hour and at time $t = 0$ the engine is turned off. At $t = 10$ the speed has dropped to 15 miles per hour. Using a hand calculator if you wish, find the velocity v at time t and thus get the displacement s when $t = 10, 20$, and ∞. Verify that your three values for s agree with the result obtained in the canoe problem of the text.

Note: 30 miles per hour equals 44 feet per second.

8. Newton's law of cooling. When the temperatures are not large enough to involve radiation, the rate at which a body cools is proportional to the difference between the temperature T of the body and the temperature T^* of the surrounding medium, which is assumed constant. This is Newton's law of cooling. An easy way to solve problems involving Newton's law is to set $y = T - T^*$. Then $dy/dt = ky$, $k = \text{const.}$, which can be solved with ease. Knowing T at various times t_0, t_1 we can find the corresponding values y_0, y_1 from $y = T - T^*$, and once y is found, we get T from $T = y + T^*$. If $T = T_0$ when $t = 0$, get a formula for T at any later time.

9. The temperature of a metal ingot fell from $120°$ to $70°$ in one hour when it was allowed to cool in air at a temperature of $20°$. How long will it take to cool to $40°$? $30°$? $20°$? Does it matter whether the given temperatures are stated in degrees C or degrees F? Use Problem 8.

10. Time of death. When a dead body is discovered in an environment of constant temperature, it is sometimes possible to estimate the time of death by Newton's law of cooling. Namely, measure the temperature T_0 at the time $t = 0$ when the body is discovered, and the temperature T_1 at a later time t_1. Extrapolate back to get the time of death t_d when the temperature was presumably $T_d = 98.6°$F. The positive quantity $|t_d|$ gives the time elapsed from the time of death to the time the body was discovered. Setting $y_i = T_i - T^*$, $i = 0, 1, d$, where T^* is the ambient temperature, obtain the formula
$$\frac{|t_d|}{t_1} = \frac{\ln(y_d/y_0)}{\ln(y_0/y_1)}$$
Hint: By Problem 8 show $(y_d/y_0) = (y_0/y_1)^{t_d/t_1}$ and take logs.

11. The temperature of a body upon discovery is $85°$F and after one hour the temperature is $80°$F. If the ambient temperature is $70°$F, about how long after death was the body found? Use Problem 10, taking $T_d = 99°$F. How much does the answer change if you take $T_d = 98.6°$F?

12. Assuming $y_0 > 0$ and $y > 0$, evaluate the integrals on the left and thus get the other two equations:
$$\int_{y_0}^y \frac{ds}{s} = \int_{x_0}^x k\,ds \qquad \frac{dy}{dx} = ky \qquad y(x_0) = y_0$$

13. If a and b are constant, the equation $dy/dt = ay + b$ is encountered in connection with the spread of rumors, Newton's law of cooling, certain types of chemical reactions, problems of population growth with immigration, and the theory of investment. If $a \neq 0$ show that the substitution $u = ay + b$ gives the differential equation on the left, hence gives y as on the right:
$$\frac{du}{dt} = au \qquad ay + b = (ay_0 + b)e^{at}$$

14. A curve $y = f(x)$ in Cartesian coordinates is such that the projection of part of the tangent on the x axis has a constant length $b > 0$, as shown in Figure 3. Geometrically, this means that an increased distance from the x axis is just compensated by an increased steepness of the curve, so that b remains constant. Obtain the differential equation $dy/dx = y/b$ for the curves of this family and deduce that $y = ce^{x/b}$.

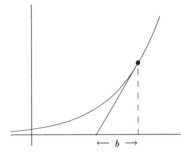

Figure 3

————————————ANSWERS————————————

3. $k = \ln 4$ 4. two inches additional   5. 32 grams, 1 year

6. $a = e^{kA}$, $b = e^{kB}$ 7. $s \ln 2 = 440(1 - 2^{-t/10})$

8. $T - T^* = (T_0 - T^*)e^{kt}$

9. About 2.32, 3.32, ∞ hours. The choice of units does not matter.

11. $t_d = \ln(29/15)/\ln(15/10) \approx 1.63$ hours; $t_d \approx 1.59$ hours. In both cases the sensible answer is *about an hour and a half.*

The heart and soul and all the poetry of Natural Philosophy are embodied in the concept of mathematical beauty.

—d'Arcy Wentworth Thompson

To myself I seem to have been only like a boy playing on the seashore, and diverting myself in now and then finding a smoother pebble or a prettier shell than ordinary, whilst the great ocean of truth lay all undiscovered before me.

—Isaac Newton

Chapter 2

METHODS OF SOLUTION

THE WORDS *linear* and *nonlinear* describe an important classification of mathematical and scientific problems. Linear problems lend themselves to comprehensive and satisfying theories, while nonlinear problems often require special devices. Both types are encountered in applications. Differential equations provide one of the few sources of significant nonlinear problems for which simple solutions are available. After setting forth the basic terminology of the subject, we give a number of elementary techniques by which differential equations can be solved.

1. Terminology

Differential equations are divided into two classes, ordinary and partial. An *ordinary differential equation* involves one independent variable and derivatives with respect to that variable. A *partial differential equation* involves more than one independent variable and corresponding partial derivatives. The order of the highest derivative in a differential equation is the *order* of the equation. For example, the first of the equations

$$\text{(1)} \qquad x^2 \frac{d^2 y}{dx^2} + 10y = 6x \frac{dy}{dx}, \qquad \frac{\partial^4 u}{\partial x^4} + \frac{\partial^4 u}{\partial y^4} = 2$$

is an ordinary differential equation of order 2 and the second is a partial differential equation of order 4. Unless the context makes it obvious that a partial differential equation is intended, the phrase "differential equation" in this book means "ordinary differential equation."

Using notation introduced in Chapter 1 we write

$$y' = \frac{dy}{dx}, \qquad y'' = \frac{d^2 y}{dx^2}, \qquad \cdots, \qquad y^{(n)} = \frac{d^n y}{dx^n}.$$

For example, $(x^5)'' = (5x^4)' = 20x^3$. In this notation the first equation (1) takes the form

$$\text{(2)} \qquad x^2 y'' + 10y = 6xy'.$$

Differential equations are usually considered on an open interval, that is, on an interval $a < x < b$ that does not contain its endpoints. Such an interval is denoted by the letter I and the statement that x is on the interval

I is written $x \in I$. A function $\phi(x)$ is said to be a *solution* of a differential equation on I if upon the substitution $y = \phi(x)$, $y^{(j)} = \phi^{(j)}(x)$ the equation is identically satisfied; that is, it holds for all $x \in I$. As an illustration, $y = x^2$ and $y = x^5$ are solutions of (2) because

$$x^2(2) + 10x^2 = 6x(2x), \qquad x^2(20x^3) + 10x^5 = (6x)(5x^4).$$

These equations hold for all values of x, hence the solutions are valid on the interval $(-\infty, \infty)$.

It was stated above that differential equations are divided into two classes, ordinary and partial. They are also divided into two other classes, linear and nonlinear. A linear equation is an equation of the form

$$p_n y^{(n)} + p_{n-1} y^{(n-1)} + \cdots + p_1 y' + p_0 y = f$$

where the coefficients p_j and f are functions of x alone, not involving y. To emphasize this point one often writes $p_j(x)$ and $f(x)$ instead of p_j and f. For example, the linear equations of order 1 and 2 are respectively

$$p_1(x)y' + p_0(x)y = f(x), \qquad p_2(x)y'' + p_1(x)y' + p_0(x)y = f(x).$$

The functions p_j and f are assumed to be defined on one and the same interval, finite or infinite. If $f = 0$, the linear equation is said to be *homogeneous*.

An equation is considered to be linear if it can be put into the above form by reversible algebraic operations. For example, the equations

$$y'' = y' + 1, \qquad y = y' \sin x, \qquad x^2(y'' + x) = xy' - y$$

are linear because they can be put into the respective forms

$$y'' - y' = 1, \qquad (\sin x)y' - y = 0, \qquad x^2 y'' - xy' + y = -x^3.$$

The second of these equations is homogeneous, the others are inhomogeneous. Examples of nonlinear equations are

$$(y')^2 = x + y, \qquad yy' = x, \qquad y' = \sin y, \qquad y' = x + e^y.$$

The most general differential equations of orders 1 and 2 are

$$F(x, y, y') = 0, \qquad F(x, y, y', y'') = 0,$$

respectively. For theoretical analysis these general equations are often assumed to be solvable for the highest derivative, thus:

$$y' = f(x, y), \qquad y'' = f(x, y, y').$$

But it is not advisable to impose this condition on the linear equation, because in important cases the coefficient of the highest-derivative term may vanish at one or more points. For example, (2) has solutions valid on $(-\infty, \infty)$, although it cannot be solved for y'' on any interval containing the origin.

Differential equations of the kind commonly encountered in applications have not just one or two solutions, but infinitely many. More specifically, an equation of order 1 usually has a family of solutions

$$y = \phi(x, c)$$

depending on a single parameter c. A second-order equation usually has a family of solutions

$$y = \phi(x, c_1, c_2)$$

depending on two parameters c_1, c_2, and so on. These parameters are like constants of integration. As an illustration, the equation $y'' = 0$ has a family of solutions

$$y = c_1 x + c_2$$

depending on two constants of integration, c_1 and c_2.

One of the simplest ways of determining the constants is to specify the value of y and some of its derivatives at a single point x_0. For an equation of order 1 there is one constant c, in general, and we need a single condition

$$y(x_0) = y_0$$

to determine c. For an equation of order 2 there are usually two constants and we need two conditions

$$y(x_0) = y_0, \qquad y'(x_0) = y_1,$$

to determine them, and so on.

The values y_j are called *initial values*, the equations $y^{(j)}(x_0) = y_j$ are called *initial conditions,* and the initial conditions together with the differential equation constitute an *initial-value problem*. For example, the initial-value problems associated with equations of orders 1 and 2 are

$$\begin{cases} y' = f(x, y) \\ y(x_0) = y_0 \end{cases} \qquad \begin{cases} y'' = f(x, y, y') \\ y(x_0) = y_0, \ y'(x_0) = y_1 \end{cases}$$

respectively. It is assumed that x_0 is on the interval I where the equation is to hold, or sometimes that x_0 is an endpoint of this interval. The reason for the term "initial value" is that in many problems x denotes the time t, and $x_0 = t_0$ is the time at which the process starts.

So far we have considered only single differential equations containing the single unknown function y. Many problems lead to two or more differential equations in two or more unknowns; in other words, they lead to a *system* of differential equations. For example,

$$\begin{cases} u' = f(x, u, v) \\ v' = g(x, u, v) \end{cases}$$

is a system of two first-order differential equations in the two unknowns u and v.

The theory of first-order systems actually embraces the entire theory of differential equations, because an equation of higher order can always be reduced to a first-order system by taking y and some of its derivatives as new unknowns. For example, the second-order equation

$$y'' = f(x, y, y')$$

is equivalent to the first-order system

$$\begin{cases} u' = v \\ v' = f(x, u, v) \end{cases}$$

with $u = y$ and $v = y'$. From this point of view the most efficient development of the subject of differential equations is to start with first-order systems. However, such an approach would seem intolerably abstract to many readers and would rob the subject of much of its charm.

EXAMPLE 1. Obtain a two-parameter family of solutions of the second-order linear equation $y'' = 6$. Integration gives $y' = 6x + c_1$, where c_1 is constant, and integrating again gives the two-parameter family

$$(3) \qquad\qquad y = 3x^2 + c_1 x + c_2, \qquad c_1, c_2 \text{ const.}$$

It is obvious that every function (3) satisfies $y'' = 6$ and almost obvious that every solution of $y'' = 6$ is in the family (3). A proof is given in Problem 11.

In general, a family of functions is the *complete solution* of a differential equation if:

1. every member of the family satisfies the equation, and

2. every solution of the equation is a member of the family.

The family of functions given by (3) is the complete solution of $y'' = 6$.

As illustrated in this example, the solution of a differential equation often involves one or more integrations, and for this reason the graph of a solution is called an *integral curve*. By (3) the integral curves of the equation $y'' = 6$ are all parabolas.

EXAMPLE 2. Find the solution of $y'' = 6$ satisfying general initial conditions

$$y(x_0) = y_0, \qquad y'(x_0) = y_1.$$

Using (3) one could determine c_1 and c_2 from the two equations $y(x_0) = y_0$, $y'(x_0) = y_1$. But a more efficient method is to note that the gist of (3) is that $y = 3x^2$ plus an arbitrary linear function. Thus one can equally well write

$$y = 3(x - x_0)^2 + a(x - x_0) + b$$

since this too is $3x^2$ plus an arbitrary linear function. The latter equation yields, by inspection, $b = y(x_0) = y_0$, $a = y'(x_0) = y_1$. Hence

$$y = 3(x - x_0)^2 + y_1(x - x_0) + y_0.$$

As explained more fully in Chapter 6, a family of solutions that can be used to satisfy arbitrarily specified initial conditions at the arbitrary point x_0 is called a *general solution*. The family (3) is a general solution of $y'' = 6$.

EXAMPLE 3. The linear homogeneous equation $y'' + y = 0$ has the obvious solutions $\cos x$ and $\sin x$. Check that

$$y = c_1 \cos x + c_2 \sin x$$

is also a solution for every choice of the constants c_1 and c_2. This follows from the calculation

$$y'' = (c_1 \cos x + c_2 \sin x)'' = (-c_1 \sin x + c_2 \cos x)' = -c_1 \cos x - c_2 \sin x = -y.$$

EXAMPLE 4. Find all constant solutions $y = c$ of

$$y'' = (y^2 - 1)\sin(y + y').$$

Since $y' = y'' = 0$ when $y = c$, we must have

$$0 = (c^2 - 1)\sin c.$$

This is satisfied by $c = 1$, $c = -1$, and $c = m\pi$ where m is an integer, and by no other values.

PROBLEMS

1. Here are twelve differential equations:

(a) $y'' + 4y = 8$ (e) $y^{iv} = 0$ (i) $x^2 y'' + 35y = 11xy'$

(b) $(y')^2 + y^2 = 1$ (f) $(y'')^2 = 4$ (j) $(y' - y)(y' - 2y) = 0$

(c) $y'' + 3 = 4y$ (g) $y^{iv} = y$ (k) $(y')^2 - 3y^2 = 2yy'$

(d) $y' = y(1 - y)$ (h) $x^2 y^{iv} = 2y''$ (l) $y' = \sin y$

Verify that the sum of the orders of these equations is 25. (The problem is worded so you can check your work without being told the answers.)

2. Exactly half of the equations in Problem 1 are linear, and two-thirds of these are both linear and homogeneous. Which are linear, and which are both linear and homogeneous?

3. Some of the equations in Problem 1 have a solution of the form $y = c$, where c is constant. Which ones are they, and what are the values of c? A solution $y = $const. is called a *stationary solution.*

4. Verify that the given function is a solution of the differential equation immediately below it:

(a) $\begin{cases} y = xe^x \\ y' = e^x + y \end{cases}$ (b) $\begin{cases} y = \tan x \\ y' = 1 + y^2 \end{cases}$ (c) $\begin{cases} y = e^x - e^{-x} \\ (y')^2 = 4 + y^2 \end{cases}$

5. Verify that the given function is a solution of the differential equation immediately below it:

(a) $\begin{cases} y = e^{7x} \\ y'' = 7y + 6y' \end{cases}$ (b) $\begin{cases} y = x^7 \\ x^2 y'' = 6xy' \end{cases}$ (c) $\begin{cases} y = xe^{-x} \\ y'' + 2y' + y = 0 \end{cases}$

6P. Verify that the equation $y'' + y' = 2y$ is satisfied by $y = e^{ax}$ if the constant a satisfies the quadratic equation $a^2 + a - 2 = 0$. Thus get the two solutions $y = e^x$, $y = e^{-2x}$. The behavior shown here is typical of all linear homogeneous differential equations with constant coefficients. That is, the substitution $y = e^{ax}$ always leads to an algebraic equation for a.

7. Equations (gjk) of Problem 1 have solutions of the form $y = e^{ax}$ where a is constant. Find the values of a.

8. Equations (bg) of Problem 1 have solutions of the form $y = \sin \omega x$ where ω is a positive constant. What are the possible values of ω?

9P. Check that $y = x^a$ satisfies $x^2 y'' = 2y$ if the constant a satisfies the equation $a^2 - a - 2 = 0$. Thus get the two solutions x^2, x^{-1}. Note that the first is valid on $(-\infty, \infty)$ but the second on $(-\infty, 0)$ or $(0, \infty)$ only. The behavior shown here is typical for a broad class of linear homogeneous equations known as equations of *Euler type*. For this class the substitution $y = x^a$ always leads to an algebraic equation for a.

10. Equations (efhi) of Problem 1 admit a solution $y = x^a$ where a is a nonzero constant. What are the possible values of a?

11. If $y'' = 6$ check that $(y' - 6x)' = 0$, and hence $y' - 6x = c_1$ where c_1 is constant. Using this, check that $(y - 3x^2 - c_1 x)' = 0$, hence $y - 3x^2 - c_1 x = c_2$ where c_2 is also constant. This shows that the solution obtained in Example 1 is the complete solution. We used the fact that if the derivative of a function is 0 on an interval then the function is constant on the interval. In the present case the interval is $(-\infty, \infty)$.

12. By repeated integration as in Example 1 show that

$$y''' = 24x \iff y = x^4 + c_1 x^2 + c_2 x + c_3, \qquad c_i \text{ const.}$$

A justification can be given as in Problem 11 above, but if the process of justification has been seen once, it need not be repeated (and you are not asked to repeat it).

13. Solve the initial-value problem $y''' = 24x$, $y(1) = y'(1) = y''(1) = 1$

(a) by determining the constants in the result of Problem 12.

(b) by writing the result of Problem 12 in the equivalent form

$$y = x^4 + a(x - 1)^2 + b(x - 1) + c, \qquad a, b, c \text{ const.}$$

and determining the constants a, b, c.

(c) Show that your two answers agree; no answer is given in the text.

14. Check that the equation $y'' + y = 0$ becomes $u' = v$, $v' = -u$ upon the substitution $u = y$, $v = y'$. Using these latter equations and the chain rule, verify that $u^2 + v^2$ is constant.

15. The functions $y = e^x$ and $y = \cos x$ cannot be solutions of one and the same first-order equation $y' = f(x, y)$ on any interval containing the origin. Why not?

─────────────────ANSWERS─────────────────

2. aceghi are linear, eghi are homogeneous

3. (a) $c = 2$ (b) $c = 1$ or -1 (c) $c = 3/4$ (d) $c = 0$ or 1 (e) any c will do
   (g) $c = 0$ (h) any c will do (ijk) $c = 0$ (l) $c = n\pi$, n any integer

7. (g) $a = 1$ or -1 (j) $a = 1$ or 2 (k) $a = -1$ or 3

8. (b) $\omega = 1$ (g) $\omega = 1$

10. (e) $a = 1$ or 2 or 3 (f) $a = 2$ (h) $a = 1$ or 4 (i) $a = 5$ or 7

15. Because $f(0, 1)$ would have to be both 1 and 0.

2. Implicit solutions

In the first-order equation

$$\frac{dy}{dx} = f(x, y),$$

the left-hand member dy/dx denotes the slope of the solution curve and the right-hand member $f(x, y)$ gives the value of the slope at the point (x, y) on the curve. It is as if we had a traffic policeman stationed at each point (x, y) directing traffic to flow along the direction specified by $f(x, y)$. From this point of view, a *solution* is a curve that obeys the law at each of its points.

The formulation by use of dy/dx breaks down when the tangent is vertical, since a vertical tangent corresponds to $dy/dx = \infty$. But the geometric interpretation remains meaningful. To deal with this matter, note that a curve in the (x, y) plane can be described, not only by $y = \phi(x)$, but also by $x = \psi(y)$. The first of the following conditions is meaningless, but the second is permissible:

$$\frac{dy}{dx} = \phi'(x) = \infty, \qquad \frac{dx}{dy} = \psi'(y) = 0$$

Thus, there is merit in treating x and y symmetrically, so that the independent variable can be either x or y.

The same conclusion is suggested when differential equations are applied to practical problems. Nature has no cognizance of coordinate systems, which merely provide a framework for the mathematical description of an underlying reality. If a problem seems intractable when we insist on a solution $y = \phi(x)$, but easy when we allow $x = \psi(y)$, it could mean that we have made an inappropriate choice of independent and dependent variables in the initial formulation. These remarks motivate the following discussion.

A first-order equation is said to be in *differential form* if it is written

(1) $$P(x, y)dx + Q(x, y)dy = 0.$$

A differentiable function $y = \phi(x)$ is a *solution* of (1) on an interval I if the substitution

$$(2) \qquad\qquad y = \phi(x), \qquad dy = \phi'(x)\,dx$$

makes the left-hand member of (1) vanish for $x \in I$. A differentiable function $x = \psi(y)$ is a *solution* of (1) on an interval J if the substitution

$$(3) \qquad\qquad x = \psi(y), \qquad dx = \psi'(y)\,dy$$

makes the left-hand member of (1) vanish for $y \in J$. The advantage of writing the equation in differential form is that it allows solutions of either type; indeed, we may have $y = \phi(x)$ at some points on the solution-curve and $x = \psi(y)$ on others.

If an equation

$$(4) \qquad\qquad F(x, y) = c, \qquad c = \text{const.}$$

can be solved for $y = \phi(x)$ or for $x = \psi(y)$ in the neighborhood of each point (x, y) satisfying (4), and if the corresponding function ϕ or ψ satisfies (1) in the sense explained above, (4) is said to give an *implicit solution* of (1). The term "implicit" distinguishes solutions (4) from the explicit solutions $y = \phi(x)$ or $x = \psi(y)$.

As a matter of terminology, we mention that the set of points satisfying an equation (4) is called the *locus* of the equation. The word "locus" is Latin for "place" and refers to the places in the (x, y) plane where the equation is true. For example, the locus defined by $x^2 + y^2 = 1$ is a circle centered at the origin.

Implicit solutions of the form $F(x, y) = c$ are commonly obtained from an equation $dF = 0$, as explained next. If $F(x, y)$ has continuous partial derivatives, an *exact differential* is an expression of the form

$$dF = \frac{\partial F}{\partial x}\,dx + \frac{\partial F}{\partial y}\,dy.$$

For example, each of the following is an exact differential:

$$d(x^2 + y^3) = 2x\,dx + 3y^2\,dy, \qquad d(x^3 y^5) = 3x^2 y^5\,dx + 5x^3 y^4\,dy$$

$$d\left(\frac{y}{x}\right) = \frac{x\,dy - y\,dx}{x^2}, \qquad x \neq 0.$$

It is an important fact that, under suitable conditions, the equations $F = \text{const.}$ and $dF = 0$ are equivalent, that is,

$$dF = 0 \iff F(x, y) = c,$$

where c is constant. A precise statement and proof of conditions under which this holds requires a deeper knowledge of two-variable calculus than is assumed at this point, and we merely give a few examples. The general theory is outlined in Chapter A.

EXAMPLE 1. The equation $y\,dx + x\,dy = 0$ can be written $d(xy) = 0$, giving the implicit solution $xy = c$ where c is constant. The integral curves form a set of rectangular hyperbolas as shown in Figure 1. There is one and only one solution through each point (x_0, y_0) other than the origin, and the solution can be written in the explicit form $y = c/x$ or $x = c/y$ as the case may be. There are two solutions through the origin, namely $y = 0$ and $x = 0$.

EXAMPLE 2. Consider the equation $x\,dy - y\,dx = 0$. If $x \neq 0$ the equation can be multiplied by $1/x^2$ and if $y \neq 0$ it can be multiplied by $1/y^2$ to give, respectively,

$$(5) \qquad \frac{x\,dy - y\,dx}{x^2} = 0 \qquad \text{or} \qquad \frac{x\,dy - y\,dx}{y^2} = 0.$$

These in turn are equivalent to

$$(6) \qquad d\left(\frac{y}{x}\right) = 0, \qquad -d\left(\frac{x}{y}\right) = 0,$$

respectively. The solutions are $y/x = c$ in the first case and $x/y = c$ in the second, where c is constant. The integral curves are shown in Figure 2. There is one and only one integral curve through each point (x_0, y_0) other than the origin, but there are infinitely many through the origin.

The factors $1/x^2$ and $1/y^2$ in (5) are called *integrating factors* because they reduce the left side to an exact differential (6) which is readily integrated. Under mild hypotheses it is seen that an equation (1) always admits an integrating factor, but construction of the factor in the general case requires prior knowledge of the solution. The usefulness of the method does

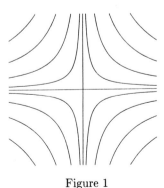

Figure 1

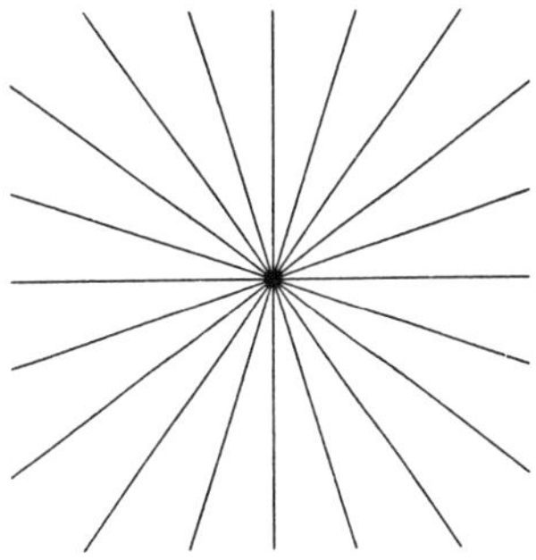

Figure 2

not depend so much on general theory as on the fact that in many problems an integrating factor can be found more or less by inspection.

EXAMPLE 3. The equation $x\,dx + y\,dy = 0$ can be written

$$\frac{1}{2}d(x^2 + y^2) = 0$$

to give the implicit solution

(7) $x^2 + y^2 = c, \qquad c \text{ const.}$

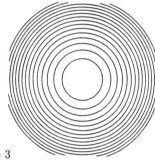

Figure 3

If $c < 0$ the locus is empty, that is, has no points. If $c = 0$ the locus consists of a single point $x = y = 0$. However, this does not give a solution, since it is not a differentiable curve of the form $y = \phi(x)$ or $x = \psi(y)$. If $c > 0$, which is the only interesting case, the integral curves are circles centered at the origin as shown in Figure 3. There is no integral curve through the origin and exactly one through every point (x_0, y_0) other than the origin.

As discussed more fully in the next section, an equation of the form

$$f(x)\,dx = g(y)\,dy$$

is said to be *separated* and a solution is given, in general, by

$$\int f(x)\,dx = \int g(y)\,dy + c, \qquad c \text{ const.}$$

The above equation $x\,dx = -y\,dy$ falls into this pattern.

EXAMPLE 4. Discuss the relationship of the three equations

(8) $\dfrac{dy}{dx} = -\dfrac{x}{y}, \qquad \dfrac{dx}{dy} = -\dfrac{y}{x}, \qquad x\,dx + y\,dy = 0.$

The first equation implies that $y = \phi(x)$ where ϕ is differentiable, and no integral curve of the equation can contain a point (x, y) where $y = 0$. The second equation requires $x = \psi(y)$ where ψ is differentiable and no integral curve of the equation can contain a point (x, y) where $x = 0$. The third equation allows solutions of either type and leads to the family of circles obtained in Example 3.

In the neighborhood of any point (x, y) of the circle we can express y as a differentiable function of x, or x as a differentiable function of y, using one of the relations

$$y = (c - x^2)^{1/2}, \quad x = (c - y^2)^{1/2}, \quad x = -(c - y^2)^{1/2}, \quad y = -(c - x^2)^{1/2}.$$

To be sure, one can get additional functions $y = \phi(x)$ by taking $y > 0$ on part
of the circle and $y < 0$ on the remainder.
However, such a function is not differentiable and does not satisfy any equation
(8). See Figure 4.

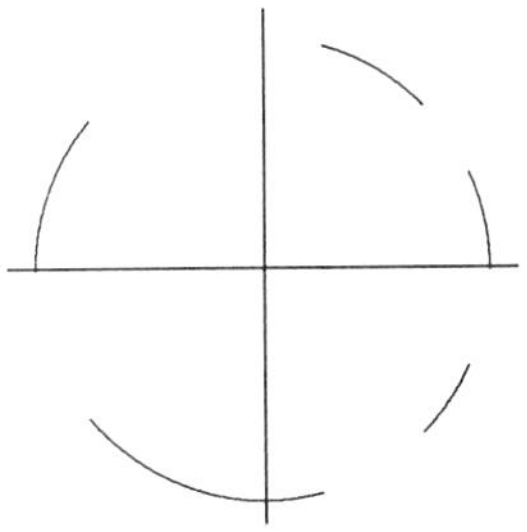

EXAMPLE 5. The position of a moving point at time t is given by

Figure 4

$$x = \sin t, \qquad y = \cos t.$$

Check that $x\,dx + y\,dy = 0$ where t, rather than x or y, is taken as the
independent variable. The proof is simple. Since $d(\sin t) = \cos t\,dt$ and
$d(\cos t) = -\sin t\,dt$, we get

$$x\,dx + y\,dy = (\sin t)(\cos t\,dt) + (\cos t)(-\sin t\,dt) = 0$$

as required. The corresponding locus is $x^2 + y^2 = 1$, which agrees with the
results of the preceding examples. Since a locus $F(x, y) = c$ can generally be
described as the path of a moving point, a locus is often called an *orbit*.

In applications it is seldom necessary to distinguish the forms

$$y = \phi(x, c), \qquad F(x, y) = c, \qquad x = \psi(y, c)$$

as carefully as was done in the examples above, and the main purpose of these
examples was to explain what an implicit solution is. Since the matter of
interpretation has now been sufficiently emphasized, the following problems
are concerned, not with the interpretation of implicit solutions, but with
methods by which they are discovered.

PROBLEMS

1P. A sum such as $du + dv + dw$ equals $d(u + v + w)$, hence $du + dv + dw = 0$
has the implicit solution $u + v + w = c$. Using this idea obtain an implicit
solution of

$$(2x + y - y^{-1})dx + (x + 2y + xy^{-2})dy = 0, \qquad y \neq 0$$

Hint: By rearrangement (and this is by far the hardest step) the equation
can be written

$$2(x\,dx + y\,dy) + (x\,dy + y\,dx) + \left(\frac{x\,dy - y\,dx}{y^2}\right) = 0$$

Results in the text now give the solution with ease.

2P. Obtain an implicit solution of $(xy)^3(x\,dy + y\,dx) = dx + dy$.

Hint: If $u = xy$ the left side is $u^3\,du$. The method illustrated here is general.

3. The following problems are to be solved as in Problems 1 or 2, but they do not require any rearrangement. Obtain implicit solutions by inspection:

(a) $d(x^2 y) = x\,dy + y\,dx$ (d) $2xy\,dy + y^2\,dx = dx + dy$

(b) $d(xy^2) = x\,dx + y\,dy$ (e) $2x\,dx + 6y^2\,dy = x\,dy + y\,dx$

(c) $x^2\,dy + 2xy\,dx = x\,dx$ (f) $(xy)^2(x\,dy + y\,dx) = 2x\,dx$

Hint: For (c), compute the left side of (a).

4. Rearrange suitably and thus get an implicit solution:

(a) $(y - 2x + 1)dx = (2y + 1 - x)dy$ (c) $(y - x^3)dx = (y^3 - x)dy$

(b) $(y + y^{-1})dx = x(y^{-2} - 1)dy$ (d) $(e^x + x^{-2}y)dx = (x^{-1} - e^{-y})dy$

5. One of the following has an integrating factor $1/x^2$ and the other has an integrating factor $1/y^2$. Decide which is which, and solve:

(a) $x\,dy = (y + x^4)dx + x^4 y^2 d(xy)$ (b) $(x^4 y^7 + y)dx = (x - x^5 y^6)dy$

6. The orbit of an implicit solution of $(3x^2 - y)dx + (3y^2 - x)dy = 0$ contains the point $(1, 1)$. Show that it also contains $(-1, 1), (1, -1), (0, 1), (1, 0)$.

7. Show that the orbits of $(x + 1)dx + (y + 1)dy = 0$ are circles. What is the radius of the orbit containing the point $(1, 1)$?

8. If a is constant, $x > 0$ as needed, and $y = \phi(x)$ is differentiable, then

$$x^{a-1}(x\,dy + ay\,dx) = d(x^a y)$$

Hence the equation $x\,dy + ay\,dx = f(x)dx$ can be solved if both sides are multiplied by x^{a-1}. Using this method solve the following explicitly, in the form $y = \phi(x, c)$:

(a) $x\,dy + 2y\,dx = 5x^3\,dx$ (b) $x\,dy - y\,dx = x^2\,dx$ (c) $x\,dy - 2y\,dx = dx$

Note that division by dx gives linear equations. The procedure illustrates a general method that will be discussed later.

9. Solve $y\,dx + 3x\,dy = 14y^4\,dy$ in the form $x = \psi(y, c)$ as follows: Interchange x and y to get an equation of the form considered in Problem 8, solve that equation in the form $y = \phi(x, c)$, and interchange x and y in the latter solution.

10. Solve Problem 9 directly, without interchanging x and y, by a procedure similar to that in Problem 8.

11. In comparing solutions of form $y = \phi(x, c)$ and $x = \psi(y, c)$ for Problems 9 and 10, which statement is nearest the truth?

(a) The first is much harder to get than the second.

(b) The second is much harder to get than the first.

(c) The difficulty is about the same.

The situation illustrated here is typical, and it is one of the reasons for introducing implicit solutions and for interchanging the roles of x and y.

─────────────────────ANSWERS─────────────────────

1. $x^2 + y^2 + xy - x/y = c$, $y \neq 0$ 2. $(xy)^4 - 4(x + y) = c$

3. (a) $x^2 y - xy = c$ (b) $2xy^2 - x^2 - y^2 = c$ (c) $2x^2 y - x^2 = c$

   (d) $xy^2 - x - y = c$ (e) $x^2 + 2y^3 - xy = c$ (f) $(xy)^3 - 3x^2 = c$

    To avoid fractions, some of the equations have been multiplied by a constant and c has been renamed; for example, $2c$ is an arbitrary constant if c is.

4. (a) $xy - x^2 - y^2 + x - y = c$ (b) $(x/y) + xy = c$, $y \neq 0$

   (c) $4xy - x^4 - y^4 = c$ (d) $(y/x) - e^x + e^{-y} = c$, $x \neq 0$

5. (a) $3(y/x) - x^3 - (xy)^3 = c$, $x \neq 0$ (b) $5(x/y) + (xy)^5 = c$, $y \neq 0$

7. $2\sqrt{2}$ 8. (a) $x^3 + cx^{-2}$, $x \neq 0$ (b) $x^2 + cx$ (c) $cx^2 - 1/2$

9. $x = 2y^4 + cy^{-3}$, $y \neq 0$

11. (a) The equation for y as function of x is of degree 7 and cannot be solved by radicals.

───

3. A survey of basic procedures

This section illustrates the use of integration to find the solution of differential equations. With that object in mind, we begin by giving a brief review of the concepts of *integration* and *solution*.

The word "integral" has two meanings. The *indefinite* integral

$$\int f(x) \, dx$$

denotes any function F such that $F' = f$. The set of all such functions is

$$\int f(x)\, dx + c, \qquad c = \text{const.}$$

On the other hand the *definite* integral

$$\int_a^b f(x)\, dx$$

is defined by a limiting process that has nothing to do with differential equations. It is true that

$$\frac{d}{dx} \int_a^x f(t)\, dt = f(x)$$

if f is continuous, but this is the fundamental theorem of calculus, not a definition.

For precision of notation the variable of integration should be distinguished from the upper limit, and that is why we used t instead of x in the integrand above. However it would be inconvenient to insist on this throughout our study of differential equations, and in simple cases in which there is no danger of confusion we agree that

$$\int_a^x f(x)\, dx = \int_a^x f(t)\, dt.$$

The symbol on the left is defined by the symbol on the right. Naturally, a similar convention holds with other letters, such as y, in place of x.

A *solution* of a differential equation

$$\frac{dy}{dx} = f(x, y)$$

is a function $y = \phi(x)$ that satisfies the equation on an interval I. To know the solution one must know both ϕ and I. In nonlinear problems of interest in applications, however, the interval of existence usually depends on the initial value $y(x_0)$ and is difficult to determine. Aside from the problem of establishing an interval of validity, it may be virtually impossible to express the solution in the explicit form $y = \phi(x)$.

Much of the difficulty disappears if we allow solutions $x = \psi(y)$ or $F(x, y) = c$. For this reason first-order equations are often restated in differential form

$$dy = f(x, y)\, dx.$$

Methods based on such a restatement do not give incorrect information, but they often give information in a more accessible form than would be the case if we insisted on the original formulation $y' = f(x, y)$. The latter requires

$y = \phi(x)$ by definition of the term "solution." The use of alternatives such as $x = \psi(y)$ or $F(x, y) = c$ is such a powerful aid in the study of nonlinear problems that no attempt to avoid it will be made in this text. One should be aware, however, that a reformulation of the differential equation is involved.

In this section and at many subsequent places in the text we make the convention that the letters

$$c,\ C,\ c_i,\ C_i, \qquad i = 0, 1, 2, 3, \cdots$$

if introduced without explanation, always denote constants.

EXAMPLE 1. Using indefinite integrals, solve the initial-value problem

$$\frac{dy}{dx} = \frac{1}{1 + x^2}, \qquad y(x_0) = y_0.$$

Integration and the initial condition give the two equations

$$y = \tan^{-1} x + c, \qquad y_0 = \tan^{-1} x_0 + c$$

respectively. The second equation gives c and the first then gives the solution,

(1) $$y = \tan^{-1} x - \tan^{-1} x_0 + y_0.$$

The integral curves form a family of parallel curves defined on $(-\infty, \infty)$. To get the graph of the curve through (x_0, y_0) all we have to do is translate the graph of

$$y = \tan^{-1} x$$

upward by the amount $y_0 - \tan^{-1} x_0$. (If this quantity is negative the "upward" translation is really downward.) See Figure 5(a).

EXAMPLE 2. Solve the preceding problem by definite integrals. Writing the equation in differential form

$$dy = \frac{1}{1 + x^2} dx$$

we note that y_0 corresponds to x_0 and y to x. Integrating between corresponding limits gives

(2) $$\int_{y_0}^{y} dy = \int_{x_0}^{x} \frac{dx}{1 + x^2} = \tan^{-1} x \Big|_{x_0}^{x} = \tan^{-1} x - \tan^{-1} x_0.$$

Since the left-hand side is $y - y_0$, this agrees with (1).

Both the above methods apply to the general equation

$$\frac{dy}{dx} = f(x), \qquad y(x_0) = y_0$$

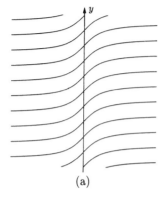

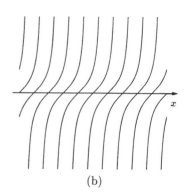

(a)   Figure 5   (b)

if f is continuous, but there is a subtle difference that is now explained. If no elementary function F such that $F' = f$ is at hand, the method of indefinite integration does not lead anywhere; it simply restates the question. But the method of definite integrals gives the explicit solution

$$y = y_0 + \int_{x_0}^x f(t)\,dt.$$

That $y(x_0) = y_0$ is obvious, and $y' = f(x)$ follows from the fundamental theorem of calculus. The same result would be obtained by writing $dy = f(x)\,dx$ and integrating between corresponding limits.

As an illustration, the solution of the initial-value problem on the left below is given by $y = S(x)$, where $S(x)$ is the integral on the right:

$$(3) \qquad \frac{dy}{dx} = \sin x^2, \qquad y(0) = 0; \qquad S(x) = \int_0^x \sin t^2\,dt.$$

There is no elementary function F such that $F'(x) = \sin x^2$, but the function $S(x)$ defined by (3) gives a perfectly good solution. It is called the *Fresnel sine integral* and is important in geometric optics.

An equation of form $g(y)\,dy = f(x)\,dx$ is said to be *separated*, and an equation that can be put into this form by reversible algebraic transformations is called a *separable equation*. A separated equation can be solved, in general, by writing

$$\int g(y)\,dy = \int f(x)\,dx + c$$

and carrying out the indicated integrations.

EXAMPLE 3. By separation of variables solve the initial-value problem

$$\frac{dy}{dx} = 1 + y^2, \qquad y(x_0) = y_0.$$

Separating variables and integrating, we get

$$\frac{dy}{1 + y^2} = dx, \qquad \tan^{-1} y = x + c, \qquad \tan^{-1} y_0 = x_0 + c$$

where the last equality follows from the initial condition. This last equality gives c and the second equation then gives the solution

$$y = \tan(x + c).$$

See Figure 5(b). The same result would be obtained by integrating between corresponding limits, thus:

(4)
$$\int_{y_0}^{y} \frac{dy}{1 + y^2} = \int_{x_0}^{x} dx.$$

Some separable equations arise so often that it is advisable to short-circuit the separation procedure by a suitable substitution. An equation of this kind is

$$y' = p(x)y$$

where p is continuous. Writing it in the form $y'/y = p$ we note that the left side is the derivative of $\ln y$. This suggests the substitution $u = \ln y$, or equivalently $y = e^u$. The result of substituting $y = e^u$ into the equation $y' = py$ is

$$e^u u' = pe^u, \qquad u' = p, \qquad u = \int p(x)\,dx + C.$$

Hence

$$y = e^u = e^{\int p(x)\,dx + C} = e^{\int p(x)\,dx} e^C = ce^{\int p(x)\,dx}$$

where $c = e^C$. Although this equation requires $c > 0$, actually the sign of c does not matter. For many purposes only a single solution is needed, as in the following example.

EXAMPLE 4. Setting $y = e^u$ obtain a positive solution of $y' = y \sin x$. The substitution gives, in succession,

$$u'e^u = e^u \sin x, \qquad u' = \sin x, \qquad u = -\cos x, \qquad y = e^{-\cos x}.$$

No additive constant is used in the equation $u = -\cos x$ because only a single solution (a particular solution) is asked for.

Many second-order differential equations can be reduced to first-order equations by an appropriate substitution. For example, an equation of form

$$y'' = f(x, y')$$

can be reduced to a first-order equation by the substitution $v = y'$. Since $y'' = (y')' = v' = dv/dx$ the first-order equation is

$$\frac{dv}{dx} = f(x, v).$$

When $v = dy/dx$ is found from this, an integration gives y.

An equation of form

$$y'' = g(y, y')$$

can also be reduced to a first-order equation, provided $v = y'$ can be expressed as a differentiable function of y. The basis of the method is the chain rule

$$y'' = \frac{dv}{dx} = \frac{dv}{dy}\frac{dy}{dx} = \frac{dv}{dy}v.$$

Using this in $y'' = g(y, y')$ we get the first-order equation

$$v\frac{dv}{dy} = g(y, v).$$

Once v is known, an integration gives y as before.

We have used the notation $v = y'$ because in many problems the independent variable is the time t, and $y' = dy/dt$ denotes the velocity. The notation v for velocity is standard in engineering and in physics. In mathematics, however, the notation $dy/dx = p$ is often used. The above transformations then become

$$\frac{d^2y}{dx^2} = \frac{dp}{dx} = \frac{dp}{dy}\frac{dy}{dx} = p\frac{dp}{dy}.$$

The expression $p\,dp/dy$ is more subtle than dp/dx, because y is thought to be a function of x in the differential equation and is treated as an independent variable in the chain rule. A mathematical justification is given in Chapter A. Here we show how the method is used.

EXAMPLE 5. When a second-order equation is reduced to an equation of first order, a solution of the latter is often called a *first integral* of the original differential equation. Obtain a first integral for the equation

$$2y'' = 3y^2.$$

Setting $p = y'$ and $y'' = p\,dp/dy$ we get, in succession,

$$2p\frac{dp}{dy} = 3y^2, \qquad 2p\,dp = 3y^2\,dy, \qquad p^2 = y^3 + c.$$

The latter equation gives the desired first integral. Since $p = dy/dx$ we have reduced the original second-order equation to a separable first-order equation,

$$\left(\frac{dy}{dx}\right)^2 = y^3 + c.$$

PROBLEMS

1P. Suppose a function $y = \phi(x)$ satisfies $dy = 4y \sin 2x \, dx$ and the initial condition $y(\pi) = e$. It is required to find $y(\pi/6)$, the value of y when $x = \pi/6$.

(a) Separate variables and integrate to get $\ln y = c - 2 \cos 2x$, $y > 0$. (Note that if you add a constant c_1 on the left and c_2 on the right when evaluating the indefinite integrals, the result is the above formula with $c = c_2 - c_1$, which is no more general.) By use of the initial condition show that $c = 3$ and then get $y(\pi/6)$.

(b) The initial condition $y(\pi) = e$ means that $x = \pi$ corresponds to $y = e$. Likewise, $x = \pi/6$ corresponds to $y = a$, where $a = y(\pi/6)$ is the unknown value that is to be found. Integrating between corresponding limits gives

$$\int_e^a \frac{dy}{y} = \int_\pi^{\frac{\pi}{6}} 4 \sin 2x \, dx$$

Evaluate the definite integrals and solve the resulting equation for a. Does this produce the same answer as found above?

(c) If $x \, dy + 3y \, dx = 0$ and $y(-\pi) = e$ you can't find $y(\pi)$. Why not?

2. Interpret in differential form and obtain implicit solutions $F(x, y) = c$ by separation of variables:

(a) $\dfrac{dy}{dx} = \dfrac{6(x^5 + x)}{5(y^4 + 1)}$ (b) $\dfrac{dy}{dx} = \dfrac{1 + \cos x}{1 + e^y}$ (c) $\dfrac{dy}{dx} = \dfrac{1}{x(4y^3 + 1)}$

3. Obtain the solution satisfying $y(0) = 1$ in explicit form $y = \phi(x)$ and give the interval or intervals of existence:

(a) $y' = y^2$ (b) $y' = \dfrac{1 + y}{1 + x}$ (c) $y' = 3(ye^x)^3$

4. Using the method of integration between limits, express the solution satisfying $y(0) = y_0$:

(a) $\dfrac{dy}{y} = \dfrac{dx}{1 + x^2}$ (b) $e^y \, dy = \dfrac{dx}{1 + x}$ (c) $(\tan y) \, dy = 2x \, dx$

5. Obtain a single positive solution by the substitution $y = e^u$:

(a) $y' = 8x^7 y$ (b) $y' = \dfrac{2xy}{x^2 + 1}$ (c) $y' = y \sec x$ (d) $y' = y \tan x$

6. Obtain a first integral by the substitution $y' = v$, $y'' = v'$:

(a) $y'y'' = x$ (b) $y'' = e^{y'}$ (c) $y'' = \dfrac{y'}{x}$ (d) $y'' = e^x (y')^2$

7. Obtain a first integral by the substitution $y' = p$, $y'' = p \, dp/dy$:

(a) $y'' = 2yy'$ (b) $y'' + y = 1$ (c) $y'' = e^y y'$ (d) $yy'' = 1$

8. Solve $y' = 1 + y^2$ by setting $y = \tan u$.

9. If x and y are interchanged, dy/dx becomes dx/dy and hence the equation solved in Example 1 becomes, after the interchange, the equation solved in Example 3. If you replace (x, y) by (y, x) and (x_0, y_0) by (y_0, x_0) in the solution of Example 1, do you get the solution of Example 3?

─────────────────────ANSWERS─────────────────────

1. (b) Yes, both give $y(\pi/6) = e^2$. (c) Because dx/x is not integrable on any interval containing the point $x = 0$, the solution $y = -e(\pi/x)^3$ satisfying $y(-\pi) = e$ cannot be continued beyond $x = 0$. The point $x = 0$ is called a *singularity* for the function $1/x$. In solving differential equations you may not in general integrate across a singularity.

2. (a) $y^5 + 5y - x^6 - 3x^2 = c$ (b) $y + e^y - x - \sin x = c$
   (c) $y^4 + y - \ln|x| = c$

3. (a) $y = 1/(1 - x)$ on $(-\infty, 1)$ or $(1, \infty)$ (b) $y = 1 + 2x$ on $(-\infty, \infty)$
   (c) $y = (3 - 2e^{3x})^{-1/2}$ on $(-\infty, a)$ where $a = (1/3)\ln(3/2)$

4. (a) $\ln y/y_0 = \tan^{-1} x$ (b) $e^y - e^{y_0} = \ln(1 + x)$ (c) $\sec y = (\sec y_0)e^{x^2}$

5. (a) $y = e^{x^8}$ (b) $y = x^2 + 1$ (c) $y = |\sec x + \tan x|$ (d) $y = |\sec x|$

6. (a) $v^2 - x^2 = c$ (b) $e^{-v} + x = c$ (c) $v = cx$ (d) $e^x + (1/v) = c$

7. (a) $p = 0$ or $p - y^2 = c$ (c) $p = 0$ or $p - e^y = c$
   (b) $p^2 + y^2 - 2y = c$ (d) $p^2 - \ln(y^2) = c$

8. It will be found that $u' = 1$, hence $u = x + c$ and $y = \tan(x + c)$.

9. Yes, and the method is valid in general.

───

4. The first-order linear equation

We shall give a systematic method of solving the linear equation

(1) $$y' + py = f$$

on a given interval I, where f and p are continuous. The method depends on construction of a nonvanishing function r such that

(2) $$r(y' + py) = (ry)'$$

(and hence such that the left side of (2) can be easily integrated). Multiplying out the left side and carrying out the differentiation on the right we see that (2) is equivalent to

$$ry' + rpy = ry' + r'y.$$

This holds if $rpy = r'y$ and hence if $r' = rp$. But the equation $r' = rp$ was solved in the preceding section, by setting $r = e^u$. The result was

(3) $$r = e^u \quad \text{where} \quad u = \int p(x)\,dx.$$

We do not add a constant because only one function r is needed here.

If (1) is multiplied by r in (3) the result is

$$r(y' + py) = rf \quad \text{or} \quad (ry)' = rf$$

where the second equation follows from $r' = rp$. An integration gives

(4) $$ry = \int r(x)f(x)\,dx + c, \qquad c = \text{const.}$$

Since the function $r = e^u$ in (3) does not vanish we can divide by r to get y.

EXAMPLE 1.   Solve the equation $y' + 2xy = f(x)$ where f is continuous. The solution is in four steps.

First step: Find $u = \int 2x\,dx = x^2$ and $r = e^u = e^{x^2}$.

Second step: Multiply the equation by r to get

$$e^{x^2}y' + e^{x^2}2xy = e^{x^2}f(x).$$

Third step: Verify that the above reduces, as it should, to

$$(e^{x^2}y)' = e^{x^2}f(x).$$

Fourth step: Integrate to get

$$e^{x^2}y = \int e^{x^2}f(x)\,dx + c.$$

Division by e^{x^2} now gives y.

EXAMPLE 2.   On an interval I where $x \neq 0$ solve

(5) $$xy' + 2y = 12x^4.$$

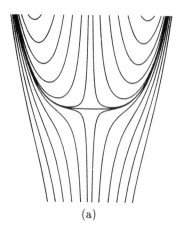

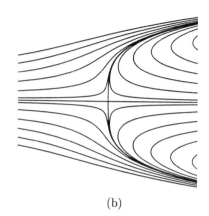

(a)   Figure 6   (b)

Since the coefficient of y' is 1 in the general theory, we divide by x to get

$$y' + \frac{2}{x}y = 12x^3.$$

The four steps of the solution are:

First step: $\int 2/x\,dx = 2\ln|x| = \ln x^2, \quad r = e^u = e^{\ln x^2} = x^2.$

Second step: $x^2 y' + 2xy = 12x^5.$

Third step: $(x^2 y)' = 12x^5.$

Fourth step: $x^2 y = 2x^6 + c, \quad y = 2x^4 + cx^{-2}.$

EXAMPLE 3. Solve the equation

(6) $y\,dx + 2x\,dy = 12y^4\,dy$

by interchanging x and y. The result of the interchange is

$$x\,dy + 2y\,dx = 12x^4\,dx.$$

Division by dx gives Equation (5), which was solved in the preceding example. Interchanging x and y in the solution found above gives a solution of (6),

(7) $x = 2y^4 + cy^{-2}, \qquad y \neq 0.$

The condition $y \neq 0$ comes from the condition $x \neq 0$ that was imposed in (5), and is clearly necessary for the formula (7) to be meaningful. Nevertheless, Equation (6) has the solutions $x = 2y^4$ and $y = 0$, both of which go through the origin. See Figure 6(a) for (5), Figure 6(b) for (6).

Since (6) is in differential form, a solution $x = \psi(y)$ is just as satisfactory as $y = \phi(x)$. An attempt to get the latter requires that we solve the sixth-degree equation

$$2y^6 - xy^2 + c = 0.$$

This is a cubic in y^2, hence it is solvable by radicals, but the resulting equation $y = \phi(x)$ is vastly more complicated than (7).

The preceding calculation illustrates a matter of personal taste with which not everyone will agree. If equation (6) is written in the form

$$y\frac{dx}{dy} + 2x = 12y^4$$

it is linear with x as the unknown function and y as the independent variable. Hence it can be solved, just as it is, by the usual method for linear equations.

For psychological reasons, the suggestion to interchange x and y is made from time to time in this text. It is only a suggestion, however, and you are not obliged to follow it.

If p and f are continuous functions of x and n is constant, the *Bernoulli equation*

$$y' + py = fy^n, \qquad y > 0$$

can be reduced to a linear equation by taking $y = u^m$ for a suitable constant m. This substitution gives $y' = mu^{m-1}u'$ by the chain rule and hence

$$mu^{m-1}u' + pu^m = fu^{mn}.$$

Dividing by u^{m-1} yields

$$mu' + pu = fu^{mn-m+1}.$$

If we choose m so that the exponent $mn - m + 1$ on the right is 0, we get a linear equation for u. The condition $m(n-1) + 1 = 0$ gives $m = 1/(1-n)$ and the final result is:

$$(8) \qquad mu' + pu = f, \qquad y = u^m, \qquad m = \frac{1}{1-n}.$$

The method breaks down if $n = 1$, but then the original equation is linear:

$$y' + (p - f)y = 0, \qquad n = 1.$$

The condition $y > 0$ ensures that the function $u = y^{1/m}$ is meaningful.

When faced with a specific Bernoulli equation, it is better to set $y = u^m$ and repeat the procedure than to memorize the result (8). The general technique of dealing with differential equations is of more value than the formula.

EXAMPLE 4.   Given that $y > 0$, reduce $y' + y = xy^3$ to a linear equation. The solution is in three steps:

First step: $y = u^m$: $mu^{m-1}u' + u^m = xu^{3m}$.

Second step: Divide by u^{m-1}: $mu' + u = xu^{2m+1}$.

Third step: $2m + 1 = 0$, $m = -\frac{1}{2}$: $-\frac{1}{2}u' + u = x$.

Multiplying the last equation by -2 we see that

$$y = u^{-1/2} \quad \text{where} \quad u' - 2u = -2x.$$

Although the method requires $y \neq 0$, the function $y = 0$ satisfies the differential equation. Note that $u > 0$ is needed to get a real y, but the basic equation $y^{-2} = u$ allows $y < 0$ as well as $y > 0$. Indeed, $-y$ satisfies the equation if y does. In general, the question whether $y > 0$ or $u > 0$ is required depends on m.

The following example illustrates a somewhat different procedure for the Bernoulli equation, which produces the correct value of m at once:

EXAMPLE 5.   If the equation $y' + y = xy^3$ of Example 4 is divided by y^3 the result is

$$y^{-3}y' + y^{-2} = x.$$

This suggests the substitution $u = y^{-2}$, which is the same as that in Example 4. Since

$$u' = -2y^{-3}y'$$

the term $y^{-3}y'$ can be expressed in terms of u' and the equation for u so obtained agrees with that found above.

PROBLEMS

1P. This problem is concerned with the two equations

$$y' + ay = f(x), \qquad y' + \frac{b}{x}y = f(x)$$

where a and b are constant, f is continuous, and in the second case $x > 0$. Both these equations were considered, with special choices f, in the foregoing pages.

(a) Taking $r = e^u$ where $u' = a$ or b/x, obtain the integrating factors

$$r = e^{ax}, \qquad r = x^b$$

(b) Verify that, when the equations are multiplied by r, they become

$$(e^{ax}y)' = e^{ax}f(x), \qquad (x^b y)' = x^b f(x)$$

(c) Integrate and solve for y to get

$$y = e^{-ax}\int e^{ax}f(x)\,dx + ce^{-ax}, \qquad y = x^{-b}\int x^b f(x)\,dx + cx^{-b}$$

2. In this and subsequent problems it is assumed that x or y is restricted to an interval on which the relevant functions are continuous. Solve:

(a) $y' + y\cos x = \cos x$ (b) $y' + y\tan x = 2\cos^2 x$ (c) $y' + y\sec x = \cos x$

3. Check that if the equation $(x^2 + 1)y' + 2xy = 3x^2$ is solved by the standard procedure for linear equations, you get the original equation back again. What is going on here? Solve the equation.

4. Find the solution satisfying $y(0) = -1$:

(a) $y' + xy = x$ (b) $y' + 2xy = e^{-x^2}$ (c) $(1 + x^3)y' + 3x^2 y = \sin x$

5. Show that the following equations become linear in y if x and y are interchanged. Thus get solutions $x = \psi(y)$ to the original equations:

(a) $(8e^y - 3x)dy = dx$ (b) $y\,dx + 2x\,dy = 6y^4\,dy$

If solutions $y = \phi(x)$ are wanted, it will be found that the first leads to a quartic equation for e^y and the second to a cubic equation for y^2.

6. Reduce to a linear equation by the substitution $y = u^m$ as in Examples 4 or 5 and check that your result agrees with the formula (8):

(a) $y' + y\cos x = xy^{4/5}$ (b) $y' + e^x y = y^{-4}\sin x$ (c) $y' + x^2 y = y^5$

7. Using (8) if you wish, reduce to a linear equation and solve completely:

(a) $x^2 y' + y^2 = xy,\; y(e) = e$ (b) $xy' + y = 2x^3 y^6,\; y(2) = (0.005)^{1/5}$

8. (a) Verify that the equation $dx + 3x\,dy = 9e^{2y}x^{2/3}\,dy$ becomes an equation of Bernoulli type if x and y are interchanged. Solve that equation and interchange x and y in the solution to get

$$x = e^{-3y}(e^{3y} + c)^3$$

(b) Find the solution satisfying $y(1) = 0$ in explicit form $y = \phi(x)$ for $x > 0$ and check by substitution into the original equation. Although we have not insisted upon it in this text, a solution obtained after a long sequence of calculations should be checked whenever the check is easy.

9P. Variation of parameters. If v is a positive solution of $v' + pv = 0$ we are going to solve $y' + py = f$ by the substitution $y = uv$. Show that this substitution yields $u' = f/v$. An integration gives v and then $y = uv$ is the solution sought.

10. (abc) Solve the equations of Problem 2 by variation of parameters.

REVIEW PROBLEMS——————————————————————————

11. In each equation $y' = dy/dx$, $y'' = d^2y/dx^2$:

(a) $y'' = (y')^2 + y$ (d) $yy'' = \sin y$ (g) $y'' + (y')^2 = 3$

(b) $y'' = (y')^2 + x$ (e) $xy'' = \sin y'$ (h) $y'' + (y')^2 = x^2$

(c) $y'' = (y' + 1)^2$ (f) $xy'' = \sin y$ (i) $y'' + y' \sin y' = 1$

Six of these can be reduced to first-order equations by the substitution $y' = v$, $y'' = dv/dx$. Five can be reduced to first-order equations by the substitution $y' = p$, $y'' = p\,dp/dy$. One can't be reduced by either substitution. Write the first-order equation if it can be found, or write "no go" if both substitutions fail.

12. Show that the solution of $2y'' = 3y^2$, $y(0) = 0$, $y'(0) = 1$ is given implicitly by

$$\int_0^y \frac{dt}{\sqrt{1 + t^3}} = x$$

The integral belongs to an important class of functions known as *elliptic integrals*. It is not expressible in terms of elementary functions.

13. Discuss the relative difficulty of solving $y'' = \sin x$, $y'' = \sin y$. No answer is provided.

14. Using the formula

$$\tan(A + B) = \frac{\tan A + \tan B}{1 - \tan A \tan B}$$

obtain the explicit solution on the right for the initial-value problem on the left:

$$y' = 1 + y^2, \quad y(x_0) = y_0 \qquad\qquad y = \frac{y_0 \cos(x - x_0) + \sin(x - x_0)}{\cos(x - x_0) - y_0 \sin(x - x_0)}$$

15. Show that the differential equation on the left has the family of solutions on the right:

$$(x^2 + 1)\frac{dy}{dx} + y^2 + 1 = 0 \qquad y = \frac{c - x}{1 + cx}$$

16. This problem pertains to the equation $y' + y = y^{a+1}$ where a is a nonzero constant. The equation has the stationary solution $y = 1$ and, if $a > -1$, the stationary solution $y = 0$. Setting these cases aside, consider the equation on an interval on which $y > 0$ and $y \neq 1$.

(a) By separating variables obtain the implicit solution

$$\int \frac{dy}{y(y^a - 1)} = x + c.$$

(b) Using Bernoulli's procedure with $n = a + 1$ obtain the explicit solution

$$y = (1 + ce^{ax})^{-1/a}$$

(c) Check both results by substituting the latter function into the integral. The calculation is simplified if you use $dy/y = d(\log y)$.

———————————ANSWERS———————————

2. (a) $y = 1 + ce^{-\sin x}$ (b) $y = \sin 2x + c\cos x$
   (c) $y = (c + x - \cos x)/(\sec x + \tan x)$

3. The left side is already in the form $(ry)'$. $y = (x^3 + c)/(x^2 + 1)$

4. (a) $y = 1 - 2e^{-x^2/2}$ (b) $y = (x - 1)e^{-x^2}$ (c) $y = (-\cos x)/(1 + x^3)$

5. (a) $x = 2e^y + ce^{-3y}$ (b) $x = y^4 + c/y^2$

6. (a) $y = u^5$ where $5u' + u\cos x = x$ (b) $y = u^{1/5}$ where $u' + 5e^x u = 5\sin x$
   (c) $y = u^{-1/4}$ where $u' - 4x^2 u = -4$.

7. (a) $y = x/\ln x$ (b) $y = (5x^3 + 5x^5)^{-1/5}$ 8. (b) $y = (1/6)\ln x$

11. (a) $p\,dp/dy = p^2 + y$ (e) $x\,dv/dx = \sin v$
    (b) $dv/dx = v^2 + x$ (f) no go
    (c) $p\,dp/dy = (p + 1)^2$ (g) $p\,dp/dy + p^2 = 3$ or $dv/dx + v^2 = 3$
    or $dv/dx = (v + 1)^2$ (h) $dv/dx + v^2 = x^2$
    (d) $yp\,dp/dy = \sin y$ (i) $p\,dp/dy + p\sin p = 1$ or $dv/dx + v\sin v = 1$

And our own elders, may God have mercy upon them, used to say, "The study of mathematics is for the mind like soap for the clothes, which washes away from them dirt and cleans the spots and stains."

—Ibn Khaldun

Chapter 3
APPLICATIONS

APPLICATIONS of differential equations can be organized by the type of differential equation or by the nature of the application. Both methods of organization are used here. We begin with some problems from biomathematics, military strategy, the theory of epidemics, and ecology, all of which lead to a first-order linear equation. Topics connected with geometry are discussed next. They include the spiral form of a snail shell, the relation between ray optics and wavefront optics in a parabolic mirror, and the equations for orthogonal trajectories. An introduction to envelopes sheds light on the behavior of solutions when uniqueness fails. Finally, we show that arc length, volume of revolution, and surface of revolution satisfy differential equations that lead immediately to the familiar integral formulas.

1. Linear equations

In a sense, most of the equations considered up to now are equivalent to linear equations, because they were reduced either to an integration problem of the form

$$(1) \qquad \frac{dy}{dx} = f(x), \qquad y = \int f(x)\,dx + c$$

or to a succession of such problems. The first-order linear equation itself was solved by two integrations and the Bernoulli equation was linearized by the substitution $y = u^m$. If x and y are interchanged, an equation

$$(2) \qquad \frac{dy}{dx} = g(y), \qquad g(y) \neq 0,$$

goes over into (1) with $f(x) = 1/g(x)$. The equations

$$(3) \qquad y'' = g(y'), \qquad y'' = g(y)$$

take the form (2) upon the substitution $v = y'$, with

$$y'' = \frac{dv}{dx} \quad \text{or} \quad y'' = v\,\frac{dv}{dy}$$

as the case may be.

The foregoing remarks summarize the highlights of Chapter 2 and include every type of equation that will be needed in Chapter 3. In this section we consider four problems that lead directly to linear equations without the special devices mentioned above.

(a) A principle of similitude

Interaction between a solid object and its environment generally takes place at the surface. For example, the rate at which a mothball evaporates is proportional to the area and so is the rate at which a fuel pellet burns. Some microscopic organisms obtain nutrients through a permeable membrane forming the boundary of the organism, and here again the rate of growth is more or less proportional to the surface area. If V is the volume of the object at time t and A its surface area, the differential equation governing processes of this kind is

$$(4) \qquad\qquad \frac{dV}{dt} = kA, \qquad k = \text{const.}$$

Suppose an object obeys the differential equation (4) and stays similar to itself as its size changes. Then, with no further assumptions, one can show that a graph of its length L versus time will necessarily be a straight line. To see why this is so, recall that for similar figures the area is proportional to the square and the volume to the cube of the linear dimensions. Thus,

$$A = aL^2, \qquad V = bL^3$$

where a and b are constant. Differentiating the second equation by the chain rule and substituting into (4) we get

$$3bL^2 \frac{dL}{dt} = kaL^2 \quad \text{hence} \quad \frac{dL}{dt} = \frac{ka}{3b}.$$

This shows that $L = mt + c$ where $m = ka/(3b)$. Thus

$$L = mt + c, \qquad A = a(mt + c)^2, \qquad V = b(mt + c)^3.$$

These represent (part of) a line, parabola, and cubic curve, respectively.

The only relevant feature of the shape of the object is given by the single constant a/b. For a sphere of diameter L or a cube of edge L the value of a/b is 6, so that for sphere-like or cube-like objects one has, approximately, $k = m/2$. The slope m is given by an experimental graph of L versus t and $m/2$ then gives the growth constant k.

(b) The square law of military strategy

Two opposing forces of x_0 soldiers and y_0 soldiers, respectively, have a military encounter in an open field without cover or concealment. This means that each soldier can see and fire upon the opposing soldiers without hindrance. The soldiers of a given force do not impede one another, so that if the size of the force is doubled its effectiveness is also doubled. If the size is tripled its effectiveness is tripled, and so on. The effectiveness of the x force is measured by the rate of decrease dy/dt of the y force and vice versa. If the soldiers of both forces have equal fighting skill, on the average, these considerations lead to the equations

$$\frac{dx}{dt} = -ky, \qquad \frac{dy}{dt} = -kx$$

where k is a constant measuring the soldiers' effectiveness.

Multiply the first of the above equations by x, the second by y, and subtract. The result is

$$x\frac{dx}{dt} - y\frac{dy}{dt} = 0 \quad \text{or} \quad \frac{d}{dt}(x^2 - y^2) = 0.$$

This gives

$$x^2 - y^2 = c$$

where c is constant.

To interpret the result, suppose the battle continues until the smaller force, say the y force, is annihilated. If the larger x force ends up with x_1 survivors and the smaller y force with $y_1 = 0$ survivors the equation

$$x_0^2 - y_0^2 = x_1^2 - y_1^2 = x_1^2$$

shows that $\sqrt{x_0^2 - y_0^2} = x_1$. Hence $\sqrt{x_0^2 - y_0^2}$ measures the numerical advantage of the superior force. This is one form of the *square law*, which was proposed by Lanchester in 1916.

If a superior force z_0 meets a force x_0 and then a second force y_0, the square law indicates that there are $\sqrt{z_0^2 - x_0^2}$ z-survivors after the first battle and $\sqrt{(z_0^2 - x_0^2) - y_0^2}$ after the second. Thus, for the encounter to be on equal terms $z_0^2 = x_0^2 + y_0^2$. This is another version of the square law.

The assumptions underlying the square law are tolerably well fulfilled in old-fashioned naval battles, in which there is no concealment and every ship can act, more or less, without hindrance from its neighbors. Application of the second form of the square law to a naval battle of historical importance is discussed in the problems.

(c) The spread of disease

Differential equations have been applied with success to the study of epidemics and contagious disease. The first important application of this kind was made by Daniel Bernoulli in 1760. Because of its historical significance and because it is carried out at a high level of sophistication, Bernoulli's line of thought will be presented in full.

The disease in question is smallpox, which for most of recorded history was a serious medical problem. The last known case was in 1978, and the conquest of smallpox is one of the major medical achievements of this century. The disease is extremely contagious, but confers virtually complete immunity on anyone who has caught it and recovered. It is this last characteristic that makes vaccination so effective, and finally made it possible to eradicate the disease. The fact that smallpox confers immunity is an essential part of Bernoulli's analysis.

Bernoulli starts with a population of people all born at time $t = 0$. It is assumed that at time t

$x(t)$ people are alive,

$y(t)$ are alive and have not yet had smallpox.

The latter group, which we refer to as the y-population, are still susceptible to the disease. Suppose susceptible persons contract smallpox at the rate a, so that the y-population decreases from this cause alone at the rate

$$\frac{dy}{dt} = -ay.$$

These people leave the y-population either by getting smallpox and recovering, or by getting it and not recovering. Denote the fraction of the latter by b where $0 < b < 1$. For example if $b = 1/4$ this means that on the average $1/4$ of those getting smallpox die from it. The population as a whole thus declines, from this cause alone, at the rate

$$\frac{dx}{dt} = -aby.$$

Finally, suppose the death rate from all other causes is a time-dependent function $d(t)$ which is the same for both the x- and y- populations. If only this cause were acting, without smallpox, we would have

$$\frac{dx}{dt} = -d(t)\,x, \qquad \frac{dy}{dt} = -d(t)\,y.$$

It is a well-accepted principle that rates add in cases such as this, so that

$$\frac{dx}{dt} = -aby - d(t)\,x, \qquad \frac{dy}{dt} = -ay - d(t)\,y.$$

These are Bernoulli's equations.

To eliminate the unknown function $d(t)$, multiply the first equation by y and the second by x and subtract. The result is

$$y\,\frac{dx}{dt} - x\,\frac{dy}{dt} = -aby^2 + axy.$$

The form of the left side suggests that $1/y^2$ is an integrating factor. If we multiply by $1/y^2$ the result is

$$\frac{d}{dt}\left(\frac{x}{y}\right) = -ab + a\frac{x}{y}.$$

Thus the ratio $z = x/y$ satisfies the linear equation

$$\frac{dz}{dt} = -ab + az, \qquad z(0) = 1.$$

The initial condition results from an assumption that just after birth no survivors have smallpox (it is generally fatal to infants) so that $x(t) = y(t)$ for small t. The steady-state solution is $z = b$. Hence

$$z = b + ce^{at}, \qquad z = b + (1-b)e^{at}$$

where the first result is obtained by adding the general solution of the homogeneous equation and the second is obtained from $z(0) = 1$.

Bernoulli estimated that $a = b = 1/8$ and hence that

$$z(20) = \frac{1}{8} + \frac{7}{8}e^{5/2} \approx 11.$$

The estimate indicates that about 9% of the population of twenty-year-olds would be at risk of smallpox. After studying mortality tables, Bernoulli recommended vaccination.

(d) Pollution

The volume of a lake is G cubic feet. An effluent that is wholly polluted runs into the lake at the rate of a cubic feet per minute and clean water runs in at the rate b cubic feet per minute. The lake drains at the rate $a + b$ cubic feet per minute so that the volume remains constant. (We assume that evaporation is, on the average, canceled by rain.) If the effluent distributes itself more or less uniformly before reaching the exit, it is required to find the concentration $P = E/G$, where $E = E(t)$ is the number of cubic feet of effluent in the lake at time t. Clearly E/G is a better measure of pollution than E itself; a cubic foot of polluted effluent is worse in the bathtub than in Lake Superior.

It is conceptually clearer to assume separate sources of water and pollutant, but the effect is the same if the incoming water itself is polluted. In any case $a/(a + b)$ measures the concentration of pollutant in the water entering the lake.

The first and most important step in the solution of this problem is to write the equation of continuity,

increase equals income minus outgo

for the amount of effluent E. In time Δt the increase is $a\Delta t$ from incoming effluent, and it remains to compute the amount lost due to the outward flow. The number of cubic feet leaving is $(a+b)\Delta t$, the concentration of effluent in the lake is E/G, and hence the $(a+b)\Delta t$ cubic feet leaving the lake contain approximately

$$\frac{E}{G}(a+b)\Delta t$$

cubic feet of effluent. (The result is only approximate because E is not constant.) By the equation of continuity,

$$(5) \qquad \Delta E \doteq a\Delta t - (a+b)\frac{E}{G}\Delta t.$$

Here $\doteq$ denotes *approximate equality* in a special sense, which is discussed in Part (e) below. Suffice it to say that a precise meaning can be given to this symbol in equations such as (5) and it is found, in general, that the error disappears completely when we pass from an equation involving $\doteq$ to a differential equation by taking limits.

In the present case dividing (5) by Δt and letting $\Delta t \to 0$ gives

$$\frac{dE}{dt} = a - (a+b)\frac{E}{G}$$

and with $E = PG$ this becomes

$$(6) \qquad G\frac{dP}{dt} = a - (a+b)P.$$

The equation is easily solved by an integrating factor, but it is more instructive to proceed as follows. Setting $P = $ const. gives the solution

$$P_\infty = \frac{a}{a+b}.$$

Add to this the general solution of the homogeneous equation to get

$$(7) \qquad P = P_\infty + ce^{-(a+b)t/G}$$

where the constant c depends on the initial condition. For any initial condition, that is, for any choice of c, Equation (7) gives

$$\lim_{t\to\infty} P(t) = P_\infty.$$

This agrees with the fact that in the long run the concentration of pollutant in the lake will be the same as the concentration $a/(a+b)$ in the sources feeding the lake. Borrowing from the terminology of electrical engineering, we say that P_∞ is the *steady-state solution* and the remaining term in (7) is a *transient* that depends on the initial conditions.

Figure 1   Figure 2

(e) Slipping of a belt

In general, an approximate equality such as $\Delta y \doteq f(x,y)\Delta x$ means

$$\Delta y = f(x,y)\Delta x + \epsilon \Delta x \quad \text{where} \quad \lim_{\Delta x \to 0} \epsilon = 0.$$

Here x is the independent variable in a given investigation and y the dependent variable. Division by Δx yields the equation on the left below, and letting $\Delta x \to 0$, we get the equation on the right:

$$\frac{\Delta y}{\Delta x} = f(x,y) + \epsilon, \qquad \frac{dy}{dx} = f(x,y).$$

What should be emphasized is that the differential equation follows as an exact mathematical consequence, even though the initial equality is only approximate. These ideas are now illustrated in connection with a problem of mechanics.

A flexible rope of negligible weight is wrapped around a rotating capstan as shown in Figure 1. A tension T_0 is exerted at one end of the rope in order to raise an anchor at the other end. The more turns around the capstan, the more the capstan helps you raise the anchor. Our problem is to find out how the tension at the anchor end depends on the angle of wrap. The same problem arises in connection with the slipping of a belt on a pully.

Figure 2 shows a small piece of rope with ends at positions specified by the angles θ and $\theta + \Delta\theta$. The statement that the rope is flexible means that the forces due to tension are directed tangentially as in the figure. If these forces have magnitudes T and $T + \Delta T$ the corresponding component of force in a direction normal to the small piece at its center is

(8)
$$\Delta N = (T + \Delta T)\sin\frac{\Delta\theta}{2} + T\sin\frac{\Delta\theta}{2}.$$

L'Hospital's rule

$$\lim_{h \to 0} \frac{\sin h}{h} = \lim_{h \to 0} \frac{\cos h}{1} = 1$$

shows that $\sin h$ is well approximated by h when h is close to 0. We use this with $h = \Delta\theta/2$. We also assume that the tension T is a continuous function of θ, so that $\Delta T \to 0$ as $\Delta\theta \to 0$. These two assumptions in (8) yield the approximate equality

$$\Delta N \doteq 2T \sin \frac{\Delta\theta}{2} \doteq T\Delta\theta, \qquad \Delta\theta \to 0.$$

If the short piece of rope were straight, the frictional force on it would be directed tangentially and would have magnitude $\mu\Delta N$, where μ is the coefficient of static or sliding friction as the case may be. It is an experimental fact that a similar result holds for the curved piece as an approximate equality, namely,

$$\Delta T \doteq \mu\Delta N, \qquad \Delta\theta \to 0.$$

These results yield

$$\Delta T \doteq \mu T\Delta\theta.$$

Dividing by $\Delta\theta$ and letting $\Delta\theta \to 0$ we are led to the differential equation on the left below, with the solution on the right:

$$\frac{dT}{d\theta} = \mu T, \qquad T = T_0 e^{\mu\theta}.$$

It is remarkable that the result depends only on the angle of wrap, θ, and is independent of the radius of the capstan.

We now give a precise meaning to the symbol $\doteq$ used in this derivation, and we show that the differential equation really follows from the physical assumptions. Namely, the two equations

$$\Delta N \doteq T\Delta\theta, \qquad \Delta T \doteq \mu\Delta N$$

are defined by

$$\Delta N = T\Delta\theta + \epsilon_1\Delta\theta, \qquad \Delta T = \mu\Delta N + \epsilon_2\Delta\theta$$

where $\epsilon_1 \to 0$ and $\epsilon_2 \to 0$ as $\Delta\theta \to 0$. The first of these equations is a consequence of the assumed continuity of T, the second is a precise formulation of our assumption regarding the behavior of frictional forces. It follows from these two relations that

$$(9) \qquad\qquad \Delta T = \mu T\Delta\theta + \epsilon\Delta\theta$$

where $\epsilon \to 0$ as $\Delta\theta \to 0$. Dividing by $\Delta\theta$ and letting $\Delta\theta \to 0$ we get the same differential equation as before. Equation (9) shows how the approximate equality $\Delta T \doteq \mu T\Delta\theta$ should be understood and also shows why this approximate equality yields the differential equation exactly.

As illustrated in this discussion, approximate equalities can often be given a precise interpretation that converts seemingly doubtful derivations into convincing demonstrations. Since the matter has now been brought to your attention, it will not be emphasized again.

PROBLEMS

1. According to the square law, if a total force c is divided into two parts a and b, so that $a + b = c$, the effective strength when the parts are used in two separate engagements is $\sqrt{a^2 + b^2}$. Show that this is least when the two parts are equal, and that the effective strength is then $c/\sqrt{2}$.

2. The Battle of Trafalgar in 1805 started an era of British supremacy on the high seas that lasted more than a hundred years and had a profound influence on the science of naval strategy. Nelson's plan of action, written before the battle, assumed 46 ships of the combined French-Spanish fleet against 40 of his own. (The numbers in the actual engagement were somewhat different, but the important thing here is Nelson's intent.) He planned to cut the other fleet exactly in half by use of 16 of his ships, and to attack the two groups of 23 one at a time by means of these 16 ships combined with another 16. The remaining eight ships were to prevent the two groups of 23 from uniting. Note that by Problem 1 a division into equal parts, rather than some other ratio, is optimum.

(a) Check that the 32 ships that Nelson planned to use successively against the two groups of 23 is, within about a half a ship, the correct number to make this part of the battle take place on equal terms.

(b) Check that if Nelson had used all 40 ships against the enemy's 46 ships in a single battle, as was naval practice up to that time, he would have had an effective disadvantage of about 23 ships.

(c) Show, however, that with his plan of $32 + 8$ against $23 + 23$, he had an effective advantage of about $5\frac{1}{2}$ ships.

3. In the pollution problem of the text suppose P has reached its steady-state level P_∞ by time $t = 0$, at which time the source of pollution is removed; in other words, a is replaced by 0. Show that P decays according to the law

$$P = P_\infty e^{-(b/G)t}$$

and hence the time for P to decay to half its initial value is

$$T = (\ln 2)\frac{\text{capacity of lake}}{\text{rate of flow}}$$

These quantities can be measured in any system of compatible units. Note that this problem is mathematically the same as the problem of half-life for radioactive decay we considered in Chapter 1. According to data taken by Rainey in 1967, the ratio multiplying $\ln 2$ is about 185 years for Lake Superior and about 2.6 years for Lake Erie.

4. Suppose a tank contains G gallons of water and that brine containing a pounds of salt per gallon flows into the tank and out again at a constant rate of b gallons per minute, starting at time $t = 0$. The mixture is kept uniform by stirring. Show that the amount $S(t)$ of salt in the tank at time $t \geq 0$ satisfies the differential equation on the left, hence is given by the formula on the right:

$$\frac{dS}{dt} = ab - \frac{b}{G}S, \qquad S(t) = aG(1 - e^{-bt/G})$$

Hint: In the time interval from t to $t + \Delta t$ the number of gallons leaving is $b\Delta t$, the concentration of the mixture in pounds per gallon is S/G, hence the number of pounds leaving is approximately $(S/G)b\Delta t$.

5. In the preceding problem suppose a block of salt in the tank dissolves at the constant rate k pounds per minute. Show that this produces an extra term k on the right side of the differential equation, and hence the effect is the same as if the concentration of the incoming brine were increased from a pounds of salt per gallon to $a + k/b$ pounds of salt per gallon. Do this result and the differential equation of Problem 4 remain valid if a, b, k are continuous functions of t, rather than constants?

6. A sailor raising a 2000 pound anchor with the aid of a capstan finds that a tension of 100 pounds is just enough to keep the rope from slipping. Suppose the rope starts to slip, so that the coefficient of friction changes from its former static-friction value of 0.4 to a value of 0.2 for sliding friction. Show that the sailor must now exert $200\sqrt{5}$ pounds of tension to stop the slipping.

7. A radioactive substance A decomposes into a new substance B, which in turn decomposes into a third substance C. Assuming that an amount A_0 of A is present initially, and no B, show that the amount of B at time t is given, in general, by

$$B(t) = aA_0 \frac{e^{-at} - e^{-bt}}{b - a}$$

where a and b are constants giving decay rates of A and B, respectively.

Hint: The rate at which B increases is equal to the rate at which B is formed from A minus the rate at which B decomposes. Thus, using the same letters for the substances and for their amounts,

$$\frac{dB}{dt} = -\frac{dA}{dt} - bB$$

Since $A = A_0 e^{-at}$ you get a linear equation for B. The result for $b = a$ can be obtained by taking the limit as $b \to a$, or also by direct solution, but the probability that $b = a$ in any realistic case is 0.

8. A capacitor of capacity C is discharged through a resistance R as shown in Figure 3. If Q is charge on the capacitor at time t, v is the voltage across the capacitor, and i is the current in the circuit, show that

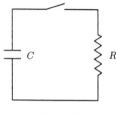

$$\frac{dQ}{dt} = -kQ, \qquad \frac{dv}{dt} = -kv, \qquad \frac{di}{dt} = -ki, \qquad k = \frac{1}{RC}$$

Figure 3

Outline of solution: We have $i = dQ/dt$ by the definition of current and $Q = Cv$ by the definition of capacity. The voltage across the resistor is iR by Ohm's law, and $iR + Q/C = 0$ since the voltage drop around the circuit must be 0.

9. The voltage across an inductor is $L\,di/dt$ where L is the inductance, and hence the current i in the RL circuit of Figure 4 satisfies

$$L\frac{di}{dt} + Ri = v_0(t), \qquad i(0) = 0$$

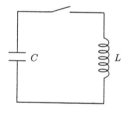

Solve in integral form if $v_0(t)$ is continuous and specialize to the case $v_0 = \text{const}$. The interest of this problem is that the constants R, L, v_0 can be so chosen that the equation exactly duplicates the equations

Figure 4

$$G\frac{dP}{dt} + (a+b)P = a, \qquad G\frac{dS}{dt} = abG - bS$$

obtained in the text and in Problem 4. Before the advent of digital computers, electrical circuits were often used as a means of analog computation.

10. We consider the equation $dy/dt = f - py$ for $t \geq 0$ where p and f are continuous functions of t and where y gives the number of individuals in a certain population at time t. Here f measures an external influence that promotes or retards growth while py represents another retarding or promoting influence proportional to p. If y_1 and y_2 are any two solutions show that $y_1(t) - y_2(t) = ce^{-P(t)}$ where P is an integral of p. Thus deduce that two distinct solutions y_1 and y_2 satisfy the condition on the left if, and only if, p satisfies the condition on the right:

$$\lim_{t \to \infty} (y_1(t) - y_2(t)) = 0 \qquad \lim_{t \to \infty} \int_0^t p(t_1)\, dt_1 = \infty$$

Also show that any three distinct solutions satisfy the identity

$$\frac{y_1(t) - y_2(t)}{y_1(t) - y_3(t)} = \text{const.}$$

────────────────ANSWERS────────────────

5. Yes.   9. $i = (1/L)e^{-Rt/L} \int_0^t e^{Rs/L} v_0(s)\, ds, \quad i = (v_0/R)(1 - e^{-Rt/L})$

2. Problems involving geometry

When differential equations are used in scientific investigations, the main ideas sometimes depend on geometry. This is to be expected, since a derivative can be interpreted geometrically as the slope of a curve. Several problems with a geometric flavor are discussed here.

(a) The logarithmic spiral

Suppose a curve $r = f(\theta)$ in polar coordinates cuts the radius at a constant angle ψ as shown in Figure 5. We want to get the equation of the curve. If $\psi = 0$ the

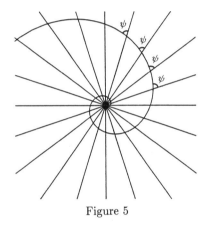

Figure 5

curve is a ray extending from the origin to ∞ and cannot be represented in the form $r = f(\theta)$. If $\psi = \pi/2$ or $-\pi/2$ the curve is a circle centered at the origin. Setting these cases aside, assume

(1) $-\pi/2 < \psi < \pi/2, \quad \psi \neq 0.$

Since ψ is constant, $\tan \psi$ is also constant. We write

(2) $\tan \psi = \dfrac{1}{k}$

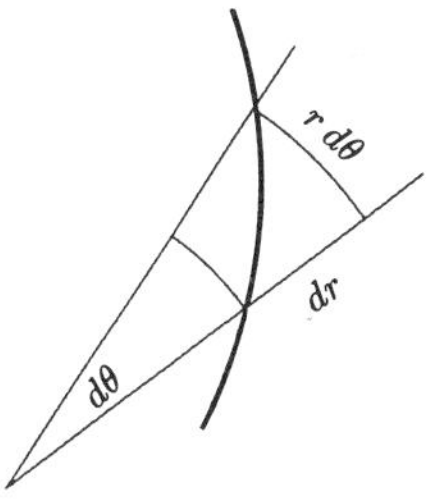

Figure 6

where the choice $1/k$ rather than k is made to show the connection with the law of exponential growth discussed in Chapter 1.

The differential triangle in Figure 6 suggests that

$$\tan \psi = \frac{r\, d\theta}{dr}$$

and by elementary calculus this formula is correct whenever the right-hand side is meaningful. Using (2) we get, in succession,

$$\frac{r\,d\theta}{dr} = \frac{1}{k}, \qquad \frac{dr}{d\theta} = kr, \qquad r = ce^{k\theta}.$$

A curve of this kind is called a *logarithmic spiral* because, apart from constants, $\theta = \ln r$. The steps are reversible and hence a curve $r = f(\theta)$ cuts the radius at a constant angle ψ satisfying (1) if, and only if, it is a logarithmic spiral. Since k is arbitrary, this gives a geometric interpretation of all processes obeying the exponential law of growth.

A logarithmic spiral looks like a snail shell, and this is not just a coincidence. A snail shell has the characteristic that it grows only at one end; the part of the shell already laid down does not change. The growth is specified by its rate in two perpendicular directions, which we take as radial and transverse. Let us suppose that both rates are proportional to the size at time t. The precise meaning of "size" will not matter, and we use the weight W as a quantity that is easily measured. If arc length in the radial direction is s_1 and in the transverse direction is s_2, our hypothesis is

(3) $$\frac{ds_1}{dt} = k_1 W, \qquad \frac{ds_2}{dt} = k_2 W, \qquad k_1, k_2 \text{ const.}$$

By a differential triangle, or also by a familiar formula of calculus, arc length in polar coordinates satisfies

$$ds^2 = dr^2 + (r\,d\theta)^2.$$

Setting $d\theta = 0$ we get $ds_1 = dr$ and setting $dr = 0$ we get $ds_2 = rd\theta$. Thus if the second equation (3) is divided by the first, the result is

$$\frac{r\,d\theta}{dr} = \frac{k_2}{k_1}.$$

Figure 7

This is the same equation as was considered in the foregoing discussion, with $k = k_1/k_2$. Hence the curve is a logarithmic spiral. Figure 7 shows a photograph of a cross-section of a nautilus shell with a best-fitting logarithmic spiral drawn beside it.

(b) The parabolic mirror

It is required to find a mirror such that light from a point source at the origin is reflected in a beam parallel to the x axis. The mirror is a surface of revolution whose profile is shown in Figure 8. The ray of light OP strikes the mirror at P and is reflected along PR. If PQ is the tangent at P and $\theta, \phi, \alpha, \beta$ are the angles indicated in the figure, then $\beta = \phi$ because vertical angles are equal and $\beta = \alpha$ by the optical law of reflection. Hence $\alpha = \phi$. Since θ and $\alpha + \phi = 2\phi$ are angles formed by parallel lines meeting a common transversal, we have $\theta = 2\phi$. This equation gives the key to our problem, and must be kept well in mind.

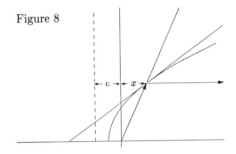

Figure 8

As is usually done in geometrical arguments, we have taken ϕ to be an *unsigned* angle. The *signed* angle ψ from the radius to the curve is measured counterclockwise and hence is negative. Thus $\psi = -\phi$. Using the formula

$$\tan \psi = \frac{r\, d\theta}{dr}$$

with $\psi = -\phi$, $\theta = 2\phi$, and $r > 0$, we get

$$-\tan \phi = \frac{2r\, d\phi}{dr}, \qquad 2 \cot \phi\, d\phi = -\frac{dr}{r}.$$

The second form is obtained by separation of variables. If $0 < \phi < \pi/2$ and $r > 0$ as in the figure, integration gives

$$2 \ln \sin \phi = -\ln r + C.$$

The left side is $\ln \sin^2 \phi$ so that, taking exponentials,

$$\sin^2 \phi = \frac{c_1}{r}, \qquad c_1 = e^C.$$

By a trigonometric identity $2 \sin^2 \phi = 1 - \cos 2\phi = 1 - \cos \theta$. Hence if the above equation is multiplied by 2 the result is

(4)
$$1 - \cos \theta = \frac{c}{r}, \qquad r = \frac{c}{1 - \cos \theta}$$

where $c = 2c_1$. This is the equation of a parabola with focus at the origin and with axis parallel to the x axis. Although we assumed $\theta > 0$ in the derivation, the final result is valid for $|\theta| < \pi/2$, $\theta \neq 0$, and gives the whole parabola. As might be expected, the lower part of the curve, not shown in the figure, is the mirror image of the upper part in the x axis.

The fact that (4) represents a parabola is known from calculus but can be verified directly as follows: Multiplying the left-hand equation (4) by r gives

$$(5) \qquad\qquad\qquad r - x = c$$

since $x = r \cos \theta$. This shows that $r = d$, where $d = x + c$ is the distance from the point (x, y) on the curve to the vertical line $x = -c$. The latter line is the directrix of the parabola and the point $r = 0$ is the focus. Since r is the distance to the focus, we have shown that

distance to focus equals distance to directrix.

Hence the curve is a parabola by definition.

The actual parabolic mirror is obtained by revolving the curve of Figure 8 about the x axis, and we take this point of view from now on. It turns out that the equation $r = x + c$ has an interesting bearing on the relation between *ray optics* and *wave-front optics*. For a point source, the rays are radial lines emanating from the point, and the wave fronts are spherical surfaces centered at the point. The main requirement is that the optical length of the paths leading from the point source to the wave front must be the same on all rays, so the waves arrive in phase. In a uniform medium, which is assumed here, the optical length need not be distinguished from the geometric length.

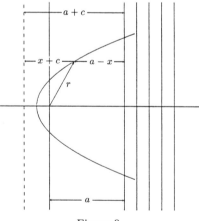

Figure 9

To apply these ideas to the parabolic mirror, let $x = a$ be a plane perpendicular to the x axis and to the right of the origin as shown in Figure 9. The path length from the focus to this plane is

$$r + (a - x) = (x + c) + (a - x) = a + c$$

which is constant. Hence, the plane $x = a$ is a wave front. The parabolic mirror has the effect of converting the *spherical waves* from the point source into *plane waves* parallel to one another. This is the physical significance of the condition $r = x + c$.

(c) Orthogonal trajectories, Cartesian

If two families of curves are such that every curve of one family intersects the curves of the other family at right angles, it is said that the two families are *orthogonal trajectories* of each other. For example, the coordinate lines

$$x = c_1, \qquad y = c_2$$

in a Cartesian coordinate system form a set of orthogonal trajectories, and so do the circles and radial lines

$$r = c_1, \qquad \theta = c_2$$

of a polar coordinate system. In geometric optics the wave fronts are the orthogonal trajectories of the rays, and when heat flows in a uniform plane sheet the lines of flow are orthogonal trajectories of the curves of constant temperature. Further examples of orthogonal trajectories are found in electrostatics, in electromagnetic theory, and in the theory of flow of ideal fluids.

Suppose a given curve in the (x, y) plane is such that the tangent at a point (x, y) on it makes an angle ϕ with the x axis. Then the orthogonal trajectory through the same point (x, y) makes an angle $\phi + \pi/2$ with the x axis. Since

$$\tan(\phi + \pi/2) = -\cot\phi = -\frac{1}{\tan\phi}$$

and since the slope of the curve is $dy/dx = \tan\phi$, we see that we should

$$\text{replace} \quad \frac{dy}{dx} \quad \text{by} \quad -\frac{dx}{dy}$$

in the differential equation for the original family to get the differential equation for the orthogonal trajectories. The process is illustrated in the following example.

EXAMPLE 1. The family of curves

$$(6) \qquad x^2 + y^2 = cx$$

represents circles tangent to the y axis as shown in Figure 10. To get the orthogonal trajectories, first differentiate and eliminate c. The result is

$$2x + 2y\frac{dy}{dx} = c$$

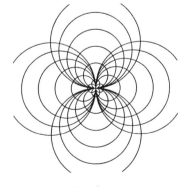

Figure 10

and substituting this value of c into the original equation gives

$$x^2 + y^2 = 2x^2 + 2xy \frac{dy}{dx}$$

which simplifies to

$$y^2 - x^2 = 2xy \frac{dy}{dx}.$$

Replace dy/dx by $-dx/dy$ to get the equation of the orthogonal trajectories,

$$y^2 - x^2 = -2xy \frac{dx}{dy}.$$

Writing in differential form and putting all terms on the left side gives

$$2xy\,dx - x^2 dy + y^2 dy = 0.$$

This equation has the integrating factor $1/y^2$. Multiplying by $1/y^2$ gives

$$d\left(\frac{x^2}{y}\right) + dy = 0, \quad \text{hence} \quad \frac{x^2}{y} + y = c.$$

The result is

$$x^2 + y^2 = cy$$

which represents a family of circles tangent to the x axis as shown in Figure 10. The following comments give useful insight:

Although the analytical steps require $x \neq 0$, $y \neq 0$, and $y^2 \neq x^2$, the final result is valid without these restrictions.

The quantity c defined by $x^2 + y^2 = cx$ is constant on the original curves of the family, but not constant on the orthogonal trajectories. That is why c must be eliminated in the first step.

Since a horizontal or vertical line that intersects a circle generally intersects it in two points, neither family can be represented in the form $y = \phi(x)$ or $x = \psi(y)$. In the study of orthogonal trajectories it is advisable to reinterpret the differential equation by use of differentials, and this was done in the calculation above.

(d) Orthogonal trajectories, polar

If a curve in polar coordinates is such that its tangent makes an angle ψ with the radius vector at a point (r, θ) of the curve, then the orthogonal trajectory through the same point (r, θ) makes an angle $\psi + \pi/2$ with the radius vector. Since

$$\tan\left(\psi + \pi/2\right) = -\cot\psi = -\frac{1}{\tan\psi}$$

and since $\tan\psi = r\,d\theta/dr$, one should

$$\text{replace} \quad \frac{r\,d\theta}{dr} \quad \text{by} \quad -\frac{dr}{r\,d\theta}$$

in the differential equation for the original family to get the differential equation for the orthogonal trajectories. We illustrate by doing Example 1 in polar coordinates.

EXAMPLE 2. If the equation $r = c\cos\theta$ is multiplied by r the result is $r^2 = cx$, which is the same as (6). Clearly

$$\frac{r\,d\theta}{dr} = \frac{c\cos\theta\,d\theta}{-c\sin\theta\,d\theta} = -\cot\theta$$

and hence the equation for the orthogonal trajectories is

$$-\frac{dr}{r\,d\theta} = -\cot\theta.$$

Separating variables and integrating give

$$\frac{dr}{r} = \cot\theta\,d\theta, \qquad \ln r = \ln\sin\theta + C.$$

Taking exponentials we get $r = c\sin\theta$ which is a family of circles tangent to the x axis. Although the analysis assumes $r > 0$, $\sin\theta > 0$, and $c = e^C > 0$, none of these conditions are needed in the final result.

(e) Envelopes

An *envelope* of a family of curves is a curve C such that every member of the family is tangent to C and such that C is tangent, at each of its points, to some member of the family. See Figure 11. At a point (x, y) on C, the values x, y, y' for C are the same as for the curve of the family that is tangent to C. This shows that, if the family satisfies a first-order differential equation

(7) $$f(x, y, y') = 0,$$

then C satisfies the same equation. Thus an envelope of a family of solutions is again a solution. It is sometimes called a *singular solution*.

Let a family of curves satisfying (7) be defined implicitly by

$$F(x, y, c) = 0$$

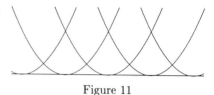

Figure 11

and suppose the family has an envelope. In many cases, an equation for the envelope can be obtained by eliminating c from the equations

(8) $F(x, y, c) = 0$ and $F_c(x, y, c) = 0$

or also by eliminating p from the equations

(9) $f(x, y, p) = 0$ and $f_p(x, y, p) = 0$.

Here the subscript denotes partial differentiation,

$$F_c = \frac{\partial F}{\partial c}, \qquad f_p = \frac{\partial f}{\partial p}.$$

Eliminating c from (8) or p from (9) yields the c *discriminant* or the p *discriminant*, respectively. The underlying theory is discussed in Chapter 14, Section 5. Here we show how (8) and (9) are used.

EXAMPLE 3. If $(y')^2 = 4y$ separation of variables gives

$$y = (x - c)^2$$

which is the family of parabolas shown in Figure 11. Here

$$F(x, y, c) = y - (x - c)^2, \qquad f(x, y, p) = p^2 - 4y$$

and the equations $F = 0$, $F_c = 0$ and $f = 0$, $f_p = 0$ are respectively

$$y = (x - c)^2, \; 2(x - c) = 0; \qquad p^2 = 4y, \; 2p = 0.$$

Either pair of equations yields the envelope $y = 0$ shown in the figure. Since $y = 0$ satisfies $(y')^2 = 4y$, it is a singular solution.

One might think that the complete solution could be obtained by adjoining the singular solution $y = 0$ to the family $y = (x - c)^2$. But this idea is entirely mistaken. The complete solution contains the two-parameter family defined by

(10) $y = (x-c_1)^2, \quad y = 0, \quad y = (x-c_2)^2$

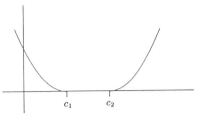

Figure 12

for $x < c_1$, $c_1 \leq x \leq c_2$, $c_2 < x$ respectively, where c_1 and c_2 are arbitrary constants with $c_1 \leq c_2$. A typical curve of this family is shown in Figure 12.

PROBLEMS

1. A family of curves is given by $y = c \cos x$. Compute y', eliminate c by division, and thus verify that the differential equations for the family and for the orthogonal trajectories are respectively

$$\frac{dy}{dx} = -y \tan x, \qquad \frac{dx}{dy} = y \tan x$$

Solve the latter equation by separating variables. Although the analysis assumes $x \neq n\pi/2$ for integral n, these values do no harm. See Figure 13.

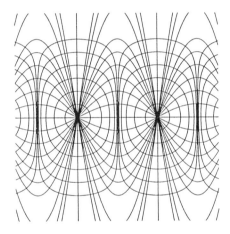

Figure 13

2. Find the orthogonal trajectories:

    (a) $x^2 + y^2 = c$ (b) $xy = c$ (c) $y^2 = x + c$ (d) $y = ce^{x^2}$

3. Check that the orthogonal trajectories for $y = ce^x$ and for $y = \ln(cx)$ are given respectively by $2x + y^2 = c$ and $2y + x^2 = c$. Note that one of these families goes into the other when x and y are interchanged, and explain how this property could have been forseen from the original equations.

4. Verify that the differential equation for orthogonal trajectories of the family of curves $x^2 = 2cy + c^2$ is the same as for the family itself. Such a family is said to be self-orthogonal.

5. Let F_1 be the family of curves whose tangents go through the origin and let F_2 be the family of curves whose normals go through the origin. It is obvious that these families must be orthogonal trajectories of each other. Why is it obvious? Find the families.

6. If the differential equations for a family of curves and for the orthogonal trajectories are written in differential form, it may happen that the one equation goes into the other when x and y are interchanged. Show that this happens for each of the following families, and thus get the orthogonal trajectories by inspection:

$$\text{(a) } x^3 - 3xy^2 = c \qquad \text{(b) } y = x + 1 + ce^x$$

7. If k is a nonzero constant, the polar equation $r = ce^{k\theta}$ represents a family of logarithmic spirals. Here k specifies the family, c specifies the curve.

(a) Verify that a differential equation for the family is given by the first equation below and hence the orthogonal trajectories satisfy the second and third equations:

$$\frac{r\,d\theta}{dr} = \frac{1}{k}, \qquad -\frac{dr}{r\,d\theta} = \frac{1}{k}, \qquad \frac{r\,d\theta}{dr} = -k$$

Comparing the first and third equations shows that the orthogonal trajectories are again logarithmic spirals, with k replaced by $-1/k$.

(b) What happens to the configuration of spirals when $k \to 0$ or $k \to \infty$?

(c) The fact that the orthogonal trajectories are logarithmic spirals follows from Part (a) of this section, without the theory of Part (d). Why does it follow?

8. Show that each pair of equations in polar coordinates describes a set of orthogonal trajectories:

$$\text{(a) } \begin{cases} r = c\sec\theta \\ r = c\csc\theta \end{cases} \qquad \text{(b) } \begin{cases} r^2 = c\sin 2\theta \\ r^2 = c\cos 2\theta \end{cases} \qquad \text{(c) } \begin{cases} r = c(1 + \cos\theta) \\ r = c(1 - \cos\theta) \end{cases}$$

Hint: Problem 8(a) is trivial if expressed in Cartesian coordinates.

9. In this problem and the next two, k is a fixed parameter that specifies the family and c varies from one curve to the next in a given family. (a) Show that the orthogonal trajectories of $|y| = c|x|^k$ are ellipses for $k > 0$ and hyperbolas for $k < 0$, with the equation $x^2 + ky^2 = c$. (b) Sketch the original family and the orthogonal trajectories for $k = 2, 1, 0, -2$.

Hint: Assume $x > 0$, $y > 0$ and drop the absolute values. The extension to other quadrants is made by symmetry.

10. If $|k| \neq 1$ show that the following families are othogonal trajectories of each other:

$$xx^k + yy^k = c, \qquad xx^{-k} - yy^{-k} = c \qquad (x > 0,\ y > 0)$$

When k is an integer, the side condition can be replaced by $x \neq 0$, $y \neq 0$.

11. Find the orthogonal trajectories of $x^k + cy^k = 1$ assuming when necessary that $x > 0$, $y > 0$ or both. What happens if $k = 0$? $k = 1$? $k = 2$?

12. This problem is harder than those above. If a and b are constant and λ is a parameter, show that the family of curves

$$\frac{x^2}{a^2 + \lambda} + \frac{y^2}{b^2 + \lambda} = 1$$

satisfies an equation, free of λ, that is unaltered when y' is replaced by $-1/y'$. This means that the family is self-orthogonal.

Suggestion: Differentiate and use the result to eliminate $x^2/(a^2 + \lambda)$ from the original equation. After clearing of fractions you will have

$$x(b^2 + \lambda) + yy'(a^2 + \lambda) = 0, \qquad -xyy' + y^2 = b^2 + \lambda$$

13. In Example 3 consider, besides the family (10) with $\infty < c_1 \leq c_2 < \infty$, the corresponding families with $c_1 = -\infty$ and with $c_2 = \infty$.

(a) By a sketch check that each of the above families is a general solution in the region defined by $y \geq 0$. That is, if $y_0 \geq 0$ there is at least one curve of the family through the point (x_0, y_0).

(b) If $y_0 \geq 0$, how many solutions are there through the point (x_0, y_0) in Example 3? How many if $y_0 > 0$ and the equation is considered in the region $y > 0$ rather than $y \geq 0$?

14. (a) Obtain the following algebraic and differential equations for the family of circles with radius 1 and centers on the x axis:

$$(x - c)^2 + y^2 = 1, \qquad (yp)^2 = 1 - y^2, \qquad p = y'$$

Obtain the envelope of the family by inspection of a graph, check by (8) and (9), and verify that the envelope satisfies the differential equation.

(b) The differential equation of Part (a) is now interpreted in differential form, so that solutions $x = \psi(y)$ as well as $y = \phi(x)$ are allowed. Sketch some of the odd-looking solutions that are obtained by going back and forth between the circles and the envelope.

(c) If $p = y'$ obtain the equation $p^2(1 - x^2) = x^2$ for the family of circles of radius 1 with centers on the y axis. Show that the p discriminant gives $x = 0$, which does not describe the envelope but does describe a locus where uniqueness fails. A locus on which two members of the family have the same tangent, as here, is called a *tac locus*.

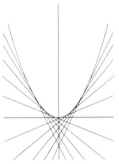

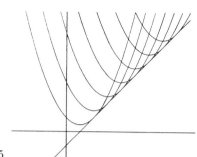

Figure 14   Figure 15

15. Obtain the equation $y + c^2 = 2cx$ for the tangent line to the parabola $y = x^2$ at $x = c$. Verify that the parabola is the envelope of the family of tangents, and discuss after the manner of Problem 14. If the tangents are regarded as light rays, the envelope is called a *caustic*. See Figure 14.

16. The family $y - c = (x - c)^2$ represents a family of congruent parabolas as shown in Figure 15. Discuss after the manner of Problem 14.

──────────────ANSWERS──────────────

1. $y^2 = \ln(\sin x)^2 + c$

2. (a) $y = cx$ (b) $x^2 - y^2 = c$ (c) $y = ce^{-2x}$ (d) $y^2 + \ln(cx) = 0$

3. It could have been forseen because the original families go into each other if you interchange x and y and replace c by $1/c$.

5. It is obvious because at any point (x, y) common to both curves the tangent to one curve has the same direction as the normal to the other. Hence they are orthogonal. The curves are $y = cx$, $x^2 + y^2 = c$.

6. (a) $y^3 - 3yx^2 = c$ (b) $x = y + 1 + ce^y$

7. (b) It becomes a set of circles centered at the origin together with a set of radial lines through the origin.
   (c) Because the original curves make a constant angle ψ with the radius, hence the orthogonal trajectories make a constant angle $\psi + \pi/2$ with the radius.

11. $x^2 + y^2 = 2x^{2-k}/(2-k) + c$ in general. If $k = 0$ the problem is nonsensical. If $k = 1$ the family is a set of lines through $(1, 0)$, hence the orthogonal trajectories are circles centered at this point. If $k = 2$ the problem changes character and the trajectories are $x^2 + y^2 = \ln x^2 + c$.

12. $(x + yy')y(y - xy') = (b^2 - a^2)yy'$

13. (b) infinitely many; two.   15. Partial answer: $4y + p^2 = 4px$

16. Partial answer: The envelope is $4y = 4x - 1$.

3. Applications to calculus

We show how the concepts of length, area, and volume as presented in elementary calculus can be developed by means of differential equations. It is taken for granted that these various quantities are numerical-valued and that they are additive. For example, if $s(P, Q)$ represents arc length along a curve from P to Q and if P, Q, R are three consecutive points on the curve as shown in Figure 16, then

Figure 16

$$s(P, R) = s(P, Q) + s(Q, R).$$

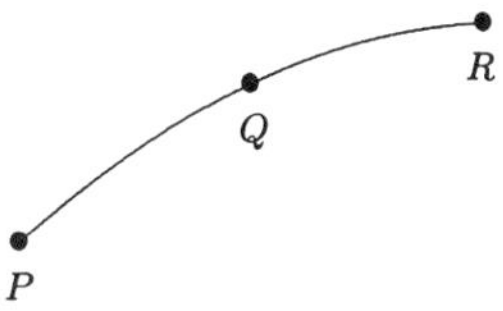

Similar relations hold for area and volume. In advanced mathematics it is said that length, area, and volume are *additive functionals*.

Our objective is to present the main ideas as simply as possible, and intuitive properties like additivity will not be stated formally as they would be in a strictly axiomatic development. In the same spirit, we take the increment Δx to be positive.

(a) Arc length

Given a curve $y = f(x)$ for $a \leq x \leq b$, let $f'(x)$ be continuous, and let $s = s(x)$ denote the arc length along the curve from its left-hand endpoint to (x, y). The main hypothesis about arc length is the following: If Δs represents arc length from the point (x, y) of the curve to $(x + \Delta x, y + \Delta y)$ and if

Figure 17

$$\Delta L = \sqrt{\Delta x^2 + \Delta y^2}$$

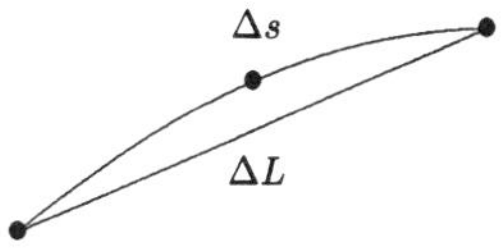

is the length of the corresponding chord, then the ratio arc/chord tends to 1. This means

(1)
$$\lim_{\Delta x \to 0} \frac{\Delta s}{\Delta L} = 1$$

with Δs and ΔL as shown in Figure 17. If $\Delta x > 0$

$$\frac{\Delta s}{\Delta x} = \frac{\Delta s}{\Delta L} \frac{\Delta L}{\Delta x} = \frac{\Delta s}{\Delta L} \frac{\sqrt{\Delta x^2 + \Delta y^2}}{\Delta x}$$

and hence

$$\frac{\Delta s}{\Delta x} = \frac{\Delta s}{\Delta L} \sqrt{1 + \left(\frac{\Delta y}{\Delta x}\right)^2}.$$

When $\Delta x \to 0$ the result is the differential equation

$$\frac{ds}{dx} = \sqrt{1 + \left(\frac{dy}{dx}\right)^2}$$

since the factor $\Delta s/\Delta L$ tends to 1 by hypothesis. Clearly $s(a) = 0$, and $s(b)$ corresponds to the arc of the curve from a to b. Integrating between corresponding limits and writing $dy/dx = f'(x)$ yields

$$s(b) = \int_a^b \sqrt{1 + f'(x)^2}\, dx.$$

Although we shall not give the proof, it can be shown conversely that if arc length is defined by this formula whenever $f'(x)$ is continuous, then arc length has all the properties used in the derivation, including the basic property (1). Thus the formula establishes *existence* of arc length, as well as a means of computation. The interest of the result is that it does not require a constructive approach by means of inscribed polygonal lines.

(b) Volume of revolution

Let the region bounded by the lines $x = a, x = b$ and the curve $y = f(x)$ be revolved about the x axis to generate a solid. We assume that $f(x)$ is positive and continuous. The volume of a right circular cylinder of radius r and height h is $\pi r^2 h$ and this formula will be used here.

Let $V(x)$ be the volume corresponding to the interval $[a, x]$ and let

$$m = \min f(t), \qquad M = \max f(t), \qquad x \le t \le x + \Delta x.$$

As a basic axiom we assume that the volume ΔV corresponding to the interval $[x, x + \Delta x]$ satisfies

(2) $$\pi m^2 \Delta x \le \Delta V \le \pi M^2 \Delta x.$$

This assumption is appropriate because the solid of which ΔV is the volume contains a cylinder of height Δx and radius m and is contained in a cylinder of height Δx and radius M. Dividing by Δx gives

(3) $$\pi m^2 \le \frac{\Delta V}{\Delta x} \le \pi M^2.$$

By continuity of f

$$\lim_{\Delta x \to 0} m = f(x), \qquad \lim_{\Delta x \to 0} M = f(x)$$

and in (3) this gives

$$\pi f(x)^2 \leq \frac{dV}{dx} \leq \pi f(x)^2.$$

Here we use the squeeze theorem of calculus, which states that if A, B, C are functions satisfying $A \leq B \leq C$, and if A and C have the same limit L, then also B has the limit L. Hence

$$\frac{dV}{dx} = \pi f(x)^2.$$

Since $V(a) = 0$, integrating between corresponding limits gives

$$V(b) = \pi \int_a^b f(x)^2 \, dx.$$

Conversely, if the volume is defined by this formula, one can deduce all the properties used in the derivation, including the formula $\pi r^2 h$ for the volume of a cylinder and the double inequality (2).

(c) Surface of revolution

The following discussion uses results from both (a) and (b) above. Let a curve $y = f(x)$ for $a \leq x \leq b$ be revolved about the x axis to generate a surface. We assume that $f(x)$ is positive and that $f'(x)$ is continuous; this implies that $f(x)$ is also continuous. If $S(x)$ is the area of (one side of) the surface swept out by the curve on the interval $[a, x]$, we want to get a differential equation for S.

Let ΔS be the area swept out by the curve on $[x, x + \Delta x]$, let Δs and ΔL be as in Part (a), and let m and M be as in Part (b). The surface giving ΔS as its area can be thought to be a ribbon, approximately, whose width is between Δs and ΔL and whose length is between $2\pi m$ and $2\pi M$. This suggests the estimate

(4) $$2\pi m \Delta L \leq \Delta S \leq 2\pi M \Delta s$$

which is now taken as a basic postulate.

Dividing by Δx gives

(5) $$2\pi m \frac{\Delta L}{\Delta x} \leq \frac{\Delta S}{\Delta x} \leq 2\pi M \frac{\Delta s}{\Delta x}.$$

We have

$$\lim_{\Delta x \to 0} \frac{\Delta L}{\Delta x} = \sqrt{1 + f'(x)^2}, \qquad \lim_{\Delta x \to 0} \frac{\Delta s}{\Delta x} = \sqrt{1 + f'(x)^2}$$

by Part (a), and

$$\lim_{\Delta x \to 0} m = f(x), \qquad \lim_{\Delta x \to 0} M = f(x)$$

by continuity as in Part (b). Using these relations and the squeeze theorem in (5) gives

$$\frac{dS}{dx} = 2\pi f(x)\sqrt{1 + f'(x)^2}$$

with a corresponding integral formula for $S(b)$. Here too, the formula yields all the properties that were used in the derivation, including the double inequality (4).

PROBLEMS

1. Area in polar coordinates. The area of a sector of angle $\Delta\theta$ in a circle of radius R is $(1/2)R^2\Delta\theta$. Consider a curve

$$r = f(\theta), \qquad a \le \theta \le b$$

in polar coordinates, where f is continuous and positive and $b - a < 2\pi$. Let $A(\theta)$ be the area swept out by the radius $0 \le r \le f(\phi)$ as ϕ varies from $\phi = a$ to $\phi = \theta$, and let m be the minimum of $f(\phi)$, M the maximum of $f(\phi)$ on the interval $\theta \le \phi \le \theta + \Delta\theta$. Explain why the double inequality

$$\frac{1}{2}m^2\Delta\theta \le \Delta A \le \frac{1}{2}M^2\Delta\theta, \qquad \Delta\theta > 0$$

is a reasonable assumption. If this is now postulated, show that

$$\frac{dA}{d\theta} = \frac{1}{2}f(\theta)^2 \quad \text{hence} \quad A(b) = \frac{1}{2}\int_a^b f(\theta)^2\, d\theta$$

Conversely, if area is defined by the integral show that you can recover the formula for area of a sector with which we started.

2. Volume by the shell method. The volume of a cylinder of radius r and height h is $\pi r^2 h$. Using this, check that the volume of a cylindrical shell of inner radius x, outer radius $x + \Delta x$ and height h lies between

$$2\pi xh\Delta x \quad \text{and} \quad 2\pi(x + \Delta x)h\Delta x, \qquad \Delta x > 0$$

Let $y = f(x)$ be continuous and positive for $a \le x \le b$, where $0 \le a < b$, and let it be required to find the volume of revolution when the region bounded by the curve, the x axis, and the lines $x = a, x = b$ is revolved about the y

axis. The shell method is used. With suitable definitions explain why the double inequality

$$2\pi x m \Delta x \le \Delta V \le 2\pi (x + \Delta x) M \Delta x, \qquad \Delta x > 0$$

is a reasonable postulate. Assuming this, deduce

$$\frac{dV}{dx} = 2\pi x f(x) \quad \text{hence} \quad V(b) = 2\pi \int_a^b x f(x)\, dx$$

Show conversely that if volume is defined by the integral, you can recover the formula for volume of a cylinder with which the discussion began.

3. A curve starts at the point $(0,1)$ and lies in the first quadrant. The arc from $(0,1)$ to the point (x,y) of the curve equals the area of the region bounded by the axes, the curve, and the ordinate.

(a) Find two distinct solutions satisfying $y(0) = 1$.

   Hint: Let $y = \cosh u$ in the integral; see Chapter 5, Section 2.

(b) Using the results (a) and the fact that arc and area are additive, find a two-parameter family of distinct solutions.

4. A curve $r = f(\theta)$ in polar coordinates starts at $(1,0)$ and lies in the first quadrant. The arc from $(1,0)$ to the point (r,θ) on the curve is twice the area swept out by the radius. Discuss after the manner of Problem 3.

5. Illustrate (3) and (4) by figures analogous to Figure 17.

───────────────────────────────**ANSWERS**───────────────────────────────

3. (a) $y = 1$, $y = \cosh x$
   (b) Follow $y = \cosh(x - c_1)$ to the point $x = c_1 \ge 0$, follow the line $y = 1$ to some point $c_2 \ge c_1$, and follow $y = \cosh(x - c_2)$ from there on.
4. Partial answer: $r = 1$, $r = \sec \theta$

Where a mathematical reasoning can be had, it's as great folly to make use of any other, as to grope for a thing in the dark, when you have a candle standing by you.

—Arbuthnot

Chapter 4

FUNDAMENTAL PRINCIPLES

FOR ANYONE involved in mathematical science, differential equations provide a tool of such power that they are impossible to ignore; and this power increases with each advance in the technology of high-speed computation. Through all the diversity of applications, a few principles and methods are encountered again and again. Among these are the principle of superposition for linear problems, theorems pertaining to existence and uniqueness of solutions, and methods of getting useful information from a differential equation without actually solving it.

1. Linearity

It is a familiar fact that, if u and v are differentiable functions and c is constant, then
$$(cu)' = cu', \qquad (u+v)' = u' + v'.$$
In terms of the differential operator $D = d/dx$ the above equations become

$$D(cu) = cDu, \qquad D(u+v) = Du + Dv.$$

These results of calculus resemble the algebraic rules

$$p(cu) = cpu, \qquad p(u+v) = pu + pv$$

for real numbers p, u, v, c.

An operator T with the two properties described above is said to be *linear*. For example, the operator D is linear. The operation of multiplying by a given function p is also linear.

For precise formulation we assume that T is defined for a class of functions y on an interval I, and that the result Ty of the operation is also a function defined on I. For example, $Dy = y'$ transforms the differentiable function y into another function y'. The class of functions y for which Ty is defined is called the *domain* of T.

Here is the formal definition of linearity:

DEFINITION OF LINEARITY. An operator T is linear if

$$T(cu) = cTu \quad \text{and} \quad T(u+v) = Tu + Tv$$

for all constants c and for all functions u, v in the domain of T.

As an illustration, let us show that the operator T defined by

$$(1) \qquad\qquad Ty = y' + py$$

is linear, where $p = p(x)$ is a given function, not involving y, defined on an open interval I. The domain of this operator consists of the functions y that are differentiable on I. In terms of T, the linear equation considered in the last chapter is $Ty = f$.

It should be emphasized that (1) describes the action of T, not only on y, but on all functions in its domain. For example,

$$Tu = u' + pu, \qquad Tv = v' + pv$$

and so on. In view of this, linearity follows from the calculation

$$T(cu) = (cu)' + p(cu) = cu' + pcu = c(u' + pu) = cTu$$

$$T(u + v) = (u + v)' + p(u + v) = u' + v' + pu + pv = Tu + Tv.$$

Here we have used the linearity of the differentiation operator D and of the operation of multiplying by p, as discussed earlier.

Sometimes the fact that an operator is linear is expressed in somewhat different notation. For example, the equations

$$(cu)'' = (cu')' = cu'', \qquad (u + v)'' = (u' + v')' = u'' + v''$$

show that the operator T defined by $Ty = y''$ is linear.

The importance of linear operators depends on the following:

GENERAL PRINCIPLE OF SUPERPOSITION. If u satisfies $Tu = f$ and v satisfies $Tv = g$, where T is a linear operator, then

$$y = c_1 u + c_2 v$$

satisfies $Ty = c_1 f + c_2 g$ for any choice of the constants c_1 and c_2.

The proof is immediate, since

$$T(c_1 u + c_2 v) = T(c_1 u) + T(c_2 v) = c_1 Tu + c_2 Tv = c_1 f + c_2 g.$$

The first two equalities follow from linearity and the third from the hypothesis $Tu = f$, $Tv = g$.

We illustrate the concept of linearity by several examples.

EXAMPLE 1. If $p = p(x)$ on I as above, show that the operator T defined by $Ty = y'' + py$ is linear. This follows from

$$T(cu) = (cu)'' + p(cu) = cu'' + cpu = c(u'' + pu) = cTu$$

$$T(u + v) = (u + v)'' + p(u + v) = u'' + v'' + pu + pv = Tu + Tv.$$

EXAMPLE 2. If $Ty = y'' + y$ the functions $u = \sin x$ and $v = \cos x$ obviously satisfy $Ty = 0$. The general principle of superposition shows then that

$$w = c_1 \sin x + c_2 \cos x \quad \text{satisfies} \quad Tw = c_1 0 + c_2 0 = 0.$$

Thus we get a two-parameter family of solutions of $y'' + y = 0$ from the particular solutions $\sin x$ and $\cos x$.

EXAMPLE 3. If $Ty = y' - 2y$ it is seen by inspection that

$$T(1) = -2, \qquad T(e^{5x}) = 3e^{5x}, \qquad T(e^{2x}) = 0.$$

The general principle of superposition shows then that the function

$$(2) \qquad y = a + be^{5x} + ce^{2x} \quad \text{satisfies} \quad Ty = -2a + 3be^{5x}$$

for any choice of the constants a, b, c. See Example 5 below.

EXAMPLE 4. Using the result of Example 3 find a general solution of

$$(3) \qquad\qquad y' - 2y = 8 + 6e^{5x}.$$

Comparing the right side of (3) with Ty in (2) we set $-2a = 8$ and $3b = 6$. By (2) the desired solution is

$$y = -4 + 2e^{5x} + ce^{2x}.$$

EXAMPLE 5. An expression of the form

$$y = c_1 u_1 + c_2 u_2 + \cdots + c_m u_m,$$

where the c_j are constant, is called a *linear combination* of the functions u_j. If each function u_j satisfies the homogeneous equation $Tu_j = 0$, show that every linear combination of the u_j satisfies the same homogeneous equation. The result is immediate, since

$$(4) \qquad\qquad Ty = c_1 Tu_1 + c_2 Tu_2 + \cdots + c_m Tu_m$$

by linearity, and each $Tu_j = 0$ by hypothesis. Hence $Ty = 0$, which is what we were to prove. The calculation depends on the fact that the basic linearity property extends to more than two summands by repetition. For example, the property for two summands can be used twice to yield

$$T((c_1u_1+c_2u_2)+c_3u_3) = T(c_1u_1+c_2u_2)+T(c_3u_3) = c_1Tu_1+c_2Tu_2+c_3Tu_3.$$

This gives (4) for $m = 3$. In a like manner

$$T((c_1u_1 + c_2u_2 + c_3u_3) + c_4u_4) = T(c_1u_1 + c_2u_2 + c_3u_3) + T(c_4u_4)$$

and using the result (4) with $m = 3$, we get (4) with $m = 4$. The process continues in an obvious manner and shows that (4) holds whenever the functions u_j are in the domain of the linear operator T.

PROBLEMS

1. In this problem T is defined by $Ty = yy''$.

(a) Check that

$$T(x^2) = 2x^2, \quad T(x^5) = 20x^8, \quad T(\sin x) = -\sin^2 x, \quad T(e^x) = e^{2x}$$

(b) Show that $T(cu) = cTu$ and $T(u + v) = Tu + Tv$ hold for all linear functions u and v, that is, all functions of form $a + bx$ where a and b are constant.

(c) Find a twice-differentiable function u such that $T(2u) \neq 2Tu$. This shows that T is not linear.

2P. Suppose u and v satisfy the equations $Tu = f$ and $Tv = f$, where T is linear. Show that the function $w = u - v$ satisfies $Tw = 0$.

3. If $u' + u\sin x = x^3 \cos 2x$ and $v' + v\sin x = x^3 \cos 2x$ show that

$$u(x) = v(x) + ce^{\cos x}$$

for some constant c. Hint: Use Problem 2 with $Ty = y' + y\sin x$.

4. Let $y = u$ and $y = v$ be two solutions of

$$y' - \frac{e^x}{1 + e^x}y = x^5 \cos x$$

that satisfy $u(0) = v(0) + 2$. Show that $u(x) = v(x) + e^x + 1$.

5. (a) If $Ty = y' + y$ check that $T(1) = 1$, $T(e^{2x}) = 3e^{2x}$, $T(e^{-x}) = 0$, hence

$$T(a + be^{2x} + ce^{-x}) = a + 3be^{2x}$$

for any constants a, b, c. Note that this last equation follows from the general superposition principle; no additional calculation is needed.

(b) If a and b are constant, solve the linear equation

$$y' + y = a + 3be^{2x}$$

by the methods of Chapter 2, and thus confirm the result (a).

6P. If u is a particular solution of $Ty = f$, where T is a linear operator, and if v is any solution of $Tv = 0$, show that $y = u + v$ satisfies $Ty = f$.

REVIEW PROBLEMS————————————————————————

7. One of the following equations is separable, two are linear in y, and one becomes linear if x and y are interchanged. Decide which is which, and solve:

(a) $y(1 + y^2)\, dx = 2(1 - xy^2)\, dy$　(c) $x(1 + y^2)\, dx = 2e^x y\, dy$

(b) $x\, dy = (8xe^{x^4} - 3y)\, dx$　　　(d) $x\, dy + (1 - 2x^2)y\, dx = 2x\, dx$

8. One of the equations in Problem 7 admits an integrating factor $1/y$. Which one? Prove your answer, of course.

————————————————**ANSWERS**————————————————

7. (a) $x(1 + y^2) = \log(y^2) + c$　(c) $e^{-x}(1 + x) + \log(1 + y^2) = c$

　　(b) $x^3 y = 2e^{x^4} + c$　　　　　(d) $xy = ce^{x^2} - 1$　　　　8. (a)

2. Existence and uniqueness

When a problem from the sciences is formulated in mathematical terms, two demands are usually imposed: The solution should exist, and it should be unique. These demands lead to the subject of existence and uniqueness theorems. The difference between the two kinds of theorems is as follows:

An existence theorem asserts that a problem has *at east* one solution.

A uniqueness theorem asserts that a problem has *at most* one solution.

Although existence theory is not slighted in this text, one can defend the view that existence is less important than uniqueness in many applications. There is nothing wrong with looking for a solution before you know it exists, and if you find it the question of existence is settled then and there.

To bring these remarks into focus, suppose we have a physical system that satisfies

(1) $$y'' + y = 0, \qquad y(0) = 0, \qquad y'(0) = 1.$$

Suppose further that we have noticed that $y = \sin x$ is a solution. Then the question of existence is settled. But as long as uniqueness is in doubt, we cannot say that $y = \sin x$ describes the behavior of the system. The most we can say is: $y = \sin x$ describes a possible behavior of the system.

This example is not artificial. The equation $y'' + y = 0$ describes vibration of a mass-spring system, the current in a circuit having capacitance and inductance, the small oscillations of a pendulum, and small oscillations of an object floating in still water, to mention four examples.

Next, suppose another physical system leads to the problem

(2) $$y'' + y = 0, \qquad y(0) = 0, \qquad y(\pi) = 0.$$

Again $y = \sin x$ is a solution, and again the question of existence is trivial. But there is a spectacular failure of uniqueness, since the system has infinitely many solutions

$$y = c \sin x, \qquad c = \text{const.}$$

Yet this system, like the other, is described by a linear second-order equation with two side conditions. This problem describes small oscillations of a stretched string or the electric field in a microwave resonant cavity. The side conditions $y(0) = 0$, $y(\pi) = 0$ are given on the boundary of the interval in which the differential equation holds, and (2) is an example of a *boundary-value problem*.

Although much of the theory of differential equations is involved with the proof of existence or uniqueness under suitable conditions, in applications a failure of existence or uniqueness can bring important information. For example, the solution $y = \tan x$ of the problem

$$y' = 1 + y^2, \qquad y(0) = 0$$

becomes infinite as $x \to \pi/2$ and does not exist on any interval containing the point $\pi/2$. Such behavior in a chemical system can mean that the reaction leads to an explosion. In the problem of free fall through a resisting medium, the fact that the solution fails to exist precisely when the velocity exceeds a certain critical value v_c means that v_c gives the terminal velocity; that is, the velocity that is approached as the time increases infinitely.

From the point of view of applications, failure of uniqueness is also significant. A point where a presumably unique solution branches into two solutions is called a *bifurcation point*. At such points the system can experience a sudden and sometimes catastrophic change in behavior. For example, when a supporting column in a building is subject to increasing axial pressure, at a certain critical pressure p_c the column suddenly buckles and becomes curved. Mathematically this means that uniqueness fails, and that besides the solution $y = 0$ corresponding to the straight column there is another solution.

When there is a failure of either existence or uniqueness, the trouble is usually caused by the fact that certain functions connected with the differential equation are not bounded. Since the concept of boundedness is important in both the theory and applications of differential equations, we give a formal definition:

DEFINITION OF BOUNDEDNESS. A function f is bounded on a region R if there is a constant M such that $|f(p)| \leq M$ for all p in R.

The precise sense of the word "region" does not matter. For functions $f(x)$ of one variable the region is usually an interval; for functions $f(x, y)$ of two variables with $p = (x, y)$ the region is usually a rectangular region in the plane, and similarly in higher dimensions. What is important is that the inequality $|f(p)| \leq M$ must hold for all values of the variables under consideration. As an example, the function $f(x) = 1/x$ is bounded for $1 \leq x \leq 2$, and also for $1 < x < \infty$, but not for $0 < x < 1$.

Let us now state an existence and uniqueness theorem for the initial-value problem

A : $\qquad y'' = f(x, y, y'), \qquad y(x_0) = y_0, \qquad y'(x_0) = y_1$

where the function $f(x, y, z)$ is defined in the following region R:

$$a_1 < x < b_1, \qquad a_2 < y < b_2, \qquad a_3 < z < b_3.$$

Here are three hypotheses on f:

B: f is continuous in R.

C: The partial derivatives $f_y = \partial f / \partial y$ and $f_z = \partial f / \partial z$ are bounded in R.

D: f is bounded in R.

The following theorem is proved in Chapter 14:

THEOREM 1. *If conditions BCD hold, the problem A has one, and only one solution for each point (x_0, y_0, y_1) in R. The solution can be extended for $x < x_0$ and for $x > x_0$ until it reaches the boundary of R.*

The concluding statement about "reaching the boundary" means: If $a < x < b$ is the longest interval on which the solution can exist while the point

$$(x, y, z) = (x, y(x), y'(x))$$

remains in R, then $(x, y(x), y'(x))$ tends to a definite point on the boundary of R as $x \to a$ through values on (a, b), and also tends to a definite point on the boundary of R as $x \to b$ through values on (a, b). The three hypotheses B,C,D make three independent contributions to the theorem and their respective roles are illustrated by the accompanying problems.

A result similar to Theorem 1 holds for the nth order initial-value problem. For example, when $n = 1$ the problem corresponding to that in the theorem is

$$y' = f(x, y), \qquad y(x_0) = y_0$$

on a region R of the form $a_1 < x < b_1$, $a_2 < y < b_2$. The only change is that f_z is omitted in condition C and the point (x_0, y_0, y_1) in the theorem is replaced by (x_0, y_0).

Although we refer to Theorem 1 and its n-dimensional extensions from time to time, the development in the first half of this book is nearly independent of these results. Hence, the proof is postponed. The following examples illustrate the limitations, as well as the strength, of general theories that do not take account of the specific form of the equations considered.

EXAMPLE 1. For the equation $y'' = 3y(y' - x)^{1/3}$ at what points of (x, y, z) space, if any, does the hypothesis of Theorem 1 fail? Setting $y' = z$ we have, in this case,

$$f(x, y, z) = 3y(z - x)^{1/3}.$$

This function is continuous for all (x, y, z) and we check to see if the partial derivatives

$$\frac{\partial f}{\partial y} = 3(z - x)^{1/3}, \qquad \frac{\partial f}{\partial z} = y(z - x)^{-2/3} = \frac{y}{(z - x)^{2/3}}$$

are bounded. The first is indeed bounded in every rectangular region R, but the second is unbounded in any region that contains a point of the plane $z = x$. Hence, one expects trouble if the solution ever gets to a value x where $y'(x) = x$. In particular, if the initial conditions are of the form

$$y(x_0) = y_0, \qquad y'(x_0) = x_0$$

the solution might very well fail to be unique even if it exists. Nonlinear equations such as that in this problem are, as a rule, extremely difficult to handle, and it would be hard to get the above conjectural information without the use of Theorem 1.

EXAMPLE 2. If p, q, r are bounded continuous functions of x on an interval (a, b) containing x_0, Theorem 1 applies to the initial-value problem

$$y'' + p(x)y' + q(x)y = r(x), \qquad y(x_0) = y_0, \qquad y'(x_0) = y_1.$$

Continuity alone suffices for existence and uniqueness, but not for the assertion about endpoint limits; see Problem 15 and Chapter 14.

To apply Theorem 1, write the equation in the form

$$y'' = r(x) - q(x)y - p(x)y'.$$

Setting $z = y'$ we get $f(x, y, z) = r(x) - q(x)y - p(x)z$, which is continuous. Since the partial derivatives

$$\frac{\partial f}{\partial y} = -q(x), \qquad \frac{\partial f}{\partial z} = -p(x)$$

are assumed bounded, the hypothesis of Theorem 1 holds. The equation in this example does not, as a rule, have elementary solutions, so again Theorem 1 gives information that would be hard to get in any other way.

EXAMPLE 3. If p and q are constant in Example 2, Theorem 1 still applies. As shown later, however, the equation in that case can be reduced to two first-order linear equations,

$$u' - au = r, \qquad v' - bv = u$$

where a and b are constant. This fact not only gives the existence and uniqueness asserted in Theorem 1 but gives additional information not contained in that theorem.

EXAMPLE 4. The equation $y' = g(y)$ satisfies the hypothesis of the one-dimensional form of Theorem 1 if g is continuous and $g'(y)$ is bounded. However, by separation of variables one can obtain a formula for the solution and show that existence and uniqueness hold under much weaker conditions.

Equations such as $y' = g(y)$ and $y'' = g(y)$ are said to be *solvable by quadrature*, because their solutions can be expressed by means of integration. Theorems pertaining to solution by quadrature are given in Chapter A. The gist of this development is that, under natural hypotheses, the methods of Chapter 2 are justified and yield the unique solutions of the equations considered there. We do not wish to impose Chapter A as a prerequisite, but we do refer to it from time to time. For example, Chapter A, Theorem 2 justifies the method of separation of variables as used in the following problems.

PROBLEMS————————————————————————————

1. At what points of the (x, y) plane does the hypothesis of the existence-uniqueness theorem (stated informally in the text) fail for the following equations? Further insight is given by Figure 1 and by Problems 2 and 3.

(a) $y' = \sin(xy)$ (b) $y' = x^{1/3}y^3$ (c) $y' = (xy)^{1/5}$

2. In Problem 1(b) verify that $y = 0$ is the only solution that exists on an interval of infinite length. Nevertheless the hypothesis of the existence-uniqueness theorem holds in every rectangular region, no matter how large.

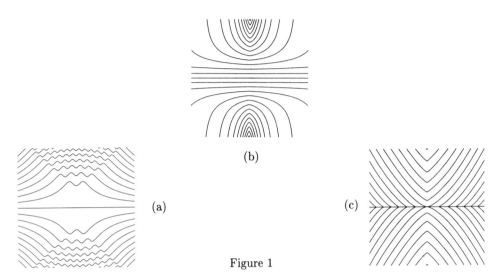

(b)

(a)

(c)

Figure 1

3. This problem requires a hand calculator. The solutions of Problem 1(c) clearly have slope 0 on the axes and yet the graphs in Figure 1(c) seem to have corners there. Near $(0,1)$ the slope y' is approximately $x^{1/5}$. Compute this for $|x| = .1, .001, .00001$ and thus explain the appearance of the graphs. Similarly show that y' is approximately $\pm(x-1)^{1/4}$ near $(1,0)$ and explore numerically. No answer is given.

4. Show that the equation $y' = (y-x)^{1/3}$ does not satisfy the hypothesis of the uniqueness theorem near the line $y = x$, yet it has a unique solution through every point of that line.

Hint: Set $u = y - x$ and separate variables.

5. At what points of (x, y, z) space does the hypothesis of Theorem 1 fail for the following equations?

(a) $y'' = (xy)^{1/3}$ (b) $y'' = (y'-y)^{1/5}$ (c) $y'' = (yy')^4$ (d) $(\sin x)y'' = e^y$

6. Check that the equation $y' = 3xy^{1/3}$ has two solutions $y = 0$, $y = x^3$ through the origin, although the right-hand side is continuous for all x and y. Why does the uniqueness theorem fail?

7. Check that the equation $xy' = y$ has infinitely many solutions through the origin, and find them. Why does the uniqueness theorem fail?

8. In what sense does the existence theorem fail for $y^2 + (y')^2 = -1$?

9. Show that the equation $y' = 3y^{2/3}$ has a solution $y = (x-c)^3$ where c is any constant, and thus construct a two-parameter family of solutions through the origin.

Hint: Pick $c_1 < 0 < c_2$. Follow a curve of the family to the point $(c_1, 0)$, then follow the x axis to the point $(c_2, 0)$, and finally follow a curve of the family beyond c_2. See Figure 2.

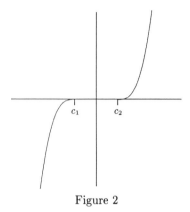

Figure 2

10. Let $f(x) = x/|x|$ for $x \neq 0$ and $f(0) = k$. It will be shown that, no matter how the constant k is chosen, the equation $dy/dx = f(x)$ has no solution on an interval containing the origin. This may serve to explain why, as a rule, a hypothesis of continuity is necessary. Since $f(x) = -1$ for $x < 0$ and $f(x) = 1$ for $x > 0$ we must have

$$y = -x + c_1, \qquad x < 0; \qquad y = x + c_2, \qquad x > 0$$

where c_1 and c_2 are constant. Existence of the derivative $y'(0)$ requires continuity of y at $x = 0$ and hence it requires $c_1 = c_2$. Taking $c_1 = c_2 = c$, show that y is not differentiable at $x = 0$. Hence y does not satisfy the differential equation at $x = 0$, no matter how $f(0)$ is defined.

11. Show that the operator T defined by $Ty = x^2 y'' - 2xy' + 2y$ is linear.

12. Show that the equation $Ty = 0$ in Problem 11 has the two solutions $y = x$, $y = x^2$ and hence the two-parameter family $y = c_1 x + c_2 x^2$. Does the hypothesis of Theorem 1 hold at $x = 0$? Can you use this two-parameter family to get at least the part of the conclusion of Theorem 1 asserting existence at $x = 0$?

13. This problem shows why differential equations are always considered on an interval rather than on a more general set. The equation $xy' + y = 0$ ought to be considered on the interval $(-\infty, 0)$ or on $(0, \infty)$. Suppose, however, that it is considered on the union U of these two intervals. Thus, U is the set of all $x \neq 0$. By separation of variables or by trying $y = x^a$ get the family of solutions $y = \phi(x, c_1, c_2)$ where

$$\phi(x, c_1, c_2) = c_1/x, \qquad x < 0; \qquad \phi(x, c_1, c_2) = c_2/x, \qquad x > 0$$

Thus show that if $x_0 \neq 0$, that is, if $x_0 \in U$, the initial-value problem

$$xy' + y = 0, \quad x \in U; \qquad y(x_0) = y_0$$

has infinitely many solutions.

14. By choosing $1/x = n\pi/2$, where n is an integer, show that the function defined for $x \neq 0$ by $y = \sin(1/x)$ oscillates infinitely often between the values $-1, 0, 1$ as $x \to 0$ and hence has no limit. See Figure 3.

15. For $x \neq 0$, check that the function $y = \sin(1/x)$ of Problem 14 satisfies $(x^2 y')' = -y/x^2$ and hence it satisfies

$$x^4 y'' + 2x^3 y' + y = 0, \qquad x \neq 0$$

This shows that the conclusion asserting existence of a limit at the boundary can fail even for linear equations if the coefficients are unbounded.

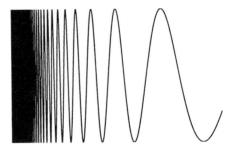

Figure 3

ANSWERS

1. (a) none　　(b) none　　(c) where $y = 0$　　(d) where $y = x$

5. (a) where $y = 0$　　(c) none

   (b) where $y = z$　　(d) where $x = n\pi$, n an integer.

6. Because f_y is unbounded near $y = 0$.

7. Because $y' = y/x$ and the coefficient $1/x$ is unbounded near $x = 0$.

8. If we solve in the form $y' = f(x, y)$ neither of the possible values for f is real, hence one would not expect real solutions. If complex solutions are allowed a modified form of the existence theorem would apply.

12. No, since if we solve for y'' the coefficients are unbounded. No, since the only initial values we can satisfy are $y(0) = 0$, $y'(0) = y_1$, which are not general.

---

3. Qualitative behavior

The equation $dy/dx = f(x, y)$ gives the slope of a solution-curve $y = \phi(x)$ at any point of this curve, even before you solve the equation. As an illustration consider the equation

$$(1) \qquad \frac{dy}{dx} = (y - 1)(y - 2)(y - 4) = g(y)$$

where $g(y)$ is defined by the product in the center. The sign of $g(y)$ can be determined by use of the following two properties:

(a) If $y > 4$ all factors are positive, hence $g(y) > 0$.

(b) $g(y)$ changes sign at the points 1,2,4 and only there.

The first of these statements is obvious and the second follows from the fact that a continuous function defined on an interval can change sign only by passing through the value 0.

If we plot the points 1,2,4 on the number line the above information is summarized as follows:

The encircled + indicates that the function is positive for $y > 4$. Starting at the right of 4 we move to the left, changing sign whenever we get to one of the points 4, 2, 1. This produces the three unencircled signs. The arrows below the line indicate that y increases when $dy/dx > 0$. If $dy/dx < 0$ then y decreases and the arrow is directed to the left. Note that the sign of $g(y)$ is also the sign of dy/dx in (1).

Equation (1) has the stationary solutions $y = 1$, $y = 2$, $y = 4$. For example if $y = 2$ then $dy/dx = 0$ and $g(y) = 0$, hence the equation is satisfied. The arrows in the figure indicate that if y is on either of the intervals $(-\infty, 1)$ or $(1, 2)$ adjacent to the stationary value $y = 1$ then y moves away from that value as x increases. For this reason the stationary solution $y = 1$ is said to be *unstable*. The stationary solution $y = 4$ is also unstable. However if y is on either of the intervals $(1, 2)$ or $(2, 4)$ adjacent to the value $y = 2$ then y moves towards the value 2 as x increases, and the stationary solution $y = 2$ is said to be *stable*.

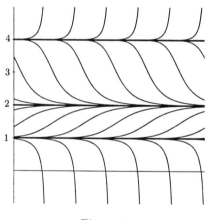

Figure 4

Typical solution-curves for (1) are shown in Figure 4. Note that the stationary solutions $y = 1, 2, 4$ divide the plane into strips, and no solution crosses from one strip into another.

The same sort of thing can be done when the polynomial has factors $(y - y_j)^m$ corresponding to repeated roots. If m is an odd integer such a factor changes sign as y passes through the value y_j, but if m is even, the factor does not change sign. Thus, the first of the factors

$$(y - 3)^5, \qquad (y - 3)^6$$

changes sign when y passes from values $y > 3$ to $y < 3$, but the second is positive in either case.

Suppose, for example, that

(2) $$\frac{dy}{dx} = (y + 1)^3 y^4 (y - 2)^7 (y - 3)^5 (y - 5)^8.$$

The relevant information is summarized as follows:

The small letters e and o above the values $-1, 0, 2, 3, 5$ indicate whether the corresponding zero is of even or odd order. For example the letter e appears above 5 because the exponent 8 on the factor $y - 5$ is even. For $y > 5$ all factors are positive, hence $g(y) > 0$, and this is indicated by the encircled $+$ sign. Starting with $y > 5$ we move to the left, changing sign whenever we pass a value marked o, but leaving the sign unchanged when we pass a value marked e. This produces the unencircled signs, $+$ or $-$. Note that the results would be the same if $g(y)$ had additional factors corresponding to complex roots, such as

$$(y^2 + 1)(y^2 + 2y + 2)(y^4 + 6).$$

Factors of this kind are always positive, hence they do not alter the pattern of signs.

The arrows below the line show the direction in which y moves on that interval as x increases. The stationary solutions $y = -1$ and $y = 3$ are unstable while $y = 2$ is stable. The stationary solutions $y = 0$ and $y = 5$ are said to be *semistable* because they have one arrow on an adjacent interval pointing toward them and one arrow pointing away from them.

An interval adjacent to a stationary solution, whose arrow points toward that solution, is called an *interval of attraction* for the solution. In this example the stationary solutions $y = 0$ and $y = 5$ each have one interval of attraction, $y = 2$ has two, and $y = -1$ and $y = 3$ have none.

The following theorem supplements the foregoing analysis:

THEOREM 1. *Let $dy/dx = g(y)$ where $g(y)$ is a real polynomial whose real zeros are $y_1, y_2, \ldots, y_m$; there may also be complex zeros but these do not affect the result. Then:*

(a) *If a solution $y = \phi(x)$ meets a stationary solution $y = y_j$, it coincides with that stationary solution on its whole interval of definition.*

(b) *If a solution $y = \phi(x)$ starts on an interval of attraction for a stationary solution y_j, then $\lim_{x \to \infty} \phi(x) = y_j$.*

The result (a) is easily proved. Since y_j is a zero of $g(y)$ we can write

$$g(y) = (y - y_j)h(y)$$

where $h(y)$ is a polynomial. Hence our solution $y = \phi(x)$ satisfies

$$\frac{dy}{dx} = (y - y_j)h(y)$$

or also

$$\frac{dy}{dx} = (y - y_j)f(x) \quad \text{where} \quad f(x) = h(\phi(x)).$$

Since $f(x)$ is continuous we can solve in the usual way and deduce

$$y - y_j = ce^{F(x)} \quad \text{where} \quad F(x) = \int f(x)dx.$$

Since $y = y_j$ at some value x_0 by hypothesis, the constant c must be 0. Hence $y = y_j$, as was to be shown.

Part (b) depends on the solution of a differential inequality by quadrature, and its proof is best postponed until you have had more experience with differential equations. A result established in Chapter A contains Part (b) as a special case.

PROBLEMS

1P. An equation $dy/dx = g(y)$ of the type considered in the text is in standard form if each factor is of the form $y - y_j$ rather than $y_j - y$.

(a) Write the following equation in standard form:

$$\frac{dy}{dx} = (8 - y)^4(2 - y)(y - 6)^3(4 - y)^2$$

(b) Plot the points 2,4,6,8 on the number line accompanied by letters e or o to indicate whether the corresponding zero is of even or odd order.

(c) If $y > 8$ all factors $y - y_j$ are positive but the function is prefixed by a minus sign (assuming you did Part (a) correctly) and hence it is negative. Starting with an encircled minus sign at $y > 8$ move to the left, changing sign when you get to a zero of odd order. Thus make a diagram similar to those in the text. In particular, classify the four stationary solutions as stable, unstable, or semistable.

(d) As $x \to \infty$, what happens to solutions $y = \phi(x)$ that pass through the points $(x, y) = (5, 3)$, $(-4, 7)$, $(15, 5)$, $(-7, 9)$? Hint: Use Theorem 1(b).

2. If $y' = dy/dx$, determine the stationary solutions, the nature of their stability, and their intervals of attraction:

(a) $y' = (y-3)(y-5)$ (b) $y' = (1-y)(y-2)^3$ (c) $y' = y(5-y)(y-4)^2$

3. Proceed as in Problem 2:

(a) $y' = (1 + y^2)(y - 3)^3(y - 5)^4$ (c) $y' = (y^2 - 1)(2 - y)(y + 4)^5$

(b) $y' = (1 - y)(3 - y)(4 - y)$ (d) $y' = (y^2 + 3y + 2)(y^4 - 16)$

4. (a) If $y = \phi(x)$ satisfies $dy/dx = g(y)$, show that $y = \phi(x - c)$ satisfies the same equation for any choice of the constant c. This means that any particular solution $y = \phi(x)$ generates a whole family of solutions by horizontal translation. If the original solution $y = \phi(x)$ lies in some strip $y_{j-1} < y < y_j$ determined by the stationary values y_j, the process gives all solutions in that strip.

(b) If x and y are interchanged, the nonlinear equation $dy/dx = g(y)$ becomes a linear equation of a very simple type, namely, $dy/dx = 1/g(x)$. Interpret the result (a) in this context.

5. (abc) Sometimes the function $g(y)$ in $dy/dx = g(y)$ is given as a graph. Since the intervals of positivity can be read from the graph, you can apply the methods of this section. A point y_0 where the graph of g crosses the y axis gives a stationary solution $y = y_0$, and a nonzero maximum or minimum of $g(y)$ gives a point of inflection on the graph of the solution. Using these remarks as a clue, plot y when $g(y)$ is as in Figure 5.

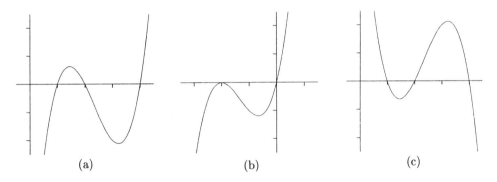

(a)   (b)   (c)

Figure 5

REVIEW PROBLEMS

Here are twelve equations, for which you are to make a list of twelve letters a,b,$\cdots$, l:

(a) $dy + 2xy\,dx = e^{-x^2}\,dx$

(b) $(1 + x)\,dy = (1 + y)\,dx$

(c) $dx + 4y^3x\,dy = y^7\,dy$

(d) $dx + x\,dy = x^3y\,dy$

(e) $y^3d(xy) = x^2\,dx$

(f) $d(xy) = x^6y^3\,dy$

(g) $d(xy) = y^2\sin y\,dy$

(h) $d(xy) = y^2\log x\,dx$

(i) $(\sin x\cos^2 y)dx = \cos^2 x\,dy$

(j) $x\,dy + y\,dx = e^{xy}\,dx$

(k) $dy + y\cos x\,dx = \cos^3 x\,dx$

(l) $x\,dx + y\,dy = \sec(x^2 + y^2)\,dy$

6. Which are separable? Cross them off your list.

7. Of those remaining, which are linear? Cross them off your list.

8. Among those remaining, which are of Bernoulli type? Cross them off.

9. Some of those remaining become linear when x and y are interchanged. Which are they? Cross them off.

10. Some of those remaining become of Bernoulli type when x and y are interchanged. Which? Cross them off.

11. If all has gone well, you should have two equations remaining that do not belong in any of the categories above. They can be solved, however, almost by inspection. Which are they, and what are their solutions?

12. Assuming that $f(x)$ is continuous for all relevant x, solve the following equations with the initial condition $y(0) = y_0$:

(a) $y' - (\tan x)y = f(x)$ (b) $y' + (\tan x)y = f(x)$

What, if any, are the limitations on the interval of existence?

13. (a) Show that the solution-curves of the initial-value problem

$$y' = 3y^{2/3}, \quad y(0) = 0$$

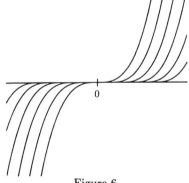

entirely fill up the region between the curve $y = x^3$ and the x axis. That is, if (x_1, y_1) is any point in this region, a solution satisfies $y(x_1) = y_1$. See Figure 6.

Figure 6

(b) Show that if the equation $y' = 3y^{2/3}$ is considered in the whole (x, y) plane, there are infinitely many solutions through each point (x_0, y_0). The two facts illustrated here are typical of what happens when uniqueness fails.

Hint for both parts: See Section 2, Problem 9.

14. (a) Show that the equation $y' = y - y^2$ becomes a linear equation upon the substitution $y = 1/u$, and thus solve it.

(b) Solve the same equation by separation of variables.

(c) Verify the equivalence of the results. No answer is given.

15. Check that the solution of $y'' + 2xy' = 0$, $y(0) = 0$, $y'(0) = 2/\sqrt{\pi}$ is

$$y = \operatorname{erf} x \quad \text{where} \quad \operatorname{erf} x = \frac{2}{\sqrt{\pi}} \int_0^x e^{-t^2} dt$$

The function $\operatorname{erf} x$ is called the *error function*; it is important in probability.

────────────ANSWERS────────────

1. (c) 2 unstable, 4 semistable, 6 stable, 8 semistable
   (d) $y \to 4$, $y \to 6$, $y \to 6$, $y \to 8$

2. Partial answer: (a) 3 stable, 5 unstable   (b) 1 unstable, 2 stable
                    (c) 0 unstable, 4 semistable, 5 stable

3. Partial answer: (a) 3 unstable, 5 semistable
                    (b) 1 stable, 3 unstable, 4 stable
                    (c) −4 unstable, −1 stable, 1 unstable, 2 stable
                    (d) −2 semistable, −1 stable, 2 unstable

4. (b) The result (a) expresses the fact that all solutions of $dy/dx = 1/g(x)$ on an interval can be obtained by adding a constant to any particular solution.

5. Partial answer: The function $g(y)$ in (a) is the same as that in (1) and the trajectories are shown in Figure 4.

6. b,i   7. a,k   8. e,h   9. c,g   10. d,f

11. j: $x + e^{-xy} = c$,   l: $\sin(x^2 + y^2) - 2y = c$

12. (a) $y = (\sec x) \int_0^x f(t) \cos t \, dt + y_0 \sec x$, $-\pi/2 < x < \pi/2$
    (b) $y = (\cos x) \int_0^x f(t) \sec t \, dt + y_0 \cos x$, $-\pi/2 < x < \pi/2$

Mathematics takes us into the region of absolute necessity, to which not only the actual world but every possible world must conform.

—Bertrand Russell

Chapter 5

APPLICATIONS

THE FACT THAT scientific problems lead to differential equations has already been illustrated in this text, and further examples are given here. Among these are problems of population growth, endangered species and harvesting in ecology, and a number of problems in mathematical physics. Our goal is not to write a treatise on ecology or physics, however, but merely to indicate the diversity of the scientific problems to which differential equations can be applied.

1. The logistic equation

The exponential law for population growth goes back to an essay that appeared anonymously in 1798, but was actually written by the English economist Thomas Malthus. His main thesis was that human misery is inevitable, because population tends to grow geometrically while the means of subsistence grows only arithmetically. From the point of view of differential equations, geometric growth means $dP/dt = kP$ and arithmetic growth means $dP/dt = k$.

In 1845 the Belgian mathematician Pierre Verhulst gave a number of reasons for thinking that the growth constant k should not remain constant indefinitely, but should diminish like $M - P$ where M is the theoretical maximum for the number P of individuals in the population. The Verhulst equation is

$$(1) \qquad \frac{dP}{dt} = k(M - P)P, \qquad P(0) = P_0$$

where k and M are positive constants. Verhulst called the solutions *logistic curves* from a Greek word meaning "skilled in computation" and (1) is often referred to as the *logistic equation*.

Three of the many uses of the logistic equation are discussed in the following problems. The first is the analysis of population growth in the study of demography. The logistic curves involve k, M, and the constant of integration c. Knowing the population P at three times one can determine these constants and thus predict future growth, including the maximum. The computation is easy when the times at which the population is known are equally spaced, and that is the case we consider.

The second application pertains to the problem of endangered species, of which the California condor is an outstanding example. For most species there is a threshold L such that if the population falls below L it cannot reproduce and becomes extinct. This means that the population drops to zero in a finite time. The situation is described by the equation

(2) $$\frac{dQ}{dt} = k(M - Q)(Q - L), \qquad 0 < L \le M.$$

If Q satisfies (2) let $P = Q - L$. Then $Q = L + P$ and hence

(3) $$\frac{dP}{dt} = k(M - L - P)P$$

which is the same as (1) with M replaced by $M - L$. Hence (1) applies to the threshold problem.

The third application is to problems of harvesting. We assume that the population satisfies the logistic equation and we consider harvesting either at a constant rate or at a rate proportional to the population size. The questions are: How much harvesting can be done without driving the population to extinction, or (which is essentially the same thing) what is the maximum rate of harvesting possible in the steady state? To see how these questions lead to the logistic equation, suppose the unharvested population satisfies (1). If the rate of harvesting is proportional to the population size there is an extra term $-kHP$, where the positive constant $\cdot H$ measures the vigor of the harvesting effort as a fraction of the growth constant k. Adding this negative term to the right-hand side of (1) gives

$$\frac{dP}{dt} = k(M - H - P)P.$$

Hence the effect of the harvesting is to change M to $M - H$. In the accompanying problems it is seen that similar considerations apply when the harvesting is at a constant rate or when the basic population equation is the threshold equation (2) rather than the logistic equation (1). In all such cases we are led back to (1) with a new choice of parameters.

The rest of this section is devoted to a study of (1). Equation (1) has the stationary solutions $P = 0$ and $P = M$. The graphs of these solutions are horizontal lines that divide the plane into three regions: the half plane $P < 0$, the strip $0 < P < M$, and the half plane $M < P$. No solution can cross either of these horizontal lines, because such a crossing would violate uniqueness; see Chapter 4, Section 3, Theorem 1.

In Problem 1 it is seen that the substitution $P = 1/p$ leads to a linear equation that can be solved more or less by inspection. The result is

(4) $$\frac{1}{P} = \frac{1}{M} + \frac{c}{e^{kMt}}.$$

The constant c is given by the initial condition as

(5) $$c = \frac{1}{P_0} - \frac{1}{M}.$$

If $P_0 > 0$ then $P(t)$ is defined for $t > 0$ and (4) gives

$$\lim_{t \to \infty} P(t) = M.$$

We can get additional information by considering the sign of P' and P'' as obtained from the differential equation

$$P' = k(M - P)P = k(MP - P^2).$$

Differentiating the second form above gives

(6) $$P'' = k(M - 2P)P'$$

by the chain rule. We mark off points $0, M/2, M$ on the number line and construct the following chart:

location of P	$(-\infty, 0)$	$(0, M/2)$	$(M/2, M)$	(M, ∞)
sign of P'	$-$	$+$	$+$	$-$
sign of $M - 2P$	$+$	$+$	$-$	$-$
sign of P''	$-$	$+$	$-$	$+$

The second line is filled in by looking at $(M - P)P$, the third is obvious, and the fourth is the product of the preceding two in accord with (6). From this chart we can read off the sense of monotonicity and concavity of the graph $y = P(t)$ and confirm the behavior shown in Figure 1. In particular, if $0 < P < M$ the chart shows that P is increasing and hence M is approached from below as an asymptotic maximum. If $M < P$, on the other hand, the chart shows that P is decreasing and M is approached from above as an asymptotic minimum.

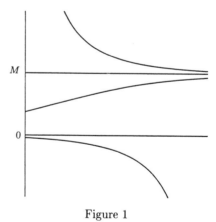

Figure 1

The last line of the table shows that P'' changes sign at $M/2$ and hence the curve has an inflection point. It is most interesting that the value of P there is half the maximum, no matter what k and c may be.

The above remarks apply to the threshold equation with $P = Q - L$ and with M replaced by $M - L$. The main new feature is that the condition $Q_0 < L$ leads to $P_0 < 0$, which we have not considered up to now. We shall show that if $Q_0 < L$ then $Q(t)$ vanishes at a certain time T. Since $Q = 0$ means that the population is extinct, the time T is called the *extinction time*.

We can get a formula for the extinction time as follows. When $Q_0 < L$ and $Q(T) = 0$ the interval

$$Q_0 \geq Q \geq 0 \quad \text{corresponds to} \quad 0 \leq t \leq T.$$

Separate variables in (2) and integrate between corresponding limits to get

$$\int_0^T k\,dt = \int_{Q_0}^0 \frac{dQ}{(M-Q)(Q-L)} = \frac{1}{M-L} \ln \frac{L-Q}{M-Q} \bigg|_{Q_0}^0.$$

The integral was evaluated by expanding the integrand in partial fractions. We summarize as follows:

EXTINCTION THEOREM. Suppose a population Q satisfies the threshold equation

$$\frac{dQ}{dt} = k(M-Q)(Q-L), \qquad Q(0) = Q_0$$

where $k > 0$ and $0 < L < M$. Suppose further that $Q_0 < L$. Then the population becomes extinct at a time T given by

$$kT = \frac{1}{M-L} \ln \left(\frac{M-Q_0}{L-Q_0} \frac{L}{M} \right).$$

Instead of taking $Q_0 = Q(0)$ we could take $Q_0 = Q(t_0)$ and the conclusion would be that the population becomes extinct at the time $t_0 + T$. The gist of the result is that *if the population ever falls below L, it goes to extinction.* That is why L is called the threshold.

The logistic equation applies not only to populations of fish and animals, including man, but also to such things as yeast plants and the growth of sunflowers. A specific numerical equation may be realistic if the population is measured in units of 1000 and the time in days, or in units of 100,000,000 and the time in decades. Changing the units changes the numerical values of the coefficients. With this thought in mind, we have chosen simple coefficients in most of the following problems.

PROBLEMS

1. The logistic equation is a Bernoulli equation with $1/(1-n) = -1$. Show that the substitution $P = 1/p$ leads to the equation on the left and hence, with $m = 1/M$ and $C = cM$, to the two formulas on the right:

$$p' + kMp = k, \qquad p = m + ce^{-kMt}, \qquad P = \frac{M}{1 + Ce^{-kMt}}$$

2. This problem illustrates essentially everything a logistic curve can do. On a single diagram of large size sketch the solutions of

$$P' = P(1 - P), \qquad t \geq 0; \qquad P(0) = -1, \quad 0, \quad \frac{1}{8}, \quad 1, \quad 2$$

One of these five curves exists only on a finite interval $0 \leq t < t_1$. Which one, and what is the value of t_1?

Suggestion: For all but one case, use the right-hand formula of Problem 1. The value of C is given by (5) as $C = (M/P_0) - 1$.

3. A population P satisfying the logistic equation is known at three equally-spaced times t_0, t_0+d, t_0+2d. Without loss of generality $t_0 = 0$; this amounts to choosing, temporarily, a new origin for the time coordinate. Thus

$$P(0) = P_0, \qquad P(d) = P_1, \qquad P(2d) = P_2$$

are known. To get a formula for $P(t)$, let

$$p = \frac{1}{P}, \qquad p_i = \frac{1}{P_i}, \qquad m = \frac{1}{M}, \qquad p(t) = m + ce^{-kMt}$$

where the last equation follows from Problem 1. With $x = e^{-kMd}$ check that

$$p_0 = m + c, \qquad p_1 = m + cx, \qquad p_2 = m + cx^2$$

and hence that

$$x = \frac{p_1 - p_2}{p_0 - p_1}, \qquad c = \frac{p_0 - p_1}{1 - x}, \qquad m = p_0 - c, \qquad kM = \frac{-\ln x}{d}$$

4. Suppose a population measured in units of 10,000,000 satisfies the logistic equation and has the values $1/7, 1/5, 1/4$ at $1850, 1900, 1950$, respectively.

(a) Show that $x = 1/2$, $c = 4$, $m = 3$, $kM = (\ln 2)/50$ in the notation of Problem 3. Thus get the formula

$$P(t) = \frac{1}{3 + 4 \cdot 2^{-s/50}}, \qquad s = t - 1850$$

(b) Check that the formula gives the right result when $t = 1850, 1900, 1950$.

(c) Show that $P(2000)/P(1800)$ is a good approximation to π.

5. The population of the United States in units of $100,000,000$ is given by the following table (courtesy Department of Commerce, Bureau of Census):

year	1820	1860	1900	1940	1980	1985
pop.	.0964	.3144	.7599	1.317	2.265	2.377

Make a corresponding table of reciprocals $p_i = 1/P_i$ and, assuming that $y = P(t)$ is a logistic curve, get two formulas, one based on $1820, 1900, 1980$, and one on $1900, 1940, 1980$.

(a) How well does the first curve match the values at 1860 and 1940?

(b) Which curve gives a better match at 1985?

(c) What maximum M is predicted by the two curves and what do you think is the cause of this astonishing discrepancy?

Note: Alaska and Hawaii are included in the last two figures but not in the first four. This makes a difference of .014 in 1980 and .016 in 1985.

6. This problem pertains to the harvesting equation $P' = k(M - H - P)P$.

(a) If $H = M$ show that $P(t) = P_0/(1 + kP_0t)$, hence $P(t) \to 0$ as $t \to \infty$.

(b) If $H > M$ show that $P(t) \to 0$ at a much faster rate than in (a), but $P(t)$ does not vanish completely at any finite time T.

(c) If $H < M$ show that $P(t) \to M - H$ as $t \to \infty$, hence the steady-state yield kHP is $kH(M - H)$.

(d) Show that $kH(M - H)$ in Part (c) is maximum when $H = M/2$. The maximum value $kM^2/4$ is called the *maximum sustainable yield*.

(e) Show that the maximum in (d) is the same as the maximum value of P' for the unharvested population when $0 < P < M$.

Hint for (b) and (c): Use the result of Problem 1 with $M - H$ instead of M.

7. This problem pertains to the threshold equation (2) with $L = M$, which is a condition of zero probability but leads to simple solutions.

(a) If $Q(0) > M$ show that $Q(t) \to M$ as $t \to \infty$.

(b) If $Q(0) < M$ find the time T at which the population becomes extinct.

8. Plot the parabola $S = (M - Q)(Q - L)$ versus Q for $0 < Q < \infty$ and show that the location and value of the maximum are given by

$$\overline{Q} = \frac{M + L}{2}, \qquad S_{\max} = \frac{(M - L)^2}{4}$$

respectively. Note that kS is the slope dQ/dt of solutions in (2).

9. A population satisfying the threshold law (2) is harvested at a constant rate kh, so that the positive constant h measures the vigor of the harvesting as a fraction of the growth constant k. This produces an extra term $-kh$ in the differential equation and a term $-h$ in the quadratic of Problem 8.

(a) If $0 < h < S_{\max}$, where $S_{\max}$ is as in Problem 8, check that the harvesting leads to a new threshold problem with values L_h, M_h that are given by the construction illustrated in Figure 2. Thus the location of the maximum is unchanged, its value is reduced by h, and the population is overexploited if

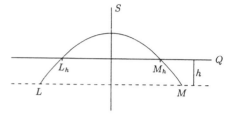

Figure 2

$$h > h^* = \frac{(M-L)^2}{4}$$

(b) The values L_h and M_h are roots of a quadratic equation. What equation? Show how to solve it graphically by intersecting the parabola of Problem 8 with a horizontal line.

(c) In analogy with Problem 6, the quantity h^* is often referred to as "the maximum sustainable yield." Check that kh gives the value $kM^2/4$ of Problem 6 if $L = 0$. Nevertheless, Problem 6 represents the better policy. Why?

10. In the notation of the preceding problems, a population of fish is simultaneously exploited by sports fishermen who catch fish at the rate kHP and by commercial fishermen who catch fish at the rate kh. If the unexploited population satisfies the threshold equation the effect of fishing is described by

$$\frac{dP}{dt} = k(M-P)(P-L) - kHP - kh$$

Explain why this problem falls within the scope of Problem 9.

---------------------------ANSWERS---------------------------

2. The first, $t_1 = \ln 2$.   4. (d) 22/7

5. (a) 0.289, 1.54   (b) The value 2.32 is not quite as good as 2.42 but the average of these two values is nearly exact.   (c) Partial answer: 2.86, 52.6

7. (b) $kMT = Q_0/(M - Q_0)$ where $Q_0 = Q(0)$

9. (b) $(M - Q)(Q - L) = h$. Draw a line of height h.   (c) Because Q becomes arbitrarily close to the semistable value given by Problem 7(a), and if it drops the least bit below this value, the population goes to zero by Problem 7(b). In Problem 6, the harvesting rate decreases as the population decreases, and this protects the population.

10. With $M^* = M + L - H$ and $h^* = ML + h$ the differential equation is

$$\frac{dP}{dt} = k(M^* - P)P - kh^*$$

This is the special case $L = 0$ of Problem 9.

2. Hyperbolic functions

This section introduces some functions that are repeatedly encountered in scientific and engineering problems. These functions are related to the equation $y'' - y = 0$ in much the same way as the trigonometric functions are related to the equation $y'' + y = 0$.

Indeed, $y = \sin x$ and $y = \cos x$ are solutions of $y'' + y = 0$ with the initial values

$$(1) \qquad\qquad y(0) = 0, \quad y'(0) = 1; \qquad y(0) = 1, \quad y'(0) = 0$$

for $\sin x$ and $\cos x$, respectively. In a like manner the functions $\sinh x$ and $\cosh x$ satisfy $y'' - y = 0$ with the same initial values (1). These functions are called the *hyperbolic sine* and *hyperbolic cosine*, respectively. For want of a better option sinh is pronounced "cinch" and a similar convention is followed for other hyperbolic functions introduced below.

To get a formula for $\sinh x$ and $\cosh x$, set $y = e^{ax}$ in $y'' - y = 0$. This procedure produces the two solutions e^x and e^{-x} and by linearity

$$(2) \qquad\qquad y = c_1 e^x + c_2 e^{-x}$$

is also a solution for any constants c_1 and c_2. The initial conditions (1) give the following equations for the constants c_1 and c_2 in (2):

$$c_1 + c_2 = 0, \ c_1 - c_2 = 1; \qquad c_1 + c_2 = 1, \ c_1 - c_2 = 0.$$

These equations can be solved by inspection, with the result

$$(3) \qquad\qquad \sinh x = \frac{e^x - e^{-x}}{2}, \qquad \cosh x = \frac{e^x + e^{-x}}{2}.$$

The logic here is that $\sinh x$ and $\cosh x$ are defined by the same initial-value problems as led to the right-hand sides in (3), and hence the solutions coincide. This reasoning depends on a uniqueness theorem that was stated without proof in Chapter 4. Alternatively, one can accept (3) as a definition.

It is readily checked that

$$(4) \qquad\qquad \frac{d}{dx} \sinh x = \cosh x, \qquad \frac{d}{dx} \cosh x = \sinh x$$

and also that

$$(5) \qquad\qquad \cosh^2 x - \sinh^2 x = 1.$$

The hyperbolic tangent and hyperbolic secant are defined, respectively, by

$$(6) \qquad\qquad \tanh x = \frac{\sinh x}{\cosh x}, \qquad \operatorname{sech} x = \frac{1}{\cosh x}.$$

By a short calculation, using (4) and (5),

(7) $$\frac{d}{dx}\tanh x = \operatorname{sech}^2 x, \qquad \frac{d}{dx}\operatorname{sech} x = -\operatorname{sech} x \tanh x$$

and also

(8) $$1 - \tanh^2 x = \operatorname{sech}^2 x.$$

One of the main uses of these functions is to simplify integrals, as explained next. If $a > 0$ it is a familiar fact that the integral

$$\int \frac{dx}{\sqrt{a^2 - x^2}}$$

is simplified by setting $x = a\sin\theta$. Namely, this substitution gives

$$\int \frac{a\cos\theta\, d\theta}{\sqrt{a^2 - a^2\sin^2\theta}} = \int \frac{a\cos\theta\, d\theta}{a\cos\theta} = \theta + c = \sin^{-1}\frac{x}{a} + c.$$

In a like manner the integral

$$\int \frac{dx}{\sqrt{a^2 + x^2}}$$

is simplified by setting $x = a\sinh u$. The result of substitution is

$$\int \frac{a\cosh u\, du}{\sqrt{a^2 + a^2\sinh^2 u}} = \int \frac{a\cosh u\, du}{a\cosh u} = u + c = \sinh^{-1}\frac{x}{a} + c$$

where $\sinh^{-1} x$ is the function inverse to $\sinh x$ as in Problem 6. An evaluation without hyperbolic functions is harder.

Anticipating a use of complex numbers that will be developed systematically in Chapter 6, we set $i = \sqrt{-1}$ so that $i^2 = -1$. Let us suppose that the function

$$C(x) = \cos ix$$

is defined in such a way that it can be differentiated by the ordinary rules, treating i as a constant. Under this assumption

$$C'(x) = -i\sin ix, \qquad C''(x) = -i^2\cos ix = \cos ix$$

and also $C(0) = \cos 0 = 1$, $C'(0) = -i\sin 0 = 0$. Hence $y = C(x)$ satisfies the same initial-value problem as was used to define $\cosh x$, namely,

$$y'' - y = 0, \qquad y(0) = 1, \qquad y'(0) = 0.$$

Thus a logical candidate for $\cos ix$ with the above properties is

$$\cos ix = \cosh x$$

and this is now taken as a definition. A similar procedure leads to the definition

$$\sin ix = i \sinh x.$$

If you are familiar with Taylor series, you will find that the substitution $t = ix$ in the Taylor series for $\sin t$ and $\cos t$ produces the above results at once. Our purpose in mentioning these equations, however, is not so much to *prove* anything as to give added insight.

PROBLEMS

1. Verify the formulas $(\sinh x)' = \cosh x$, $(\cosh x)' = \sinh x$.

2. If $z = (\cosh x)^2 - (\sinh x)^2$ show that $z' = 0$, hence $z = $ const., and hence $z = 1$. This gives (5). Check that division by $(\cosh x)^2$ yields (8).

3. Verify the formulas $(\tanh x)' = \text{sech }^2 x$, $(\text{sech } x)' = -\text{sech } x \tanh x$.

4. A function $f(x)$ is even if $f(-x) = f(x)$, odd if $f(-x) = -f(x)$. Which of the functions $\sinh x$, $\cosh x$, $\tanh x$, $\text{sech } x$ are even and which are odd?

5. Check that the graphs of the four functions in Problem 4 have the general appearance shown in Figure 3.

6. The equation $y = \sinh^{-1} x$ means $x = \sinh y$, and $y = \tanh^{-1} x$ means $x = \tanh y$. Set $e^y = z$, $e^{-y} = 1/z$ in these equations, solve for z and compute $y = \ln z$ to get

$$\text{(a) } \sinh^{-1} x = \ln(x + \sqrt{1 + x^2}) \qquad \text{(b) } \tanh^{-1} x = \frac{1}{2} \ln \frac{1 + x}{1 - x}$$

for $|x| < \infty$ and $|x| < 1$, respectively. Why is the minus sign for the radical excluded in the first formula?

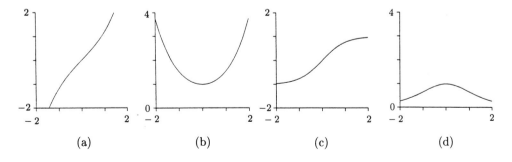

(a)　　　　　　(b)　　　　　　(c)　　　　　　(d)

Figure 3

7. If a is a positive constant set $x = a \tan \theta$ in the first integral and $x = a \tanh u$ in the second to establish the formulas

$$\int \frac{dx}{a^2 + x^2} = \frac{1}{a} \tan^{-1} \frac{x}{a} + c, \qquad \int \frac{dx}{a^2 - x^2} = \frac{1}{a} \tanh^{-1} \frac{x}{a} + c$$

The first is valid without restriction, the second for $|x| < a$.

8. If a is a positive constant and f is continuous, the equations

$$\frac{dy}{dx} = (a^2 + y^2)f(x), \qquad |y| < \infty; \qquad \frac{dy}{dx} = (a^2 - y^2)f(x), \qquad |y| < a$$

lead to integrals of the type considered in Problem 7 upon separation of variables. Guided by this, set $y = a \tan u$ in the first equation and $y = a \tanh u$ in the second to obtain the solutions

$$y = a \tan(aF(x) + c), \qquad y = a \tanh(aF(x) + c)$$

where F is any indefinite integral of f.

9. If a is a positive constant and f is continuous, show that the equation on the left below is solved by the two equations on the right:

$$\frac{dy}{dx} = \sqrt{a^2 + y^2}f(x), \qquad y = a \sinh u, \qquad u = \int f(x)\, dx + c$$

10. As seen in Figure 3, the equation $y = \cosh x$ has no solution x if $y < 1$, one if $y = 1$, and two solutions if $y > 1$. For this reason the theory of the inverse function requires more care than for the sinh or tanh. The equation $y = \cosh^{-1} x$ means $x = \cosh y$ and $y \geq 0$. Show that

$$\cosh^{-1} x = \ln(x + \sqrt{x^2 - 1}), \qquad x \geq 1$$

11. If $a > 0$ the first of the following formulas was derived in the text. Assuming $x > a > 0$ set $x = a \cosh u$ and derive the second:

$$\int \frac{dx}{\sqrt{a^2 + x^2}} = \sinh^{-1} \frac{x}{a} + c, \qquad \int \frac{dx}{\sqrt{x^2 - a^2}} = \cosh^{-1} \frac{x}{a} + c$$

12. Using the results of Problems 7 and 11 with $a = 1$ show that

$$\frac{d}{dx} \tanh^{-1} x = \frac{1}{1 - x^2} \qquad \frac{d}{dx} \sinh^{-1} x = \frac{1}{\sqrt{1 + x^2}} \qquad \frac{d}{dx} \cosh^{-1} x = \frac{1}{\sqrt{x^2 - 1}}$$

In the first case $|x| < 1$, in the second x is unrestricted, and in the third $x > 1$.

13. Clearly $x = \cos t$, $y = \sin t$ give parametric equations for the circle on the left below. Check that $x = \cosh t$, $y = \sinh t$ give parametric equations for one branch of the hyperbola on the right:

$$x^2 + y^2 = 1, \qquad x^2 - y^2 = 1$$

This is the reason why $\sin t$, $\cos t$ are called *circular functions* and $\sinh t$, $\cosh t$ are called *hyperbolic functions*.

14. Give reasonable definitions for the hyperbolic cotangent, $\coth x$, and the hyperbolic cosecant, $\operatorname{csch} x$, and establish formulas for their derivatives. The chances are that these functions will not be used in this book.

————————————————————ANSWERS————————————————————

    4. oeoe   6. Because $x - \sqrt{1 + x^2} < 0$ for all x and the logarithm of a negative number is undefined.

14. Partial answer: $\coth x = (\cosh x)/(\sinh x)$, $\operatorname{csch} x = 1/\sinh x$

---

3. A hyperbolic substitution

The logistic equation considered in Section 1 is a nonlinear equation of first order that is linearized by the substitution $P = 1/u$. Here we discuss two nonlinear equations of the second order that are linearized by the substitution $y' = \sinh u$. The first pertains to a chain hanging under gravity, the second to a pursuit problem.

(a) The loaded chain

We shall find the equation for the curve formed by a flexible chain bearing a load under gravity. As shown in Figure 4, the coordinates are chosen so that $x = 0$ at the lowest point and s denotes arc as measured along the chain. The height of the chain at x is $y = \phi(x)$, where $\phi(x)$ is to be found. Let

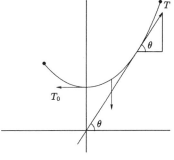

Figure 4

$$w_0 = \text{weight density of chain}, \qquad f(x) = \text{loading function}.$$

The constant w_0 gives the weight per foot of chain and $f(x)$ gives the load-density in pounds per foot as measured along the x axis. We assume $f(x)$ is continuous and nonnegative. The reason for assuming a chain rather than a cable is that we want the chain to be perfectly flexible. Mathematically, this means that the force due to tension is directed tangentially at every point.

The equation of the chain will be obtained from the fact that the portion of chain between 0 and x is in static equilibrium. Let the tension at the lowest point be T_0 and the tension at point (x, y) of the chain be T. Equating vertical forces gives

$$(1) \qquad T \sin \theta = \int_0^s w_0 \, ds + \int_0^x f(x) \, dx$$

since the weight of load-plus-chain must be balanced by the vertical component of force due to T. The first integral in (1) is $w_0 s$ and is so expressed in the sequel. Equating horizontal forces gives

$$T \cos \theta = T_0.$$

If we divide the first of the above equations by the second and recall that

$$\tan \theta = \text{slope of curve} = \frac{dy}{dx},$$

the result is

$$T_0 \frac{dy}{dx} = w_0 s + \int_0^x f(x) \, dx.$$

We now differentiate with respect to x, using the formula

$$\frac{ds}{dx} = \sqrt{1 + \left(\frac{dy}{dx}\right)^2} = \sqrt{1 + (y')^2}.$$

The result is

$$(2) \qquad T_0 y'' = w_0 \sqrt{1 + (y')^2} + f(x), \qquad y'(0) = 0.$$

The initial condition $y'(0) = 0$ expresses the fact that the slope of the chain is 0 at its lowest point.

(b) The unloaded chain

If $f(x) = 0$ in (2) the substitution $p = y'$ gives

$$T_0 \frac{dp}{dx} = w_0 \sqrt{1 + p^2}.$$

Separation of variables leads to an integral that is easily evaluated by setting $p = \sinh u$ and hence we make this substitution in the differential equation itself. The result is

$$T_0 \cosh u \, \frac{du}{dx} = w_0 \sqrt{1 + \sinh^2 u} = w_0 \cosh u.$$

Hence $T_0 \, du/dx = w_0$. Solving for u and substituting into $p = \sinh u$ we get

$$p = \sinh\left(\frac{w_0}{T_0}x + c_1\right).$$

The initial condition $p(0) = 0$ shows that $c_1 = 0$ and, since $p = dy/dx$, an integration gives

(3) $$y = \frac{T_0}{w_0} \cosh \frac{w_0}{T_0}x + c.$$

The constant of integration can be taken as 0 if the x axis is located in such a way that the height of the chain at its lowest point is T_0/w_0. The curve so obtained is called a *catenary* from the Latin *catena*, chain.

(c) A pursuit problem

A boat A moves along the y axis with constant speed a. It is required to find the path of a second boat B that moves in the left-hand half of the (x, y) plane with constant speed b and always points directly at A. See Figure 5. At time t minutes after A is at $(0,0)$ we shall have A at $(0, at)$ and B at (x, y), say. Since the line AB is tangent to the path of B the slope of this line equals the slope of the path, so that

$$\frac{dy}{dx} = \frac{y - at}{x - 0}.$$

Writing $y' = dy/dx$ we get $xy' - y = -at$, and differentiating with respect to x gives

(4) $$(xy' - y)' = xy'' = -a\frac{dt}{dx}.$$

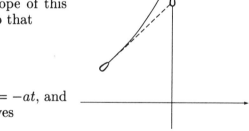

Figure 5

Since $ds/dt = b$, where s is the arc length on the trajectory of B, we have

(5) $$\frac{dt}{dx} = \frac{dt}{ds}\frac{ds}{dx} = \frac{1}{b}\sqrt{1 + (y')^2}.$$

Substituting (5) into (4) gives

$$xy'' = -k\sqrt{1 + (y')^2}, \qquad k = \frac{a}{b}.$$

As might be expected on physical grounds, the path of B depends only on the ratio of speeds k and not on the speeds themselves.

If $y' = p$ the equation becomes

$$xp' = -k\sqrt{1 + p^2}.$$

Separation of variables leads to an integral that is easily evaluated by setting $p = \sinh u$, and hence we let $p = \sinh u$ in the differential equation as it stands. The result is

$$x(\cosh u)u' = -k \cosh u, \quad \text{or} \quad xu' = -k.$$

If $c < 0$, so that c and x have the same sign, the solution of $du/dx = -k/x$ can be written

$$u = k \ln \frac{c}{x} = \ln \left(\frac{c}{x}\right)^k.$$

This gives simple expressions for e^u and e^{-u} and hence for $\sinh u$. Recalling that $dy/dx = p$ and that $p = \sinh u$, we get

$$\frac{dy}{dx} = \frac{1}{2}\left(\left(\frac{c}{x}\right)^k - \left(\frac{x}{c}\right)^k\right).$$

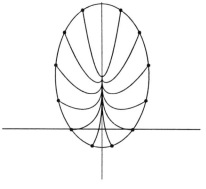

Figure 6

From this, y is obtained by integration. If the boat starts at (x_0, y_0) the initial conditions are

$$y(x_0) = y_0, \qquad y'(x_0) = \frac{y_0}{x_0}$$

where the second condition results from the fact that the boat is pointing toward the origin at time $t = 0$.

When dt/dx was introduced in (4) we assumed on physical grounds that x is strictly increasing. Hence the inverse function $t = \psi(x)$ for $x = \phi(t)$ is strictly increasing too. If the boat starts in the right half of the plane instead, then x is strictly decreasing, both dt/dx and ds/dx in (5) are negative, and a minus sign must be used. The trajectories starting in the left and right half planes are mirror images of each other as shown in Figure 6.

PROBLEMS

1. A flexible weightless chain supports a uniform roadway weighing w pounds per foot. If the chain is symmetric about the y axis and has its lowest point at the origin, show that it forms a parabola with equation $y = wx^2/(2T_0)$.

2. If the x axis is chosen so that $c = 0$ in the equation (3) for a hanging chain, show that the tension at any point is given by $T = w_0 y$.

Hint: $T = T_0 \sec \theta$ and $dy/dx = \tan \theta$.

3. Uniform rods are suspended from a chain in such a way that when equilibrium is reached the rods are equally spaced in the x direction and their ends are all on the x axis. This corresponds to the choice $f(x) = w_1 y$ in the text, where w_1 is a positive constant. Show that there is a solution in which the chain forms a catenary $y = (1/k) \cosh kx$ and find the constant k.

4. Problems 4, 5, and 6 range from difficult to very difficult. Solve the pursuit problem of the text under the hypothesis that A is at $(0,0)$ and B is at $(x_0, 0)$ at time $t = 0$. It is advisable to distinguish the cases $k = 1$ and $k \neq 1$. If $k < 1$ at what point and when does B overtake A? If $k = 1$ how close can B get to A?

5. Lotka's problem. Assuming $b > a$ suppose both boats start at the same time, A at the origin and B at the point P. Find the locus of P from which B catches A in the fixed time T.

6. It is desired to find the path of a small boat in a wide river with uniform current if the boat has constant speed relative to the water and always heads towards a fixed point on the bank. Show that this problem is mathematically equivalent to the pursuit problem of the text, and discuss in detail.

───────────────────ANSWERS───────────────────

3. $T_0 k^2 = k w_0 + w_1$. This has a unique positive solution for w_0 and w_1 positive, as seen by solving it or by plotting the parabola $y = T_0 k^2$ together with the line $y = k w_0 + w_1$.

4. $T = |x_0| b / (b^2 - a^2)$, $y = aT$;   almost within $|x_0|/2$

5. An ellipse with one focus at the origin and with center at the point of meeting, $(0, aT)$. It is obvious from the start that the points $(0, aT \pm bT)$ are on the locus, hence the major axis has length $2bT$. Thus the locus is the set of points (x_0, y_0) such that the sum of its distances from $(0,0)$ and $(0, 2aT)$ is $2bT$. See Figure 6.

6. Partial answer: Let a be the speed of the boat and $b = ra$ the speed of the river. Choose coordinates so that the y axis is the left bank, the boat heads toward the origin and the direction of flow is toward $y = +\infty$. Then $2y = x((cx)^{-r} - (cx)^r)$, $c > 0$. When $r = 1$ the boat follows a parabolic path that approaches, but does not reach, a point $(0, y_0)$, $y_0 > 0$.

4. Physical problems

The first of the topics considered here is an elaboration of the following question: If a stone is dropped from height h, with what speed does it hit the ground? One could find the position at any time t and differentiate to get the speed, but it is more efficient to reason as follows: Let m be the mass of the stone and g the acceleration of gravity. Then the weight of the stone is mg and the work needed to raise it through the height h is mgh. If it drops from rest and hits with velocity v_0, the gain in kinetic energy is $(1/2)mv_0^2$. Equating the gain in kinetic energy to the work needed to raise the stone gives

$$\frac{1}{2}mv_0^2 = mgh \quad \text{or} \quad |v_0| = \sqrt{2gh}.$$

(a) Conservation of energy

Suppose an object of mass m moves in a straight line under a force $F(s)$, where s is displacement from the origin and F is a continuous function. It is required to deduce a first-order equation for the velocity $v = ds/dt$ as a function of s. Newton's law

$$(\text{mass})(\text{acceleration}) = \text{force}$$

gives

$$(1) \qquad\qquad ms'' = F(s)$$

where the primes denote differentiation with respect to the time t.

We shall use the transformation $y' = v$, $y'' = v\,dv/dy$ that was introduced in Chapter 2. In the present case $y = s$, $v = ds/dt$, and the result is

$$mv\frac{dv}{ds} = F(s) \quad \text{or} \quad mv\,dv = F(s)\,ds.$$

Let us suppose that this holds on an interval $s_0 \leq s \leq s_1$, that $v = v_0$ when $s = s_0$, and that $v = v_1$ when $s = s_1$. Integrating between corresponding limits yields

$$(2) \qquad\qquad \frac{1}{2}mv^2\Big|_{v_0}^{v_1} = \int_{s_0}^{s_1} F(s)\,ds.$$

This is the *principle of conservation of energy* as applied to the equation (1). From Chapter A, Theorem 7 it follows that one can go in the opposite direction, if (2) is given for variable upper limits $v_1 = v$ and $s_1 = s$, and deduce (1) at any point where $v \neq 0$. Thus Newton's law is equivalent to the principle of conservation of energy on any interval on which the velocity does not vanish and the force F is a continuous function of displacement s.

To get a physical interpretation, recall that in moving a distance against a constant force the work is given by

$$\text{work} = (\text{force})(\text{distance}).$$

Hence $F(s)\Delta s$ is the work done in moving through a distance Δs against the force $F(s)$, and by a limit process with which you are probably familiar, the integral on the right of (2) represents the work done in moving from s_0 to s_1 against the variable force F. To interpret the left side, $(1/2)mv^2$ represents the *kinetic energy* due to the fact that the object is in motion. Then the principle of conservation of energy says:

$$(\text{change in kinetic energy}) = (\text{work done by the force}).$$

(b) Escape velocity

A rocket is fired vertically upwards and its motor is turned off as soon as it is high enough so that air resistance can be neglected — about 70 miles. If its velocity when the motor is turned off is v_0, what value v_0 just ensures that the rocket will not return to earth? This problem will be solved here by use of the principle of conservation of energy. The velocity needed to escape the earth's gravitational field is called the *escape velocity*.

Let m be the mass of the rocket and m_e the mass of the earth. By Newton's law of gravitation the force on the rocket due to gravity is

$$F(r) = -\frac{Gmm_e}{r^2}$$

where r is the distance to the earth's center and G is a constant, called the gravitational constant, whose value need not concern us in this problem. The minus sign is used because the positive direction is upwards, away from the earth's center.

The force at the earth's surface is $-mg$ and hence, if r_e is the radius of the earth,

$$\frac{-Gmm_e}{r_e^2} = -mg.$$

This gives G and shows that the force at a distance $r \geq r_e$ from the earth's center is

$$F = -mg \left(\frac{r_e}{r}\right)^2.$$

As a check and an aid to memory, note that the expression is of the form k/r^2, as it should be, and it gives the right value when $r = r_e$.

Although the rocket's coasting trip starts somewhat above the earth's surface, so air resistance can be neglected, we do the problem as if it starts at $r = r_e$. The work needed to raise the rocket to an unlimited height is

$$\int_{r_e}^{\infty} mg \left(\frac{r_e}{r}\right)^2 dr = \lim_{r \to \infty} mgr_e^2 \left(-\frac{1}{r} + \frac{1}{r_e}\right) = mgr_e.$$

By the principle of conservation of energy, this should be just enough to equal the kinetic energy $(1/2)mv_0^2$. Hence

$$\frac{1}{2}mv_0^2 = mgr_e, \quad \text{or} \quad v_0 = \sqrt{2gr_e}.$$

Comparing with the formula $v_0 = \sqrt{2gh}$ mentioned at the beginning of this section, we see that the escape velocity is the same as would be attained in a free fall through a height $h = r_e$ in a constant gravitational field equal to that at the earth's surface.

For numerical aproximation let us ignore the extra height needed to get above the atmosphere and let us take $g \approx 32$ ft/sec^2, $r_e \approx 4000$ miles. These values give

$$v_0^2 = 2gr_e \approx \left(\frac{64 \text{ ft}}{\text{sec}^2}\right)(4000 \text{ mi})\left(\frac{1 \text{ mi}}{5280 \text{ ft}}\right).$$

Hence the escape velocity is about 7 miles per second.

(c) Maximum height

Suppose that a stone is thrown vertically upward with velocity v_0 and that air resistance is not neglected. What is the maximum height attained? The point of this problem is that besides the force $-mg$ on the stone due to gravity there is a retarding force that depends on the velocity v. For viscous damping the force is $k|v|$ where k is constant. For power-law damping it is $k|v|^a$ where k and a are constant. The special case kv^2 is called *Newtonian damping*.

Here we assume, more generally, that the retarding force is a continuous function $k(v) \geq 0$. If s is the height of the stone measured positively upward and m is its mass, Newton's law gives

$$ms'' = -mg - k(v).$$

The minus sign is used on $-k(v)$ because v is positive as long as the stone is moving upward, and the resistance force has a sign opposite to the sign of v. This equation is of the form $s'' = g(s')$, and not $s'' = g(s)$ as in the previous discussion. Nevertheless, we shall use the same substitution $s'' = v \, dv/ds$. The result is

$$m \, v \frac{dv}{ds} = -mg - k(v), \quad \text{hence} \quad \frac{mv \, dv}{mg + k(v)} = -ds.$$

Since the stone is thrown upward with velocity v_0 we have $v = v_0$ when $s = 0$. Also $v = 0$ at the highest point. Denoting the maximum height by h and integrating between corresponding limits, we get

(3)
$$\int_0^{v_0} \frac{mv \, dv}{mg + k(v)} = \int_h^0 -ds = h.$$

This is an explicit formula for the maximum height h.

It is of some interest to compare this result with the formula $v_0 = \sqrt{2gh}$ for free fall through a distance h. If $k(v) = 0$ the formula (3) gives $v_0^2/(2g) = h$, which is consistent with the former result. Thus, in the absence of air resistance, we get the same pair of values (h, v_0) if a stone is dropped from height h and hits the ground with speed v_0, or if it is thrown upward with speed v_0 and attains the maximum height h.

Most people would consider this equivalence to be rather obvious. However, such a result is never to be expected in the presence of air resistance. To see why, suppose you make a parachute jump from 10,000 feet and land with the same speed $|v_0|$ you would acquire in jumping off a ten-foot building. If you now reverse the process, and jump upwards with speed $|v_0|$, you would only go back the 10 feet to the top of the building (and not the 10,000 feet to your starting point). Still less, if you are dragging the parachute behind you.

(d) The Torricelli-Borda law

Evangelista Torricelli is known mainly for his invention of the barometer in 1643. However he also stated a law of fluid flow under gravity when a tank is drained through a hole in its bottom. Namely, the exit speed from the hole is the same as would be attained by a free fall through a vertical distance equal to the head. Here the "head" means the depth y of the hole beneath the surface of the water. Since a stone dropped from height h hits the ground with speed $\sqrt{2gh}$, the Torricelli law gives

(4)
$$\text{(exit speed at depth } y) = \sqrt{2gy}.$$

It was observed by Torricelli's contemporary Borda that the fluid stream has a smaller diameter than the aperture, so that the effective size of the latter is diminished. Specifically, if the cross-section area of the exit pipe is

a the effective area is ba where the *Borda constant* b is usually between $1/2$ and 1. For water, b is about 0.6. We summarize as follows:

(5)   (effective aperture size) $= ba,$ $0 < b \leq 1.$

It is a general principle of physics that the rate of flow in a pipe is

(6)   (rate of flow)=(effective cross section)(speed of fluid flow).

The units depend on the units used to measure speed and cross-section area. For example, if the speed is in feet per second and the area in square feet, the rate of flow is in cubic feet per second. To see why (6) is reasonable, let the speed of flow be S and the effective cross-section area A. In time Δt a cylinder of water of length $S\Delta t$ and of base-area A exits from the pipe, and hence the change in volume is $\Delta V = AS\Delta t$. This gives $dV/dt = AS$, which is what the principle says. When the principle is combined with (4) and (5) the result is the *Torricelli-Borda law*

(7)   (rate of flow) $= (ba)\sqrt{2gy} = k\sqrt{y},$ $k = ab\sqrt{2g}.$

Note that a single constant k takes account of the size of the drain pipe as measured by a, of the nature of the fluid as measured by b, and even of the acceleration of gravity, g. (If the tank were drained on the moon, g and hence k would change accordingly.)

(e) Draining a tank

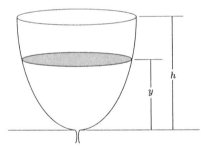

Figure 7 shows a water tank that is to be drained by a pipe at its bottom. We assume that the tank has no pits or depressions, so that all the water has access to the drain. Also the diameter of the exit pipe is negligible compared to the size of the tank. As the depth of the water decreases the rate of flow through the pipe also decreases, and

Figure 7

the rate approaches zero when the depth approaches zero. Hence it is not immediately obvious that the tank will drain in finite time. Our problem is to show that the time is finite, and to get a formula for the time in terms of the geometry of the tank. The key to the solution is provided by the Torricelli-Borda law.

Let $A(x)$ denote the cross-section area of the tank at height x above the bottom. We assume that $A(x)$ is positive and continuous. By elementary calculus (volume by slicing) the volume of water when the depth is y is

$$V(y) = \int_0^y A(x)\,dx.$$

This gives

$$\frac{d}{dt}V(y) = \frac{dV}{dy}\frac{dy}{dt} = A(y)\frac{dy}{dt}$$

where the first equality follows from the chain rule and the second from the fundamental theorem of calculus. Since the volume is decreasing, its rate of increase dV/dt is negative and is given by (7) as $-k\sqrt{y}$. Hence

$$-k\sqrt{y} = A(y)\frac{dy}{dt} \qquad \text{or} \qquad -k\,dt = \frac{A(y)}{\sqrt{y}}\,dy.$$

The value $t = 0$ corresponds to the initial depth h and $y = 0$ corresponds to the time T for emptying the tank. Integrating between corresponding limits gives a formula of remarkable simplicity and generality,

$$(8) \qquad\qquad kT = \int_0^h \frac{A(y)}{\sqrt{y}}\,dy.$$

PROBLEMS

If nothing is said about air resistance, air resistance is to be neglected. For numerical evaluation $g = 32\,\text{ft/sec}^2$.

1. A stone is thrown vertically upward with velocity 8 feet per second at time $t = 0$. Find the velocity and position at any time t. Find the time at which the velocity is 0 and show that the height is then maximum. Verify that the maximum height agrees with the principle of conservation of energy, that is, with the formula $mgh = mv_0^2/2$.

2. An iceboat weighing 500 pounds is driven by a wind that exerts a force of 25 pounds. Five pounds of this force are expended in overcoming frictional resistance. What speed will the boat acquire at the end of 30 seconds if it starts from rest?

Hint: The relation between weight w and mass m is $w = mg$.

3. (a) What constant acceleration is needed to run the hundred-yard dash in 10 seconds?

(b) If you and your frictionless rocket-powered skateboard together weigh 160 pounds, what constant thrust of the rocket will do the same job?

4. A brick is set moving over ice with initial speed v_0. If the coefficient of friction between the brick and the ice is μ, how long will it be before the brick stops? By integrating $ds = v\,dt$ obtain the total distance s traveled and verify that the work μmgs done by frictional forces equals the loss in kinetic energy. This illustrates the fact that conservation of energy can apply to dissipative systems.

5. A particle of mass m slides down an inclined plane making an angle θ with the horizontal. If friction is neglected the component of force in the direction of motion is $F = mg \sin\theta$. Hence the equation of motion is $m\, d^2r/dt^2 = mg \sin\theta$ where r is the distance along the inclined plane. If a number of particles are released from rest at the same point and time on a number of inclined planes as suggested in Figure 8, show that at any later time T the particles all lie on a circle. This result was known to Galileo before Newton's laws were available.

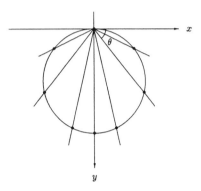

Figure 8

Suggestion: Regard (r, θ) as polar coordinates, multiply by r and convert to Cartesian coordinates.

6. If the coefficient of friction in the above problem is μ there is an extra term $-mg\mu \cos\theta$ that tends to retard the motion. Find the locus of the various particles at any subsequent time.

7. In this problem x is measured horizontally and y vertically with the positive direction upward. At time $t = 0$ a golf ball is hit with speed v_0 in a direction making an angle $\theta > 0$ with the horizontal. Thus the horizontal and vertical components of the initial velocity are respectively

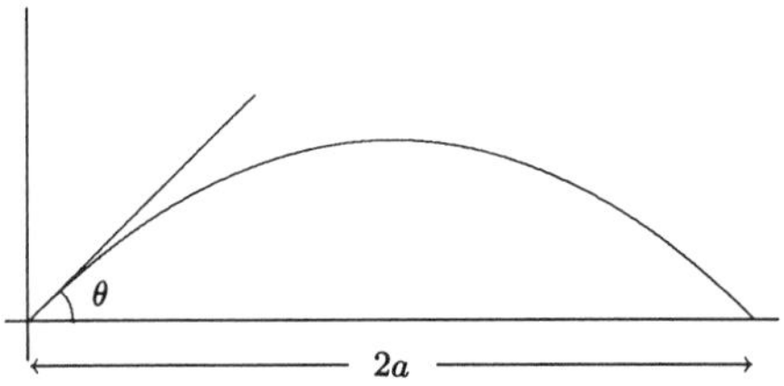

$$v_1 = v_0 \cos\theta, \qquad v_2 = v_0 \sin\theta$$

Figure 9

It is an important principle of physics that x and y act independently of each other in situations such as this. Hence if (x, y) gives the position of the golf ball at time t and m is its mass, the equations of motion are

$$mx'' = 0, \qquad my'' = -mg$$

(a) If the golf ball starts at the origin show that

$$x = v_1 t \qquad y = v_2 t - (1/2)gt^2 \qquad y = x \tan\theta - \frac{g}{2v_0^2}x^2 \sec^2\theta$$

The third equation holds for $\theta \neq \pi/2$ and shows that the orbit is a parabola in that case.

(b) Show that y is maximum when $gt = v_2$. Conclude that the vertex of the parabola is at (a, b) where $ag = v_1 v_2$, $2bg = v_2^2$.

(c) By setting $y = 0$ check that the distance to the point where the ball lands is twice the value a, which equals $(v_0^2/g) \sin 2\theta$. See Figure 9.

8. A building with vertical walls stands on a horizontal plain. A stone is thrown horizontally out of a window at height H feet above the plain and is D feet from the building when it lands; see Figure 10. Show that the speed with which it was thrown is approximately $4D/\sqrt{H}$ feet per second.

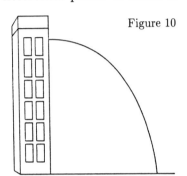

Figure 10

9. Water is flowing from a 2-inch horizontal pipe running full. Find the rate of discharge in cubic feet per minute if the jet of water strikes the ground 4 feet beyond the end of the pipe when the pipe is 2 feet above the ground.

10. An ancient water clock called a clepsydra is so designed that the surface descends at a constant rate as the water runs out through a hole in the bottom. This design allows use of a uniform scale to mark the hours. Find the cross-section area $A(y)$ of the tank as a function of y, and conclude that, if the tank is a volume of revolution with vertical axis, its profile is given by $y = cx^4$.

Hint: In the notation of Part 4(e), $T(h) = (\text{const.})h$. Differentiate.

11. For the tank of Part 4(e) let $A(y) = cy^n$ where c and n are positive constants. Compute kT, V and kT/V and thus show that

$$kT = \frac{V}{\sqrt{h}} \frac{2n+2}{2n+1}$$

where V is the volume of the tank. Find n for the four tanks illustrated in Figure 11, given that the curve in the last two figures is a parabola.

 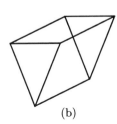 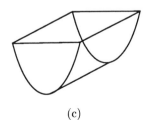 

(a)   (b)   (c)   (d)

Figure 11

12. Two tanks are both right circular cylinders of height L and diameter L and both are initially full. The first stands on end and drains through a hole in its flat base. The second lies on its side, with its axis horizontal, and drains through a hole in the lowest part of the curved side. If the holes are the same size, which tank do you think will drain faster? Check your conjecture by finding the ratio of the two times, first/second.

13. Show that the time T for maximum height in the problem leading to (3) is given by

$$T = \int_0^{v_0} \frac{m\,dv}{mg + k(v)}$$

14. You drop a wrench into an abandoned mine shaft and 16/3 seconds later you hear the noise it made when it hit the bottom.

(a) If $g = 32$ feet per second per second and the speed of sound is 1200 feet per second how deep is the shaft?

(b) Does it make a significant difference if you use the more accurate values 32.17, 1125 for the acceleration of gravity and the speed of sound, respectively?

15. An anti-aircraft gun standing on a broad horizontal plane has the constant muzzle velocity v_0. Its angle of elevation θ can be set at any value from 0 to $\pi/2$, and the gun swivels so that it can point in any direction. Describe the region that is within the range of the gun. See Figure 12.

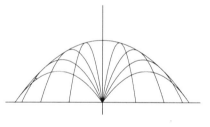

Hint: The Cartesian equation of the trajectories is given in Problem 7. Write in terms of $c = \tan\theta$ and find the envelope.

Figure 12

――――――――――――――――ANSWERS――――――――――――――――

1. $v = 8(1 - 4t)$, $s = 8t(1 - 2t)$, $v = \pm 8\sqrt{1 - s}$ 2. l92/5 ft/sec

3. (a) 6 ft/sec^2 (b) Since this is $(3/16)g$, and since 160 pounds is the force associated with g, the rocket force is $(3/16)(160) = 30$ lbs.

4. $t_0 = v_0/(\mu g)$, $s_0 = v_0^2/(2\mu g)$

5. $r = 2c\sin\theta$ where $4c = gT^2$. Hence $r^2 = 2cy$, which is a circle.

6. $(|x| + \mu c)^2 + (y - c)^2 = c^2(1 + \mu^2)$ where $4c = gT^2$. This represents two circular arcs through the points $(0,0)$ and $(0, 2c)$. The arcs make an angle θ with the horizontal at $(0,0)$ where $\tan\theta = \mu$. At smaller angles the particle does not move.

9. About 15 cubic feet per minute. "Running full" means that the effective pipe size is 2 inches after taking account of the Borda constant.

10. $A(y) = (\text{const})\sqrt{y}$ 11. $n = 2, 1, 1/2, 1$ 12. $3\pi/8$

14. (a) 400 feet (b) No. The new answer is 398.7 feet, which corresponds to an error of about 0.006 seconds in the time. This is well beyond the probable accuracy of the measurement. Similar considerations motivate our use of simple values, such as $g = 32$, in other problems.

15. The envelope is the parabola through the three points $(0, v_0^2/2g)$ and $(\pm v_0^2/g, 0)$ that describe the maximum height and the maximum horizontal range, respectively. If you revolve the envelope about a vertical axis, the region is bounded by the parabolic surface so obtained and by the plane $y = 0$.

Not only the movements of the heavenly host must be determined by observation and elucidated by mathematics, but whatsoever else can be expressed by a number and defined by natural law.

—d'Arcy Wentworth Thompson

Nature first made things in her own way, and then made human reason skillful enough to be able to understand, but only by hard work, some part of her secrets.

—Galileo

As far as the laws of mathematics refer to reality they are not certain, and as far as they are certain, they do not refer to reality.

—Albert Einstein

Chapter 6
LINEAR EQUATIONS I

SECOND-ORDER linear equations with constant coefficients are discussed here by means of general principles that are valid in a broader context. Among these are the principle of superposition and the principle of the complementary function. An interesting feature of the analysis is that many problems are best solved by use of complex-valued functions, especially when the solution exhibits an oscillatory behavior. The bridge between real and complex solutions is provided by a third general principle, the principle of equating real and imaginary parts.

1. Basic principles

Numerous problems in mechanical and electrical engineering lead to an equation of the form

$$(1) \qquad y'' + py' + qy = f(t), \qquad t \in I$$

where f is continuous on a given interval I, the coefficients p and q are constant, and the primes denote differentiation with respect to t. If we define an operator T by

$$(2) \qquad Ty = y'' + py' + qy$$

then T is linear; that is,

$$(3) \qquad T(cu) = cTu, \qquad T(u+v) = Tu + Tv$$

for any constant c and for all functions u and v in the domain of T. In the present case the domain of T consists of functions twice differentiable on I. The equation $T(cu) = cTu$ follows from

$$(cu)'' + p(cu)' + q(cu) = cu'' + pcu' + qcu = c(u'' + pu' + qu).$$

The proof that $T(u+v) = Tu + Tv$ is similar.

Since T is linear, we can use the following principle to get a family of solutions of $Ty = 0$ from two solutions:

PRINCIPLE OF SUPERPOSITION. Let T be any linear operator. If $Tu = 0$ and $Tv = 0$, then $T(c_1u + c_2v) = 0$ for any constants c_1 and c_2.

The proof is immediate because $T(c_1u+c_2v) = c_1Tu+c_2Tv$ by linearity and Tu and Tv are 0 by hypothesis.

In order to get the particular solutions u and v, note that, if s is any constant, the function e^{st} satisfies

$$(e^{st})' = se^{st}, \qquad (e^{st})'' = s^2e^{st},$$

and so on. Hence substitution in (2) gives

$$T(e^{st}) = e^{st}(s^2 + ps + q)$$

when the common factor e^{st} is factored out. This equation is used in the following example.

EXAMPLE 1.   By setting $y = e^{st}$ find two solutions of $y'' + 3y' + 2y = 0$. The suggested substitution gives

$$Ty = e^{st}(s^2 + 3s + 2) = e^{st}(s+1)(s+2)$$

and $Ty = 0$ if $s = -1$ or $s = -2$. The solutions are e^{-t}, e^{-2t}.

The equation
$$s^2 + ps + q = 0$$

is called the *characteristic equation* of the operator T defined by (2). The gist of the above method is that the homogeneous equation $Ty = 0$ has a solution $y = e^{st}$ if, and only if, s satisfies the characteristic equation.

Most people would think that two solutions e^{-t} and $2e^{-t}$ are more or less the same, while e^{-t} and e^{-2t} are really different. This intuitive feeling is given a mathematical significance by the following definition:

DEFINITION. On a given interval two functions u and v are linearly dependent if one of them is a constant multiple of the other, and they are linearly independent if neither is a constant multiple of the other.

According to this definition e^{-t} and $2e^{-t}$ are linearly dependent on every interval and e^{-t} and e^{-2t} are linearly independent on every interval. The concept of linear dependence is discussed in Section 6, where it is seen that linearly independent solutions of the homogeneous equation $Ty = 0$ can be used to satisfy arbitrary initial conditions. Instead of developing theory at this point, we give an example.

EXAMPLE 2.   Find the solution in Example 1 satisfying arbitrary initial conditions
$$y(0) = y_0, \qquad y'(0) = y_1.$$

Since the independent solutions e^{-t} and e^{-2t} were found in Example 1, the superposition principle gives the solution

$$y = c_1 e^{-t} + c_2 e^{-2t}$$

for any constants c_1 and c_2. The initial conditions hold if

$$c_1 + c_2 = y_0, \qquad -c_1 - 2c_2 = y_1.$$

Adding the two equations we get $c_2 = -(y_0 + y_1)$ and the first equation then gives c_1. The result is

$$y = (2y_0 + y_1)e^{-t} - (y_0 + y_1)e^{-2t}.$$

So far we have considered only the homogeneous equation, that is, the equation with $f = 0$. The case of general f can often be handled by a trial solution, much as we used the trial solution $y = e^{st}$ to discuss the homogeneous equation. Since the trial solution involves coefficients that are initially undetermined , this method is called the *method of undetermined coefficients.*

EXAMPLE 3. If the constant a does not have one of two exceptional values, show that the equation

$$y'' + 3y' + 2y = 7e^{at}$$

has a solution of the form $y = Ae^{at}$ where A is constant. The function Ae^{at} satisfies the equation if

$$Ae^{at}(a^2 + 3a + 2) = 7e^{at}.$$

Dividing by e^{at} shows that the equation holds if, and only if,

$$A(a^2 + 3a + 2) = 7.$$

This can be solved for A unless a has one of the values -1 or -2 for which the expression in parentheses vanishes. In the following section we shall see what to do about that case.

EXAMPLE 4. If a, b, c are constant and $a^2 \neq b^2$ find a particular solution of

$$y'' + a^2 y = c \sin bt.$$

Here we try $y = A \sin bt$ where A is constant. The result is

$$(-b^2 + a^2)A \sin bt = c \sin bt, \quad \text{hence} \quad A = \frac{c}{a^2 - b^2}.$$

In Section 4 we shall see what to do if $a^2 = b^2$.

The following result enables us to get a family of solutions of the inhomogeneous equation $Ty = f(t)$ from a particular solution:

PRINCIPLE OF THE COMPLEMENTARY FUNCTION. Let u be a particular solution of $Tu = f$, where T is a linear operator, and let v be any solution of $Tv = 0$. Then $T(u + v) = f$, and every solution of $Ty = f$ can be obtained in this way.

That $T(u + v) = f$ is evident because $T(u + v) = Tu + Tv$ by linearity and $Tu = f$ and $Tv = 0$ by hypothesis. The meaning of the second assertion is that if $Ty = f$, then y can be written in the form $y = u + v$, where v satisfies the homogeneous equation $Tv = 0$. To see this, define $v = y - u$. Then $y = u + v$ and $Tv = T(y - u) = Ty - Tu = f - f = 0$.

EXAMPLE 5. In three steps, we shall solve the initial-value problem

$$(4) \qquad y'' - 4y = 8t - 16, \qquad y(0) = 4, \; y'(0) = 2.$$

First step: Solve the homogeneous equation $Ty = 0$. The substitution $y = e^{st}$ into the equation $y'' - 4y = 0$ gives $s^2 - 4 = 0$. Hence $s = 2$ or -2, and the superposition principle gives the family of solutions

$$v = c_1 e^{2t} + c_2 e^{-2t}, \qquad Tv = 0.$$

Second step: Find a particular solution u of the inhomogeneous equation $Ty = f$. Since the right-hand side has the form $at + b$ we try $y = At + B$. Substitution into the differential equation (4) gives

$$-4(At + B) = 8t - 16.$$

Hence, by equating coefficients, $-4A = 8, -4B = -16$. The result is

$$u = 4 - 2t, \qquad Tu = f.$$

Third step: Use the principle of the complementary function to form the family of solutions

$$(5) \qquad y = 4 - 2t + c_1 e^{2t} + c_2 e^{-2t}, \qquad Ty = f.$$

The constants are determined by the initial conditions as follows:

$$y(0) = 4: \qquad 4 + c_1 + c_2 = 4, \qquad c_1 + c_2 = 0$$
$$y'(0) = 2: \qquad -2 + 2c_1 - 2c_2 = 2, \qquad c_1 - c_2 = 2.$$

The values $c_1 = 1$, $c_2 = -1$ can be read from the last two equations and (5) gives the solution,

$$y = 4 - 2t + e^{2t} - e^{-2t}.$$

We conclude with some statements that have not been proved at this point but give useful insight. On an interval I let u and v be linearly independent solutions of $Ty = 0$ and let w be a particular solution of $Ty = f$. Then the family of functions

$$y = c_1 u + c_2 v + w$$

is both the *general solution* and the *complete solution* of $Ty = f$. It is a general solution because the constants c_i can be chosen so as to satisfy arbitrary initial conditions

$$y(t_0) = y_0, \qquad y'(t_0) = y_1$$

at any specified point $t_0 \in I$. It is the complete solution because every solution of $Ty = f$ is contained in the family. Since any general solution coincides with the complete solution, one can speak of "the general solution" rather than "a general solution." The function $c_1 u + c_2 v$ is called *the complementary function* because when added to any particular solution of $Ty = f$, it produces the complete solution of this equation.

PROBLEMS

1P. Show that the functions u and v are linearly dependent on an interval I if, and only if, there are constants c_1 and c_2, not both 0, such that

$$c_1 u(t) + c_2 v(t) = 0, \qquad t \in I$$

2P. In this problem a and b are constant and I is an interval. Show that e^{at} and e^{bt} are linearly independent on I if $a \neq b$, and e^{at} and te^{at} are linearly independent on I for all a.

3. Obtain two linearly independent solutions of the form e^{st} where s is constant:

(a) $y'' = 9y$ (b) $y'' = 4y'$ (c) $y'' - y' = 2y$ (d) $y'' + 5y' + 6y = 0$

4P. This problem pertains to the equation $y'' + 2y = 3y'$.

(a) Obtain two linearly independent solutions by trying $y = e^{st}$.

(b) Write the general solution by use of the superposition principle.

(c) Thus find the solution satisfying $y(0) = y'(0) = 1$.

5. Solve the equation subject to the initial conditions below it:

(a) $y'' - y = 0$ (b) $y'' + 7y' + 12y = 0$ (c) $y'' + 7y' + 10y = 0$

$y(0) = 1,\ y'(0) = 0$ $y(0) = -1,\ y'(0) = 4$ $y(0) = 12,\ y'(0) = -3$

6. Obtain a particular solution by use of a trial function of the form Ae^{st}, $A\sin st$, $A\cos st$, or $At + B$, whichever is appropriate:

(a) $y'' + 3y = 4\sin 2t$ (c) $y'' - y = t + 1$ (e) $y'' = y' + y + 10e^{3t}$

(b) $y'' + y' + y = 9e^{t}$ (d) $y'' - y = 10\cos t$ (f) $y'' + t = y' + y + 4$

7. Obtain a particular solution under the assumption that the constant a does not have certain exceptional values. What are these exceptional values?

(a) $y'' + y = \sin at$ (c) $y'' + y = e^{-at}$ (e) $y'' + 9y' + 14y = e^{at}$

(b) $y'' - y' = e^{at}$ (d) $y'' - 9y = \cos at$ (f) $y'' + 8y' + 12y = e^{-at}$

8P. The *rest solution* of a differential equation is the solution that satisfies null initial conditions at $t = 0$. We are going to find the rest solution of

$$Ty = y'' + y = 3\sin 2t + 3 + 4e^{t}$$

(a) Solve $Ty = 3\sin 2t$, $Ty = 3$, $Ty = 4e^{t}$ by use of appropriate trial solutions. The general superposition principle of Chapter 4 then yields a particular solution of the given equation.

(b) Adding the complementary function gives the two-parameter family

$$y = c_1\cos t + c_2\sin t - \sin 2t + 3 + 2e^{t}$$

Determine the constants so that $y(0) = y'(0) = 0$.

9. Obtain the rest solution:

(a) $y'' + y = 2e^{t}$ (c) $y'' + y' = 2e^{-2t} + 4e^{t}$

(b) $y'' - y = 4\sin t$ (d) $y'' + 9y = 15\sin 2t + 36e^{-3t}$

10. Solve, subject to the given initial conditions:

(a) $y'' + 3y' + 2y = 24e^{-5t} - 4,$ $y(0) = 0,\ y'(0) = -9$

(b) $y'' - y' - 2y = -8e^{-2t} - 2t - 3,$ $y(0) = 1,\ y'(0) = 9$

11P. The operator $D = d/dt$ denotes differentiation; for example,

$$D\sin t = \cos t, \qquad De^{2t} = 2e^{2t}$$

Give a reasonable interpretation of each of the following expressions, and decide whether they are equal. No answer is provided:

$$(D + 2)(D - 3)t^{3}, \qquad (D - 3)(D + 2)t^{3}$$

─────────────────────────ANSWERS─────────────────────────

1. If $u = cv$ then $c_1 u + c_2 v = 0$ with $c_1 = 1 \neq 0$ and $c_2 = -c$.
   If $c_1 u + c_2 v = 0$ and $c_1 \neq 0$ then $u = cv$ with $c = -c_2/c_1$.

3. (a) e^{3t}, e^{-3t} (b) 1, e^{4t} (c) e^{-t}, e^{2t} (d) e^{-2t}, e^{-3t} 4. e^t

5. (a) $\cosh t$ (b) $-e^{-4t}$ (c) $19e^{-2t} - 7e^{-5t}$

6. (a) $-4\sin 2t$ (b) $3e^t$ (c) $-t - 1$ (d) $-5\cos t$ (e) $2e^{3t}$ (f) $t - 5$

7. (a) $(\sin at)/(1 - a^2)$, $a \neq 1, -1$ (d) $-(\cos at)/(a^2 + 9)$, no exceptions
   (b) $e^{at}/(a^2 - a)$, $a \neq 1, 0$ (e) $e^{at}/(a^2 + 9a + 14)$, $a \neq -7, -2$
   (c) $e^{-at}/(a^2 + 1)$, no exceptions   (f) $e^{-at}/(a^2 - 8a + 12)$, $a \neq 6, 2$

8. $c_1 = -5$,   9. (a) $e^t - \cos t - \sin t$ (c) $e^{-2t} + 2e^t - 3$
   $c_2 = 0$ (b) $-2\sin t + 2\sinh t$ (d) $3\sin 2t + 2e^{-3t} - 2\cos 3t$

10. (a) $2e^{-5t} + e^{-t} - e^{-2t} - 2$ (b) $1 + t + 4\sinh 2t$

2. Operator notation

The study of linear differential equations becomes easier if we introduce a special symbol $D = d/dt$ for the operation of differentiation. This symbol should be distinguished from the symbol d used in connection with differentials. Thus

$$D(t^3) = 3t^2, \qquad d(t^3) = 3t^2 dt.$$

Since

(1) $$D(cu) = cDu, \qquad D(u + v) = Du + Dv$$

for all differentiable functions u, v and for any constant c, the operator D is linear. Its usefulness stems from additional properties which indicate that D behaves much like a number. For example, if we define powers by

$$D^0 u = u, \quad D^m u = u^{(m)} = \frac{d^m u}{dt^m} \qquad (m = 1, 2, \cdots)$$

whenever u has m derivatives, then

(2) $$D^m D^n = D^{m+n}, \qquad (D^m)^n = D^{mn} \qquad (m, n = 0, 1, 2, \cdots).$$

Operator equations such as these mean that the operators on each side have the same domain and do the same thing to functions in their domain. Since

(2) is an immediate consequence of the definitions, we do not give a formal proof.

An operator expression such as

$$P(D) = D^2 + pD + q$$

is defined by prescribing its domain and stating what it does to any function in its domain. In this case the domain consists of all functions twice differentiable on some interval I. For such functions, by definition,

(3) $(D^2 + pD + q)u = D^2u + pDu + qu = u'' + pu' + qu.$

Although (3) resembles the linearity property (1), it is merely a definition while (1) is a theorem about differentiation.

If a and b are constant a short calculation based on (1) and (3) gives

$$(D + a)(D + b)u = D^2u + (a + b)Du + abu$$

for any twice-differentiable function u. Since the right-hand side is unchanged when a and b are permuted, that is, when (a, b) is replaced by (b, a), the left-hand side must have the same property. This shows that

$$(D + a)(D + b) = (D + b)(D + a), \qquad a, b \text{ const.}$$

The gist of these remarks is that polynomials $P(D)$ can be manipulated much as if D were a number, provided the coefficients are constant. That this provision is essential is shown in Problem 1.

There is an interesting connection between the polynomial $P(D)$ and the characteristic equation introduced in Section 1. If s is constant,

$$De^{st} = se^{st}, \qquad D^2e^{st} = s^2e^{st},$$

and so on. Hence

$$(D^2 + pD + q)e^{st} = (s^2 + ps + q)e^{st}.$$

The result is summarized by the equation

(4) $P(D)e^{st} = P(s)e^{st},$

which shows that the characteristic equation for the operator $T = P(D)$ can be obtained by writing s instead of D.

EXAMPLE 1. Find two linearly independent solutions by inspection:

(5) $(D - 4)(D + 5)y = 0.$

The characteristic equation is $(s - 4)(s + 5) = 0$ with roots $s = 4$, $s = -5$. This yields the solutions

$$e^{4t}, \qquad e^{-5t}.$$

In numerous investigations one requires trial solutions of the form $e^{kt}u$ where u is a sufficiently differentiable function and k is constant. By the rule for differentiating a product

$$D(e^{kt}u) = e^{kt}Du + ke^{kt}u = e^{kt}(D + k)u.$$

Hence if b is constant

$$(D - b)e^{kt}u = e^{kt}(D + k - b)u.$$

Thus, the factor e^{kt} can be moved to the left of the operator $D - b$ provided D is replaced by $D + k$. Repetition gives a similar result for two factors,

$$(D - a)(D - b)e^{kt}u = (D - a)e^{kt}(D + k - b)u = e^{kt}(D + k - a)(D + k - b)u.$$

This is important enough to state as a theorem:

EXPONENTIAL SHIFT THEOREM. Let k be constant and let

$$P(D)y = (D - a)(D - b)y$$

where a and b are constant. Then a factor e^{kt} can be moved across the operator $P(D)$ provided D is replaced by $D + k$. In symbols

$$P(D)e^{kt}u = e^{kt}P(D + k)u$$

for all u in the domain of $P(D)$.

As our first application of the shift theorem let us see what to do about the case of repeated roots. Here the differential equation is

$$(D - a)^2 y = 0, \qquad a = \text{const.}$$

and the trial solution $y = e^{st}$ yields only the single solution e^{at} with a corresponding family

$$y = e^{at}c.$$

Using a method suggested by the French-Italian mathematician Lagrange, we replace c by a function u and try

$$y = e^{at}u.$$

By the shift theorem with $k = a$

$$(D - a)^2 e^{at} u = e^{at}(D + a - a)^2 u = e^{at} D^2 u.$$

The desired equation $(D - a)^2 y = 0$ holds if $D^2 u = 0$ and hence if u is a linear function of t. Thus we get the family of solutions

$$y = (c_1 + c_2 t)e^{at}.$$

In particular, the choices $(c_1, c_2) = (1, 0)$ and $(0, 1)$ yield two linearly independent solutions

$$e^{at}, \qquad te^{at}.$$

As a second illustration, the shift theorem shows how to get a particular solution of an equation such as

(6) $$(D - 3)^2 y = 16e^{3t}.$$

The trial solution $y = Ae^{3t}$ does not work because 3 is a root of the characteristic equation and hence this choice of y gives $Ty = 0$. Instead let us try $y = e^{3t} u$ and use the shift theorem with $k = 3$. The result is

$$(D - 3)^2 e^{3t} u = e^{3t}(D + 3 - 3)^2 u = e^{3t} D^2 u = 16e^{3t}$$

where the last equality is imposed because we want a solution of (6). Clearly $D^2 u = 16$ does the job. Integrating twice gives $u = 8t^2$ and hence

$$y = 8t^2 e^{3t}.$$

We drop the constants of integration because only a particular solution is wanted.

As a third illustration let us find a particular solution of

$$(D - 1)(D + 2)y = 12e^t.$$

The trial solution $y = Ae^t$ is not appropriate because $s = 1$ is a root of the characteristic equation, so that this choice makes the left-hand side vanish. Instead let us set $y = e^t u$ and use the shift theorem with $k = 1$. The result is

$$(D - 1)(D + 2)e^t u = e^t(D + 1 - 1)(D + 1 + 2)u = 12e^t.$$

This reduces to

$$D(D + 3)u = 12 \quad \text{or} \quad (D + 3)Du = 12.$$

The second form is preferable because with $v = Du$ it becomes

$$(D + 3)v = 12.$$

This has the obvious solution $v = 4$. Since $v = Du$ an integration gives u and the final result is

$$v = 4, \qquad u = 4t, \qquad y = 4te^t.$$

Although we assumed $P(D)$ to be in factored form when proving the theorem, the following example shows that $P(D)$ need not be factored when the theorem is used.

EXAMPLE 2. By the trial function $y = e^{-2t}u$ obtain a particular solution:

$$(D^2 + 2D + 8)y = 4(1 + 4t)e^{-2t}.$$

The suggested substitution gives

$$(D^2 + 2D + 8)e^{-2t}u = 4(1 + 4t)e^{-2t}$$

or, by the shift theorem with $k = -2$,

$$e^{-2t}((D - 2)^2 + 2(D - 2) + 8)u = 4(1 + 4t)e^{-2t}.$$

Canceling the factor e^{-2t} and simplifying, we get

$$(D^2 - 2D + 8)u = 4(1 + 4t).$$

The trial function $u = at + b$ gives

$$-2a + 8(at + b) = 4 + 16t, \quad \text{hence} \quad 8a = 16, \quad -2a + 8b = 4.$$

The final result is $u = 2t + 1, \quad y = (2t + 1)e^{-2t}.$

PROBLEMS

1. Show that the equation $(D + t)(D + 1) = (D + 1)(D + t)$ fails if both sides are applied to the function $u(t) = 1$.

2. Write the general solution by inspection:

(a) $(D - 1)^2 y = 0$ (b) $(D + 1)(D - 2)y = 0$ (c) $(D + 4)^2 y = 0$

3P. We are going to find the general solution of

$$(D + 1)(D - 2)y = (9t^2 + 6t - 3)e^{2t}$$

(a) Set $y = e^{2t}u$ and use the shift theorem to get the equivalent equation

$$(D + 3)Du = 9t^2 + 6t - 3$$

(b) Set $Du = At^2 + Bt + C$ and find first A, then B, and finally C.

(c) Finish the job.

4. Find the general solution:

(a) $(D-1)^2 y = 80(t^3+1)e^t$ (c) $(D-1)(D+2)y = (18t^2 - 12t + 3)e^{-2t}$

(b) $(D^2-9)y = (2+12t)e^{3t}$ (d) $(D-\pi)^2 y = 120t^4 e^{\pi t}$

5. Solve $Dy = t^3 e^t$ by the shift theorem and thus integrate $t^3 e^t$ without using the formula for integration by parts.

6. Obtain a particular solution without factoring $P(D)$, as in Example 2:

(a) $(D^2+D+1)y = te^t$ (b) $(D^2-2D+8)y = e^t \sin 2t$ (c) $e^t(D^2+1)y = t^2$

7P. If a, b, c, k are constant and u is a three-times differentiable function, explain why

$$(D-a)(D-b)(D-c)e^{kt}u = e^{kt}(D+k-a)(D+k-b)(D+k-c)u$$

8P. With a, b, c as in Problem 7, explain why

$$(D-a)(D-b)(D-c) = (D-c)(D-b)(D-a)$$

───────────────ANSWERS───────────────

2. (a) $(c_1 + c_2 t)e^t$ (b) $c_1 e^{-t} + c_2 e^{2t}$ (c) $(c_1 + c_2 t)e^{-4t}$
3. $y = (t^3 - t + c_1)e^{2t} + c_2 e^{-t}$
4. (a) $(4t^5 + 40t^2 + c_1 t + c_2)e^t$ (c) $(c_1 - t - 2t^3)e^{-2t} + c_2 e^t$
     (b) $(t^2 + c_1)e^{3t} + c_2 e^{-3t}$ (d) $(4t^6 + c_1 t + c_2)e^{\pi t}$
5. $\int t^3 e^t dt = (t^3 - 3t^2 + 6t - 6)e^t + c$
6. (a) $(t-1)e^t/3$ (b) $(e^t \sin 2t)/3$ (c) $e^{-t}(t+1)^2/2$
7. Because you can move e^{kt} to the left of each factor one at a time, as was done for two factors in the text.
8. Because an interchange of two adjacent factors does no harm, and you can get from $(D-a)(D-b)(D-c)$ to $(D-c)(D-b)(D-a)$ by three interchanges of adjacent factors, as follows:

$$(D-b)(D-a)(D-c), \quad (D-b)(D-c)(D-a), \quad (D-c)(D-b)(D-a)$$

3. Complex numbers

If you try to solve $y'' + y = 0$ by setting $y = e^{st}$ you are led to

(1) $$s^2 + 1 = 0.$$

This equation has no solution in real numbers because the square of a real number cannot be negative. But if you adjoin a symbol i to the real number system, which satisfies the equation

$$i^2 = -1$$

by definition, you can construct a system of *complex numbers* in which (1) has a solution.

Complex numbers are an indispensable aid in the study of oscillations of mechanical and electrical systems, in electrostatics, in the theory of fluid flow, and in other branches of physics and engineering. They greatly increase the scope of the methods we have introduced up to now and, as seen presently, they lead to a complete solution of the general linear equation with constant coefficients. In the next section we indicate some of the reasons why complex numbers are so useful. Here we explain what a complex number is.

A complex number is a symbol $a + ib$ where a and b are real, and this condition (that a and b are real) is always understood when we write $a + ib$ without explanation. The number a is called the *real part* and b the *imaginary part* of $a + ib$, in symbols

$$\text{Re}(a + ib) = a, \qquad \text{Im}(a + ib) = b.$$

One could define the imaginary part to be ib but this is not as convenient.

With two exceptions, the algebra of complex numbers is just like the algebra of real numbers, provided we agree to replace i^2 by -1 whenever the occasion arises. The first exception involves the notion of equality . By definition,

$$a + ib = c + id \quad \text{means} \quad a = c \text{ and } b = d.$$

Thus, a *single* complex equation leads to *two* real equations. We emphasize this point because it might cause confusion if you are seeing complex numbers for the first time.

The second exceptional feature of complex numbers involves the notion of inequality. Since the square of a complex number can be negative, there is no relation of inequality between complex numbers that has the familiar properties of this relation as applied to real numbers. However, inequalities can be expressed by means of the symbol for *absolute value*,

$$|a + ib| = \sqrt{a^2 + b^2}.$$

If the complex number $A = a + ib$ is associated with the point (a, b) in the (x, y) plane then A is the distance from the representative point of A to the origin and $|A - B|$ is the distance from the representative point for A to that for B. The geometric interpretation suggests the inequality

(2) $$|A + B| \leq |A| + |B|$$

and an algebraic proof is outlined in Problem 12.

A *complex-valued function* is defined by $w = u + iv$ where u and v are real-valued functions of the real variable t. The domain of w is the set of points t where both $u(t)$ and $v(t)$ are defined. Familiar concepts such as the notions of limit, continuity, derivative and integral extend in a natural way from u and v to w. For example, w is differentiable if, and only if, both u and v are, and in that case the derivative is $w' = u' + iv'$. The phrase "complex function" means the same as "complex-valued function."

With the aid of (2) one can extend the theory of limits from real to complex functions and the proofs remain virtually unchanged. As a consequence, familiar formulas for differentiating sums, products, and so on, continue to hold for complex-valued functions of a real variable t. One can also develop the calculus of complex functions from that of real functions without repeating the proofs, as illustrated in Problems 13 and 14. These results are not deep and are taken for granted in the sequel.

PROBLEMS

1P. To solve the equation $A(2 + 3i) = 6 - 7i$ you can do this:

$$A = \frac{6 - 7i}{2 + 3i}\frac{2 - 3i}{2 - 3i} = \frac{12 - 14i - 18i - 21}{4 + 9}$$

Express A in the form $x + yi$ by completing the calculation and verify that it satisfies the given equation.

2. Assuming $|a + ib| \neq 0$ in (c), express A in the form $x + yi$ and check:

  (a) $A(1 - i) = 4 + 6i$ (b) $A(5 - 2i) = 29(3 + 4i)$ (c) $A(a + bi) = c + di$

3P. Write the real and imaginary parts by inspection:

    (a) $(2 + 3i)(4 - i)$ (b) $(1 + i)(c + is)$ (c) $(a + ib)(c + is)$

4P. Check that $|\cos t + i \sin t| = 1$, $-\infty < t < \infty$.

5P. In this problem we assume that e^{it} can be so defined that its derivative is ie^{it}, and also $e^{i0} = e^0 = 1$.

(a) Using the quotient rule show that the derivative of the following function is 0 for all t. How do you know the denominator does not vanish?

$$\phi(t) = \frac{e^{it}}{\cos t + i \sin t}$$

(b) Hence $\phi(t) = c$, a constant; see Problem 14. Determine c by setting $t = 0$ and thus establish the *Euler-de Moivre formula*

$$e^{it} = \cos t + i \sin t$$

(c) What does the formula say if $t = \pi$?

6P. If the complex number $z = x + iy$ is correlated with the point (x, y) in the Cartesian plane, show that polar coordinates

$$x = r \cos \theta, \qquad y = r \sin \theta$$

give $z = re^{i\theta}$ when you use the result of Problem 5. This is called the *polar form* of the complex number z.

7P. Let ω be real. Writing $C = c_1 - ic_2 = Ae^{-i\phi}$, show that

$$c_1 \cos \omega t + c_2 \sin \omega t = \mathrm{Re}\,(Ce^{i\omega t}) = A \cos(\omega t - \phi)$$

8. Using the result of Problem 5, show that the equation $e^{i(a+b)} = e^{ia}e^{ib}$ is equivalent to the familiar formulas for $\sin(a + b)$ and $\cos(a + b)$.

9P. For $z = x + iy$ we define $e^z = e^{x+iy} = e^x e^{iy}$ where e^{iy} is given by Problem 5 with $t = y$. Check that $|e^z| = e^x$ and

$$e^{z_1} e^{z_2} = e^{z_1 + z_2}$$

Hint: $e^{x_1} e^{iy_1} e^{x_2} e^{iy_2} = e^{x_1} e^{x_2} e^{iy_1} e^{iy_2}$. Use properties of the real exponential function together with Problem 8.

10P. The *conjugate* of a complex number z is denoted by $\bar{z}$ and is obtained by changing this sign of its imaginary part. Thus,

$$z = x + iy \iff \bar{z} = x - iy$$

(a) Check that $\overline{z_1 + z_2} = \bar{z}_1 + \bar{z}_2$, $\overline{z_1 z_2} = \bar{z}_1 \bar{z}_2$, $|z|^2 = z\bar{z}$.

(b) Explain the following proof that $|z_1 z_2| = |z_1||z_2|$:

$$|z_1 z_2|^2 = z_1 z_2 \overline{z_1 z_2} = z_1 z_2 \bar{z}_1 \bar{z}_2 = z_1 \bar{z}_1 z_2 \bar{z}_2 = |z_1|^2 |z_2|^2$$

11. The Schwarz inequality, special case. Apply the inequality $|\text{Re } z| \leq |z|$ to $z = (a + ib)(c - id)$ and deduce

$$|ac + bd| \leq \sqrt{a^2 + b^2}\sqrt{c^2 + d^2}$$

12. Fill in the details and deduce $|A + B| \leq |A| + |B|$:

$$|A + B|^2 = |A|^2 + 2\text{Re}A\overline{B} + |B|^2 \leq (|A| + |B|)^2$$

13. At any point where w_1' and w_2' exist, use familiar results from the calculus of real functions to get

$$(w_1 w_2)' = w_1 w_2' + w_2 w_1'$$

14. Let w be continuous for $a \leq t \leq b$ and let $w'(t) = 0$ for $a < t < b$. Show that w is constant for $a \leq t \leq b$.

Hint: By definition, the hypothesis on $w = u + iv$ gives corresponding conditions on u and v. Use a familiar result from the calculus of real functions.

──────────────ANSWERS──────────────

2. Self-checking.   **3.** (a) 11, 10   (b) $c - s$, $c + s$ (c) $ac - bs$, $bc + as$
5. (a) By Problem 4.   (c) $e^{i\pi} = -1$

───────────────────────────────────

4. Complex solutions

Since the equation $y'' + y = 0$ leads to the values $s = i$ or $-i$ upon the substitution $y = e^{st}$, it is natural to ask whether the corresponding solutions

$$e^{it}, \qquad e^{-it}$$

can be given an interpretation that justifies this substitution. Such an interpretation is possible and has the utmost importance for differential equations.

In Section 6, Theorem 4 it is shown that every solution of $y'' + y = 0$ must have the form

$$y = c_1 \cos t + c_2 \sin t$$

and this result is taken for granted here. (If complex-valued solutions are in question the constants c_1 and c_2 are complex.) Since we want the function e^{it} to be a solution, it follows that

(1) $$e^{it} = c_1 \cos t + c_2 \sin t.$$

Let us suppose that e^{it} can be differentiated by the usual rules, so that by differentiation of (1)

(2) $$i e^{it} = -c_1 \sin t + c_2 \cos t.$$

Setting $t = 0$ in (1) and (2) gives $c_1 = 1$, $c_2 = i$. Hence the only possibility, subject to our requirements, is

(3) $$e^{it} = \cos t + i \sin t.$$

It is not hard to show, conversely, that if e^{it} satisfies (3), then e^{it} has all the properties used in the derivation. Hence we accept (3) as a definition, which is completed by setting

$$e^{s+it} = e^s e^{it} = e^s(\cos t + i \sin t).$$

That z in e^z acts like an exponent was shown in Section 3, Problem 9.

Equation (3) is known as the *Euler-de Moivre formula* after the British mathematician de Moivre and the Swiss mathematician Euler (pronounced "oiler") who discovered it independently. If you know about Taylor series you will recognize that the substitution $x = it$ in the series for e^x produces (3) at once. However, this argument is not a proof of (3), but it is only another way of motivating the definition.

EXAMPLE 1. By setting $y = e^{st}$ obtain two linearly independent solutions of

$$y'' + 4y' + 13y = 0.$$

The suggested substitution leads to the equation

$$s^2 + 4s + 13 = 0$$

which has the roots

$$s = \frac{-4 \pm \sqrt{16 - 52}}{2} = -2 \pm 3i.$$

Using (3) with t replaced by $3t$ or $-3t$, we get the solutions

(4) $$e^{(-2\pm 3i)t} = e^{-2t} e^{\pm 3it} = e^{-2t}(\cos 3t \pm i \sin 3t).$$

The center expression shows that the ratio of the solutions is e^{6it}, which is not constant. Hence the solutions are independent.

Since the coefficients of the equation leading to (4) are real, it will be seen shortly that the real and imaginary parts of any complex solution are again solutions. Hence we get the linearly independent real solutions

$$e^{-2t} \cos 3t, \qquad e^{-2t} \sin 3t$$

by taking real and imaginary parts of the expression with the plus sign on the right of (4), and the pair

$$e^{-2t}\cos 3t, \qquad -e^{-2t}\sin 3t$$

from the expression with the minus sign. Thus the minus sign gives nothing new.

When a linear operator T is applied to a complex function $x + iy$ it is assumed that Tx and Ty are meaningful and, by linearity,

$$T(x + iy) = Tx + iTy.$$

However, this is not necessarily a decomposition into real and imaginary parts, because Tx and Ty on the right might not be real. If Tu is real for every real function u in the domain of T the operator T is said to be *real*. As an illustration, the operator T defined by

$$Tz = z'' + pz' + qz$$

is real if the constants p and q are real.

Much of the usefulness of complex solutions depends on the following:

PRINCIPLE OF EQUATING REAL PARTS. Let $z = x + iy$ be a complex solution of $Tz = f$, where T is a real linear operator. Then $Tx = \operatorname{Re} f$ and $Ty = \operatorname{Im} f$.

The proof is simple. By linearity $Tz = Tx + iTy$ and this equals f by hypothesis. Hence the real part of f is Tx and the imaginary part is Ty, as stated in the theorem. In the special case in which $Tz = 0$ the principle says that x and y satisfy the homogeneous equation, a fact that was used in Example 1.

EXAMPLE 2. Find a solution of each equation:

$$(5) \qquad x'' + 3x' + 2x = \cos t, \qquad y'' + 3y' + 2y = \sin t.$$

Since $\cos t$ and $\sin t$ are the real and imaginary parts of e^{it} we consider, instead of (5), the equation

$$z'' + 3z' + 2z = e^{it}.$$

The trial solution $z = Ae^{it}$ satisfies this equation if

$$Ae^{it}(i^2 + 3i + 2) = e^{it}$$

or $A(1 + 3i) = 1$. This gives

$$A = \frac{1}{1 + 3i}\frac{1 - 3i}{1 - 3i} = \frac{1 - 3i}{10}$$

and $z = Ae^{it}$ becomes $z = (.1 - .3i)(\cos t + i \sin t)$. The real and imaginary parts of $z = x + iy$ can be found by inspection with the result

$$x = .1 \cos t + .3 \sin t, \qquad y = .1 \sin t - .3 \cos t.$$

These functions satisfy (5).

The problem could be solved without using complex numbers by means of the trial solution

$$y = A \cos t + B \sin t.$$

However this procedure leads to two pairs of simultaneous equations for the constants A and B and the details are harder.

EXAMPLE 3. Obtain a solution of

(6) $$y'' + y = 5e^t \sin t.$$

Since the right-hand side is the imaginary part of $5e^t e^{it} = 5e^{(1+i)t}$ we consider, instead of (6),

(7) $$z'' + z = 5e^{(1+i)t}.$$

The trial solution $z = Ae^{(1+i)t}$ leads to the equation

$$A((1 + i)^2 + 1) = 5 \quad \text{or} \quad (1 + 2i)A = 5.$$

By a brief calculation $A = 1 - 2i$ and the solution of (7) is

$$z = (1 - 2i)e^{(1+i)t} = (1 - 2i)e^t(\cos t + i \sin t).$$

The imaginary part $y = e^t(\sin t - 2 \cos t)$ satisfies (6).

This problem could be solved by use of the trial solution

(8) $$y = e^t(A \cos t + B \sin t)$$

but the calculations would be harder.

For an equation of the form $Ty = e^{at} \cos bt$ or $e^{at} \sin bt$, one can use the shift theorem to eliminate the factor e^{at} and thus avoid the complex exponent $a + bi$. This approach is illustrated in the following example.

EXAMPLE 4. By the trial function $y = e^t u$ obtain a particular solution:

$$(D^2 + 4D + 7)y = e^t(17\cos t + 5\sin t).$$

By the shift theorem with $k = 1$ it is found that

$$(D^2 + 6D + 12)u = 17\cos t + 5\sin t.$$

This equation can be solved with equal ease by trying $u = A\cos t + B\sin t$ or by consideration of the complex equation

$$(D^2 + 6D + 12)z = e^{it}.$$

It is left for you to verify that either method gives

$$u = \cos t + \sin t, \quad \text{hence} \quad y = e^t(\cos t + \sin t).$$

PROBLEMS————————————————————————————

1P. For the equation $z'' - 10z' + 26z = 0$ obtain the complex solution

$$z = e^{(5+i)t} = e^{5t}e^{it} = e^{5t}(\cos t + i\sin t)$$

by trying $z = e^{st}$. Thus get two linearly independent real solutions, and from this, get the general solution in real form. What if you use the root $5 - i$ instead of $5 + i$?

2. Find the general solution in real form:

   (a) $y'' + 9y = 0$ (b) $y'' - 2y' + 2y = 0$ (c) $(D^2 - 8D + 25)y = 0$

3. (a) For the equation $z'' + 4z' + 3z = 130e^{2it}$ show that the trial solution $z = Ae^{2it}$ leads to $(-1 + 8i)A = 130$. Solve for A and thus obtain

$$z = (-2 - 16i)(\cos 2t + i\sin 2t)$$

(b) Using the result (a) obtain particular solutions of

$$x'' + 4x' + 3x = 130\cos 2t, \qquad y'' + 4y' + 3y = 130\sin 2t$$

4. By use of an equivalent complex equation or by a trial solution of the form $A\sin st + B\cos st$, as you prefer, obtain particular solutions of:

(a) $x'' - 3x' + 2x = 20\cos 2t, \quad y'' - 3y' + 2y = 20\sin 2t$

(b) $x'' + x' + 17x = 17\cos 4t, \quad y'' + y' + 17y = 17\sin 4t$

5. As in Example 3 or 4, or by a trial function of the form suggested by (8), obtain a particular solution:

(a) $x'' + 2x' + 5x = 20e^t \cos 2t,$ $y'' + 2y' + 5y = 20e^t \sin 2t$

(b) $x'' - x' - 6x = 102e^{3t} \cos 3t,$ $y'' - y' - 6y = 102e^{3t} \sin 3t$

6. Solve $z' = e^{(a+ib)t}$, separate into real and imaginary parts, obtain the formula on the left, and answer the question on the right:

$$\int e^{at} \cos bt \, dt = e^{at} \frac{a \cos bt + b \sin bt}{a^2 + b^2} + c, \qquad \int e^{at} \sin bt \, dt = \ ?$$

7P. Let ω be a nonzero real constant. Solve $(D + i\omega)(D - i\omega)z = e^{i\omega t}$ and thus show that the equations below have the indicated solutions:

$$x'' + \omega^2 x = \cos \omega t, \quad x = \frac{t \sin \omega t}{2\omega}; \qquad y'' + \omega^2 y = \sin \omega t, \quad y = -\frac{t \cos \omega t}{2\omega}$$

───────────────ANSWERS───────────────

1. $e^{5t}(c_1 \cos t + c_2 \sin t)$. The root $5 - i$ gives $e^{5t}(c_1 \cos t - c_2 \sin t)$, which is the same family in different notation.

2. (a) $c_1 \cos 3t + c_2 \sin 3t$ (c) $e^{4t}(c_1 \cos 3t + c_2 \sin 3t)$
   (b) $e^t(c_1 \cos t + c_2 \sin t)$

3. $-2 \cos 2t + 16 \sin 2t,$ $-16 \cos 2t - 2 \sin 2t$

4. (a) $-3 \sin 2t - \cos 2t, \ 3 \cos 2t - \sin 2t$
   (b) $\cos 4t + 4 \sin 4t,$ $-4 \cos 4t + \sin 4t$

5. (a) $e^t(\cos 2t + 2 \sin 2t),$ $e^t(-2 \cos 2t + \sin 2t)$
   (b) $e^{3t}(-3 \cos 3t + 5 \sin 3t),$ $e^{3t}(-5 \cos 3t - 3 \sin 3t)$

5. Existence and uniqueness

Throughout this section the letters p, q, a, b, k, A_j, B_j denote real or complex constants. We write

$$Ty = y'' + py' + qy = (D - a)(D - b)y$$

where the form on the right is obtained by factoring the characteristic equation; see Problem 9. The factorization yields:

THEOREM 1.   *Let f be a continuous real or complex-valued function on an open interval I containing the point t_0. Then the following initial-value problem has one, and only one, solution:*

$$Ty = f \text{ on } I, \qquad y(t_0) = y_0, \qquad y'(t_0) = y_1.$$

The proof is simple. Setting $u = (D - b)y$ shows that the equation

$$(D - a)(D - b)y = f$$

is equivalent to the two equations

$$(D - a)u = f, \qquad (D - b)y = u.$$

Since $u = y' - by$, the initial conditions for u can be read off from those for y. The problem in the theorem is thus reduced to a succession of two first-order problems:

$$u' - au = f, \qquad u(t_0) = y_1 - by_0,$$

$$y' - by = u, \qquad y(t_0) = y_0.$$

By Chapter A, Theorem 1 there is one, and only one solution u. After u is determined the same theorem shows there is one, and only one solution y. This completes the proof.

For important classes of functions f one can go beyond a statement of mere existence and say something about the form of the solution. This information is an aid in solving the equation. If you know that a solution has a certain form, for example

$$B_0 e^t \quad \text{or} \quad B_0 + B_1 t + B_2 t^2 \quad \text{or} \quad B_0 \cos t + B_1 \sin t,$$

the constants B_j can be found by substition. Many examples of this method were given in the foregoing pages. It will now be explored more fully, using the same ideas that were used in the proof of Theorem 1.

Let us begin with the equation

$$(D - a)y = A(t), \qquad a \neq 0$$

where A is a polynomial of degree n. We claim that the equation has a solution B which is also a polynomial of degree n. To see why, consider the special case $n = 3$, for which the equation is

(1) $$(D - a)y = A_0 + A_1 t + A_2 t^2 + A_3 t^3.$$

The trial function

$$y = B_0 + B_1 t + B_2 t^2 + B_3 t^3$$

satisfies (1) if, and only if, the polynomial

$$-aB_0 + (B_1 - aB_1 t) + (2B_2 t - aB_2 t^2) + (3B_3 t^2 - aB_3 t^3)$$

agrees with the polynomial on the right of (1). We now borrow a result from Chapter 21, Section 3, which states that two polynomials have the same value on an interval if, and only if, corresponding coefficients are equal. Equating coefficients of t^3, t^2, t and 1, in that order, we get

$$-aB_3 = A_3, \ 3B_3 - aB_2 = A_2, \ 2B_2 - aB_1 = A_1, \ B_1 - aB_0 = A_0.$$

The first of these equations gives B_3, the second then gives B_2, the third then gives B_1, and the last gives B_0. A similar procedure works when $A(t)$ is a polynomial of any degree and shows that there is one, and only one, polynomial solution $y = B(t)$ in that case. Furthermore, the degree of B is the same as that of A.

Suppose next that our equation is

$$(D - a)(D - b)y = A(t), \qquad a \neq 0, \ b \neq 0.$$

We can write it in the form

(2) $$(D - a)v = A(t), \qquad (D - b)y = v.$$

If A is a polynomial, the above result shows that the first equation (2) has a unique polynomial solution v. Now that we know this, the same result shows that the second equation has a unique polynomial solution y. In both cases the degree is the same as that of A.

If $a \neq 0$ but $b = 0$ the equation has the form

$$(D - a)Dy = A(t).$$

The above results give $Dy = B$ where B is a polynomial of the same degree as A. Integrating to get y raises the degree by 1. If the constant of integration is taken to be zero, y has the form $tC(t)$ where C is a polynomial of the same degree as A. In a like manner, if $a = b = 0$ the equation is $D^2y = A$, and two integrations yield a solution of the form $t^2C(t)$.

In the first case considered above, 0 is not a root of the characteristic equation

$$P(s) = (s - a)(s - b) = 0.$$

In the second case 0 is a simple root, and in the third case the root is double. It is said that 0 is a root of multiplicity 0, 1, or 2 in the three cases respectively.

If the equation is

$$Ty = A(t)e^{kt}$$

we set $y = ue^{kt}$ and use the exponential shift theorem. The resulting equation for u is like that for y, except that its characteristic polynomial is $P(s + k)$. A moment's thought shows that 0 is a root of $P(s + k) = 0$

if, and only if, k is a root of $P(s) = 0$. Furthermore, the multiplicities are the same. The above remarks are summarized as follows:

THEOREM 2.   *Let A be a polynomial. Then the equation $Ty = A(t)e^{kt}$ has a solution of the form*

$$B(t)e^{kt} \quad or \quad tB(t)e^{kt} \quad or \quad t^2B(t)e^{kt}$$

if k is a root of the characteristic equation $P(s) = 0$ of multiplicity 0, 1, or 2 respectively. In each case B is a polynomial of the same degree as A. When $k = \alpha + i\beta$ we get corresponding results for equations of the form

$$(3) \qquad Ty = A(t)e^{\alpha t}\cos\beta t, \qquad Ty = A(t)e^{\alpha t}\sin\beta t.$$

The concluding statement follows from the principal of equating real parts; see also Chapter 8, Section 3, Theorem 4.

EXAMPLE 1.   For $y'' - 3y' = 1 + 9t^2$ the appropriate trial solution is

$$y = t(B_0 + B_1t + B_2t^2).$$

The extra factor t is needed because 0 is a root of the characteristic equation.

EXAMPLE 2.   For $y'' - 4y' + 4y = (6t - 2)e^{2t}$ the appropriate trial solution is

$$y = t^2(B_0 + B_1t)e^{2t}.$$

The factor t^2 is needed because $k = 2$ is a root of multiplicity 2.

EXAMPLE 3.   For $y'' - y' - 2y = (6t - 2)e^{2t}$ the trial solution is the same as in Example 2, except that the factor t^2 is replaced by t. In this case $k = 2$ is a root of multiplicity 1.

EXAMPLE 4.   For $y'' + y' + 2y = 2\sin t$ the trial solution is

$$y = B_0\cos t + B_1\sin t.$$

EXAMPLE 5.   For $y'' + 4y = 4\cos 2t$ the trial solution is

$$y = t(B_0\cos 2t + B_1\sin 2t).$$

The factor t is needed because $k = 2i$ is a root of the characteristic equation.

EXAMPLE 6.   For $y'' - 6y' + 25y = (2 + 16t)e^{3t}\cos 4t$ the trial solution is

$$y = te^{3t}((B_0 + B_1t)\cos 4t + (B_2 + B_3t)\sin 4t).$$

The factor t is needed because $k = 3 + 4i$ is a root of the characteristic equation. Note that two different polynomials are required for the sine term and for the cosine term.

REVIEW PROBLEMS────────────────────────────────

1. Write two linearly independent real solutions:

(a) $y'' + 3y' = 54y$ (d) $y'' + 6y = 5y'$ (g) $y'' + 6y' + 10y = 0$

(b) $y'' + y = 2y'$ (e) $y'' + 4y = 4y'$ (h) $y'' + 10y' + 29y = 0$

(c) $y'' + 5y = 4y'$ (f) $y'' + 50y = 2y'$ (i) $y'' + 12y' + 35y = 0$

2. This problem pertains to the equation $y'' + 4y' + ay = 0$ where a is a real constant. Obtain the particular solutions

$$e^{-t} \qquad te^{-2t} \qquad e^{-2t}\cos t$$

for $a = 3, 4, 5$ respectively. You may find it advisable to solve for general a and then specialize.

3. This problem pertains to the equation $y'' + 2by' + 2y = 0$ where b is a real constant. For $b^2 = 1, 2, 3$, respectively, obtain the particular solutions

$$e^{-bt}\cos t \qquad e^{-bt}t \qquad e^{-bt}e^t$$

4. Get a particular solution of the first three equations in simplest possible form and thus get a solution of the fourth equation, where a, b, c are any constants:

$$y'' - y' + 7y = \cos 2t, \qquad y'' - y' + 7y = \sin 2t, \qquad y'' - y' + 7y = e^{2t}$$

$$y'' - y' + 7y = 13(a\cos 2t + b\sin 2t) + 9ce^{2t}$$

5. You are to get a particular solution for each of the two equations

$$x'' + 2x' + 2x = 2e^{-t}\cos t, \qquad y'' + 2y' + 2y = 2e^{-t}\sin t$$

Do either (a) or (b), as you prefer: (a) Try $te^{-t}(A\cos t + B\sin t)$.

(b) Set $z = e^{(-1+i)t}u$ and use the shift theorem with $k = -1 + i$ in the equivalent complex equation

$$(D + 1 + i)(D + 1 - i)z = 2e^{(-1+i)t}$$

6. Write the appropriate form for a trial solution. No answer is provided:

(a) $y'' + y = t^3$ (d) $y'' + y = \sin 2t$ (g) $y'' + 2y' + 2y = t^2e^t\cos t$

(b) $y'' = t^3 + t^2$ (e) $y'' + y = \sin t$ (h) $y'' - 2y' + 2y = t^2e^t\sin t$

(c) $y'' - y' = te^t$ (f) $y'' - y = t^3e_t$ (i) $y'' - 6y' + 9y = t^5e^{3t}$

7P. If all constants are real, solve $Ty = e^{\alpha t}(A_0 \cos \beta t + A_1 \sin \beta t)$ by use of

$$Tz = e^{(\alpha + i\beta)t}, \qquad z = u + iv$$

8. (abcdef) Using whatever method seems best to you, obtain a particular solution of the equations in Examples 1,2,3,4,5,6.

9. Here we solve the general quadratic equation with complex coefficients.

(a) Show that $z^2 + 2rz + q = 0$ is equivalent to $(z + r)^2 = r^2 - q$, hence it can be solved as soon as you find the two square roots of $c = r^2 - q$.

(b) If $w = u + iv$ satisfies $w^2 = c$, check that

$$u^2 + v^2 = |c|, \qquad u^2 - v^2 = \text{Re } c, \qquad 2uv = \text{Im } c$$

The first two equations determine u^2 and v^2 and the signs of u and v are chosen so that uv has the same sign as Im c. Show then that $w^2 = c$.

————————————ANSWERS————————————

1. (a) e^{-9t}, e^{6t} (d) e^{3t}, e^{2t} (g) $e^{-3t} \cos t, e^{-3t} \sin t$

     (b) e^t, te^t (e) e^{2t}, te^{2t} (h) $e^{-5t} \cos 2t, e^{-5t} \sin 2t$

     (c) $e^{2t} \cos t, e^{2t} \sin t$ (f) $e^t \cos 7t, e^t \sin 7t$ (i) e^{-5t}, e^{-7t}

4. $a(3 \cos 2t - 2 \sin 2t) + b(2 \cos 2t + 3 \sin 2t) + ce^{2t}$

5. $te^{-t} \sin t, \quad -te^{-t} \cos t$ 7. $y = A_0 u + A_1 v$

8. (a) $-t(1 + t + t^2)$ (b) $t^2(t - 1)e^{2t}$ (c) $3y = t(3t - 4)e^{2t}$

     (d) $\sin t - \cos t$ (e) $t \sin 2t$ (f) $4y = te^{3t}(\cos 4t + (1 + 4t) \sin 4t)$

6. The Wronskian and linear dependence

Throughout this section u, v and all constants can be complex. The letter I denotes an open interval of the t-axis and T is defined by

$$Ty = y'' + py' + qy.$$

Except in Theorem 5, both p and q are constant.

The *Wronskian* of two differentiable functions u and v is, by definition,

$$W = \begin{vmatrix} u & v \\ u' & v' \end{vmatrix} = uv' - vu'$$

where the symbol in the center is a 2-by-2 determinant. Determinants of this form were introduced by the Polish mathematician Wronski in 1811, hence the name. We write $W(t)$ or $W(u,v)$ to emphasize dependence on t or on the functions u, v, as the case may be. For example

$$W(\cos t, \sin t) = \begin{vmatrix} \cos t & \sin t \\ -\sin t & \cos t \end{vmatrix} = \cos^2 t + \sin^2 t = 1.$$

The Wronskian gives a simple criterion for linear dependence. On any interval on which $v \neq 0$ the functions u and v are linearly dependent if, and only if, the ratio u/v is constant. In that case

(1) $$\frac{d}{dt}\left(\frac{u}{v}\right) = \frac{vu' - uv'}{v^2} = 0$$

provided the derivatives exist. Comparing W with the numerator of the fraction in the center, we see that $W(u,v) = 0$. Conversely, if $W(u,v) = 0$ on an interval the above calculation shows that the derivative of u/v is 0 and hence u/v is constant.

The assumption that $v \neq 0$ is essential for the second part of this argument but not for the first. If $u = cv$ on an interval, the functions being differentiable, then

$$u = cv, \qquad u' = cv'.$$

Here the second relation is obtained by differentiating the first. If we multiply the first by v', the second by v, and subtract, we get

$$uv' - u'v = cvv' - cv'v = 0.$$

This says that $W(u,v) = 0$. The same conclusion follows if $v = cu$, and hence it follows whenever u and v are linearly dependent. Thus we have established:

THEOREM 1. *Let u and v be differentiable on I.*

(a) *If u and v are linearly dependent, then $W(u,v) = 0$ on I.*

(b) *If $W(u,v) = 0$ and $v \neq 0$ on I, then u and v are linearly dependent.*

For example, if a and b are unequal constants, real or complex,

$$W(e^{at}, e^{bt}) = \begin{vmatrix} e^{at} & e^{bt} \\ ae^{at} & be^{bt} \end{vmatrix} = e^{(a+b)t}(b - a).$$

Since this is not 0 at any t, Theorem 1 indicates that the functions e^{at} and e^{bt} are linearly independent on every interval.

It is seen in Problem 3 that the condition $W(u, v) = 0$ on an interval does not, in general, ensure that the functions u and v are linearly dependent. The following theorem shows, however, that a result of this kind does hold if u and v are solutions of $Ty = 0$:

THEOREM 2.   *Let u and v satisfy the equation $Ty = 0$ on I, and suppose that their Wronskian satisfies $W(t_0) = 0$ at some point $t_0 \in I$. Then u and v are linearly dependent on I, and hence $W(t) = 0$ for all $t \in I$.*

The proof is simple. Since $W(t_0) = 0$ it is known from linear algebra that you can choose constants c_i, not both 0, such that

$$c_1 u(t_0) + c_2 v(t_0) = 0$$
$$c_1 u'(t_0) + c_2 v'(t_0) = 0.$$

(Note that $W(t_0)$ is the determinant of the system.) Hence the function

$$y = c_1 u(t) + c_2 v(t)$$

satisfies null initial conditions at t_0. The uniqueness theorem of Section 5 now gives $y = 0$ on I, which shows that u and v are linearly dependent. The equation $W(t) = 0$ follows from Theorem 1.

An obvious consequence of Theorem 2 is:

THEOREM 3.   *If u and v are solutions of the homogeneous equation $Ty = 0$ on I, either $W(t) = 0$ at every point of I or $W(t) = 0$ at no point of I.*

The Wronskian has an interesting application to initial-value problems. If the homogeneous equation $Ty = 0$ has solutions u and v on an interval I then $y = c_1 u + c_2 v$ is also a solution for any constants c_1 and c_2. To satisfy prescribed initial conditions

$$y(t_0) = y_0, \qquad y'(t_0) = y_1$$

at $t_0 \in I$ we must determine the constants c_1 and c_2 in such a way that

(2)
$$c_1 u(t_0) + c_2 v(t_0) = y_0$$
$$c_1 u'(t_0) + c_2 v'(t_0) = y_1.$$

It is known from linear algebra that the system can be solved for arbitrary choices of y_0 and y_1 if, and only if, the coefficient determinant is not zero. This determinant is $W(t_0)$. By Theorem 3, the condition $W(t_0) \neq 0$ for solvability of the initial-value problem does not depend on t_0.

These remarks give information about the possible form of a solution of $Ty = 0$. On a given interval I let u and v be particular solutions of $Ty = 0$ such that $W(u, v) \neq 0$ and let y be any solution whatever. If $t_0 \in I$ we can choose constants c_1 and c_2 such that (2) holds with $y_0 = y(t_0)$ and $y_1 = y'(t_0)$. This means that the given function y and the special function

$$\tilde{y} = c_1 u + c_2 v$$

both satisfy $Ty = 0$ with the same initial conditions at t_0. But the uniqueness theorem of the last section says that there can be *at most one* solution with prescribed initial conditions. Hence $y = \tilde{y}$. In view of Theorem 3, we summarize as follows:

THEOREM 4. *On I let $Ty = 0$ have two solutions u, v such that $W(u, v) \neq 0$ at some point $t \in I$. Then every solution of $Ty = 0$ can be written in the form $y = c_1 u + c_2 v$ and the constants c_j can be chosen so as to satisfy arbitrarily specified initial conditions $y(t_0) = y_0$, $y'(t_0) = y_1$ at any point $t_0 \in I$. In other words, the family $c_1 u + c_2 v$ is both the complete solution and the general solution of $Ty = 0$.*

The only reason for assuming p and q constant in the foregoing discussion is so that we can use the existence-uniqueness theorem of Section 5. It is seen in Chapter 10 that uniqueness holds when p and q are any continuous functions of t, and existence is established under the same hypothesis in Chapter 14. By inspection of the proofs, we obtain the following:

THEOREM 5. *Theorems 2, 3, and 4 remain valid if $p = p(t)$ and $q = q(t)$ are continuous functions of t on I.*

PROBLEMS

1. Using Theorem 1 or the definition of linear dependence, whichever you prefer, test the following pairs of functions for linear dependence:

(a) e^t, e^{2t} (c) $(t + 1)^2, t^2$ (e) $t \sin t, t \cos t$

(b) e^{t+1}, e^t (d) te^t, e^t (f) $e^t \sin t, e^t \cos t$

2P. (a) If u'' and v'' exist, check that $(uv' - u'v)' = uv'' - vu''$.

(b) Let $Tu = 0$ and $Tv = 0$ on an interval. Multiply the second of these equations by u, the first by v, and subtract. Using the result of Part (a) deduce that the Wronskian W satisfies $W' + pW = 0$. This is *Abel's identity*, which was discovered by the Norwegian mathematician Niels Henrik Abel in 1826. It does not matter whether p is constant or depends on t.

(c) Obtain two linearly independent solutions of $y'' - 10y' + 74y = 0$ and verify that their Wronskian satisfies $W = ce^{10t}$.

(d) Deduce Theorem 3 from Abel's identity.

3. By considering $u(t) = t^3$, $v(t) = |t|^3$ show that the hypothesis $v \neq 0$ cannot be dropped in Theorem 1(b).

4. On the interval $0 < t < 3$ let $u(t) = 0$ except that $u(1) = 1$, and let $v(t) = 0$ except that $v(2) = 1$. Show that u and v are linearly independent on $0 < t < 3$.

5. Let u and v be continuous and linearly independent on an interval I. Suppose w is defined on I and has only finitely many zeros.

(a) Show that wu, wv are linearly independent on I.

(b) You can't use the Wronskian in this problem. Why not?

(c) Show that the result can fail if u and v are not continuous.

Hint for (c): Use u and v from Problem 4.

6. Using Problem 5 with $w(t) = t^m$, $m = 0, 1, 2, 3, \cdots$, show with little calculation that the following pairs of functions are linearly independent on every interval:

(a) $t^m \cos \omega t$, $t^m \sin \omega t$, $\omega \neq 0$ (b) $t^m e^{at}, t^m e^{bt}$, $a \neq b$.

7. Three functions u, v, w are linearly independent on an interval I if the condition
$$c_1 u(t) + c_2 v(t) + c_3 w(t) = 0, \qquad t \in I$$
can hold only when all constants $c_i = 0$. Show that the three functions e^t, e^{2t}, e^{-t} are linearly independent on the interval $(-\infty, \infty)$.

Hint: If $c_1 e^t + c_2 e^{2t} + c_3 e^{-t} = 0$, $-\infty < t < \infty$, multiply this equation by e^{-2t} and let $t \to \infty$. The result is $c_2 = 0$.

8. Linear independence of four or more functions is defined by analogy with Problem 7. Show that the four functions $\sin t$, $\cos t$, $t \sin t$, $t \cos t$ are linearly independent on the interval $(-\infty, \infty)$.

Hint: Let $t = 2n\pi$ or $2n\pi + \pi/2$ and $n \to \infty$

──────────────────────**ANSWERS**──────────────────────

1. The pair (b) is linearly dependent on every interval, all others are linearly independent on every interval.

5. (b) Because the functions are not assumed differentiable.

───

> Mathematics is the standard of objective truth for all intellectual endeavors.
>
> —Hermann Weyl

Chapter 7

APPLICATIONS

THE DISPLACEMENT of a mass suspended from a spring and the voltage across the condenser in a series circuit both satisfy a second-order linear differential equation with constant coefficients. Analysis of either problem leads to the concepts of underdamped and overdamped oscillation, resonance and beats, and phase lag and gain. Next we derive the three laws of Kepler for planetary motion, and we conclude with a discussion of the elementary functions familiar from calculus. Although these various topics are entirely different from one another, each of them leads to a second-order equation with constant coefficients and all but the last make essential use of complex numbers.

1. Mechanical and electrical systems

If a mass m is attached to one end of a suspended spring, it will produce an elongation s which, according to Hooke's law, is proportional to the force mg. Thus

$$F = ks = mg$$

where k is the stiffness constant for the spring. If an additional force is applied to produce an additional extension y, and is then removed, the mass-spring system will start to oscillate; see Figure 1. The problem is to determine the subsequent motion. Here, and throughout this chapter, we assume that the spring does not go slack or exceed its elastic limit.

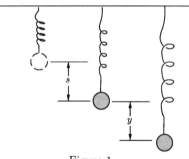

Figure 1

The forces acting on the mass are the force of gravity mg downward and the tension T in the spring that acts in a direction opposite to the force of gravity. We take the positive direction as downward. Since T is the tension when the elongation is $s + y$, Hooke's law gives $T = k(s + y)$. The resultant downward force due to the combined action of the spring and gravity is therefore

$$F = mg - k(s + y) = -ky$$

when we recall that $mg = ks$. Here y is the displacement from the equilibrium position. By Newton's second law

$$(\text{mass})(\text{acceleration}) = \text{force}$$

we get $my'' = -ky$ where, as throughout this chapter, primes denote differentiation with respect to t. Thus

$$my'' + ky = 0.$$

This differential equation describes free oscillations of the undamped mass-spring system.

Suppose next that the mass is in a resisting medium in which the damping is proportional to the velocity y'. Since the resistance opposes the motion, the damping acts in a direction opposite to the velocity. Hence the resistive force is $-ry'$, where r is a positive constant. By Newton's law $my'' = -ky - ry'$ and hence

$$my'' + ry' + ky = 0.$$

This is the equation for free oscillations of the damped mass-spring system.

Next, let the point of support vibrate in accordance with some law that gives the displacement of the top of the spring as a function of time, say $x = x(t)$ where x is measured positively downward. Since the elongation of the spring is diminished by the amount x, the term ky in the foregoing discussion must be replaced by $k(y - x)$. The result is

(1) $$my'' + ry' + ky = kx$$

when the term involving x is transferred to the right-hand side. This equation describes forced oscillation of the damped mass-spring system. The result for the undamped spring is obtained by setting $r = 0$ and for free oscillations by setting $x = 0$. Thus (1) summarizes the preceding discussion.

We shall exhibit a striking analogy between the mass-spring system and a corresponding electrical circuit. This analogy makes it possible to apply the methods of electrical engineering to a variety of mechanical problems.

Let a source of electromotive force be connected in series with a capacitor, an inductor, and a resistor, as shown in Figure 2. If the current is i the voltage drop across the resistance R is iR by Ohm's law and the drop across the inductance L is

$$L\frac{di}{dt}$$

by the law of Faraday. The voltage equation is therefore

(2) $$Li' + Ri + v = v_0(t)$$

where v is the voltage across the capacitor and $v_0(t)$ is the voltage at the source.

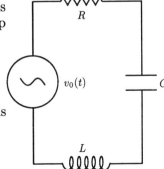

Figure 2

The charge Q on a capacitor plate satisfies $Q = Cv$ where C is a constant called the *capacitance*. Since the current i is the rate of flow of charge by definition, we have $i = Q' = Cv'$. Substitution into (2) gives

$$LCv'' + RCv' + v = v_0(t)$$

or, dividing by C and setting $S = 1/C$,

(3) $$Lv'' + Rv' + Sv = Sv_0(t).$$

The constant $S = 1/C$ is called the *elastance*.

Upon comparing (1) and (3) we see an exact correspondence between the variables

$$v, \ L, \ R, \ S$$

of the circuit problem and the variables

$$y, \ m, \ r, \ k$$

of the mechanical problem. The displacement $x(t)$ of the top of the spring corresponds to the voltage $v_0(t)$ of the source. Correspondences of this kind form the basis for analog computers, in which mechanical systems are simulated by electrical circuits. More important is the fact that the whole technology of transmission-line theory, with its concepts of impedance, transfer function, and so on can be applied to mechanical systems. In fact, from the point of view of communications engineering the mass-spring system is a *transducer* that transforms an input x into an output y.

2. Undamped unforced oscillations

If the equation

(1) $$my'' + ky = kx$$

for undamped oscillations is divided by m the result is

(2) $$y'' + \omega^2 y = \omega^2 x(t)$$

where $\omega^2 = k/m$. This equation is considered here under the assumption that $\omega > 0$. We begin with the case of unforced oscillations, for which $x = 0$.

(a) Simple harmonic motion

According to Chapter 6 the general solution of the equation

(3) $$y'' + \omega^2 y = 0$$

is given by

(4) $$y = c_1 \cos \omega t + c_2 \sin \omega t$$

or also by

(5) $$y = A \cos(\omega t - \phi), \qquad A \geq 0.$$

The equivalence of these two forms is seen most easily by writing

$$y = \mathrm{Re}\ Ce^{i\omega t}, \qquad C = c_1 - ic_2 = Ae^{-i\phi}$$

but can also be verified by real-variable methods. The expression (4) is convenient for satisfying prescribed initial conditions while (5) makes it easy to plot the graph of the solution.

Motion described by any one of the equivalent equations (3), (4), and (5) is called *simple harmonic motion*. The *amplitude* is A, the *phase* is ϕ, and the *period* of the motion is

$$T = \frac{2\pi}{\omega}.$$

This is the smallest positive constant such that, for all t,

$$y(t + T) = y(t).$$

In other words, T is the smallest time for one complete oscillation. The *frequency*

$$f = \frac{1}{T} = \frac{\omega}{2\pi}$$

gives the number of oscillations per second and ω is the *angular frequency* in radians per second. The notation T for period and f for frequency is standard. One must not, however, confuse this use of the letters T, f with their use to denote an operator T or a forcing function $f(t)$ in a differential equation.

(b) The phase plane

If we set $v = y'$, $y'' = v\,dv/dy$ as in Chapter 2, the equation

$$my'' + ky = 0$$

becomes $mv\,dv + ky\,dy = 0$ after multiplication by dy. An integration gives

(6) $$\frac{1}{2}mv^2 + \frac{1}{2}ky^2 = E$$

where E is constant. This equation expresses the principle of conservation of energy. The term $(1/2)\,mv^2$ represents the kinetic energy in the moving mass and the term $(1/2)\,ky^2$ gives the work

$$W = \int_0^y F\,dy = \int_0^y ky\,dy$$

needed to stretch the spring by the amount y. Hence this term represents the potential energy. Since there is no damping and no forcing, one would expect the total
energy, kinetic plus potential, to remain constant. That expectation is confirmed by the
analysis.

The (y, v) plane is called the *phase plane*
and a plot of (y, v) in this plane as t varies is
called an *orbit*. By equation (6) the orbits in
simple harmonic motion are ellipses. A given
ellipse corresponds to a given energy-level E.
As shown in Figure 3, the orbits obtained for
equal increments of E, such as $E = 1, 2, 3, \cdots$,
are not equally spaced, but crowd together as
the orbits get larger. When $\omega = 1$ the ellipses
are circles, and the configuration has the same
mathematical structure as a diagram of Fresnel
zones in geometrical optics.

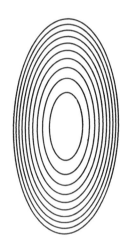

Figure 3

PROBLEMS————————————————————————————

1P. Review Chapter 6, Section 4, including the P problems.

2P. (a) The general solution of $y'' + \omega^2 y = 0$ is $c_1 \cos \omega t + c_2 \sin \omega t$. Determine c_1 and c_2 from the initial conditions $y(0) = y_0$, $y'(0) = y_1$ and thus get

$$y = y_0 \cos \omega t + y_1 \frac{\sin \omega t}{\omega}$$

(b) Let I be an interval containing the point t_0. A *fundamental solution* of a second-order homogeneous linear equation $Ty = 0$ at t_0 is a pair of functions u, v such that $Tu = Tv = 0$ on I and

$$u(t_0) = 1,\ u'(t_0) = 0; \qquad v(t_0) = 0,\ v'(t_0) = 1$$

If (u, v) is a fundamental solution show that $y = y_0 u + y_1 v$ satisfies

$$Ty = 0, \qquad y(t_0) = y_0, \qquad y'(t_0) = y_1$$

3. A force of 900 dynes stretches a spring 1 cm. A mass of 100 grams is suspended from the spring and allowed to oscillate.

(a) Find the differential equation of the motion.

(b) Using Problem 5, obtain displacement versus time if the mass is pulled down 2 cm from its equilibrium position and gently released.

(c) Obtain displacement versus time if the mass is projected downward from its equilibrium position with speed 9 centimeters per second.

4. (a) A mass M is hung on a spring of stiffness k and allowed to come to equilibrium. The total length of the spring is now y_0. If an additional mass m increases the length to y_1, show that $mg = k(y_1 - y_0)$ independently of the mass M.

(b) Two equal masses m are suspended from a spring of stiffness k. After equilibrium is reached one of the masses falls off. Describe the motion of the remaining mass.

Hint: Measure y from the equilibrium position of a single mass and use the result of Part (a).

5. A concrete block suspended by a cable oscillates vertically with frequency f. When a known mass m_1 is added the frequency drops to a new value f_1. Show that

$$m = \frac{f_1^2}{f^2 - f_1^2} m_1, \qquad k = \frac{(2\pi f f_1)^2}{f^2 - f_1^2} m_1$$

6. (a) Explain how you can determine the phase-plane ellipse for a solution of $y'' + \omega^2 y = 0$ from the initial conditions $y(0)$ and $y'(0)$.

(b) If y and z are two solutions satisfying

$$y(t_0) = a, \ y'(t_0) = b, \ z(t_1) = c, \ z'(t_1) = d$$

what conditions on a, b, c, d ensure that both solutions have the same orbit in phase space?

(c) Give the equation of the phase-plane orbit for $y'' + 9y = 0$, $y(0) = 2$, $y'(0) = -3$, and sketch.

(d) In an artificial mass-spring system in which the spring has a negative stiffness constant $k = -1$, the displacement satisfies $y'' = y$. Sketch the phase-plane orbits. What if $|y(t_0)| = |y'(t_0)|$ at some value t_0?

7P. An important generalization of $y'' + y = 0$ is $y'' + g(y) = 0$ where g is continuous. As in the text, set $v = y'$ and obtain the equation

$$\frac{1}{2}v^2 + G(y) = E, \qquad G(y) = \int_0^y g(\eta)\, d\eta$$

Hence the orbits in the phase plane can be found by intersecting the surface

$$z = \frac{1}{2}v^2 + G(y)$$

with the plane $z = E$ and then projecting onto the (y, v) plane.

8. In many cases the orbits of Problem 7 form closed curves and the corresponding solutions are periodic in t; the period is the time needed to go once around the orbit. Given a closed orbit, let α be the minimum and β the maximum of y on it. Check that the period is given, in general, by

$$T = \sqrt{2} \int_{\alpha}^{\beta} \frac{dy}{\sqrt{E - G(y)}}$$

Hint: Set $v = dy/dt$ and separate variables. The time to traverse half the orbit (from $y = \alpha$ to $y = \beta$) is $T/2$. If the integral diverges it takes infinitely long to traverse the orbit and the period is infinite.

9. In the special case $y'' + \omega^2 y = 0$ check that the formula of Problem 7 yields the correct result, $T = 2\pi/\omega$.

10. This is difficult. (a) For $y'' + y = y^3$ and (b) for $y'' + y^3 = y$, find the phase-plane orbits. Solutions with $v = 0$ are called stationary. Check that each equation has the stationary solutions $y = 0$, $y = -1$, $y = 1$. Choose E so that the stationary point is on the orbit and plot these three orbits; use a hand calculator if you wish. Also plot orbits for two other well-chosen values of E and thus confirm the behavior shown in Figure 4. A systematic method of analyzing orbits near a stationary point is given in Chapter 20.

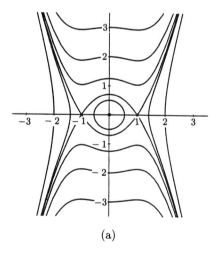

(a)

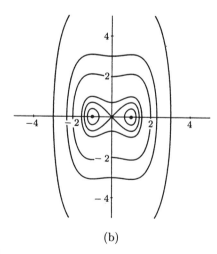

(b)

Figure 4

ANSWERS

3. (a) $y'' + 9y = 0$ (b) $y = 2\cos 3t$ (c) $y = 3\sin 3t$

4. (b) $y = (mg/k)\cos t\sqrt{k/m}$

6. (a) If $v = y'$ then $v^2 + \omega^2 y^2$ has the same value at t as at $t = 0$, since it is independent of t. Hence $v^2 + \omega^2 y^2 = y'(0)^2 + \omega^2 y(0)^2$.

    (b) $b^2 - d^2 = \omega^2(c^2 - a^2)$ (c) $v^2 + 9y^2 = 45$ (d) $v^2 - y^2 = 2E$, $v = \pm y$

10. (a) $2v^2 + 2y^2 - y^4 = 4E$ (b) $2v^2 - 2y^2 + y^4 = 4E$

---

3. Undamped forced oscillations

In the equation

$$(1) \qquad\qquad y'' + \omega^2 y = \omega^2 x(t), \qquad \omega > 0$$

the forcing-function $x(t)$ is often a sinusoidal input,

$$(2) \qquad\qquad x(t) = A_0 \cos \omega_0 t \quad \text{or} \quad x(t) = B_0 \sin \omega_0 t,$$

where A_0, B_0, and ω_0 are real constants with $\omega_0 > 0$. Part of the importance of this choice stems from the fact that the output of an alternating-current generator is, as a rule, sinusoidal. But more fundamental is the fact that an arbitrary periodic input of period T can generally be approximated by sums of the form

$$(3) \qquad \frac{1}{2}a_0 + a_1 \cos \omega_1 t + b_1 \sin \omega_1 t + \cdots + a_n \cos \omega_n t + b_n \sin \omega_n t$$

where $T\omega_k = 2\pi k$, $k = 1, 2, 3, \cdots$. If we know the response to the particular inputs (2), the general principle of superposition gives the response to (3) and hence, approximately, to an arbitrary periodic input $f(t)$. A discussion of the underlying theory can be found in Chapter 18.

(a) Sinusoidal input

Suppose, then, that $x(t)$ in (1) is given by either expression (2). If $\omega_0 \neq \omega$ as now assumed, the trial solution

$$y = A \cos \omega_0 t \qquad \text{or} \qquad y = B \sin \omega_0 t$$

gives $A(-\omega_0^2 + \omega^2) = \omega^2 A_0$. By this and by a similar result for B,

$$(4) \qquad\qquad A = \frac{\omega^2}{\omega^2 - \omega_0^2} A_0 \qquad B = \frac{\omega^2}{\omega^2 - \omega_0^2} B_0.$$

The ratio A/A_0 of output to input amplitude is called the *gain* of the system. This ratio can be read off from (4), where it is seen that the same value is obtained for B/B_0. Note that the gain is large if ω_0 is close to ω.

When ω_k is different from ω for all relevant k, the response to the input (3) is given by a series of the same form with coefficients

$$\frac{a_0}{2}, \quad \frac{\omega^2 a_k}{\omega^2 - \omega_k^2}, \quad \frac{\omega^2 b_k}{\omega^2 - \omega_k^2}.$$

This follows from (4) together with the general superposition principle. Addition of the complementary function

$$c_1 \cos \omega t + c_2 \sin \omega t$$

gives the complete solution, but in realistic problems there is always at least a small amount of dissipation, and the complementary function tends to zero as $t \to \infty$. The effect of dissipation is discussed later. Anticipating the result of this discussion, we omit the complementary function here.

If a particular frequency ω_k is close to ω the corresponding term in the series is strongly emphasized and the output may look entirely different from the input. This behavior is now illustrated numerically.

(b) An example

Let $x(t)$ be the function of period 2π defined for $|t| \leq \pi$ by

$$x(t) = \frac{\pi}{2} - |t|.$$

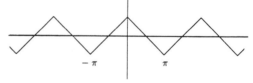

Figure 5

The graph of this function is the saw-tooth wave illustrated in Figure 5. In the theory of Fourier series it is shown that

$$x(t) = \frac{4}{\pi} \left(\frac{\cos t}{1^2} + \frac{\cos 3t}{3^2} + \frac{\cos 5t}{5^2} + \cdots \right)$$

and that if the series is carried only to n terms the error is at most $1/(4n-2)$. The above discussion indicates that if $x(t)$ is approximated by such a finite sum, the steady-state response to the approximation is given by the same number of terms in

$$y(t) = \frac{4\omega^2}{\pi} \left(\frac{\cos t}{1^2(\omega^2 - 1^2)} + \frac{\cos 3t}{3^2(\omega^2 - 3^2)} + \frac{\cos 5t}{5^2(\omega^2 - 5^2)} + \cdots \right).$$

When ω is close to an odd integer the corresponding term tends to dominate. For example if $\omega \approx 3$ one would expect

$$y(t) \approx \frac{4}{\pi} \frac{1}{\omega^2 - 9} \cos 3t.$$

The response for several values of ω is shown in Figure 6.

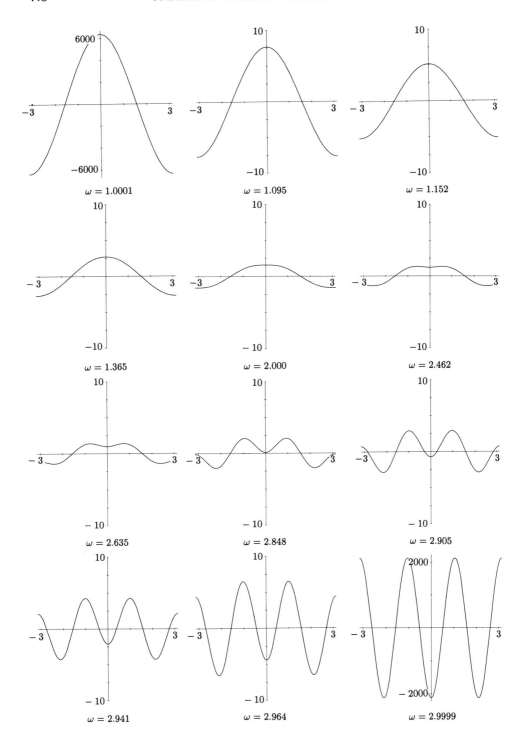

Figure 6

(c) Beats

Assuming $\omega_0 \neq \omega$, we are going to find the rest solution of

$$y'' + \omega^2 y = A \cos \omega_0 t \quad \text{where} \quad A = \omega_0^2 - \omega^2.$$

Since any solution can be multiplied by a constant, the choice of A involves no loss of generality. The general solution is

$$y = c_1 \cos \omega t + c_2 \sin \omega t - \cos \omega_0 t.$$

The initial conditions $y(0) = y'(0) = 0$ give $c_1 = 1$ and $c_2 = 0$, so that

(5) $$y = \cos \omega t - \cos \omega_0 t.$$

Physically, this represents the response of a system with the natural frequency ω to a forcing function of frequency ω_0, if the system is initially at rest.

By a trigonometric identity the function (5) can be written

$$y = 2 \sin \frac{\omega_0 + \omega}{2} t \, \sin \frac{\omega_0 - \omega}{2} t.$$

If ω_0 is close to ω, and both are large, the second sine term is slowly varying in comparison to the first and the curve has the general appearance shown in Figure 7. The low-frequency factor $\sin(\omega_0 - \omega)t/2$ induces an *amplitude modulation* on the high-frequency factor $\sin(\omega_0 + \omega)t/2$. The resulting modulated oscillation is an example of the phenomenon of *beats*.

When a tuning fork is sounded simultaneously with a piano string at nearly the same frequency, the beats allow you to tell how far apart the two notes are even if you are tone-deaf; that is, if your ear is sensitive only to

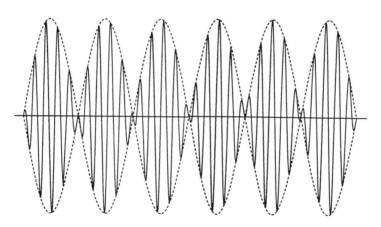

Figure 7

intensity and not to pitch. The slower the beats, the more nearly the string is in tune. This technique is used by piano tuners.

The phenomenon of beats has an interesting bearing on electronic communication and detection. The year 1902 saw the invention of the *heterodyne method* in which a circuit at the receiver generates a signal that differs only slightly in frequency from the high-frequency signal of the sender. The difference produces a frequency in the audible range. As a spin-off from radar technology, in 1946 the heterodyne method was used to detect one-centimeter microwave radiation from the sun. That experiment marked a major step in radio astronomy and microwave spectroscopy.

(d) Modulation

The phenomenon of amplitude modulation illustrated in Figure 7 is important in radio transmission. A carrier wave of high frequency can be modulated symmetrically by one signal as shown in Figure 8, or by two different signals as suggested in Figure 9. Further development of this technique makes it possible to transmit several messages with a single carrier. A mathematical description is given by the product of a high-frequency carrier wave

$$c(t) = A \cos(\omega t - \phi)$$

with a comparatively slowly varying signal $s(t)$. The signal changes A to $s(t)A$, and that is the reason for the term "amplitude modulation."

In *frequency modulation* it is ω that is altered and in *phase modulation*, ϕ is altered. The point we want to make is that a good part of this highly developed and sophisticated technology has its roots in the simple phenomenon illustrated in Figure 7.

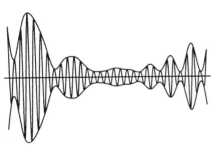

Figure 8

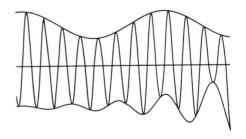

Figure 9

(e) Resonance

When $\omega_0 = \omega$ it is said that the system is *resonant*. Thus, resonance means that the angular frequency ω_0 of the forcing function agrees with the natural frequency ω of the oscillating system. From results obtained in Chapter 6 the two equations

$$x'' + \omega^2 x = \omega^2 A_0 \cos \omega t, \qquad y'' + \omega^2 y = \omega^2 B_0 \sin \omega t$$

have the particular solutions

(6)
$$x = \frac{1}{2} A_0 \omega t \sin \omega t, \qquad y = -\frac{1}{2} B_0 \omega t \cos \omega t$$

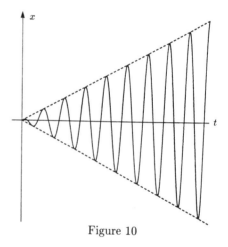

Figure 10

respectively. Here it is convenient to let both x and y denote responses, in agreement with the notation $z = x + iy$. Thus, x must not be confused with the input-function x introduced earlier. The oscillatory factor $\sin \omega t$ has the expected behavior but the amplitude is proportional to t and is unbounded as $t \to \infty$. See Figure 10.

The phenomenon of resonance has profound importance in engineering and physical situations. The failure of the Tacoma bridge was explained by some authorities on the basis of forced vibrations, and there are instances of the collapse of buildings induced by the rhythmic swaying of dancing couples. The failure of propellor shafts and of airplane wings has been attributed to resonance, and accidents have been caused by resonance in connection with power steering of automobiles. See also, Joshua 6:20.

PROBLEMS

1.   (a) If $|\omega|$ does not equal 1, 2, or 3 obtain the following particular solution for the equation $y'' + \omega^2 y = \sin t + \sin 2t + \sin 3t$:

$$y = \frac{\sin t}{\omega^2 - 1} + \frac{\sin 2t}{\omega^2 - 4} + \frac{\sin 3t}{\omega^2 - 9}$$

(b) This problem is best done with a hand calculator. Sketch a graph showing the approximate behavior of the input and of the particular solution found in Part (a) when $\omega = 1.01$, 2.01, and 2.99. (Four graphs.)

2. When placed on a rubber pad, a motor compresses the pad 1/24 inch. Find the frequency of natural vibrations of the system and thus show that there is a risk of uncontrolled oscillations when the motor reaches a speed of about 900 revolutions per minute.

Hint: $mg = k(1/24)(1/12)$ where $g \approx 32$, and $\omega^2 = k/m$.

3. As you and your new Ferrari together are being lowered by a crane, your friend notices that it sinks 6 inches from the time the tires just touch the ground.

(a) Check that the frequency of free oscillations when you're on the road is about $4/\pi$ cycles per second.

(b) On your way to school you traverse a washboard section of road in which the pavement forms a sinusoid of low amplitude and period 70; that is, the distance from one maximum to the next is 70 feet. At what speed will you be shaken so badly that you might lose control of the car?

Note: 70 is about 22π and 30 miles per hour is 44 feet per second.

4. This problem pertains to the equations

$$y'' + \omega^2 y = \cos \omega_0 t, \qquad y'' + \omega^2 y = \sin \omega_0 t$$

where ω_0 is close to but different from ω. The choice $A_0 = B_0 = 1/\omega^2$ in (4) yields the particular solutions

$$\frac{\cos \omega_0 t}{\omega^2 - \omega_0^2}, \qquad \frac{\sin \omega_0 t}{\omega^2 - \omega_0^2}$$

(a) As $\omega_0 \to \omega$ show that one of the intial conditions $y(0)$ or $y'(0)$ becomes infinite. (The one that becomes infinite is different for the two solutions.)

(b) Check that the following are particular solutions for which the initial conditions remain finite as $\omega_0 \to \omega$:

$$\frac{\cos \omega_0 t - \cos \omega t}{\omega^2 - \omega_0^2}, \qquad \frac{\sin \omega_0 t - \sin \omega t}{\omega^2 - \omega_0^2}$$

(c) By l'Hospital's rule show that the limit as $\omega_0 \to \omega$ in (b) gives the resonant solutions (6) with $A_0 = B_0 = 1/\omega^2$.

───────────────**ANSWERS**───────────────

2. $f = 48/\pi$ cycles per second.   **3.** (b) About 60 miles per hour.

4. Viscous damping

The equation obtained for free damped oscillations in Section 1 can be written

(1) $$y'' + 2by' + a^2 y = 0$$

where a and b are positive constants. In the case of a mass-spring system

$$2b = \frac{r}{m}, \qquad a^2 = \frac{k}{m}$$

whereas if the equation refers to an electrical circuit then

$$2b = \frac{R}{L}, \qquad a^2 = \frac{1}{LC}.$$

The following discussion is worded in terms of the mass-spring system but the analysis applies to both cases.

The characteristic polynomial $P(s) = s^2 + 2bs + a^2$ associated with (1) has roots

$$r_1 = -b + \beta, \quad r_2 = -b - \beta \quad \text{where} \quad \beta = \sqrt{b^2 - a^2}.$$

It will prove instructive to interpret the behavior of the solution in the three cases $b > a$, $b = a$, $b < a$ for which β is real and nonzero, or $\beta = 0$, or β is purely imaginary, respectively.

(a) Nonoscillatory behavior

When $\beta \neq 0$ the roots are distinct and the general solution is

(2) $$y = c_1 e^{(-b+\beta)t} + c_2 e^{(-b-\beta)t}.$$

If $b > a$ then β is real and the equation $\beta^2 = b^2 - a^2$ shows that $|\beta| < b$. Hence both of the quantities $-(b - \beta)$ and $-(b + \beta)$ in the exponents of (2) are negative. This is the *overdamped* case. The retarding force is so great that no oscillation can occur and the solution decays exponentially as $t \to \infty$. Figure 11 shows the behavior of several curves with a common initial value but several initial velocities. The latter is the slope at the left-hand endpoint of the curves.

When $a = b$ the characteristic polynomial is

$$s^2 + 2bs + b^2 = (s + b)^2$$

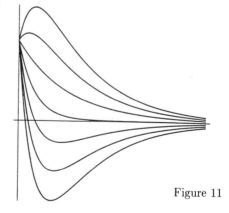

Figure 11

and the two roots are both equal to $-b$. By Chapter 6 the solution is

$$y = e^{-bt}(c_1 + c_2 t), \qquad a = b.$$

This is the *critically damped* case, so called because the motion becomes oscillatory if the damping is reduced by an arbitrarily small amount. In the design of instruments such as a voltmeter one wants the response to be as fast as possible without oscillation. This leads to critical damping as a useful design feature. Aside from that application, however, the case $a = b$ has little practical importance, because it is unstable with respect to small changes of the coefficients. A minute change of a or b moves the system out of the critically-damped case and into one of the other two cases. The general appearance of the curves when $a = b$ is similar to that when $a < b$.

(b) Oscillatory behavior

When $b < a$ the roots of the characteristic equation are complex and the solution is oscillatory. If ω is defined by

$$\omega^2 = a^2 - b^2, \qquad \omega > 0,$$

comparison with $\beta^2 = b^2 - a^2$ shows that $\beta = \pm i\omega$. Hence the solution can be written in the form

$$y = e^{-bt}(c_1 \cos \omega t + c_2 \sin \omega t)$$

or also in the form

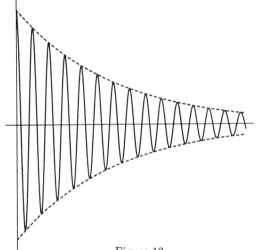

Figure 12

(3) $$y = e^{-bt} A \cos(\omega t - \phi)$$

where $C = c_1 - ic_2 = Ae^{-i\phi}$. This is the *underdamped* case. The first expression is useful for satisfying initial conditions and the second for plotting the graph of the solution. See Figure 12.

According to (3) the motion consists of a cosine factor with period $2\pi/\omega$ multiplied by the damping factor e^{-bt}. Since the cosine factor oscillates between -1 and 1, the graph of y versus t oscillates between the curves

$$y = -Ae^{-bt} \quad \text{and} \quad y = Ae^{-bt}.$$

It touches the lower curve at the values

$$\omega t = \phi + \pi, \quad \phi + 3\pi, \quad \phi + 5\pi, \quad \cdots$$

where the cosine is -1 and it touches the upper curve at the values halfway between these.

(c) Forced oscillations

The sinusoidal input discussed in Section 2 leads to the two equations

(4) $x'' + 2bx' + a^2 x = a^2 \cos \omega_0 t,$ $y'' + 2by' + a^2 y = a^2 \sin \omega_0 t$

where, without loss of generality, the amplitudes A_0 and B_0 have been taken equal to 1. As in the previous section, here too x denotes a response rather than an input. These equations can be solved by use of

(5) $z'' + 2bz' + a^2 z = a^2 e^{i \omega_0 t}.$

The trial solution $z = C e^{i \omega_0 t}$ in (5) leads to $C(-\omega_0^2 + 2bi\omega_0 + a^2) = a^2$, which gives the particular solution

(6) $z = \dfrac{a^2}{a^2 - \omega_0^2 + 2ib\omega_0} e^{i \omega_0 t}.$

Since the roots of the characteristic equation have negative real parts, as seen above, the complementary function tends to 0 as $t \to \infty$. This means that every other solution has the same behavior as the solution found above, in the long run. It is said, briefly, that the equation has a well-defined *steady-state* behavior, and that the effect of the initial conditions is a *transient*.

When $b = 0$ the complementary function does not tend to 0 and the steady-state response to an input of period T is interpreted as follows:

(a) Introduce a small dissipation $b > 0$.

(b) Keep only the terms of the response that do not tend to 0 as $t \to \infty$.

(c) Now let $b \to 0$ in the result (b).

Further discussion of these matters can be found in Chapter 8, Section 2.

(d) Phase lag and gain

Let the denominator in (6) be written in polar form $Re^{i\phi}$. The equation

(7) $a^2 - \omega_0^2 + 2bi\omega_0 = R(\cos \phi + i \sin \phi)$

holds if R is the magnitude of the expression on the left and ϕ is the phase angle. By a short calculation

(8) $R^2 = (a^2 - \omega_0^2)^2 + 4b^2 \omega_0^2,$ $\tan \phi = \dfrac{2b\omega_0}{a^2 - \omega_0^2}$

where $R \geq 0$ and $0 < \phi < \pi$. The inequality $0 < \phi < \pi$ follows from the equation $2b\omega_0 = R \sin \phi$, which shows that $\sin \phi > 0$.

Let us define G by

$$(9) \qquad G = \frac{a^2}{R} = \frac{a^2}{\sqrt{(a^2 - \omega_0^2)^2 + 4b^2\omega_0^2}}.$$

Equations (6), (7), and (9) give $z = Ge^{i(\omega_0 t - \phi)}$ and the corresponding real solutions are

$$x = G\cos(\omega t - \phi), \qquad y = G\sin(\omega t - \phi).$$

The steady-state response has the same angular frequency ω as the impressed oscillation, but there is a phase lag ϕ given by (8). When $\omega_0 < a$ the angle ϕ is between 0 and $\pi/2$, and it approaches 0 as $\omega_0 \to 0$. Thus, the system tends to follow a slow oscillation in phase. However if $\omega_0 > a$ then ϕ is between $\pi/2$ and π, and ϕ tends to π as $\omega_0 \to \infty$. This shows that when the impressed oscillation is rapid, the response is almost out of phase with it. As $\omega_0 \to a$ the phase lag tends to $\pi/2$. This corresponds to the condition for near-resonance, as seen next.

Since the input amplitudes are 1 and the response amplitudes are G the coefficient G is a measure of the gain of the system. When ω_0 is large G is approximately a^2/ω_0^2, which is small. Thus a rapid impressed oscillation does not produce much response. But if $\omega_0 = a$ then $G = a/(2b)$, and this can be dangerously large when the damping term b is close to 0. As $b \to 0$ we are led to the condition of resonance discussed in Section 3, in which the amplitude of the response tends to ∞ with t.

PROBLEMS

1. In the overdamped case check that the conditions $y(0) = y_0$, $y'(0) = y_1$ lead to the equations

$$c_1 + c_2 = y_0, \qquad c_1 - c_2 = (y_1 + by_0)\beta^{-1}$$

Thus obtain the solution $y = y_0 u + y_1 v$ where

$$u = e^{-bt}\left(\cosh \beta t + b\frac{\sinh \beta t}{\beta}\right), \qquad v = e^{-bt}\frac{\sinh \beta t}{\beta}$$

Note that (u, v) gives a fundamental solution at $t = 0$. The same applies to u and v in Problem 2.

2. (a) Using $e^{i\omega t} = \cos \omega t + i \sin \omega t$, check that

$$\cosh i\omega t = \cos \omega t, \qquad \frac{\sinh i\omega t}{i\omega} = \frac{\sin \omega t}{\omega}$$

(b) Setting $\beta = i\omega$ in the result of Problem 1, obtain the following formulas for the underdamped case:

$$u = e^{-bt}\left(\cos\omega t + b\frac{\sin\omega t}{\omega}\right), \qquad v = e^{-bt}\frac{\sin\omega t}{\omega}$$

3. (a) For the critically damped case $a = b$, obtain the fundamental solution

$$u = e^{-bt}(1 + bt), \qquad v = e^{-bt}t$$

(b) Show that the result of Part (a) can be obtained by letting $\beta \to 0$ in the result of Problem 1, and also by letting $\omega \to 0$ in the result of Problem 2.

4. (a) A force of 3380 dynes extends a spring 2 cm. A mass of 10 grams is suspended from one end of the spring, pulled down 5 cm from its equilibrium position, and gently released. Find a differential equation and a formula for the motion if resistance is neglected.

(b) Solve Part (a) under the hypothesis that the motion is viscously damped. It is given that a resistance of 100 dynes corresponds to a speed of 1 cm per second.

5. What value of the resistance in Problem 4 would make the motion critically damped, and what is the motion in that case?

6. (a) A force of 120 pounds extends a spring one inch. A weight of 320 pounds is suspended from the spring, pulled down 1 foot below its equilibrium position, and gently released. Find a differential equation and a formula for the motion when resistance is neglected.

Hint: One inch is $1/12$ foot. The weight in pounds is related to the mass in newtons by the formula $w = mg$. Take $g = 32$ feet per second per second.

(b) Solve the problem of Part (a) if there is viscous damping of 260 pounds corresponding to 1 foot per second.

7. What value of the resistance in Problem 6 would give critical damping and what is the motion in that case?

8. Logarithmic decrement. Let $y(t) = Ae^{-bt}\cos(\omega t - \phi)$ where A, b and ω are positive constants. If $T = 2\pi/\omega$ is the period of the cosine factor show that

$$\frac{y(t+T)}{y(t)} = e^{-bT}$$

independently of t. Show that successive maxima y_k, y_{k+1} of the curve y versus t are spaced by a time interval T and hence the foregoing equation also gives y_{k+1}/y_k. You need not compute the maximum values y_k. The quantity $\delta = \log(y_k/y_{k+1}) = bT$ is called the *logarithmic decrement*. It allows estimation of the damping constant b from observation of the motion. Since b depends on the medium in which the motion takes place, whether liquid or gaseous, the logarithmic decrement is useful for measuring viscosity.

9. A body suspended from the end of a viscously damped spring makes 60 oscillations per minute and after one minute the amplitude is halved. Find the differential equation of the motion.

10. Show that $G > 1$ if, and only if, $\omega_0^2 + 4b^2 < 2a^2$. This is the condition for amplification, that is, for the steady-state response to have a larger amplitude than the input.

11. If $s = \omega_0 - a$ and $|s|$ is small, show that

$$4G^2 \approx \frac{a^2}{s^2 + b^2}$$

Sketch the corresponding approximate graph of G^2 versus s for fixed $b > 0$ and show that G^2 assumes half its maximum value when $s = b$. Hence if G^2 is plotted versus ω_0, the full width at half maximum is approximately $2b$. The *full width at half maximum* has been given the misleading name of "half width." The half width embodies a relation betweeen frequency response and dissipation that is useful in the analysis of microwave and acoustic resonators and in the measurement of dielectric materials.

12. In the RLC circuit considered in Section 1 suppose $L = 0.1$ henry, $C = 40$ microfarads, and $R = 80$ ohms. Find the differential equation for the voltage v and its general solution. Hint: Compatible units for R, L, C and v are ohms, henrys, farads, and volts, respectively. A microfarad is a millionth of a farad.

13. In the RLC circuit considered in Section 1 suppose $L = 1/8$ henry, $C = 50$ microfarads, $R = 125$ ohms. Find a differential equation for the voltage and its general solution.

14. If i is the current in the series circuit considered in Section 1, differentiate the voltage equation with respect to t, use $i = dQ/dt = C dv/dt$, and thus show that

$$Li'' + Ri' + \frac{1}{C}i = v_0'(t)$$

15. In Problem 14 let $v_0(t) = v_0 e^{j\omega t}$ where v_0 is constant and $j = \sqrt{-1}$. Show that the substitution $i = Ae^{j\omega t}$ yields a solution $i = i_0(t)$ satisfying

$$v_0(t) = i_0(t)Z \quad \text{where} \quad Z = R + j\left(\omega L - \frac{1}{\omega C}\right)$$

In electrical enginering $\sqrt{-1}$ is denoted by j to avoid confusion with the current i. The number Z is called the *complex impedance*. The equation $v_0(t) = Zi_0(t)$ is a sophisticated generalization of Ohm's law, to which it reduces when $L = 0$ and $\omega C = \infty$. Complex numbers were introduced into circuit analysis by the American electrical engineer Charles Steinmetz. His mathematical innovations led to major advances in the theory of alternating-current machinery and were responsible, in part, for the early rise of the General Electric Company to a position of industrial dominance.

─────────────────ANSWERS─────────────────

4. (a) $y'' + 169y = 0$, $\quad y = 5\cos 13t$
   (b) $y'' + 10y' + 169y = 0$, $\quad y = (5/12)(12\cos 12t + 5\sin 12t)e^{-5t}$
5. $r = 260$, $\quad y = 5(1 + 13t)e^{-13t}$
6. (a) $y'' + 144y = 0$, $\quad y = \cos 12t$
   (b) $y'' + 26y' + 144y = 0$, $\quad y = (1/5)(9e^{-8t} - 4e^{-18t})$
7. $r = 240$, $\quad y = (1 + 12t)e^{-12t}$
9. $y'' + y' \log 4 + ((120\pi)^2 + (\log 2)^2)y = 0$ if t is measured in minutes
12. $v'' + 800v' + 250000v = 0$, $v = Ae^{-400t}\cos(300t - \phi)$
13. $v'' + 1000v' + 160000v = 0$, $v = c_1 e^{-200t} + c_2 e^{-800t}$

5. Kepler's laws of planetary motion

Newton's law of gravitation states that two mass points attract each other with a force proportional to the product of the masses and inversely proportional to the square of the distance between them. If the masses are m_1 and m_2, and the distance is r, then the magnitude of the force is

$$\frac{Gm_1 m_2}{r^2}$$

where G is a proportionality constant called the *gravitational constant*.

There is a close connection between Newton's law of gravitation and three laws of planetary motion that were discovered empirically by the German mathematical astronomer and mystic Johannes Kepler (1571–1630).

Because of its importance in the history of science, this connection will be developed in detail. We use complex numbers and second-order linear differential equations as presented in Chapter 6.

Kepler's first law states that the planets move in ellipses with the sun at one focus. The second law states that the radius from a planet to the sun sweeps out equal areas in equal times. The third law states that the square of the time for a single traverse of an orbit is proportional to the cube of the mean distance to the sun. Here the term "mean distance" means the average of the maximum and minimum distances.

To deduce these laws from Newton's law of gravitation, we need three simplifications. Newton showed that a thin uniform shell attracts a particle at an external point as if all the mass of the shell is concentrated at its center. By integration, he then deduced a similar result for a solid sphere, provided the density depends only on the distance to the center. A second integration gives a corresponding result for two spheres. Thus, if the planet and sun are thought to be uniform spheres, they can be replaced by points, of the same mass, at their centers. This is the first simplification.

The second simplification is that we consider the sun and a single planet in isolation, neglecting the influence of the other planets. This approximation was made in Newton's early work and is correct within the accuracy of Kepler's observations.

Finally, we assume that the mass of the planet is so small compared to that of the sun that the sun is not moved by the planet's attraction. This is the third simplification.

Let us begin by outlining a proof that the (x, y, z) orbit of the planet lies in a plane. As long as the planet does not fall into the sun, the differential equations of the motion satisfy the hypothesis of a uniqueness theorem for systems established in Chapter 14. By taking the (x, y) plane to be the plane through the sun, the planet, and the vector v_0 that gives the planet's velocity at some specified time $t = t_0$, we shall get a solution in which the third-coordinate function z is identically 0. Uniqueness shows that this is the sole solution, hence the orbit is plane.

We place the center of the sun at the origin and the center of the planet at (x, y). The components of force on the planet due to the sun's attraction are denoted by mf_x and mf_y where m is the planet's mass. Newton's law of motion

$$\text{(mass) (acceleration)} = \text{(force)}$$

yields $mx'' = mf_x$ and $my'' = mf_y$ or, dividing by m,

$$x'' = f_x, \qquad y'' = f_y.$$

When the second equation is multiplied by i and added to the first we get

$$w'' = f \quad \text{where} \quad w = x + iy, \quad f = f_x + if_y.$$

It is convenient to introduce polar coordinates (r, θ) so that

$$x = r\cos\theta, \qquad y = r\sin\theta, \qquad w = re^{i\theta}.$$

Since the force f is directed toward the sun it has the form $Re^{i\theta}$ where R is a real-valued function of r and θ. The equation $w'' = f$ is now

(1) $$(re^{i\theta})'' = Re^{i\theta}.$$

A field in which the force is directed toward a fixed point is said to be *central*. Equation (1) is the equation for plane motion in a central force field.

The left side of (1) is evaluated by differentiating the equation

$$(re^{i\theta})' = (r' + ir\theta')e^{i\theta}.$$

We substitute the result into (1) and divide by $e^{i\theta}$ to get

$$(r' + ir\theta')' + i\theta'(r' + ir\theta') = R.$$

Since R is real, equating real and imaginary parts gives the basic equations for planetary motion:

(2) $$r'' - r(\theta')^2 = R, \quad (r\theta')' + r'\theta' = 0.$$

If the second equation (2) is multiplied by r it becomes $(r^2\theta')'=0$. Hence

(3) $$r^2\theta' = C$$

where C is constant. The case $C = 0$ means that the orbit is a straight line and is excluded from now on.

Writing (3) in the form

$$\frac{1}{2}r^2 \, d\theta = \frac{C}{2} \, dt$$

and integrating between corresponding limits yields

(4) $$\int_{\theta_1}^{\theta_2} \frac{1}{2}r^2 \, d\theta = \frac{C}{2} \int_{t_1}^{t_2} dt.$$

This is Kepler's second law. The integral on the left describes the area swept out by the radius vector and that on the right measures the elapsed time. The equation requires the planet to move faster when close to the sun, so that the then-shorter radius will still sweep out equal areas in equal times. The planet can be thought to be falling toward or receding from the sun. Kepler's second law agrees with the intuitive idea that a falling body moves faster the farther it falls.

What has been shown is that Kepler's second law holds for motion in any central force field. We now use the further information that the magnitude of the force is inversely proportional to r^2; more specifically, we set $R = -k/r^2$

where k is a positive constant. The minus sign is used because the force is directed toward the origin. In view of (3), the first equation (2) is now

$$(5) \qquad\qquad r'' - \frac{C^2}{r^3} = -\frac{k}{r^2}.$$

We set $s = 1/r$ and we take θ instead of t as the independent variable. Then (3) gives $\theta' = Cs^2$ and by the chain rule

$$r' = \frac{dr}{dt} = -\frac{1}{s^2}\frac{ds}{dt} = -\frac{1}{s^2}\frac{ds}{d\theta}\frac{d\theta}{dt} = -C\frac{ds}{d\theta}.$$

In a like manner,

$$r'' = \frac{dr'}{dt} = \frac{dr'}{d\theta}\frac{d\theta}{dt} = \left(-C\frac{d^2 s}{d\theta^2}\right)Cs^2 = -\frac{C^2}{r^2}\frac{d^2 s}{d\theta^2}.$$

Substitution into (5) and multiplying by $-r^2/C^2$, we get

$$(6) \qquad\qquad \frac{d^2 s}{d\theta^2} + s = \alpha \quad \text{where} \quad \alpha = \frac{k}{C^2}.$$

The methods of Chapter 6 give

$$(7) \qquad\qquad s = \alpha - \beta\cos(\theta - \theta_0)$$

where β and θ_0 are constant. Since $s = 1/r$ this is the polar equation of a conic with one focus at the origin; see Problem 2. In the case of a planet the orbit is bounded, hence an ellipse. Thus we have established Kepler's first law.

     Kepler's third law depends on the fact that $R = -k/r^2$ where the constant k depends only on the mass of the sun, not on the mass of the planet. (Recall that m canceled out in our derivation of $w'' = f$.) By (4) the time T for a complete revolution satisfies

$$\frac{1}{2}CT = \int_0^{2\pi} \frac{1}{2}r^2 \, d\theta = A$$

where A is the area of the elliptical region enclosed by the orbit. Using this and $k = \alpha C^2$ from (6) we get

$$(8) \qquad\qquad T^2 = \frac{4A^2}{C^2} = \frac{4A^2\alpha}{k}.$$

In Problem 3 it is seen that

$$(9) \qquad\qquad kT^2 = (2\pi)^2 a^3$$

where the semimajor axis a of the ellipse is the mean of the maximum and minimum distances from the planet to the sun. Since k is the same for all planets, this is Kepler's third law.

The foregoing discussion illustrates an interesting contrast between theoretical and empirical modes of procedure. Kepler's remarkable investigations, sustained by a deep faith in the mathematical harmony of nature, were extended over a period of thirty years and one of his many notebooks contains over 800 pages of calculations in small handwriting. Nevertheless his work reduces to three statements of fact about the solar system while Newton's applies to an enormous diversity of problems throughout the universe.

PROBLEMS

In satellite problems the satellite is so high that air resistance can be neglected and the satellite is revolving around the earth rather than the sun. We assume that the satellite does not hit the earth and we retain the assumption $C \neq 0$ mentioned in the text.

1. If the earth is the attractor, instead of the sun, the constant k is given by $k = g(r_e)^2$ where g is the acceleration of gravity at the earth's surface and r_e is the radius of the earth. Equating gravitational and centrifugal force gives $mv^2/a = mk/a^2$ for a satellite in a circular orbit of radius a. Obtain the equation

$$T = 2\pi \frac{a}{r_e} \sqrt{\frac{a}{g}}$$

for the period and check that this is consistent with Kepler's third law.

2. Let L be a fixed line in the plane and F a fixed point not on L. The focus-directrix definition of a conic is

$$\text{(distance to F)} = \epsilon(\text{distance to L})$$

where ϵ is the eccentricity. If $\theta_0 = 0$ and $\beta > 0$ in (7), check that (7) gives $r\alpha - \beta x = 1$. Solve for r and deduce that the locus is a conic with one focus at the origin, with directrix given by $x = -1/\beta$, and with eccentricity β/α. Since we can choose the x axis in any direction, the conditions on θ and β involve no loss of generality.

3. The condition $\alpha < 0$ in (7) corresponds to a force of repulsion and $0 < \alpha \leq \beta$ gives a parabola or hyperbola. Assuming $0 < \beta < \alpha$, let the ellipse have $2a$, $2b$, $2c$ as major axis, minor axis, and distance between foci respectively. Show that

$$a = \frac{\alpha}{\alpha^2 - \beta^2} \qquad b^2 = \frac{1}{\alpha^2 - \beta^2} \qquad c = \frac{\beta}{\alpha^2 - \beta^2}$$

and hence that (9) follows from (8). Hint: Get a and c by a sketch, using $\theta - \theta_0 = 0$ or π. Then get b from $b^2 = a^2 - c^2$ and A from $A = \pi ab$.

4. (a) If a particle moves in a central force field, deduce from (3) that its speed $v = |w'|$ is related to $s = 1/r$ by

$$\left(\frac{v}{C}\right)^2 = \left(\frac{ds}{d\theta}\right)^2 + s^2$$

(b) In Part (a) let $s = \alpha - \beta \cos\theta$ where $s = 1/r$ and α and β are positive constants. Obtain the following equation and interpret by the law of cosines:

$$v = C(\alpha^2 - 2\alpha\beta\cos\theta + \beta^2)^{1/2}$$

5. Set $C^2 = k/\alpha$ and deduce from Problems 4 and 3, respectively, that

$$\frac{2}{r} - \frac{v^2}{k} = \frac{\alpha^2 - \beta^2}{\alpha} = \frac{1}{a}$$

6. A satellite is launched so that it has speed v_0 at height r_0 above the earth's center.

(a) Show that the orbit is an ellipse, a parabola, or hyperbola when $2k - r_0 v_0^2$ is positive, zero, or negative, respectively.

(b) If the orbit is an ellipse show that the time for return depends only on (r_0, v_0) and not on the direction of launch.

Hint: For (a) use Problem 5 and the formula $\gamma = \beta/\alpha$ from Problem 2. For (b) use Problem 5 and Kepler's third law.

7. If a coasting satellite has speed v when it is at a distance r from the earth's center, the potential energy is the work needed to raise it to the height r and the kinetic energy is $mv^2/2$.

(a) By the principle of conservation of energy, give a simple proof that the left-hand expression in Problem 5 is constant.

(b) Show that the result of Problem 6(a) agrees with what one would expect from the escape velocity; see Chapter 5, Section 4.

8. If the earth were a uniform sphere the gravitational force on a particle of mass m at a distance $r \le r_e$ from its center would be directed toward the center and would have magnitude mgr/r_e. Imagine a straight tube extending from a point P on the earth's surface to another point Q on the surface. A particle is released from rest in the tube at some point R so that it slides back and forth in the tube without air resistance or friction. Show that the period of oscillation is independent of P, Q, R and is the same as the period of a satellite in an orbit of radius r_e.

9. (a) Show that Kepler's third law is equivalent to the statement that a plot of $\log a$ versus $\log T$ will be a straight line of slope $2/3$.

(b) Using a hand calculator, verify the result of Part (a) by reference to the following table. The letters are abbreviations for the planets Mercury, Venus, Earth, Mars, Jupiter, Saturn, Uranus, Neptune, and Pluto. The mean distance a is given in millions of miles and the period T in years:

Planet	M	V	E	M	J	S	U	N	P
a	36.0	67.2	92.9	142	483	886	1780	2790	3670
T	.241	.615	1.00	1.88	11.9	29.5	84.0	165	248

10. Open-ended problem. Historically, Newton started from Kepler's laws and deduced the law of gravity as a necessary consequence. Reinterpret the equations of the text with this as the primary goal. It will be found that Kepler's second law ensures that the force is central and directed towards the sun. The first law ensures that the force for a given planet is inversely proportional to r^2, and the third law shows that the constant k in the text is independent of the planet's mass.

6. The elementary functions

When the differential equation

$$\frac{dy}{dt} = t^n$$

was first encountered more than three centuries ago, the solution was quickly seen to be a constant times a power of t, provided n is an integer different from -1. However the case $n = -1$ resisted solution for a number of years, until it was realized that this case calls for a function of an entirely different type. In more recent times the problems of mathematical physics have also led to new functions such as Bessel functions. The main properties of these functions are often deduced from the differential equations they satisfy.

Here we show how this line of thought applies to the functions

$$(1) \qquad\qquad e^t, \qquad \log t, \qquad t^m, \qquad \sin t, \qquad \cos t$$

familiar from calculus. In particular, we shall deduce the main properties of e^t and t^m without prior knowledge of the theory of exponents, and the main properties of $\sin t$ and $\cos t$ without prior knowledge of the geometry of the circle.

The functions that will turn out to be identical with those in (1) are denoted respectively by

(2) $$E(t), \qquad L(t), \qquad P_m(t), \qquad S(t), \qquad C(t).$$

Differential equations such as

$$\frac{dE}{dt} = E(t), \qquad \frac{dL}{dt} = \frac{1}{t}, \qquad \frac{d^2 S}{dt^2} + S = 0$$

are motivated by familiar properties of the functions (1), and that is the reason for calling (1) to your attention. The notation (2), however, provides a safeguard against taking other properties of these functions for granted. Since we want to deduce these properties from differential equations, it is important not to assume them at the outset.

The existence and uniqueness theorems stated without proof in Chapter 4 will be used here. They indicate that the functions $E(t)$, $S(t)$, and $S(t)$ introduced below exist for all t, and $L(t)$ and $P_m(t)$ exist for $t > 0$.

(a) The exponential

The function $y = E(t)$ is defined by the initial-value problem

(3) $$E'(t) = E(t), \qquad E(0) = 1.$$

Let $x = E(t + s)$ where s is constant. Then by (3)

$$\frac{dx}{dt} = E'(t + s) = E(t + s) = x, \qquad x(0) = E(s).$$

On the other hand, let $y = E(t)E(s)$. Then, again by (3),

$$\frac{dy}{dt} = E'(t)E(s) = E(t)E(s) = y, \qquad y(0) = E(s).$$

Thus x and y satisfy the same initial-value problem. By uniqueness $x = y$, hence

(4) $$E(t + s) = E(t)E(s).$$

Since e^t satisfies the same initial-value problem (3) we must have $e^t = E(t)$, and the above equation yields the main property of the exponential function:

$$e^{t+s} = e^t e^s.$$

(b) The logarithm

By definition, $L(t)$ is the solution of the initial-value problem

$$(5) \qquad \frac{dL}{dt} = \frac{1}{t}, \qquad L(1) = 0, \quad t > 0.$$

Let $x = L(st)$ where s is a positive constant. Then by the chain rule and (5) with st replacing t

$$\frac{dx}{dt} = L'(st)s = \frac{1}{st}s = \frac{1}{t}, \qquad x(1) = L(s).$$

On the other hand let $y = L(s) + L(t)$. Then by (5)

$$\frac{dy}{dt} = L'(t) = \frac{1}{t}, \qquad y(1) = L(s).$$

Since x and y satisfy the same initial-value problem we conclude that $y = x$, or in other words

$$(6) \qquad L(st) = L(s) + L(t).$$

The function $\log t$ satisfies the initial-value problem (5). Hence $\log t = L(t)$ and (6) yields the main property of the logarithm,

$$\log(st) = \log s + \log t.$$

(c) Powers

The function $P_m(t)$ is defined by the initial-value problem

$$(7) \qquad \frac{dP_m}{dt} = \frac{mP_m(t)}{t}, \qquad P_m(1) = 1, \quad t > 0.$$

Although this differential equation may seem more mysterious than those above, it is satisfied by t^m.

If $x = P_m(P_n(t))$ then by the chain rule and (7)

$$\frac{dx}{dt} = P'_m(P_n(t))P'_n(t) = \frac{mP_m(P_n(t))}{P_n(t)}\frac{nP_n(t)}{t} = \frac{mnx}{t}.$$

Here (7) has been used twice, once for P_m and once for P_n. If $y = P_{mn}(t)$ then by (7) with mn replacing m

$$\frac{dy}{dt} = \frac{mny}{t}.$$

Since $x(1) = y(1) = 1$, we see that x and y satisfy the same initial-value problem. Hence $x = y$ or in other words

$$P_m(P_n(t)) = P_{mn}(t).$$

It is left for you to verify by a similar but simpler argument that, if $s > 0$ and $t > 0$,

$$P_{m+n}(t) = P_m(t)P_n(t), \qquad P_m(st) = P_m(s)P_m(t).$$

These three equations yield the principal laws of exponents, namely,

$$(t^n)^m = t^{mn}, \qquad t^{m+n} = t^m t^n, \qquad (st)^m = s^m t^m.$$

Note that m and n can be any positive numbers; they need not be integers.

(d) The trigonometric functions

Let functions $C(t)$ and $S(t)$ be defined by the differential equations

$$C'' + C = 0, \qquad S'' + S = 0$$

with the initial conditions

$$C(0) = 1, \quad C'(0) = 0; \qquad S(0) = 0, \quad S'(0) = 1$$

respectively. The function

$$F = (S' - C)^2 + (S + C')^2$$

satisfies

$$F' = 2(S' - C)(S'' - C') + 2(S + C')(S' + C'')$$

$$= 2(S' - C)(-S - C') + 2(S + C')(S' - C) = 0$$

where the second equality follows from $S'' = -S$, $C'' = -C$ and the third follows by inspection; it is not necessary to multiply out. The equation $F' = 0$ shows that F is constant. By the initial conditions for S and C

$$F(0) = (1 - 1)^2 + (0 + 0)^2 = 0.$$

This shows that the constant value of F must be 0. Since the two terms in F, being squares, are nonnegative, these terms are 0 separately. Thus

(8) $$S' = C, \qquad C' = -S$$

or, in conventional notation,

$$\frac{d}{dt}\sin t = \cos t, \qquad \frac{d}{dt}\cos t = -\sin t.$$

These are the main differentiation formulas of trigonometry.

If we set $x(t) = S(s+t)$ where s is constant, the chain rule gives

$$x'' + x = S''(s+t) + S(s+t) = 0.$$

Here the second equality follows from $S'' + S = 0$ at the argument $s+t$. Now comes an important point. In view of the initial conditions, the Wronskian of the functions S and C at the origin is

$$W(C,S) = \begin{vmatrix} C & S \\ C' & S' \end{vmatrix} = \begin{vmatrix} 1 & 0 \\ 0 & 1 \end{vmatrix} = 1, \qquad t = 0.$$

Therefore Chapter 6, Section 6, Theorem 4 indicates that every solution of the equation $x'' + x = 0$ can be expressed in terms of S and C. Applying this to the function x above we get the first equation below, where c_1 and c_2 are constant:

$$S(s+t) = c_1 C(t) + c_2 S(t), \qquad C(s+t) = -c_1 S(t) + c_2 C(t).$$

In view of (8), the second equation follows by differentiation of the first.

The constants c_1 and c_2 are found by setting $t = 0$. The result is

$$S(s+t) = S(s)C(t) + C(s)S(t), \qquad C(s+t) = -S(s)S(t) + C(s)C(t).$$

In conventional notation these are the familiar formulas

$$\sin(s+t) = \sin s \cos t + \cos s \sin t, \qquad \cos(s+t) = \cos s \cos t - \sin s \sin t.$$

PROBLEMS

1. Show that $C(-t) = C(t)$ and $S(-t) = -S(t)$.

Hint: $x = C(t)$ and $y = C(-t)$ satisfy the same second-order initial-value problem.

2. (a) Show that $P_m(t)P_n(t)$ satisfies the same initial-value problem as $P_{m+n}(t)$ and thus deduce one of the three laws of exponents.

(b) Similarly consider the equation $P_m(st) = P_m(s)P_m(t)$ where s is a positive constant.

3. Show that the function $G = S^2 + C^2$ is constant and, setting $t = 0$, deduce that the value is 1. The formula $\sin^2 t + \cos^2 t = 1$ is equivalent to the Pythagorean theorem.

4. According to Problem 3 the point $x = C(t)$, $y = S(t)$ lies on the circle $x^2 + y^2 = 1$. Show that the arc of the circle from $t = 0$ to t is t, and hence t can be identified with radian measure. This problem gives the key to the relation of the functions C and S to the geometry of the circle.

Hint: $C' = -S$ and $S' = C$.

5. The following result is needed in Problem 6. Fill in the details:

$$E(t)E(s) = E(t + s) \Rightarrow E(t)E(-t) = 1 \Rightarrow E(t) \neq 0 \Rightarrow E(t) > 0$$

6. Relations among the functions. We shall establish the four relations

$$\ln(e^t) = t, \qquad e^{\ln t} = t, \qquad \ln t^m = m \ln t, \qquad (e^t)^m = e^{mt}$$

The first and fourth hold without restriction, the second and third for $t > 0$.

(a) If $x = L(E(t))$ deduce $x' = 1$, $x(0) = 0$, hence $x = t$.

(b) If $y = E(L(t))$ for $t > 0$ deduce $y' = y/t$, $y(1) = 1$, hence $y = t$.

(c) If $z = L(P_m(t))$ for $t > 0$ deduce $z' = m/t$, $z(1) = 0$, hence $z = mL(t)$.

(d) If $z = P_m(E(t))$ deduce $z' = mz$, $z(0) = 1$ hence $z = E(mt)$.

One cannot escape the feeling that mathematical formulas have an independent existence and intelligence of their own, that they are wiser than we are.

—Heinrich Hertz

I can wait 100 years for a reader, the Lord God also had to wait 6,000 years for the discoverer of His works.

—Johannes Kepler

Chapter 8
LINEAR EQUATIONS II

THIS CHAPTER gives a comprehensive development of the theory of linear differential equations with constant coefficients. The concept of asymptotic stability allows us to get the approximate long-time behavior by use of any particular solution without solving the characteristic equation. For broad classes of forcing-functions f the form of particular solutions can be precisely specified, and this is done here by means of techniques that are computationally effective. For general f, both existence and uniqueness are deduced from the fundamental theorem of algebra.

1. Techniques of solution

Consider the equation

$$Ty = y^{(n)} + p_{n-1}y^{(n-1)} + \cdots + p_1y' + p_0y = f$$

on an open interval I, where the coefficients p_j are real or complex constants. The domain of the operator T defined by the first equality consists of the functions that are n times differentiable on I. We write, as in Chapter 6,

$$Ty = (D^n + p_{n-1}D^{n-1} + \cdots + p_1D + p_0)y = P(D)y.$$

Let us show that T is linear. Since D is linear it is readily checked that

$$D^j(u+v) = D^ju + D^jv, \qquad j = 0, 1, 2, \cdots$$

when u and v are j-times differentiable. Multiplying by p_j, we get

$$p_jD^j(u+v) = p_jD^ju + p_jD^jv.$$

If this is summed on j from $j = 0$ to $j = n$ the result is

$$T(u+v) = Tu + Tv.$$

The proof that $T(cu) = cTu$ is similar.

Since T is linear, all the principles established for general linear operators in Chapter 6 hold for it, and we need not repeat the proofs. The action of T on exponential functions e^{st} is also similar to that in the earlier discussion, as seen next.

If s is constant then

$$D^j e^{st} = s^j e^{st}, \qquad j = 0, 1, 2, \cdots$$

If this equation is multiplied by p_j and the results added we get

$$P(D)e^{st} = P(s)e^{st}$$

where $P(s)$ is the *characteristic polynomial*,

$$P(s) = s^n + p_{n-1}s^{n-1} + \cdots + p_1 s + p_0.$$

Hence $y = e^{st}$ satisfies the homogeneous equation $Ty = 0$ if, and only if, s is a root of the *characteristic equation $P(s) = 0$*.

Just as one can solve the characteristic equation by factoring $P(s)$, one can solve the differential equation by factoring $P(D)$. If

$$a_1, a_2, \cdots, a_n$$

are the roots of $P(s) = 0$ counted with their multiplicities, then

$$P(D) = (D - a_1)(D - a_2) \cdots (D - a_n).$$

The action of $P(D)$ on functions in its domain is independent of the order in which the factors are written. To see this, note that the equation

$$(D - a)(D - b) = (D - b)(D - a)$$

obtained in Chapter 6, Section 2 shows that an interchange of adjacent factors does not affect the result. However, any permutation of the factors can be obtained by successive interchanges of adjacent factors, as is easily proved. Hence the permutation leaves the operator unaltered.

The following example illustrates the correspondence between $P(D)$, $P(s)$, and Te^{st}.

EXAMPLE 1. Find all solutions of the form $y = e^{st}$:

(1) $$(D - 4)(D^2 - 1)(D^2 + 1)y = 0.$$

The characteristic equation is

$$(s - 4)(s^2 - 1)(s^2 + 1) = 0$$

with roots 4, 1, -1, i, $-i$. These roots give the solutions

$$e^{4t}, \qquad e^t, \qquad e^{-t}, \qquad e^{it}, \qquad e^{-it}.$$

The last two functions could be replaced by

$$\cos t, \qquad \sin t.$$

This follows from the principle of equating real parts, or also from the equations

$$\cos t = \frac{e^{it} + e^{-it}}{2}, \qquad \sin t = \frac{e^{it} - e^{-it}}{2i}.$$

By the principle of superposition

$$y = c_1 e^{4t} + c_2 e^t + c_3 e^{-t} + c_4 \cos t + c_5 \sin t$$

is a solution for any choice of the constants c_j. It is seen later that this is both the complete solution and the general solution.

As in Chapter 6, here too the problem of multiple roots is taken care of by the following:

EXPONENTIAL SHIFT THEOREM. If u is in the domain of $P(D)$ and k is constant, then

(2) $$P(D)e^{kt}u = e^{kt}P(D+k)u.$$

The proof is simple. In the equation

$$P(D)e^{kt}u = (D - a_1)(D - a_2) \cdots (D - a_n)e^{kt}u$$

you can move the factor e^{kt} to the left of $(D - a_n)$ if D in this factor is replaced by $D + k$. Similarly you can move it to the left of the next factor, and so on. After n steps e^{kt} is at the left of every factor $(D - a_j)$ and the result is (2).

If some particular root $a = a_j$ is of multiplicity m, so that $P(D)$ has a factor

$$(D - a)^m,$$

the exponential shift theorem shows that the corresponding solutions of the homogeneous equation are

$$e^{at}, \qquad te^{at}, \qquad t^2 e^{at}, \qquad \cdots, \qquad t^{m-1}e^{at}.$$

A formal proof is given in Section 2. Here we illustrate by an example.

EXAMPLE 2. Obtain an appropriate set of solutions for each root of the characteristic equation:

(3) $$(D - 1)^3(D + 1)(D^2 + 1)^2 y = 0.$$

The trial solution $y = e^t u$ and the shift theorem with $k = 1$ lead to

$$e^t D^3(D + 2)((D + 1)^2 + 1)^2 u = 0.$$

The equation can be written

$$(D + 2)(D^2 + 2D + 2)^2 D^3 u = 0$$

and holds if $D^3 u = 0$. This gives

$$u = c_0 + c_1 t + c_2 t^2, \qquad y = (c_0 + c_1 t + c_2 t^2)e^t.$$

The functions y obtained for various choices of c_j are solutions corresponding to the root $s = 1$ of multiplicity 3. Note that y is a linear combination of the three functions

$$e^t, \qquad te^t, \qquad t^2 e^t.$$

This family of three functions is an equally good description of the solutions corresponding to $s = 1$.

Once the logic of the method is understood, it is no longer necessary to use the shift theorem. Instead, you can get the answer by inspection. Since the characteristic equation for (3) has the root -1 with multiplicity 1 and the roots i, $-i$ each with multiplicity 2, the corresponding solutions

$$e^{-t}, \qquad e^{it}, \qquad te^{it}, \qquad e^{-it}, \qquad te^{-it}$$

can be written with ease. If a real form is desired one can replace the pair e^{it}, e^{-it} by the pair $\cos t$, $\sin t$ to get

$$\cos t, \qquad t \cos t, \qquad \sin t, \qquad t \sin t.$$

By the principle of superposition any linear combination

$$(c_0 + c_1 t + c_2 t^2)e^t + c_3 e^{-t} + (c_4 + c_5 t)\cos t + (c_6 + c_7 t)\sin t$$

is also a solution. Note that there are eight constants c_j, corresponding to the fact that the equation is of order eight. It is seen later that this expression gives both the complete and the general solution.

So far we have considered only the homogeneous equation, for which $f = 0$. Techniques for solving the inhomogeneous equation are similar to those in Chapter 6, as illustrated next.

EXAMPLE 3. If the constant a is not a root of the characteristic equation, obtain a particular solution of

$$(4) \qquad\qquad P(D)y = e^{at}.$$

The substitution $y = Ae^{at}$ with A constant gives

$$P(D)Ae^{at} = AP(D)e^{at} = AP(a)e^{at} = e^{at}$$

where the last equality is stated because we want to solve (4). Hence $A = 1/P(a)$ and

$$y = \frac{e^{at}}{P(a)}.$$

EXAMPLE 4. Obtain a particular solution of each equation:

(5) $(D^2 + 3D + 2)^2 x = \cos t,$ $(D^2 + 3D + 2)^2 y = \sin t.$

Since $\cos t$ and $\sin t$ are the real and imaginary parts of e^{it} we set $z = x + iy$ and consider, instead of (5),

$$(D^2 + 3D + 2)^2 z = e^{it}.$$

By Example 3 with $a = i$ a solution is

$$z = \frac{e^{it}}{(-1 + 3i + 2)^2} = \frac{e^{it}}{(1 + 3i)^2} = \frac{e^{it}}{-8 + 6i}.$$

Since $64 + 36 = 100$ this gives

$$z = \frac{e^{it}}{-8 + 6i} \frac{-8 - 6i}{-8 - 6i} = (-.08 - .06i)(\cos t + i \sin t).$$

The solutions of (5) are the real and imaginary parts,

$$x = -.08 \cos t + .06 \sin t,   y = -.08 \cos t - .06 \sin t.$$

Example 5. Obtain a particular solution of

$$(D^2 + 1)(D - 3)^2 y = 12e^{3t}(10t + 1).$$

Substituting $y = e^{3t}u$ and using the shift theorem with $k = 3$, we get

$$e^{3t}((D + 3)^2 + 1)D^2 u = 12e^{3t}(10t + 1)$$

or, setting $v = D^2 u$,

$$(D^2 + 6D + 10)v = 120t + 12.$$

The trial solution $v = At + B$ gives v, integrating twice gives u, and the result is

$$v = 12t - 6,   u = 2t^3 - 3t^2,   y = (2t^3 - 3t^2)e^{3t}.$$

EXAMPLE 6.   If ω is a nonzero real constant, obtain a solution of each equation:

(6) $$(D^2 + \omega^2)^4 x = \cos \omega t, \qquad (D^2 + \omega^2)^4 y = \sin \omega t$$

Results of this kind are useful in the study of resonance in oscillating systems. We set $z = x + iy$ and consider, instead of (6), the equation

(7) $$(D + i\omega)^4 (D - i\omega)^4 z = e^{i\omega t}.$$

If $z = e^{i\omega t} u$ the shift theorem with $k = i\omega$ yields

$$e^{i\omega t}(D + 2i\omega)^4 D^4 u = e^{i\omega t}.$$

We set $D^4 u = A$, where A is constant, and are led to

$$(D + 2i\omega)^4 A = 1.$$

Since $DA = 0$, $D^2 A = 0$, and so on, the equation reduces without detailed calculation to
$$(2i\omega)^4 A = 1 \quad \text{or} \quad 16\omega^4 A = 1.$$

This gives A, the equation $D^4 u = A$ gives u, and the result is

$$z = \frac{1}{384} \left(\frac{t}{\omega} \right)^4 (\cos \omega t + i \sin \omega t).$$

The real and imaginary parts satisfy the two equations (6).

Initial conditions are introduced with the aid of the complementary function as in Chapter 6. Initial conditions will not be emphasized here, however, because they are handled more easily by the Laplace transform, and because they are often irrelevant to the steady-state behavior. These matters are taken up in Section 2 of this chapter and in Chapter 21.

PROBLEMS

1. This problem and Problem 2 are to be done, as far as possible, by inspection. No proof is required. Obtain n distinct solutions of the form e^{kt} where n is the order of the equation:

(a) $(D - 1)(D - 2)(D - 3)y = 0$ (c) $(D^3 + 9D)(D^4 - 16)y = 0$

(b) $(D^2 - 9)(D^2 + 3D + 2)y = 0$ (d) $(D^2 + 4D + 5)(D^2 + 2D + 5)y = 0$

2. Obtain n distinct solutions of the form $t^j e^{kt}$ where n is the order of the equation:

(a) $(D-1)^3(D-2)^2(D-3)y = 0$ (d) $(D^2+1)^4y = 0$

(b) $(D^2-1)^3(D^2-7D+10)^2y = 0$ (e) $(D^3+4D^2)^3y = 0$

(c) $(D^2-D)^2(D^2+4D+13)^3y = 0$ (f) $(D^2+2D+2)^4y = 0$

3. Obtain a particular solution by trying $y = Ae^{st}$ with $s = -2$ or $s = 1$:

(a) $(D^4+D^3+D^2+D+1)y = 33e^{-2t}$ (b) $(D^4-2)^5y = 5e^t$

4. Obtain a real solution by use of $Ae^{i\omega t}$ or $A\cos\omega t + B\sin\omega t$, whichever you prefer, with $\omega = 1$ or $\omega = 2$:

(a) $(D^2+2D+2)(D^2+2)^4x = \cos t$, $\quad$ $(D^2+2D+2)(D^2+2)^4y = \sin t$

(b) $(D^2+D+1)^2(D^2+1)x = \cos 2t$, $\quad$ $(D^2+D+1)^2(D^2+1)y = \sin 2t$

5. Obtain a solution of $(D^2-1)^4(D^3+1)^5y = 3e^t$ by the trial solution At^4e^t or by writing the equation in the form

$$(D^3+1)^5(D+1)^4(D-1)^4y = 3e^t$$

and using the shift theorem with $k = 1$. Hint for second method: If $D^4u = A$ is constant then $D^jA = 0$ for $j \geq 1$ and hence

$$((D+1)^3+1)^5(D+2)^4A = (1+1)^52^4A$$

6. Obtain a real solution of each of the two equations

$$(D^2+1)^2(D^2+D+1)x = \cos t, \qquad (D^2+1)^2(D^2+D+1)y = \sin t$$

by trying $y = t^2(A\cos t + B\sin t)$ or by consideration of the complex equation

$$(D^2+D+1)(D+i)^2(D-i)^2z = e^{it}$$

7. Show that the five solutions obtained in Example 1 are linearly independent on the interval $(0, \infty)$. That is, if the following equation holds for $0 < t < \infty$, all five constants c_j are 0:

$$c_1e^{4t} + c_2e^t + c_3e^{-t} + c_4\cos t + c_5\sin t = 0$$

Hint: Read Problems 7 and 8 of Chapter 6, Section 6.

8. Show that the eight real solutions obtained in Example 2 are linearly independent on the interval $(1, \infty)$.

9. Check that the functions te^{st}, t^2e^{st}, t^3e^{st}, $\cdots$ asssociated with a multiple root s can be obtained by differentiating e^{st} with respect to s. This illustrates a phenomenon that pervades the theory of differential equations. It will be seen again in connection with the Frobenius method for handling multiple roots in power-series solutions.

────────────────────ANSWERS────────────────────

1. (a) $k = 1, 2, 3$ (c) $k = 0, 3i, -3i, 2, -2, 2i, -2i$
   (b) $k = 3, -3, -1, -2$ (d) $k = -2 + i, -2 - i, -1 + 2i, -1 - 2i$

2. (a) $(j, k) = (0, 1),\ (1, 1),\ (2, 1),\ (0, 2),\ (1, 2),\ (0, 3)$
   (b) $(j, k) = (0, \pm 1),\ (1, \pm 1),\ (2, \pm 1),\ (0, 2),\ (1, 2),\ (0, 5),\ (1, 5)$
   (c) $(j, k) = (0, 0),\ (1, 0),\ (0, 1),\ (1, 1),\ (0, -2\pm 3i),\ (1, -2\pm 3i),\ (2, -2\pm 3i)$
   (d) $(j, k) = (0, \pm i),\ (1, \pm i),\ (2, \pm i),\ (3, \pm i)$
   (e) $(j, k) = (0, 0),\ (1, 0),\ (2, 0),\ (3, 0),\ (4, 0),\ (5, 0),\ (0, -4),\ (1, -4),\ (2, -4)$
   (f) $(j, k) = (0, -1 \pm i),\ (1, -1 \pm i),\ (2, -1 \pm i),\ (3, -1 \pm i)$

3. (a) $3e^{-2t}$ (b) $-5e^t$

4. (a) $x = .2\cos t + .4\sin t$, $y = -.4\cos t + .2\sin t$
   (b) $507x = -5\cos 2t + 12\sin 2t$, $507y = -12\cos 2t - 5\sin 2t$

5. $y = 2^{-12}t^4e^t$ 6. $8x = -t^2\sin t$, $8y = t^2\cos t$

──

2. Asymptotic stability

Throughout the rest of this chapter an operator T is defined by

$$T = (D^n + p_{n-1}D^{n-1} + \cdots + p_1D + p_0) = (D - a_1)(D - a_2)\cdots(D - a_n)$$

where the p_j and a_j are real or complex constants. It is not assumed that the a_j are distinct. If $P(s)$ has a zero $a_j = a$ of multiplicity m, then

$$P(s) = Q(s)(s - a)^m$$

where $Q(s)$ is a polynomial that does not vanish when $s = a$. The corresponding operator equation is

$$P(D) = Q(D)(D - a)^m.$$

To solve $P(D)y = 0$ we set $y = e^{at}u$ and use the shift theorem with $k = a$, getting

(1) $$P(D)y = Q(D)(D - a)^m e^{at}u = e^{at}Q(D + a)D^m u.$$

The right-hand side is 0 if $D^m u = 0$, hence if u is a polynomial of degree at most $m - 1$. This shows that the expression

$$y = c(t)e^{at}$$

satisfies $Ty = 0$ when $c(t)$ is any polynomial of the form

$$c(t) = c_0 + c_1 t + \cdots + c_{m-1} t^{m-1}.$$

The same sort of thing can be done for each root a_j of the characteristic equation. Let the roots be so numbered that the *distinct* roots are

$$a_1, \, a_2, \, \cdots, \, a_r$$

with multiplicities $m_1, m_2, \cdots, m_r$ respectively. Thus

$$(2) \qquad\qquad m_1 + m_2 + \cdots + m_r = n$$

where n is the degree of $P(s)$. The above method shows that the homogeneous equation $Ty = 0$ has the solution

$$(3) \qquad\qquad y = c_1(t)e^{a_1 t} + c_2(t)e^{a_2 t} + \cdots + c_r(t)e^{a_r t}$$

where $c_j(t)$ is an arbitrary polynomial of degree $m_j - 1$. Such a polynomial has m_j coefficients and hence, by (2), the number of arbitrary constants in the solution (3) is n.

It turns out that (3) is both the complete solution and the general solution of $Ty = 0$. This can be proved by mathematical induction, as illustrated for other results of a similar nature in Section 4. Another proof, which depends on general ideas and requires little calculation, can be found in Chapter 11, Section 2. The fact that (3) is the complete solution of $Ty = 0$ is needed in the Corollary at the end of this section.

In practical investigations one is often interested in the long-time behavior, that is, in the behavior of solutions as $t \to \infty$. The long-time behavior is also referred to as the *steady state*. The following informal discussion may give some idea of what the subject is about and why it is important.

For some equations the behavior as $t \to \infty$ depends strongly on the initial conditions. For example, if the equation is

$$y' - y = 5$$

the particular solution $y = -5$ is overwhelmed by the term ce^t in the complementary function. To be sure, the initial conditions could by accident make $c = 0$, but in any realistic case $c \neq 0$ is a virtual certainty. For this equation there is no well-defined steady state.

For the equation $y' + y = 5$, however, the particular solution $y = 5$ is almost unaltered, for large t, by the addition of the complementary function ce^{-t}. Every solution tends to 5 as a limit and, indeed, with exponential rapidity. Since every solution tends to 5, it is said that the solution $y = 5$ is *asymptotically stable*.

In problems of circuit analysis the steady state is ordinarily attained in a small fraction of a second. In other areas—ecology, economics, biology—the effect of the initial conditions generally persists for a long time, but still the concepts of "asymptotic stability" and "steady-state solution" retain their usefulness. The rest of this section is devoted to a precise formulation and elaboration of these ideas, starting with the following:

DEFINITION. A particular solution u of the equation $Ty = f$ is asymptotically stable if every solution v of this equation satisfies

$$\lim_{t \to \infty} |u(t) - v(t)| = 0.$$

The concept of asymptotic stability is illustrated in Example 1.

EXAMPLE 1. To get a particular solution u of the equation

$$(D^4 + 4D^3 + 6D^2 + 5D + 2)y = 260 \sin 2t$$

replace the right-hand member by $260e^{2it}$ and try $u = Ae^{2it}$. This leads to

$$(16 - 32i - 24 + 10i + 2)A = 260 \quad \text{or} \quad (-6 - 22i)A = 260.$$

By a short calculation $A = -3 + 11i$ and hence

$$u = \text{Im } (-3 + 11i)(\cos 2t + i \sin 2t) = 11 \cos 2t - 3 \sin 2t.$$

If v is any other solution, the principle of the complementary function gives

$$(4) \qquad v(t) = u(t) + c_1 e^{a_1 t} + c_2 e^{a_2 t} + c_3 e^{a_3 t} + c_4 e^{a_4 t}$$

where the a_j are the roots of the characteristic equation $P(s) = 0$. In Problem 2 it is found that each a_j satisfies

$$\text{Re } a_j < 0, \quad \text{hence} \quad \lim_{t \to \infty} e^{a_j t} = 0.$$

This shows that $v(t) = u(t) + \varepsilon(t)$ where $\lim_{t \to \infty} \varepsilon(t) = 0$. Thus $v(t) - u(t) \to 0$ as $t \to \infty$ so that u is asymptotically stable.

A moment's thought shows that if *any* solution of $Ty = f$ is asymptotically stable then *every* solution is asymptotically stable and every solution is equally valid as a description of the steady state. Since we can often find a particular solution of $Ty = f$ without solving the characteristic equation, the notion of asymptotic stability leads to great simplification in practice.

The following theorem shows that the question of asymptotic stability depends only on the homogeneous equation, not on f:

PRINCIPLE OF ASYMPTOTIC STABILITY. Let u be a solution of Tu = f. Then u is asymptotically stable if, and only if, every solution of the homogeneous equation Ty = 0 tends to 0 as t → ∞.

For proof, let v be any solution of $Tv = f$ and let $w = u - v$. By linearity $Tw = Tu - Tv = f - f = 0$. If *every* solution of the homogeneous equation tends to 0, then w does, and this shows that u is asymptotically stable. On the other hand if some solution w of the homogeneous equation does not tend to 0 we could pick $v = u + w$. Then $Tv = f$, but $u - v$ does not tend to 0. This shows that u was not asymptotically stable and completes the proof.

Equation (3) together with the result of Problem 1 yields the following:

COROLLARY. The solutions of Ty = f are asymptotically stable if, and only if, all roots of the characteristic equation have negative real parts.

Under the hypothesis of the corollary any particular solution of $Tu = f$ can be used to describe the steady state, and the complementary function v represents a *transient* that depends on the initial conditions. The transient decays exponentially with time, and that is the reason why initial conditions are not emphasized in this chapter. Separation of a solution $y = u + v$ into a steady-state part u and a transient v will be seen again in connection with the Laplace transform.

PROBLEMS————————————————————————————

1P. If m is a positive integer, $s = \sigma + i\tau$, and $\delta = -\sigma/m$ check that

$$\left| t^m e^{st} \right| = t^m e^{\sigma t} = (te^{-\delta t})^m, \qquad t \geq 0$$

Evaluate the limit of $t/e^{\delta t}$ by l'Hospital's rule and deduce that

$$\lim_{t \to \infty} \left| t^m e^{st} \right| = 0, \qquad \text{Re } s < 0, \quad m = 0, 1, 2, \cdots$$

Show conversely that Re s must be negative if the limit is 0.

2. Check that the characteristic polynomial in Example 1 has the following factorization, and hence that the zeros a_j satisfy Re $a_j < 0$:

$$P(s) = (s^2 + 3s + 2)(s^2 + s + 1)$$

3. Let u and v be two particular solutions of $Tu = f(t)$, $Tv = f(t)$ defined for $t \geq 0$. Suppose further that all roots a_i of the characteristic equation satisfy Re $a_j \leq 0$. If all roots of the form $i\beta$ are simple show that $|u(t) - v(t)|$ is bounded for $t \geq 0$. Show also that if some root of the form $i\beta$ is not simple, there are solutions u and v for which $u - v$ is not bounded. (A *simple root* has multiplicity 1.)

4. The solutions of $Tx = A\cos\omega t$ and of $Ty = B\sin\omega t$ are said to be *resonant* if $i\omega$ is a root of the characteristic equation. Show that the equation on the left has the resonant solution on the right:

$$(D+3)(D^2+1)^5 x = 640\cos t, \qquad x = \frac{t^5}{5!}(-2\cos t + 6\sin t)$$

The point is that this solution grows like t^5 while the complementary function has the order of t^4 at most. Thus any particular solution gives the dominant behavior for large t, even though the solutions are not asymptotically stable. Hint: Consider $(D+3)(D+i)^5(D-i)^5 z = 640e^{it}$ and use the shift theorem.

5. Show that the equation on the left has the resonant solution on the right:

$$(D+2)(D^2+1)^3 y = 240\sin t, \qquad t^3(2\cos t + \sin t)$$

6. For the equation $(D^2+1)(D^2+4)^2 x = 288\cos\omega t$ show that the resonant solutions when $\omega = 1$ and $\omega = 2$ have the following leading terms; that is, terms of highest power in t:

$$16t(\sin t + c_1\cos 2t + c_2\sin 2t), \qquad 3t^2\cos 2t$$

The resonant solution $16t\sin t$ for $\omega = 1$ does not describe the leading term, because the complemenatry function has terms of the same order. But the solution $3t^2\cos 2t$ for $\omega = 2$ falls into the pattern of Problem 4.

7. Let $P(s) = s^2 + p_1 s + p_0 = (s - r_1)(s - r_2)$ be a quadratic with real coefficients p_j and real or complex zeros r_j. Show that Re $r_1 < 0$ and Re $r_2 < 0$ both hold if, and only if, $p_1 > 0$ and $p_0 > 0$ both hold. Results such as this make it possible to decide about asymptotic stability without solving the characteristic equation.

8. This problem requires knowledge of higher-order determinants. The *Routh-Hurwitz criterion* gives a necessary and sufficient condition for all the zeros of a real polynomial

$$b_0 s^n + b_1 s^{n-1} + b_2 s^{n-2} + \cdots + b_{n-1} s + b_n, \qquad b_0 > 0$$

to have negative real parts. The criterion is as follows. Set $b_j = 0$ for $j > n$ and form the determinant

$$D_n = \begin{vmatrix} b_1 & b_0 & 0 & 0 & 0 & 0 & \cdots & 0 \\ b_3 & b_2 & b_1 & b_0 & 0 & 0 & \cdots & 0 \\ b_5 & b_4 & b_3 & b_2 & b_1 & b_0 & \cdots & 0 \\ \vdots & \vdots & \vdots & \vdots & \vdots & \vdots & \ddots & \vdots \\ b_{2n-1} & b_{2n-2} & b_{2n-3} & b_{2n-4} & b_{2n-5} & b_{2n-6} & \cdots & b_n \end{vmatrix}$$

Then all of the roots have negative real parts if, and only if, D_n and all its principle minors

$$b_1 \qquad \begin{vmatrix} b_1 & b_0 \\ b_3 & b_2 \end{vmatrix}, \qquad \begin{vmatrix} b_1 & b_0 & 0 \\ b_3 & b_2 & b_1 \\ b_5 & b_4 & b_3 \end{vmatrix}, \qquad \begin{vmatrix} b_1 & b_0 & 0 & 0 \\ b_3 & b_2 & b_1 & b_0 \\ b_5 & b_4 & b_3 & b_2 \\ b_7 & b_6 & b_5 & b_4 \end{vmatrix}, \qquad \cdots$$

are positive. Check that the determinant for the polynomial in Example 1 is the following determinant D_4 and show that it satisfies the Routh-Hurwitz criterion:

$$D_4 = \begin{vmatrix} 4 & 1 & 0 & 0 \\ 5 & 6 & 4 & 1 \\ 0 & 2 & 5 & 6 \\ 0 & 0 & 0 & 2 \end{vmatrix}$$

---

3. Undetermined coefficients

Let $A(t)$ be a polynomial. In Chapter 6, Section 5 it was seen that the equation

$$(D - a_1)y = A(t), \qquad a_1 \neq 0$$

has one, and only one, polynomial solution $B(t)$, and the degree of B is the same as that of A. It was also seen that repetition gives a similar result for

$$(D - a_1)(D - a_2)y = A(t), \qquad (D - a_1)(D - a_2)(D - a_3)y = A(t),$$

and so on, provided the constants a_j are all different from 0. Thus we get the following:

THEOREM 1. *Let $A(t)$ be a polynomial and suppose $P(0) \neq 0$. Then the equation $Ty = A(t)$ has one, and only one, polynomial solution $B(t)$, and the degree of B is the same as that of A.*

We now consider the equation $Tz = e^{kt}A(t)$ where k is a real or complex constant and A is a polynomial. The substitution $z = e^{kt}u$ and the shift theorem give the equivalent equation

$$P(D + k)u = A(t).$$

The condition $P(0) \neq 0$ of Theorem 1 holds for this equation if $P(k) \neq 0$. Thus we get:

THEOREM 2. *Let k be a real or complex constant, let $A(t)$ be a polynomial, and suppose $P(k) \neq 0$. Then the equation $Tz = e^{kt}A(t)$ has one, and only one, solution of the form $z = e^{kt}B(t)$ where B is a polynomial. The degree of B is the same as that of A.*

Assume next that k is a zero of $P(s)$ of multiplicity m, so that

$$P(s) = Q(s)(s - k)^m$$

where Q is a polynomial. Taking $P(D) = Q(D)(D - k)^m$ and using the shift theorem as above, we are led to the equation

$$Q(D + k)D^m u = A(t).$$

Simce m is the multiplicity of k we know that $Q(k) \neq 0$. Hence Theorem 1 shows that there is one, and only one, solution such that $D^m u = B(t)$, a polynomial. The function u is found from B by integrating m times. Exactly one of the functions obtained by integrating B has the form $t^m C(t)$, where $C(t)$ is a polynomial of the same degree as B. Namely, such a function $t^m C(t)$ is obtained when, and only when, the constants of integration are set equal to 0. These considerations give the following:

THEOREM 3. *Suppose the constant k in Theorem 2 is a zero of $P(s)$ of multiplicity m. Then the equation $Tz = e^{kt}A(t)$ has one, and only one, solution of the form $z = t^m C(t)e^{kt}$ where $C(t)$ is a polynomial. The degree of C is the same as that of A.*

If $k = a + ib$ is complex the polynomial C in Theorem 3 is also, in general, complex. With $C = F + iG$ the solution given by Theorem 3 is

$$z = t^m e^{at}(F + iG)(\cos bt + i \sin bt).$$

Setting $z = x + iy$ and equating real and imaginary parts yields:

THEOREM 4. *Let $A(t)$ be a real polynomial, let T be real, and let $k = a+ib$ be a zero of $P(s)$ of multiplicity m. Then the equations*

$$Tx = A(t)e^{at}\cos bt, \qquad Ty = A(t)e^{at}\sin bt$$

have solutions of the form

$$x = t^m e^{at}(F\cos bt - G\sin bt), \qquad y = t^m e^{at}(G\cos bt + F\sin bt)$$

where $F + iG$ is a polynomial of the same degree as A.

Equations of the type considered in Theorems 1–4 have a host of applications in a branch of engineering known as systems science, which embraces the theory of mechanical and electrical circuits and certain aspects of the theory of feedback and control. Since these equations are the subject of a vast literature, it is of considerable interest that so much can be said about the form of their solutions. But the exponential shift theorem is usually preferable for calculation. As in the above proofs, it leads to the correct form automatically and reduces complicated cases to the case covered by Theorem 1.

EXAMPLE 1. What form do you expect for a particular solution?

$$(D-1)^3(D-2)y = 3t^2, \qquad (D-8)^4Dy = te^{7t}, \qquad (D+1)(D^2+4)^2y = \sin t$$

Since 0 is not a root of the first characteristic equation, nor 7 a root of the second, nor i a root of the third, the form is similar to that of the right-hand side. Namely, Theorems 1, 2, and 4 give the respective trial solutions

$$a + bt + ct^2, \qquad (a + bt)e^{7t}, \qquad a\sin t + b\cos t.$$

EXAMPLE 2. What form do you expect for a particular solution now?

$$D^3(D - 2)y = 3t^2, \qquad (D - 7)^4y = te^{7t}, \qquad (D + 1)(D^2 + 1)^2y = \sin t$$

Here the characteristic equations have a root 0 of multiplicity 3, a root 7 of multiplicity 4, and a root i of multiplicity 2, respectively. By Theorems 3 and 4 the trial solutions of Example 1 must be multiplied respectively by t^3, t^4, and t^2.

PROBLEMS

1P. Review Chapter 6, Section 5, including the examples.

2. What form do you expect for particular solutions of the following equations?

(a) $(D^4 + D^2)y = t^2$ (b) $(D^2 - 1)^3y = te^t$ (c) $(D^2 + 1)^4y = t\sin t$

3. For an equation of the form $Ty = Ae^{kt}$, where A and k are constant, a useful strategy is to try $y = Be^{kt}$ with constant B. If $P(k) \neq 0$ the process gives an immediate solution. If $P(k) = 0$ you have not wasted your time, because you now know that $D - k$ is a factor of $P(D)$. As an illustration, consider the following equation where $C = 5$ or $C = 6$:

$$(D^4 + 2D^3 + 3D + C)y = 7e^{-2t}$$

(a) Try $y = Be^{-2t}$ and show that this gives an immediate solution if $C = 5$, and it guarantees that $P(D)$ is divisible by $D + 2$ if $C = 6$.

(b) By division get the other factor when $C = 6$ and solve by the shift theorem. Or use the shift theorem directly, without factoring $P(D)$.

4. Obtain a particular solution (a) when $C = 2$ and (b) when $C = 1$:

$$(D^4 + 2D^3 + 2D^2 + 2D + C)y = 8e^{-t}$$

5. By two different procedures, we are going to get a particular solution of the equation
$$Tu = u'' + 6u' + 12u = 17\cos t + 5\sin t$$

(a) Try $u = a\cos t + b\sin t$ and equate coefficients.

(b) Solve $Tz = e^{it}$ by trying $z = Be^{it}$. If $z = x + iy$, show that $u = 17x + 5y$ agrees with the value found in Part (a). Since this provides a check, no answer is given.

6. This problem pertains to the double integral

$$y = \int_0^t \int_0^r se^{as} \sin bs \, ds \, dr$$

where a and b are real nonzero constants.

(a) By considering $D^2z = te^{(a+ib)t}$ reduce the problem to a simpler form.

(b) Solve completely when $a = b = 1$.

7. This is difficult. Formulate and solve an analog of Problem 6 when the factor s in the integrand is replaced by s^2.

──────────────────ANSWERS──────────────────

2. (a) $t^2(at^2 + bt + c)$ (b) $t^3(at + b)e^t$ (c) $t^4((a + bt)\sin t + (c + dt)\cos t)$
3. (a) $y = -7e^{-2t}$ $P(D) = (D^3 + 3)(D + 2)$ (b) $5y = -7te^{-2t}$
4. (a) $8e^{-t}$ (b) $2t^2e^{-t}$ 6. (b) $2y = e^t(\cos t + \sin t - t\cos t) - 1 - t$
7. When $a = b = 1$, $2y = e^t(2t - t^2)\cos t + e^t(2t - 3)\sin t + t$

4. Existence and uniqueness

We now give an existence and uniqueness theorem for the equation $Ty = f$ on an open interval I on which f is continuous. The discussion is simpler if initial conditions are specified at $t = 0$ rather than at t_0. Hence we assume that I contains 0 and that

$$(1) \qquad y(0) = y_0, \qquad y'(0) = y_1, \qquad \cdots, \qquad y^{(n-1)}(0) = y_{n-1}.$$

Since $t - t_0$ could be introduced as a new variable, this simplification involves no loss of generality.

We begin with the equation

$$(2) \qquad\qquad (D - a)u = f$$

where a is constant. If (2) is multiplied by e^{-at} we get

$$D(e^{-at}u) = e^{-at}f(t)$$

and integration yields

$$(3) \qquad\qquad u(t) = e^{at} \int_0^t e^{-as} f(s)\, ds + ce^{at}$$

where $c = u(0)$ is constant. Conversely, differentiating (3) yields (2).

The expression on the right of (3) involves two parameters, a and c. We denote this expression by $T(a, c)f$, so that

$$T(a, c)f = e^{at} \int_0^t e^{-as} f(s)\, ds + ce^{at}.$$

What we have shown is that u is a solution of $(D - a)u = f$ if, and only if,

$$u = T(a, c)f, \qquad c = u(0).$$

In a like manner, we can solve the initial-value problem

$$(4) \qquad (D - a_2)(D - a_1)y = f, \qquad y(0) = y_0,\ y'(0) = y_1$$

by setting $u_1 = (D - a_1)y$. Then (4) holds if, and only if,

$$y = T(a_1, c_1)u_1 \quad \text{and} \quad u_1 = T(a_2, c_2)f$$

where $c_1 = y(0)$ and $c_2 = u_1(0)$. Since $u_1 = y' - a_1 y$ we have

$$c_2 = y_1 - a_1 y_0.$$

This shows that the constants c_i are uniquely determined by the initial conditions and hence (4) has one, and only one, solution. The solution is given by the explicit formula

$$y = T(a_1, c_1) T(a_2, c_2) f$$

with c_j as above. The process continues in an obvious manner and the final result is summarized as follows:

EXISTENCE AND UNIQUENESS THEOREM. Let f be continuous on an open interval I containing the point $t = 0$. Then the initial-value problem

$$Ty = f, \qquad y(0) = y_0, \qquad y'(0) = y_1, \qquad \cdots, \qquad y^{(n-1)}(0) = y_{n-1}$$

has one, and only one, solution valid on I. The solution is given by the formula

$$y = T(a_1, c_1) T(a_2, c_2) \cdots T(a_n, c_n) f$$

where the a_j are the roots of the characteristic equation and where the constants c_j are uniquely determined by the initial conditions.

The most significant part of this theorem is the second statement, which gives an algorithm for getting the solution. Being uniform and iterative, the algorithm is well suited for automatic computation. It also gives information about the nature of the solution that is not contained in a mere statement of existence and uniqueness.

The existence theorem has a close connection with the following algebraic theorem, which was first proved by Carl Friedrich Gauss:

FUNDAMENTAL THEOREM OF ALGEBRA. If P is a nonconstant polynomial with real or complex coefficients, the equation $P(s) = 0$ has at least one root.

It is because of this theorem that a polynomial of degree $n \geq 1$ can be factored into linear factors; see Chapter 21, Section 3, Theorem 2. If the leading coefficient is 1, as it is for $P(s)$, the factorization has the form

$$P(s) = (s - a_1)(s - a_2) \cdots (s - a_n).$$

Since the factorization is what makes the foregoing proof possible, the existence theorem is, essentially, a consequence of the fundamental theorem of algebra.

In conclusion, we mention that the main theorem of this section, and other results of a similar nature, can be established with ease by a method of proof known as *mathematical induction*. Since you are probably familiar with mathematical induction, only a brief summary is given here.

The usual setting for mathematical induction is a sequence of statements $S(n)$ depending on the positive integer n. An inductive proof of these statements requires the following two steps:

(a) $S(1)$ is true.

(b) If $S(n-1)$ is true then $S(n)$ is true.

The conclusion is that $S(n)$ is true for all positive integers n.

The assumption (b) that $S(n-1)$ is true is called the *induction hypothesis*. With the induction hypothesis you can assume not only $S(n-1)$ but

$$S(1), \ S(2), \ \cdots, \ S(n-1).$$

Also the starting point need not be $n = 1$ as in Step (a) but can be, for example, $n = 2$ or $n = 0$. These variants are logically equivalent to the principle of induction as formulated above.

Proof by induction is illustrated in the following example.

EXAMPLE 1. If the a_j are unequal constants, real or complex, show that any n functions $e^{a_j t}$ are linearly independent on any given interval I. That is, if

(5) $$c_1 e^{a_1 t} + c_2 e^{a_2 t} + \cdots + c_n e^{a_n t} = 0, \qquad t \in I$$

the constants c_j are all 0.

Here $S(n)$ is the statement that (5) implies $c_j = 0$ for $j = 1, 2, \cdots, n$. In particular, $S(1)$ is the assertion that

$$c_1 e^{a_1 t} = 0, \qquad t \in I \quad \Rightarrow \quad c_1 = 0.$$

This is true, since an exponential cannot vanish. Suppose then that $n \geq 2$ and that $S(n-1)$ is known to be true. If (5) is divided by $e^{a_n t}$ the result is

(6) $$c_1 e^{b_1 t} + c_2 e^{b_2 t} + \cdots + c_{n-1} e^{b_{n-1} t} + c_n = 0$$

where $b_j = a_j - a_n$. Differentiation with respect to t gives a similar relation involving only the first $n-1$ constants $b_j c_j$. By the induction hypothesis $b_j c_j = 0$, hence $c_j = 0$ for $j = 1, 2, \cdots, n-1$. Equation (5) now gives $c_n = 0$, completing the proof.

PROBLEMS

1. Assuming u and v sufficiently differentiable, prove by mathematical induction that

$$D^n(u + v) = D^n u + D^n v$$

2. If u is sufficiently differentiable and k is constant prove by mathematical induction that

$$D^n e^{kt} u = e^{kt}(D + k)^n u$$

3. Write the result of Problem 2 with j in place of n, multiply by p_j, and sum from 0 to n to get the exponential shift theorem. The advantage of this proof over that in Section 1 is that this proof does not presuppose the fundamental theorem of algebra.

4. If T is linear, u_j are in the domain of T, and c_j are constants, prove by mathematical induction that

$$T(c_1 u_1 + c_2 u_2 + \cdots + c_n u_n) = c_1 T u_1 + c_2 T u_2 + \cdots + c_n T u_n$$

The discussion in Chapter 4, Section 1, Example 5 amounts to an informal proof by mathematical induction and you are asked here to give a formal proof.

5. Taking $S(n)$ to be the statement: "The theorem is true for equations of order n" prove the main theorem of this section by mathematical induction.

Outline of solution: Statement $S(1)$ is established in the text. Taking $n > 1$ write the equation in the form $Q(D)(D - a_n)y = f$ where $Q(D)$ is a polynomial in D of degree $n-1$. Let $u = (D-a_n)y$. Since $u(t) = y'(t) - a_n y(t)$ you can read off the $n-1$ initial conditions for u from the n initial conditions for y. The induction hypothesis shows then that u is uniquely determined. Once you know this, y is uniquely determined too.

An elegantly executed proof is a poem in all but the form in which it is written.

—Morris Kline

Chapter 9

APPLICATIONS

MANY PROBLEMS lead to a system of two or more differential equations in two or more unknowns. Such systems can often be reduced to a single differential equation of higher order and solved as in the previous chapter. Using this technique we analyze the transfer of energy between coupled pendulums, the behavior of a radioactive chain in which the decay products are themselves radioactive, and the diffusion of a solute between compartments separated by a membrane. Similar methods apply to boundary-value problems in which the side conditions are specified not just at one point but at two. As examples, we consider the buckling of a supporting column in a building and the stability of a rotating propeller shaft.

1. A class of linear systems

Let S and T be two linear differential operators of orders m and n, respectively, with constant coefficients. This means

$$S = D^m + p_{m-1}D^{m-1} + \cdots + p_1 D + p_0$$

$$T = D^n + q_{n-1}D^{n-1} + \cdots + q_1 D + q_0$$

where the p_j and q_j are constant. The domain of ST consists of the functions that have $m + n$ derivatives on some interval I, and this is also the domain of TS. Furthermore

$$ST = TS$$

in the sense that each side does the same thing to any function in the common domain. This is true because S and T can be written in factored form. As an illustration, suppose

$$S = (D - a_1)(D - a_2), \qquad T = (D - b_1)(D - b_2)(D - b_3).$$

Then ST is the product of these linear factors in one order and TS is the product of the same linear factors in another order. Since the order does not affect the result, as was shown in Chapter 8, we conclude that $ST = TS$.

These remarks lead to a powerful method of solving simultaneous differential equations, which is discussed from a general point of view later in this book. Here we consider the special case

(1) $$Sx = ay, \qquad Ty = bx$$

where a and b are nonzero constants. If at least one of the operators S or T actually involves differentiation, it can be shown that all solutions of (1) have derivatives of all orders, and that fact is taken for granted here.

To solve (1), operate on the first equation with T and use the second equation to get

$$TSx = T(ay) = aTy = abx.$$

Thus the system becomes

(2) $$TSx = abx, \qquad ay = Sx.$$

Since the first equation (2) involves x alone we can get x by the methods of Chapter 8 and the second equation then gives y.

Conversely, if x and y satisfy (2), then

$$aTy = T(ay) = TSx = abx$$

and, dividing by a, we get $Ty = bx$. This shows that any solution of (2) is also a solution of (1).

EXAMPLE 1. Solve the system

(3) $$x' = x + y, \qquad y' = 6x$$

where, as throughout this section, the primes denote differentiation with respect to t. When the system is written in the form

$$(D - 1)x = y, \qquad Dy = 6x$$

the general procedure gives

$$D(D - 1)x = Dy = 6x, \quad \text{hence} \quad (D^2 - D - 6)x = 0.$$

Since $D^2 - D - 6 = (D - 3)(D + 2)$ the result is

$$x = c_1 e^{3t} + c_2 e^{-2t}.$$

The value of y is now found from the first equation (3):

$$y = x' - x = 2c_1 e^{3t} - 3c_2 e^{-2t}.$$

EXAMPLE 2. Solve

(4) $$x'' = 2x + 3y, \qquad y'' = 2x + y.$$

Here the standard form is

$$(D^2 - 2)x = 3y, \qquad (D^2 - 1)y = 2x.$$

and the general procedure gives

$$(D^2 - 1)(D^2 - 2)x = 3(D^2 - 1)y = 6x.$$

Hence

$$(D^4 - 3D^2 - 4)x = 0 \quad \text{or} \quad (D^2 - 4)(D^2 + 1)x = 0.$$

This yields

$$x = c_1 e^{2t} + c_2 e^{-2t} + c_3 \cos t + c_4 \sin t$$

and y is obtained from the first equation (4).

EXAMPLE 3.   The solution of Example 2 has four arbitrary constants. Using this fact, try to impose the four initial conditions

$$x(0) = x'(0) = x''(0) = 0, \qquad y(0) = y_0.$$

Setting $t = 0$ in the first equation (4) yields

$$x''(0) = 2x(0) + 3y(0)$$

and hence the initial values of x'', x, and y cannot all be assigned arbitrarily. In the present case the initial values for x give $y(0) = 0$. Hence there is no solution unless $y_0 = 0$. If $y_0 = 0$ a short calculation shows that there are infinitely many solutions,

$$x = c(e^{2t} - e^{-2t} - 4 \sin t), \qquad 3y = 2c(e^{2t} - e^{-2t} + 6 \sin t).$$

The following theorem gives conditions under which the initial conditions determine the solution uniquely, so that the phenomenon noted in the preceding example does not arise:

THEOREM 1.   *Let S and T be linear operators, with constant coefficients, of orders m and n respectively. Suppose $m \geq 1$ and $n \geq 1$ and suppose a and b are nonzero constants. Then the system*

$$Sx = ay, \qquad Ty = bx$$

has one, and only one, solution satisfying the initial conditions

$$x(t_0) = x_0, \ x'(t_0) = x_1, \ \cdots, x^{(m-1)}(t_0) = x_{m-1}$$

$$y(t_0) = y_0, \ y'(t_0) = y_1, \ \cdots, y^{(n-1)}(t_0) = y_{n-1}.$$

The solution exists on $(-\infty, \infty)$ and x and y have derivatives of all orders at every point.

A result similar to Theorem 1 holds for the inhomogeneous system

$$Sx = ay + f, \qquad Ty = bx + g$$

and for a more general system considered in Problem 7.

Theorem 1 is true because the system can be written as a first-order system in $m + n$ unknowns, as illustrated in the following example. A development that contains Theorem 1 as a special case is in Chapter 14.

EXAMPLE 4. Write as a first-order system in five variables z_i and interpret the initial conditions $z_i(t_0) = a_i$ in terms of the original variables x and y:

(5) $$x'' - x = 6y, \qquad y''' + 3y'' + 2y' + 4y = 3x.$$

With

$$z_1 = x, \qquad z_2 = x', \qquad z_3 = y, \qquad z_4 = y', \qquad z_5 = y''$$

the system (5) becomes

$$z_1' = z_2, \qquad z_3' = z_4, \qquad z_4' = z_5$$

$$z_2' = z_1 + 6z_3, \qquad z_5' = 3z_1 - 3z_5 - 2z_4 - 4z_3.$$

In a scheme such as this the first row specifies the meaning of the variables and the last row embodies the given differential equations. The initial conditions $z_i(t_0) = a_i$ for $i = 1, 2, 3, 4, 5$ give

$$x(t_0) = a_1, \; x'(t_0) = a_2, \; y(t_0) = a_3, \; y'(t_0) = a_4, \; y''(t_0) = a_5.$$

This agrees with the form of the initial conditions in Theorem 1.

PROBLEMS

1P. (a) Check that the system $x' = 2(y - x)$, $y' = 3x - y$ can be written

$$(D + 2)x = 2y, \qquad (D + 1)y = 3x$$

(b) Operate on the first equation with $D + 1$ and use the second to get

$$(D + 1)(D + 2)x = 6x, \quad \text{hence} \quad (D^2 + 3D - 4)x = 0$$

(c) Find x and then determine $2y$ from the first of the original equations.

2. Solve as in Problem 1:

(a) $x' = y$, $y' = -x$ (b) $x' = y$, $y' = x$ (c) $x'' = y$, $y'' = x$

3. Solve as in Problem 1:

(a) $x' = 3x - 2y$, $y' = 2x - y$ (d) $x' + y = 6x$, $y' - 5x = 4y$

(b) $x' = 3x + 4y$, $y' = 2x + y$ (e) $x'' = 3x + 2y$, $y'' = -4x - 3y$

(c) $x' = 5x + 9y$, $y' = -x + 11y$ (f) $x'' = x + 2y$, $y'' = -2x - 3y$

4. Solve the following system subject to the initial conditions $x'(0) = y'(0) = 0$ and to the additional conditions given below:

$$x'' = 4y - 7x, \qquad y'' = -3y + 3x$$

(a) $x(0) = 2$, $y(0) = 3$ (b) $x(0) = -2$, $y(0) = 1$ (c) $x(0) = 0$, $y(0) = 0$

5.   We are going to solve the inhomogeneous initial-value problem

$$x' + 2x - 2y = 16t, \qquad y' - 3x + y = 5e^t, \qquad x(0) = -1, \qquad y(0) = -14$$

(a) Check that $(D + 2)x = 2y + 16t$, $(D + 1)y = 3x + 5e^t$.

(b) Operate on the first of these equations with $D + 1$ and use the second to get an equation in x alone:

$$(D^2 + 3D - 4)x = 10e^t + 16t + 16$$

(c) Finish the job.

6. Let $x' + 2x - 2y = 16t$, $y' - 3x + y = 5e^t$, $x(0) = -3$ and $x'(0) = -2$.

(a) Find $y(0)$, $y'(0)$, and $y''(0)$ without solving either differential equation.

(b) Check by solving completely.

7.   Let P, Q, R be linear operators with constant coefficients like those in Theorem 1, and let a be a nonzero constant. We are going to solve the system

$$Px = ay + f, \qquad Qx + Ry = g$$

where f and g are sufficiently differentiable functions on a given interval I. It is assumed that $RP + aQ$ is not identically 0.

(a) Operate on the first equation with R and eliminate Ry by the second equation to get

$$(RP + aQ)x = ag + Rf, \qquad ay = Px - f$$

As in the discussion leading to Theorem 1, you can find x from the first equation and y is then given uniquely by the second.

(b) Show that if (x, y) satisfies the system obtained in Part (a) then (x, y) also satisfies the original system.

8. The system $(D+1)x+(D+6)y = 1$, $(2D+4)x+(D+15)y = 4$ does not fall into the pattern of Problem 7, because neither x nor y is given explicitly in terms of the other. However, if the first equation is subtracted from the second the result is $(D+3)x+9y = 3$. Apply Problem 7 to this equation and one of the others, and solve completely. Problems 7 and 8 together illustrate a method of great scope and power, which is often more efficient than other procedures.

————————————————ANSWERS————————————————

1. $x = c_1 e^{-4t} + c_2 e^t$, $2y = -2c_1 e^{-4t} + 3c_2 e^t$
2. (a) $x = c_1 \sin t + c_2 \cos t$, $y = c_1 \cos t - c_2 \sin t$
   (b) $x = c_1 e^t + c_2 e^{-t}$, $y = c_1 e^t - c_2 e^{-t}$
   (c) $x = c_1 e^t + c_2 e^{-t} + c_3 \cos t + c_4 \sin t$, $y = c_1 e^t + c_2 e^{-t} - c_3 \cos t - c_4 \sin t$
3. (a) $x = (c_1 + c_2 t)e^t$, $2y = (2c_1 - c_2 + 2c_2 t)e^t$
   (b) $x = c_1 e^{5t} + c_2 e^{-t}$, $2y = c_1 e^{5t} - 2c_2 e^{-t}$
   (c) $x = (c_1 + c_2 t)e^{8t}$, $9y = (3c_1 + c_2 + 3c_2 t)e^{8t}$
   (d) $x = e^{5t}(c_1 \cos 2t + c_2 \sin 2t)$, $y = e^{5t}((c_1-2c_2)\cos 2t + (2c_1+c_2)\sin 2t)$
   (e) $x = c_1 e^t + c_2 e^{-t} + c_3 \sin t + c_4 \cos t$
       $y = -c_1 e^t - c_2 e^{-t} - 2c_3 \sin t - 2c_4 \cos t$
   (f) $x = (c_1 + c_2 t)\cos t + (c_3 + c_4 t)\sin t$
       $y = (c_4 - c_1 - c_2 t)\cos t - (c_2 + c_3 + c_4 t)\sin t$
4. (a) $x = 2\cos t$, $y = 3\cos t$ (b) $x = -2\cos 3t$, $y = \cos 3t$ (c) $x = y = 0$
5. $x = 2te^t + 6e^{-4t} - 4t - 7$, $y = (1+3t)e^t - 6e^{-4t} - 12t - 9$
6. (a) $-4, 0, -1$
8. $x = 1 + c_1 \sin 3t + c_2 \cos 3t$, $3y = (c_2 - c_1)\sin 3t - (c_2 + c_1)\cos 3t$

---

2. Applications of systems

Here we show how linear systems arise in the study of mechanics. Applications to other disciplines are given in the accompanying problems.

(a) A coupled mass-spring system

Let masses m_1 and m_2 be suspended in tandem from springs of stiffness coefficients k_1 and k_2 as shown in Figure 1. If x and y denote the displacements of the masses m_1 and m_2 from their equilibrium positions, a procedure similar to that in Chapter 5, Section 1 gives the two equations

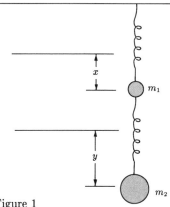

Figure 1

$$m_1 x'' = k_2(y - x) - k_1 x, \quad m_2 y'' = -k_2(y - x)$$

when friction is negligible. Upon division by m_1 and m_2 these equations become

$$x'' = c(y - x) - ax, \quad y'' = -b(y - x)$$

or also

$$x'' + (a + c)x = cy, \quad y'' + by = bx$$

where $a = k_1/m_1$, $b = k_2/m_2$, and $c = k_2/m_1$. If we define operators S and T by

$$S = D^2 + a + c, \quad T = D^2 + b$$

the equations take the form

$$Sx = cy, \quad Ty = bx.$$

This is a system of two simultaneous differential equations for the displacements x and y.

To find y let us operate on the second equation with S and substitute for Sx the value given by the first equation. The result is $STy = bSx = bcy$ or, written in full,

$$(D^2 + a + c)(D^2 + b)y = bcy.$$

Multiplying out yields

(1) $$(D^4 + (a + b + c)D^2 + ab)y = 0.$$

After we have solved this fourth-order equation for y the equation

(2) $$bx = y'' + by$$

gives x. There will be four arbitrary constants, since the equation for y is of order 4 and no new constants are introduced in the calculation of x. By Theorem 1 of the last section, the constants enable us to satisfy the initial conditions

$$x(0) = x_0, \quad x'(0) = x_1, \quad y(0) = y_0, \quad y'(0) = y_1.$$

These conditions are physically reasonable, since they specify the initial position and velocity of each mass.

(b) Natural frequencies

The characteristic equation

$$s^4 + (a + b + c)s^2 + ab = 0$$

associated with (1) is a quadratic in s^2. The discriminant is positive because

$$(a + b + c)^2 > (a + b)^2 \geq 4ab.$$

(The latter inequality follows from $(a - b)^2 \geq 0$.) Since the discriminant is positive the roots s^2 must be real and distinct, and they are negative because the coefficients are positive. Thus

$$s = \pm i\omega_1, \qquad s = \pm i\omega_2$$

where ω_i are real and distinct. The corresponding solution is

$$y = c_1 \cos \omega_1 t + c_2 \sin \omega_1 t + c_3 \cos \omega_2 t + c_4 \sin \omega_2 t$$

and the equation (2) gives a similar expression for x.

The analysis indicates that the motion of each mass is a combination of two simple harmonic motions of angular frequencies ω_1 and ω_2. These are the *natural frequencies* of the system.

(c) Coupled pendulums

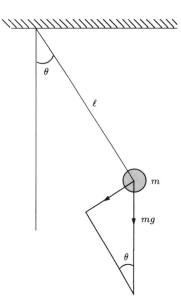

Figure 2 shows an oscillating pendulum of length ℓ and mass m. It is assumed that the bob can be regarded as a mass-point and that the rod joining it to the point of support is of negligible weight. Air resistance and friction at the pivot are also neglected.

Let θ be the angle in radians between the pendulum and the vertical as shown in the figure. The corresponding arc is $s = \ell\theta$ and the tangential acceleration is $s'' = \ell\theta''$. The force due to gravity on the pendulum bob is mg and the tangential component of this force is $mg \sin\theta$ as shown by the small triangle in the figure. Hence the equation of motion is

Figure 2

$$m\ell\theta'' = -mg \sin\theta \quad \text{or} \quad \ell\theta'' = -g \sin\theta.$$

The minus sign is needed because the force of gravity is directed downwards and tends to move the pendulum back to the equilibrium position. The normal component of the force and the centrifugal force are cancelled by the reaction of the supporting rod.

Since l'Hospital's rule gives

$$\lim_{\theta \to 0} \frac{\sin \theta}{\theta} = \lim_{\theta \to 0} \frac{\cos \theta}{1} = 1$$

we have $\sin \theta \approx \theta$ when $|\theta|$ is small. Hence the equation for small oscillations is

$$\ell\theta'' = -g\theta.$$

The motion is simple harmonic motion with angular frequency

$$\omega = \sqrt{\frac{g}{\ell}}.$$

Suppose next that we have two pendulums each with parameters (m, ℓ). As shown in Figure 3 they are both supported by the same taut metal strip, so that a displacement of one of the pendulums twists the strip and influences the other pendulum. We assume that the rods supporting the pendulums are attached rigidly to the strip and have some stiffness.

If the pendulums did not influence each other the equations for small oscillations would be

$$\ell\theta'' = -g\theta, \qquad \ell\phi'' = -g\phi$$

where ϕ is the angle corresponding to θ for the second pendulum. However, when one pendulum twists the strip from which both are suspended, this tends to move the other pendulum in the same direction. If Hooke's law is assumed we are led to the equations

$$\ell\theta'' = -g\theta + k(\phi - \theta), \qquad \ell\phi'' = -g\phi + k(\theta - \phi)$$

where k is a positive constant. See Figure 4, which shows a view looking along the supporting strip. With $a = g/\ell$ and $b = k/\ell$ division by l gives

(3) $$\theta'' = -a\theta + b(\phi - \theta), \qquad \phi'' = -a\phi + b(\theta - \phi).$$

These equations describe small oscillations of identical coupled pendulums in the absence of friction.

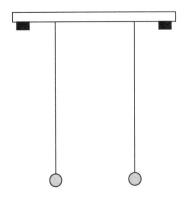

Figure 3

Figure 4

(d) Transfer of energy

The system considered above exhibits the following astonishing behavior: If you set one pendulum to swinging, its oscillations gradually diminish until they almost vanish and only the other pendulum is oscillating. Then the first pendulum takes over again, and so on, back and forth between the two. The behavior is so surprising that it is worthwhile to build a model and see it in action.

To understand why the pendulums act in this way, let us solve the system (3). Addition or subtraction of the equations gives, respectively,

$$(\theta + \phi)'' = -a(\theta + \phi), \qquad (\theta - \phi)'' = -(a + 2b)(\theta - \phi).$$

With

$$\omega_1^2 = a, \qquad \omega_2^2 = a + 2b$$

the solutions are

$$(4) \qquad \theta + \phi = c_1 \cos \omega_1 t + c_2 \sin \omega_1 t, \qquad \theta - \phi = c_3 \cos \omega_2 t + c_4 \sin \omega_2 t.$$

The presence of four constants c_j agrees with the fact that appropriate initial conditions are of the form

$$\theta(0) = \theta_0, \qquad \theta'(0) = \theta_1, \qquad \phi(0) = \phi_0, \qquad \phi'(0) = \phi_1.$$

According to (4) the variables $\theta + \phi$ and $\theta - \phi$ execute independent harmonic oscillations and are effectively decoupled. Expressions with this property are called *canonical variables* and are an important aid in the study of vibrating systems.

With the second pendulum at rest in the equilibrium position, let us displace the first pendulum and release it gently. The initial values are then

$$\theta_0 = 1, \qquad \theta_1 = 0, \qquad \phi_0 = 0, \qquad \phi_1 = 0.$$

(The choice $\theta_0 = 1$ involves no loss of generality, since any solution (θ, ϕ) can be multiplied by a constant). These initial values in (4) yield

$$\theta + \phi = \cos \omega_1 t, \qquad \theta - \phi = \cos \omega_2 t.$$

Adding the equations gives θ and subtracting them gives ϕ. By use of trigonometric identities the results can be written

$$(5) \qquad \theta = \cos \frac{\omega_2 + \omega_1}{2} t \cos \frac{\omega_2 - \omega_1}{2} t, \qquad \phi = \sin \frac{\omega_2 + \omega_1}{2} t \sin \frac{\omega_2 - \omega_1}{2} t.$$

To interpret these equations let us suppose that k is very small, so that b is almost 0 and ω_2 is only slightly larger than ω_1. Then both expressions

(5) involve a product of a rapidly oscillating factor with a slowly oscillating factor, but the latter is a cosine in the first case and a sine in the second. The magnitude of the cosine is maximum when that of the sine is zero and vice versa. That is why the behavior is as described at the beginning of this discussion. See Figure 5. The problem of coupled pendulums resembles the problem of beats as discussed in Chapter 7, but it has a good deal more subtlety and sophistication.

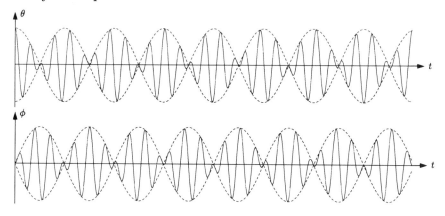

Figure 5

PROBLEMS

1. A two-compartment problem. Figure 6 shows a tank with two compartments separated by a permeable membrane. The volumes of the compartments are V_1 and V_2, independent of time. At time t a quantity Q_1 of dissolved substance is in the first compartment and a quantity Q_2 in the second. The concentration is $C_i = Q_i/V_i$. The rate of diffusion through the membrane is proportional to the difference in concentrations on the two sides, and the direction is from higher to lower concentration. Thus

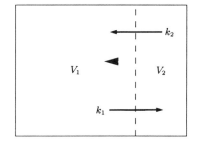

Figure 6

$$\frac{dQ_1}{dt} = k_1(C_2 - C_1), \qquad \frac{dQ_2}{dt} = k_2(C_1 - C_2)$$

where the constants k_i measure the diffusivity of the membrane in the two directions.

(a) Divide the first equation by k_1, the second by k_2, and add to deduce

$$C_1(t)\frac{V_1}{k_1} + C_2(t)\frac{V_2}{k_2} = C_1(0)\frac{V_1}{k_1} + C_2(0)\frac{V_2}{k_2}$$

(b) With $Q_i = C_i V_i$, divide the first equation by V_1, the second by V_2, and subtract to get an equation for $C_1 - C_2$. Thus deduce

$$C_1(t) - C_2(t) = (C_1(0) - C_2(0))e^{-kt} \quad \text{where} \quad k = \frac{k_1}{V_1} + \frac{k_2}{V_2}$$

2. A radioactive chain. Let P, Q, R denote the amounts of three radioactive substances at time t. Suppose $P \to Q \to R$ in the sense that the decay product of P is Q and the decay product of Q is R. This means

$$\frac{dP}{dt} = -aP, \qquad \frac{dQ}{dt} = aP - bQ, \qquad \frac{dR}{dt} = bQ - cR$$

where a, b, c are positive constants related to the half-lives of P, Q, R respectively. The process starts with an amount I of the first substance, so that $P(0) = I$, and none of the others has yet been produced. If a, b, c are all unequal, show that at time t

$$\frac{R}{I} = \frac{ab}{(a-b)(a-c)}e^{-at} + \frac{ab}{(b-c)(b-a)}e^{-bt} + \frac{ab}{(c-a)(c-b)}e^{-ct}$$

Hint: Find first P, then Q, then R.

3. Isolated coupled masses. Figure 7 shows two freight cars joined by a spring and constrained to move on a straight frictionless track. Since the track cancels the force due to gravity there is no external force on the system as a whole, and in this sense the system is isolated. The position and mass of the cars are x_1, m_1 for the first and x_2, m_2 for the second, with $x_1 \le x_2$. We assume Hooke's law for the spring whether it is in tension or compression, so that the stretching beyond the equilibrium length ℓ_0 can be either positive or negative. This stretching is not equal to the *length* of the spring $x_2 - x_1$, however, but to the *elongation* $x_2 - x_1 - \ell_0$. We take the positive direction to the right. Thus the effect of the spring on the first car is to pull it to the right, and on the second car, to pull it to the left. These considerations lead to the following system where k is the spring constant:

$$m_1 \frac{d^2 x_1}{dt^2} = k(x_2 - x_1 - \ell_0), \qquad m_2 \frac{d^2 x_2}{dt^2} = -k(x_2 - x_1 - \ell_0)$$

(a) By adding the equations show that the center of mass

$$\bar{x} = \frac{m_1 x_1 + m_2 x_2}{m_1 + m_2}$$

moves with constant velocity. Similar behavior is found for

Figure 7

more general mass systems whenever the total external force on the system is 0.

(b) The quantity $u = x_2 - x_1 - \ell_0$ represents the elongation of the spring at time t. Check that

$$m_1 x_1'' = ku, \qquad m_2 x_2'' = -ku, \qquad (x_2 - x_1)'' = u''$$

(c) Deduce from Part (b) that u executes simple harmonic motion according to the equation

$$mu'' + ku = 0 \quad \text{where} \quad \frac{1}{m} = \frac{1}{m_1} + \frac{1}{m_2}$$

4. Coupled electrical circuits. The currents i_1 and i_2 in the two coupled circuits shown in Figure 8 satisfy the following differential equations:

$$M \frac{d^2 i_2}{dt^2} + L_1 \frac{d^2 i_1}{dt^2} + R_1 \frac{di_1}{dt} + \frac{i_1}{C_1} = 0$$

$$M \frac{d^2 i_1}{dt^2} + L_2 \frac{d^2 i_2}{dt^2} + R_2 \frac{di_2}{dt} + \frac{i_2}{C_2} = 0$$

Here M is the *mutual inductance* between the inductors associated with L_1 and L_2. We assume $M > 0$, so the circuits are actually coupled.

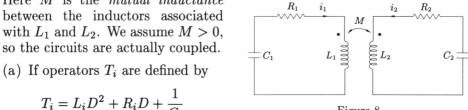

Figure 8

(a) If operators T_i are defined by

$$T_i = L_i D^2 + R_i D + \frac{1}{C_i}$$

check that the above equations become

$$M D^2 i_2 = -T_1 i_1, \quad M D^2 i_1 = -T_2 i_2$$

(b) Operate on the second equation of Part (b) with $M D^2$ and use the first equation to show that

$$(M^2 D^4 - T_2 T_1) i_1 = 0$$

Since $T_1 T_2 = T_2 T_1$, the same result holds for i_2.

(c) Assuming $M^2 < L_1 L_2$, analyze the case $R_1 = R_2 = 0$. In particular, show that the characteristic equation reduces to a quadratic in s^2 in which all the coefficients have the same sign and the discriminant is

$$\left(\frac{L_1}{C_2} - \frac{L_2}{C_1} \right)^2 + \frac{4M^2}{C_1 C_2} > 0$$

Hence the response is purely oscillatory with two natural frequencies ω_1 and ω_2, as in Part (b) of the text.

3. Boundary-value problems

In all examples discussed above the independent variable has been the time t, and arbitrary constants in the solutions have been determined by prescribing the value and some of the derivatives at a single time, t_0. We now present two applications in which the independent variable is a space variable x and the side conditions are specified not just at one point but at two. Problems of this type are known as *boundary-value problems*. They have great importance in mathematics and in several branches of mathematical physics.

(a) The Bernoulli-Euler law

Consider a uniform horizontal beam under the action of vertical loads. We assume that all force-vectors acting on the beam lie in a vertical plane containing the central axis. Choose the x axis along the central axis of the beam in the undeformed state and the positive y axis downward as shown in Figure 9. Under the action of the external forces the beam is bent and its central axis is deformed. The deformed axis, shown by the dashed line in the figure, is called the *elastic curve* and it is an important problem in the theory of elasticity to determine its shape.

According to the *Bernoulli-Euler law*, a beam made of a material that obeys Hooke's law deforms in such a way that the curvature K of the elastic curve at a given point is proportional to the bending moment M at that point. More specifically,

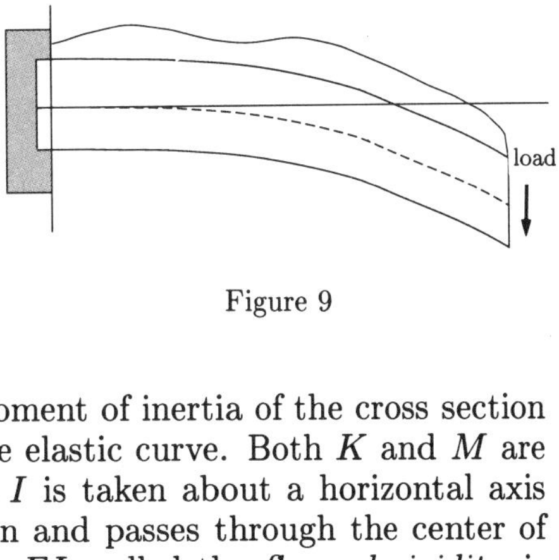

$$(1) \quad K = \frac{y''}{(1 + (y')^2)^{3/2}} = \frac{M}{EI}.$$

Figure 9

where E is Young's modulus, I is the moment of inertia of the cross section of the beam, and y is the ordinate of the elastic curve. Both K and M are measured at x. The moment of inertia I is taken about a horizontal axis that lies in the plane of the cross section and passes through the center of mass of the cross section. The quantity EI, called the *flexural rigidity*, is assumed constant in the sequel.

When the beam is a structural element like a girder in a building, one would expect the deflection to be small, so that the slope y' is also small. If the term $(y')^2$ in (1) is neglected the result is the linear equation

$$(2) \qquad y'' = \frac{M}{EI}.$$

This is, by definition, the equation for small deflection of an elastic beam under loads as described above. Surprising as it may seem, the following analysis, based on (2), gives results that are mathematically exact when applied to (1).

(b) The Euler column

It is known from experiments that a long rectilinear rod subjected to the action of axial compressive forces is compressed and retains its initial shape as long as the compressive forces do not exceed a certain critical value. Upon gradual increase of the compressive load P a value $P = P_1$ is reached when the rod buckles suddenly and becomes curved. The deflection of a rod so compressed is extremely sensitive to minute changes of the load and increases rapidly as P increases. Since the rod is initially straight, analysis of the onset of this *instability* or *buckling* phenomenon can be based on the equation for small deflections.

Thus, consider a rod of uniform cross section and length ℓ compressed by the forces P applied to its ends. Initially the rod is straight but after the critical load P_1 is reached it becomes curved as shown in Figure 10. We denote the deflection of its central line at point x by y and we note that, as the process just begins, y is arbitrarily small. Hence (2) can be used. The bending moment in this case is $M = -Py$, since y is the distance from (x, y) to the line of action of the force. Thus

$$(3) \qquad\qquad y'' + k^2 y = 0, \qquad k^2 = \frac{P}{EI}$$

where the primes denote differentiation with respect to x. Equation (3) must be solved subject to the conditions

$$(4) \quad y(0) = 0, \qquad y(\ell) = 0$$

since the ends of the rod remain on the axis.

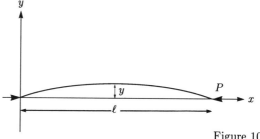

Figure 10

The *boundary-value problem* described by Equations (3) and (4) is quite different from the initial-value problems we have considered hitherto. Since the conditions (4) are given at two points, it is not obvious that there is any solution other than the trivial solution $y = 0$. We shall see, however, that for suitable choices of the parameter k, nontrivial solutions do exist.

The general solution of (3) is

$$y = c_1 \cos kx + c_2 \sin kx$$

and on imposing the two conditions (4) we get

$$c_1 = 0, \qquad c_2 \sin k\ell = 0.$$

The choice $c_1 = c_2 = 0$ gives $y = 0$ corresponding to the rectilinear shape of the rod. If the rod does not remain straight we must have $c_2 \neq 0$, hence $\sin k\ell = 0$, and therefore

$$k = \frac{n\pi}{\ell}, \qquad n = 0, 1, 2, 3, \cdots.$$

The choice $n = 0$ gives $y = 0$. If $n = 1$ then $k = \pi/\ell$ and on recalling the definition of k from (3) we see that

$$P_1 = EI\frac{\pi^2}{\ell^2}.$$

This is the *critical*, or *Euler* load. The choice $n = 2, 3, \cdots$ gives other critical loads $P_2, P_3, \cdots$.

(c) Rotating shaft

If a long shaft such as a propeller shaft is allowed to rotate, its initially rectilinear shape is preserved only if the angular speed of rotation ω does not exceed a certain critical value ω_1. On approaching the speed ω_1 the shaft starts pulsating and its shape changes. On further increase of the speed another critical value ω_2 is reached when the shaft starts beating and its shape changes again, and so on. This phenomenon can be explained by calculations similar to those used in determining the Euler load.

Let us suppose that the shaft is supported by bearings at $x = 0$ and $x = \ell$ and is rotating with angular speed ω. An element of length Δs of the shaft experiences the centrifugal force

$$F\Delta s \doteq \rho\,\Delta s\,\omega^2 y$$

where ρ is the density per unit length of shaft and y is the deflection at point x. The inequality is only approximate because y is not constant. However in the limit we find that

(5) $$F = \rho\omega^2 y$$

is the force per unit length of shaft distributed along its length. For small deflections we need not distinguish the arc s from the coordinate x.

It is shown in books on the strength of materials that when forces F acting on a rod are normal to its axis then $F = M''$ where the bending moment M is given by the Bernoulli-Euler law $M = EIy''$. Combining these remarks we get

$$y^{(iv)} = \frac{F}{EI}.$$

The substitution of F from (5) gives the following equation for the rotating shaft:

$$(6) \qquad y^{(iv)} = k^4 y, \qquad k^4 = \frac{\rho \omega^2}{EI}.$$

Throughout the above analysis we have assumed small deflection, but this is permissible for the same reason as in the discussion of Part (b). Namely, we are interested in the deformation only when it has just barely started.

Since the characteristic equation for (6) is $s^4 = k^4$ its roots are

$$s = \pm k, \qquad s = \pm ki$$

and the general solution is

$$(7) \qquad y = c_1 \cosh kx + c_2 \sinh kx + c_3 \cos kx + c_4 \sin kx.$$

If the bearings are short, that is, if they extend only a small distance along the axis of the shaft, the bending moment M at the bearings as well as the deflection y is almost 0. Since M is proportional to y'' the initial conditions are then

$$(8) \qquad y(0) = 0, \ y''(0) = 0, \qquad y(\ell) = 0, \ y''(\ell) = 0.$$

Substitution of the first two equations (8) into (7) yields

$$c_1 + c_3 = 0, \qquad c_1 - c_3 = 0.$$

Hence $c_1 = c_3 = 0$ and

$$y = c_2 \sinh kx + c_4 \sin kx.$$

The conditions (8) at ℓ now yield

$$c_2 \sinh k\ell + c_4 \sin k\ell = 0, \qquad c_2 \sinh k\ell - c_4 \sin k\ell = 0.$$

This is a linear system for c_2 and c_4 with determinant $-2 \sinh k\ell \sin k\ell$. Since the determinant must be 0 for a nontrivial solution, we conclude that

$$k\ell = n\pi, \qquad n = 1, 2, 3, \cdots.$$

Upon recalling the value of k from (6) we see that the first critical speed is

$$\omega_1 = \frac{\pi^2}{\ell^2} \sqrt{\frac{EI}{\rho}}.$$

The critical speeds $\omega_2, \omega_3, \cdots$ are determined by taking $n = 2, 3, \cdots$.

PROBLEMS

1. If a wire is stretched between two fixed points, as in a piano or harp, analysis of its vibrations leads to the boundary-value problem

$$y'' + ky = 0, \qquad y(0) = y(\ell) = 0$$

where k is constant and ℓ is the length of the wire.

(a) Check that the general solution of the differential equation is

$$c_1 \cos ax + c_2 \sin ax, \qquad c_1 + c_2 x, \qquad c_1 \cosh ax + c_2 \sinh ax$$

when $k = a^2$ is positive, $k = 0$, or $k = -a^2$ is negative, respectively.

(b) Show that in two of these cases the sole solution is $y = 0$ but in the remaining case there is a nontrivial solution for certain values of k. What are these values?

2. In this problem the independent variable is the time t and primes denote differentiation with respect to t. An oscillating mass-spring system satisfies $my'' + ky = 0$ where m and k are positive constants. It is observed that $y = 0$ at $t = 0$ and at a later time T. What are the possible values of the ratio k/m?

3. Solve, if possible, subject to the boundary conditions $y(0) = y(\pi) = 0$:

(a) $y'' - y' = 0$ (e) $y'' + y = 0$ (i) $y'' + y = x^2 - \pi x + 2$

(b) $y'' - y' = 2x$ (f) $y'' + y = e^x$ (j) $y'' + y = x^2 - 3x + 2$

(c) $4y'' + y = 0$ (g) $y'' - 2y' + 2y = 0$ (k) $y'' - 4y' + 5y = 8 \sin x$

(d) $4y'' + y = 1$ (h) $y'' - 2y' + 2y = 2$ (l) $y'' - 4y' + 5y = 8e^x$

4. Fredholm alternative. This problem sheds light on Problem 3 above. Let $Ty = y'' + p_1 y' + p_0 y$ where p_1 and p_0 are constant, let u and v be linearly independent solutions of $Ty = 0$, and let f be continuous on the interval $[a, b]$. Consider the two problems

$$Ty = 0, \quad y(a) = y(b) = 0; \qquad Ty = f, \quad y(a) = y(b) = 0$$

Explain why the first problem has a nontrivial solution if, and only if, the determinant

$$\begin{vmatrix} u(a) & v(a) \\ u(b) & v(b) \end{vmatrix}$$

is 0, while the second has a solution for arbitrary f if, and only if, the above determinant is not 0. Hence exactly one of the following statements is true:

(1) The first problem has a nontrivial solution.

(2) The second problem has a solution for all f.

This dichotomy is known as the *Fredholm alternative* after the Swedish mathematician Ivar Fredholm, who proved a similar result for equations of far greater generality.

5. This problem and the next pertain to the situation illustrated in Figure 11. In the case of a uniformly loaded beam the moment $M(x)$ is a quadratic function of x and hence, for small deflection, $y^{(iv)} = $ const. By choice of units we can make the length of the beam 1, and we then multiply y by a constant so that the equation is $y^{(iv)} = 24$.

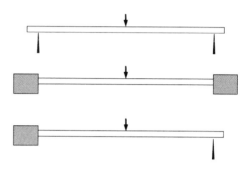

(a) If $y(0) = y(1) = 0$ show that
$y = x^4 + ax^3 + bx^2 - (1+a+b)x$
where a and b are constant.

Figure 11

(b) If the ends are pivoted y satisfies $y''(0) = y''(1) = 0$ in addition to the boundary conditions of Part (a). Find y.

(c) If the ends are clamped y satisifes $y'(0) = y'(1) = 0$ in addition to the boundary conditions of Part (a). Find y.

(d) Show that the maximum deflection when the ends are clamped is $1/5$ as much as when they are pivoted. Since this information gives a check, no further answer is provided.

6. Suppose the beam in Problem 5 has the end $x = 0$ clamped and the end $x = 1$ pivoted. Find y and find the point x at which the deflection is maximum. What is the approximate value of the maximum deflection? The true value is given in the answers.

7. If the bearings in the propeller problem keep the shaft horizontal as it enters the bearings, the boundary conditions are

$$y(0) = y'(0) = 0, \qquad y(\ell) = y'(\ell) = 0$$

(a) Show that the constant k giving the critical speeds now satisfies the transcendental equation $\cos k\ell \cosh k\ell = 1$.

(b) Plot the curves $h = \cos k\ell$, $h = \operatorname{sech} k\ell$ on a single diagram and deduce that the equation of Part (a) has infinitely many solutions k.

——————————————————ANSWERS——————————————————

1. $k = (n\pi/l)^2$, $n = 1, 2, 3, \cdots$ 2. $n^2\pi^2/T^2$, $n = 1, 2, 3, \cdots$

3. (a) $y = 0$ (g) $y = ce^x \sin x$

   (b) $y = \pi(\pi + 2)(e^x - 1)/(e^\pi - 1) - x^2 - 2x$ (h) no solution

   (c) $y = 0$ (i) $y = x(x - \pi) + c \sin x$

   (d) $y = 1 - \cos(x/2) - \sin(x/2)$ (j) no solution

   (e) $y = c \sin x$ (k) no solution

   (f) no solution   (l) no solution

6. $2y = 2x^4 - 5x^3 + 3x^2$, $2^4 x = 15 - \sqrt{33}$, $2^{13} y_{max} = 117 + 165\sqrt{33}$

The function of mathematics in providing answers is often less important than its function in providing understanding.

—Ben Noble

It has come to pass, I know not how, that Mathematics and Logic, which ought to be the handmaids of Physic, nevertheless presume on the strength of the certainty which they possess to exercise dominion over it.

—Francis Bacon

Mathematics weaves a robe of wonderful colors to clothe the unseen nakedness of nature.

—Eric Temple Bell

Chapter 10
LINEAR EQUATIONS III

SECOND-ORDER linear equations with variable coefficients do not have elementary solutions, in general, and lead to new functions. Some of these, for example the Bessel functions, are defined by use of the differential equations that they satisfy. We present a variety of techniques for getting a complete solution of the inhomogeneous equation from a single nonvanishing solution of the homogeneous equation. These methods extend to equations of higher order. We also give qualitative results that depend on conditions at an interior maximum or minimum and apply to second-order equations only. The chapter concludes with an introduction to differential inequalities and their bearing on uniqueness for initial-value problems.

1. Some useful identities

The general linear equation has the form

$$(1) \qquad p_n(x)y^{(n)} + p_{n-1}(x)y^{(n-1)} + \cdots + p_1(x)y' + p_0(x) = f(x).$$

This differs from most of the linear equations discussed hitherto in that the coefficients need not be constant. We use x rather than t because equations with nonconstant coefficients are often encountered in boundary-value problems, in which the variable is a space variable rather than the time.

If (1) is written in the form $Ty = f$, a discussion virtually identical to that for constant coefficients shows that T is linear. Hence the general principle of superposition, the principle of equating real parts, and all the other results established for linear operators in the foregoing discussion can be used for (1).

When $n = 2$ the equation is of the second order. That case will be emphasized here because it is typical of the general case and includes many equations of mathematical physics.

If $n = 2$ and $p_2(x) \neq 0$ on I we can divide by $p_2(x)$ and write the equation in the form

$$(2) \qquad Ty = y'' + p(x)y' + q(x)y = f(x), \qquad x \in I.$$

Until further notice, the operator T will be defined by (2), the functions p, q, and f are assumed continuous on I, and

$$P(x) = \int p(x)\, dx, \qquad W(x) = \begin{vmatrix} u & v \\ u' & v' \end{vmatrix}.$$

Thus, P is an indefinite integral of p and $W = W(u, v)$ is the Wronskian of the functions u and v.

This section is devoted mainly to an exposition of the following theorem:

THEOREM 1. *Suppose the equation $Tu = 0$ has a solution u that does not vanish on I. Then:*

(a) *A second solution v of $Tv = 0$, independent of u, is given by*

$$\left(\frac{v}{u}\right)' = \frac{ce^{-P(x)}}{u^2}, \qquad c \neq 0.$$

(b) *A particular solution of the inhomogeneous equation $Tw = f$ is given by $w = uz$ where*

$$(e^P u^2 z')' = e^P uf.$$

(c) *The general solution of $Ty = f$ is $y = c_1 u + c_2 v + w$.*

We say "the general solution" rather than "a general solution" because it is seen later that the general solution coincides with the complete solution, hence is unique. The point of the theorem is that P, v, z are determined by integration, that is, by quadrature. The theorem shows that the second-order linear equation $Ty = f$ can be solved by quadrature if a single nonvanishing solution of the homogeneous equation $Tu = 0$ is known.

Since $Tv = 0$ gives $T(cv) = 0$ for every constant c, the choice $c = 1$ in Theorem 1(a) involves no loss of generality. We include c nevertheless because it allows us to choose the *simplest* function v without changing its name. For example, suppose $c = 1$ gives $v = 23x/135$. If we stated Theorem 1 with $c = 1$ we would be obliged to introduce another letter, not v, for the solution x that we want to use. Here we get rid of the unwanted factor by taking $c = 135/23$.

(a) An identity involving the Wronskian

Starting with

$$Tv = v'' + pv' + qv, \qquad Tu = u'' + pu' + qu$$

multiply the first equation by u, the second by v, and subtract. This gives

$$uTv - vTu = (uv'' - vu'') + p(uv' - vu').$$

It was seen in Chapter 6 that

$$uv'' - vu'' = W'$$

where W is the Wronskian; the easy proof will not be repeated here. Hence

$$(3) \qquad uTv - vTu = W' + pW.$$

As a first application, suppose $Tu = Tv = 0$. Then

$$W' + pW = 0$$

and, solving for W, we get *Abel's identity*

$$W(x) = ce^{-P(x)}.$$

It follows that $W(x) = 0$ at every $x \in I$, or $W(x) = 0$ at no $x \in I$.

Instead of $Tu = Tv = 0$ as above, let us now assume that $Tu = 0$ and $u \neq 0$ on I. The identity (3) shows that

$$Tv = 0 \iff W' + pW = 0$$

and hence, if v is chosen so as to satisfy Abel's identity

$$(4) \qquad uv' - vu' = ce^{-P},$$

v is a solution of $Tv = 0$. Equation (4) is a first-order linear equation for v and admits the integrating factor $1/u^2$. In fact, if we divide by u^2, the left side of (4) becomes $(v/u)'$ and we get the equation in Theorem 1(a). The solutions are independent because the quotient v/u is not constant.

EXAMPLE 1. The trial solution $y = x^m$ shows that the equation

$$(5) \qquad x^2 y'' - 13xy' + 49y = 0, \qquad x > 0$$

has a solution $u = x^7$. To get a second solution, compute

$$p(x) = \frac{-13}{x}, \qquad P(x) = -13 \ln x, \qquad e^{-P(x)} = x^{13}.$$

Theorem 1(a) with $u = x^7$ and $c = 1$ gives

$$\frac{v}{u} = \int x^{13} x^{-14} \, dx = \ln x.$$

Hence $v = x^7 \ln x$. The constant of integration was ignored because only a single function v is wanted. The general solution of (5) is

$$x^7 (c_1 + c_2 \ln x)$$

and it is here that we introduce the constants.

(b) An identity involving a product

Another useful identity is obtained by setting $y = uz$ where u and z are twice-differentiable functions on I. Since

$$y = uz, \qquad y' = uz' + zu', \qquad y'' = uz'' + 2u'z' + zu''$$

the result of substitution into the operator T is

$$T(uz) = (uz'' + 2u'z' + zu'') + p(uz' + zu') + quz.$$

The terms with z as a factor together give zTu, so that

(6) $$T(uz) = uz'' + (2u' + pu)z' + zTu.$$

By means of (6) one can solve the inhomogeneous equation $Tw = f$ when a single nonvanishing solution u of the homogeneous equation is known. Let $Tu = 0$ and let a solution of the general equation $Tw = f$ be sought in the form $w = uz$. In (6) the term zTu drops out, since $Tu = 0$, and the condition $Tw = f$ becomes

$$uz'' + (2u' + pu)z' = f.$$

This is a first-order linear equation in the unknown z'. It is seen in Problem 10 that an integrating factor is ue^P. If we multiply by ue^P the result is

$$z''u^2e^P + e^P(2uu' + pu^2)z' = ue^P f$$

or, by use of $P' = p$,

(7) $$(z'u^2e^P)' = ue^P f.$$

An integration gives $z'u^2e^P$ and, if $u \neq 0$, a second integration gives z. Thus we have established Part (b) of Theorem 1. Part (c) follows from the theory of the Wronskian as presented in Chapter 6, so that the proof of Theorem 1 is now complete.

EXAMPLE 2.   Taking $u = x^7$ express a solution of the following equation by means of integration:

$$x^2y'' - 13xy' + 49y = x^2 f(x), \qquad x > 0.$$

The formulas in Theorem 1 assume that the leading term in Ty is y''. With this in mind, the factor x^2 was introduced on the right to give $f(x)$ after the equation is divided by x^2.

That $Tu = 0$ was verified in Example 1 where it was also seen that

$$e^{P(x)} = e^{-13\ln x} = x^{-13}.$$

Hence the equation (7) for z becomes

$$(x^{-13}x^{14}z')' = x^{-13}x^7 f(x).$$

After simplification we integrate and divide by x to get

(8) $$z' = \frac{1}{x}\int \frac{f(x)}{x^6}\,dx.$$

A second integration gives z, and $w = uz$ is the solution sought. As an illustration, it is left for you to verify that the choice $f(x) = x^m$ gives, aside from constants of integration,

$$w = \frac{x^{m+2}}{(m-5)^2}, \qquad m \ne 5; \qquad w = \frac{1}{2}x^7(\ln x)^2, \qquad m = 5.$$

The first of these could have been found by the trial solution $w = Ax^{m+2}$ but the second is less obvious.

(c) Elimination of a term

Let us again use (6), but instead of $Tu = 0$ we now set

(9) $$2u' + pu = 0.$$

In this case (6) gives $T(uz) = uz'' + zTu$ and, if $w = uz$, the equation $Tw = f$ reduces to

(10) $$uz'' + zTu = f.$$

The advantage is that (10) lacks the term z'. Since $P' = p$, Equation (9) gives

$$u = e^{-P/2}, \qquad u' = e^{-P/2}\left(-\frac{p}{2}\right), \qquad u'' = e^{-P/2}\left(-\frac{p'}{2} + \frac{p^2}{4}\right).$$

(We do not introduce a constant of integration for u because only a single solution is wanted.) By a short calculation

$$Tu = e^{-P/2}\left(q - \frac{p'}{2} - \frac{p^2}{4}\right).$$

Substitution of $u = e^{-P/2}$ and of Tu into (10) yields the following theorem:

THEOREM 2. *Let p be differentiable and let*

$$P = \int p(x)\, dx, \qquad Q = q - \frac{p'}{2} - \frac{p^2}{4}, \qquad z = e^{P/2} w.$$

Then the following equations are equivalent. That is, if either holds, both do:

(11) $$z'' + Qz = e^{P/2} f, \qquad w'' + pw' + qw = f.$$

EXAMPLE 3. The *Hermite equation* is

$$Ty = y'' - 2xy' + 2my = 0$$

where m is a constant parameter. To remove the second term from this equation we take $p = -2x$, $q = 2m$ and hence

$$e^{P(x)} = e^{-x^2}, \qquad Q(x) = 2m + 1 - x^2.$$

By Theorem 2 the equation for z, which lacks the second term, is

(12) $$z'' + (2m + 1 - x^2)z = 0.$$

If $H_m(x)$ satisfies the Hermite equation then the function

$$z = e^{-x^2/2} H_m(x)$$

satisfies (12) and conversely.

In conclusion, we mention that the quantity e^P yields

$$e^{-P} = \frac{1}{e^P}, \qquad e^{P/2} = \sqrt{e^P}.$$

The first of these is needed in Abel's equation for the Wronskian and also in the result of Theorem 1(a). The second occurs twice in Theorem 2. Because of its many uses, computation of e^P is suggested as an initial step in several of the following problems.

PROBLEMS

In this and subsequent sections, the functions p, q, f, g introduced without explanation are continuous on the relevant intervals.

1P. Review the discussion of the Wronskian in Chapter 6, Section 6.

2P. Here are some expressions Sy, containing a constant parameter m, that are frequently used in applied mathematics. The operator is denoted by S instead of T because the leading term in Sy is not y''. We give first the name of the mathematician with whom the equation $Sy = 0$ is identified, then Sy, then the most common notation for certain solutions of $Sy = 0$, and finally e^P for the associated operator T. You are asked to verify the given value of e^P and to state Abel's equation for the Wronskian $W(u,v)$ when $Su = Sv = 0$. Since division is needed to go from $Sy = 0$ to the standard form $Ty = 0$, assume $x > 0$ in (ad) and $-1 < x < 1$ in (bc).

(a) Laguerre	$xy'' + (1-x)y' + my$		L_m	$e^P = xe^{-x}$
(b) Legendre	$(1-x^2)y'' - 2xy' + m(m+1)y$		P_m	$e^P = 1 - x^2$
(c) Chebychev	$(1-x^2)y'' - xy' + m^2y$		T_m	$e^P = \sqrt{1-x^2}$
(d) Bessel	$x^2y'' + xy' + (x^2 - m^2)y$		$J_{\pm m}$	$e^P = x$

3. If $u = 1$ the equation of Theorem 1 (a) is $v' = ce^{-P}$. Apply this to get linearly independent solutions $(1, v)$ of:

(a) $xy'' + 2y' = 0$ (b) $xy'' + y' = 0$ (c) $y'' \sin x = y' \cos x$

4P. This problem pertains to the operator $T = x^{-2}S$ for $x > 0$ where

$$Sy = x^2y'' - 4xy' + 4y, \qquad e^P = x^{-4}, \qquad x > 0$$

(a) Verify the above formula for e^P and write Abel's equation for the Wronskian W of two solutions of $Ty = 0$.

(b) Check that $u = x$ satisfies $Su = 0$ and find an independent solution v by $(v/x)' = ce^{-P(x)}/x^2$. The latter equation comes from Theorem 1.

(c) Show that the Wronskian of your two solutions satisfies Abel's equation as found in Part (a).

(d) Using $u = x$, express a solution of $Sy = x^2f(x)$ by two integrations. Since Part (e) asks for a check, no answer is provided.

(e) Compute the solution (d) when $f(x) = x^m$ and verify that it satisfies $Sy = x^{m+2}$. Two values of m are exceptional and lead to logarithms.

5. With α and β constant, this problem pertains to the operator $T = x^{-2}S$ where for $x > 0$

$$Sy = x^2y'' + (1 - \alpha - \beta)xy' + \alpha\beta y, \qquad e^P = x^{1-\alpha-\beta}$$

(a) Verify the above equation for e^P.

(b) For what values of the constant m, if any, does the trial solution $y = x^m$ satisfy $Sy = 0$?

(c) Obtain two linearly independent solutions of $Ty = 0$ when $\alpha = \beta$.

6. When $m = 1$ each equation $Su = 0$ in Problem 2 (abc) has a solution of the form $u = x + c$ where c is constant. Find c and show that a second solution v is

(a) $(x - 1) \displaystyle\int \frac{e^x}{x(x-1)^2} \, dx$ (b) $1 - x \tanh^{-1} x$ (c) $\sqrt{1 - x^2}$

Hint: The integral (a) is not elementary. For (b) note that

$$\frac{1}{x^2(1 - x^2)} = \frac{1}{x^2} + \frac{1}{1 - x^2}$$

and integrate the second term by setting $x = \tanh u$. Part (c) leads to an integral that is best evaluated by setting $x = \cos \theta$.

7. Let $Tu = 0$ and $u \neq 0$ where T is the Hermite operator in Example 3.

(a) Obtain the two equations

$$\left(\frac{v}{u}\right)' = \frac{e^{x^2}}{u^2}, \qquad (e^{-x^2} u^2 z')' = e^{-x^2} u f$$

The first is the recipe of Theorem 1 for a solution of $Tv = 0$ and the second is the recipe for a solution of $Ty = f$.

(b) In Part (a) obtain the explicit formulas

$$v(x) = u(x) \int \frac{e^{x^2}}{u(x)^2} \, dx, \qquad z'(x) = \frac{e^{x^2}}{u(x)^2} \int e^{-x^2} u(x) f(x) \, dx$$

An integration gives z and then $y = uz$ satisfies $Ty = f$.

8. Remove the second term from Bessel's equation, Problem 2(d), and thus show that the functions on the left below satisfy the equation on the right:

$$\sqrt{x} J_m(x), \qquad \sqrt{x} J_{-m}(x); \qquad x^2 z'' + (x^2 - m^2 + \tfrac{1}{4})z = 0$$

9. (abc) This problem has rather unattractive solutions and no answer is provided. Remove the second term from equations (abc) of Problem 2.

10. For simplicity let $u > 0$. If the equation $uz'' + (2u' + pu)z' = f$ is divided by u the result is the first-order linear equation

$$w' + \left(\frac{2u'}{u} + p\right) w = \frac{f}{u}$$

where $w = z'$. By Chapter 2 an integrating factor is e^g where

$$g(x) = \int \left(\frac{2u'}{u} + p \right) dx = 2 \ln u + P$$

Check that $e^g = u^2 e^P$ and hence an integrating factor for the original equation is ue^P.

11. If $Ty = y'' + qy$, so that the second term is missing, check that Abel's equation for the Wronskian, the procedure for getting a second solution v from u, and the method for solving $Ty = f$ take the following simple form:

$$W(u, v) = c, \qquad \left(\frac{v}{u} \right)' = \frac{c}{u^2}, \qquad (u^2 z')' = uf$$

12. Discuss Problem 11 as applied to $y'' + y = f(x)$. No answer is provided, but see Problem 7 of the following section.

————————————ANSWERS————————————

2. $W = c/e^P$ **3.** (a) $1/x$ for $x \neq 0$ (b) $\ln|x|$ for $x \neq 0$ (c) $\cos x$
4. (a) $W = cx^4$ (e) $y = x^{m+2}/(m+1)(m-2)$, $-(x \ln x)/3$, $(x^4 \ln x)/3$
5. (b) $m = \alpha$ or β (c) x^α, $x^\alpha \ln x$

---

2. Variation of parameters

When a nonvanishing solution of the homogeneous equation $Tu = 0$ is known, the inhomogeneous equation $Ty = f$ can be solved by two integrations. This is the main result of the foregoing section. It was discovered by Lagrange that if two linearly independent solutions of the homogeneous equation are known, the inhomogeneous equation can be solved by a single integration. Lagrange's procedure extends to equations of order n and represents a major advance in the theory of differential equations.

The basis of the method is easily described. Let u and v be a pair of linearly independent solutions of $Ty = 0$ and form the expression $y = au + bv$. If a and b are constant parameters, this represents the general solution of $Ty = 0$. Lagrange solves $Ty = f$ by choosing a trial solution of this same form, but with a and b functions of x rather than constants. Since the formerly constant parameters are now variables, the method is called the method of *variation of parameters*.

A similar idea has already been used in this book. For example, in studying equations with constant coefficients, we replaced the expression ce^{kt}, which occurs in connection with the homogeneous equation, by ue^{kt}. In the preceding section the constant c in the solution cu of the homogeneous equation was replaced by v to get the trial solution $y = uv$.

(a) The Lagrange procedure

Suppose then that $Tu = 0$, $Tv = 0$, and

$$(1) \qquad\qquad y = au + bv$$

where a and b are differentiable functions of x. Differentiation gives

$$y' = (au' + bv') + (a'u + b'v).$$

Now comes an important point. The calculation of y'' will be materially simplified if a and b are so chosen that the second expression in parentheses vanishes. Accordingly we set

$$(2) \qquad\qquad a'u + b'v = 0$$

and get $y' = au' + bv'$. Hence

$$y'' = au'' + bv'' + a'u' + b'v'$$

$$p: \qquad y' = au' + bv'$$

$$q: \qquad y = au + bv$$

where the second and third equations have been repeated for convenience. The symbols p and q at the left mean that you are to multiply the second equation by p, the third by q, and add the three equations. Since

$$Ty = y'' + py' + qy$$

and similarly for Tu and Tv, the result of addition is

$$Ty = aTu + bTv + a'u' + b'v' = a'u' + b'v'.$$

The second equality holds because $Tu = Tv = 0$ by hypothesis.

Thus (2) and the desired equation $Ty = f$ are

$$a'u + b'v = 0$$

$$a'u' + b'v' = f.$$

This is a linear system in the unknowns (a', b'). By Cramer's rule

$$\begin{vmatrix} u & v \\ u' & v' \end{vmatrix} a' = \begin{vmatrix} 0 & v \\ f & v' \end{vmatrix}, \qquad \begin{vmatrix} u & v \\ u' & v' \end{vmatrix} b' = \begin{vmatrix} u & 0 \\ u' & f \end{vmatrix}.$$

The coefficient determinant is the Wronskian $W(u, v)$, so that

$$a' = \frac{-fv}{W(u, v)}, \qquad b' = \frac{fu}{W(u, v)}.$$

An integration gives a and b and (1) then gives the solution,

(3) $$y = u(x) \int \frac{-fv}{W(u, v)}\, dx + v(x) \int \frac{fu}{W(u, v)}\, dx.$$

Since the constants of integration produce an added term $c_1 u + c_2 v$ the result is actually the *general* solution of $Ty = f$.

EXAMPLE 1. As an illustration let us solve

$$x^2 y'' - 2xy' + 2y = x^2 f(x), \qquad x \neq 0.$$

The side condition $x \neq 0$ means that the equation is considered on an interval I that does not contain the origin. It is easily checked that a pair of linearly independent solutions and their Wronskian is

$$u = x, \qquad v = x^2, \qquad W(u, v) = x^2.$$

To reduce the equation to standard form we must divide by x^2. The factor x^2 has been inserted on the right so that the notation after division will be consistent with that in the general theory. Lagrange's formula (3) gives

$$y = -x \int f(x)\, dx + x^2 \int \frac{f(x)}{x}\, dx.$$

It is left for you to verify that the choice $f(x) = x^m$ with m constant yields the solutions

$$-x \ln x, \qquad x^2 \ln x, \qquad \frac{x^{m+2}}{m(m + 1)}$$

for $m = -1$, for $m = 0$, and for other values of m, respectively. The general solution when $f(x) = x^m$ is obtained by adding $c_1 x + c_2 x^2$ to these particular solutions.

(b) Green's function: initial values

The basic formula (3) has several applications. Let x_0 be any point on the interval I where the differential equation is to hold and let the indefinite integrals defining a and b be replaced by definite integrals. Thus

(4) $$a(x) = \int_{x_0}^{x} \frac{-f(t)v(t)}{W(t)}\, dt, \qquad b(x) = \int_{x_0}^{x} \frac{f(t)u(t)}{W(t)}\, dt$$

where $W(t) = W(u(t), v(t))$. Since the coefficients $u(x)$ and $v(x)$ in (3) are constant with respect to the integration variable t, they can be moved under the integral sign. The result is the formula

(5) $$y(x) = \int_{x_0}^{x} \frac{u(t)v(x) - u(x)v(t)}{u(t)v'(t) - u'(t)v(t)} f(t)\, dt$$

when the Wronskian $W(t)$ is replaced by its definition.

To interpret this formula let us recall the equations

(6) $$y(x) = a(x)u(x) + b(x)v(x), \qquad y'(x) = a(x)u'(x) + b(x)v'(x)$$

with which our discussion began. Since (4) gives $a(x_0) = b(x_0) = 0$, it follows that

$$y(x_0) = 0, \qquad y'(x_0) = 0.$$

Thus, (5) gives the rest solution of $Ty = f$ as referred to the initial point x_0.

EXAMPLE 2. For the equation

$$x^2 y'' - 2xy' + 2y = x^2 f(x), \qquad x \neq 0$$

considered in Example 1, substitution of $u = t, v = t^2$ into (5) yields the rest solution

$$y(x) = x \int_{x_0}^{x} (x - t)\frac{f(t)}{t}\, dt.$$

For example, if $f(t) = t \sin t$ partial integration gives

$$\int_{x_0}^{x} (x - t) \sin t\, dt = (x - t)(-\cos t)\Big|_{t=x_0}^{x} - \int_{x_0}^{x} \cos t\, dt.$$

Evaluating this expression and multiplying by x we get

$$y(x) = x(x - x_0)\cos x_0 - x(\sin x - \sin x_0).$$

The differential equation and initial conditions are readily verified.

The result (5) can be written

$$y(x) = \int_{x_0}^{x} G_0(x,t)f(t)\,dt$$

where

(7) $$G_0(x,t) = \frac{u(t)v(x) - u(x)v(t)}{u(t)v'(t) - u'(t)v(t)}.$$

The subscript 0 suggests that the formula is concerned with initial values at x_0 and serves to distinguish G_0 from another function, G, that occurs in connection with boundary-value problems.

(c) Green's function: boundary values

We denote the basic interval I by $[x_0, x_1]$ where $x_0 < x_1$. In the previous pages we gave several examples of boundary-value problems in which the conditions are specified not just at one point but at two. Three typical boundary conditions are

$$y(x_0) = y(x_1) = 0; \qquad y'(x_0) = y'(x_1) = 0; \qquad y(x_0) = y'(x_1) = 0.$$

All three of these are included in

$$\alpha_0 y(x_0) + \beta_0 y'(x_0) = 0, \qquad \alpha_1 y(x_1) + \beta_1 y'(x_1) = 0$$

where α_j and β_j are constant.

It is a remarkable fact that if u satisfies the boundary conditions at x_0 and v satisfies them at x_1, we can arrange matters so that y satisfies them at *both* x_0 and x_1. The only restriction is the basic condition $W(u,v) \neq 0$ which is always needed.

To see how this is done recall the conditions (6), which we repeat for convenience:

(8) $$y(x) = a(x)u(x) + b(x)v(x), \qquad y'(x) = a(x)u'(x) + b(x)v'(x).$$

We want to choose a and b so that y behaves like $u(x)$ near $x = x_0$ and like $v(x)$ near $x = x_1$. To this end let

$$a(x_1) = 0, \qquad b(x_0) = 0.$$

Recalling the earlier formulas for a' and b' we set

$$a(x) = \int_{x}^{x_1} \frac{v(t)}{W(t)} f(t)\,dt, \qquad b(x) = \int_{x_0}^{x} \frac{u(t)}{W(t)} f(t)\,dt.$$

Then a' and b' have the correct values and the result is the formula

$$(9) \qquad y(x) = \int_{x_0}^{x} \frac{u(t)v(x)}{W(t)} f(t)\, dt + \int_{x}^{x_1} \frac{u(x)v(t)}{W(t)} f(t)\, dt.$$

To verify the boundary conditions use (8) and get

$$y(x_0) = a(x_0)u(x_0), \qquad y'(x_0) = a(x_0)u'(x_0).$$

If u satisfies the conditions at x_0 then

$$\alpha_0 y(x_0) + \beta_0 y'(x_0) = a(x_0)(\alpha_0 u(x_0) + \beta_0 u'(x_0)) = 0$$

and hence y satisfies them too. The behavior of v at x_0 does not matter because $b(x_0) = 0$. In a like manner, if v satisfies the boundary conditions at x_1 then y does also, since $a(x_1) = 0$.

If we define

$$(10) \qquad G(x,t) = \frac{u(t)v(x)}{W(t)}, \qquad G(x,t) = \frac{u(x)v(t)}{W(t)}$$

for $x_0 \le t \le x$ and $x \le t \le x_1$ respectively, Equation (9) becomes

$$(11) \qquad y(x) = \int_{x_0}^{x_1} G(x,t) f(t)\, dt.$$

The functions G_0 and G are called *Green's functions* after George Green, the self-taught son of an English miller. In a long-neglected paper on electricity and magnetism (1828) Green gave formulas of this general type for partial differential equations.

EXAMPLE 3. Let us solve the boundary-value problem

$$y'' + y = f(x), \qquad y'(0) = 0, \qquad y(\pi) = 0.$$

Independent solutions of the homogeneous equation satisfying the first and second boundary conditions, respectively, are

$$u(x) = \cos x, \qquad v(x) = \sin x.$$

The Wronskian is $W(u,v) = 1$ and by (10) we have

$$G(x,t) = \cos t \sin x, \qquad G(x,t) = \cos x \sin t$$

according as $t \le x$ or $t \ge x$. The solution is given by (11), but for numerical evaluation it is better to use (9). For example, when $f(t) = 1$ we get

$$y(x) = \int_{0}^{x} \cos t \sin x\, dt + \int_{x}^{\pi} \cos x \sin t\, dt = 1 + \cos x.$$

PROBLEMS————————————————————————————————

1. The sole purpose of this problem is to help you understand the text. Do not use the formulas (3), (5), or (9). Instead, use the trial function below the equation and solve by the procedure that led to (3). In the last case $x > 0$:

(a) $y'' + y = 1$ (b) $y'' - y' = 1$ (c) $xy'' - y' = 1$

$a(x)\sin x + b(x)\cos x$ $a(x) + b(x)e^x$ $a(x) + x^2 b(x)$

2P. The Lagrange procedure is one of those rare procedures that is easier to carry out in general than in special cases. The reason is that if you have specific functions like $x \ln x$ or $e^x \sin x$ for u and v, you will be led to compute u', u'', v', v'' when you follow the pattern of Problem 1. It is a waste of time to find these derivatives, however, because they drop out of the calculation.

For this reason you are not asked to derive the basic formulas anew in the following problems. Instead, you are now asked to study the derivation of the equations

$$a'u + b'v = 0, \qquad a'u' + b'v' = f$$

leading to (3) so thoroughly that you can write it without the aid of the text.

3. Here and in subsequent problems of similar nature it is assumed that $x > 0$. Using $(u, v) = (x, x^4)$ in the first case and $(x^7, x^7 \ln x)$ in the second, solve by (3). Evaluate when $f(x) = 1$ and verify that the resulting solution satisfies the differential equation. The formulas for (u, v) come from Problem 4 and Example 1 of Section 1:

(a) $x^2 y'' - 4xy' + 4y = x^2 f(x)$ (b) $x^2 y'' - 13xy' + 49y = x^2 f(x)$

4. If α and β are constant parameters, solve (a) when $\alpha \neq \beta$ and (b) when $\alpha = \beta$. Take $(u, v) = (x^\alpha, x^\beta)$ in the first case and $(x^\alpha, x^\alpha \ln x)$ in the second. These values were found in Section 1, Problem 5:

$$x^2 y'' + (1 - \alpha - \beta)xy' + \alpha\beta y = x^2 f(x)$$

5. This problem depends on the formulas (5) and (9). Using the functions $(1, \cos x)$ from Section 1, Problem 3(c), find (a) the solution satisfying $y(\pi/2) = y'(\pi/2) = 0$ and (b) the solution satisfying $y(\pi/2) = y(\pi) = 0$. Evaluate when $f(x) = \sin x$ and check that the result satisfies both the differential equation and the initial or boundary conditions:

$$y'' \sin x - y' \cos x = f(x) \sin x$$

Hint for (b): By consideration of $c_1 + c_2 \cos x$ construct u satisfying the first boundary condition and v satisfying the second.

6. For the equation $x^2 y'' - 4xy' + 6y = x^2 f(x)$ find the solution satisfying

$$\text{(a) } y(1) = y'(1) = 0 \qquad \text{(b) } y(1) = 0, \ y(2) = y'(2)$$

Evaluate when $f(x) = x^2$ and verify that the result satisfies both the differential equation and the initial or boundary conditions.

Hint: Solve the homogeneous equation by trying $y = x^m$.

7. If ω is a positive constant, obtain the formula on the right for the solution of the initial-value problem on the left:

$$y'' + \omega^2 y = f(x), \quad y(x_0) = y'(x_0) = 0; \quad y = \frac{1}{\omega} \int_{x_0}^{x} \sin \omega (x - t) f(t) \, dt$$

8. Show that the boundary conditions $y(0) = y(\pi) = 0$ can be imposed in Problem 7 if, and only if, $\omega \neq$ integer.

Hint: For the boundary condition $u(0) = 0$ it will be found that $u = \sin \omega x$ is, essentially, the only choice. You need a function $v = c_1 \cos \omega x + c_2 \sin \omega x$, vanishing at π, which is not a multiple of u.

9. If the expression Q of Section 1, Theorem 2 is a positive constant ω^2, Theorem 1 of that section reduces the inhomogeneous equation to

$$z'' + \omega^2 z = g(x), \qquad g(x) = e^{P(x)/2} f(x)$$

Using this show that the general solution of $y'' + py' + qy = 0$ is

$$y = (c_1 \cos \omega x + c_2 \sin \omega x) e^{-P(x)/2}$$

10. Show that the Bessel equation $Ty = 0$ satisfies the condition $Q =$ const of the preceding problem if, and only if, $m = \pm 1/2$. Thus show that the general solution when $m = \pm 1/2$ is

$$x^{-1/2}(c_1 \cos x + c_2 \sin x)$$

If T is the Bessel operator with $m = 1/2$ the methods of this section yield an elementary solution for $Ty = f$, that is, a solution by quadrature. It is seen later that the result of this problem also gives

$$J_{1/2}(x) = \sqrt{\frac{2}{\pi x}} \sin x, \qquad J_{-1/2}(x) = \sqrt{\frac{2}{\pi x}} \cos x$$

11. Section 1, Problem 2 gives $W(u,v) = c/x$ if u, v are solutions of the Bessel equation. For the Bessel functions $J_{\pm m}$ it can be shown that this result takes the specific form

$$J_m(x)J'_{-m}(x) - J_{-m}(x)J'_m(x) = \frac{-2}{\pi x}\sin \pi m, \qquad x > 0$$

and hence the functions are linearly independent unless m is an integer. Assuming $x > 0$, $x_0 > 0$ and m not an integer, obtain the formula

$$y(x) = \frac{\pi}{2\sin \pi m} \int_{x_0}^{x} (J_m(x)J_{-m}(t) - J_m(t)J_{-m}(x)) \frac{g(t)}{t}\, dt$$

for rest solutions of the inhomogeneous Bessel equation:

$$x^2 y'' + xy' + (x^2 - m^2)y = g(x), \qquad y(x_0) = y'(x_0) = 0$$

12. The inhomogeneous Legendre equation with $m = 1$ and the inhomogeneous Chebychev equation with $m = 1$ are, respectively,

(a) $(1 - x^2)y'' - 2xy' + 2y = g(x)$ \qquad (b) $(1 - x^2)y'' - xy' + y = g(x)$

Using the results of Section 1, Problem 6, solve these equations.

13. If $p = 0$ show that $G(x, t) = G(t, x)$.

————————————ANSWERS————————————

1. (a) $c_1 \cos x + c_2 \sin x + 1$ \quad (b) $c_1 + c_2 e^x - x$ \quad (c) $c_1 + c_2 x^2 - x$
3. See answer to Problem 4.
4. (a) $(\alpha - \beta)y = x^\alpha \int x^{1-\alpha}f(x)\, dx - x^\beta \int x^{1-\beta}f(x)\, dx$
   (b) $y = x^\alpha \ln x \int x^{1-\alpha}f(x)\, dx - x^\alpha \int x^{1-\alpha}f(x)\ln x\, dx$
5. (a) $y = \int_{\pi/2}^{x}(\cos t - \cos x)(f(t)/\sin t)\, dt$
      $= \sin x - 1 - (x - \pi/2)\cos x$ when $f(x) = \sin x$
   (b) $y = \int_{\pi/2}^{x} \cos t(1+\cos x)(f(t)/\sin t)\, dt + \int_{x}^{\pi} \cos x(1+\cos t)(f(t)/\sin t)\, dt$
      $= \sin x + (\pi - x - 1)\cos x - 1$ when $f(x) = \sin x$
6. (a) $\int_{1}^{x}(x/t)^2(x - t)f(t)\, dt = x^2(x - 1)^2/2$ when $f(x) = x^2$
   (b) $\int_{1}^{x} (x/t)^2 (1 - t)f(t)\, dt + \int_{x}^{2} (x/t)^2 (1 - x)f(t)\, dt$
      $= x^2(1 - x)(3 - x)/2$ when $f(x) = x^2$
   We took $u = x^2 - x^3$ and $v = x^2$ to satisfy the boundary conditions.
12. Partial answer: If $u = x$ and if v is the second solution, $W(u, v) = 1/(x^2 - 1)$ in the first case and $-1/\sqrt{1 - x^2}$ in the second. Once you know this all you have to do is substitute u, v, W into (3), taking $f(x) = g(x)/(1 - x^2)$.

3. Qualitative behavior

By considering the points where a function attains a maximum or minimum one can get a good deal of information about the solutions of a differential equation without solving it. This procedure is illustrated here. We begin with a review of terminology, most of which is already familiar.

A point x is *interior* to an interval I if x is in I but is not an endpoint. As an illustration, if I is defined by one of the inequalities

(1) $1 < x \leq 3,$ $0 \leq x < \infty,$ $a \leq x \leq b,$ $0 < x$

the interior is defined by the corresponding inequality

$$1 < x < 3, \qquad 0 < x < \infty, \qquad a < x < b, \qquad 0 < x.$$

An interval is *finite* or *bounded* if its length is finite and *closed* if it contains its endpoints. For example, the first interval (1) is bounded but not closed, the second is closed but not bounded, the third is both bounded and closed, and the fourth is neither bounded nor closed.

Let w be a real-valued function defined on the interval I. It is said that w has a *maximum* at c if $c \in I$ and

$$w(c) \geq w(x), \qquad x \in I.$$

The maximum is interior if c is an interior point of I and it is positive if $w(c) > 0$. The term "negative interior minimum" is defined similarly.

If the inequality $w(c) \geq w(x)$ is required to hold only in some open interval $J \subset I$, with $x \in J$ and $c \in J$, the maximum is said to be *local*. From now on we use the terms *maximum* and *minimum* to mean local maximum or local minimum, respectively. Figure 1 shows a function with a positive interior maximum, a function with a negative interior minimum, a function with both, and a function with neither.

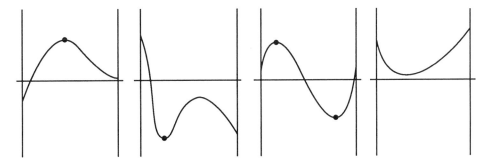

Figure 1

The following theorem of advanced calculus ensures existence of a maximum and minimum in all cases of interest here:

THEOREM A. *A continous real-valued function on a bounded closed interval attains its maximum and minimum on the interval.*

At an interior maximum or minimum c a differentiable function w satisfies $w'(c) = 0$. If w is twice differentiable there is the additional condition $w''(c) \leq 0$ at a maximum, $w''(c) \geq 0$ at a minimum. Thus, at a positive interior maximum a twice-differentiable function w satisfies

$$(2) \qquad\qquad w(c) > 0, \qquad w'(c) = 0, \qquad w''(c) \leq 0.$$

At a negative interior minimum it satisfies

$$(3) \qquad\qquad w(c) < 0, \qquad w'(c) = 0, \qquad w''(c) \geq 0.$$

In the rest of this section we show how (2) and (3) are used to get information about the solutions of differential equations. We adopt the convention that $a < b$ and that y is continuous on the closed interval

$$a \leq x \leq b \quad \text{or} \quad a \leq x < \infty,$$

as the case may be, which is implicitly specified in the statement of the problem. The function y is also twice (or at least once) differentiable on the interior, but this follows from other hypotheses and need not be separately postulated. Our first example illustrates an application to uniqueness.

EXAMPLE 1. Show that the sole solution of the following boundary-value problem is $y = 0$:

$$y'' + e^x y' = (x^2 + 1)y, \qquad y(a) = y(b) = 0.$$

If y is not identically 0 there must be a positive maximum or a negative minimum at some point c, $a < c < b$. At this point $y'(c) = 0$ and the differential equation becomes

$$y''(c) = (c^2 + 1)y(c).$$

Either condition (2) or (3) leads to a contradiction.

Equations (2) and (3) together constitute the *maximum principle* in its broadest sense. In a narrower sense, the term "maximum principle" is used to describe an estimate for y on $a < x < b$ from knowledge of the boundary values $y(a)$ and $y(b)$. Here is an example.

EXAMPLE 2. On an interval (a, b) let

$$(\cos x)^2 y'' + x^2(1 - x)y' = 2y, \qquad |y(a)| \leq m, \qquad |y(b)| \leq m$$

where m is a positive constant. Show that $|y(x)| < m$ for $a < x < b$.

If $y(x) \geq m$ at any point of (a, b) then y has a positive interior maximum, and if $y(x) \leq -m$ at any point of (a, b) then y has a negative interior minimum. Either condition leads to a contradiction as in Example 1.

It is a consequence of Rolle's theorem that if $y' > 0$ on an interval then y is strictly increasing. The relation $y' > 0$ is an example of a *differential inequality*. Differential inequalities form a major subfield of the modern theory of differential equations and are used systematically in the next section. Example 3 shows that monotonicity of y can sometimes be deduced from a second-order differential inequality $Ty \geq 0$.

EXAMPLE 3. Show that a function y satisfying the following conditions must be strictly increasing:

$$e^x y'' + y' \sin x - (1 + x)y \geq 0, \qquad x > 0; \qquad y(0) \geq 0, \qquad y'(0) > 0.$$

If the conclusion fails there are points x_1 and x_2 such that

$$0 < x_1 < x_2, \qquad y(x_1) \geq y(x_2).$$

It is obvious from a sketch that y attains a positive maximum at an interior point c of the interval $0 \leq x \leq x_2$. (An analytic proof can be given by use of Theorem A.) At the maximum we have $y'(c) = 0$, the differential inequality becomes

$$e^c y''(c) - (1 + c)y(c) \geq 0$$

and (2) leads to a contradiction.

A *zero* of a function y is a point where $y(x) = 0$. In general, a solution of a second-order linear differential equation cannot have zeros too close together unless it is identically zero. This fact is illustrated in Example 4.

EXAMPLE 4. Let y satisfy $y(0) = y(x_1) = 0$ where $x_1 > 0$ and also

$$y'' + e^{-x}y' = y \sin x, \qquad x > 0.$$

Show that $x_1 > \pi$ unless y is identically zero.

Suppose that $x_1 \leq \pi$ and that y is not identically 0 for $0 < x < x_1$. Then there is a positive maximum or negative minimum at an interior point c of this interval. At c the differential equation becomes $y''(c) = y(c) \sin c$ and either condition (2) or (3) leads to a contradiction. Hence $y = 0$ on the interval $0 < x < x_1$. This gives $y(x_0) = y'(x_0) = 0$ and then $y = 0$ by the uniqueness theorem for second-order initial-value problems.

We now give an application that involves only the first derivative. The functions $\sin x$ and $\cos x$ are linearly independent solutions of $y'' + y = 0$. These functions have the remarkable property that their zeros interlace. That is, if one of them has consecutive zeros x_1 and x_2 then the other is not zero at x_1 or x_2 but has exactly one zero between x_1 and x_2. The following theorem generalizes this result:

THEOREM 1. *Let u and v be differentiable functions on an interval I and suppose the Wronskian $W(x) = W(u, v)$ does not vanish at any point of I. Then the zeros of u and v interlace.*

For proof, let x_1 and x_2 be consecutive zeros of one of the functions, which without loss of generality we take to be u. If $v(x_1) = 0$ then the first row of the Wronskian is 0 at x_1, and this makes $W(x_1) = 0$ contrary to hypothesis. Similarly for x_2. Hence v does not vanish at x_1 or x_2.

If v does not vanish anywhere on the interval (x_1, x_2) then the function $y = u/v$ is differentiable for $x_1 \leq x \leq x_2$ and vanishes at the ends of this interval. By Rolle's theorem $y'(c) = 0$ at some value c. This gives $W(c) = 0$, which again contradicts the hypothesis. Hence v must vanish at some point of (x_1, x_2).

Finally, v cannot have two or more zeros between x_1 and x_2. If it did, the function u would have to vanish somewhere between them, contradicting the fact that x_1 and x_2 are consecutive zeros of u. This completes the proof.

It should be observed that Rolle's theorem, upon which Theorem 1 depends, is really an application of Theorem A and of the fact that $y'(c) = 0$ at an interior maximum. The same applies to other consequences of Rolle's theorem, some of which were already used at earlier points of this text.

To illustrate Theorem 1, consider the Bessel equation

$$x^2 y'' + xy' + (x^2 - m^2)y = 0, \qquad x > 0.$$

If m is not an integer it is shown later in this book that the solutions $J_m(x)$ and $J_{-m}(x)$ are linearly independent and that both have infinitely many zeros. Linear independence gives the hypothesis $W \neq 0$ of Theorem 1. Hence the zeros of either of these functions separate those of the other.

The main importance of maximum principles, by far, is that they apply to nonlinear problems. We conclude with a simple but significant theorem that illustrates this. The notation $f(x, p, q \uparrow)$ means that f is weakly increasing with respect to q, in other words

$$f(x, p, q_1) \geq f(x, p, q_2) \quad \text{for} \quad q_1 \geq q_2.$$

The notation $g(x, q \uparrow)$ is defined similarly. In both cases, the inequality is to hold for all relevant values of the variables.

The *inner normal derivative* u_ν of a function on $a \leq x \leq b$ is defined by

$$u_\nu(a) = \frac{du}{dx}\bigg|_a, \qquad u_\nu(b) = -\frac{du}{dx}\bigg|_b.$$

In the following theorem, the constants ϵ and δ measure the equation error and the boundary error, respectively, and the conclusion gives an estimate for the error in the solution:

THEOREM 2. *With $f(x, p, q \uparrow)$ and $g(x, q \uparrow)$ let*

$$Tu = u - f(x, u', u''), \qquad Ru = u - g(x, u_\nu).$$

Then $|Tu - Tv| \leq \epsilon$ and $|Ru - Rv| \leq \delta$ implies $|u - v| \leq \max(\epsilon, \delta)$.

The letter R is for the German Rand, meaning "boundary." It is a long-standing convention in theorems of this kind that inequalities involving the differential operator T hold for $a < x < b$, those involving the boundary operator R hold at a and b, and the conclusion holds for $a \leq x \leq b$.

    For proof, suppose $w = u - v$ has a maximum of value $m > \max(\epsilon, \delta)$. If the maximum is interior the conditions $w = m$, $w' = 0$, $w'' \leq 0$ yield

$$Tu = u - f(x, v', u'') \geq u - f(x, v', v'') = m + v - f(x, v', v'') > \epsilon + Tv.$$

This is a contradiction. If the maximum occurs on the boundary then $w_\nu \leq 0$ at that point, hence $u_\nu \leq v_\nu$, and this gives another contradiction,

$$Ru = u - g(x, u_\nu) \geq u - g(x, v_\nu) = m + v - g(x, v_\nu) > \delta + Rv.$$

Since the hypothesis is symmetric in u and v, Theorem 2 follows.

PROBLEMS————————————————————————————————

1P. Numerous applications require a nontrivial solution of

$$w'' + \lambda q w = 0, \qquad w(a) = w(b) = 0$$

where q is a positive function of x and λ is constant. (A *nontrivial solution* is a solution other than $w = 0$.) If λ is negative show by (2) and (3) that no nontrivial solution exists.

2. If $m > 0$ show that the Bessel function $J_m(x)$ cannot have a positive maximum or negative minimum on the interval $0 < x < m$. Given $J_m(0) = 0$ for $m > 0$, deduce that the first positive zero x_m of $J_m(x)$ satisfies $x_m > m$.

3. Let $w(0) = 0$ and, for all x, $e^{\cos x}w'' - x^2w + x^3 = 0$. If $y = w - x$ show that y cannot have a positive maximum or negative minimum at any value $x = c$. Show also that the sign of w'' is the same as the sign of $w - x$. Using these two facts sketch the graph of w for several values of $w'(0)$.

4. Let $(\cosh x)y'' + (\cos x)y' = (1+x^2)y$ for $a < x < b$ and let $y(a) = y(b) = 1$. As in example 4, show that $0 < y(x) < 1$ for $a < x < b$.

5. Let $y'' + pxy' = py$ where $p(x)$ is continuous. Check that $y = x$ is a solution and let v be any solution other than $v = 0$. Deduce from Theorem 1 that v can have at most two zeros, x_1 and x_2, and in this case $x_1x_2 < 0$. Finally, choose p in such a way that $v = 1 - x^2$ is a solution, thus showing that two zeros with opposite signs can actually occur.

6. Let $y'' + p(x)(\sin x)y' + y = p(x)(\cos x)y$ where p is continuous. Check that $y = \sin x$ is a solution and deduce that every solution, except $y = 0$, must have exactly one zero on every interval $n\pi \le x < (n+1)\pi$, $n = 0, \pm 1, \pm 2, \cdots$.

7. (a) Check that the result of Example 1 remains valid, and the proof is virtually unchanged, if the coefficient e^x is replaced by any function p and $x^2 + 1$ is replaced by any positive function q.

(bcdef) Similarly generalize Examples 2,3,4 and Problems 3 and 4.

8. Let $p(x,s)$, $q(x,s)$ $r(x,s)$ be real-valued functions such that

$$q(x,0) > 0, \quad r(x,0) \ge 0 \qquad a < x < b$$

Show that no solution of the following equation can have a positive interior maximum or a negative interior minimum:

$$r(x,y')y'' + p(x,y')y' = q(x,y')y$$

9. (abcdef) Apply Problem 8 to Examples 1, 2, 3, and 4 and to Problems 3 and 4.

10. Open-ended problem. In Theorem 2 let u be an exact solution of $Tu = \tau(x)$, $Ru = \rho(x)$ and let v be an approximation thereto. Discuss the implications of the theorem as regards uniqueness, stability, and numerical estimation.

4. Differential inequalities and uniqueness

Here we establish a theorem that yields uniqueness for solutions of linear equations with bounded coefficients. Similar methods apply to broad classes of nonlinear equations and even to nonlinear systems.

We shall need the following properties of *absolute value*, which hold no matter whether the numbers a, b are real or complex:

$$|a + b| \leq |a| + |b|, \qquad |ab| = |a||b|, \qquad |a| = 0 \iff a = 0.$$

Repeated use of the first two of these gives similar results for three or more numbers a, b, c, $\cdots$. As an illustration, the equation

$$u''' + pu'' + qu' + ru = 0$$

yields $|u'''| \leq |p||u''| + |q||u'| + |r||u|$. If $|p|, |q|$ and $|r|$ are all $\leq K$, where K is constant, we get

(1) $$|u'''| \leq K(|u''| + |u'| + |u|).$$

Equation (1) is a differential inequality, as is the main hypothesis in the following theorem:

THEOREM 1. *Let I be an open interval containing the point x_0, let M be a positive constant, and let w be a real-valued function satisfying the two conditions $|w'(x)| \leq Mw(x)$ for $x \in I$ and $w(x_0) = 0$. Then $w = 0$.*

For proof, suppose, contrary to the conclusion, that $w(x_1) > 0$ at some point $x_1 \in I$. We take $x_1 > x_0$; the case $x_1 < x_0$ is similar. We have

$$\frac{d}{dx} e^{-Mx} w(x) = e^{-Mx}(w'(x) - Mw(x)) \leq 0.$$

Hence $e^{-Mx} w(x)$ is decreasing and, since $x_1 > x_0$,

$$e^{-Mx_1} w(x_1) \leq e^{-Mx_0} w(x_0) = 0.$$

Multiplication by e^{Mx_1} gives $w(x_1) \leq 0$, contrary to our initial assumption. This completes the proof.

As a first application of Theorem 1 we establish:

THEOREM 2. *Let $p(x)$ and $q(x)$ be bounded on an open interval I containing the point x_0. Then the following conditions imply $u(x) = 0$ for $x \in I$:*

$$u'' + p(x)u' + q(x)u = 0, \qquad x \in I; \qquad u(x_0) = 0, \qquad u'(x_0) = 0.$$

Instead of assuming p and q bounded, we could assume that they are continuous on I. By Theorem A of the previous section, they are bounded on every closed subinterval J contained in I. Theorem 2 gives $u(x) = 0$ on J and, since J is arbitrary, this implies $u(x) = 0$ on I. A similar remark applies to all theorems of this general structure and shows that boundedness is more general than continuity. The hypothesis of boundedness is also more elementary, in that it avoids Section 3, Theorem A.

For proof of Theorem 2, let $|p| \leq K$ and $|q| \leq K$ where K is constant. The differential equation gives

$$(2) \qquad |u''| \leq K(|u| + |u'|)$$

and this suffices. For simplicity we assume that u is real; the case of complex u is taken up in the problems. Let

$$w = (u)^2 + (u')^2$$

and differentiate to get

$$|w'| = |2uu' + 2u'u''| \leq 2|u||u'| + 2|u'||u''|.$$

Estimating $|u''|$ by (2) yields

$$(3) \qquad |w'| \leq 2|u||u'| + 2|u'|K(|u| + |u'|).$$

Now comes an important point. Since squares are nonnegative we have $w \geq u^2$ and $w \geq (u')^2$, hence

$$|u| \leq \sqrt{w}, \qquad |u'| \leq \sqrt{w}.$$

Using these estimates in (3) yields $|w'| \leq (2 + 4K)w$. Since $w(x_0) = 0$ by the initial conditions, Theorem 1 with $M = 2 + 4K$ gives $w = 0$. The fact that $|u| \leq \sqrt{w}$ now gives $u = 0$, completing the proof.

Theorem 2 yields the basic uniqueness theorem for the initial-value problem

$$Ty = y'' + p(x)y' + q(x)y = f(x), \qquad y(x_0) = y_0, \qquad y'(x_0) = y_1$$

when p and q are bounded. Namely, if z is another solution then $u = y - z$ satisfies the hypothesis of Theorem 2 and the conclusion $u = 0$ gives $y = z$.

The only use we made of the differential equation was to get the differential inequality (2). Thus, $u = 0$ if (2) holds and u and u' vanish at some point of I. This formulation is no more complicated than Theorem 2, but it yields uniqueness for broad classes of nonlinear initial-value problems as seen later. The analogous result for equations of order n is as follows:

THEOREM 3.   *On an open interval I let u be a real- or complex-valued function satisfying the two conditions*

$$|u^{(n)}| \leq K(|u| + |u'| + \cdots + |u^{(n-1)}|), \qquad \text{for all } x \in I$$

$$u(x_0) = u'(x_0) = \cdots = u^{(n-1)}(x_0) = 0, \qquad \text{for some } x_0 \in I$$

where K is constant. Then $u(x) = 0$ on I.

Again the proof is simple. Assuming that u is real, let

$$w = (u)^2 + (u')^2 + \cdots + (u^{(n-1)})^2.$$

By differentiation

(4) $$w' = 2uu' + 2u'u'' + \cdots + 2u^{(n-1)}u^{(n)}.$$

Since squares are nonnegative we have

$$w \geq |u^{(j)}|^2, \quad \text{hence} \quad |u^{(j)}| \leq \sqrt{w}, \qquad j = 0, 1, \cdots, n-1.$$

The differential inequality in Theorem 3 now gives

$$|u^{(n)}| \leq K(\sqrt{w} + \sqrt{w} + \cdots + \sqrt{w}) = Kn\sqrt{w}.$$

Taking absolute values in (4) we get

$$|w'| \leq 2w + 2w + \cdots + 2w + 2wKn = 2(n-1)w + 2Knw.$$

Hence $|w'| \leq M|w|$ with $M = 2(n-1)+2Kn$. Since the initial conditions give $w(x_0) = 0$ Theorem 1 gives $w = 0$, the inequality $|u| \leq \sqrt{w}$ gives $u = 0$, and this completes the proof when u is real. The result is extended to complex u in Problem 2.

For readers familiar with the notion of a Lipschitz condition, Theorem 3 has an interesting bearing on the initial-value problem

$$y^{(n)} = f(x, y, y', \cdots, y^{(n-1)}), \qquad y^{(j)}(x_0) = y_0, \qquad j = 0, 1, \cdots, n-1.$$

If y and z are two solutions, and if f admits the Lipschitz constant K, then $u = y - z$ satisfies the hypothesis of Theorem 3. Hence $u = 0$ and $y = z$. This is a classical uniqueness theorem for nth-order nonlinear equations. In Chapter 14 it is seen that the Lipschitz condition holds if f has bounded partial derivatives. Hence the special cases $n = 1$ and $n = 2$ yield the assertions pertaining to uniqueness in Chapter 4, Section 2.

REVIEW PROBLEMS————————————————————————

1. With $x_0 \in I$ and K constant, let u be a complex-valued function satisfying the two conditions

$$|u'(x)| \le K|u(x)|, \quad x \in I; \quad u(x_0) = 0$$

Let $w = |u|^2 = u\bar{u}$ where the absolute value pertains to complex numbers and the bar denotes the complex conjugate. Check that $|u'| \le K|u|$ gives

$$|w'| \le 2K|u|^2 = 2Kw$$

and hence that $w = 0$ follows from Theorem 1 with $M = 2K$.

2. If u is a complex-valued function in Theorem 3 we define

$$w = |u|^2 + |u'|^2 + \cdots + |u^{(n-1)}|^2$$

where $|u^{(j)}|^2 = u^{(j)}\bar{u}^{(j)}$. Show that you get the same differential inequality for w as in the real case, hence the same conclusion.

3. Obtain a uniqueness theorem for a general third-order initial-value problem by combining (1) with Theorem 3.

4. Verify that $(\sin x)/x$, e^x, x satisfy the following equations, respectively, and thus obtain a general solution. Part (c) leads to a tricky integral:

(a) $\qquad xy'' + 2y' + xy = 0$ $\hfill (x > 0)$
(b) $\qquad (2x - x^2)y'' + (x^2 - 2)y' + (2 - 2x)y = 0 \qquad 0 < x < 2$
(c) $\qquad (2x - 1)y'' - 4xy' + 4y = 0 \qquad\qquad 2x > 1$

5. If u and v satisfy $(ry')' + qy = 0$ show that $rW(u, v) = c$.

Hint: Repeat the derivation of Abel's equation, keeping the expressions (ru') and (rv') as inviolable units.

6. Explain why $y'' + py' + qy = f$ can be solved by quadrature, in general, on any interval on which one of the following holds:

(a) $p(x) + xq(x) = 0$ $\qquad$ (b) $1 + q(x) = p(x)$ $\qquad$ (c) $2q + qx^2 = px$

7. (a) Taking $u = e^h$ show that $Ty = f$ can be solved by quadrature if

$$h'' + (h')^2 + ph' + q = 0$$

(b) Check that the following equation has a solution of the form $y = e^{ax^2}$, find a, and obtain the formula on the right for a second solution:

$$y'' + (x + x^2)y' + (x^3 + 1)y = 0, \qquad y = e^{-x^2/2} \int e^{x^2/2 - x^3/3} \, dx$$

8. This pertains to $Ty = (1 - x)y'' + xy' - y = g(x)$. Given that x satisfies $Ty = 0$, (a) find a second solution of $Ty = 0$, (b) solve the inhomogeneous equation by the method of Lagrange and (c) evaluate when $g(x) = (1 - x)^2$. Part (a) leads to a tricky integral.

9. (a) Check that Green's function for $y'' = f(x)$, $y(0) = y(1) = 0$ is

$$G(x, t) = t(x - 1), \quad t < x; \quad G(x, t) = x(t - 1), \quad x < t$$

(b) Using $G(x, t)$ compute y when $f(x) = x^m$, $m \geq 0$ and verify that the expected properties hold. No answer is given.

10. For the problem $y'' - 2y' + y = f(x)$:

(a) Find the solution satisfying $y(x_0) = y'(x_0) = 0$.

(b) Find Green's function for the boundary conditions $y(0) = y'(1) = 0$.

──────────────ANSWERS──────────────

4. (a) $(c_1 \cos x + c_2 \sin x)/x$ (b) $c_1 e^x + c_2 x^2$ (c) $c_1 x + c_2 e^{2x}$
6. Partial answer: Try $u = x^m$ and $u = e^{ax}$.
8. (c) $y = c_1 x + c_2 e^x + x^2 + 1$
10. (a) $y = \int_{x_0}^x (x - t)e^{x-t} f(t)\, dt$
    (b) $2G(x, t) = (x - 2)te^{x-t}$ for $0 < t < x$;
        $2G(x, t) = (t - 2)xe^{x-t}$ if $x < t < 1$.

────────────────────────

Mathematics is a vast adventure in ideas; its history reflects some of the noblest thoughts of countless generations.

—Dirk Struik

Mathematics may be defined as the subject in which we never know what we are talking about, nor whether what we say is true.

—Bertrand Russell

Chapter 11
LINEAR EQUATIONS IV

THE CONCEPTS of uniqueness, linear dependence, general solution and complete solution are developed for linear equations of arbitrary order with variable coefficients. The homogeneous case is emphasized because a general solution of the homogeneous equation yields a general solution of the inhomogeneous equation by variation of parameters. Much of our work up to this point can be regarded as a preview of topics that are here presented from a comprehensive point of view.

This chapter assumes knowledge of determinants to the extent developed in Chapter E, Parts (abc).

1. Mathematical foundations

Throughout this chapter the letters c_j denote constants; p_j, u_j, y, and f are real- or complex-valued functions on a given interval I; and the operator T is defined by

$$Ty = y^{(n)} + p_{n-1}y^{(n-1)} + \cdots + p_1 y' + p_0 y.$$

Our results apply to complex functions because the algebra of linear equations and determinants is the same for complex numbers as for real.

In the preceding pages the theory of linear differential equations was developed for two special but significant cases:

(i) nth-order equations with constant coefficients.

(ii) 2nd-order equations with variable coefficients.

Now we consider the general case,

(iii) nth-order equations with variable coefficients.

The basic vocabulary was already introduced in the earlier discussion, where we defined such notions as general solution, linear dependence, and initial values. For clarity, however, these terms are defined again here.

(a) Uniqueness

A *solution* of the equation $Ty = f$ is a function $y = y(x)$ that satisfies the equation on I. Associated with the concept of solution are the *initial conditions*

$$(1) \qquad y(x_0) = y_0, \qquad y'(x_0) = y_1, \qquad \cdots, \qquad y^{(n-1)}(x_0) = y_{n-1}$$

where x_0 is a given point of I. The following uniqueness theorem is one of the main results in the theory:

THEOREM 1. *If the coefficients p_k are bounded, the equation $Ty = f$ has at most one solution satisfying the initial conditions (1).*

For proof, let z be another solution, so that $Tz = f$. With $u = y - z$ linearity gives

$$Tu = T(y - z) = Ty - Tz = f - f = 0$$

which is equivalent to

$$u^{(n)} = -p_{n-1}u^{(n-1)} - \cdots - p_1 u' - p_0 u.$$

If $|p_k| \leq K$ for $k = 0, 1, 2, \cdots, n-1$ taking absolute values yields

$$|u^{(n)}| \leq K(|u^{(n-1)}| + \cdots + |u'| + |u|).$$

Since y and z satisfy the same initial conditions, the initial conditions for u are 0. Chapter 10, Section 4, Theorem 3 now gives $u = 0$. Hence $y = z$, completing the proof.

(b) The Wronskian

If $Tu_j = 0$ for $j = 1, 2, \cdots, n$ the function

(2) $$y = c_1 u_1 + c_2 u_2 + \cdots + c_n u_n$$

satisfies $Ty = 0$ by the superposition principle. A family of solutions such as this is a *general solution* of $Ty = 0$ if the constants c_j in it can be chosen so as to satisfy arbitrarily specified initial conditions at the arbitrarily specified point $x_0 \in I$. If the family (2) contains every solution of the equation $Ty = 0$, it is the *complete solution*.

Substitution of (2) into the initial-value equations (1) yields

(3) $$c_1 u_1^{(j)}(x_0) + c_2 u_2^{(j)}(x_0) + \cdots + c_n u_n^{(j)}(x_0) = y_j$$

for $j = 0, 1, 2, \cdots, n-1$. This is a system of n linear equations in the n unknowns c_j. The coefficient determinant is the *Wronskian* $W(x_0)$ where, by definition,

$$W = \begin{vmatrix} u_1 & u_2 & \cdots & u_n \\ u_1' & u_2' & \cdots & u_n' \\ \vdots & \vdots & \vdots & \vdots \\ u_1^{(n-1)} & u_2^{(n-1)} & \cdots & u_n^{(n-1)} \end{vmatrix}$$

For example, when $n = 3$ the linear system is

$$c_1 u_1(x_0) + c_2 u_2(x_0) + c_3 u_3(x_0) = y_0$$

$$c_1 u_1'(x_0) + c_2 u_2'(x_0) + c_3 u_3'(x_0) = y_1$$

$$c_1 u_1''(x_0) + c_2 u_2''(x_0) + c_3 u_3''(x_0) = y_2$$

and its determinant is

$$W(x_0) = \begin{vmatrix} u_1(x_0) & u_2(x_0) & u_3(x_0) \\ u_1'(x_0) & u_2'(x_0) & u_3'(x_0) \\ u_1''(x_0) & u_2''(x_0) & u_3''(x_0) \end{vmatrix}$$

The prescription of initial values is not the only problem that leads to the Wronskian. The Wronskian is also encountered in the theory of linear dependence. By definition, the functions u_j are *linearly dependent on I* if there are constants c_j, not all 0, such that

(4) $$c_1 u_1(x) + c_2 u_2(x) + \cdots + c_n u_n(x) = 0, \qquad x \in I.$$

If no such constants exist, the functions are *linearly independent on I*. The qualifying phrase "on I" is understood whether mentioned or not, and may be omitted in the following discussion.

Differentiation of (4) yields the equations

(5) $$c_1 u_1^{(j)}(x) + c_2 u_2^{(j)}(x) + \cdots + c_n u_n^{(j)}(x) = 0$$

for $j = 0, 1, 2, \cdots, n - 1$. This is a homogeneous linear system in the unknowns c_j. Both systems (3) and (5) have essentially the same determinant, $W(x_0)$ in the first case and $W(x)$ in the second. For example, when $n = 3$ the system (5) is the 3 by 3 system displayed above with x_0 replaced by x and with $y_j = 0$.

(c) The Fredholm alternative

The main fact we need about linear systems pertains to a system

$$a_{i1} c_1 + a_{i2} c_2 + \cdots + a_{in} c_n = y_i, \qquad i = 0, 1, \cdots, n - 1$$

of n equations in n unknowns c_j. It states that *exactly one* of the following alternatives holds:

(i) the system has a unique solution for all y_j, or

(ii) the homogenous system has a nontrivial solution.

Furthermore, (i) happens if the coefficient determinant is not zero, and (ii) if it is zero. We refer to these statements collectively as the *Fredholm alternative*, after the Swedish mathematician Ivar Fredholm who formulated similar properties in a more general setting.

We shall use the Fredholm alternative to establish the following theorem:

THEOREM 2. *Let $Tu_j = 0$ for $j = 1, 2, \cdots, n$ and let $W(x)$ be the Wronskian of the functions u_j. Then:*

(a) *The expression $c_1 u_1 + c_2 u_2 + \cdots + c_n u_n$ is a general solution of $Ty = 0$ if, and only if, $W(x) \neq 0$ holds at every point of I.*

(b) *The functions u_j are linearly independent if $W(x) \neq 0$ holds at a single point of I.*

The initial-value problem is solvable at x_0 if the c_j can be determined so as to satisfy (3) for arbitrary choices of the y_j. According to the Fredholm alternative this can be done if, and only if, $W(x_0) \neq 0$. If x_0 as well as y_j is arbitrary we conclude that $W(x) \neq 0$ for every $x \in I$. In other words the functions u_j produce a general solution by way of (2) if, and only if, $W(x) \neq 0$ holds at every point of I. This gives Part (a) of Theorem 2.

On the other hand if the functions are linearly dependent then the system (4) has a nontrivial solution c_j at every value $x \in I$. By the Fredholm alternative we must have $W(x) = 0$. Thus, the functions u_j can be linearly dependent on I only if $W(x) = 0$ at every point of I. This gives Part (b) of the theorem.

For most functions u_j it is to be expected that $W(x) = 0$ at some points of I and $W(x) \neq 0$ at others. Solutions of $Ty = 0$, however, exhibit the following surprising behavior:

THEOREM 3. *On a given interval I let u_j be solutions of $Tu_j = 0$ for $j = 1, 2, \cdots, n$ where T has bounded coefficients p_j. Suppose that $W(x_0) = 0$ at some point $x_0 \in I$. Then the functions u_j are linearly dependent and $W(x) = 0$ at every $x \in I$.*

It follows, in particular, that either $W(x) = 0$ at every $x \in I$, or $W(x) = 0$ at no $x \in I$.

The proof uses both the Fredholm alternative and uniqueness. By the Fredholm alternative we can choose constants c_j, not all 0, in such a way that the function

$$(6) \qquad y = c_1 u_1 + c_2 u_2 + \cdots + c_n u_n$$

satisfies null initial conditions at the point x_0. The solution given by the zero function 0 satisfies these same null conditions, hence by uniqueness $y = 0, x \in I$. This shows that the u_j are linearly dependent, and $W(x) = 0$ follows from the remarks above.

Another consequence of uniqueness together with the Fredholm alternative is:

THEOREM 4.   Let u_j satisfy $Tu_j = 0$ on I for $j = 1, 2, \cdots, n$, where T has bounded coefficients p_j. Suppose further that $W(x_0) \neq 0$ at some point $x_0 \in I$. Then the expression (6) is the complete solution of $Ty = 0$.

The concluding statement means that every solution of $Ty = 0$ can be written in the form (6).

For proof, note that the constants c_j can be chosen so that y in (6) satisfies arbitrary initial conditions at x_0. This follows from the Fredholm alternative, since $W(x_0) \neq 0$. If z is any solution of $Tz = 0$, let us determine the c_j in such a way that the initial conditions for y at x_0 agree with those for z. Since $Ty = 0$ by the superposition principle and $Tz = 0$ by hypothesis, uniqueness gives $y = z$. Hence the arbitrary solution z is in the family (6), which shows that (6) is the complete solution.

PROBLEMS———————————————————————————————

Assume all functions sufficiently differentiable to justify the calculations.

1. If $u_1 = x^2 + 2x$, $u_2 = x^3 + x$, $u_3 = 2x^3 - x^2$ find constants c_1 and c_2 such that $u_3 = c_1 u_1 + c_2 u_2$. What does this indicate about the Wronskian? Verify your answer by direct computation.

2. Check for linear dependence by any method:

(a) $(x + 1)^2$, $(x - 1)^2$, $3x$ (d) e^{ix}, $\cos x$, $\sin x$ (g) 1, x, x^2, x^3

(b) $x^2 - 2x$, $3x - 1$, $\sin x$ (e) x, x^2, x^3 (h) e^{-x}, e^x, 1, $\sinh x$

(c) 1, $\sin x$, $\cos x$ (f) e^x, xe^x, $x^2 e^x$ (i) e^x, e^{2x}, e^{3x}, e^{4x}

3. Reduction of order. If u is a solution of $Tu = 0$ that is nowhere zero on a given interval I, the substitution $y = uv$ reduces the equation $Ty = f$ on I to an equation of lower order in the unknown v'. The reduction depends on formulas for differentiating a product such as

$$(uv)''' = u'''v + 3u''v' + 3u'v'' + uv'''$$

By this method reduce the equation $y''' + py'' + qy' + ry = f$ to

$$uv''' + (3u' + pu)v'' + (3u'' + 2pu' + qu)v' = f$$

4. Reduce the following equation to one of second order in v' by the choice (a) $u = x$ or (b) $u = x^2$ or (c) $u = x^3$ in Problem 3. Only the answer (a) is provided:

$$x^3 y''' - 3x^2 y'' + 6xy' - 6y = g(x), \qquad x > 0$$

5. Write an equation for $(uv)^{(j)}$ involving binomial coefficients and use it to generalize Problem 3. No answer is provided.

6. Let $\delta_{ij} = 1$ for $i = j$ and $\delta_{ij} = 0$ for $i \neq j$. This symbol is called the *Kronecker* delta after the German mathematician Leopold Kronecker. A *fundamental solution* of a linear homogeneous equation $Ty = 0$ is a set of solutions u_i satisfying the initial conditions

$$u_i^{(j-1)} = \delta_{ij}, \qquad i, j = 1, 2, \cdots, n$$

For example, when $n = 3$ the conditions have the following appearance:

$$u_1(x_0) = 1 \qquad\qquad u_2(x_0) = 0 \qquad\qquad u_3(x_0) = 0$$

$$u_1'(x_0) = 0 \qquad\qquad u_2'(x_0) = 1 \qquad\qquad u_3'(x_0) = 0$$

$$u_1''(x_0) = 0 \qquad\qquad u_2''(x_0) = 0 \qquad\qquad u_3''(x_0) = 1$$

Let y_j be given constants and let $\{u_i\}$ be a fundamental solution. Show that the function on the left below satisfies the given initial conditions:

$$y = y_0 u_1 + y_1 u_2 + \cdots + y_{n-1} u_n, \qquad y^{(j)}(x_0) = y_j, \qquad j = 0, 1, \cdots, n - 1$$

7. (a) Apply the following equation to $W(xe^x, x^2 e^x, x^3 e^x)$:

$$W(uv_1, uv_2, \cdots, uv_n) = u^n W(v_1, v_2, \cdots, v_n)$$

(bc) If u and v_k have $n - 1$ derivatives, derive the above equation (b) for $n = 3$ and (c) in general.

8. Let I be an open interval containing $x = 0$. Explain why the functions x, x^2, x^3 cannot be solutions of one and the same linear homogeneous equation $y''' + py'' + qy' + ry = 0$, $x \in I$ with bounded coefficients, though they all satisfy $y^{iv} = 0$.

───────────────────────────ANSWERS───────────────────────────

1. $W = 0$ 2. (adh) are dependent, the rest independent.

4. (a) $x^4 v''' = g(x)$ 7. (a) $(xe^x)^3 W(1, x, x^2) = 2(xe^x)^3$

8. Because the Wronskian vanishes at $x = 0$ but is not identically 0.

2. Basis

Let the functions u_1, u_2, $\cdots u_n$ be solutions of the homogeneous equation $Ty = 0$. These functions are said to form a *basis* if every solution of $Ty = 0$ can be written in the form

$$(1) \qquad\qquad y = c_1 u_1 + c_2 u_2 + \cdots + c_n u_n$$

and this representation is unique. We shall establish:

THEOREM 1. *Let u_j for $j = 1, 2, \cdots, n$ be n linearly independent solutions of $Tu_j = 0$ where T has bounded coefficients. Then the u_j form a basis for the solutions of $Ty = 0$.*

Since the functions u_j are linearly independent, Theorem 3 of the previous section shows that $W(x_0) \neq 0$ at each $x_0 \in I$. Theorem 4 of that section then shows that every solution can be written in the form (1). To see why the representation is unique, suppose there is another such representation with constants d_j instead of c_j. By subtraction

$$(c_1 - d_1)u_1 + (c_2 - d_2)u_2 + \cdots + (c_n - d_n)u_n = 0.$$

Since the u_j are linearly independent each coefficient $c_j - d_j = 0$. Hence $d_j = c_j$, which shows that the new representation is the same as the old. This completes the proof.

Instead of assuming that the u_j are linearly independent we could make the stronger assumption that $W(x) \neq 0$ for $x \in I$. In that case the conclusion of Theorem 1 remains valid with no hypothesis on the coefficients p_j. Proof of this remarkable fact is outlined in Problem 4 at the end of the following section.

(a) Constant coefficients

Let us show that there is a basis consisting only of elementary functions when the coefficients p_j are constant. We know from Chapter 8 that the equation $Ty = 0$ has n distinct solutions of the form

$$x^j e^{q_j x}$$

where the j are nonnegative integers and the q_j are real or complex constants. We shall show that these n solutions are linearly independent on every interval I and therefore form a basis.

To this end, suppose a linear combination of these functions is 0 on I. Let the *distinct* exponents q_j be

$$r_1, r_2, \cdots, r_k$$

and group terms with the same exponent. The result is an equation

(2) $\qquad c_1(x)e^{r_1 x} + c_2(x)e^{r_2 x} + \cdots + c_k(x)e^{r_k x} = 0, \qquad x \in I,$

where the c_j are polynomials. The total number of constants is n. Thus n is the sum of the numbers $d_j + 1$ where d_j is the degree of c_j.

It will be established by induction on n that all $c_j = 0$. This is obvious if $n = 1$, since there is only a single term ce^{rx}. Suppose then that the result is known for $n - 1$. To get it for n, multiply (2) by $e^{-r_1 x}$. The result is

(3) $\qquad c_1(x) + c_2(x)e^{s_2 x} + \cdots + c_k(x)e^{s_k x} = 0, \qquad x \in I,$

where $s_j = r_j - r_1$. The s_j, like the r_j, are all unequal, and none of them equals the first exponent, 0, associated with c_1. If we differentiate (3) with respect to x we get an equation with $n - 1$ constants to which the induction hypothesis applies. Hence all the coefficients are 0 and in particular $c_1'(x) = 0$. This shows that c_1 is constant. A similar argument applies to each c_j and shows that each c_j is constant. However, the case of constant coefficients c_j was already settled in Chapter 8, Section 4. This completes the proof.

(b) The Euler-Cauchy equation

There are only a few interesting classes of nth-order linear equations for which a basis can be expressed by means of elementary functions. One of these is the class of equations with constant coefficients, as we have seen. Another is the class of equations of the form

$$x^n y^{(n)} + p_{n-1} x^{n-1} y^{(n-1)} + \cdots + p_1 x y' + p_0 y = 0, \qquad x > 0$$

where the p_j are constant. An equation of this form is often called *Cauchy's equation* though such equations were studied earlier by Euler. The coefficients x^j exactly compensate for the lowering of the exponent by differentiation, hence the substitution $y = x^\rho$ leads to a polynomial equation for ρ. If this polynomial equation has n distinct roots ρ_j the functions x^{ρ_j} so obtained form a basis. Thus every solution of the equation is uniquely representable in the form

$$y = c_1 x^{\rho_1} + c_2 x^{\rho_2} + \cdots + c_n x^{\rho_n}.$$

For complex roots $\rho = \sigma + i\tau$ we write

$$x^\rho = x^{\sigma + i\tau} = x^\sigma x^{i\tau} = x^\sigma e^{i\tau \ln x}.$$

If the coefficients p_j are real, taking real and imaginary parts gives the two solutions

$$x^\sigma \cos(\tau \ln x), \qquad x^\sigma \sin(\tau \ln x).$$

These correspond to the complex root $\sigma + i\tau$.

If ρ is a repeated root of multiplicity m the corresponding family of solutions is

$$x^\rho, \qquad x^\rho \ln x, \qquad x^\rho (\ln x)^2, \quad \cdots, \quad x^\rho (\ln x)^{m-1}.$$

The family of n distinct functions obtained by this process again forms a basis. These results are easy to prove because the Euler-Cauchy equation can be reduced to an equation with constant coefficients as in the following problems.

PROBLEMS

1. The Euler-Cauchy equation. If $Ty = x^3 y''' + p_2 x^2 y'' + p_1 x y' + p_0 y$ for $x > 0$ and ρ is constant, check that

$$Tx^\rho = (\rho(\rho - 1)(\rho - 2) + p_2 \rho(\rho - 1) + p_1 \rho + p_0)x^\rho$$

Hence $Tx^\rho = 0$ holds if ρ satisfies a certain cubic equation. Also $Ty = Ax^\rho$ has a solution of the form $y = Bx^\rho$ if ρ is not a root of that cubic equation.

2. (a) If $x = e^t$, $D_x = d/dx$, $D_t = d/dt$ show that sufficiently differentiable functions defined by $y(x) = z(t)$ satisfy

$$D_x y = e^{-t} D_t z, \qquad D_x^2 y = e^{-2t} D_t (D_t - 1)z,$$
$$D_x^3 y = e^{-3t} D_t (D_t - 1)(D_t - 2)z,$$

and so on. Hint: $dy/dx = (dy/dt)/(dx/dt)$, hence $D_x = e^{-t} D_t$.

(b) With $D = D_t$ in Part (a) show that the Euler-Cauchy equation for $n = 3$ can be reduced to the following; note the resemblance to the expression obtained in Problem 1:

$$(D(D - 1)(D - 2) + p_2 D(D - 1) + p_1 D + p_0)z = f(e^t)$$

3. By the result of Problems 1 or 2 obtain two linearly independent solutions:

(a) $x^2 y'' + 4xy' = 54y$ (c) $x^2 y'' + 5y = 3xy'$ (e) $x^2 y'' + 4y = 3xy'$

(b) $x^2 y'' + y = xy'$ (d) $x^2 y'' + 6y = 4xy'$ (f) $x^2 y'' + 50y = xy'$

4. Reduce to an equation with constant coefficients by setting $x = e^t$ and thus obtain a particular solution:

(a) $x^2 y'' + 4xy' + 2y = \cos(\ln x)$ (b) $x^2 y'' + xy' + y = 5x \sin(\ln x)$

(c) $x^2 y'' + 5xy' + 7y = 17x \cos(\ln x) + 5x \sin(\ln x)$

5. Obtain three linearly independent solutions of $x^3 y''' + xy' = y$ by the method of Problem 2.

6. Show that for arbitrary n the results of Problems 1 and 2 are respectively

$$(\rho(\rho - 1) \cdots (\rho - n + 1) + \cdots + p_2\rho(\rho - 1) + p_1\rho + p_0)x^\rho = Tx^\rho$$

$$(D(D - 1) \cdots (D - n + 1) + \cdots + p_2 D(D - 1) + p_1 D + p_0)z = f(e^t)$$

7. The method of annihilators. In this problem S and T are linear differential operators with constant coefficients. Consider the differential equation $Ty = f$ where f is a function such that S *annihilates* f; that is, $Sf = 0$. For example, we could have

$$f(x) = Ax^m e^{ax}, \qquad S = (D - a)^{m+1}, \qquad D = \frac{d}{dx}$$

since the results of Chapter 8 give $Sf = 0$. Show that all solutions of $Ty = f$ are among the solutions of the homogeneous equation $STy = 0$. By results of Part (a) the form of the latter is known, hence the form of possible solutions of $Ty = f$ is also known. The process thus gives an alternative approach to the method of undetermined coefficients.

8. (a) Using the method of Problem 7 obtain the general form of a particular solution of

$$(D^2 + 1)(D - 3)^2 y = 12e^{3x}(10x + 1)$$

Outline of solution: If f denotes the right-hand side then $(D - 3)^2 f = 0$. Hence any solution of the inhomogeneous equation must satisfy

$$(D^2 + 1)(D - 3)^4 y = 0$$

By the theorem established in Part (a) of the text the only possibility is

$$y = c_1 \cos x + c_2 \sin x + (c_3 + c_4 x + c_5 x^2 + c_6 x^3)e^{3x}$$

Discarding the terms that satisfy the homogeneous equation, and hence do not help, you are left with the trial solution

$$y = x^2(c_5 + c_6 x)e^{3x}$$

(b) This is difficult. Complete the solution without using the shift theorem. The latter condition is imposed because if you have the shift theorem you can solve the problem without any of the theory illustrated here.

————————————ANSWERS————————————

3. (a) x^{-9}, x^6 (c) $x^2 \cos(\ln x)$, $x^2 \sin(\ln x)$ (e) x^2, $x^2 \ln x$

   (b) x, $x \ln x$ (d) x^3, x^2 (f) $x \cos(7 \ln x)$, $x \sin(7 \ln x)$

4. (a) $.1 \cos(\ln x) + .3 \sin(\ln x)$ (b) $x \sin(\ln x) - 2x \cos(\ln x)$
                (c) $x \cos(\ln x) + x \sin(\ln x)$

5. $c_1 x + c_2 x \ln x + c_3 x (\ln x)^2$

8. Replace t by x in Chapter 8, Section 1, Example 5.

-----

3. Variation of parameters

In Chapter 10 we presented the method of Lagrange for solving the second-order equation $Ty = f$ by using two linearly independent solutions of the homogeneous equation $Ty = 0$. Lagrange extended his method to equations of order n, as described now. Let

$$(1) \qquad c_1 u_1 + c_2 u_2 + \cdots + c_n u_n$$

be a general solution of $Ty = 0$. Since it is a general solution, results of Section 1 show that the Wronskian does not vanish anywhere on I. Replacing the constants c_j by functions v_j, we seek a solution of $Ty = f$ in the form

$$(2) \qquad y = u_1 v_1 + u_2 v_2 + \cdots + u_n v_n.$$

To determine the v_j, differentiate (2) and set the sum of the terms involving the differentiated functions v_i' equal to 0. Differentiate again, using the fact that the first step leads to simplification, and again set the sum of terms involving the functions v_i' equal to 0. Continue until you have $n-1$ equations. The nth equation is obtained by inserting the simplified derivatives in the equation $Ty = f$. Using Cramer's rule, you get the formula

$$y(x) = \sum_{i=1}^{n} u_i(x) \int \frac{W_i(u_1, u_2, \cdots, u_n)}{W(u_1, u_2, \cdots, u_n)} f(x) \, dx$$

provided f is continuous. Here W is the Wronskian and W_i is the determinant obtained from W when the ith column is replaced by a column whose entries, from top to bottom, are $(0, 0, \cdots, 0, 1)$. For $n = 2$ the formula reduces, as it should, to the result obtained in Chapter 10, Section 2.

In the following problems we derive *Abel's identity*, which states that

$$W(x) = W(x_0) e^{-P(x)} \quad \text{where} \quad P(x) = \int_{x_0}^{x} p_{n-1}(t) \, dt.$$

Here x_0 is any point of I and it is assumed that $p_{n-1}(x)$ is continuous so we can form the integral. This formula gives $W(x)$ in terms of $W(x_0)$. For example, if the point $x_0 = 0$ or $x_0 = 1$ is on I it suffices to find $W(0)$ or $W(1)$ and the formula gives $W(x)$ without laborious computation.

EXAMPLE 1. Let us see how the Lagrange procedure works out for the third-order equation

$$(3) \qquad Ty = y''' + p_2 y'' + p_1 y' + p_0 y = f.$$

It is assumed that the coefficients p_j are bounded, that f is continuous, and that u_1, u_2, u_3 form a linearly independent set of solutions of the homogeneous equation.

Starting with

$$p_0: \qquad y = u_1 v_1 + u_2 v_2 + u_3 v_3$$

we differentiate and get

$$p_1: \qquad y' = u_1' v_1 + u_2' v_2 + u_3' v_3 + (u_1 v_1' + u_2 v_2' + u_3 v_3').$$

As a first condition on the v_j the expression in parentheses is set equal to zero. A second differentiation then gives

$$p_2: \qquad y'' = u_1'' v_1 + u_2'' v_2 + u_3'' v_3 + (u_1' v_1' + u_2' v_2' + u_3' v_3').$$

Again the expression in parenthesis is set equal to zero and a third differentiation gives

$$y''' = u_1''' v_1 + u_2''' v_2 + u_3''' v_3 + (u_1'' v_1' + u_2'' v_2' + u_3'' v_3').$$

The letters p_0, p_1, p_2 at the left mean that the first equation is to be multiplied by p_0, the second by p_1, and the third by p_2. These results are then added to the fourth equation. Using the definition of T as given in (3) we get

$$Ty = (Tu_1)v_1 + (Tu_2)v_2 + (Tu_3)v_3 + (u_1'' v_1' + u_2'' v_2' + u_3'' v_3').$$

By hypothesis $Tu_j = 0$ for $j = 1, 2, 3$. Hence the equations obtained by setting the above parentheses equal to zero, together with the equation $Ty = f$, yield the system

$$u_1 v_1' + u_2 v_2' + u_3 v_3' = 0$$
$$u_1' v_1' + u_2' v_2' + u_3' v_3' = 0$$
$$u_1'' v_1' + u_2'' v_2' + u_3'' v_3' = f.$$

This is a system of three equations in the three unknowns v_1', v_2', v_3'. Since the coefficient determinant is the Wronskian, which is not zero, we can solve by Cramer's rule. Namely,

$$v_1' \begin{vmatrix} u_1 & u_2 & u_3 \\ u_1' & u_2' & u_3' \\ u_1'' & u_2'' & u_3'' \end{vmatrix} = \begin{vmatrix} 0 & u_2 & u_3 \\ 0 & u_2' & u_3' \\ f & u_2'' & u_3'' \end{vmatrix} = \begin{vmatrix} 0 & u_2 & u_3 \\ 0 & u_2' & u_3' \\ 1 & u_2'' & u_3'' \end{vmatrix} f$$

where the third expression shows the connection with the general formula stated above. This equation and those for v_2' and v_3' are

$$W v_1' = \begin{vmatrix} u_2 & u_3 \\ u_2' & u_3' \end{vmatrix} f, \qquad W v_2' = \begin{vmatrix} u_3 & u_1 \\ u_3' & u_1' \end{vmatrix} f, \qquad W v_3' = \begin{vmatrix} u_1 & u_2 \\ u_1' & u_2' \end{vmatrix} f.$$

Division by the Wronskian W gives v_j' and an integration gives v_j.

PROBLEMS

1. Here and below, f is continuous. Consider

$$x^3 y''' - 3x^2 y'' + 6xy' - 6y = x^3 f(x), \qquad x > 0$$

(a) By trying $y = x^m$ find the solutions x, x^2, x^3 for the homogeneous equation. Verify that the Wronskian of these three functions is $2x^3$ and that $W(x) = cx^3$ agrees with the integral formula in the text.

(b) Taking $u_1 = x$, $u_2 = x^2$, $u_3 = x^3$ in Example 1 check that the result of that example gives

$$2v_1' = x f(x), \qquad v_2' = -f(x), \qquad 2x v_3' = f(x)$$

and hence the inhomogeneous equation has the solution

$$y = \frac{x}{2} \int x f(x)\, dx - x^2 \int f(x)\, dx + \frac{x^3}{2} \int \frac{f(x)}{x}\, dx$$

(c) Let $x^3 f(x) = x^m$ assuming, for simplicity, that $m \neq 1, 2, 3$. Check that the general formula produces the same result as the trial solution $y = Ax^m$.

2. Discuss as in Problem 1. Since the process incorporates a check, no answer is provided:

(a) $x^3 y''' - x^2 y'' = x^3 f(x)$ (c) $x^3 y''' + 2x^2 y'' - 2xy' = x^3 f(x)$

(b) $x^3 y''' + 3x^2 y'' = x^3 f(x)$ (d) $x^3 y''' + x^2 y'' - 2xy' + 2y = x^3 f(x)$

3. This takes a good deal of time. Apply the method of Example 1 to

$$x^3 y''' + xy' - y = 24x \ln x, \qquad x > 0$$

with $u_1 = x$, $u_2 = x \ln x$, $u_3 = x(\ln x)^2$; by Section 2, Problem 5 these functions satisfy the homogeneous equation. It will be found that

$$v_1 = 3(\ln x)^4 \qquad v_2 = -8(\ln x)^3 \qquad v_3 = 6(\ln x)^2 \qquad y = x(\ln x)^4$$

4. Let $Tu_j = 0$ and $W(x) \neq 0$ both hold for $x \in I$. Prove (a) for $n = 3$ and (b) for arbitrary n that the u_j form a basis. The interest of this result is that nothing is assumed about the coefficients p_j.

Hint for (a): Let $Ty = 0$. Since $W(x) \neq 0$ there are functions v_j such that

$$y^{(j)} = u_1^{(j)} v_1 + u_2^{(j)} v_2 + u_3^{(j)} v_3, \qquad j = 0, 1, 2$$

Cramer's rule gives a formula for the v_j which shows that they are differentiable. Referring to the equations in Example 1 deduce that

$$0 = u_1^{(j)} v_1' + u_2^{(j)} v_2' + u_3^{(j)} v_3', \qquad j = 0, 1, 2$$

Since the coefficient determinant is $W(x)$, which is not 0, the sole solution is $v_1' = v_2' = v_3' = 0$. Hence the v_j are constant and y is a linear combination of the u_j.

5. The expanded form of a 3 by 3 determinant is

$$\begin{vmatrix} u_1 & u_2 & u_3 \\ v_1 & v_2 & v_3 \\ w_1 & w_2 & w_3 \end{vmatrix} = \sum \pm u_i v_j w_k$$

where the sign $\pm$ is chosen according to a definite rule and where there are six terms in the sum. We assume that the elements of this determinant are differentiable functions. Using the equation

$$(uvw)' = u'vw + uv'w + uvw'$$

with $u = u_i, v = v_j, w = w_k$ explain why the derivative of the determinant is obtained by differentiating rows one at a time and adding the results. The same result, with essentially the same proof, holds for determinants of order n. In that case there are n factors in the product corresponding to uvw and $n!$ terms in the sum.

6. Let W be the Wronskian of three functions u, v, w each of which is three times differentiable on an interval I. Deduce from Problem 5 that dW/dx is obtained by differentiating the elements u'', v'', w'' of the bottom row and leaving the other rows unchanged.

7. If the three functions u, v, w satisfy the third-order equation

$$Ty = y''' + py'' + qy' + ry = 0$$

use first Problem 6 and then $Tu = Tv = Tw = 0$ to get

$$\frac{dW}{dx} = \begin{vmatrix} u & v & w \\ u' & v' & w' \\ -pu'' - qu' - ru & -pv'' - qv' - rv & -pw'' - qw' - rw \end{vmatrix}$$

Using row operations simplify the bottom row of the determinant and thus obtain Abel's equation, $dW/dx = -pW$.

8. If p is continuous, show that the result of Problem 7 yields Abel's equation for $n = 3$ as stated in the text.

9. Generalize Problems 6 and 7 to determinants of arbitrary order, n.

---

4. Summary

As in Section 1 let

$$Ty = y^{(n)} + p_{n-1}y^{(n-1)} + \cdots + p_1 y' + p_0 y$$

where the p_j are defined on an open interval I. Suppose further that

$$Tu_j = 0, \qquad x \in I, \qquad j = 1, 2, \cdots, n.$$

Several results of this chapter pertain to one or another of the following five statements:

A: $y = c_1 u_1 + c_2 u_2 + \cdots + c_n u_n$ is a general solution of $Ty = 0$ on I.

B: $W(x) \neq 0$ for every $x \in I$.

C: $W(x_0) \neq 0$ for at least one $x_0 \in I$.

D: The functions $u_1, u_2, \cdots, u_n$ are linearly independent on I.

E: $y = c_1 u_1 + c_2 u_2 + \cdots + c_n u_n$ is the complete solution of $Ty = 0$ on I.

Without some condition on the coefficients, ABCDE are not equivalent. That is, some of these statements can be true while others are false. For example, consider the equation

$$y'' + p(x)y' + q(x) = 0, \qquad p(0) = q(0) = 0.$$

If $p(x) = 0$ and $q(x) = -20/x^2$ for $x \neq 0$, the functions x^5 and $|x|^5$ satisfy the equation for all x. These functions are linearly independent on any open interval containing the origin, although their Wronskian is 0 for every value of x. Hence Condition D does not imply B or C even if $p_{n-1} = 0$.

As another illustration let $p(x) = -8/x$, $q(x) = 20/x^2$ for $x \neq 0$. In this case the equation has the solutions u, v where $u(x) = x^4$ for all x and $v(x) = x^4$ for $x \leq 0$ but $v(x) = x^5$ for $x \geq 0$. The Wronskian $W(x)$ of u and v is 0 for $x \leq 0$ and nonzero for $x > 0$. This shows that B and C may not be equivalent even in the case $n = 2$.

Despite these examples, the following statements are true:

(a) A and B are equivalent and either of them implies CDE.

(b) If p_{n-1} is bounded then ABC are equivalent.

(c) If all p_j are bounded then ABCD are equivalent.

(d) If all coefficients are continuous then ABCDE are equivalent.

To get B $\Rightarrow$ E in (a) you will need Section 3, Problem 4, and the implication C $\Rightarrow$ B in (b) depends on Chapter 10, Section 4, Theorem 1. Other aspects of (abc) follow with ease from results of this chapter.

To prove (d) we pass from continuity to boundedness as explained in Chapter 10, Section 4. By (a)

(1) $$A \Rightarrow B \Rightarrow C \Rightarrow D \Rightarrow E$$

and, except for one statement, everything needed for (d) has now been established. The exception is the implication E $\Rightarrow$ A. This depends on an existence theorem which asserts that the equation has n linearly independent solutions on I. Since the complete solution E must contain these linearly independent solutions, A follows. The existence theorem is established in Chapter 14. Together with (1), the statement E $\Rightarrow$ A closes the ring of implications and yields the desired equivalence. For example, if we want to show that D $\Rightarrow$ C we go around the ring as follows:

$$D \Rightarrow E \Rightarrow A \Rightarrow B \Rightarrow C.$$

What is provable should not be believed in science without proof.

—J. W. R. Dedekind

Chapter 12

SERIES SOLUTIONS I

EVEN WHEN they do not have elementary solutions, differential equations can often be solved by means of power series. The procedure is similar to the use of undetermined coefficients for polynomial solutions, except that there are infinitely many coefficients. Equating corresponding powers leads to a recurrence relation that expresses any given coefficient in terms of its predecessors. Thus they can be determined successively, one after another. In cases commonly encountered a formula for the general term can be surmised from the first few terms and proved by mathematical induction. The solution is then obtained as an infinite series.

This chapter and the next assume knowledge of power series to the extent developed in Chapter B.

1. Fundamental principles

Unless the contrary is specifically stated, the word "power series" means "power series about the point $c = 0$." In inequalities of the form $|x| < r$ we agree that r denotes a positive number. This convention is easy to remember, because if $r \leq 0$ the inequality $|x| < r$ is satisfied by no value of x.

The following theorem shows why power series are useful in the solution of differential equations:

THEOREM 1. *Let y be a formal power-series solution of the differential equation*

(1) $$p_2 y'' + p_1 y' + p_0 y = 0$$

where the p_j are polynomials. Suppose further that the series y has a positive radius of convergence r. Then the function $y(x)$ defined by the sum of the series satisfies the differential equation for $|x| < r$.

The proof is simple. For $|x| < r$ the series on the left of (1) converges to

$$p_2(x)y''(x) + p_1(x)y'(x) + p_0(x)y(x)$$

where the primes denote the ordinary derivative as in calculus. Since y is a formal solution by hypothesis, each coefficient in the series on the left of (1) is 0; this is what is meant by a "formal solution." Hence

$$p_2(x)y''(x) + p_1(x)y'(x) + p_0(x) = \sum_{j=0}^{\infty} 0x^j = 0, \qquad |x| < r$$

which is what we wanted to show.

The argument can be generalized. Let y be a formal power-series solution of

$$p_n y^{(n)} + p_{n-1} y^{(n-1)} + \cdots + p_1 y' + p_0 = f$$

where all p_j and f have convergent power-series expansions for $|x| < r$. Suppose the series giving the formal solution y also converges for $|x| < r$. Then the function $y(x)$ defined by the sum of this series satisfies the differential equation for $|x| < r$. The proof is virtually identical to that given above.

EXAMPLE 1.   To solve the equation

(2) $$y'' + y = xy'$$

we try the substitution

$$y = a_0 + a_1 x + a_2 x^2 + \cdots + a_j x^j + \cdots.$$

The general terms in the series for y, y', and y'' are respectively

$$a_j x^j, \qquad j a_j x^{j-1}, \qquad j(j-1) a_j x^{j-2}.$$

This yields the following table for the expressions that occur in the differential equation:

Expression:	y''	y	xy'
General term:	$j(j-1)a_j x^{j-2}$	$a_j x^j$	$j a_j x^j$
Coefficient of x^n:	$(n+2)(n+1)a_{n+2}$	a_n	$n a_n$

The second row gives the general term in the expansion of the entries in the first row and the third row is the coefficient of x^n in these expansions. It is obtained by the choices $j = n+2$, $j = n$, $j = n$ in the respective entries of the second row.

The equation $y + y'' = xy'$ holds in the sense of equality for formal power series if, and only if, the coefficients of x^n on both sides agree. Equating coefficients of x^n yields the *recurrence formula*

$$(n+2)(n+1)a_{n+2} + a_n = n a_n.$$

The formula allows us to determine all the a_j, starting from any prescribed values a_0 and a_1. To see how this is done, solve for a_{n+2}:

(3) $$a_{n+2} = \frac{n-1}{(n+2)(n+1)} a_n.$$

Setting $n = 0,\ 2,\ 4\ \cdots$ we get, in succession,

$$a_2 = -\frac{1}{2}a_0, \qquad a_4 = \frac{1}{3\cdot 4}a_2 = -\frac{1}{4!}a_0, \qquad a_6 = \frac{3}{6\cdot 5}a_4 = -\frac{3}{6!}a_0,$$

and so on. Thus the even powers of x in the formal series give

$$(4) \qquad a_0\left(1 - \frac{1}{2!}x^2 - \frac{1}{4!}x^4 - \frac{1\cdot 3}{6!}x^6 - \frac{1\cdot 3\cdot 5}{8!}x^8 - \frac{1\cdot 3\cdot 5\cdot 7}{10!}x^{10} - \cdots\right).$$

The way the above series was constructed shows that it is a formal solution. We now investigate the convergence.

Since the series has only even powers of x, the ratio of two consecutive terms is

$$\left|\frac{a_{2(n+1)}x^{2(n+1)}}{a_{2n}x^{2n}}\right| = \left(\frac{|a_{2n+2}|}{|a_{2n}|}\right)|x^2|.$$

Using (3) with n replaced by $2n$ we get

$$\frac{a_{2n+2}}{a_{2n}} = \frac{2n - 1}{(2n + 2)(2n + 1)}.$$

This tends to 0 as $n \to \infty$. Hence the above ratio tends to 0 and the series converges for all x. By Theorem 1, its sum satisfies the differential equation.

Note the way we avoided use of the general term of the series. It would be inefficient to work with the general term because the formula (3) already gives exactly what is needed. In fact, if you apply the ratio test directly to the series, you will find yourself going backward and getting (3) again.

We now use (3) when n is odd. Setting $n = 1$ in (3) yields $a_3 = 0$ and the formula then gives $a_5 = 0$, $a_7 = 0$, and so on. Thus the odd powers of x in the formal series reduce to the single term $a_1 x$, all other terms being 0. It is easily checked that $y = a_1 x$ satisfies the differential equation.

If $u(x)$ denotes the function multiplying a_0 in (4) and if $v(x) = x$, our solution is

$$(5) \qquad\qquad y = a_0 u(x) + a_1 v(x).$$

The Wronskian of u and v at 0 is 1 and hence these functions are linearly independent. It follows from Chapter 10 that (5) is both the general solution and the complete solution. The latter statement means that every solution can be represented in the form (5), and hence *every solution has a power-series expansion*. This fact was by no means obvious when we started our investigation.

For many purposes it is advantageous to define $a_j = 0$ for $j < 0$, while keeping the convention that $0x^j = 0$ for all x. Then

$$\sum_{j=0}^{\infty} a_j x^j = \sum_{j=-\infty}^{\infty} a_j x^j, \qquad a_{-1} = 0,\ a_{-2} = 0,\ \cdots.$$

As illustrated in the following example, this interpretation is to be used whenever you encounter a term a_j with negative j.

EXAMPLE 2. Airy's equation. The equation $y'' = xy$ was first analyzed by the English astronomer Sir George Airy and is named after him. We shall find solutions u and v with initial values

$$u(0) = 1, \qquad u'(0) = 0; \qquad v(0) = 0, \qquad v'(0) = 1.$$

By superposition, the solution satisfying arbitrary initial values is then

$$y = y_0 u(x) + y_1 v(x) \quad \text{where} \quad y_0 = y(0), \ y_1 = y'(0).$$

We begin with the table

Expression:	y	xy	y''
General term:	$a_j x^j$	$a_j x^{j+1}$	$j(j-1)a_j x^{j-2}$
Coefficient of x^n:	——	a_{n-1}	$(n+2)(n+1)a_{n+2}$

The equation $y'' = xy$ then yields the recurrence formula

(6) $$(n+2)(n+1)a_{n+2} = a_{n-1}.$$

If $y = u$ the initial values give $a_0 = 1$, $a_1 = 0$. Setting $n = 0$ in (6) we get

$$2a_2 = a_{-1} = 0$$

since $a_j = 0$ for negative j as explained above. The initial conditions give $a_1 = 0$. Each use of (6) increases the subscript on a by 3. Consequently

$$0 = a_1 = a_4 = a_7 = \cdots, \qquad 0 = a_2 = a_5 = a_8 = \cdots.$$

Thus the only nonzero a_j are those in which $j = 3n$, a multiple of 3. Setting $n + 2 = 3, \ 6, \ 9, \ \cdots$, we get $n - 1 = 0, \ 3, \ \cdots$ and hence

$$3 \cdot 2a_3 = a_0, \qquad 6 \cdot 5a_6 = a_3, \qquad 9 \cdot 8a_9 = a_6, \qquad \cdots.$$

Successive determination of the a_j starting with $a_0 = 1$ yields

$$a_0 = 1, \qquad a_3 = \frac{1}{3 \cdot 2}, \qquad a_6 = \frac{1}{6 \cdot 5 \cdot 3 \cdot 2}, \qquad \cdots$$

Upon recalling that the other $a_j = 0$ we get the series

$$u(x) = 1 + \frac{x^3}{3 \cdot 2} + \frac{x^6}{6 \cdot 5 \cdot 3 \cdot 2} + \frac{x^9}{9 \cdot 8 \cdot 6 \cdot 5 \cdot 3 \cdot 2} + \cdots.$$

For the solution v we have $a_0 = 0$, $a_1 = 1$ and a similar process yields

$$v(x) = x + \frac{x^4}{4 \cdot 3} + \frac{x^7}{7 \cdot 6 \cdot 4 \cdot 3} + \frac{x^{10}}{10 \cdot 9 \cdot 7 \cdot 6 \cdot 4 \cdot 3} + \cdots.$$

It is easily checked by (6) that both series converge for all x; see Problem 5. Hence, both satisfy the differential equation.

The following example differs from the preceding in that the equation is of the first order and has an arbitrary constant m as one of its coefficients. That the procedure works for equations of arbitrary order was indicated in the remarks following Theorem 1.

EXAMPLE 3.   Binomial theorem. We shall find a series solution for the initial-value problem

$$(1 + x)y' = my, \qquad y(0) = 1.$$

The table now is

Expression:	y	y'	xy'
General term:	$a_j x^j$	$j a_j x^{j-1}$	$j a_j x^j$
Coefficient of x^n:	a_n	$(n+1)a_{n+1}$	$n a_n$

Since $y' = my - xy'$ we get the recurrence formula

$$(7) \qquad (n+1)a_{n+1} = ma_n - na_n \quad \text{or} \quad a_{n+1} = \left(\frac{m-n}{n+1} \right) a_n.$$

Successive use of this formula, starting with $a_0 = 1$, yields

$$a_0 = 1, \qquad a_1 = m, \qquad a_2 = \frac{m(m-1)}{2!}, \qquad a_3 = \frac{m(m-1)(m-2)}{3!},$$

and so on, and the resulting series is the binomial series. By (7) the ratio of two successive terms satisfies

$$\left| \frac{a_{n+1}x^{n+1}}{a_n x^n} \right| = |x| \left| \frac{a_{n+1}}{a_n} \right| = |x| \left| \frac{m-n}{n+1} \right|.$$

As $n \to \infty$ the limit is $|x|$. Hence the series converges for $|x| < 1$, diverges for $|x| > 1$, and the radius of convergence is 1.

By separation of variables the solution is

$$y = (1 + x)^m.$$

Equating this to the series solution found above we get the *binomial theorem*,

$$(1 + x)^m = 1 + mx + \frac{m(m-1)}{2!}x^2 + \frac{m(m-1)(m-2)}{3!}x^3 + \cdots, \qquad |x| < 1.$$

The interest of the method is that it yields the binomial theorem for all real exponents m.

If m is a positive integer in Example 3 the series terminates at x^m; that is, all terms $a_j x^j$ for $j > m$ are 0. Hence, although the series solution converges only for $|x| < 1$ in general, when m is a positive integer it reduces to a polynomial and converges for all x. In the next section we discuss a second-order equation that shows similar behavior.

Having presented a number of examples, we now return to the general case. If $p_2(0) \neq 0$ in Theorem 1, one can divide the equation by $p_2(x)$ for small $|x|$ and reduce the equation to the standard form in which the coefficient of y'' is 1. The following theorem holds:

THEOREM 2.   (a) *Suppose the equation $y'' + p(x)y' + q(x)y = 0$ has two power-series solutions $u(x)$ and $v(x)$ for $|x| < r_1$. Suppose further that the Wronskian W of u and v satisfies $W(0) \neq 0$. Then p and q have power-series expansions on some interval $|x| < r$.*

(b) *Conversely, if p and q have power-series expansions for $|x| < r$, there exist solutions u and v with $W(0) \neq 0$ that have power-series expansions at least for $|x| < r$.*

Whenever one says that a function "has a power-series expansion," one means that the series converges on some interval and that the function is given by its sum. For example the hypothesis of (a) means that $u(x)$ and $v(x)$ are the sums of convergent power series at least for $|x| < r_1$ and that they both satisfy the equation on this interval. The conclusion says that $p(x)$ and $q(x)$ are sums of convergent power series on some possibly smaller interval $|x| < r$.

For proof of Part (a), write the equations satisfied by u and v in the form

$$pu' + qu = -u'', \qquad pv' + qv = -v''.$$

This can be regarded as a linear system in the two unknowns p, q. Solving by Cramer's rule we get

$$p = \frac{u''v - v''u}{W}, \qquad q = \frac{u'v'' - u''v'}{W}$$

where W is the Wronskian. Since $W(0) \neq 0$ these expressions have convergent power-series expansions on some interval $|x| < r$, where $r > 0$. This completes the proof.

A similar result holds for the linear equation of order n. Given n solutions u_i for such an equation, we can write

$$p_0 u_i + p_1 u_i' + \cdots + p_{n-1} u_i^{(n-1)} = -u_i^{(n)}, \qquad i = 1, 2, \cdots, n.$$

and an analog of Part (a) follows at once.

Proof of Part (b) is harder and is postponed. We mention, however, that this part also extends to equations of order n.

PROBLEMS

1P. In this and subsequent problems assume a power series for y with coefficients a_j as in the text. Here we are going to solve the equation $y'' = xy'$ by power series.

(a) Make a table showing the general term for y', y'', xy'. Thus get the coefficient of x^n in the series for y'' and xy' and from this deduce the recursion formula

$$a_{n+2} = \frac{n}{(n+2)(n+1)} a_n$$

(b) Taking first $n = 0, 2, 4, \cdots$ and then $n = 1, 3, 5, \cdots$ obtain

$$y = a_0 + a_1 \left(x + \frac{x^3}{3!} + \frac{3 \cdot 1}{5!} x^5 + \frac{5 \cdot 3 \cdot 1}{7!} x^7 + \cdots \right)$$

The solution is valid for all x. Why?

2. Solve by series and express your answer in terms of elementary functions:

(a) $y' = y$ (b) $y' = -2y$ (c) $y'' = y$ (d) $y'' = -y$ (e) $y' = 2xy$

3. If a_j are the coefficients of a power-series solution, obtain the recursion formula shown at the right of the equation:

(a) $x(1-x)y' = y$ $(n-1)a_n = (n-1)a_{n-1}$

(b) $(x^2 + 1)y'' = 2y$ $(n+2)a_{n+2} = (2-n)a_n$

(c) $(x^2 - 1)y'' + 4xy' + 2y = 0$ $a_{n+2} = a_n$

(d) $y'' = xy' + y$ $(n+2)a_{n+2} = a_n$

(e) $(1+x)y' = 3y$ $(n+1)a_{n+1} = (3-n)a_n$

(f) $y'' + y = xy'$ $(n+2)(n+1)a_{n+2} = (n-1)a_n$

(g) $(x^2 + x)y' = (2x+1)y$ $(n-1)a_n = (3-n)a_{n-1}$

(h) $xy' = (2x^2 + 1)y$ $(n-1)a_n = 2a_{n-2}$

(i) $y'' = x^2 y$ $(n+2)(n+1)a_{n+2} = a_{n-2}$

4. (abcdefghi) Obtain series solutions for the equations in Problem 3.

5. Let m be a positive integer. An *m-step recursion formula* gives a_{j+m} in terms of a_j, starting from some definite value a_k. Usually one starts at a_0 or a_1.

(a) Write the following recursion relations in terms of $j = n - 1$, n, $n - 2$, as the case may be, and thus convince yourself that these are 1-step, 2-step and 4-step formulas respectively:

$$na_n = a_{n-1} \qquad (n^2 + 1)a_{n+2} = (2n)^2 a_n \qquad 16na_{n+2} = (n+1)a_{n-2}$$

(b) Starting with a_k, the m-step formula gives a_{k+m}, a_{k+2m}, and so on. Hence it determines the coefficients in the series

$$a_k x^k + a_{k+m} x^{k+m} + a_{k+2m} x^{k+2m} + \cdots + a_{k+nm} x^{k+nm} + \cdots$$

Show that the limit relation below yields the formula on the right for the radius of convergence of this series:

$$(j = k + nm), \qquad \lim_{j \to \infty} \left| \frac{a_{j+m}}{a_j} \right| = L, \qquad r = \frac{1}{\sqrt[m]{L}}$$

The equation $(j = k+nm)$ means that j has the form $k+nm$ as $n \to \infty$. This avoids coefficients a_j that are not in the above series and may be zero.

(c) How should you interpret the cases $L = 0$, $L = \infty$?

(d) Check that the formula of Part (b) yields $r = \infty$, $1/2$, 2 in the three examples of Part (a) unless the series terminates (in which case the limit condition of Part (b) is meaningless and $r = \infty$).

6. (abcdefghi) Assuming that the starting value is chosen so that the series does not terminate, use the formula of Problem 5 to find the radius of convergence of the power-series solutions in Problem 3. Also verify agreement with the result of Theorem 2(b) in those cases in which this theorem applies.

7. Problems (abcd) are special cases of the equations of Legendre, Hermite, Chebychev, and Laguerre respectively. Give a recursion formula for the coefficients of a series solution. Taking $(a_0, a_1) = (1, 0)$ or $(0, 1)$, as the case may be, use your formula to obtain a polynomial solution, and check the latter by substitution into the differential equation. (Three things to do):

(a) $(1 - x^2)y'' + 12y = 2xy'$ (c) $(1 - x^2)y'' + 9y = xy'$

(b) $y'' + 8y = 2xy'$ (d) $xy'' + 2y = (x - 1)y'$

8. (a) Solve the equation $x^2y'' - 2xy' + 2y = 0$ by trying $y = x^s$ where s is constant and thus show that the hypothesis "$W(0) \neq 0$" in Theorem 2(a) cannot be replaced by "u and v are linearly independent for $|x| < r_1$."

(b) Under the hypothesis of Theorem 2(b) does every solution y necessarily have a power-series expansion for $|x| < r$? Why or why not?

9. The recursion formula $(n+1)a_{n+1} = (m-n)a_n$ with the initial condition $a_0 = 1$ was obtained for the binomial coefficients a_n in Example 3. Prove by mathematical induction that

$$a_n = \frac{m(m-1)(m-2)\cdots(m-n+1)}{n!}, \qquad n \geq 1$$

Although the matter will not be emphasized, a similar procedure applies to nearly all the examples and problems in this chapter. Only when the recursion formula is too complicated to guess the general term a_n does the method of induction fail. In such cases one gets a few terms only, relying on general theorems for assurance that the full series would converege.

───────────────────────────────ANSWERS───────────────────────────────

1. By Theorem 1 and the ratio test.
2. Partial answer: (a) $a_0 e^x$ (b) $a_0 e^{-2x}$ (c) $a_0 \cosh x + a_1 \sinh x$
   (d) $a_0 \cos x + a_1 \sin x$ (e) $a_0 e^{x^2}$
4. (a) $a_1 x/(1-x)$

   (b) $a_0(1+x^2) + a_1\left(x + \frac{1}{3}x^3 - \frac{1}{3\cdot 5}x^5 + \frac{1}{5\cdot 7}x^7 - \frac{1}{7\cdot 9}x^9 + \cdots\right)$

   (c) $(a_0 + a_1 x)/(1-x^2)$

   (d) $a_0 e^{x^2/2} + a_1\left(x + \frac{x^3}{3} + \frac{x^5}{3\cdot 5} + \frac{x^7}{3\cdot 5\cdot 7} + \cdots\right)$ (e) $a_0(1+x)^3$

   (f) $a_0\left(1 - \frac{x^2}{2!} - \frac{1}{4!}x^4 - \frac{3\cdot 1}{6!}x^6 - \frac{5\cdot 3\cdot 1}{8!}x^8 - \cdots\right) + a_1 x$

   (g) $a_1(x+x^2)$ (h) $a_1 x e^{x^2}$

   (i) $a_0\left(1 + \frac{x^4}{3\cdot 4} + \frac{x^8}{3\cdot 4\cdot 7\cdot 8} + \cdots\right) + a_1\left(x + \frac{x^5}{4\cdot 5} + \frac{x^9}{4\cdot 5\cdot 8\cdot 9} + \cdots\right)$

5. (c) If $L = 0$ it converges for all x, if $L = \infty$ it converges only for $x = 0$.
6. (a) 1 (b) 1 (c) 1 (d) ∞ (e) terminates (f) ∞ (g) terminates (h) ∞ (i) ∞
7. (a) $(n+2)(n+1)a_{n+2} = a_n(n-3)(n+4)$, $x - (5/3)x^3$
   (b) $(n+2)(n+1)a_{n+2} = 2(n-4)a_n$, $1 - 4x^2 + (4/3)x^4$
   (c) $(n+2)(n+1)a_{n+2} = (n+3)(n-3)a_n$, $x - (4/3)x^3$
   (d) $(n+1)^2 a_{n+1} = (n-2)a_n$, $1 - 2x + (1/2)x^2$

8. (a) Partial answer: x, x^2. Note however that if p and q are bounded, linear independence on an interval implies $W(x) \neq 0$ at every point of the interval, hence it gives the needed condition $W(0) \neq 0$.

   (b) Yes, because every solution can be written in the form $y = c_1 u + c_2 v$.

---

2. Four special equations

Legendre's equation

$$(1 - x^2)y'' - 2xy' + m(m+1)y = 0$$

was introduced in Chapter 10; it has many applications in mathematics and mathematical physics. To solve this equation by power series we make the following table:

y''	$-x^2 y''$	$-2xy'$	$m(m+1)y$
$j(j-1)a_j x^{j-2}$	$-j(j-1)a_j x^j$	$-2ja_j x^j$	$m(m+1)a_j x^j$
$(n+2)(n+1)a_{n+2}$	$-n(n-1)a_n$	$-2na_n$	$m(m+1)a_n$

The differential equation says that the sum of the expressions in the first row is 0. Hence the sum of those in the third row must also be 0. Moving all terms in a_n to the right yields

$$(n+2)(n+1)a_{n+2} = -a_n(m^2 + m - n^2 - n)$$

or equivalently,

(1) $$(n+2)(n+1)a_{n+2} = -a_n(m-n)(m+n+1).$$

The coefficients for even n are determined by a_0 as follows:

$$a_2 = -a_0 \frac{m(m+1)}{2 \cdot 1}, \quad a_4 = -a_2 \frac{(m-2)(m+3)}{4 \cdot 3}, \quad a_6 = -a_4 \frac{(m-4)(m+5)}{6 \cdot 5}$$

and so on. It is left for you to determine the a_{2j} from this and to carry out a similar calculation for odd n starting with $n = 1$. The result is

$$y = a_0 \left(1 - \frac{m(m+1)}{2!}x^2 + \frac{m(m-2)(m+1)(m+3)}{4!}x^4 - \cdots\right)$$

$$+ a_1 \left(x - \frac{(m-1)(m+2)}{3!}x^3 + \frac{(m-1)(m-3)(m+2)(m+4)}{5!}x^5 + \cdots\right).$$

Since (1) gives

$$\lim_{n\to\infty} \left| \frac{a_{n+2}}{a_n} \right| = \lim_{n\to\infty} \left| \frac{(m-n)(m+n+1)}{(n+2)(n+1)} \right| = 1$$

the radius of convergence, in general, is 1; see Problem 5 of the last section. However, if m is a positive integer or 0, the choice $n = m$ in (1) gives $a_{m+2} = 0$, and the same holds for subscripts $m+4$, $m+6$ and so on. Thus the series terminates. Hence the coefficient of a_0 or a_1 reduces to a polynomial of degree m according as m is even or odd.

Choosing a_0 or a_1 for $m = 0, 1, 2, \cdots$ in such a way that the value of the polynomial is 1 when $x = 1$, we get a sequence of polynomials P_m as follows:

$$1, \qquad x, \qquad \frac{3}{2}x^2 - \frac{1}{2}, \qquad \frac{5}{2}x^3 - \frac{3}{2}x, \qquad \frac{35}{8}x^4 - \frac{15}{4}x^2 + \frac{3}{8}, \qquad \cdots.$$

These are the *Legendre polynomials*. For any integer $m \geq 0$ the Legendre polynomial $P_m(x)$ is the unique polynomial of degree m that satisfies Legendre's equation with parameter m and also satisfies $P_m(1) = 1$.

In the following problems we discuss three other equations that also lead to important classes of polynomials.

PROBLEMS

1. This pertains to *Hermite's equation* $y'' - 2xy' + 2my = 0$.

(a) Obtain a recursion formula for a power-series solution, find the corresponding radius of convergence, and verify agreement with Theorem 2(b) of Section 2. Taking $y(0) = 1$, $y'(0) = 0$, or $y(0) = 0$, $y'(0) = 1$, get

(b) $$y = 1 - \frac{2m}{2!}x^2 + \frac{2^2 m(m-2)}{4!}x^4 - \frac{2^3 m(m-2)(m-4)}{6!}x^6 + \cdots$$

(c) $$y = x - \frac{2(m-1)}{3!}x^3 + \frac{2^2(m-1)(m-3)}{5!}x^5 - \cdots$$

2. When m is a nonnegative integer the series in Problem 1(b) or 1(c) terminates. The resulting polynomials, multiplied by a constant to give 2^m as the leading coefficient, are the *Hermite polynomials* $H_m(x)$. Verify that the first five Hermite polynomials are

$$1, \qquad 2x, \qquad 4x^2 - 2, \qquad 8x^3 - 12x, \qquad 16x^4 - 48x^2 + 12$$

3. This pertains to *Chebychev's equation* $(1 - x^2)y'' - xy' + m^2 y = 0$.

Part (a) is the same as in Problem 1. For Parts (bc) obtain the solutions

(b) $y = 1 - \dfrac{m^2}{2!}x^2 + \dfrac{m^2(m^2 - 2^2)}{4!}x^4 - \dfrac{m^2(m^2 - 2^2)(m^2 - 4^2)}{6!}x^6 + \cdots$

(c) $y = x - \dfrac{m^2 - 1^2}{3!}x^3 + \dfrac{(m^2 - 1^2)(m^2 - 3^2)}{5!}x^5 - \cdots$

4. When m is a nonnegative integer the series in Problem 3(b) or 3(c) terminates. The resulting polynomials, multiplied by a constant to give 2^{m-1} as the leading coefficient, are the *Chebychev polynomials* $T_m(x)$ for $m \geq 1$, and by convention $T_m(x) = 1$ when $m = 0$. Verify that the first five are

$$1, \qquad x, \qquad 2x^2 - 1, \qquad 4x^3 - 3x, \qquad 8x^4 - 8x^2 + 1$$

5. This pertains to *Laguerre's equation* $xy'' + (1 - x)y' + my = 0$.

(a) Obtain a recursion formula for a series solution, determine the corresponding radius of convergence, and discuss the applicability of Section 2, Theorem 2.

(b) Obtain the following solution:

$$y = 1 - mx - \frac{m(1 - m)}{(2!)^2}x^2 - \frac{m(1 - m)(2 - m)}{(3!)^2}x^3 - \cdots$$

6. If m is a nonnegative integer in Problem 5 the series terminates. The resulting polynomials, multiplied by a constant to give $y(0) = m!$, are the *Laguerre polynomials* $L_m(x)$. Verify that the first four are

$$1, \qquad 1 - x, \qquad 2 - 4x + x^2, \qquad 6 - 18x + 9x^2 - x^3$$

7. It can be shown that the Legendre polynomials $P_m(x)$ satisfy

$$(1 - 2xh + h^2)^{-1/2} = P_0(x) + P_1(x)h + P_2(x)h^2 + \cdots + P_m(x)h^m + \cdots$$

The function on the left is called the *generating function* of the sequence $\{P_m(x)\}$. Verify this equation through terms in h^4.

Hint: Taking $Q = 2hx - h^2$, expand $(1 - Q)^{-1/2}$ by the binomial theorem and collect powers of h.

8. For $m = 0, 1, 2, 3$ verify *Rodrigues' formula*

$$2^m m! P_m(x) = \frac{d^m}{dx^m}(x^2 - 1)^m$$

9. (a) Check that the equation $z'' + (2m+1-x^2)z = 0$ leads to a three-term recursion formula for the coefficients of a power-series solution, and that the determination of z by this method presents a difficult task.

(b) Show however that the substitution $z = ye^{-x^2/2}$ leads to the Hermite equation for y, and from this z can be written by inspection.

10. The generating function for the Hermite polynomials is

$$e^{2xt-t^2} = H_0(x) + H_1(x)t + \frac{H_2(x)}{2!}t^2 + \frac{H_3(x)}{3!}t^3 + \cdots$$

Verify this equation through terms in t^3 in two ways:

(a) Expand e^u, set $u = 2xt - t^2$, and collect terms.

(b) Multiply the series for e^{2xt} by that for e^{-t^2}.

11. Verify the following Rodrigues-type equation for $m = 1, 2, 3$:

$$H_m(x) = (-1)^m e^{x^2} \frac{d^m}{dx^m} e^{-x^2}$$

12. (a) The Chebychev polynomials satisfy $T_m(\cos\theta) = \cos m\theta$. Verify this for $m = 0, 1, 2, 3$.

(b) Also verify the following for $m = 1, 2, 3$:

$$2T_m(x) = (x + \sqrt{x^2 - 1})^m + (x - \sqrt{x^2 - 1})^m$$

(c) For arbitrary m, derive the result of Part (a) from that of Part (b). Hint:

$$2\cos m\theta = 2\text{Re } e^{im\theta} = (\cos\theta + i\sin\theta)^m + (\cos\theta - i\sin\theta)^m$$

--------------------------------ANSWERS--------------------------------

1. (a) $(n+2)(n+1)a_{n+2} = 2(n-m)a_n$, $r = \infty$
3. (a) $(n+2)(n+1)a_{n+2} = (n^2 - m^2)a_n$, $r = 1$
5. (a) $(n+1)^2 a_{n+1} = (n-m)a_n$, $r = \infty$. Theorem 2(b) does not apply but Theorem 2(a) shows there cannot be a second power-series solution making $W(0) \neq 0$ for the pair.

3. Convergence

The theory of convergence has two aspects. The first involves only a few elementary facts about limits, while the second depends on deeper properties of the real number system. Examples of the first aspect given by the following, which are immediate consequences of the definitions:

(i) The general term of a convergent series tends to 0.

(ii) The operation denoted by $\sum$ is linear.

Item (ii) means that

$$\sum_{j=0}^{\infty} ca_j = c \sum_{j=0}^{\infty} a_j, \qquad \sum_{j=0}^{\infty}(a_j + b_j) = \sum_{j=0}^{\infty} a_j + \sum_{j=0}^{\infty} b_j$$

provided the series on the right-hand side of these equations are convergent.

The second aspect of the theory of convergence is most easily formulated in terms of sequences that are bounded and increasing. A sequence $\{s_n\}$ is *bounded* if $|s_n| \le M$ for some constant M and *increasing* if $s_n \ge s_{n-1}$. Both conditions are required for $n = 1, 2, 3, \cdots$. Note that the word "increasing" is used in a weak sense. If $s_n > s_{n-1}$ for all n the sequence is said to be *strictly increasing*.

As an illustration, consider the decimal representation for an irrational number such as π. The representation gives π as limit of the sequence

$$3, \quad 3.1, \quad 3.14, \quad 3.141, \quad 3.1415, \quad 3.14159, \quad \cdots.$$

Each term of the above sequence is between 3 and 4, and any term beyond the first is at least equal to its predecessor. Hence the sequence is bounded and increasing. The fact that the sequence actually defines a real number is a consequence of the following:

COMPLETENESS AXIOM. Every bounded increasing sequence has a limit.

This is the property of the real number system that underlies the deeper aspects of the theory of convergence.

We shall use the completeness axiom to establish the comparison test for real series, which is as follows:

(iii) If $0 \le a_j \le b_j$ and $\displaystyle\sum_{j=0}^{\infty} b_j$ converges, then $\displaystyle\sum_{j=0}^{\infty} a_j$ converges.

The proof follows from

$$s_n = \sum_{j=0}^{n} a_j \le \sum_{j=0}^{n} b_j \le \sum_{j=0}^{\infty} b_j.$$

The condition $a_j \geq 0$ shows that $\{s_n\}$ is increasing and the above calculation shows that $\{s_n\}$ is bounded. Hence s_n has a limit by the completeness axiom.

If the first of the following series converges the second is said to be *absolutely convergent*:

$$\sum_{j=0}^{\infty} |a_j|, \quad \sum_{j=0}^{\infty} a_j.$$

We shall establish:

(iv) An absolutely convergent series is convergent.

For proof, let $a_j = u_j + iv_j$. Since

$$0 \leq u_j + |a_j| \leq 2|a_j|$$

the comparison test shows that the series with general term $u_j + |a_j|$ converges. Hence the series with general term

$$u_j = (u_j + |a_j|) - |a_j|$$

also converges. A similar argument applies to v_j and this completes the proof. Items (iii) and (iv) together yield the following comparison test for complex series:

(v) If $|a_j| \leq b_j$ and $\displaystyle\sum_{j=0}^{\infty} b_j$ converges, then $\displaystyle\sum_{j=0}^{\infty} a_j$ converges.

Surprising as it may seem, item (v) provides a basis for the general existence theory of differential equations and systems, both linear and nonlinear. Its usefulness is not restricted to linear equations that have power-series solutions, though it will be applied only to such equations in this chapter.

(a) The radius of convergence

We allow complex coefficients a_j and we also replace the variable x by a complex variable z. This increase in generality gives added insight with no increase in complication.

The following theorem is basic in the sequel:

THEOREM 1. *Suppose the power series*

$$\sum_{j=0}^{\infty} a_j z^j$$

converges at some real or complex value $z_0 \neq 0$. *Then there is a constant* M *such that*

$$|a_j| \leq \frac{M}{|z_0|^j}, \qquad j = 0, 1, 2, 3, \cdots.$$

For proof, note that the general term $a_j z_0^j$ tends to 0. Hence there is an integer N such that

$$|a_j z_0^j| \leq 1, \qquad j > N.$$

This gives $|a_j| \leq 1/|z_0|^j$ for $j > N$. To take care of the cases $j \leq N$, as well as those for $j > N$, let M be the largest of the numbers

$$1, \qquad |a_0|, \qquad |a_1||z_0|, \qquad \cdots, \qquad |a_N||z_0|^N.$$

Then $|a_j| \leq M/|z_0|^j$ for all j and this completes the proof.

Under the hypothesis of Theorem 1, if $|z| < |z_0|$ we have

$$|a_j z^j| \leq M \left| \frac{z}{z_0} \right|^j$$

and the series converges absolutely by comparison with a geometric series. Hence *if a power series converges at some value $z_0 \neq 0$ then it converges for all z such that $|z| < |z_0|$.* The locus $|z| = |z_0|$ represents a circle in the complex plane and the locus $|z| < |z_0|$ represents the disk bounded by this circle. We shall now explain why a power series has a disk of convergence in the complex plane, analogous to the interval of convergence discussed in the foregoing pages.

If the series converges only for $z = 0$, or if it converges for all z, there is nothing to do. Suppose, however, that the series converges for some value $z_0 \neq 0$ and diverges for some other value z_1. We let r denote the *largest value* such that the series converges for all $|z| < r$. Then r is at least equal to $|z_0|$, since it was seen above that convergence at z_0 implies convergence for $|z| < |z_0|$. We also have $r \leq |z_1|$, since the series diverges at z_1. These two remarks together show that $0 < r < \infty$.

The series converges for $|z| < r$ by the definition of r, but it cannot converge at any value z_2 with $|z_2| > r$. If it did, it would converge for $|z| < |z_2|$, and the presumably largest value r could be replaced by the larger value $|z_2|$. Thus the series converges for $|z| < r$ and diverges for $|z| > r$, and that is what we wanted to show. The number r is called the *radius of convergence*, with the convention that $r = 0$ if the series converges for $z = 0$ only and $r = \infty$ if the series converges for all z.

The only gap in the argument is the proof that a largest value r exists. Without belaboring the matter let us indicate how existence of r can be established by use of decimal notation. To illustrate the main ideas suppose the radius of convergence will finally turn out to be π. First we pick the largest integer, r_0, such that the series converges for $|z| < r_0$. In the case we are considering this integer will be 3. Next we pick the largest number r_1 among

$$3.0, \ 3.1, \ 3.2, \ 3.3, \ 3.4, \ 3.5, \ 3.6, \ 3.7, \ 3.8, \ 3.9$$

such that the series converges for $|z| < r_1$. In our case this number will be 3.1. Next we consider the largest number r_2 from

$$3.10, \ 3.11, \ 3.12, \ 3.13, \ 3.14, \ 3.15, \ 3.16, \ 3.17, \ 3.18, \ 3.19$$

such that the series converges for $|z| < r_2$. In our case this number will be 3.14. The process continues in an obvious manner and gives r as an infinite decimal.

The concept of "disk of convergence" sheds light on the behavior of power series. For example, the series

$$\frac{1}{1 + x^2} = 1 - x^2 + x^4 - x^6 + \cdots$$

converges for $|x| < 1$ by the ratio test and it diverges for $|x| \geq 1$ because the general term does not tend to 0. When $1/(1 + x^2)$ is considered as a function of the *real* variable x it has derivatives of all orders at every value of x, and there seems to be no reason why the series should diverge when $|x| \geq 1$. But if x is replaced by a *complex* variable z the divergence is explained by the fact that the denominator vanishes at $z = i$. The disk of convergence cannot contain the point $z = i$, hence the radius of convergence cannot exceed 1.

(b) Proof of convergence

We consider the equation

$$y'' + p(x)y' + q(x)y = 0,$$

taking x as a real variable because we do not want to discuss the derivative with respect to a complex variable z. Our goal is to prove Section 1, Theorem 2(b), which is repeated for convenience:

THEOREM 2.   *If p and q have power-series expansions for $|x| < r$ where $r > 0$, there is a pair of solutions u and v, with $W(0) \neq 0$, that have power-series expansions at least for $|x| < r$.*

By Chapter 10, Section 1, Theorem 2, the transformation

$$P = \int p(x)\, dx, \qquad Q = q - \frac{p'}{2} - \frac{p^2}{4}, \qquad y = e^{P/2}w$$

gives an equation for w that lacks the term w', namely

$$w'' + Qw = 0.$$

The properties of power series stated in Chapter B imply that P, Q, and $e^{P/2}$ have power-series expansions at least for $|x| < r$. Also, if we show that w has such an expansion, the same is true of y.

The net result of the above discussion is that we can confine attention to the case $p = 0$. Suppose, then, that

$$-y'' = q(x)y$$

where $q(x)$ has a power-series expansion with radius of convergence $r > 0$. Setting

$$q(x) = \sum_{j=0}^{\infty} q_j x^j, \qquad y(x) = \sum_{j=0}^{\infty} a_j x^j$$

we get the recursion formula

$$-(n+2)(n+1)a_{n+2} = q_0 a_n + q_1 a_{n-1} + \cdots + q_n a_0.$$

With n replaced by $n-1$ this becomes

(1) $$-(n+1)n a_{n+1} = q_0 a_{n-1} + q_1 a_{n-2} + \cdots + q_{n-1} a_0.$$

The coefficients a_0 and a_1 can be prescribed arbitrarily and the formula then gives a_n for $n \geq 2$. Thus the series is uniquely determined as a formal power series.

We shall show that the series converges for $|x| < r$, hence has radius of convergence at least r. To this end let x be given with $|x| < r$ and choose R so that $|x| < R < r$. The series for $q(x)$ converges whenever $|x| < r$, hence it converges at $x = R$. Theorem 1 gives

(2) $$|q_j| \leq \frac{B}{R^j}, \qquad j = 0, 1, 2, \cdots$$

for some constant B. Our objective is to get a similar relation for a_j, namely

(3) $$|a_j| \leq \frac{A}{R^j}, \qquad j = 0, 1, 2, \cdots$$

where A is constant. Once this is done the comparison test shows that the series for y converges at x, since $|x| < R$. But x was an *arbitrary* real number satisfying $|x| < r$. Hence the series converges for all $|x| < r$, as claimed.

The inequality (3) will be established by mathematical induction. Let us first see under what conditions (3) can be obtained for $j = n+1$ if it is known for $j \leq n$. By (2) and (3) we have

$$|q_j a_{n-j-1}| \leq \frac{B}{R^j} \frac{A}{R^{n-j-1}} = \frac{BA}{R^{n-1}}$$

for $0 \leq j \leq n - 1$. In the recursion formula (1) this gives

$$(n+1)n|a_{n+1}| \leq n\frac{BA}{R^{n-1}}$$

and therefore

$$|a_{n+1}| \leq \frac{BR^2}{n+1}\frac{A}{R^{n+1}}.$$

As soon as $n + 1 \geq BR^2$ this produces the desired inequality,

$$|a_{n+1}| \leq \frac{A}{R^{n+1}}.$$

To get started we pick an integer $N > BR^2$ and we let A be the largest of the numbers

$$|a_0|, \quad |a_1|R, \quad |a_2|R^2, \quad \cdots, \quad |a_N|R^N.$$

The inequality

$$|a_j| \leq \frac{A}{R^j}$$

holds for $j = 0, 1, 2, \cdots, N$ by choice of A and it holds for $j > N$ by the above discussion of the recursion formula. Since a_0 and a_1 are arbitrary, the condition $W(0) \neq 0$ is easily fulfilled, and this completes the proof.

(c) Singularities

When x is replaced by a complex variable z as in Part (b) above, the radius of convergence of the power series for a given function can often be obtained by inspection of the function. We shall show how this is done, and shall also show how it applies to differential equations.

A complex-valued function f of the complex variable z is *analytic* at a point z_0 if it has a convergent power-series expansion about z_0. It is analytic in a disk $|z - z_0| < r$ if it is analytic at every point of the disk. The following theorem, which we accept without proof, is established in a branch of mathematics known as the theory of functions of a complex variable:

THEOREM A. *The function $f(z)$ has a series expansion about $z = c$ with radius of convergence at least r if, and only if, $f(z)$ is analytic at every point of the disk $|z - c| < r$.*

To get at the meaning of this theorem, let us define a *singularity* of $f(z)$ to be a point z_0 at which f fails to be analytic. For example, the function

$$\frac{z^2 + 4}{(z + 2)(z^2 + 1)}$$

has singularities at -2, i, $-i$ where the denominator vanishes. These are the only singularities, since results stated in Chapter B indicate that this function has a power-series expansion about every other point. Theorem A can now be reworded as follows:

THEOREM 3. *The radius of convergence of the power series for $f(z)$ about $z = c$ is the distance from c to the nearest singularity of f.*

If c is real, a moment's thought shows that the disk of convergence $|z - c| < r$ in the complex plane leads to the same value r as that associated with the interval of convergence $|x - c| < r$ on the real axis. When combined with Theorem 2, this fact gives information about the interval of convergence for power-series solutions $y(x)$ when we know the location of the singularities of the coefficients in the complex plane.

EXAMPLE 1. Solutions of the following differential equation are expressed as power series about the point $x = 0$. What radius of convergence is to be expected?

$$(x + 2)(x^2 + 1)(x^2 - 4x + 5)y'' + (x^2 + 4)y = 0.$$

If this is written in standard form $y'' + q(x)y = 0$ the function q has, as sole singularities, the points

$$-2, \ i, \ -i, \ 2 + i, \ 2 - i$$

where the denominator vanishes. As seen in Figure 1, the distance from 0 to the nearest singularity is 1 and hence the expected radius of convergence for the solutions is 1. We say "expected" because at least this radius is guaranteed by Theorem 2, but the actual radius is sometimes larger. See Problem 1.

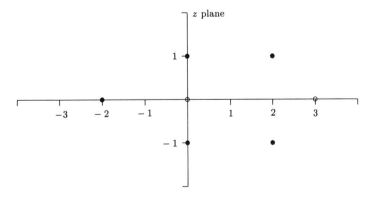

Figure 1

PROBLEMS————————————————————————————————————

1. Check that the equation $(2x - 1)y'' - 4xy' + 4y = 0$ has the linearly independent solutions x, e^{2x} although the expected radius of convergence for power-series solutions expanded about $x = 0$ is only $1/2$.

2P. Solutions of the differential equation in Example 1 are expanded about the point $x = 3$. Show that the expected radius of convergence is $\sqrt{2}$.

Hint: By Figure 1 the singularities nearest to $x = 3$ are $2 + i$ and $2 - i$.

3. What is the expected radius of convergence for series solutions of the following equations if the expansion is about $x = 0$? $x = 4$? $x = -1$?

(a) $(x^2 + 1)y'' = y$ (c) $x(x^2 + 6x + 9)y'' + y' = xy$

(b) $(x^2 - 9)y'' = y'$ (d) $(x^2 + 2x + 2)(x^2 + 4)y'' = (xy)'$

4. Suppose the equation $y'' + py' + qy = 0$ has solutions u and v that have power-series expansions about $x = 0$ with respective radii of convergence A and B. Suppose further that the distance from the origin to the nearest real or complex zero of the Wronskian $W(u, v)$ is C. Outline a proof that p and q have power-series expansions about $x = 0$ with radius of convergence at least equal to the smallest of the three numbers A, B, C.

5. Apply the result of Problem 4 to Problem 1.

6. In the examples and problems given hitherto the recurrence relation involved only two terms, for example a_n and a_{n-1} or a_n and a_{n-2}. For the equation

$$(2 - x)y'' + (x - 1)y' - y = 0$$

show that the substitution of a power series yields the three-term recurrence formula

$$2(n + 2)(n + 1)a_{n+2} - (n + 1)^2 a_{n+1} + (n - 1)a_n = 0$$

Convince yourself that it is hard to get a_n starting with arbitrary values a_0 and a_1. Show however that the initial conditions

$$y(0) = 1, \qquad y'(0) = 1; \qquad y(0) = -1, \qquad y'(0) = 1$$

lead to simple calculations and thus express the complete solution in terms of elementary functions.

7. Leibniz' rule for differentiating a product is

$$(uv)^{(n)} = uv^{(n)} + nu'v^{(n-1)} + \frac{n(n - 1)}{2!}u''v^{(n-2)} + \cdots$$

We shall apply this to Airy's equation $y'' - xy = 0$.

(a) Taking $n \geq 1$, differentiate n times to obtain

$$y^{(n+2)} - xy^{(n)} - ny^{(n-1)} = 0, \quad \text{hence} \quad y^{(n+2)}(c) = cy^{(n)}(c) + ny^{(n-1)}(c)$$

The differential equation itself gives $y''(c) = cy(c)$.

(b) Get the first six terms of the Taylor expansion for y about $x = c$.

(c) When $c = 0$ the recurrence formula is $y^{(n+2)}(0) = ny^{(n-1)}(0)$. Verify that the power series about $x = 0$ so obtained agrees with the discussion of Airy's equation in Section 1.

8. In the preceding problem $f^{(n)}(c)$ is considered to be the primary unknown. This amounts to use of $b_n = n!a_n$ instead of a_n, so that the series for y has the form

$$y = \sum_{j=0}^{\infty} \frac{b_n}{n!}(x - c)^n$$

Check that the coefficient of $(x - c)^n$ after differentiating any number of times is obtained by merely increasing the subscript on b.

9. Open-ended problem. Are the methods of Problems 7 and 8 simpler than those used elsewhere in this chapter? Justify your answer by examples.

───────────────────────**ANSWERS**───────────────────────

3. (a) 1, $\sqrt{17}$, $\sqrt{2}$ (b) 3, 1, 2   (c) 0, 4, 1   (d) $\sqrt{2}$, $2\sqrt{5}$, 1

5. The value $r = \infty$ given by Problem 4 agrees with that in Problem 1.

6. $c_1 e^x + c_2(1 - x)$. Note that this yields the complete solution of the three-term recurrence formula. A complicated recurrence formula can often be simplified by appropriate initial conditions.

7. (b) Partial answer: The coefficient of $(x - c)^5$ is $(4c\alpha + c^2\beta)/5!$
     where $\alpha = y(c)$, $\beta = y'(c)$

───

A demonstration without rigor is not a demonstration.

—Henri Poincaré

Chapter 13

SERIES SOLUTIONS II

Instead of assuming that the coefficients p and q of the differential equation $y'' + py' + qy = 0$ have power-series expansions about $x = 0$, as in the previous chapter, we assume only that $xp(x)$ and $x^2 q(x)$ have such expansions. Under this weaker assumption the solutions are generally of the form $x^s A(x)$ where s is a constant and where $A(x)$ has a power-series expansion. It turns out that s must satisfy a certain quadratic equation, called the indicial equation, and the coefficients are then determined by recursion. Introduction of the factors x, x^2, x^s as described above seems to be only a modest change. Nevertheless it leads to a dramatic increase in the scope of the theory.

1. The Frobenius normal form

Throughout this chapter, the phrase "power-series expansion", unless further qualified, means "convergent power-series expansion on some interval $|x| < r$." Since $x - x_0$ can be taken as a new variable, the assumption that our expansions are about the point $x = 0$ involves no loss of generality,

Let us consider the general second-order linear equation

$$(1) \qquad p_2(x)y'' + p_1(x)y' + p_0(x)y = 0$$

where the coefficients have power-series expansions. We assume $p_2(x) \neq 0$ except possibly at $x = 0$. Multiplying the equation by $1/p_2(x)$ yields

$$y'' + p(x)y' + q(x)y = 0$$

where

$$p(x) = \frac{p_1(x)}{p_2(x)}, \qquad q(x) = \frac{p_0(x)}{p_2(x)}.$$

If $p_2(0) = 0$ neither function is defined at $x = 0$ but we agree that

$$p(0) = \lim_{x \to 0} \frac{p_1(x)}{p_2(x)}, \qquad q(0) = \lim_{x \to 0} \frac{p_0(x)}{p_2(x)}$$

provided the limits exist. This definition is motivated by a desire to make the functions p and q continuous at $x = 0$. When the continuous functions so obtained have power-series expansions, $x = 0$ is called an *ordinary point*. A sufficient, but by no means necessary, condition for 0 to be an ordinary point is that $p_2(0) \neq 0$.

Instead of multiplying by $1/p_2(x)$ we could multiply by $x^2/p_2(x)$. The result is

(2) $$x^2 y'' + P(x)xy' + Q(x)y = 0$$

where

$$P(x) = \frac{xp_1(x)}{p_2(x)}, \qquad Q(x) = \frac{x^2 p_0(x)}{p_2(x)}.$$

If $p_2(0) = 0$ neither function is defined for $x = 0$, but we agree that

$$P(0) = \lim_{x \to 0} \frac{xp_1(x)}{p_2(x)}, \qquad Q(0) = \lim_{x \to 0} \frac{x^2 p_0(x)}{p_2(x)}$$

provided the limits exist. The factors x and x^2 greatly improve the behavior at the origin, and it is quite possible for P and Q to have power-series expansions when p and q do not. In that case the origin is said to be a *regular singular point*. The fact that (2) is an important and natural class of equations will become clear in the course of the following discussion. First we illustrate the definitions by an example.

EXAMPLE 1. Classify the singularity at the origin:

$$xy'' + (\sin x)y = 0, \quad x^2 y'' + (\sin x)y' + (\cos x)y = 0, \quad x^2 y'' + y' = 0.$$

If the left-hand equation is divided by x the coefficient of y becomes

$$q(x) = \frac{\sin x}{x} = 1 - \frac{x^2}{3!} + \frac{x^4}{5!} - \cdots$$

since, by our convention regarding continuity, $q(0) = 1$. The origin is an ordinary point.

The coefficient of y' in the second equation can be written

$$\sin x = xP(x) \quad \text{where} \quad P(x) = \frac{\sin x}{x}.$$

Taking $P(0) = 1$ gives a power series for $P(x)$ as seen above. Since $Q(x) = \cos x$ also has a series expansion, the origin is a regular singular point. It is not an ordinary point because neither $x^{-2} \sin x$ nor $x^{-2} \cos x$ has a power-series expansion.

In the third case the coefficient of y' is $(1/x)x$. The function $P(x) = 1/x$ does not have a series expansion and the origin is an example of an *irregular singular point*. This means that the equation has the form (1) with power-series coefficients as described above, but that the origin is neither an ordinary point nor a regular singular point.

The theory of (2) is due mainly to the German mathematicians Carl Friedrich Gauss, Bernhard Riemann, Georg Frobenius, and Lazarus Fuchs. Gauss and Riemann initiated the investigation by profound study of certain second-order equations (1812, 1857). The general equation of order n was dealt with by Fuchs (1866), and his methods were later simplified by Frobenius (1873). It was Frobenius who emphasized the convenience of introducing the factors x^2 and x, and (2) is sometimes said to be in *Frobenius normal form*. Without intending to disparage prior claims, we refer to (2) as "the Frobenius equation."

An example of an equation in Frobenius normal form is the Euler-Cauchy equation

$$x^2 y'' + P_0 x y' + Q_0 y = 0, \qquad x > 0$$

where P_0 and Q_0 are real constants. The Euler-Cauchy equation and its nth-order generalization were discussed in Chapter 11. All you need to know here is that the substitution

$$y = x^s, \qquad y' = s x^{s-1}, \qquad y'' = s(s-1)x^{s-2}$$

yields the quadratic equation

$$\text{(3)} \qquad\qquad s(s-1) + P_0 s + Q_0 = 0$$

for the constant s. This fact was illustrated in Chapter 2 and can be verified by inspection.

If (3) has two roots λ and μ the process gives the solutions

$$\text{(4)} \qquad\qquad u = x^\lambda, \qquad v = x^\mu.$$

When a root $s = \sigma + i\tau$ is complex we define x^s for $x > 0$ by

$$x^s = (e^{\ln x})^{\sigma + i\tau} = e^{\sigma \ln x} e^{i\tau \ln x} = x^\sigma (\cos(\tau \ln x) + i \sin(\tau \ln x)).$$

The principle of equating real parts gives corresponding real solutions

$$x^\sigma \cos(\tau \ln x), \qquad x^\sigma \sin(\tau \ln x).$$

Conversely, if λ and μ are prescribed real or complex numbers we can define P_0 and Q_0 in such a way that

$$s(s-1) + P_0 s + Q_0 = (s - \lambda)(s - \mu).$$

Then (3) has roots λ, μ and the Euler-Cauchy equation has the solutions (4). Equating coefficients yields the formulas

$$P_0 = 1 - (\lambda + \mu), \qquad Q_0 = \lambda\mu.$$

These equations show that P_0 and Q_0 are real if the prescribed roots λ and μ are real, and also if they are complex conjugates.

The point we want to make is that solutions of the Euler-Cauchy equation can have a great variety of behaviors near $x = 0$. The following pairs of linearly independent solutions (u, v) are obtained by suitable choices of P_0 and Q_0:

$$(x, x^2), \qquad (x^{-1}, x^{-2}), \qquad (x^{1/5}, x^{1/3}), \qquad (x^{-1}, x^2).$$

The first shows that both solutions can have power-series expansions, the second shows that both solutions can be unbounded near 0, the third shows that both solutions can be continuous yet such that neither has a power-series expansion, and the fourth shows that one solution can be unbounded while the other has a power-series expansion.

Similar behavior is found for solutions of the general Frobenius equation, which is repeated for convenience:

(5) $$x^2 y'' + P(x) x y' + Q(x) y = 0.$$

When x is near 0, Equation (5) is approximately

(6) $$x^2 y'' + P(0) x y' + Q(0) y = 0.$$

This is the Euler-Cauchy equation with $P_0 = P(0)$ and $Q_0 = Q(0)$. The corresponding quadratic polynomial

(7) $$I(s) = s(s - 1) + sP(0) + Q(0)$$

is called the *indicial polynomial* of (5), the equation $I(s) = 0$ is called the *indicial equation*, and the roots λ, μ of the indicial equation are called the *exponents* at the regular singular point $x = 0$. Equation (6) retains its relevance as a means of approximation when P and Q are merely continuous at $x = 0$, without having power-series expansions. If P and Q do have power-series expansions, as assumed here, the quantities

$$P(0) = P_0, \qquad Q(0) = Q_0$$

are the constant terms in these expansions.

The following theorem shows that solutions of (5) near the origin behave much like solutions of (6):

THEOREM 1. *Suppose P and Q have power-series expansions for $|x| < r$. Let (λ, μ) be the roots of the indicial equation, so labeled that $\lambda \geq \mu$ when both are real. Then the Frobenius equation has a solution*

(8) $$u(x) = x^s \sum_{j=0}^{\infty} a_j x^j, \qquad a_0 \neq 0$$

with $s = \lambda$. If $\lambda - \mu$ is not an integer there is a second solution $v(x)$ of this form with $s = \mu$. In both cases the series converges at least for $|x| < r$ and the given function satisfies the differential equation at least for $0 < x < r$.

It is best to postpone the proof until you have had some experience with the Fuchs-Frobenius procedure. We shall, however, make some general remarks.

Even though we have assumed P and Q to be real, we must allow complex coefficients a_j. This is true because the coefficients a_j are almost always complex if the roots of the indicial equation are complex. The situation is similar to that for the equation

$$y'' + ay' + by = 0, \qquad a, b \text{ const.,}$$

where the substitution $y = e^{st}$ may lead to complex solutions even if a and b are real.

Since the exponent s can be increased at will, the assumption $a_0 \neq 0$ involves no loss of generality. For example, if $a_0 = a_1 = 0$ but $a_2 \neq 0$ the series starts as follows:

$$a_2 x^{s+2} + a_3 x^{s+3} + \cdots.$$

This has the same form as the original series, with the role of a_0 taken by a_2 and with s replaced by $s+2$. When $a_0 \neq 0$, as assumed here, it will be found that s has to be a root of the indicial equation. That is why the indicial equation is important.

The differential equation is considered for $0 < x < r$ rather than for $|x| < r$ because the factors x^λ and x^μ can be infinite at $x = 0$ or, if finite, can be complex when $x < 0$ even if λ and μ are real. Examples are $1/x$ and $x^{5/2}$. If we replace x^λ and x^μ by

$$|x|^\lambda, \qquad |x|^\mu$$

respectively, we get a solution valid for $-r < x < 0$ and hence, in some cases, a solution valid for $|x| < r$. See Problem 5.

If there are two solutions as described in Theorem 1, one for $s = \lambda$ and one for $s = \mu$, with $\lambda \neq \mu$, the two solutions are linearly independent on the interval $(0, r)$. Hence the expression

$$c_1 u(x) + c_2 v(x)$$

yields both the complete solution and the general solution on this interval. This follows from the fact that for x positive and close to zero the ratio $u(x)/v(x)$ is defined and not constant.

We now illustrate the procedure by an example.

EXAMPLE 2. For the equation

(9) $$x^2 y'' + (x^2 - 20)y = 0$$

we have $P(x) = 0$ and $Q(x) = x^2 - 20$. Hence the indicial polynomial (7) is

$$(10) \qquad I(s) = s(s-1) - 20 = (s-5)(s+4)$$

and the exponents are $\lambda = 5$, $\mu = -4$.

To solve Equation (9) we try a series of the form

$$y = x^s \sum_{j=0}^{\infty} a_j x^j = \sum_{j=0}^{\infty} a_j x^{s+j}.$$

The series is easily differentiated and the general term is given by the following table:

y	y'	y''
$a_j x^{s+j}$	$(s+j)a_j x^{s+j-1}$	$(s+j)(s+j-1)a_j x^{s+j-2}$

This yields another table for the expressions in the differential equation:

$x^2 y''$	$x^2 y$	$-20y$
$(s+j)(s+j-1)a_j x^{s+j}$	$a_j x^{s+j+2}$	$-20a_j x^{s+j}$
$(s+n)(s+n-1)a_n$	a_{n-2}	$-20a_n$

The second row gives the general term in the expansion of the entries of the first row, and the third row is the coefficient of x^{s+n} in these expansions. It is obtained by the choices $j = n$, $j = n-2$, $j = n$ in the respective entries of the second row.

Setting the coefficient of x^{s+n} equal to 0 in (9), we get the recursion formula

$$((s+n)(s+n-1) - 20)\, a_n + a_{n-2} = 0.$$

Since $s(s-1) - 20 = I(s)$ by (10), this can be written

$$I(s+n)a_n = -a_{n-2}.$$

When $n = 0$ we get $I(s)a_0 = -a_{-2} = 0$ and hence $I(s) = 0$. The two roots $\lambda = 5$ and $\mu = -4$ were already found above. If $s = 5$ the recursion formula becomes

$$n(n+9)a_n = -a_{n-2}.$$

It is easily checked that $a_n = 0$ when n is odd and that the a_n with even n yield the solution

$$a_0 x^5 \left(1 - \frac{x^2}{2 \cdot 11} + \frac{x^4}{(4 \cdot 2)(13 \cdot 11)} - \frac{x^6}{(6 \cdot 4 \cdot 2)(15 \cdot 13 \cdot 11)} + \cdots \right).$$

The factor x^5 comes from x^s with $s = 5$.

If we set $s = -4$, which is the other root μ, the recurrence formula becomes

$$n(n-9)a_n = -a_{n-2}.$$

By a short calculation

$$y = a_0 x^{-4}\left(1 + \frac{x^2}{2 \cdot 7} + \frac{x^4}{(4 \cdot 2)(5 \cdot 7)} + \frac{x^6}{(6 \cdot 4 \cdot 2)(3 \cdot 5 \cdot 7)} + \cdots\right).$$

The pattern is clear from the recursion formula. Notice, however, that the signs alternate as soon as $n > 9$.

Since $\lambda - \mu = 9$ this example shows that the condition $\lambda - \mu \neq$ integer in Theorem 1 is not always necessary for the existence of two linearly independent solutions of the form (8).

PROBLEMS

1P. The notation in this and subsequent problems is the same as in the text. For the equation $x^2 y'' = (6 + 2x)y$:

(a) Find $P(0)$ and $Q(0)$ by inspection and check that the indicial polynomial is $I(s) = s^2 - s - 6 = (s-3)(s+2)$.

(b) Make a table showing the coefficient of x^{s+n} in $x^2 y''$, y, and xy and thus get the recurrence formula $I(s+n)a_n = 2a_{n-1}$.

(c) The choice $n = 0$ makes $I(s) = 0$, hence $s = 3$ or -2. If $s = 3$ show that $n(n+5)a_n = 2a_{n-1}$, which gives

$$y = a_0 x^3 \left(1 + \frac{2x}{1! \cdot 6} + \frac{(2x)^2}{2! \cdot 6 \cdot 7} + \frac{(2x)^3}{3! \cdot 6 \cdot 7 \cdot 8} + \cdots\right)$$

(d) If $s = -2$ show that $n(n-5)a_n = 2a_{n-1}$. This equation for $n = 1, 2, 3, 4, 5$ makes $a_0 = 0$, contrary to our basic assumption. Why did the method fail to give a second solution?

2P. Multiply $(x + x^2)y'' + (2x + 1)y' + (5x + 4)y = 0$ by $x^2/(x + x^2)$ to get the Frobenius normal form

$$x^2 y'' + \frac{2x+1}{x+1}xy' + \frac{5x^2 + 4x}{x+1}y = 0$$

Thus show that the indicial polynomial is $I(s) = s^2$. Now that you know $I(s)$, which is better for finding a series solution: the original form of the equation, or the normal form?

3. There are four things to do in each part: (i) Find the indicial polynomial $I(s)$. (ii) Derive the recurrence formula given at the right of the equation. (iii) Determine s by setting $n = 0$ and specialize the recurrence relation accordingly. (iv) Get one or two solutions when $a_0 = 1$. For two problems, which you will recognize when you come to them, the recurrence formula has been written with $n - 1$ replacing n.

(a) $x^2 y'' = xy$ $\qquad\qquad$ $I(s+n)a_n = a_{n-1}$

(b) $xy'' + y' = xy$ $\qquad\qquad$ $I(s+n)a_n = a_{n-2}$

(c) $xy'' = xy' + 2y$ $\qquad\qquad$ $I(s+n)a_n = (s+n+1)a_{n-1}$

(d) $x^2 y'' = (6 + 2x^2)y$ $\qquad\qquad$ $I(s+n)a_n = 2a_{n-2}$

(e) $x^2 y'' + 4xy' + 2y = 0$ $\qquad\qquad$ $I(s+n)a_n = 0$

(f) $4x^2 y'' = (4x - 1)y$ $\qquad\qquad$ $I(s+n)a_n = a_{n-1}$

(g) $x^2 y'' + 6xy' = (x^2 - 6)y$ $\qquad\qquad$ $I(s+n)a_n = a_{n-2}$

(h) $x^2 y'' + xy' = (1 + x^2)y$ $\qquad\qquad$ $I(s+n)a_n = a_{n-2}$

(i) $2x(1-x)y'' + (1-8x)y' = 4y$ $\quad$ $I(s+n)a_n = (s+n)(s+n+1)a_{n-1}$

(j) $4x^2 y'' = (4x^2 - 1)y$ $\qquad\qquad$ $I(s+n)a_n = a_{n-2}$

(k) $2xy'' + y' = 2xy' + y$ $\qquad\qquad$ $I(s+n)a_n = (s+n-\frac{1}{2})a_{n-1}$

(l) $2x^2 y'' + y = xy'$ $\qquad\qquad$ $I(s+n)a_n = 0$

(m) $9x^2 y'' = (3x - 2)y$ $\qquad\qquad$ $I(s+n)a_n = \frac{1}{3}a_{n-1}$

(n) $xy'' = (6 + 5x)y' + 5y$ $\qquad\qquad$ $I(s+n)a_n = 5(s+n)a_{n-1}$

(o) $16x^2 y'' = (8x^2 - 3)y$ $\qquad\qquad$ $I(s+n)a_n = \frac{1}{2}a_{n-2}$

(p) $3x(1-x)y'' + (1-7x)y' = y$ $\quad$ $I(s+n)a_n = (s+n)(s+n-\frac{2}{3})a_{n-1}$

(q) $4x^2 y'' + 4xy' = (4x^2 + 1)y$ $\qquad$ $I(s+n)a_n = a_{n-2}$

4. Since $I(s + n)$ is essentially n^2 for large n the radius of convergence corresponding to the recursion relations in Problem 3 can be determined more or less by inspection. In all but two cases the radius is infinite. What are the two exceptions and what is the radius of convergence in those cases?

5. In this problem $t = -x$, $y(x) = z(t)$, and λ, μ are the exponents associated with the Euler-Cauchy equation. Check that the Euler-Cauchy equation for y as a function of x yields the same equation for z as a function of t. Since $t = -x = |x|$ for $x < 0$ this shows that $|x|^\lambda$, $|x|^\mu$ are solutions on $(-\infty, 0)$. If λ and μ are real, what conditions ensure that these solutions are valid on $(-\infty, \infty)$?

6. Change of variable. If P and Q have convergent power-series expansions about x_0, we say that x_0 is a regular singular point for the equation

$$(x - x_0)^2 y'' + P(x)(x - x_0)y' + Q(x)y = 0$$

Let $t = x - x_0$ and $y(x) = z(t)$ to obtain

$$t^2 z'' + P(x_0 + t)tz' + Q(x_0 + t)z = 0$$

where the primes on z denote differentiation with respect to t. Check that this is in the Frobenius normal form with $t = 0$ as a regular singular point, and that the series on the left below for z yields the series on the right for y:

$$z = t^\lambda \sum_{j=0}^{\infty} a_j t^j, \qquad y = (x - x_0)^\lambda \sum_{j=0}^{\infty} a_j (x - x_0)^j$$

————————————ANSWERS————————————

1. Because $\lambda - \mu = 5$, an integer.
2. The original form, because its coefficients are polynomials while those in the normal form involve infinite series. In theory, nonterminating series do no harm; but in practice they complicate the recurrence formulas.
3. (a) general term $nx^n/(n!)^2$ when $s = 1$, no second solution
   (b) general term $x^{2n}/(2^n n!)^2$ when $s = 0$, no second solution
   (c) $(x + x^2/2)e^x$ when $s = 1$, no second solution.
   (d) The following when $s = 3$ and $s = -2$ respectively:

$$x^3 \left(1 + \frac{x^2}{1! \, 7} + \frac{x^4}{2! \, 7 \cdot 9} + \frac{x^6}{3! \, 7 \cdot 9 \cdot 11} + \cdots \right)$$

$$x^{-2} \left(1 - \frac{x^2}{3 \cdot 1!} + \frac{x^4}{3 \cdot 2!} + \frac{x^6}{3 \cdot 3! \cdot 1} + \frac{x^{10}}{3 \cdot 5! \cdot 1 \cdot 3 \cdot 5} + \cdots \right)$$

   (e) $1/x$, $1/x^2$
   (f) general term $x^{n+1/2}/(n!)^2$ when $s = 1/2$, no second solution
   (g) $(\cosh x)/x^3$, $(\sinh x)/x^3$
   (h) general term $x(x/2)^{2n}/((n+1)! \, n!)$ when $s = 1$, no second solution
   (i) the following when $s = 0$ and $s = 1/2$, respectively:

$$1 + \frac{2!}{1}(2x) + \frac{3!}{1 \cdot 3}(2x)^2 + \frac{4!}{1 \cdot 3 \cdot 5}(2x)^3 + \cdots,$$

$$x^{1/2} \left(1 + \frac{5}{2}x + \frac{7 \cdot 5}{4 \cdot 2}x^2 + \frac{9 \cdot 7 \cdot 5}{6 \cdot 4 \cdot 2}x^3 + \cdots \right)$$

   (j) general term $x^{2n+1/2}/(2^n n!)^2$, no second solution   (k) e^x and

$$x^{1/2} \left(1 + \frac{2x}{3} + \frac{(2x)^2}{3 \cdot 5} + \frac{(2x)^3}{3 \cdot 5 \cdot 7} + \cdots \right)$$

(l) x, $x^{1/2}$ (m) The following for $s = 1/3$ and $s = 2/3$, respectively:

$$x^{1/3}\left(1 + \frac{x}{1!\,2} + \frac{x^2}{2!\,2\cdot 5} + \frac{x^3}{3!\,2\cdot 5\cdot 8} + \cdots\right)$$

$$x^{2/3}\left(1 + \frac{x}{1!\,4} + \frac{x^2}{2!\,4\cdot 7} + \frac{x^3}{3!\,4\cdot 7\cdot 10} + \cdots\right)$$

(n) $x^7 e^{5x}$, no second solution

(o) The following for $s = 1/4$ and $3/4$ respectively:

$$x^{1/4}\left(1 + \frac{x^2}{(2)(3)} + \frac{x^4}{(2\cdot 4)(3\cdot 7)} + \frac{x^6}{(2\cdot 4\cdot 6)(3\cdot 7\cdot 11)} + \cdots\right)$$

$$x^{3/4}\left(1 + \frac{x^2}{(2)(5)} + \frac{x^4}{(2\cdot 4)(5\cdot 9)} + \frac{x^6}{(2\cdot 4\cdot 6)(5\cdot 9\cdot 13)} + \cdots\right)$$

(p) $1/(1-x)$, $x^{2/3}/(1-x)$ (q) $x^{-1/2}\sinh x$, $x^{-1/2}\cosh x$

4. $r = 1$ in (ip)   5. $\lambda \geq 2$, $\mu \geq 2$

---

2. Equations with parameters

In the examples and problems of the preceding section we considered only equations having specific numerical coefficients, such as 2 or -5. Now we consider equations in which the coefficients depend on numerical parameters such as a, b, c, m. One equation of this kind is discussed in the text and others are taken up in the accompanying problems.

The *Bessel equation* is

(1) $$x^2 y'' + xy' + (x^2 - m^2)y = 0, \qquad m \geq 0$$

where m is a real constant. This is in the Frobenius normal form with

$$P(x) = 1, \qquad Q(x) = x^2 - m^2.$$

Since $P(0) = 1$ and $Q(0) = -m^2$ the indicial polynomial is

$$I(s) = s(s-1) + P(0)s + Q(0) = s^2 - m^2,$$

the indicial equation is $s^2 = m^2$, and the exponents are $\lambda = m$, $\mu = -m$. The condition $m \geq 0$ ensures $\lambda \geq \mu$.

The general terms in the series for y, y', and y'' are, respectively,

$$a_j x^{s+j}, \qquad (s+j)a_j x^{s+j-1}, \qquad (s+j)(s+j-1)x^{s+j-2}a_j.$$

This yields the following table for the functions appearing in the Bessel equation:

$x^2 y''$	xy'	$x^2 y$	$-m^2 y$
$(s+j)(s+j-1)a_j x^{s+j}$	$(s+j)a_j x^{s+j}$	$a_j x^{s+j+2}$	$-m^2 a_j x^{s+j}$
$(s+n)(s+n-1)a_n$	$(s+n)a_n$	a_{n-2}	$-m^2 a_n$

The second row gives the general terms in the corresponding series and the third row gives the coefficients of x^{s+n}. These coefficients are obtained by the respective choices $j = n$, n, $n-2$, n in the second row. By (1) and the last line of the table we get the recurrence formula

$$(s+n)^2 a_n + a_{n-2} - m^2 a_n = 0.$$

Since $I(s) = s^2 - m^2$ the coefficient of a_n is $I(s+n)$ and hence

(2) $$I(s+n)a_n = -a_{n-2}.$$

The value $n = 0$ in (2) shows, as expected, that s must be one of the roots m or $-m$ of the indicial equation. The choice $s = m$ yields

$$I(s+n) = I(m+n) = (m+n)^2 - m^2 = n(2m+n)$$

and the recurrence formula (2) becomes

$$a_n = -\frac{a_{n-2}}{n(2m+n)}, \qquad n \geq 1.$$

Taking $n = 1, 3, 5, \cdots$ shows that $a_n = 0$ for all odd n. The values $n = 2, 4, 6, \cdots$ yield

$$a_2 = -\frac{a_0}{2(2m+2)}, \qquad a_4 = -\frac{a_2}{4(2m+4)}, \qquad a_6 = -\frac{a_4}{6(2m+6)},$$

and so on. Our solution is therefore

$$y = a_0 x^m \left(1 - \frac{x^2}{2(2m+2)} + \frac{x^4}{4 \cdot 2(2m+4)(2m+2)} - \cdots \right)$$

where the initial factor x^m is x^s with $s = m$. A solution for the exponent $s = -m$ is obtained by writing $-m$ in place of m provided the denominators so obtained do not vanish. This holds if, and only if, m is not a positive integer.

The terms can be simplified by factoring out 2 from each factor in the denominator and combining it with x in the numerator. The result is

$$(3) \qquad y = a_0 x^m \left(1 - \frac{(x/2)^2}{1!(m+1)} + \frac{(x/2)^4}{2!(m+2)(m+1)} - \cdots \right).$$

Assuming for the moment that m is a positive integer, we can write this in the form

$$y = a_0 x^m \left(\frac{m!}{m!} - \frac{m!(x/2)^2}{1!(m+1)!} + \frac{m!(x/2)^4}{2!(m+2)!} - \cdots \right).$$

If we now choose $a_0 = 1/(2^m m!)$ we get

$$y = \left(\frac{x}{2} \right)^m \left(\frac{1}{m!} - \frac{(x/2)^2}{1!(m+1)!} + \frac{(x/2)^4}{2!(m+2)!} - \cdots \right).$$

This function is denoted by $J_m(x)$ and is called *the Bessel function of order* m. Thus, by definition,

$$(4) \qquad J_m(x) = \left(\frac{x}{2} \right)^m \sum_{j=0}^{\infty} (-1)^j \frac{(x/2)^{2j}}{j!(m+j)!}.$$

A graph of $J_0(x)$ is given in Chapter 18, Section 3, Figure 6.

The Bessel equation is meaningful even when m is not a positive integer, and so is the series (3). It is natural to inquire whether we can define $m!$ for arbitrary real m and thus give a meaning to the series (4). In Chapter 22, Section 4 it is seen that such an extension is in fact possible. The extended definition of $m!$ agrees with the familiar definition when m is a positive integer and it satisfies the equation

$$(5) \qquad (m+1)\frac{1}{(m+1)!} = \frac{1}{m!}$$

for all m without exception. The universal validity of (5) depends on the following equation, which holds by definition:

$$(6) \qquad \frac{1}{m!} = 0 \quad \text{when} \quad m = -1, -2, -3, \cdots.$$

The definition is appropriate because $|m!|$ tends to ∞ when m tends to a negative integer.

Equation (5) shows that (4) remains a valid solution for all m, and hence $J_m(x)$ and $J_{-m}(x)$ both satisfy the Bessel equation. If m is not an integer it is easily checked that the two functions J_m and J_{-m} are linearly independent on $(0, \infty)$ and hence

$$y = c_0 J_m(x) + c_1 J_{-m}(x)$$

gives both the general solution and the complete solution of the Bessel equation on this interval. If $m = 0$, however, we obtain only the single function $J_0(x)$. Another exceptional case arises when m is a positive integer. In this case the series (3), with m replaced by $-m$, has a zero denominator and is meaningless. However the series (4) for the Bessel function remains well defined and satisfies

(7) $$J_{-m}(x) = (-1)^m J_m(x), \qquad m = 1, 2, 3, \cdots.$$

For proof, write $-m$ in place of m in (4). Since (6) shows that $1/(j-m)! = 0$ when $j < m$, the result is

$$J_{-m}(x) = \sum_{j=0}^{\infty} (-1)^j \frac{(x/2)^{2j-m}}{j!(j-m)!} = \sum_{j=m}^{\infty} (-1)^j \frac{(x/2)^{2j-m}}{j!(j-m)!}.$$

The truth of (7) becomes evident when this series and the one for $J_m(x)$ are both written in expanded form.

It should be noticed that the exceptional cases $m = 0, 1, 2, \cdots$ make the value of $\lambda - \mu = 2m$ an integer, and Theorem 1 of Section 4 excludes this case when affirming existence of a second solution. We shall presently see that a second solution can indeed be found when $\lambda - \mu$ is an integer. Sometimes, but not always, the second solution involves $\ln x$.

We conclude with some historical remarks. The Bessel function of order $1/2$ was introduced by Jakob Bernoulli in a letter to Leibniz (1703). The function $J_0(x)$ occurs in a memoir by Daniel Bernoulli on the oscillation of heavy chains (1732). The functions $J_m(x)$ for integral m were used by Euler in connection with the theory of vibrations of a circular membrane (1766). In a paper on planetary motion the systematic study of Bessel functions was initiated by the German astronomer F. W. Bessel in 1824, hence the name. A complete development of the theory of Bessel functions would occupy more than one book the size of this one.

PROBLEMS

1. The Bessel equation with $m = 1/2$ is

$$x^2 y'' + xy' + \left(x^2 - \frac{1}{4} \right) y = 0$$

In this case $\lambda - \mu = m - (-m) = 2m = 1$, an integer. Using the method but not the result in the text, obtain series solutions that you will recognize as

$$a_0 x^{-1/2} \sin x \quad \text{and} \quad a_0 x^{-1/2} \cos x$$

for $s = 1/2$ and $s = -1/2$ respectively. In Chapter 22 it is shown that $(-1/2)! = \sqrt{\pi}$, and $(1/2)!$ can then be found from $m! = m(m-1)!$. Thus verify that the choice $a_0 = 1/(2^m m!)$ used in the text yields

$$J_{1/2}(x) = \sqrt{\frac{2}{\pi x}} \sin x, \qquad J_{-1/2}(x) = \sqrt{\frac{2}{\pi x}} \cos x$$

2. Generalization of Section 1, Problem 3(afm). This pertains to the equation $x^2 y'' = (a^2 - a + bx)y$ where a and b are constants. Obtain the solution

$$y = x^a \left(1 + \frac{bx}{1 \cdot 2a} + \frac{(bx)^2}{2!(2a)(2a+1)} + \frac{(bx)^3}{3!(2a)(2a+1)(2a+2)} + \cdots \right)$$

and another solution with a replaced by $1 - a$. It is assumed that $2a$ is not 0 or a negative integer in the first case and that $2(1 - a)$ is not 0 or a negative integer in the second. Hence you get two linearly independent solutions unless $2a$ is an integer.

3. Generalization of Section 1, Problem 3(djo). This pertains to the equation $x^2 y'' = (a^2 - a + bx^2)y$ where a and b are constants. Obtain the solution

$$x^a + \frac{x^{a+2}(b/2)}{1!(2a+1)} + \frac{x^{a+4}(b/2)^2}{2!(2a+1)(2a+3)} + \frac{x^{a+6}(b/2)^3}{3!(2a+1)(2a+3)(2a+5)} + \cdots$$

and another solution with a replaced by $1 - a$. It is assumed that $2a \neq -1, -3, -5, \cdots$ in the first case and that $2a \neq 3, 5, 7, \cdots$ in the second. If $2a = 1$ the two solutions coincide. Hence you get two linearly independent solutions unless $2a$ is an odd integer.

4. Generalization of Section 1, Problem 3(bhq). A modification of the Bessel equation is
$$x^2 y'' + x y' = (x^2 + m^2)y, \qquad m \geq 0$$

where m is constant. Show that the theory parallels that for the Bessel function as developed in the text, with the sole exception that the terms of the series do not alternate in sign.

5. Generalization of Section 1, Problem 3(ck). The *confluent hypergeometric equation* is $xy'' + ay' = xy' + by$ where a and b are constants.

(a) For $s = 0$ obtain the solution

$$y = 1 + \frac{b}{a}\frac{x}{1!} + \frac{b(b+1)}{a(a+1)}\frac{x^2}{2!} + \frac{b(b+1)(b+2)}{a(a+1)(a+2)}\frac{x^3}{3!} + \cdots$$

(b) Taking $s = 1 - a$ show that another solution is

$$y = x^{1-a}G(2 - a, b + 1 - a, x)$$

where $G(a, b, x)$ denotes the above series. It is assumed that a is not 0 or a negative integer in the first case and that $2 - a$ is not 0 or a negative integer in the second.

6. Generalization of Section 1, Problem 3(ip). The *hypergeometric equation* is

$$x(1 - x)y'' + (c - (a + b + 1)x)y' - aby = 0, \qquad a, b, c \text{ const.}$$

Discuss (a) when $a = 1$ and (b) when $b = c$. No answer is provided but see Problem 7.

7. (a) For the hypergeometric equation of Problem 6 obtain the following solution corresponding to the exponent $s = 0$:

$$y = 1 + \frac{ab}{1!c}x + \frac{(a+1)a(b+1)b}{2!(c+1)c}x^2 + \frac{(a+2)(a+1)a(b+2)(b+1)b}{3!(c+2)(c+1)c}x^3 + \cdots$$

(b) The above series is called the *hypergeometric series* and is denoted by $F(a, b, c, x)$. Check that a solution with $s = 1 - c$ is given by

$$y = x^{1-c}F(a - c + 1, b - c + 1, 2 - c, x)$$

It is assumed that c is not 0 or a negative integer in Part (a) and that c is not a positive integer exceeding 1 in Part (b).

8. The *gamma function* is defined by $\Gamma(p) = (p-1)!$. If the series in Problem 7(a) is multiplied by a suitable constant show that it can be written

$$y = \sum_{n=0}^{\infty} \frac{\Gamma(a + n)\Gamma(b + n)}{\Gamma(1 + n)\Gamma(c + n)}x^n$$

9. Show (a) for $m = 1$ and (b) in general that

$$\frac{d}{dx}(x^m J_m(x)) = x^m J_{m-1}(x), \qquad \frac{d}{dx}(x^{-m} J_m(x)) = -x^{-m}J_{m+1}(x)$$

10. Deduce from Problem 9 that

$$J_{m-1}(x) - J_{m+1}(x) = 2J_m'(x), \qquad J_{m-1}(x) + J_{m+1}(x) = \frac{2m}{x}J_m'(x)$$

11. The generating function of the sequence $\{J_n(x)\}$ is

$$e^{(x/2)(h-h^{-1})} = \sum_{n=-\infty}^{\infty} h^n J_n(x)$$

Verify that the coefficient of h^0 on the right is indeed $J_0(x)$.

---

3. Exponents differing by an integer

As in the preceding discussion $I(s)$ is the indicial polynomial and λ, μ are the corresponding exponents, with $\lambda \geq \mu$ when both are real. We denote by u a solution with exponent λ; namely, a solution of the form $x^\lambda A(x)$ where $A(x)$ has a power-series expansion and $A(0) \neq 0$. The phrase "second solution" means a solution $y = v$ that is valid on some interval $0 < x < r$ and is such that (u, v) are linearly independent. In the sequel it is left for you to verify that the ratio of the alleged second solution to the first is not constant and hence that the two are, in fact, independent.

If $\lambda = \mu$, there is only one exponent and the method of the preceding sections gives only one solution. There may also be difficulty if $\lambda - \mu$ is a positive integer. This difficulty comes from the recursion formula as explained next.

In all examples and problems discussed hitherto the recursion formula had the form

$$I(n + s)a_n = c_n a_{n-1} \quad \text{or} \quad I(n + s)a_n = c_n a_{n-2}$$

where c_n were known constants. A similar result holds in the general case, except that the right-hand side can involve all a_j from $j = 0$ to $n - 1$. The general case is taken up later in connection with the problem of convergence. What we want to emphasize now is that if $I(n + s) = 0$ for some positive integer n, the formula fails to give a_n. In fact it leads to a contradiction unless the right-hand side happens to be 0 for the corresponding value n.

Let us recall that the equation for $n = 0$ requires $s = \lambda$, where λ is a root of the indicial equation. This follows from our basic hypothesis that $a_0 \neq 0$. If also $I(n + s) = 0$ we conclude that $n + s = \mu$, the other root. Hence $\mu - \lambda = n$. This equation cannot hold if λ and μ are complex, nor if $\lambda \geq \mu$. But it always holds for some n if $\lambda - \mu$ is a positive integer and we try to get a solution with exponent μ by interchanging the roles of λ and μ.

Regardless of the value of $\lambda - \mu$, one can obtain a second solution by the method of Chapter 10. Define a linear operator T by

(1) $$Ty = x^2 y'' + P(x)xy' + Q(x)y, \qquad x > 0.$$

If u and W are twice differentiable, a short calculation gives

$$T(uW) = WTu + x^2(2u'W' + uW'') + xuPW'.$$

This is essentially the product identity of Chapter 10 and will be used in the same way here. Let u be a solution of $Tu = 0$ and let a second solution be sought in the form $v = uW$. Since $Tu = 0$ the product identity gives

$$x^2(2u'W' + uW'') + xuPW' = 0.$$

If we set $W' = w$ and divide by x^2uw the equation becomes

$$(2) \qquad \frac{w'(x)}{w(x)} = -2\frac{u'(x)}{u(x)} - \frac{P(x)}{x}.$$

It is assumed that x is confined to an interval on which x, $u(x)$, and $w(x)$ do not vanish.

Equation (2) involves various powers of x, including a term $P(0)/x$. This term produces a logarithm when (2) is integrated. To isolate it, we write

$$\frac{P(x)}{x} = \frac{P(x) - P(0)}{x} + \frac{P(0)}{x}$$

and then we define a power series $R(x)$ by

$$(3) \qquad R'(x) = \frac{P(x) - P(0)}{x}, \qquad x \neq 0; \qquad R(0) = 0.$$

Thus, if the coefficients of $P(x)$ are P_j, we get $R(x)$ as follows: Take the terms $P_1 x + P_2 x^2 + \cdots$, divide by x, and integrate. The result is

$$R(x) = P_1 x + P_2 \frac{x^2}{2} + P_3 \frac{x^3}{3} + P_4 \frac{x^4}{4} + \cdots.$$

The indicial polynomial is

$$I(s) = s(s-1) + P(0)s + Q(0) = (s - \lambda)(s - \mu)$$

and hence, equating coefficients, $P(0) = 1 - \lambda - \mu$. Equation (3) now yields

$$\frac{P(x)}{x} = \frac{P(0)}{x} + R'(x) = \frac{1 - \lambda - \mu}{x} + R'(x)$$

and (2) takes the form

$$(4) \qquad \frac{w'(x)}{w(x)} = \frac{\lambda + \mu - 1}{x} - 2\frac{u'(x)}{u(x)} - R'(x).$$

This equation is easily integrated to give the following theorem:

THEOREM 1. *In the above notation, if $u(x)$ is a nonvanishing solution of $Ty = 0$ a second solution is*

$$v(x) = u(x) \int w(x)\, dx \quad \text{where} \quad w(x) = \frac{x^{\lambda+\mu-1}}{(u(x))^2} e^{-R(x)}.$$

It should be noticed that u need not be associated with the exponent λ, but can equally well be associated with μ. This remark is helpful when, as is often the case, the latter solution is simpler than the former.

EXAMPLE 1. For the equation $x^2 y'' + P_0 xy' + Q_0 = 0$ the coefficient $P(x)$ is constant, hence $R(x) = 0$, and $u(x) = x^\lambda$ in Theorem 1 gives

$$y = x^\lambda \int w(x)\, dx \quad \text{where} \quad w(x) = x^{\mu-\lambda-1}.$$

If $\mu \neq \lambda$ we get the expected result $y = cx^\mu$ where c is constant. But if $\mu = \lambda$ we have $w(x) = 1/x$ and the second solution is $y = x^\lambda \ln x$. This was obtained by an entirely different method in Chapter 11.

We now use Theorem 1 to get information about a second solution when we have a first solution

$$u(x) = x^\lambda A(x), \qquad A(0) \neq 0$$

corresponding to λ. Substitution into the formula of Theorem 1 gives

$$w(x) = x^{\mu-\lambda-1} C(x) \quad \text{where} \quad C(x) = A(x)^{-2} e^{-R(x)}.$$

Since $A(x)$ and $R(x)$ have power-series expansions, and $A(0) \neq 0$, the results of Chapter B show that $C(x)$ also has a power-series expansion,

(5) $$C(x) = A(x)^{-2} e^{-R(x)} = \sum_{j=0}^{\infty} c_j x^j.$$

Theorem 1 involves the integral $W(x)$ of the function

$$w(x) = x^{\mu-\lambda-1} C(x) = \sum_{j=0}^{\infty} c_j x^{\mu-\lambda+j-1}.$$

We take the constant of integration to be 0. If $\lambda - \mu$ is different from j for every j the exponent -1 does not arise and

$$W(x) = x^{\mu-\lambda} \sum_{j=0}^{\infty} \frac{c_j}{\mu - \lambda + j} x^j.$$

But if $\lambda - \mu = n$, an integer, the term for $j = n$ is c_n/x and integrates to give $c_n \ln x$. In this case

$$W(x) = x^{\mu - \lambda} \sideset{}{'}\sum_{j=0}^{\infty} \frac{c_j}{\mu - \lambda + j} x^j + c_n \ln x$$

where the prime on the sum means that the term for $j = n$ is omitted. Multiplying by $u(x) = x^{\lambda} A(x)$ yields the following:

THEOREM 2. Let $u(x) = x^{\lambda} A(x)$ be a solution for the exponent λ. Then there is a second solution of the form

$$v(x) = x^{\mu} B(x) + cu(x) \ln x$$

where $B(x)$ has a convergent power-series expansion. The constant c is 0 if $\lambda - \mu$ is not an integer. But if $\lambda - \mu = n$, an integer, then c is the coefficient of x^n in the series for $C(x)$ in (5).

It is easily checked that $c_0 \neq 0$ and hence, if $\lambda = \mu$, there is always a logarithmic term. But there may or may not be such a term when $\lambda - \mu$ is a positive integer. The following example illustrates a case, already familiar, in which the logarithmic term is absent.

EXAMPLE 2. For the Bessel equation $\lambda - \mu = 2m$ and $R(x) = 0$. It was seen in Section 2 that $J_m(x)$ has the form $x^m A(x)$ where $A(x)$ has only even powers of x in its series expansion. Hence $C(x) = 1/A(x)^2$ also has only even powers of x; in other words, if $n = 2m$, where $2m$ is odd, then $c_n = 0$. Theorem 2 now shows that the log term is missing. This agrees with Section 2, where it was found that the general solution of Bessel's equation is

$$c_1 J_m(x) + c_2 J_{-m}(x)$$

unless m itself (and not only $2m$) is an integer.

The foregoing calculations give a method of approximation near the singular point. Without loss of generality let $A(0) = 1$, so that $A = 1 + h$ where

$$h(x) = a_1 x + a_2 x^2 + a_3 x^3 + \cdots.$$

We can get $A^{-2} = (1 + h)^{-2}$ by substituting this into the binomial theorem,

$$(6) \qquad (1 + h)^{-2} = 1 - 2h + 3h^2 - 4h^3 + \cdots.$$

If the series expansion of $P(x)$ has coefficients P_j then

$$R(x) = P_1 x + P_2 \frac{x^2}{2} + P_3 \frac{x^3}{3} + \cdots$$

and the series for e^{-R} can be found by substituting this into

(7) $$e^{-R} = 1 - R + \frac{R^2}{2!} - \frac{R^3}{3!} + \cdots.$$

By Theorem 1, a second solution $v(x)$ is determined as follows:

$$C(x) = A(x)^{-2}e^{-R(x)}, \quad W(x) = \int x^{\mu-\lambda-1}C(x)\, dx, \quad v(x) = A(x)x^{\lambda}W(x).$$

If a logarithmic term is needed, it will be produced automatically in the calculation of W.

EXAMPLE 3. In Problem 3(c) of Section 1 it was seen that the equation

$$xy'' = xy' + 2y$$

has the exponents $\lambda = 1$, $\mu = 0$ and that a solution corresponding to λ is

$$u(x) = x(1 + h) \text{ where } h(x) = \frac{3}{2}x + x^2 + \cdots.$$

We call this a *partial series* since it has only the first few terms. By (6)

$$(1 + h)^{-2} = 1 - 2(\frac{3}{2}x + x^2) + 3(\frac{3}{2}x)^2 + \cdots$$

aside from terms of order x^3. Thus, to the same approximation,

$$A(x)^{-2} = (1 + h)^{-2} = 1 - 3x + \frac{19}{4}x^2 + \cdots.$$

In the present case $P(x) = R(x) = -x$ and $C = A^{-2}e^{-R}$ satisfies

$$C(x) = (1 - 3x + \frac{19}{4}x^2 + \cdots)(1 + x + \frac{1}{2}x^2 + \cdots).$$

This gives the series for $C(x)$ and then

$$w(x) = x^{\mu-\lambda-1}C(x) = x^{-2}(1 - 2x + \frac{9}{4}x^2 + \cdots).$$

Integrating $w(x)$ and multiplying by $u(x)$ yields the desired second solution,

$$v(x) = u(x)(-x^{-1} - 2\ln x + \frac{9}{4}x + \cdots).$$

The logarithmic term is $-2u(x)\ln x$. If we keep this term, but use the partial series for u in the remaining terms, the final result is

$$v(x) = -1 - \frac{3}{2}x + \frac{5}{4}x^2 + \cdots - 2u(x)\ln x.$$

We have carried only enough terms to illustrate the principles. If many terms are desired, calculations such as this should be done on a computer, not by hand.

PROBLEMS

1P. If $xy'' + (3 - 2x)y' = 2y$ check that the exponents are $0, -2$ and obtain the recurrence formulas

$$(n+2)a_n = 2a_{n-1}, \qquad n \geq 1; \qquad n(n-2)a_n = 2(n-2)a_{n-1}, \qquad n \geq 0$$

for $s = 0$ and $s = -2$ respectively. Find series solutions with a_0 arbitrary in the first case, a_0 and a_2 arbitrary in the second. By comparing with the Taylor series for e^{2x}, express your solutions in terms of elementary functions. What happens if you divide the factor $n - 2$ out of the second recurrence formula?

2. Using Theorem 1 to get a second solution, solve:

(a) $x^2 y'' - 13xy' + 49y = 0$ (b) $x(1 - x)y'' + (1 - 3x)y' = y$

3. This pertains to the Bessel equation $x^2 y'' + xy' + (x^2 - m^2)y = 0$ where $m \geq 0$ and x is confined to interval I on which $x > 0$ and $J_m(x) \neq 0$.

(a) Taking $u = J_m$, $P = 1$, $R = 0$ in Theorem 1, show that the first equation below yields a second solution on I even if m is 0 or a positive integer:

$$v(x) = J_m(x) \int \frac{dx}{x J_m(x)^2}, \qquad v(x) = c_1 J_m(x) + c_2 J_{-m}(x)$$

(b) If m is not an integer explain why there are constants c_1, c_2 such that the second equation holds on I.

4. Divide the second equation of Problem 3 by $J_m(x)$, differentiate, and thus deduce it from Abel's identity.

5. It was seen in Section 1, Problem 3(c) that the equation considered in Example 3 has the closed-form solution

$$u(x) = x\left(1 + \frac{x}{2}\right)e^x$$

By Theorem 1 obtain a second solution

$$v(x) = x\left(1 + \frac{x}{2}\right)e^x \int x^{-2}\left(1 + \frac{x}{2}\right)^{-2} e^{-x}\, dx$$

Expand both solutions in series to enough terms to verify agreement with the result of Example 3.

6. Let $Ty = x^2 y'' + P(x)xy + Q(x)y$ as in the text. If $Tu = 0$ check that

$$T(u(x)\ln x) = 2xu'(x) + (P(x) - 1)u(x)$$

7. To solve Example 3 by the result of Theorem 2, take

$$u = x + \frac{3}{2}x^2 + x^3 + \cdots, \qquad v = b_0 + b_1 x + b_2 x^2 + \cdots + cu\ln x$$

The function u is the solution used in the example and the expression for v is given by Theorem 2 with $\mu = 0$. Substitute into $Tv = 0$, equate coefficients, and thus get the recurrence formulas

$$2b_0 = c, \qquad 2b_2 - 3b_1 = -\frac{7}{2}c, \qquad 4b_2 = \frac{7}{2}c$$

(Computation of $T(cu\ln x)$ is facilitated by, or can be checked by, the result of Problem 6.) Taking $b_0 = 4$ to avoid fractions, get the solution

$$v(x) = 4 + 14x + 7x^2 + \cdots + 8u(x)\ln x$$

Is this result consistent with that of Example 3? Why or why not?

8. Here are five equations from Section 1, Problem 3 for which we did not get a second solution. Using the partial series for u given at the right of the equation, get a partial series for a second solution as in Example 3 or as in Problem 7. Answers are given for the method of Example 3:

(a) $xy'' = y$ $\qquad\qquad\qquad x\left(1 + \dfrac{x}{2} + \dfrac{x^2}{12} + \cdots\right)$

(b) $4x^2 y'' = (4x - 1)y$ $\qquad\quad x^{1/2}\left(1 + x + \dfrac{x^2}{4} + \cdots\right)$

(c) $x^2 y'' + xy' = (1 + x^2)y$ $\qquad x\left(1 + \dfrac{x^2}{8} + \cdots\right)$

(d) $4x^2 y'' = (4x^2 - 1)y$ $\qquad\quad x^{1/2}\left(1 + \dfrac{x^2}{4} + \cdots\right)$

(e) $xy'' = (6 + 5x)y' + 5y$ $\qquad x^7\left(1 + 5x + \dfrac{25x^2}{2} + \cdots\right)$

9. In this problem you will need both $P_0 = 1 - \lambda - \mu$ and $Q_0 = \lambda\mu$.

(a) Substitute $y = x^\lambda(1 + a_1 x + \cdots)$ into the general Frobenius equation

$$x^2 y'' + (P_0 + P_1 x + \cdots)xy' + (Q_0 + Q_1 x + \cdots)y = 0$$

keep only the leading terms, and thus get $a_1 = -(Q_1 + P_1\lambda)/(\lambda - \mu + 1)$.

(b) When $\lambda - \mu = 1$ find c_1 corresponding to the function $u(x)$ in Part (a) and deduce that the logarithmic term in the second solution is

$$\left(Q_1 - \frac{P_0 P_1}{2}\right) u(x) \ln x$$

In particular, the logarithmic term is missing if, and only if, $P_0 P_1 = 2Q_1$.

10. This assumes familiarity with Section 2, Problem 7. The exponents $0, 1 - c$ of the hypergeometric equation differ by 1 if $c = 0$ or $c = 2$. Referring to Problem 9 show that the logarithmic term in a second solution for these two cases, respectively, is

$$-abx F(a + 1, b + 1, 2, x) \ln x, \qquad -(1 - a)(1 - b)F(a, b, 2, x) \ln x$$

11. If $A(0) = 1$ the Cauchy product yields $A(x)^2$ in the form

$$A(x)^2 = 1 + \alpha_1 x + \alpha_2 x^2 + \alpha_3 x^3 + \cdots$$

The reciprocal $A(x)^{-2} = 1 + \beta_1 x + \beta_2 x^2 + \cdots$ can be computed by equating coefficients in

$$(1 + \beta_1 x + \beta_2 x^2 + \cdots)(1 + \alpha_1 x + \alpha_2 x^2 + \cdots) = 1$$

Check that

$$\beta_1 = -\alpha_1, \qquad \beta_2 = -(\beta_1 \alpha_1 + \alpha_2), \qquad \beta_3 = -(\alpha_1\beta_2 + \alpha_2\beta_1 + \alpha_3)$$

This gives an alternative approach to the problem of computing $C(x)$.

12. Taking $A(0) = 1$ and using (6) or Problem 11 as you prefer, show that

$$A(x)^{-2} = 1 - 2a_1 x + (3a_1^2 - 2a_2)x^2 + (-4a_1^3 + 6a_1 a_2 - 2a_3)x^3 + \cdots$$

Also show, in the notation of the text, that

$$e^{-R(x)} = 1 - P_1 x + (P_1^2 - P_2)\frac{x^2}{2} + (3P_1 P_2 - P_1^3 - 2P_3)\frac{x^3}{6} + \cdots$$

The Cauchy product gives $C(x)$, leading to an additional term in calculations like those in Example 3.

——————————————————ANSWERS——————————————————

1. $(a_0/2)x^{-2}(e^{2x} - 1 - 2x)$, $a_0 x^{-2}(1 + 2x) + (a_2/2)x^{-2}(e^{2x} - 2x - 1)$. If you divide by $n - 2$ you get $a_2 = 2a_0$ although in reality a_2 is arbitrary. In the present case this does no harm, since the extra information is already given by the first solution. However, in some cases you may be led to a contradiction, or you may lose the second solution. When finding a second solution in the case $\lambda - \mu =$ integer, do not divide by a common factor until you have passed all values n for which it vanishes.

2. (a) x^7, $x^7 \ln x$ (b) $1/(1 - x)$, $(\ln x)/(1 - x)$. This is the special case $a = b = c = 1$ of the hypergeometric equation.

3. (b) Because the expression on the left satisfies Bessel's equation, as shown in Part (a). If m is not an integer the expression on the right gives the complete solution of this equation.

7. Yes, because if v is the result from Example 3 and w the result obtained in this problem, $4v + w = 8u$ within the accuracy of the calculations.

8. For convenience in printing the following solutions have been multiplied by -2, -4, -16, -4, -42, respectively:

(a) $2 + x - x^2 + \cdots + 2u(x) \ln x$

(b) $x^{1/2}(8x + 3x^2 + \cdots) - 4u(x) \ln x$

(c) $x^{-1}(8 + x^2 + \cdots) + 4u(x) \ln x$

(d) $x^{-1/2}(x^2 + \cdots) - 4u(x) \ln x$

(e) $6 - 5x + 5x^2 + \cdots$.
The coefficient of the log term is $c_7 = -5^7/7!$, which is beyond the scope of the calculation.

4. The method of Frobenius

In principle, the methods of the previous section give the second solution v explicitly as soon as the first solution u is known. In practice, however, the series for v is difficult to determine. Another shortcoming is that the procedure gives little information about the length of the interval on which the second solution is valid.

An alternative approach due to Frobenius gives the correct interval and often leads to simpler calculations. In this approach the exponent s is not set equal to λ or μ but is left as a variable. The corresponding series

(1)
$$y(x, s) = x^s \sum_{j=0}^{\infty} a_j(s)x^j$$

yields $Ty(x, s) = a_0(s)I(s)x^s$ when substituted into the differential equation $Ty = 0$. The term on the right arises from $j = 0$ and is not zero unless

$I(s) = 0$ (or $a_0(s) = 0$, a possibility we exclude). The terms for $j \geq 1$ do give 0, however, as can be seen from the recurrence formula.

The basic idea is to differentiate with respect to s. If the power-series factor in $y(x, s)$ converges for $|x| < r$, then

$$(2) \qquad T\frac{\partial}{\partial s}y(x, s) = \frac{\partial}{\partial s}Ty(x, s), \qquad 0 < x < r.$$

Though details will not be given here, this equation can be proved by the theory of uniform convergence as presented in Chapter F.

Suppose now that λ is a double root of $I(s)$. With $a_0(s) = 1$ we get

$$Ty(x, s) = (s - \lambda)^2 x^s$$

and hence by (2)

$$T\frac{\partial}{\partial s}y(x, s) = 2(s - \lambda)x^s + (x - \lambda)^2 x^s \ln x.$$

Setting $s = \lambda$ yields a second solution.

If $\lambda - \mu = n$, a positive integer, we choose $a_0(s) = s - \mu$ and obtain

$$Ty(x, s) = (s - \lambda)(s - \mu)^2 x^s.$$

Differentiating with respect to s and setting $s = \mu$ again yields a second solution. The results are summarized as follows:

THEOREM 1. *Let $P(x)$ and $Q(x)$ have power-series expansions for $|x| < r$.*

(a) *If $\lambda = \mu$ and $y(x, s)$ starts with $a_0(s) = 1$, a second solution is*

$$\frac{\partial}{\partial s}y(x, s)\Big|_{s=\lambda}.$$

(b) *If $\lambda - \mu$ is a positive integer and $y(x, s)$ starts with $a_0(s) = s - \mu$, a second solution is*

$$\frac{\partial}{\partial s}y(x, s)\Big|_{s=\mu}.$$

(c) *With $y(x, s)$ as in (1) we have, in either case,*

$$\frac{\partial}{\partial s}y(x, s) = x^s \sum_{j=0}^{\infty} a_j'(s)x^j + x^s \ln x \sum_{j=0}^{\infty} a_j(s)x^j, \qquad 0 < x < r.$$

The concluding statement is suggested by formal differentiation and can be proved by uniform convergence. It shows that the solutions have the same form as those in Section 3, Theorem 2, and gives the additional information that they are valid for $0 < x < r$.

EXAMPLE 1.   The Bessel equation of order 0 is $x^2 y'' + xy' + x^2 y = 0$. Substituting

$$y(x, s) = \sum_{j=0}^{\infty} a_j(s) x^{s+j}$$

with $a_0(s) = 1$ we get

(3) $(s + n)^2 a_n(s) = -a_{n-2}(s), \qquad n \geq 1.$

Equation (3) is not imposed for $n = 0$ because that would make $s = 0$, a restriction we want to avoid.

It is easily seen that $a_j(s) = 0$ for $j = 1, 3, 5, \cdots$ and that the remaining $a_j(s)$ are uniquely determined provided, as we now suppose, s is not a negative integer. (Actually, s will be close to 0.) Equating the logarithm of the absolute values in (3) we get

$$2 \ln |s + n| + \ln |a_n(s)| = \ln |a_{n-2}(s)|$$

and differentiation yields

$$\frac{2}{s+n} + \frac{a_n'(s)}{a_n(s)} = \frac{a_{n-2}'(s)}{a_{n-2}(s)}.$$

We introduce the abbreviations $a_n = a_n(0)$, $b_n = a_n'(0)$ and now, after the differentiation, we set $s = 0$. The result is

$$\frac{b_n}{a_n} - \frac{b_{n-2}}{a_{n-2}} = -\frac{2}{n}.$$

If you write this for $n = 2, 4, 6, \cdots, 2k$ and add, you will find that the series on the left telescopes to give

$$\frac{b_{2k}}{a_{2k}} = -h_k$$

where h_k is the kth partial sum of the *harmonic series*:

$$h_k = 1 + \frac{1}{2} + \frac{1}{3} + \cdots + \frac{1}{k}, \qquad k \geq 1.$$

The equation $n^2 a_n = -a_{n-2}$ obtained from (3) with $s = 0$ is the same as that for the coefficients in $J_0(x)$ and yields

$$a_{2n} = \frac{(-1)^n}{2^{2n}(n!)^2}, \qquad a_{2n+1} = 0.$$

Setting $s = 0$ in Theorem 1, Part (c) we obtain the desired second solution,

$$y = J_0(x) \ln x - \sum_{j=1}^{\infty} \frac{(-1)^j (x/2)^{2j}}{(j!)^2} h_j.$$

We conclude with the following elementary theorem, which sheds further light on the Fuchs-Frobenius theory:

THEOREM 2. *Let λ and μ be unequal real or complex numbers. Let $A(x)$ and $B(x)$ be any two power series that have positive radii of convergence and satisfy $A(0) \neq 0$, $B(0) \neq 0$. Then on a sufficiently small interval $0 < x < r$ there is one, and only one, equation in the Frobenius normal form that has*

$$c_1 x^{\lambda} A(x) + c_2 x^{\mu} B(x)$$

as its general solution.

The proof of Theorem 2 is not difficult and is outlined in Problem 4.

Theorem 2 shows why the Frobenius normal form is singled out for special study. It may seem strange to start with the solution and deduce the form of the equation, but that is the way the subject started. Historically, the Frobenius normal form and its nth-order generalization were derived by requiring certain properties of the solutions. It was shown first that these properties lead to a differential equation in the Frobenius normal form and second, that the solutions of such an equation always have the specified properties. This discovery represented a major advance in the theory of differential equations.

PROBLEMS————————————————————————————

1. This problem is like a guided tour over difficult terrain. The objective is to illustrate Theorem 1(b) for the the Bessel equation with $m = 1$. Proceeding as in Example 1, obtain the recurrence formula

$$((s+n)^2 - 1)a_n(s) = -a_{n-2}(s), \qquad n \geq 1$$

You should you take $a_0(s) = s + 1$. Why? Check that $a_{2n+1}(s) = 0$ and

$$a_{2n}(s) = \frac{(-1)^n}{(s+3)^2(s+5)^2 \cdots (s+2n-1)^2(s+2n+1)}, \qquad n \geq 2$$

You can assume that the factors are positive. Why? Compute $\ln |a_{2n}(s)|$, differentiate with respect to s, and get

$$\frac{a'_{2n}(s)}{a_{2n}(s)} = -\frac{2}{s+3} - \frac{2}{s+5} - \cdots - \frac{2}{s+2n-1} - \frac{1}{s+2n+1}, \qquad n \geq 2$$

By the recurrence formula deduce $a_2(s) = 1/(s+3)$ for $s \neq -1$. You should use the same formula when $s = -1$. Why? In the notation of Example 1, and with $h_0 = 0$ by definition, set $s = -1$ to get

$$\frac{a'_{2n}(-1)}{a_{2n}(-1)} = -\frac{1}{2}(h_n + h_{n-1}), \qquad n \geq 1$$

Check that Theorem 1(c) with $s = \mu = -1$ yields the second solution

$$y = \frac{1}{x}\left(1 - \sum_{n=1}^{\infty}(h_n + h_{n-1})\frac{(-1)^n(x/2)^{2n}}{n!(n-1)!}\right) + 2\frac{\ln x}{x}\sum_{n=1}^{\infty}\frac{(-1)^n(x/2)^{2n}}{n!(n-1)!}$$

This agrees with Theorem 2 of the preceding section. Why?

2. Solve Example 1 by differentiating $\ln|a_{2n}(s)|$ as in Problem 1.

3. Solve Problem 1 by getting a recurrence formula for $a'_n(s)/a_n(s)$ as in Example 1. Hint: You will find it advisable to take $n \geq 3$ in general and to use the special formula $a_2(s) = 1/(s+3)$ separately as in Problem 1. A series that you will encounter can be expressed in terms of h_n by use of the partial fraction expansion

$$\frac{2(n-1)}{n(n-2)} = \frac{1}{n} + \frac{1}{n-2}$$

4. Proof of Theorem 2. Write the differential equation in the form

$$P(x)xy' + Q(x)y = -x^2 y''$$

substitute $u = x^\lambda A(x)$ for y, and divide by x^λ. Your result should be of the form

$$P(x)(xA'(x) + \lambda A(x)) + Q(x)A(x) = -U(x)$$

where $U(x)$ has a convergent power-series expansion. In a similar manner, the function $v = x^\mu B(x)$ leads to

$$P(x)(xB'(x) + \mu B(x)) + Q(x)B(x) = -V(x)$$

where $V(x)$ has a convergent power-series expansion. This is a system of two linear equations in the two unknowns $P(x), Q(x)$. Verify that the coefficient determinant $D(x)$ satisfies $D(0) = (\lambda - \mu)A(0)B(0) \neq 0$ and finish the proof.

———————————————ANSWERS———————————————

1. Because the second exponent is $\mu = -1$; because s will be close to -1; because you want $a_2(s)$ to be differentiable at $s = -1$; because the right-hand term is $-J_1(x)\ln x$.

4. As a check: $U(x) = x^2 A''(x) + 2\lambda x A'(x) + \lambda(\lambda - 1)A(x)$

5. Convergence

Let a linear operator T be defined by

$$Ty = x^2 y'' + P(x)xy' + Q(x)y$$

where P and Q have convergent power-series expansions

$$P(x) = \sum_{j=0}^{\infty} P_j x^j, \qquad Q(x) = \sum_{j=0}^{\infty} Q_j x^j.$$

If s is constant, a short calculation gives

(1) $\qquad Tx^s = C(x,s)x^s, \qquad C(x,s) = s(s-1) + P(x)s + Q(x).$

The series expansions for P and Q yield

(2) $\qquad C(x,s) = c_0(s) + c_1(s)x + c_2(s)x^2 + c_3(s)x^3 + \cdots$

with $c_0(s) = s(s-1) + P_0 s + Q_0$ and

(3) $\qquad\qquad c_j(s) = sP_j + Q_j, \qquad j \geq 1.$

Note that $c_0(s) = I(s)$, the indicial polynomial. This fact is used below.

These equations give a convenient method of finding the formal power-series expansion of Ty where, as elsewhere in this chapter,

$$y(x) = a_0 x^s + a_1 x^{s+1} + a_2 x^{s+2} + a_3 x^{s+3} + \cdots, \qquad a_0 \neq 0.$$

By (1) with s replaced by $s, s+1, s+2, \cdots$ we get

$$T(a_0 x^s) = a_0 C(x,s)x^s$$

$$T(a_1 x^{s+1}) = a_1 C(x, s+1)x^{s+1}$$

$$T(a_2 x^{s+2}) = a_2 C(x, s+2)x^{s+2}$$

and so on. The coefficients of x^{s+n} in these expressions are, respectively,

$$a_0 c_n(s), \quad a_1 c_{n-1}(s+1), \quad a_2 c_{n-2}(s+2), \quad \cdots$$

and hence the coefficient of x^{n+s} in the formal expansion of Ty is

$$a_0 c_n(s) + a_1 c_{n-1}(s+1) + a_2 c_{n-2}(s+2) + \cdots + a_n c_0(s+n).$$

Let us recall that $c_0(s) = I(s)$, the indicial polynomial. Setting the above coefficient equal to 0 gives $a_0 I(s) = 0$ and

$$(4) \quad -I(s+n)a_n = a_0 c_n(s) + a_1 c_{n-1}(s+1) + \cdots + a_{n-1} c_1(s+n-1)$$

for $n \geq 1$. The choice $n = 0$ shows, as expected, that s must be one of the roots λ or μ of the indicial equation; we assume $\lambda \geq \mu$ when both are real. The discussion of Section 3 shows that $I(s+n) \neq 0$ if $s = \lambda$, and also if $s = \mu$ provided $\lambda - \mu$ is not an integer. Only such cases will be considered here. What is important is that s is constant, either λ or μ, and that $I(s+n)$ does not vanish for any positive integer n. Since $I(s+n)$ behaves like n^2 for large n, and does not vanish, there is a positive constant K such that

$$(5) \quad |I(s+n)| \geq Kn^2, \qquad n = 1, 2, 3, \cdots.$$

This inequality is used below.

We shall establish Theorem 1 of Section 1, which is repeated for convenience:

THEOREM 1. *Under the above conditions suppose the series for P and Q both converge for $|x| < r$, where $r > 0$. Then the series $A(x)$ associated with the solution $y = x^s A(x)$ also converges for $|x| < r$.*

For proof, let x with $|x| < r$ be given and choose r_0 and R so that

$$|x| < r_0 < R < r.$$

By Chapter 12, Section 4, Theorem 1 there are constants B_1 and B_2 such that

$$|P_j| \leq \frac{B_1}{R^j}, \qquad |Q_j| \leq \frac{B_2}{R^j}.$$

Equation (3) yields

$$c_{n-j}(s+j) = (s+j)P_{n-j} + Q_{n-j}, \qquad j = 0, 1, \cdots, n-1$$

and hence

$$|c_{n-j}(s+j)| \leq \frac{|s+j|B_1 + B_2}{R^{n-j}}.$$

This yields an inequality of the form

$$(6) \qquad |c_{n-j}(s+j)| \leq \frac{Bn}{R^{n-j}}$$

for $n \geq 1$ and for the above range of j, where B is a positive constant.

If we set $\theta = r_0/R$ then $0 < \theta < 1$ and the formula for the sum of a geometric series yields

(7)
$$\sum_{j=0}^{n-1} \theta^{n-j} \leq \frac{1}{1-\theta}.$$

It will be seen that (7) yields

(8)
$$|a_n| \leq \frac{A}{r_0{}^n}$$

for some constant A. The hypothesis $|x| < r_0$ then shows that the series $A(x)$ converges by comparison with the geometric series.

First let us see under what conditions (8) can be obtained for n if (8) is known for all values up to $n - 1$. The recursion formula (4) is

$$-I(s+n)a_n = \sum_{j=0}^{n-1} a_j c_{n-j}(s+j).$$

By (8) and (6) with $R = r_0/\theta$

$$|a_j c_{n-j}(s+j)| \leq \frac{A}{r_0{}^j} \frac{Bn}{R^{n-j}} = \frac{ABn\theta^{n-j}}{r_0{}^n}.$$

The inequality (7) now yields

$$|I(n+s)||a_n| \leq \frac{Bn}{1-\theta} \frac{A}{r_0{}^n}.$$

Dividing by $I(s+n)$ and using (5) we get

$$|a_n| \leq \frac{B}{K(1-\theta)n} \frac{A}{r_0{}^n}.$$

This yields (8) at the value n, as desired, provided $n \geq N$ where N is sufficiently large. Having fixed N, let A be the largest of the numbers

$$|a_0|, \qquad |a_1|r_0, \qquad |a_2|r_0{}^2, \qquad \cdots, \qquad |a_N|r_0{}^N.$$

The inequality $|a_n| \leq A/r_0{}^n$ holds for $n \leq N$ by choice of A and it holds for $n > N$ as seen above. This completes the proof.

REVIEW PROBLEMS————————————————————————

1. The origin is said to be an *accidental singular point* for a second-order linear differential equation if it is not a regular point and yet the equation admits two linearly independent solutions, both of which have power-series expansions. By trying an ordinary power series as in Chapter 12, show that the following have accidental singular points at the origin:

(a) $x^2y'' - 7xy' + 7y = 0$ (b) $x^2y'' - (4 - x)xy' + (4 - x)y = 0$

(c) $x^2y'' - (4 - ax^2)xy' + (6 + bx^2)y = 0$, a, b const.

2. As mentioned in Chapter 12, an m-step recursion formula gives a_{m+n} in terms of a_n. Show that the following equations lead to a one-step and two-step recursion formula, respectively, for the coefficients of a generalized power-series solution. The letters a, b, c, d, e, f denote constants:

(a) $x^2y''(ax + b) + xy'(cx + d) + (ex + f)y = 0$

(b) $x^2y''(ax^2 + b) + xy'(cx^2 + d) + (ex^2 + f)y = 0$

3. Show that the equations in Problem 2 both lead to the same indicial equation, and find it.

4. The equation $xy'' + (1 - 2x)y' = (1 - x)y$ has exponents $\lambda = \mu = 0$ and leads to a three-term recurrence formula. Get a simple solution by a suitable choice of initial values and use it to find a second solution involving a logarithm.

5. (abc) Open-ended problems. If a and b are constant the equations

(a) $x^2y'' + ay = (x + a)xy'$ (b) $xy'' + (a - bx)y' = by$

have the elementary solutions $x^a e^x$ and $x^{1-a}e^{bx}$ respectively. Obtain a second solution in as many ways as you can.

(c) Solve $x^m y'' = y'$ as in Chapter 2 and discuss the behavior when the origin is an irregular singular point, as well as when it is a regular singular point.

————————————————ANSWERS————————————————

1. (a) x, x^7 (b) $x, \dfrac{x^4}{3 \cdot 0!} - \dfrac{x^5}{4 \cdot 1!} + \dfrac{x^6}{5 \cdot 2!} - \dfrac{x^7}{6 \cdot 3!} + \cdots$

(c) $x^2 - \dfrac{2a + b}{2!}x^4 + \dfrac{(4a + b)(2a + b)}{4!}x^6 - \cdots,$

$x^3 - \dfrac{3a + b}{3!}x^5 + \dfrac{(5a + b)(3a + b)}{5!}x^7 - \cdots$

3. $s(s - 1)b + sd + f = 0$ **4.** $e^x, e^x \ln x$

Chapter 14

EXISTENCE THEORY

THE PRINCIPAL existence and uniqueness theorems for differential equations are established in this chapter. With the aid of vector notation, it is found that the results for a first-order equation $y' = f(t, y)$ extend easily to systems of simultaneous differential equations. Important special cases include linear systems, which form the topic of Chapter 15, and equations of higher order. A brief discussion of bifurcation sheds light on the problem of uniqueness and on the theory of envelopes.

This chapter assumes knowledge of uniform convergence to the extent developed in Chapter F.

1. Introduction

In Sections 1, 2, and 3 of this chapter, all variables and functions are real. If $y = \phi(t)$ satisfies the initial-value problem

$$(1) \qquad \frac{dy}{dt} = f(t, y), \qquad y(t_0) = y_0$$

then $f(t, y)$ gives the slope of the curve $y = \phi(t)$ at any point (t, y) on it. For broad classes of functions f, we can get an idea of the shape of this curve in the following way: Draw a line segment through the point (t_0, y_0) with slope $f(t_0, y_0)$ and proceed a short distance along this segment to a new point (t_1, y_1). Draw a line segment through (t_1, y_1) with slope $f(t_1, y_1)$, proceed a short distance along it to a new point (t_2, y_2), and so on. The result is a polygonal line, called the *Euler polygon*, which is built up of short segments. The general appearance of the Euler polygon in a typical case is shown in Figure 1.

Suppose now that $|f| \leq M$ where M is constant. In this case the segments in the Euler polygon have slopes between $-M$ and M and the polygon lies between two lines of slopes $\pm M$ through the point (t_0, y_0) as shown in Figure 1. It is natural to surmise that the solution has the same property. This surmise can be proved, without using the Euler polygon and without assuming that the solution is unique. We shall establish the following:

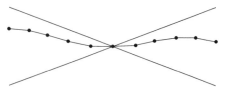

Figure 1

THEOREM 1. *Let I be an interval containing the point t_0, and let M be constant. Suppose $y = \phi(t)$ satisfies $|y'| \leq M$ for $t \in I$ and $y(t_0) = y_0$. Then*

$$|y - y_0| \leq M|t - t_0|, \qquad t \in I.$$

With $y = \phi(t)$ and $y_0 = \phi(t_0)$ the mean-value theorem gives

$$|y - y_0| = |(t - t_0)\phi'(c)| = |t - t_0||\phi'(c)|$$

where c is between t and t_0. The hypothesis $|y'| \leq M$ yields $|\phi'(c)| \leq M$ and Theorem 1 follows.

A function differentiable on an open interval can have an extremely complicated behavior at the endpoints. For example, the functions

$$\sin\frac{1}{t}, \qquad \frac{1}{t}\sin\frac{1}{t}$$

are both differentiable on the open interval $t > 0$, yet oscillate with extraordinary violence as t approaches 0, the second even more than the first. The following theorem shows that functions with a bounded derivative cannot behave in this way:

THEOREM 2. *On a finite interval $a < t < b$ let $|\phi'(t)| \leq M$ where M is constant. Then $\phi(t)$ has a limit as $t \to a+$ and also as $t \to b-$.*

The notation $t \to a+$ means that $t > a$ throughout the limiting process, and $t \to b-$ means that $t < b$ throughout the limiting process. These restrictions are needed to keep t on the interval (a, b) where $\phi(t)$ is defined.

For proof, it follows from Theorem 1 that $|\phi(t)|$ is bounded and hence the function $F(t) = \phi(t) + Mt$ is also bounded. The condition $|\phi'(t)| \leq M$ gives

$$F'(t) = \phi'(t) + M \geq 0,$$

which shows that $F(t)$ is increasing. Since $F(t)$ is both bounded and increasing, the completeness axiom of Chapter 12, Section 3 ensures that the limit

(2) $$\lim_{t \to b-} F(t) = L$$

exists. (The completeness axiom is stated for sequences in Chapter 12 but the result used here is an easy consequence.) The equation

$$\lim_{t \to b-} \phi(t) = \lim_{t \to b-} (F(t) - Mt) = L - Mb$$

now shows that $\phi(t)$ has a limit as $t \to b-$. The proof for $t \to a+$ is similar.

We used the Euler polygon as a point of departure for Theorem 1. The concept of iteration is introduced next, and the Euler polygon again provides a convenient point of departure. Let the successive points where the segments of the polygon join be

(3) $(t_0, y_0),$ $(t_1, y_1),$ $(t_2, y_2),$ $\cdots,$ $(t_n, y_n),$ $\cdots.$

For definiteness, we go a constant horizontal distance h from one point to the next, where h is a small positive number. The Euler construction is then formalized by the equations

(4) $t_{n+1} = t_n + h,$ $y_{n+1} = y_n + h f(t_n, y_n).$

Each point is determined from its predecessor by a uniform procedure, and the process as a whole is an example of an *iterative scheme*. Many methods in numerical analysis depend on iterative schemes such as this, with various refinements to improve the behavior of the errors $\phi(t_n) - y_n$.

In this chapter we use iteration in a somewhat different form. Instead of constructing a sequence of points (t_n, y_n) as in (3) we construct a sequence of functions $\{y_n(t)\}$ satisfying

$$y_1(t_0) = y_0,   y_2(t_0) = y_0,   \cdots,   y_n(t_0) = y_0,   \cdots$$

and

$$y_1' = f(t, y_0),   y_2' = f(t, y_1),   \cdots,   y_{n+1}' = f(t, y_n),   \cdots.$$

Under suitable hypotheses it will be shown that the sequence converges and yields a solution of the differential equation $y' = f(t, y)$. Note that each of the above equations can be solved by the methods of Chapter 2. Namely,

$$y_{n+1}(t) = y_0 + \int_{t_0}^{t} f(s, y_n(s)) \, ds.$$

If the integrand is continuous, as it will be in the sequel, the fundamental theorem of calculus gives

$$y_{n+1}'(t) = f(t, y_n(t))$$

and the condition $y_n(t_0) = y_0$ is also satisfied. Thus the y_j are determined one at a time, starting with the known value y_0. This process is called *Picard iteration* after the French mathematician Émile Picard, who used it in 1893 to develop a comprehensive existence theory for differential equations.

One of the main requirements for convergence is that the function $f(t, y)$ should not change too much when y is changed. In the Euler polygon, for example, it is desirable to have the successive values y_{n+1} and y_n in some sense close to each other. Since these values depend on y_n and y_{n-1} there is a cumulative effect. Rapid change of $f(t, y)$ can make y_n very different from y_{n-1}, and this difference may be further exaggerated by the effect of f at the next stage. Instead of converging to a definite curve $y = \phi(t)$ as we take smaller and smaller values of h in (4), the sequence of Euler polygons in such cases generally diverges. Similar remarks apply to the Picard method.

An appropriate restriction on f is given by the following definition:

LIPSCHITZIAN FUNCTIONS. The function $f(t, y)$ is Lipschitzian with respect to y in a region R if there is a constant K such that

$$(5) \qquad |f(t, y) - f(t, z)| \le K|y - z|, \qquad (t, y) \in R, \ (t, z) \in R.$$

The region R can be any set of points, but no harm is done if you consider R to be a familiar region such as a rectangle or disk. What is really needed is that (5) shall hold for all values of the variables under consideration.

As an illustration, the function $\sin y$ satisfies

$$\sin y - \sin z = (y - z)\cos c$$

for some c between y and z. This follows from the mean-value theorem of differential calculus. Hence $\sin y$ is Lipschitzian in the whole (t, y) plane and the Lipschitz constant K can be taken as 1. An extension of this reasoning yields the following theorem:

THEOREM 3.   *Let $f(t, y)$ be such that $f_y = \partial f/\partial y$ is bounded in a rectangular region R. Then f is Lipschitzian with respect to y in R.*

For proof, let $|f_y(t, y)| \le K$ in R where K is constant, and consider $f(t, y)$ to be a function of y alone. The mean-value theorem for functions of one variable yields

$$|f(t, y) - f(t, z)| = |(y - z)f_y(t, c)| \le |y - z|K$$

where c is between y and z. (The value of c depends on t, y, and z but that does not matter.) The inequality at the right follows from the hypothesis $|f_y| \le K$ and yields the conclusion; we assume, naturally, that (t, y) and (t, z) lie in R. A similar argument applies to more general regions.

Many functions of importance in applications are Lipschitzian but do not fall within the scope of Theorem 3. One of the simplest of these is $|y|$, which is Lipschitzian though not differentiable at $y = 0$. The function $|y|$ can be thought of as obtained by joining two Lipschitzian functions, $-y$ and y, at the origin. The following theorem applies to cases of this kind:

THEOREM 4.   *Suppose $f(t, y)$ is defined in the strip $c < y < d$ and is Lipschitzian with respect to y in each of the strips*

$$c < y \le c_1, \quad c_1 \le y \le c_2, \quad \cdots, \quad c_{n-1} \le y \le c_n, \quad c_n \le y < d.$$

Then $f(t, y)$ is Lipschitzian with respect to y for $c < y < d$.

If the theorem holds when there are only two strips, a moment's thought shows that repeated use of this fact yields the whole theorem. Also the Lipschitz constants K_1 and K_2 for the two strips can be replaced by

$$K = \max(K_1, K_2)$$

without losing the hypothesis. Thus it suffices to prove Theorem 1 when there are two strips with the same constant K for each.

This reduced version of Theorem 4 has a simple geometric intererpretation. At any fixed t the quotient

$$\frac{f(t, y) - f(t, z)}{y - z}, \qquad y \ne z$$

represents the slope of the chord joining two points on the graph of $f(t, y)$ versus y. The Lipschitz condition says that the magnitude of the slope never exceeds K. The reduced version of Theorem 4 asserts that if two curves with this property are joined end to end, the extended curve so obtained has the same property. The truth of this is made plausible by Figure 2 and a formal proof is outlined in Problem 5.

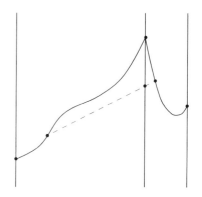

Figure 2

The following theorem from Chapter 10, Section 4 is needed in this chapter and is repeated for convenience:

THEOREM 5.   *With M constant, let $|w'(t)| \le M|w(t)|$ hold on an open interval I containing the point t_0. Then $w(t_0) = 0 \Rightarrow w(t) = 0$, $t \in I$.*

PROBLEMS——————————————————————————

1. If $p(t)$ is bounded for $a < t < b$, show that $p(t)y$ is Lipschitzian with respect to y in the strip $a < t < b$, $-\infty < y < \infty$. This is the reason why Lipschitzian functions are often met in the study of linear differential equations.

2. Is the function $|\sin y|$ Lipschitzian? Why or why not?

3. In a region in which $|y| \leq M$, where M is constant, show that the function y^2 satisfies a Lipschitz condition with constant $K = 2M$. Show however that if y is unrestricted the function y^2 is not Lipschitzian. This failure of the Lipschitz condition is reflected in the behavior of the initial-value problem

$$y' = y^2, \qquad y(0) = y_0 > 0$$

Namely, the solution $y = y_0/(1 - ty_0)$ tends to ∞ as $t \to 1/y_0$.

4. Show that the function $y^{2/3}$ is not Lipschitzian in any strip $|y| < h$ containing the origin. This fact is reflected in the behavior of

$$y' = y^{2/3}, \quad y(0) = 0$$

Namely, there are two solutions, $y = 0$ and $y = (t/3)^3$.

5. Prove the reduced form of Theorem 4 as mentioned in the text. Hint: Let $c < y \leq c_1 \leq z < d$ and use

$$f(t, y) - f(t, z) = f(t, y) - f(t, c_1) + f(t, c_1) - f(t, z)$$

6P. For the equation $y' = 2ty^2$, $y(0) = 1$, Picard's formula is

$$y_{n+1}(t) = 1 + \int_0^t 2sy_n(s)^2 \, ds, \qquad y_0 = 1.$$

(a) Check that $y_1(t) = 1 + t^2$ and find $y_2(t)$.

(b) Solve by separation of variables and compare the series expansion of the exact solution with the result obtained in Part (a).

7. Using Picard's method obtain the iterate y_j specified at the right of the problem:

(a) $y' = 1 + ty$ $y(0) = 0$ y_3 (d) $y' = y$ $y(0) = 1$ y_n

(b) $y' = 1 + ty$ $y(0) = 6$ y_2 (e) $y' = 1 + y^2$ $y(0) = 0$ y_3

(c) $y' = t + y$ $y(0) = -1$ y_{n-1} (f) $y' = 1 + ty^2$ $y(0) = 0$ y_3

8. Assuming that $y = \phi(t)$ has a continuous derivative, deduce Theorem 1 from the equation

$$\phi(t) - \phi(t_0) = \int_{t_0}^t \phi'(s) \, ds$$

9. If $w(t_0) = 0$ is not assumed in Theorem 5, show that

$$|w(t_0)|e^{-M|t-t_0|} \leq |w(t)| \leq |w(t_0)|e^{M|t-t_0|}, \qquad t \in I.$$

Hint: If $w(t) = 0$ at any point of I the conclusion follows from Theorem 5. Hence you can assume, without loss of generality, that $w(t) > 0$ on I. Set $w = e^y$ and use Theorem 1.

───────────────────────ANSWERS───────────────────────

2. Yes, by Theorem 4.

6. (a) $y_2(t) = 1 + t^2 + t^4 + t^6/3$ (b) $y = 1/(1 - t^2) = 1 + t^2 + t^4 + \cdots$.

7. (a) $t + t^3/3 + t^5/15$ (d) $1 + t + t^2/2! + \cdots + t^n/n!$
   (b) $6 + t + 3t^2 + t^3/3 + 3t^4/4$ (e) $t + t^3/3 + 2t^5/15 + t^7/63$
   (c) $-1 - t + t^n/n!$ (f) $t + t^4/4 + t^7/14 + t^{10}/160$

---

2. A local existence theorem

If $|f(t, y)| \le M$, where M is constant, every solution of the initial-value problem

$$y' = f(t, y), \qquad y(t_0) = y_0$$

lies in the region defined by

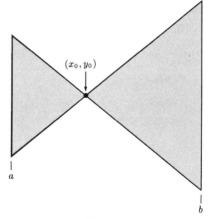

$$(1) \qquad |y - y_0| \le M|t - t_0|.$$

This follows from Section 1, Theorem 1. In general, the condition $|f(t, y)| \le M$ does not hold for all t nor does the solution exist for all t. We therefore supplement the inequality (1) by $a < t < b$ where a and b are constants satisfying $a < t_0 < b$. The region defined by these two conditions is called a *standard region* and is denoted by

Figure 3

$$R(t_0, y_0, a, b, M) = \{(t, y) \; : \; a < t < b, \; |y - y_0| \le M|t - t_0|\}.$$

A standard region is shown in Figure 3, and you may find it helpful to glance at this figure from time to time as you read this section.

It is desirable to have the same constant M in the definition of the standard region as in the inequality $|f(t, y)| \le M$. This is accomplished as follows. We start with a region—say a disk or rectangle—in which $|f(t, y)| \le M$, and then we associate a standard region with parameter M to any specified point (t_0, y_0) in it. The procedure is illustrated in Figure 4.

This section is devoted to a proof of the following:

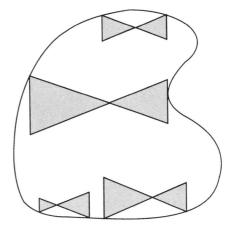

Figure 4

LOCAL EXISTENCE THEOREM. In $R = R(t_0, y_0, a, b, M)$ let f be a continuous function that is Lipschitzian with respect to y and satisfies the inequality $|f(t, y)| \leq M$. Then the initial-value problem

$$y' = f(t, y), \qquad y(t_0) = y_0$$

has a solution $y = \phi(t)$ for $a < t < b$ and the graph of y lies in R.

The theorem is *local* because in most cases a suitable standard region is available only in the immediate neighborhood of the point (t_0, y_0). A global theorem is given later.

For proof, we use Picard iteration

$$y_1'(t) = f(t, y_0), \quad y_2'(t) = f(t, y_1(t)), \quad \cdots, \quad y_{n+1}'(t) = f(t, y_n(t)), \quad \cdots$$

with the initial value $y_n(t_0) = y_0$ in each case. Here t ranges over the interval $a < t < b$. The graph of the constant function $y = y_0$, $a < t < b$ clearly lies in R. If, for a specified n, the graph of y_n lies in R then

$$|f(t, y_n(t))| \leq M, \qquad a < t < b.$$

Hence $|y_{n+1}'(t)| \leq M$. Section 1, Theorem 1 yields

$$|y_{n+1}(t) - y_0| \leq M|t - t_0|, \qquad a < t < b$$

which shows that the graph of y_{n+1} also lies in R. By mathematical induction the graph of every y_j lies in R, and this is the first step of the proof.

For the second step, we assume $t_0 \leq t < b$; the case $a < t \leq t_0$ is similar. Recall that Picard's procedure gives

$$(2) \qquad\qquad y_{n+1}(t) = y_0 + \int_{t_0}^{t} f(s, y_n(s)) \, ds.$$

If we choose a constant M_0 so that

$$|f(t, y_0)| \leq M_0, \quad a < t < b$$

Equation (2) with $n = 0$ yields

$$(3) \qquad |y_1(t) - y_0| \leq \int_{t_0}^{t} M_0 ds = M_0(t - t_0).$$

This completes the second step of the proof. The constant M_0 could be replaced by M but M_0 gives a better estimate for the rate of convergence of y_n. It also has some theoretical advantages as seen later.

We now take $n \geq 1$. If (2) is written with $n - 1$ instead of n and the results subtracted we get first

$$y_{n+1}(t) - y_n(t) = \int_{t_0}^{t} (f(s, y_n(s)) - f(s, y_{n-1}(s)))\, ds$$

and then, with the Lipschitz constant K,

$$(4) \qquad |y_{n+1}(t) - y_n(t)| \leq \int_{t_0}^{t} K |y_n(s) - y_{n-1}(s)|\, ds.$$

This follows from the Lipschitz condition, which is available because the graphs of the y_j all lie in R. The inequality (3) with t replaced by s is

$$|y_1(s) - y_0| \leq M_0(s - t_0).$$

Substituting into (4) when $n = 1$ yields

$$|y_2(t) - y_1(t)| \leq K \int_{t_0}^{t} M_0(s - t_0)\, ds = K M_0 \frac{(t - t_0)^2}{2!}.$$

We will now prove that, more generally,

$$(5) \qquad |y_n(t) - y_{n-1}(t)| \leq M_0 K^{n-1} \frac{(t - t_0)^n}{n!}.$$

Indeed, if (5) holds for a specified value n, we write (5) with t replaced by s. Substitution into (4) yields

$$|y_{n+1}(t) - y_n(t)| \leq K \int_{t_0}^{t} M_0 K^{n-1} \frac{(s - t_0)^n}{n!}\, ds$$

or also

$$(6) \qquad |y_{n+1}(t) - y_n(t)| \leq M_0 K^n \frac{(t - t_0)^{n+1}}{(n + 1)!},$$

which is (5) with n replaced by $n + 1$. Hence (5) holds by mathematical induction. This is the third step.

After these preliminaries the proof is easily completed. Since

$$|y'_{n+1}(t) - y'_n(t)| = |f(t, y_n(t)) - f(t, y_{n-1}(t))| \le K|y_n(t) - y_{n-1}(t)|$$

by the Lipschitz condition, Equation (5) gives

$$(7) \qquad |y'_{n+1}(t) - y'_n(t)| \le M_0 K^n \frac{(t - t_0)^n}{n!}.$$

Recalling that $t_0 \le t < b$, we replace $t - t_0$ by $b - t_0$ in the right-hand sides of (6) and (7). The resulting bounds are independent of t. By the M test for sequences (Chapter F) the sequences $\{y_n\}$ and $\{y'_n\}$ are uniformly convergent on the interval $t_0 \le t < b$. By Chapter F, Theorem 2 there is a function $y = \phi(t)$ defined on this interval such that simultaneously

$$\lim_{n \to \infty} y_n(t) = \phi(t), \qquad \lim_{n \to \infty} y'_n(t) = \phi'(t).$$

The conditions $y(t_0) = y_0$, $|y - y_0| \le M|t - t_0|$ are inherited from the corresponding conditions for y_n. The second of these shows that the graph of $y = \phi(t)$ is in R. The equation

$$\lim_{n \to \infty} f(t, y_n(t)) = f(t, \phi(t))$$

holds because f is continuous, or also because

$$|f(t, y_n(t)) - f(t, \phi(t))| \le K|y_n(t) - \phi(t)|.$$

Hence letting $n \to \infty$ in $y'_{n+1}(t) = f(t, y_n(t))$ yields

$$\phi'(t) = f(t, \phi(t)), \qquad t_0 \le t < b.$$

This completes the proof.

PROBLEMS

1. Rectangular region. Suppose f is continuous and Lipschitzian with respect to y in the rectangular region defined by

$$|t - t_0| < h, \qquad |y - y_0| < k, \qquad h, k > 0 \text{ const.}$$

Suppose further that $f(t, y) \le M$ in this region. Deduce from the local existence theorem that the initial-value problem considered in that theorem has a solution at least over the interval

$$|t - t_0| \le \min\left(h, \frac{k}{M}\right).$$

2. Problems 2, 3, and 4 pertain to the main theorem of this section. Let f be continuous and Lipschitzian with respect to y for $a < t < b$ and for $|y| < \infty$. Let the hypothesis that $|f(t, y)| \leq M$ be replaced by the weaker hypothesis that $f(t, y_0)$ is bounded for $a < t < b$. Does the solution $y = \phi(t)$ necessarily exist over the whole interval $a < t < b$? Why or why not?

3. Why do the limits $\lim_{t \to a+} \phi(t)$ and $\lim_{t \to b-} \phi(t)$ both exist?

4. Uniqueness. Suppose there are two solutions y and z in the theorem, valid for $a < x < b$ and such that the graphs of both solutions lie in R. Deduce from the Lipschitz condition that $|y' - z'| \leq K|y - z|$ and deduce from this that $y = z$. Hint: Use Section 1, Theorem 5.

5. A counterexample. Let $f(t, y) = 1$ in the region defined by $|y| \leq |t|$ and let f be so defined outside this region that

$$f(t, t + t^3) = 1 + 3t^2$$

Show that the problem $y' = f(t, y)$, $y(0) = 0$ has two solutions. What does this have to do with Problem 4?

————————————————ANSWERS————————————————

2. Yes, because all the functions y_n exist for $a < t < b$ automatically, and it was to ensure this that we needed the condition $|f(t, y)| \leq M$. The other estimates involve only K and M_0 and remain unchanged.
3. By Section 1, Theorem 2. It is left for you to explain why this theorem is applicable.
5. Partial answer: It shows that the hypothesis that the graphs of both solutions lie in R cannot be dropped.

---

3. A global existence theorem

We begin by discussing some terms and conventions that are used in many parts of mathematics. A *region* is a nonempty set of points; for present purposes, a set of points (t, y) in the plane. Regions are often described by inequalities, the phrase "the set of all points (t, y) such that" being understood. For example one can speak of the half plane $t > 0$, the strip $|y| < 1$, the disk $t^2 + y^2 < 1$, and so on.

A *neighborhood* of a point (t_0, y_0) is a disk

(1) $(t - t_0)^2 + (y - y_0)^2 < r^2$

or a square

(2) $|t - t_0| < r,$ $|y - y_0| < r,$

where in both cases r is a positive constant. It will not matter which definition of neighborhood is used, and we have allowed both to emphasize this
fact. A given point (t_0, y_0) is an *interior point* of R if some neighborhood
of (t_0, y_0) is wholly contained in R. The point (t_0, y_0) is a *boundary point* of
R if every neighborhood of (t_0, y_0) contains at least one point of R and at
least one point that does not belong to R. You should convince yourself that
these definitions agree with the intuitive notions of interior and boundary
when applied to familiar regions such as a rectangle or disk.

A region R is said to be *bounded* if it is contained in some neighborhood
and *open* if all of its points are interior. For example the disk (1) and square
(2) are bounded and open. If $a < b$, the strip $a < t < b$ is open but not
bounded.

Let $f(t, y)$ be defined in an open region R containing the point (t_0, y_0).
It is said that $y = \phi(t)$ is a *solution* of the initial-value problem

(3) $y' = f(t, y),$ $y(t_0) = y_0$ (in R)

if the following three conditions hold:

(i) $\phi(t)$ is defined on an open interval $a < t < b$ containing t_0.

(ii) $\phi(t_0) = y_0$ and $\phi'(t) = f(t, \phi(t)),$ $a < t < b$.

(iii) $(t, \phi(t)) \in R,$ $a < t < b$.

Condition (i) is imposed so that the graph of the solution will not consist of
two or more separate pieces. The importance of having the solution defined
on an interval has been often mentioned in this text and need not be emphasized again. The meaning of condition (ii) is obvious, and condition (iii)
says that the graph of the solution lies in R.

The above solution is said to be *maximal* if (a, b) is the longest interval
on which any solution satisfying (i), (ii), and (iii) can exist. This condition
is imposed so that our solution will be unique. For example, suppose we
consider the initial-value problem

$$y' = 1,   y(0) = 0$$

in the (t, y) plane. Somebody might claim that uniqueness fails, because the
solution $y = t$ defined for $-2 < t < 3$ is different from the solution $y = t$ on
the interval $-1 < t < 6$. These solutions are indeed different, since they have
different domains. But neither of them is maximal. The maximal solution is
$y = t$ on the interval $-\infty < t < \infty$ and it is unique.

After these preliminaries we can state the following theorem:

GLOBAL EXISTENCE THEOREM. In a bounded open region R let $f(t, y)$ be continuous, bounded, and Lipschitzian with respect to y. Let (t_0, y_0) be a point of R. Then the initial-value problem (3) has one, and only one, maximal solution $y = \phi(t)$. If $\phi(t)$ is defined on the interval $a < t < b$ then the point $(t, \phi(t))$ tends to a boundary point of R as $t \to a+$ and also as $t \to b-$.

The concluding statement means that the solution extends in both directions from (t_0, y_0) to the boundary of R. See Figure 5.

Except for the last step involving maximality of the solution, the proof is rather easy. If we place a standard region at (t_0, y_0) as shown in Figure 5 the local existence theorem gives a solution on some interval $a < t < b$, and the graph of the solution lies in R. We show that any solution with these two properties is unique. In fact, let two such solutions be y and z, so that the graphs lie in R and

$$y' = f(t, y), \qquad z' = f(t, z), \qquad a < t < b; \qquad y(t_0) = z(t_0) = y_0.$$

Subtraction yields

$$|y' - z'| = |f(t, y) - f(t, z)| \le K|y - z|$$

where K is the Lipschitz constant for f. The function $y - z$ vanishes at t_0 and hence $y = z$ by Section 1, Theorem 5.

Next we establish existence of one-sided limits at a and b. The solution $y = \phi(t)$ satisfies $|y'| \le M$, $a < t < b$ where M is a bound for $|f(t, y)|$ in R. By Section 1, Theorem 2 the limits

$$\lim_{t \to a+} \phi(t) = A, \qquad \lim_{t \to b-} \phi(t) = B$$

both exist. Hence the point $(t, \phi(t))$ tends to a definite point as t approaches the end of the interval, (a, A) in the first case and (b, B) in the second.

All that remains is to show that these points are on the boundary of R, provided the solution is maximal. We give the proof in outline only. Suppose,

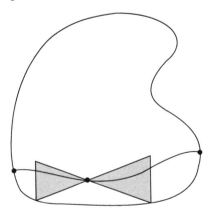

Figure 5

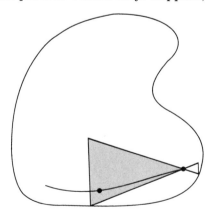

Figure 6

for contradiction, that (b, B) is an interior point of R. This means that the solution ends as shown in Figure 6. Placing a standard region at the point (b, B) we can use the local existence theorem to prolong the solution. Hence the interval (a, b) can be extended to the right and is not maximal. Similarly, if (a, A) is interior to R the interval can be extended to the left.

The part of the theorem asserting existence requires only continuity of the function f. This deep result is essentially due to the Italian mathematician Guiseppe Peano, though his proof (1886) was inadequate. The proof is difficult and will not be given in this book. We mention, however, that Peano's theorem is usually deduced from the Euler polygon rather than from Picard iteration. As was already seen in Chapter 4, Section 2, the assertions of uniqueness and of existence of the limits at the endpoints can fail if f is only assumed continuous.

PROBLEMS

1P. An error estimate. Suppose an exact solution $y = \phi(t)$ and the Picard approximations $y_n(t)$ all exist over one and the same interval $a < t < b$ as in the local theorem of the last section. Show that

$$|y(t) - y_n(t)| \le \frac{M_0}{K} \left(\frac{H^{n+1}}{(n+1)!} + \frac{H^{n+2}}{(n+2)!} + \cdots \right)$$

for $a < t < b$ where $H = K|t - t_0|$.

Hint: See the error bound in the M-test for sequences as given in Chapter F. The error estimates given here and in Problem 2 apply without change to the vector case considered in Section 4.

2. Simpler estimates. Factor out $H^{n+1}/(n+1)!$ from the series of Problem 1 and deduce the following estimates for $E_n = |y(t) - y_n(t)|$. It is assumed that $a < t < b$ and in the second case that $H < n + 2$:

$$E_n \le \frac{M_0}{K} \frac{H^{n+1}}{(n+1)!} e^H, \qquad E_n \le \frac{M_0}{K} \frac{H^{n+1}}{(n+1)!} \frac{n+2}{n+2-H}$$

3. Assuming $f(t, y)$ continuous for all relevant values of t and y, let an operator T be defined by

$$Ty = y_0 + \int_{t_0}^{t} f(s, y(s)) \, ds$$

A *fixed point* of T is a function y such that $y = Ty$. Explain why a fixed point yields a solution of the initial-value problem $y' = f(t, y)$, $y(t_0) = y_0$.

4. Solve $y' = y + e^t$, $y(0) = 1$ and verify that the solution is a fixed point for the operator T associated with this equation by the procedure of Problem 3. Since the problem is self-checking, no answer is provided.

5. This problem is unusual in that all you have to do is read it. The function f is said to be a *contraction* if it satisfies a Lipschitz condition with constant $K < 1$. In a like manner, the operator T is a contraction if in some suitable sense

$$|Ty - Tz| \leq K|y - z|$$

where $K < 1$. A useful and versatile theorem due to the Polish mathematician Stefan Banach says that, under certain general conditions, a contraction operator always has a fixed point. As suggested by Problem 3, Banach's theorem is applicable to differential equations. However, the error after n iterations generally has the form

$$|y - y_n| \leq M\frac{K^n}{1 - K}$$

where $K < 1$ is the contraction constant. Problem 6 gives one of the reasons why Picard iteration has been preferred here.

6. In a specific but typical case, suppose the errors associated with the procedure of the text and with the contraction principle are, respectively,

$$|y - y_n| \leq \frac{200}{(n + 1)!}, \qquad |y - y_n| \leq \frac{100(0.8)^n}{1 - 0.8}$$

(a) Using a hand calculator find the errors for $n = 1, 2, 5, 10, 20$.
   Answer when $n = 20$: $(3.9)10^{-18}$, 5.8.

(b) Show that 9 and 70 iterations are needed to make the respective errors fall below 10^{-4}, and that 16 and 152 iterations are needed to make them fall below 10^{-12}.

---

4. Vector notation

We shall introduce a condensed notation that greatly facilitates the analysis of systems of differential equations. Its usefulness will become clear in the course of the following discussion. First, however, we explain the notation. An ordered m-tuple of numbers

$$y = (y^1, y^2, \cdots, y^m)$$

is called a *vector* with *coordinates* y^j and *dimension m*. The superscript j on the jth coordinate y^j is not an exponent, and must also be distinguished from the superscript (j) that would denote the jth derivative. The set of all m-dimensional vectors is denoted by R^m. For example, R^2 is the set of pairs (y^1, y^2) and can be thought of as a plane. The point (t, y) of the foregoing discussion was in R^2. Here, however, (t, y) will be a point of R^{m+1}.

Two vectors are said to be equal if corresponding coordinates agree. For example,

$$(y^1, y^2, y^3) = (z^1, z^2, z^3) \quad \text{means} \quad y^1 = z^1, \; y^2 = z^2, \; y^3 = z^3.$$

The zero vector, denoted by 0, is the vector with all coordinates 0.

Two vectors are added by adding corresponding coordinates; for example,

$$(y^1, y^2, y^3) + (z^1, z^2, z^3) = (y^1 + z^1, y^2 + z^2, y^3 + z^3).$$

Two vectors *cannot be multiplied or divided*. But a vector can be multiplied by a real number λ according to the rule

$$\lambda(y^1, y^2, \cdots, y^m) = (\lambda y^1, \lambda y^2, \cdots, \lambda y^m).$$

We define $-x = (-1)x$. If $\lambda \neq 0$ division by λ means multiplication by $1/\lambda$. Aside from the fact that two vectors cannot be multiplied or divided, the algebra of vectors is much like the algebra of real numbers, and properties such as those listed in Problem 1 will be used without further comment.

The *magnitude* of a vector can be defined in many ways. The definition

$$|y| = \max(|y^1|, |y^2|, \cdots, |y^m|)$$

simplifies the theoretical development and is convenient in numerical analysis. This definition will be used here.

We shall establish the important inequality

(1) $$|y + z| \leq |y| + |z|$$

which states that the magnitude of the sum of two vectors does not exceed the sum of the magnitudes. For proof, note that for each index j

$$|(y + z)^j| = |y^j + z^j| \leq |y^j| + |z^j| \leq |y| + |z|.$$

Taking the maximum of the left side with respect to j, we get (1).

It is said that $y = \phi(t)$ is a *vector-valued function* if:

(i) each coordinate is a real-valued function, $y^j = \phi^j(t)$,

(ii) All functions ϕ^j have the same domain of definition, and

(iii) the dimension m of the vector is independent of t.

The limit of a vector-valued function is defined by taking the limit of each coordinate function. For example, if the limits on the right-hand side exist,

$$\lim_{t \to a} (\phi^1(t), \phi^2(t)) = (\lim_{t \to a} \phi^1(t), \lim_{t \to a} \phi^2(t)).$$

A vector-valued function ϕ is continuous if each coordinate function ϕ^j is continuous. In Problem 3 it is seen that $|\phi|$ is continuous if ϕ is continuous, and this fact will be used here.

A vector-valued function ϕ is differentiable if each coordinate function is differentiable and then, by definition,

$$\frac{d}{dt} \phi(t) = \left(\frac{d}{dt} \phi^1(t), \ \frac{d}{dt} \phi^2(t), \ \cdots, \ \frac{d}{dt} \phi^m(t) \right).$$

Integration is defined similarly. For example, if $m = 3$ and the coordinate functions are integrable, then, by definition,

$$\int_a^b \phi(t)\,dt = \left(\int_a^b \phi^1(t)\,dt, \ \int_a^b \phi^2(t)\,dt, \ \int_a^b \phi^3(t)\,dt \right).$$

One of the most important properties of the integral states that, if $a \geq b$ and ϕ is continuous for $a \leq t \leq b$, then

$$(2) \qquad \left| \int_a^b \phi(t)\,dt \right| \leq \int_a^b |\phi(t)|\,dt.$$

The proof follows from

$$\left| \int_a^b \phi^j(t)\,dt \right| \leq \int_a^b |\phi^j(t)|\,dt \leq \int_a^b |\phi(t)|\,dt.$$

Taking the maximum of the left side with respect to j, we get (2).

A sequence $\{s_n\}$ of vectors is a succession of vectors

$$s_1, \quad s_2, \quad s_3, \quad \cdots, \quad s_n, \quad \cdots$$

each of which has the same dimension m. The sequence converges to s if $\{s_n^j\}$ converges to s^j for $j = 1, 2, \cdots, m$. The convergence of a sequence of vector-valued functions is defined similarly. Such a sequence converges uniformly to $s(t)$ on a given interval I if the convergence of each sequence of coordinate functions, $\{s_n^j(t)\}$, is uniform on I.

EXAMPLE 1. On a given interval I suppose a sequence of vector-valued functions s_n satisfies

$$(3) \qquad |s_{n+1}(t) - s_n(t)| \leq M_n, \qquad t \in I$$

where the M_n are constants. Suppose further that the series with general term M_n converges. Then $\{s_n\}$ converges uniformly to a vector-valued function s on I and

$$(4) \qquad |s(t) - s_n(t)| \le \sum_{j=n+1}^{\infty} M_j, \qquad t \in I, \ n \ge 1.$$

This is true because $|y^j| \le |y|$ for every vector. Hence the coordinate functions satisfy

$$|s_{n+1}^j(t) - s_n^j(t)| \le M_n, \qquad j = 1, 2, \cdots, m$$

for $t \in I$. Uniform convergence of the sequence of coordinate functions now follows from the Weierstrass M test; see Chapter F. The M test gives an error estimate like (4) for the jth coordinate functions and, taking the maximum over j, we get (4).

EXAMPLE 2. If the functions s_n in Example 1 are differentiable and their derivatives $s_n'(t)$ satisfy an inequality like (3) with the same condition on M_n, then

$$\lim_{n \to \infty} s_n'(t) = s'(t), \qquad t \in I.$$

This follows by applying the result of Example 1 to $\{s_n'\}$ as well as $\{s_n\}$ and using Chapter F, Theorem 2.

EXAMPLE 3. Let M be constant. Suppose a vector-valued function $y = \phi(t)$ satisfies

$$|y'| \le M, \qquad y(t_0) = y_0$$

on an open interval I containing the point t_0. Then

$$(5) \qquad |y - y_0| \le M|t - t_0|, \qquad t \in I.$$

To see this, note that the differential inequality for y yields the same differential inequality for each coordinate y^j. By Section 1, Theorem 1

$$|y^j - y_0^j| \le M|t - t_0|, \qquad t \in I.$$

Taking the maximum of the left side with respect to j, we get (5).

EXAMPLE 4. Let M be constant and let the vector-valued function ϕ satisfy $|\phi'(t)| \le M$ on a bounded interval (a, b). Then the derivative of each coordinate function ϕ^j is bounded by this same constant M. Section 2, Theorem 2 shows that each $\phi^j(t)$ has a limit as $t \to a+$ and as $t \to b-$, and this gives existence of the one-sided limits

$$\lim_{t \to a+} \phi(t) \quad \text{and} \quad \lim_{t \to b-} \phi(t).$$

EXAMPLE 5.   Let K be constant. Suppose a vector-valued function w satisfies

$$|w'| \leq K|w|, \qquad w(t_0) = 0$$

on an open interval I containing the point t_0. Then $w(t) = 0$ for $t \in I$. As seen in Problem 5, this follows from Section 1, Theorem 5.

The gist of the foregoing results is that vector-valued functions and real-valued functions have similar analytic properties. The similarity is so great that we can use the proofs in Sections 2 and 3 without change in the vector case. For example, the function $f(t, y)$ is Lipschitzian with respect to y if there is a constant K such that

$$|f(t, y) - f(t, z)| \leq K|y - z|, \qquad t \in R^1,\ y \in R^m,\ z \in R^m.$$

If this holds when (t, y) and (t, z) are restricted to a region R of R^{m+1}, it is said that $f(t, y)$ is Lipschitzian with respect to y in R. A standard region is defined by the two inequalities

$$a < t < b, \qquad |y - y_0| \leq M|t - t_0|$$

where t_0 is a given point on (a, b) and where y and y_0 are in R^m. The local existence theorem of Section 2 can now be proved just as before.

To state a global theorem, we define a *neighborhood* of a point (t_0, y_0) in R^{m+1} to be

$$\{(t, y) : t \in R^1,\ y \in R^m,\ |t - t_0| < r,\ |y - y_0| < r\}$$

where r is a positive constant. This corresponds to the square neighborhood introduced in Section 3. The definitions of interior point, boundary point, and open region parallel those for the case $m = 1$. A global theorem then follows from the local theorem just as it did in Section 3.

One of the advantages of vector notation is that complex equations in R^m can usually be reduced to real equations in R^{2m}. Without belaboring this matter, we illustrate it in the case of a complex system with $m = 2$. Let u, v, g, h be complex-valued functions such that

$$u' = g(t, u, v), \qquad v' = h(t, u, v).$$

where t is real. This is a two-dimensional system associated with the complex two-dimensional vectors (u, v) and (g, h). Writing

$$u = y^1 + iy^2, \qquad v = y^3 + iy^4, \qquad f = f^1 + if^2, \qquad g = f^3 + if^4$$

and separating into real and imaginary parts, we get a real four-dimensional system $y' = f(t, y)$ for the real vector $y = (y^1, y^2, y^3, y^4)$. Namely,

$$(y^1)' = f^1(t, y), \qquad (y^2)' = f^2(t, y), \qquad (y^3)' = f^3(t, y), \qquad (y^4)' = f^4(t, y).$$

The foregoing remarks are summarized as follows:

FUNDAMENTAL THEOREM. The local and global theorems of Sections 2 and 3 remain valid for real vector equations. They can be applied to complex vector equations if the latter are written in real form by doubling the dimension.

PROBLEMS

1. In this problem x, y, z denote vectors of dimension m, and λ, μ denote real numbers. Establish two of the following equations with $m = 2$, three with $m = 3$, and the rest with m arbitrary:

$$x + 0 = x \quad x + y = y + x \quad x + (y + z) = (x + y) + z \quad x + (-x) = 0$$
$$\lambda(x + y) = \lambda x + \lambda y \quad (\lambda + \mu)x = \lambda x + \mu x \quad (\lambda \mu)x = \lambda(\mu x) \quad |\lambda x| = |\lambda||x|$$

2. Show that $\big| |x| - |y| \big| \le |x - y|$. Hint: $x = (x - y) + y$.

3. If ϕ is a vector-valued function deduce from Problem 2 that $|\phi|$ is continuous at any point where ϕ is continuous.

4. In m-dimensional Euclidean geometry, the length of a vector is given by

$$\|y\|^2 = (y^1)^2 + (y^2)^2 + \cdots + (y^m)^2, \qquad \|y\| \ge 0$$

Show that $|y| \le \|y\| \le \sqrt{m}|y|$.

5. In this problem $y = \phi(t)$ is a differentiable vector-valued function defined on an open interval I containing the point t_0. It is assumed that $|y'| \le K|y|$ where K is constant. Show that the real-valued function $W = \|y\|^2$ satisfies $|W'| \le 2mKW$, where $\|y\|$ is as in Problem 4. Thus establish the result stated in Example 5.

6. Let $y = \phi(t)$ be a vector-valued function. Show that

$$-|\phi'(t)| \le |\phi(t)|' \le |\phi'(t)|$$

holds at any point where ϕ and $|\phi|$ are both differentiable. This is the key to the deeper study of differential inequalities for vector-valued functions. However there is a technical difficulty, because differentiability of ϕ does not ensure that $|\phi|$ is differentiable. That is why the method of Problem 5 is preferred here.

Hint: Problem 2 gives $|\phi(t + h)| - |\phi(t)| \le |\phi(t + h) - \phi(t)|$.

7. We illustrate Picard's method in connection with

$$(y^1)' = 2ty^1 + y^2, \qquad (y^2)' = y^1y^2 + 1, \qquad y^1(0) = 1, \qquad y^2(0) = 0$$

Substitute $y_0 = (1, 0)$ to obtain $y_1 = (t^2 + 1, t)$. Thus obtain

$$4y_2 = (2t^4 + 6t^2 + 4, t^4 + 2t^2 + 4t)$$

The complexity increases rapidly as the process continues. One of the merits of the vector method is that it avoids detailed calculations such as this.

8. The definitions in the text give $0x = 0$ and $1x = x$ for all vectors x. Show that these equations follow from the algebraic properties stated in Problem 1. When the theory of vector spaces is developed abstractly, properties such as those in Problem 1 are postulated and others are deduced from them. With this abstract approach, the dimension of a vector space can be, and often is, infinite.

9. The Lipschitz condition gives $|y' - y'_n| \le K|y - y_{n-1}|$. Using this and a vector analog of Section 3, Problem 2 deduce for $n \ge 1$ that

$$|y'(t) - y'_n(t)| \le \frac{M_0}{K} \frac{H^n}{n!} e^H, \qquad H = K|t - t_0|$$

---

5. Concluding remarks

Throughout this section $I = (a, b)$ denotes a bounded open interval containing the point t_0 and

$$a < t_0 < b, \qquad y \in R^m, \qquad z \in R^m, \qquad f(t, y) \in R^m.$$

Suppose the following three conditions hold for $t \in I$, $y \in R^m$, $z \in R^m$:

A. $f(t, y)$ is continuous.

B. $|f(t, y) - f(t, z)| \le K|y - z|$ where K is constant.

C. $|f(t, y_0)| \le M_0$ where M_0 is constant.

Then for $a < t < b$ the integrals defining the Picard iterates exist automatically and we need not introduce a standard region as in the local theorem. This is clear by inspection of the formulas. The rest of the discussion goes through without change, except for existence of limiting values at the ends of the interval. This poses a problem because we do not have a bound M for

$|f(t,y)|$. However, the condition A together with $y_1' = f(t, y_0)$ shows that y_1' is bounded, and the result of Section 4, Problem 9 with $n = 1$ then shows that y' is bounded. Hence the limits exist.

(a) Linear systems

We shall apply the above results to the linear system

$$(1) \qquad (y^j)' = p_{j1}y^1 + p_{j2}y^2 + \cdots + p_{jm}y^m + q_j$$

where $j = 1, 2, \cdots, m$. The coefficients p_{ij} and q_j are assumed to be real-valued functions not involving y and defined on I.

Let us inquire what conditions on these coefficients ensure the above hypotheses ABC. With

$$(2) \qquad f^j(t, y) = p_{j1}(t)y^1 + p_{j2}(t)y^2 + \cdots + p_{jm}(t)y^m + q_j(t)$$

the linear system is $y' = f(t, y)$. Condition A holds if the p_{ij} and q_j are continuous on I and this hypothesis is now assumed.

The Lipschitz condition B holds if each coefficient p_{ij} is bounded. To see why, let

$$|p_{ij}(t)| \leq N, \qquad t \in I, \quad i, j = 1, 2, \cdots, m$$

where N is constant. We write (2) with y and with z, and subtract. The result is

$$|f^j(t, y) - f^j(t, z)| \leq N(|y_1 - z_1| + |y_2 - z_2| + \cdots + |y_m - z_m|).$$

This yields the first inequality below, and the second follows from the first:

$$|f^j(t, y) - f^j(t, z)| \leq mN|y - z|, \qquad |f(t, y) - f(t, z)| \leq mN|y - z|.$$

If, in addition, $|q_j| \leq M$ then by (2) with $y = y_0$

$$|f^j(t, y_0)| \leq mN|y_0| + M, \quad \text{hence} \quad |f(t, y_0)| \leq mN|y_0| + M.$$

This yields Condition C with $M_0 = mN + M$. Thus we have established the following theorem, which provides a foundation for Chapter 15:

THEOREM 1. *If the coefficients p_{ij} and q_j are continuous and bounded for $a < t < b$ then the the linear system (1) has one, and only one, maximal solution $y = \phi(t)$. The solution is valid on the whole interval $a < t < b$ and the limits $\lim\limits_{t \to a+} \phi(t)$ and $\lim\limits_{t \to b-} \phi(t)$ exist.*

(b) Higher-order equations

It was already mentioned in Chapter 2 that a higher-order differential equation can be reduced to a first-order system. This matter is now discussed more fully. Let the equation be

$$(3a) \qquad u^{(m)} = g(t, u, u', u'', \cdots, u^{(m-1)})$$

with the initial conditions

$$(3b) \qquad u(t_0) = u_0, \qquad u'(t_0) = u_1, \qquad \cdots, \qquad u^{(m-1)}(t_0) = u_{m-1}.$$

Here u and g are real-valued and, as elsewhere in this book, the superscript in parenthesis denotes differentiation. It is supposed that the solution u is defined on a bounded open interval $a < t < b$ containing the point t_0. With

$$y^1 = u, \qquad y^2 = u', \qquad y^3 = u'', \qquad \cdots, \qquad y^m = u^{(m-1)}$$

the differential equation is equivalent to the system

$$(y^1)' = y^2, \ (y^2)' = y^3, \cdots, (y^{m-1})' = y^m, \ (y^m)' = g(t, y^1, y^2, \cdots, y^m).$$

The real-valued function on the right is denoted by $g(t, y)$. We can write (3a) in the form

$$(4) \qquad y' = f(t, y), \quad y(t_0) = y_0$$

where the vector-valued function f is defined by

$$f^1(t, y) = y^2, \quad f^2(t, y) = y^3, \quad \cdots, \quad f^{m-1}(t, y) = y^m, \quad f^m(t, y) = g(t, y).$$

The coordinates of y_0 are given by u_j in (3b).

We now inquire what conditions on g are needed to ensure that the fundamental theorem of Section 4 applies to the system (4). Let R denote a bounded open region of R^{m+1}. In view of the above correlation between y and the derivatives $u^{(j)}$ it is said that the solution $u = \psi(t)$ lies in R if

$$(t, \psi(t), \psi'(t), \psi''(t), \cdots, \psi^{(m-1)}(t)) \in R, \qquad a < t < b.$$

Here ψ is real valued, in contrast to the vector-valued function ϕ introduced elsewhere.

Continuity of g in R ensures continuity of f in R, and that is the first hypothesis we need. The real-valued function g is said to be Lipschitzian with respect to y in the region R if there is a constant L such that

$$|g(t, y) - g(t, z)| \leq L|y - z|, \qquad (t, y) \in R, \ (t, z) \in R.$$

It is left for you to verify that when this holds, the vector-valued function f introduced above satisfies

$$|f(t,y) - f(t,z)| \leq K|y - z| \quad \text{where} \quad K = \max(1, L).$$

These remarks lead to the following:

THEOREM 2. *Let R be a bounded open region containing the point (t_0, y_0). Suppose g is continuous, Lipschitzian with respect to y, and bounded in R. Then the initial-value problem (3) has one, and only one, maximal solution $u = \psi(t)$ that lies in R and passes through the given point (t_0, y_0). If (a, b) is the interval on which this maximal solution exists, then the expression*

$$(t, \psi(t), \psi'(t), \cdots, \psi^{(m-1)}(t))$$

tends to a point on the boundary of R as $t \to a+$ and also as $t \to b-$.

We must still establish existence of the limits as asserted in the last sentence. We consider b; the proof for a is similar. Equation (3a) shows that $u^{(m)}$ is bounded, hence $u^{(m-1)}(t)$ has a limit for $t \to b-$ by Section 1, Theorem 2. Existence of this limit shows that $u^{(m-1)}(t)$ is bounded for $t < b$ and near b, and we conclude that $u^{(m-2)}(t)$ has a limit at $b-$. Continuation of this process gives the concluding statement in Theorem 2.

We supplement Theorem 2 by giving a sufficient condition for g to be Lipschitzian with respect to y. A *rectangular region* is a region in R^{m+1} defined by inequalities of the form

$$a < t < b, \quad a^1 < y^1 < b^1, \quad a^2 < y^2 < b^2, \cdots, a^m < y^m < b^m.$$

The following holds:

THEOREM 3. *In a rectangular region R let $g(t, y)$ have bounded partial derivatives with respect to each y^j, $j = 1, 2, \cdots, m$. Then g is Lipschitzian with respect to y in R.*

When combined with Theorem 1, this yields the existence and uniqueness theorems as stated in Chapter 4. The proof of Theorem 3 is not difficult and is outlined in Problem 1.

(c) Bifurcation

Although Theorem 3 is formulated for rectangular regions, it has a broader scope than this restriction would indicate. To see why, let us return to the general vector system $y' = f(t, y)$ considered in Section 4. Suppose two solutions y and z lying in R agree at some value t_1 but do not agree at some other value t_2. Without loss of generality $t_2 > t_1$. Starting at t_2 we diminish t until it first happens that the solutions agree. (That a *first value* exists can be deduced from the completeness axiom for real numbers but is taken for granted here.) If the point so obtained is t_0 then u and v have a common value y_0 at t_0, but they do not agree throughout any open interval containing t_0, no matter how short. A point $(t_0, y_0) \in R$ with this property is said to be a *bifurcation point*. What we have shown is that, if uniqueness fails, there is always at least one bifurcation point.

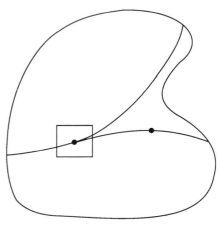

Figure 7

Since (t_0, y_0) is interior to R, the uniqueness problem can be studied in an arbitrarily small rectangular region centered at (t_0, y_0). If uniqueness holds in such a rectangle, the fact that (t_0, y_0) is a bifurcation point leads to a contradiction. The gist of the argument is made clear by Figure 7.

The above principles apply to the general vector equation and show that uniqueness is essentially a local property. In the context of Theorem 2, we get the following result from Theorem 3:

THEOREM 4.   *Suppose every point $(t_0, y_0) \in R$ is interior to a rectangle in which the partial derivatives $\partial g/\partial y^j$ are bounded. Then the problem (3) has no bifurcation point in R.*

The hypothesis of Theorem 4 holds if the partial derivatives are continuous in R. This follows from an n-dimensional version of Chapter 10, Section 3, Theorem A.

(d) Singular solutions and envelopes

By definition, a *singular solution* is a solution such that every point of its graph is a bifurcation point. For the scalar case, $m = 1$, a singular solution is essentially the same as an envelope of a family of solutions, and a brief discussion is given now. Familiarity with Chapter 3, Section 2, is assumed, and in agreement with the notation used there, we take x rather than t as the independent variable. Without going into elaborate detail, we want to explain where the two equations $f_p = 0$, $F_c = 0$ for the envelope come from.

Suppose that f has continuous partial derivatives with respect to each of its three arguments and that the solutions of $f(x, y, y') = 0$ form a family of curves that has an envelope. Near any point (x_0, y_0) at which $f_p \neq 0$, a three-variable analog of Chapter A, Theorem E states that the equation

$f(x, y, p) = 0$ can be solved for $p = g(x, y)$ where g, like f, has continuous partial derivatives. By Theorem 4 the resulting differential equation

$$y' = g(x, y)$$

can have only one solution through (x_0, y_0). But through a point on the envelope, there are, in general, two solutions: the envelope, and the member of the family to which the envelope is tangent. This failure of uniqueness explains why $f_p = 0$ is to be expected on the envelope.

Next let a family of curves C_c be given by $F(x, y, c) = 0$ and suppose the family has an envelope E. The envelope is tangent to C_c at some point $(\xi(c), \eta(c))$ and is expressed in the parametric form $x = \xi(c), y = \eta(c)$. If C_c is given as $x = \sigma(s)$, $y = \tau(s)$, the defining equations are

(5) $$F(\xi(c), \eta(c), c) = 0, \qquad F(\sigma(s), \tau(s), c) = 0.$$

Let us now assume that $F, \xi, \eta, \sigma, \tau$ have continuous first derivatives and that the curves C_c are smooth, in the sense that σ' and τ' do not vanish simultaneously. Differentiating (5) by the chain rule yields

(6) $$F_x \xi' + F_y \eta' + F_c = 0, \qquad F_x \sigma' + F_y \tau' = 0.$$

Since the envelope is tangent to C_c at $(\xi(c), \eta(c))$, the tangent vectors (ξ', η') and (σ', τ') are parallel there. Thus

$$\xi' = \lambda \sigma', \qquad \eta' = \lambda \tau'$$

where λ is some scalar. The second equation (6) now shows that the first reduces to $F_c = 0$.

(e) Remarks on notation

To distinguish vectors from real numbers one often uses boldface type. For example, the vectors $x, y, 0$ in the foregoing discussion could be written $\mathbf{x}, \mathbf{y}, \mathbf{0}$. We have avoided this notation because we want to stress the analogy between the cases $m = 1$ and $m > 1$. The fact that most of the equations for $m = 1$ and for $m > 1$ are identical in appearance is a major part of the message. It is characteristic of twentieth-century mathematics to use familiar symbols in a new way, thus achieving greater generality with no change in notation and no increase in complication.

Of course one should be aware of the distinction between real numbers and vectors, whether or not the distinction is built into the notation. For this reason we have added Problem 7.

In this chapter the coordinates of a vector were distinguished by superscripts. Superscript notation is standard in such fields as differential geometry, elasticity, and relativity. It was used here so that subscripts could be reserved for the purpose of distinguishing one vector from another.

When equations involve exponents as well as vectors, the superscript notation can lead to confusion. For example one might have x^2 and t^2 together, where 2 is a coordinate index in the first case and an exponent in the second. Except for a brief return to superscript notation in Chapter 24, we denote vector coordinates by subscripts in the rest of the text.

PROBLEMS

1. Theorem 3 is proved by changing variables one at a time. For example if $m = 2$ the expression $g(t, y) - g(t, z)$ is written

$$g(t, y^1, y^2) - g(t, z^1, y^2) + g(t, z^1, y^2) - g(t, z^1, z^2)$$

The mean-value theorem for functions of one variable now shows that g is Lipschitzian with constant $K = 2M$, where M is a bound for the two partial derivatives. Generalize (a) to $m = 3$ and (b) to arbitrary m.

2. The method used to reduce (3) to a first-order system applies to systems of higher-order equations. In general the dimension m of the first-order system equals the sum of the orders of the individual equations. Reduce the following to a first-order system of dimension $m = 9$. No answer is provided:

$$u'' = uvw''', \qquad v''' = v' + u' + w, \qquad w^{iv} = (ww'')^2 + (v'u^2) + v^2 w'''$$

3. In the preceding problem what initial conditions for u, v, w lead to the standard initial condition $y(t_0) = y_0$ for your first-order system? No answer is provided but see Chapter 9, Section 1, Example 4.

4. Uniqueness without vectors. Suppose u and v both satisfy the initial-value problem (3) on an open interval I and suppose further that g is Lipschitzian with the constant K. By subtraction deduce an inequality for $w = u - v$ to which Chapter 10, Section 4, Theorem 3 applies. Thus get $u = v$.

5. This problem and the next pertain to the scalar case $m = 1$. Let $F(x, y, c) = 0$ and $G(x, y) = 0$ describe, respectively, a family of curves $\{C_c\}$ and its envelope E.

(a) Check that the family of curves given by $F(x, y, c)G(x, y) = 0$ contains E, each C_c, and the differentiable curves that can be obtained by switching from E to C_c or from C_c to E at the point of tangency.

(b) If each C_c satisfies a first-order equation $f(x, y, y') = 0$ explain why any differentiable function $y = \phi(x)$ that satisfies the above equation also satisfies $f(x, y, y') = 0$.

6. (abc) We consider the family $y = (x - c)^2$ of solutions of $(y')^2 = 4y$ as discussed in Chapter 3, Section 2. The envelope $y = 0$ cannot be obtained by any choice of c, and sometimes this property (that it cannot be obtained by assigning a value to c) is used as a definition of the term "singular solution." Discuss that definition as applied to the following:

$$y(x - c)^2 = y^2 \qquad (cx - 1)^2 = c^2 y \qquad (c^2 + y^2)(cx - 1)^2 = c^2 y(c^2 + y^2)$$

Partial answer: For (a) see Problem 5. In (b) the omitted curve is $y = x^2$. In (c) the curve $y = x^2$ is still omitted but the envelope $y = 0$ is included.

7. Make a copy of Sections 4 and 5 of this chapter, including the problems; the publisher agrees that this is not a violation of copyright. The proofreader's sign for boldface is a wavy underline. On your copy mark everything in boldface that should be so marked, and nothing that should not be.

In most sciences one generation tears down what another has built and what one has established another undoes. In mathematics alone each generation builds a new story to the old structure.

—Hermann Hankel

It has long been a complaint against mathematicians that they are hard to convince. But it is a far greater disqualification both for philosophy and for the affairs of life to be too easily convinced, to have too low a standard of proof.

—J. S. Hill

The 19th Century, which prides itself upon the invention of steam and evolution, might have a more legitimate title to fame from the discovery of pure mathematics.

—Bertrand Russell

Chapter 15
MATRICES AND LINEAR SYSTEMS

A LINEAR homogeneous system of n differential equations in n unknowns can be described by $dy/dt = Ay$ where A is an n-by-n matrix and y is an n-dimensional column vector. If $n=1000$ the system represents 1000 scalar equations in 1000 unknowns. Nevertheless, it behaves much like the single scalar equation obtained when $n = 1$. The general theory is surprisingly easy, and it is only when we turn to specific examples that the underlying complexity becomes manifest. As examples we consider first-order systems in two variables and an important class of higher-order systems with constant coefficients.

This chapter assumes knowledge of linear algebra to the extent developed in Chapters CDE.

1. Linear independence

The following notation is used in Chapters 15, 16, and 17. By J we mean a bounded open interval containing the point t_0. Depending on context,

$$A, \; Y, \; F, \; U, \; V,$$

denote matrices or matrix-valued functions defined on J. However, C, K, and E are constant matrices, independent of t. All of these are square matrices of order n. If the meaning is clear from context, we sometimes say "matrix" instead of "matrix-valued function" and similarly for vectors. Unless the contrary is stated, it will not matter whether the elements of our vectors and matrices are real or complex.

The jth column of a square matrix is systematically denoted by the corresponding letter with a subscript. Thus,

$$A_j, \; Y_j, \; F_j, \; U_j, \; V_j, \; C_j, \; K_j, \; E_j, \; I_j$$

represent n-dimensional column vectors, the jth columns of the corresponding matrices A, Y, F, $\cdots$. We also use y, f, g, u, v to denote n-dimensional column vectors or vector-valued functions defined on J, while c and k are constant n-dimensional column vectors.

A matrix-valued function such as $A = A(t)$ is said to be continuous, bounded, or differentiable if each element a_{ij} in it is continuous, bounded, or differentiable, respectively. Differentiation and integration are performed elementwise; thus,

$$\frac{d}{dt}(a_{ij}) = \left(\frac{da_{ij}}{dt}\right), \qquad \int (a_{ij}) \, dt = \left(\int a_{ij} \, dt\right).$$

We shall be concerned with the equation $y' = Ay + f$, or more explicitly

$$\frac{dy}{dt} = A(t)y(t) + f(t), \qquad t \in J.$$

This is a condensed way of writing a linear system of differential equations in the n coordinates of y. For example, if y_j and f_j are real-valued functions defined on J, the system on the left below is equivalent to the matrix equation on the right:

$$y_1' = a_{11}y_1 + a_{12}y_2 + f_1 \qquad \frac{d}{dt}\begin{pmatrix} y_1 \\ y_2 \end{pmatrix} = \begin{pmatrix} a_{11} & a_{12} \\ a_{21} & a_{22} \end{pmatrix}\begin{pmatrix} y_1 \\ y_2 \end{pmatrix} + \begin{pmatrix} f_1 \\ f_2 \end{pmatrix}.$$
$$y_2' = a_{21}y_1 + a_{22}y_2 + f_2$$

The equation $y' = Ay + f$ is the same as $Ty = f$ where

$$Ty = y' - Ay.$$

The domain of the operator T so defined consists of n-dimensional vector-valued functions differentiable on J. It is easily checked that T is linear. That is, if u and v are in the domain of T, and α is any scalar, then

$$T(u + v) = Tu + Tv, \qquad T(\alpha u) = \alpha Tu.$$

Proof of the second equation is left for you. The first follows from

$$T(u+v) = (u+v)' - A(u+v) = u' + v' - Au - Av = u' - Au + v' - Av = Tu + Tv.$$

By mathematical induction, this extends to three or more summands.

Since T is linear, the general principle of superposition, the principle of the complementary function, and other results established for linear operators in the foregoing pages apply to T. The following theorem indicates that the behavior of the equation $Ty = f$ is also similar to that in the scalar case:

THEOREM 1. *Let A and f be continuous and bounded for $t \in J$. Then for each k the initial-value problem*

$$y'(t) = A(t)y(t) + f(t), \qquad t \in J; \qquad y(t_0) = k$$

has one, and only one, solution. The solution exists on the whole interval J and $y(t)$ has a limit as t approaches either end of the interval.

Theorem 1 is a reformulation of Chapter 14, Section 5, Theorem 1.

The assertion of uniqueness in Theorem 1 requires only that A be bounded on every closed subinterval contained in J, and this holds if $A(t)$

is continuous. (See Chapter 10, Section 3, Theorem A.) Existence of the solution on every closed subinterval of J, and hence on J, also follows from continuity alone. But the concluding statement about endpoint limits of $y(t)$ requires that A and f be bounded on J.

These matters were emphasized in Chapters 11 and 14 and will not be emphasized again here. In this chapter we drop statements about existence of endpoint limits and we make the following assumption once and for all:

BASIC ASSUMPTION. The functions A, F and f are continuous on J.

The subject of matrix differential equations is one of those rare disciplines in which specific examples tend to be much more difficult than the general theorems they are supposed to illustrate. An early preoccupation with examples obscures the essential simplicity of the theory and does not do justice to its scope and power. In Sections 1 and 2 the basic results are presented without interruption. Later we give many examples, starting with the case $n = 2$ in Section 3.

(a) Linear dependence of vectors

The vectors Y_j are said to be *linearly dependent* if there are real or complex constants c_j, not all zero, such that

$$(1) \qquad\qquad c_1 Y_1 + c_2 Y_2 + \cdots + c_n Y_n = 0.$$

If no such constants exist the vectors are linearly independent. To interpret this condition, let the coordinates of the vector Y_j be y_{ij}. When $n = 3$ the condition for linear dependence takes the form

$$c_1 \begin{pmatrix} y_{11} \\ y_{21} \\ y_{31} \end{pmatrix} + c_2 \begin{pmatrix} y_{12} \\ y_{22} \\ y_{32} \end{pmatrix} + c_3 \begin{pmatrix} y_{13} \\ y_{23} \\ y_{33} \end{pmatrix} = \begin{pmatrix} 0 \\ 0 \\ 0 \end{pmatrix}.$$

These are equivalent to three linear homogeneous equations in the three unknowns c_1, c_2, c_3, namely

$$y_{11} c_1 + y_{12} c_2 + y_{13} c_3 = 0$$
$$y_{21} c_1 + y_{22} c_2 + y_{23} c_3 = 0.$$
$$y_{31} c_1 + y_{32} c_2 + y_{33} c_3 = 0$$

The coefficient matrix of this system is the matrix Y whose columns are the vectors Y_j. There is a nontrivial solution c_j if, and only if, the determinant $|Y|$ is 0. Hence the vectors are dependent when $|Y| = 0$, independent when $|Y| \neq 0$.

A similar result holds in general. For arbitrary n, Equation (1) leads to the following system of n linear equations in the n unknowns c_j:

$$(2) \qquad \sum_{j=1}^{n} y_{ij} c_j = 0, \qquad i = 1, 2, \cdots, n.$$

These are equivalent to $Yc = 0$ where c is the column vector with elements c_j. Hence there is a nontrivial solution c if, and only if, $|Y| = 0$.

The column vector y is said to be a linear combination of the Y_j if there exist constants c_j such that

$$(3) \qquad y = c_1 Y_1 + c_2 Y_2 + \cdots + c_n Y_n.$$

The above discussion shows that (3) is equivalent to $y = Yc$. Hence there is a solution c, for every choice of y, if and only if $|Y| \neq 0$. In that case the solution c is unique.

If every vector y can be represented uniquely in the form (3) the Y_j form a *basis*; more specifically, they form a basis for real n-dimensional space R^n when the vectors and constants are real, and for complex n-dimensional space C^n when they are complex. The above discussion can now be summarized as follows:

THEOREM 2. *The vectors Y_j are linearly dependent if and only if $|Y| = 0$. They form a basis for R^n or C^n, as the case may be, if and only if $|Y| \neq 0$.*

EXAMPLE 1. Test for linear dependence the four vectors on the left below:

$$\begin{pmatrix} 1 \\ 3 \\ 4 \\ 3 \end{pmatrix} \begin{pmatrix} 2 \\ 6 \\ 6 \\ 4 \end{pmatrix} \begin{pmatrix} 3 \\ 6 \\ 8 \\ 5 \end{pmatrix} \begin{pmatrix} 4 \\ 9 \\ 10 \\ 6 \end{pmatrix}, \qquad \begin{vmatrix} 1 & 2 & 3 & 4 \\ 3 & 6 & 6 & 9 \\ 4 & 6 & 8 & 10 \\ 3 & 4 & 5 & 6 \end{vmatrix}$$

It is seen in Chapter E that the determinant on the right is 0. Hence the vectors are dependent.

(b) Linear dependence of vector-valued functions

We now extend the notion of linear dependence from vectors Y_j to vector-valued functions $Y_j(t)$. Such functions are linearly dependent on J if constants c_j, not all 0, can be found satisfying

$$c_1 Y_1(t) + c_2 Y_2(t) + \cdots + c_n Y_n(t) = 0, \qquad t \in J.$$

When no such constants exist the functions are linearly independent on J. For dependence, not only must the vectors $Y_j(t)$ be linearly dependent at each $t \in J$, but the *same* constants c_j must do the job at every t. This is an extremely restrictive condition. Nevertheless, we shall establish the following:

THEOREM 3. *Let Y_j be n solutions of $y' = A(t)y$ on J. Suppose further that $|Y(t_0)| = 0$ at some point $t_0 \in J$. Then the functions $Y_j(t)$ are linearly dependent on J and $|Y(t)| = 0$ for $t \in J$.*

By Theorem 2 the vectors $Y_j(t_0)$ are linearly dependent. Hence we can find constants c_j, not all 0, such that the function

$$(4) \qquad y(t) = c_1 Y_1(t) + c_2 Y_2(t) + \cdots + c_n Y_n(t)$$

satisfies $y(t_0) = 0$. Linearity gives $y' = Ay$ and uniqueness then gives $y(t) = 0$ for $t \in J$. A second use of Theorem 2 completes the proof.

The functions Y_j in Theorem 3 form a *basis* for solutions of $Ty = 0$ if every solution of this equation can be represented uniquely in the form (4). We shall establish:

THEOREM 4. *Let Y_j be n vector-valued functions satisfying $TY_j = 0$ on J. Then the Y_j form a basis for solutions of $Ty = 0$ if, and only if, $|Y(t_0)| \neq 0$ at some point $t_0 \in J$. In that case $|Y(t)| \neq 0$ at every point $t \in J$.*

For proof, note that (4) is the same as $y(t) = Y(t)c$. If $|Y(t_0)| \neq 0$ there is one, and only one, vector c such that

$$y(t_0) = Y(t_0)c.$$

By the uniqueness part of Theorem 1 we have $y(t) = Y(t)c$ on J and hence the Y_j form a basis. On the other hand if $|Y(t_0)| = 0$ the above discussion of (4) shows that the representation of the zero solution, $y = 0$, would not be unique. Hence the Y_j do not form a basis. The concluding statement follows from Theorem 3.

If the columns of Y are linearly independent solutions of $y' = Ay$ then Y is called a *fundamental matrix* for $y' = Ay$. By (4) and Theorem 4, a fundamental matrix yields both the general solution and the complete solution of $y' = Ay$. Further discussion is given in the following section.

PROBLEMS

1P. If α is constant check that the vectors on the left below are linearly dependent, but that the vector-valued functions on the right are linearly independent on every interval J, no matter how short:

$$\begin{pmatrix} \alpha \\ \alpha \end{pmatrix}, \begin{pmatrix} 2 \\ 2 \end{pmatrix}; \qquad \begin{pmatrix} t \\ t \end{pmatrix}, \begin{pmatrix} 2 \\ 2 \end{pmatrix}$$

2. The vectors $(t, 2t)^T$ and $(t, t^2)^T$ cannot be solutions of a 2-by-2 system $y' = Ay$, with $A(t)$ bounded, on any interval containing the point $t = 0$ or $t = 2$. Why not? Confirm this by finding the unique matrix A for which $y' = Ay$ has those solutions.

3. (abcde) For what values of the constant α, if any, are the columns of the following matrices linearly dependent?

$$\begin{pmatrix} 1 & 0 & 0 \\ 0 & 1 & \alpha \\ 0 & 0 & 1 \end{pmatrix} \begin{pmatrix} 1 & 1 & 0 \\ 3 & 1 & 1 \\ 4 & 2 & \alpha \end{pmatrix} \begin{pmatrix} 2 & 1 & 2 \\ 4 & 2 & 5 \\ 2 & 1 & \alpha \end{pmatrix} \begin{pmatrix} 1 & 1 & 0 \\ 1 & \alpha & 1 \\ -6 & 0 & \alpha \end{pmatrix} \begin{pmatrix} \alpha & -1 & 1 \\ 3 & \alpha & 1 \\ 3 & 1 & \alpha \end{pmatrix}$$

4. In Example 1 express the fourth vector as a linear combination of the other three.

───────────────────ANSWERS───────────────────

2. Because the vectors are dependent at $t = 0$ and $t = 2$, but are independent, as functions of t, on every interval. See Theorem 3.

$$A(t) = \frac{1}{t(t-2)} \begin{pmatrix} t-2 & 0 \\ -2t & -2+2t \end{pmatrix}$$

3. none, 1, all, $\{3, -2\}$, $\{-1, 0, 1\}$ 4. $V_4 = V_3 + V_2 - V_1$

───

2. Matrix differential equations

The linear operator $Ty = y' - Ay$ introduced in Section 1 has a natural extension from vectors to matrices. For example, the equation on the right below follows, *by definition*, from those on the left:

$$T\begin{pmatrix} y_1 \\ y_2 \end{pmatrix} = \begin{pmatrix} f_1 \\ f_2 \end{pmatrix}, \qquad T\begin{pmatrix} z_1 \\ z_2 \end{pmatrix} = \begin{pmatrix} g_1 \\ g_2 \end{pmatrix}, \qquad T\begin{pmatrix} y_1 & z_1 \\ y_2 & z_2 \end{pmatrix} = \begin{pmatrix} f_1 & g_1 \\ f_2 & g_2 \end{pmatrix}.$$

In a like manner, by definition,

$$TY = T(Y_1, Y_2, \cdots, Y_n) = (TY_1, TY_2, \cdots, TY_n).$$

The domain of T now consists of matrix-valued functions Y differentiable on J. This dual interpretation of T causes no difficulty. When we want to emphasize the distinction, we speak of $Ty = y' - Ay$ as the vector case and of $TY = Y' - AY$ as the matrix case.

In the matrix case linearity takes the form

$$(1) \quad T(U+V) = TU + TV, \qquad T(UC) = (TU)C, \qquad T(Uc) = (TU)c$$

where U, V are n by n matrix-valued functions in the domain of T. Proof of the first equation is left for you. The second follows from

$$T(UC) = (UC)' - A(UC) = U'C - (AU)C = (U' - AU)C = (TU)C.$$

Instead of requiring C to be square as in our notational conventions, we could let C be any constant matrix with n rows. The choice $C = c$, a column vector of dimension n, then gives the third equation (1) as a special case of the second.

If we have n vector-valued functions F_j, with corresponding solutions Y_j, the equations

$$(2) \qquad\qquad TY_1 = F_1, \qquad TY_2 = F_2, \qquad \cdots, \qquad TY_n = F_n$$

yield $T(Y_1, Y_2, \cdots, Y_n) = (TY_1, TY_2, \cdots, TY_n) = (F_1, F_2, \cdots, F_n)$. Hence $TY = F$. Conversely, $TY = F$ is equivalent to n vector equations, one for each column of Y and F. Applying Section 1, Theorem 1 to each equation (2), we get the following:

THEOREM 1. *For each K, the following initial-value problem has one, and only one, solution.*

$$Y'(t) = A(t)Y(t) + F(t), \qquad t \in J; \qquad Y(t_0) = K$$

(a) Fundamental solutions. Variation of parameters

A *fundamental solution* of the equation $TY = 0$ is a solution $Y = U$ satisfying $|U(t_0)| \neq 0$ at some point t_0. A fundamental solution of $TY = 0$ is also a fundamental matrix for $Ty = 0$. By Section 1, Theorem 3 the condition $|U(t_0)| \neq 0$ implies $|U(t)| \neq 0$ on all of J, and hence the inverse $U(t)^{-1}$ of a fundamental matrix U exists at every point $t \in J$. This fact allows us to satisfy arbitrary initial conditions as in the following example.

EXAMPLE 1. If U is a fundamental matrix for $Ty = 0$ (that is, a funda-mental solution of $TY = 0$) solve the initial-value problems

$$y' = Ay, \quad y(t_0) = k; \qquad Y' = AY, \quad Y(t_0) = K.$$

In the first case let $y = Uc$. By linearity $y' = Ay$ and the initial condition holds if

$$U(t_0)c = k, \quad \text{hence if } c = U(t_0)^{-1}k.$$

The second case is similar, with c replaced by C and k by K. Thus the desired solutions are respectively

$$y(t) = U(t)U(t_0)^{-1}k, \qquad Y(t) = U(t)U(t_0)^{-1}K.$$

Another use of the fundamental solution is to solve the nonhomogeneous equation $Ty = f$. Instead of $y = Uc$ as above, we set $y = Uv$ where v is differentiable. Substituting $y = Uv$ into the equation $y' = Ay + f$ yields

$$Uv' + U'v = AUv + f, \qquad Uv' + AUv = AUv + f, \qquad Uv' = f.$$

Here the second equation follows from $U' = AU$ and the third is evident. The third equation gives $v' = U^{-1}f$, integration gives v, and the solution is $y = Uv$. In symbols,

$$(3) \qquad\qquad y(t) = U(t) \int U(t)^{-1}f(t)\, dt.$$

The constant of integration c implied in (3) produces an added term $U(t)c$. Since this satisfies the homogeneous equation, all choices of c are permissible. As in analogous cases presented earlier, this procedure is called the method of *variation of parameters*.

EXAMPLE 2. The *rest solution* of $Ty = f$ associated with the point t_0 is the solution satisfying $y(t_0) = 0$. To express this solution by a definite integral, all we have to do is to set $v(t_0) = 0$ in the above discussion. The result is

$$(4) \qquad y(t) = U(t) \int_{t_0}^{t} U(s)^{-1}f(s)\, ds \qquad \text{(rest solution).}$$

(b) Liouville's equation

The *trace* of a matrix A is the sum of its diagonal elements,

$$\text{trace } A = a_{11} + a_{22} + \cdots + a_{nn}.$$

The following theorem of Liouville is a far-reaching generalization of Abel's identity for the Wronskian, to which it reduces when a single linear equation of order n is written as a first-order system:

THEOREM 2. *If $Y'(t) = A(t)Y(t)$ for $t \in J$ then for $t \in J$ and $t_0 \in J$*

$$(5ab) \qquad |Y(t)|' = |Y(t)|\, \text{trace } A(t), \qquad |Y(t)| = |Y(t_0)|e^{\int_{t_0}^{t} \text{trace } A(s)\, ds}$$

For proof, suppose first that $|Y(t_0)| = 0$ at some point $t_0 \in J$. Then Section 1, Theorem 3 gives $|Y(t)| = 0$ on J and (5) is true, both sides being 0. We therefore assume $|Y(t)| \neq 0$ on J. There are two cases.

Case 1. Suppose that $Y(t_0) = I$ at a specified point $t_0 \in J$. We shall show that (5a) holds at t_0. Since $Y(t_0) = I$ we have $Y_j(t_0) = I_j$, and the equation $Y_j' = AY_j$ yields $Y_j' = AI_j$ at t_0. From $AI = A$ follows $AI_j = A_j$, so that

(6) $$Y_j = I_j, \qquad Y_j' = A_j \qquad \text{at } t = t_0.$$

Let us recall that the determinant $|Y| = |Y_1\ Y_2\ \cdots\ Y_n|$ can be differentiated as follows: Replace Y_j by Y_j' and add the results. In view of (6) this gives $|Y|'$ as a sum of determinants D_j, where D_j is obtained from

$$|I_1\ I_2\ \cdots\ I_j\ \cdots\ I_n|$$

by writing A_j instead of I_j. It is easily checked that $D_j = a_{jj}$ and this gives Equation (5a) at t_0.

Case 2. Let t_0 be a specified point at which (5a) is to be verified, and suppose $|Y(t_0)| \neq 0$. If $C = Y(t_0)^{-1}$ the function $U(t) = Y(t)C$ satisfies

$$U' = AU, \qquad U(t_0) = I.$$

Case 1 yields $|U|' = (\text{trace } A)\, |U|$ at t_0, or in other words

$$\frac{d}{dt}|Y(t)C| = (\text{trace } A)\, |Y(t)C|, \qquad t = t_0.$$

Since $|YC| = |Y||C|$, division by $|C|$ yields (5a) at t_0. However, t_0 was an arbitrary point of J. Hence (5a) holds on J and this gives (5b).

PROBLEMS

1P. Show that the equation $T(CU) = CTU$ fails unless A and C commute.

2P. The terminology and algebra of complex numbers extends in a natural way to matrices and the extension will be taken for granted in this book. Check that the complex equation

$$(X + iY)' = (A + iB)(X + iY) + (F + iG)$$

is equivalent to the two real equations on the left below, and that these can be written (by block multiplication) in the matrix form on the right:

$$\begin{cases} X' = AX - BY + F \\ Y' = BX + AY + G \end{cases} \qquad \begin{pmatrix} X' \\ Y' \end{pmatrix} = \begin{pmatrix} A & -B \\ B & A \end{pmatrix} \begin{pmatrix} X \\ Y \end{pmatrix} + \begin{pmatrix} F \\ G \end{pmatrix}$$

3P. Give a sensible formulation of the principle of equating real parts for solutions of matrix differential equations and show that it follows from the special case $B = 0$ in Problem 2. No answer is provided.

4. If U is a fundamental solution and C is nonsingular show that UC is also a fundamental solution.

5. Theorem 1 (uniqueness) shows that (4) must give the same result no matter what fundamental solution U is used. If U is replaced by UC with C nonsingular as in Problem 4, show that C drops out of (4).

6. If V is a fundamental solution show that the following satisfies $U(t_0) = I$ and is also a fundamental solution: $U(t) = V(t)V(t_0)^{-1}$.

7. With U as in Problem 6, deduce from (4) that the following equation yields the unique solution of $y' = Ay + f$, $y(t_0) = k$:

$$y(t) = U(t)\left(\int_{t_0}^{t} U(s)^{-1}f(s)\,ds + k\right)$$

8. Show that the method of variation of parameters applies almost without change to $TY = F$; you merely set $Y = UV$ instead of $y = Uv$.

9. If $Y' = AY$ and $n = 2$ check that $|Y|' = D_1 + D_2$ where

$$D_1 = \begin{vmatrix} a_{11}y_{11} + a_{12}y_{21} & a_{11}y_{12} + a_{12}y_{22} \\ y_{21} & y_{22} \end{vmatrix} = a_{11}|Y|$$

Proceed similarly for D_2 and deduce Liouville's theorem when $n = 2$.

10. This pertains to Case 1 of Liouville's theorem when $n = 4$. Write out the following determinants in displayed form and verify that the sum is trace A. The proof for arbitrary n is similar:

$$|A_1 \ I_2 \ I_3 \ I_4| + |I_1 \ A_2 \ I_3 \ I_4| + |I_1 \ I_2 \ A_3 \ I_4| + |I_1 \ I_2 \ I_3 \ A_4|$$

11. Referring to (5), let J be the interval (a, ∞) and let

$$R(t) = \text{Re}\left(\int_{t_0}^{t} (a_{11}(s) + a_{22}(s) + \cdots + a_{nn}(s))\,ds\right)$$

If all solutions of $Y'(t) = A(t)Y(t)$ are (a) bounded or (b) tend to 0 as $t \to \infty$, show (a) that $R(t) \le$ const. or (b) that $R(t) \to -\infty$ as $t \to \infty$.

3. Two-dimensional systems

In this section $n = 2$. Hence

$$A = \begin{pmatrix} a_{11} & a_{12} \\ a_{21} & a_{22} \end{pmatrix}, \qquad Y = \begin{pmatrix} y_{11} & y_{12} \\ y_{21} & y_{22} \end{pmatrix}$$

and y, f, c, k, u, v are respectively

$$\begin{pmatrix} y_1 \\ y_2 \end{pmatrix}, \quad \begin{pmatrix} f_1 \\ f_2 \end{pmatrix}, \quad \begin{pmatrix} c_1 \\ c_2 \end{pmatrix}, \quad \begin{pmatrix} k_1 \\ k_2 \end{pmatrix}, \quad \begin{pmatrix} u_1 \\ u_2 \end{pmatrix}, \quad \begin{pmatrix} v_1 \\ v_2 \end{pmatrix}.$$

The three equations

$$y' = Ay + f, \qquad y' = Ay, \qquad Y' = AY$$

are mutually related as follows: If u and v are two linearly independent solutions of the second equation, the 2-by-2 matrix $U = (u, v)$ with u, v as columns yields a fundamental solution of the third equation. Then

(1) $\qquad y = U(t)U(t_0)^{-1}k \quad$ satisfies $\quad y' = Ay, \qquad y(t_0) = k$

and, by variation of parameters,

(2) $\qquad y = U(t) \displaystyle\int U(t)^{-1} f(t)\, dt \quad$ satisfies $\quad y' = Ay + f.$

Thus the main problems are solved as soon as we have two linearly independent solutions of $y' = Ay$. In expanded form the equation $y' = Ay$ is

(3) $\qquad y_1' = a_{11}y_1 + a_{12}y_2, \qquad y_2' = a_{21}y_1 + a_{22}y_2.$

To simplify notation, let $y_1 = \phi$. Then (3) gives

(4) $\qquad a_{12}y_2 = \phi' - a_{11}\phi, \quad y_2' = a_{21}\phi + a_{22}y_2.$

Under suitable hypotheses the first of these equations allows us to eliminate y_2 from the second, thus getting an equation of order 2 for ϕ alone. The procedure is illustrated in the following examples and is discussed from a general point of view in the problems.

(a) Constant coefficients

Here, in Part (a), the coefficients a_{ij} are constant. To eliminate y_2 from (4), multiply the second equation by a_{12} and substitute for $a_{12}y_2$ the expression given by the first equation. We get

$$(\phi' - a_{11}\phi)' = a_{12}a_{21}\phi + a_{22}(\phi' - a_{11}\phi)$$

and then $\phi'' - (a_{11} + a_{22})\phi' + (a_{11}a_{22} - a_{21}a_{12})\phi = 0$. This can be written

(5) $$\phi'' - (\text{trace } A)\,\phi' + (\det A)\,\phi = 0.$$

As in Chapter 6, the substitution $\phi = e^{st}$ yields

(6) $$s^2 - (\text{trace } A)\,s + \det A = 0.$$

From this we get two linearly independent solutions of (5) and Equation (4) gives corresponding values for y_2. Equation (6) is called the *characteristic equation* of A or of the system (3).

EXAMPLE 1. Consider the linear system on the left below with matrix A on the right:

$$\begin{aligned} y_1' &= 12y_1 + 5y_2 \\ y_2' &= -6y_1 + y_2 \end{aligned} \qquad A = \begin{pmatrix} 12 & 5 \\ -6 & 1 \end{pmatrix}.$$

The trace of A is 13, its determinant is 42, and the characteristic equation is

$$s^2 - 13s + 42 = (s - 6)(s - 7) = 0.$$

Corresponding solutions ϕ are $c_1 e^{6t}$, $c_2 e^{7t}$. With $\phi = y_1$ the system gives $5y_2 = \phi' - 12\phi$ and the result is

$$y_1 = c_1 e^{6t}, \qquad 5y_2 = -6c_1 e^{6t}; \qquad y_1 = c_2 e^{7t}, \qquad y_2 = -c_2 e^{7t}.$$

We take $c_2 = 1$ and (to avoid fractions) $c_1 = 5$. Thus we get the following independent solutions u, v and fundamental matrix U:

$$u = \begin{pmatrix} 5e^{6t} \\ -6e^{6t} \end{pmatrix}, \quad v = \begin{pmatrix} e^{7t} \\ -e^{7t} \end{pmatrix}; \qquad U(t) = \begin{pmatrix} 5e^{6t} & e^{7t} \\ -6e^{6t} & -e^{7t} \end{pmatrix}.$$

EXAMPLE 2. Solve the system of Example 1 subject to the initial conditions $y(0) = k$. The inverse of the fundamental matrix found above can be written by inspection. The solution $y = U(t)U(0)^{-1}k$ satisfying $y(0) = k$ is

$$\begin{pmatrix} y_1 \\ y_2 \end{pmatrix} = \begin{pmatrix} 5e^{6t} & e^{7t} \\ -6e^{6t} & -e^{7t} \end{pmatrix} \begin{pmatrix} -1 & -1 \\ 6 & 5 \end{pmatrix} \begin{pmatrix} k_1 \\ k_2 \end{pmatrix}.$$

EXAMPLE 3. This system leads to a double root λ:

$$y_1' = 6y_1 + y_2 \atop y_2' = -y_1 + 8y_2 \, , \qquad A = \begin{pmatrix} 6 & 1 \\ -1 & 8 \end{pmatrix}.$$

The trace of A is 14, its determinant is 49, and the characteristic equation is

$$s^2 - 14s + 49 = (s - 7)^2 = 0.$$

Corresponding values of $\phi = y_1$ are e^{7t}, te^{7t}. The first equation of the system gives $y_2 = \phi' - 6\phi$. In Problem 1 you are asked to finish the calculation and to get the following fundamental matrix U together with its inverse U^{-1}:

$$(7) \qquad U(t) = e^{7t} \begin{pmatrix} 1 & t \\ 1 & t+1 \end{pmatrix}, \qquad U(t)^{-1} = e^{-7t} \begin{pmatrix} t+1 & -t \\ -1 & 1 \end{pmatrix}.$$

EXAMPLE 4. Solve $y' = Ay + f$ with A as in Example 3. By (2) the solutions are given by

$$\begin{pmatrix} y_1 \\ y_2 \end{pmatrix} = e^{7t} \begin{pmatrix} 1 & t \\ 1 & t+1 \end{pmatrix} \int e^{-7t} \begin{pmatrix} t+1 & -t \\ -1 & 1 \end{pmatrix} \begin{pmatrix} f_1(t) \\ f_2(t) \end{pmatrix} dt.$$

EXAMPLE 5. This system leads to complex λ:

$$y_1' = y_1 + y_2 \atop y_2' = -4y_1 + y_2 \, , \qquad A = \begin{pmatrix} 1 & 1 \\ -4 & 1 \end{pmatrix}.$$

Here the characteristic equation is $s^2 - 2s + 5 = 0$ with roots $1 \pm 2i$. The root $1 + 2i$ gives

$$y_1 = \phi = e^{(1+2i)t}, \qquad y_2 = \phi' - \phi = 2ie^{(1+2i)t}.$$

A corresponding complex vector is

$$\begin{pmatrix} e^{(1+2i)t} \\ 2ie^{(1+2i)t} \end{pmatrix} = e^t \begin{pmatrix} e^{2it} \\ 2ie^{2it} \end{pmatrix} = e^t \begin{pmatrix} \cos 2t + i \sin 2t \\ -2 \sin 2t + 2i \cos 2t \end{pmatrix}.$$

Separating into real and imaginary parts, we get the two vector solutions on the left below and these yield the fundamental matrix on the right:

$$u = e^t \begin{pmatrix} \cos 2t \\ -2\sin 2t \end{pmatrix}, \quad v = e^t \begin{pmatrix} \sin 2t \\ 2\cos 2t \end{pmatrix}; \qquad U = e^t \begin{pmatrix} \cos 2t & \sin 2t \\ -2\sin 2t & 2\cos 2t \end{pmatrix}.$$

(b) An example with variable coefficients

On an interval not containing the origin, we shall find a fundamental matrix for

$$(8) \qquad t^2 y_1' = -2ty_1 + 4y_2, \qquad ty_2' = -2ty_1 + 5y_2.$$

With $y_1 = \phi$ the first equation (8) gives $4y_2 = t^2\phi' + 2t\phi$. Multiply the second equation (8) by 4, substitute $4y_2$ as found above, and divide by t. The result is

$$(t^2\phi' + 2t\phi)' = -8\phi + 5(t\phi' + 2\phi).$$

In Problem 1 you are asked to show that this reduces to an equation of Euler type and to solve by setting $\phi(t) = t^m$. If all goes well, you will get the two solutions $u/2$ and v where

$$(9) \qquad\qquad u = \begin{pmatrix} 2 \\ t \end{pmatrix}, \qquad v = \begin{pmatrix} t^2 \\ t^3 \end{pmatrix}.$$

A fundamental matrix U and its inverse are

$$(10) \qquad U(t) = \begin{pmatrix} 2 & t^2 \\ t & t^3 \end{pmatrix}, \qquad U(t)^{-1} = \frac{1}{t^3} \begin{pmatrix} t^3 & -t^2 \\ -t & 2 \end{pmatrix}.$$

Since trace $A = 3/t$ and $|U(t)| = t^3$, Liouville's equation holds. This gives a check on the correctness of the solution.

EXAMPLE 6. Find the solution of (8) satisfying $y(1) = k$. By (1) and (10), the desired solution is

$$\begin{pmatrix} y_1 \\ y_2 \end{pmatrix} = \begin{pmatrix} 2 & t^2 \\ t & t^3 \end{pmatrix} \begin{pmatrix} 1 & -1 \\ -1 & 2 \end{pmatrix} \begin{pmatrix} k_1 \\ k_2 \end{pmatrix}.$$

EXAMPLE 7. Express the solution as an indefinite integral:

$$y_1' = -2t^{-1}y_1 + 4t^{-2}y_2 + f_1(t), \qquad y_2' = -2y_1 + 5t^{-1}y_2 + f_2(t).$$

The factors t^2 and t in (8) have been divided out to put the problem into the form $y' = Ay + f$. By (10) and (2),

$$y = \begin{pmatrix} 2 & t^2 \\ t & t^3 \end{pmatrix} \int \frac{1}{t^3} \begin{pmatrix} t^3 & -t^2 \\ -t & 2 \end{pmatrix} \begin{pmatrix} f_1(t) \\ f_2(t) \end{pmatrix} dt.$$

PROBLEMS———————————————————————————

1P. Fill in the details omitted in the derivation of (7) and (10).

2. (a) Evaluate y in Example 2 if $k = (1, -1)^T$.

(b) In Example 4 get the rest solution when $f(t) = (t, 1)^T e^{7t}$.

(c) In Example 5 find the solution satisfying $y(0) = k$.

(d) In Example 5 write the solutions of $y' = Ay + f$ using indefinite integrals and evaluate explicitly if $f = e^t(\alpha, 4\beta)^T$ where α and β are constant.

(e) Check that the result of Example 6 gives u in (9) when $k = (2, 1)^T$ and v when $k = (1, 1)^T$.

(f) Taking the constants of integration as 0, evaluate the formula in Example 7 when $f(t) = (t^m, t^{m+1})^T$ with $t > 0$ and m constant.

3. For the system $y_1' = 3y_1 + 2y_2$, $y_2' = -2y_1 - y_2$ find the unique fundamental matrix satisfying $U(0) = I$ and verify.

4P. Canonical variables. By Section 1, the complete solution of the equation in Example 1 is $y_1 = 5c_1 e^{6t} + c_2 e^{7t}$, $y_2 = -6c_1 e^{6t} - c_2 e^{7t}$.

(a) Eliminate e^{6t} by forming $6y_1 + 5y_2$ and similarly eliminate e^{7t}.

(b) Thus conclude that the variables $x_1 = 6y_1 + 5y_2$, $x_2 = y_1 + y_2$ satisfy $x_1' = 7x_1$, $x_2' = 6x_2$. The first equation involves x_1 alone and the second involves x_2 alone. The equations in x_i are said to be *decoupled* and the x_i are called *canonical variables*. See also Chapter 16.

5. Generalize Problem 4 to $y' = Ay$, where A is any constant 2-by-2 matrix such that its characteristic equation has two distinct roots λ, μ.

6. Here are three matrices A. Using Problem 10 if you wish, get a fundamental matrix for the equation $y' = Ay$ on a suitably restricted interval J and verify by use of Liouville's equation. Problem (c) leads to a nonelementary integral:

(a) $\dfrac{1}{2t^3} \begin{pmatrix} t^2 & 3 \\ -t^4 & 9t^2 \end{pmatrix}$ 　　(b) $\begin{pmatrix} 1 & e^{3t} \\ -2e^{-3t} & 1 \end{pmatrix}$ 　　(c) $\dfrac{1}{t^2} \begin{pmatrix} 2t & -2 \\ t^2 + 2t & -2 \end{pmatrix}$

7. In Problem 6(b) obtain the following solution for $y' = Ay + f$:

$$y(t) = \begin{pmatrix} e^{3t} & e^{2t} \\ 2 & e^{-t} \end{pmatrix} \int \begin{pmatrix} -e^{-3t} & 1 \\ 2e^{-2t} & -e^t \end{pmatrix} \begin{pmatrix} f_1(t) \\ f_2(t) \end{pmatrix} dt$$

8. Complex coefficients. Let a, b, p, q be real-valued functions defined on J and let $z = p + iq = e^w$, $h = a + ib$.

(a) Show that the following system is equivalent to $z' = hz$ and, if w is differentiable, to $w' = h$:

$$\begin{pmatrix} p' \\ q' \end{pmatrix} = \begin{pmatrix} a & -b \\ b & a \end{pmatrix} \begin{pmatrix} p \\ q \end{pmatrix}$$

(b) If z satisfies $z' = hz$ so does iz. What does this say about solutions of the corresponding 2-by-2 real system?

(c) By this method solve $y_1' = 7y_1 + 9y_2$, $y_2' = -9y_1 + 7y_2$.

9. Here are three matrices A. Using the method of Part (b) or Problem 8, whichever is simpler, get a fundamental matrix for the equation $y' = Ay$ on a suitably restricted interval J. Verify by use of Liouville's equation:

(a) $\dfrac{1}{t}\begin{pmatrix} 1 & -t \\ t & 1 \end{pmatrix}$ (b) $\begin{pmatrix} -\tan t & -1 \\ 1 & -\tan t \end{pmatrix}$ (c) $\dfrac{1}{1+t^2}\begin{pmatrix} t & -1 \\ 1 & t \end{pmatrix}$

10. Suppose that $a_{12}(t)$ does not vanish at any $t \in J$ and suppose further that a_{11} and a_{12} are differentiable on J. Show that (4) yields

$$\phi'' - (a_{11} + a_{22})\phi' + \begin{vmatrix} a_{11} & a_{12} \\ a_{21} & a_{22} \end{vmatrix}\phi = a_{12}\left(\frac{a_{11}}{a_{12}}\right)'\phi + \frac{a_{12}'}{a_{12}}\phi'$$

─────────────────────ANSWERS─────────────────────

2. (a) $(e^{7t}, -e^{7t})^T$ (b) $6y = e^{7t}(6t^2 - t^3,\ 6t + 3t^2 - t^3)^T$

   (c) $2y = e^t(2k_1\cos 2t + k_2\sin 2t,\ -4k_1\sin 2t + 2k_2\cos 2t)^T$

   (d) Taking the constants of integration as zero, $y = e^t(\beta, -\alpha)^T$. The method of undetermined coefficients would have been easier.

   (f) $(m-1)y = t^m(t, t^2)^T$, $\quad m \neq 1$; $\qquad y = (\ln t)(t^2, t^3)^T$, $\quad m = 1$

3. Partial check: The first row of $U(t)$ is $e^t(1 + 2t, 2t)$.

6. With $\phi'(t) = e^{2/t}$ in (c), possible answers are:

   (a) $\begin{pmatrix} 3t & t^2 \\ t^3 & t^4 \end{pmatrix}$ (b) $\begin{pmatrix} e^{2t} & e^{3t} \\ e^{-t} & 2 \end{pmatrix}$ (c) $\begin{pmatrix} 1 & \phi(t) \\ t & t\phi(t) - (t^2/2)e^{2/t} \end{pmatrix}$

8. (b) If $(p, q)^T$ is a solution so is $(-q, p)^T$.

   (c) $y_1 = e^{7t}(c_1\cos 9y + c_2\sin 9t)$, $y_2 = e^{7t}(-c_1\sin 9y + c_2\cos 9t)$

9. The following in cases (abc) respectively:

   $t\begin{pmatrix} \cos t & -\sin t \\ \sin t & \cos t \end{pmatrix}$ $(\cos t)\begin{pmatrix} \cos t & -\sin t \\ \sin t & \cos t \end{pmatrix}$ $\begin{pmatrix} 1 & -t \\ t & 1 \end{pmatrix}$

4. A class of higher-order systems

In this section the operator method of Chapter 9 will be combined with the matrix methods of Chapters CDE. We consider a system

$$\sum_{j=1}^{n} T_{ij} y_j = f_i, \qquad i = 1, 2, \cdots, n$$

where each operator T_{ij} is a polynomial in $D = d/dt$ with constant coefficients. Thus,

$$T_{ij} = P_{ij}(D).$$

The order of T_{ij} is the degree of the corresponding polynomial $P_{ij}(s)$. A constant nonzero polynomial has degree 0. However, the 0 polynomial $P(s) = 0$ has no degree, and the corresponding operator $T = 0$ has no order.

If $T(D)$ denotes the operator matrix (T_{ij}) the system can be written in the compact form

$$(1) \qquad\qquad T(D)y = f.$$

The *characteristic polynomial* and *characteristic operator* are, respectively,

$$P(s) = |P_{ij}(s)|, \qquad P(D) = |T(D)|.$$

Equation (1) resists an attempt to get by with minimum smoothness conditions on f and y, and no such attempt will be made here. The equation also becomes intractable if the operator $P(D) = 0$. With these thoughts in mind, we introduce the following:

BASIC HYPOTHESIS. The operator $P(D)$ has order $m \geq 0$ and f and y have enough derivatives to justify the calculations.

Since each operator T_{ij} is linear, the matrix operator $T(D)$ is also linear. Hence the principle of superposition, the principle of the complementary function, and other results established for linear operators in the foregoing pages apply to this case. As explained in Chapter E, a substantial part of the algebraic theory of Chapters CDE carries over to $T(D)$, and the proofs are the same. The main restriction is that we must not divide by an expression containing D.

In particular, a suitable interpretation of Cramer's rule holds. For example let $n = 2$, so that the system and its coefficient determinant are

$$\begin{matrix} T_{11}y_1 + T_{12}y_2 = f_1 \\ T_{21}y_1 + T_{22}y_2 = f_2 \end{matrix}, \qquad P(D) = \begin{vmatrix} T_{11} & T_{12} \\ T_{21} & T_{22} \end{vmatrix}$$

Then

$$P(D)y_1 = \begin{vmatrix} f_1 & T_{12} \\ f_2 & T_{22} \end{vmatrix}, \qquad P(D)y_2 = \begin{vmatrix} T_{11} & f_1 \\ T_{21} & f_2 \end{vmatrix}$$

The first determinant is, by definition,

$$\begin{vmatrix} f_1 & T_{12} \\ f_2 & T_{22} \end{vmatrix} = -\begin{vmatrix} T_{12} & f_1 \\ T_{22} & f_2 \end{vmatrix} = -T_{12}f_2 + T_{22}f_1.$$

The second is $T_{11}f_2 - T_{21}f_1$. In general if one column of a determinant contains functions f_j and the remaining columns contain operators, the cofactor of f_j is written *on the left* of f_j when the determinant is expanded, no matter where f_j may be located in the determinant before expansion.

For proof of Cramer's rule, it suffices to examine either of the proofs in Chapter E. Only the last step, division by the determinant, is excluded.

EXAMPLE 1. For the system

$$(D-1)y_1 + (D-2)y_2 = e^t, \qquad (2D^2 + 2D + 2)y_1 + (3D^2 + 3D + 2)y_2 = 4t$$

it is easily checked that $P(D) = (D^2 + 1)(D + 2)$ and hence

$$(D^2 + 1)(D + 2)y_1 = \begin{vmatrix} e^t & D-2 \\ 4t & 3D^2 + 3D + 2 \end{vmatrix} = 8e^t + 8t - 4.$$

The methods of Chapter 8 give

$$y_1 = \frac{4}{3}e^t + 4t - 8 + c_1 e^{-2t} + c_2 \cos t + c_3 \sin t.$$

In a like manner

$$(2) \qquad (D^2 + 1)(D + 2)y_2 = \begin{vmatrix} D-1 & e^t \\ 2D^2 + 2D + 2 & 4t \end{vmatrix} = 4 - 4t - 6e^t$$

and, again by Chapter 8,

$$(3) \qquad y_2 = e^t + 2t - 1 + c_4 e^{-2t} + c_5 \cos t + c_6 \sin t.$$

Although the above results contain a good deal of information, the problem is far from solved. The trouble is that the six constants c_j are not independent. To complete the solution along these lines we would have to substitute each expression into the original system and express three of the constants in terms of the other three. Except in the simplest cases this technique is very inefficient, and a better method is presented next.

The systematic theory of (1) was developed by the British algebraist Chrystal in 1895, and a simplified version of Chrystal's theory will be given here. It depends on an elementary observation concerning the two equations

(a) $$T_{i1}y_1 + T_{i2}y_2 + \cdots + T_{in}y_n = f_i$$

(b) $$T_{j1}y_1 + T_{j2}y_2 + \cdots + T_{jn}y_n = f_j.$$

Let $Q = Q(D)$ be a polynomial in D with constant coefficients. If Equation (a) is multiplied on the left by Q and subtracted from (b) the result is

(c) $$(T_{j1} - QT_{i1})y_1 + (T_{j2} - QT_{i2})y_2 + \cdots + (T_{jn} - QT_{in})y_n = f_j - Qf_i.$$

We can now state the following:

LEMMA. Equations (ab) and (ac) have exactly the same solutions.

For proof, we already deduced (c) from (ab) and hence (ac) follows from (ab). On the other hand if (ac) is given, the terms in (c) involving Q as a factor reduce to 0 by (a), and then (c) reduces to (b). This shows that (ac) implies (ab) and completes the proof.

EXAMPLE 2. The augmented matrix in the system of Example 1 is

$$\begin{pmatrix} D - 1 & D - 2 & e^t \\ 2D^2 + 2D + 2 & 3D^2 + 3D + 2 & 4t \end{pmatrix}$$

If $2D^2 + 2D + 2$ is divided by $D - 1$ the result is $2D + 4$ with remainder 6, thus:

$$2D^2 + 2D + 2 = (2D + 4)(D - 1) + 6.$$

After substituting this in the above matrix, we subtract $2D + 4$ times the first row from the second row. This yields the equivalent system

$$\begin{pmatrix} D - 1 & D - 2 & e^t \\ 6 & D^2 + 3D + 10 & 4t - 6e^t \end{pmatrix}.$$

Since the second equation gives y_1 directly in terms of y_2, we can already complete the solution without introducing any extraneous constants of integration. Nevertheless we carry the calculation one step further.

Multiply the first row by 6 (to avoid fractions) and then subtract $D - 1$ times the second row from the first. The result is

$$\begin{pmatrix} 0 & -(D^3 + 2D^2 + D + 2) & 6e^t - 4 + 4t \\ 6 & D^2 + 3D + 10 & 4t - 6e^t \end{pmatrix}.$$

The first row yields the equation (2) for y_2 found in Example 1 and has the same three-parameter family of solutions (3). What is new is that the second row yields y_1 directly, without additional constants of integration. Namely,

$$6y_1 = 4t - 6e^t - (D^2 + 3D + 10)y_2.$$

Substitution of y_2 from Example 1 to get y_1 is routine and is omitted.

We can always use the procedure of Example 2 to reduce the matrix to row-echelon form. Let T_{i1} be the operator of lowest order in the first column of $T(D)$ (or one of these if there are several). Long division yields equations of the form

$$T_{j1} = Q_{j1}T_{i1} + R_{j1}$$

where Q_{j1} corresponds to the quotient and R_{j1} to the remainder. Thus each R_{j1} is either 0 or has a lower order than T_{i1}. By the lemma we can subtract Q_{j1} times the ith row from the jth row and remove the term $Q_{j1}T_{i1}$ from T_{j1}. This is done for each $j \neq i$ for which $T_{j1} \neq 0$.

Each operation subtracts a multiple of one row from another row and hence, by the lemma, it does not change the set of solutions. Results of Chapter E show that the coefficient determinant is also unchanged.

Repetition, with new choices of i, finally reduces the first column to a form in which one element is a nonzero operator S_{11} and all others are 0. We rearrange rows so that S_{11} is at the top. A similar reduction is applied to the matrix of order $n - 1$ obtained when the first row and column are omitted, and so on. After n steps the coefficient matrix is in row echelon form.

For example, when $n = 3$ the reduced coefficient matrix $S(D)$ and the corresponding reduced system would be

$$\begin{pmatrix} S_{11} & S_{12} & S_{13} \\ 0 & S_{22} & S_{23} \\ 0 & 0 & S_{33} \end{pmatrix}, \qquad \begin{aligned} S_{11}y_1 + S_{12}y_2 + S_{13}y_3 &= g_1 \\ S_{22}y_2 + S_{23}y_3 &= g_2 \\ S_{33}y_3 &= g_3 \end{aligned}$$

Here g is derived from f and S_{ij} from T_{ij} by the reduction process described above. We solve the reduced system by the methods of Chapter 8, starting with the third equation and working upwards.

The number of constants in the solution of the reduced system is equal to the order of the operator $S_{11}S_{22}S_{33}$. Indeed, if the order of S_{jj} is m_j, we introduce m_3 constants at the first step, then m_2 additional constants in the second step, and m_1 additional constants in the last step. Note that the above product is $|S_{ij}|$, which is the determinant of the reduced system. This equals $\pm P(D)$, where the minus is needed to account for rearrangment of the rows.

These informal remarks could be developed into a full proof of Chrystal's theorem, which is as follows:

THEOREM 1. Let $P(D) = |T(D)|$ be an operator of order m. Then the system $T(D)y = f$ has a set of solutions with m independent parameters. These can be found from an equivalent system that is in row-echelon form and has characteristic operator $P(D)$ or $-P(D)$.

PROBLEMS————————————————————————————————————

1. Find $P(D)$ and thus predict the expected number of constants in the complete solution. Also predict the form of solutions of the corresponding homogeneous system (or of the system itself, if it is already homogeneous):

(a) $\begin{cases} (D-4)y_1 - y_2 = e^{2t} \\ y_1 + (D-2)y_2 = 4e^{3t} \end{cases}$

(d) $\begin{cases} (D+3)y_1 + (D-7)y_2 = f_1 \\ (D+4)y_1 + (D-6)y_2 = f_2 \end{cases}$

(b) $\begin{cases} (D^4 - 10)y_1 = 9y_2 \\ (D^4 - 7)y_2 = 6y_1 \end{cases}$

(e) $\begin{cases} (D+3)y_1 + (D-7)y_2 = -7 \\ (D+5)y_1 + (D-6)y_2 = -6 \end{cases}$

(c) $\begin{cases} D^2 y_1 + Dy_2 = 0 \\ Dy_1 + (D+1)y_2 = 0 \end{cases}$

(f) $\begin{cases} (D^8 + D^2 + 1)y_1 + D^6 y_2 = 0 \\ (D^4 + D^2)y_1 + (D^2 + 1)y_2 = 0 \end{cases}$

2. (abcdef) Solve the systems in Problem 1.

3. This problem is due to Chrystal. Proceed as in Problems 1 and 2:

$$(D^2 + 1)y_1 + (D^2 + D + 1)y_2 = t \qquad Dy_1 + (D+1)y_2 = e^t$$

4. Solve this harder problem, which is also due to Chrystal:

$$\begin{pmatrix} D & D & D-1 \\ D-1 & D^2 & (D-1)^2 \\ D+1 & D^3 & D^2 - 1 \end{pmatrix} \begin{pmatrix} y_1 \\ y_2 \\ y_3 \end{pmatrix} = \begin{pmatrix} 0 \\ 0 \\ 0 \end{pmatrix}$$

Hint: Use the first equation to eliminate y_2 from the second and third equations. By one more step, eliminate y_3.

5. If α is a constant parameter, solve:

$$y_1' - y_1 + y_2' - 2y_2 = e^{-t} \qquad y_1' - 3y_1 + y_2' - 2\alpha y_2 = e^{-t} - 2$$

6. This problem is difficult. Solve

$$\begin{pmatrix} D^4 + 1 & 0 & 2D^2 + 2 \\ D^2 - 1 & 2D & 0 \\ 0 & D^6 + D^2 & D^5 - D \end{pmatrix} \begin{pmatrix} y_1 \\ y_2 \\ y_3 \end{pmatrix} = \begin{pmatrix} 0 \\ 0 \\ 0 \end{pmatrix}$$

Hint: One of the two values of $\sqrt{i}$ is $e^{i\pi/4}$.

7. Asymptotic stability. Within the framework of our basic hypothesis, the theory of asymptotic stability (Chapter 8, Section 2) extends to systems $T(D)y = f$. After giving suitable definitions, show that the solutions are asymptotically stable if, and only if, all the roots of the characteristic equation $P(s) = 0$ have negative real parts.

────────────────────ANSWERS────────────────────

1. The order of $P(D)$ is respectively 2, 8, 3, 0, 1, 4. This is the number of independent constants as predicted by Chrystal's theorem. If $T(D)z = 0$, each z_j satisfies $P(D)z_j = 0$ by Cramer's rule. That gives the general form. For confirmation, see answers to Problem 2.

2. (a) $y_1 = (c_1 + c_2 t + 2t^2)e^{3t}$, $y_2 = y_1' - 4y_1 - e^{2t}$

   (b) $y_1 = c_1 e^t + c^2 e^{-t} + c^3 \sin t + c_4 \cos t + c_5 e^{2t} + c_6 e^{-2t} + c_7 \sin 2t + c_8 \cos 2t$,

       $9y_2 = y_1^{(4)} - 10y_1$

   (c) $y_1 = c_1 + c_2 t + c_3 t^2$, $y_2 = 2c_3 - c_2 - 2c_3 t$

   (d) $10y_1 = f_1' - 6f_1 - f_2 + 7f_2$, $10y_2 = -f_1' - 4f_1 + f_2' + 3f_2$

   (e) $y_1 = c_1 e^{-17t}$, $y_2 = 1 - 2c_1 e^{-17t}$

   (f) $y_1 = y_2 = (c_1 + c_2 t)\cos t + (c_3 + c_4 t)\sin t$

3. $y_1 = 1 + t - 3e^t$, $y_2 = 2e^t - 1$

4. Partial answer: $y_1 = c_1 + c_2 e^t + c_3 e^{(1-\sqrt{2})t} + c_4 e^{(1+\sqrt{2})t}$,

   $(D-1)y_3 = (-D^2 + D - 1)y_1$, $Dy_2 = -Dy_1 - (D-1)y_3$

5. $y_1 = 1 + (1-\alpha)y_2$. For $\alpha = 2, 5/2, 3$, other, y_2 is respectively:

   $-1 - e^{-t}$, $(c_1 - t)e^{-t} - 1$, $c_1 - t + e^{-t}$, $\alpha_1 + \alpha_2 e^{-t} + c_1 e^{\beta t}$

   where $\alpha_1 = 1/(\alpha - 3)$, $\alpha_2 = 1/(2\alpha - 5)$, $\beta = (\alpha - 3)/(\alpha - 2)$.

6. $y_3 = c_1 + c_2 e^t + c_3 e^{-t} + c_4 \cos t + c_5 \sin t$. With $\alpha = \sqrt{2}/2$,

   $y_2 = -c_1 t + c_6 + (c_7 e^{-\alpha t} + c_8 e^{\alpha t})\cos(\alpha t) + (c_9 e^{-\alpha t} + c_{10} e^{\alpha t})\sin(\alpha t)$

   $y_1 = (D^2 + 1)(Dy_2 - y_3)$

──

> From a drop of water, a logician could infer the possibility
> of an Atlantic or a Niagara, without having seen or heard
> of one or the other.
>
>   —Sherlock Holmes

Chapter 16

EIGENVALUES AND EIGENVECTORS

$\mathbf{W}$HEN the matrix A is constant, the equation $y' = Ay$ has nontrivial solutions of the form $ce^{\lambda t}$, as one might expect from the scalar case. These solutions lead to characteristic values λ_j, characteristic vectors E_j, and to canonical variables. The latter are related to y by a linear change of variables, $y = Ex$, and they satisfy an uncoupled system that can be solved by inspection. Canonical variables play a major role in engineering, economics, mechanics, and indeed in all fields that make extensive use of linear systems with constant coefficients.

Prerequisites: Chapters CDE and Chapter 15, Section 1.

1. Introduction

In this chapter $A = (a_{ij})$ is a constant n-by-n matrix. As we know from Chapter 1, the scalar equation $y' = Ay$ for $n = 1$ has solutions of the form $ce^{\lambda t}$ where $\lambda = A$ and c is an arbitrary constant. It turns out that the vector equation $y' = Ay$ for $n \geq 2$ also has solutions of the form $ce^{\lambda t}$, but the vector c is no longer arbitrary and (unless $c = 0$) the exponent λ must be the root of a certain polynomial equation. Although our objective is to solve a differential equation, the computation of c and λ involves algebra rather than calculus.

Let us try, then, to choose λ and a vector $c \neq 0$ so that the function $y = ce^{\lambda t}$ satisfies $y' = Ay$. Substitution leads successively to

$$\lambda ce^{\lambda t} = Ace^{\lambda t}, \qquad Ac = \lambda c, \qquad (A - \lambda I)c = 0.$$

There is a nonzero solution c if, and only if, $|A - \lambda I| = 0$. This is the same as $P(\lambda) = 0$ where, by definition,

$$P(s) = |sI - A|.$$

It is easily seen that $P(s)$ is a polynomial in s. Since it behaves like $|Is| = s^n$ for large s, the leading term is s^n.

The above expression $P(s)$ is the *characteristic polynomial* of A, the equation $P(s) = 0$ is the *characteristic equation*, a solution λ of this equation is a *characteristic value*, and a nonzero vector c such that $Ac = \lambda c$ is a *characteristic vector*. The conditions are summarized by:

$$|\lambda I - A| = 0, \qquad Ac = \lambda c, \qquad c \neq 0.$$

Characteristic vectors and values are often called *eigenvectors* and *eigenvalues*, respectively. These words are hybrids of English and German, and following German usage, "ei" rhymes with π.

The following notation will be used throughout this chapter and the next. Since $P(s)$ has degree n there are n eigenvalues λ_j, not necessarily distinct. We denote corresponding eigenvectors by $E_j = (e_{ij})$. Thus, E_j is not 0 and satisfies each of the equivalent equations

$$(A - \lambda_j I)E_j = 0, \qquad AE_j = \lambda_j E_j, \qquad (\lambda_j I - A)E_j = 0.$$

By definition, the n by n matrix with columns E_j is E and the diagonal matrix with diagonal elements λ_j is Λ. When computing a particular eigenvector E_j, we use e_i as an abbreviation for e_{ij}.

The principal features of the theory are already present in the case $n = 2$, and this case is discussed next. When $n = 2$ the matrix $A = (a_{ij})$ is of order 2 and

$$\lambda I - A = \begin{pmatrix} \lambda & 0 \\ 0 & \lambda \end{pmatrix} - \begin{pmatrix} a_{11} & a_{12} \\ a_{21} & a_{22} \end{pmatrix} = \begin{pmatrix} \lambda - a_{11} & -a_{12} \\ -a_{21} & \lambda - a_{22} \end{pmatrix}.$$

By a short calculation the characteristic equation $|\lambda I - A| = 0$ is

$$\lambda^2 - (\text{trace } A)\lambda + (\det A) = 0.$$

Since the equation is quadratic there are two roots λ_j, not necessarily distinct. In this case

$$E_1 = \begin{pmatrix} e_{11} \\ e_{21} \end{pmatrix}, \quad E_2 = \begin{pmatrix} e_{12} \\ e_{22} \end{pmatrix}, \quad E = \begin{pmatrix} e_{11} & e_{12} \\ e_{21} & e_{22} \end{pmatrix}, \quad \Lambda = \begin{pmatrix} \lambda_1 & 0 \\ 0 & \lambda_2 \end{pmatrix}.$$

The matrix Λ is unique aside from the order in which the λ_i are written, but E is not unique, because any of its columns can be multiplied by a nonzero constant. For example, $5E_1$ is an eigenvector if E_1 is.

From the definitions it follows that the functions $E_1 e^{\lambda_1 t}$ and $E_2 e^{\lambda_2 t}$ each satisfy $y' = Ay$. If $\lambda_1 \neq \lambda_2$, these solutions are linearly independent, hence

$$U(t) = \begin{pmatrix} E_1 e^{\lambda_1 t} & E_2 e^{\lambda_2 t} \end{pmatrix}$$

is a fundamental matrix. The corresponding general solution is

$$y = \alpha_1 E_1 e^{\lambda_1 t} + \alpha_2 E_2 e^{\lambda_2 t}$$

where, as also below, the α_j denote real or complex constants. This illustrates one use of eigenvectors.

Another use depends on the identity $AE = E\Lambda$, which is verified as follows:

$$\begin{pmatrix} a_{11} & a_{12} \\ a_{21} & a_{22} \end{pmatrix} \begin{pmatrix} e_{11} & e_{12} \\ e_{21} & e_{22} \end{pmatrix} = \begin{pmatrix} \lambda_1 e_{11} & \lambda_2 e_{12} \\ \lambda_1 e_{21} & \lambda_2 e_{22} \end{pmatrix} = \begin{pmatrix} e_{11} & e_{12} \\ e_{21} & e_{22} \end{pmatrix} \begin{pmatrix} \lambda_1 & 0 \\ 0 & \lambda_2 \end{pmatrix}.$$

Here the first equality results from $AE_j = \lambda_j E_j$ and the second is evident. If the vectors E_1 and E_2 are linearly independent, as we now assume, then $|E| \neq 0$, and we can make the change of variables

$$y = Ex, \qquad x = E^{-1}y.$$

With $y = Ex$ and $AE = E\Lambda$ the equation $y' = Ay$ gives, in succession,

$$y' = Ay, \qquad Ex' = AEx, \qquad Ex' = E\Lambda x, \qquad x' = \Lambda x.$$

The last equation is equivalent to $x_1' = \lambda_1 x_1$, $x_2' = \lambda_2 x_2$ and hence, by inspection,

$$x_1 = \alpha_1 e^{\lambda_1 t}, \qquad x_2 = \alpha_2 e^{\lambda_2 t}.$$

It is said that the original system $y' = Ay$ is uncoupled by the change of variable $y = Ex$ and the new variables x_i are called *canonical variables*.

The success of the foregoing methods depends on the condition $|E| \neq 0$ which in turn depends on linear independence of the vectors E_1, E_2. To complete our discussion of the case $n = 2$ we show that, if the eigenvalues are distinct, the eigenvectors are independent.

Suppose, then, that $\lambda_1 \neq \lambda_2$ and that

$$(1) \qquad\qquad \alpha_1 E_1 + \alpha_2 E_2 = 0.$$

We want to show that $\alpha_1 = \alpha_2 = 0$. To this end multiply (1) on the left by A. Since $AE_j = \lambda_j E_j$ the result is

$$(2) \qquad\qquad \alpha_1 \lambda_1 E_1 + \alpha_2 \lambda_2 E_2 = 0.$$

If (1) is multiplied by λ_2 and (2) is then subtracted we get

$$\alpha_1 (\lambda_2 - \lambda_1) E_1 = 0.$$

This gives $\alpha_1 = 0$ and then $\alpha_2 = 0$ by (1).

EXAMPLE 1.   The matrix on the left below has the characteristic equation on the right:

$$A = \begin{pmatrix} 12 & 5 \\ -6 & 1 \end{pmatrix}, \qquad \begin{vmatrix} \lambda - 12 & -5 \\ 6 & \lambda - 1 \end{vmatrix} = \lambda^2 - 13\lambda + 42 = 0.$$

Hence $\lambda = 6$ and $\lambda = 7$. If $\lambda = 6$ the equation $(A - \lambda I)E_1 = 0$ reduces to

$$\begin{pmatrix} 6 & 5 \\ -6 & -5 \end{pmatrix} \begin{pmatrix} e_1 \\ e_2 \end{pmatrix} = \begin{pmatrix} 0 \\ 0 \end{pmatrix}.$$

This is equivalent to $6e_1 + 5e_2 = 0$. (We get only one equation because the determinant is 0. That was the reason for taking $\lambda = 6$.) We can use any nonzero value for e_1 and the equations then give e_2. To avoid fractions we choose $e_1 = 5$. The resulting characteristic vector E_1 and solution $u = E_1 e^{\lambda t}$ are on the left below:

$$E_1 = \begin{pmatrix} 5 \\ -6 \end{pmatrix}, \qquad u = \begin{pmatrix} 5 \\ -6 \end{pmatrix} e^{6t}; \qquad E_2 = \begin{pmatrix} 1 \\ -1 \end{pmatrix}, \qquad v = \begin{pmatrix} 1 \\ -1 \end{pmatrix} e^{7t}.$$

It is left for you to verify that the choice $\lambda = 7$ yields $e_1 + e_2 = 0$ for the corresponding eigenvector E_2, and to get E_2 and v as above.

The matrix E and its inverse are, respectively,

$$E = \begin{pmatrix} 5 & 1 \\ -6 & -1 \end{pmatrix}, \qquad E^{-1} = \begin{pmatrix} -1 & -1 \\ 6 & 5 \end{pmatrix}.$$

As a check, we write A, E, Λ side by side, thus:

$$\begin{pmatrix} 12 & 5 \\ -6 & 1 \end{pmatrix} \begin{pmatrix} 5 & 1 \\ -6 & -1 \end{pmatrix} \begin{pmatrix} 6 & 0 \\ 0 & 7 \end{pmatrix}.$$

The equation $AE = E\Lambda$ is easily verified and confirms the calculations.

EXAMPLE 2. Canonical variables x in Example 1 are given by

$$\begin{pmatrix} y_1 \\ y_2 \end{pmatrix} = \begin{pmatrix} 5 & 1 \\ -6 & -1 \end{pmatrix} \begin{pmatrix} x_1 \\ x_2 \end{pmatrix}, \qquad \begin{pmatrix} x_1 \\ x_2 \end{pmatrix} = \begin{pmatrix} -1 & -1 \\ 6 & 5 \end{pmatrix} \begin{pmatrix} y_1 \\ y_2 \end{pmatrix}.$$

The equation $x' = \Lambda x$ is in this case $x_1' = 6x_1$, $x_2' = 7x_2$. As a check, we compute x_1' by use of the original system,

$$y_1' = 12y_1 + 5y_2, \qquad y_2' = -6y_1 + y_2.$$

The equation $x_1 = -(y_1 + y_2)$ yields

$$x_1' = -y_1' - y_2' = -(12y_1 + 5y_2) - (-6y_1 + y_2) = -6y_1 - 6y_2 = 6x_1.$$

Verification that $x_2' = 7x_2$ is similar and is left for you.

EXAMPLE 3. For the problem

$$A = \begin{pmatrix} 6 & 1 \\ -1 & 8 \end{pmatrix}, \qquad \begin{vmatrix} \lambda - 6 & -1 \\ 1 & \lambda - 8 \end{vmatrix} = \lambda^2 - 14\lambda + 49 = 0$$

there is only one characteristic value, $\lambda = 7$, and the equation for a corresponding characteristic vector is

$$\begin{pmatrix} -1 & 1 \\ -1 & 1 \end{pmatrix} \begin{pmatrix} e_1 \\ e_2 \end{pmatrix} = \begin{pmatrix} 0 \\ 0 \end{pmatrix}.$$

The sole solution is $e_1 = \alpha$, $e_2 = \alpha$ where α is constant, and there is no way to construct a fundamental matrix from this.

EXAMPLE 4. The following leads to complex eigenvalues $1 \pm 2i$:

$$\begin{pmatrix} 1 & 1 \\ -4 & 1 \end{pmatrix}, \qquad \begin{vmatrix} \lambda - 1 & -1 \\ 4 & \lambda - 1 \end{vmatrix} = \lambda^2 - 2\lambda + 5 = 0.$$

It is left for you to verify that, when $\lambda = 1 + 2i$, the equation $(A - \lambda I)E_1 = 0$ reduces to $e_2 = 2ie_1$. The form of the final solution depends on the choice of e_1, whether real or complex. For example, $e_1 = 1$ leads to the result of Chapter 15, Section 3, Example 5. But $e_1 = i$ gives the complex solution on the left below, which yields the real fundamental solution on the right:

$$\begin{pmatrix} i \\ -2 \end{pmatrix} e^t (\cos 2t + i \sin 2t), \qquad e^t \begin{pmatrix} -\sin 2t & \cos 2t \\ -2\cos 2t & -2\sin 2t \end{pmatrix}.$$

PROBLEMS

1. Using the method of eigenvectors, find a fundamental matrix in real form for the equation $y' = Ay$ associated with the following matrices A. Comparison with the answers is facilitated if you number real λ_j so $\lambda_1 < \lambda_2$. In (hjl) we have taken e_2 real, though other choices may be better:

(a) $\begin{pmatrix} -5 & 3 \\ -4 & 2 \end{pmatrix}$ (d) $\begin{pmatrix} 3 & 4 \\ 2 & 1 \end{pmatrix}$ (g) $\begin{pmatrix} -6 & 1 \\ 5 & -2 \end{pmatrix}$ (j) $\begin{pmatrix} -3 & 1 \\ -5 & -7 \end{pmatrix}$

(b) $\begin{pmatrix} 2 & -1 \\ -4 & -1 \end{pmatrix}$ (e) $\begin{pmatrix} 6 & 2 \\ 3 & 7 \end{pmatrix}$ (h) $\begin{pmatrix} 2 & 1 \\ -8 & 6 \end{pmatrix}$ (k) $\begin{pmatrix} -5 & 7 \\ 3 & -1 \end{pmatrix}$

(c) $\begin{pmatrix} 1 & 4 \\ 8 & 5 \end{pmatrix}$ (f) $\begin{pmatrix} 1 & 3 \\ -2 & -4 \end{pmatrix}$ (i) $\begin{pmatrix} 7 & -1 \\ 3 & 3 \end{pmatrix}$ (l) $\begin{pmatrix} -1 & 4 \\ -10 & -5 \end{pmatrix}$

2. (abcdefgik) Obtain canonical vectors x in Problem 1 and verify as in Example 2. (We omit (hjl) because complex canonical variables for real differential equations are seldom used.)

3P. If $P(s) = s^3 + p_2 s^2 + p_1 s + p_0$ is the characteristic polynomial for a 3-by-3 matrix $A = (a_{ij})$, verify that $p_2 = -\text{trace } A$, $p_0 = -\det A$, and

$$p_1 = \begin{vmatrix} a_{11} & a_{12} \\ a_{21} & a_{22} \end{vmatrix} + \begin{vmatrix} a_{11} & a_{13} \\ a_{31} & a_{33} \end{vmatrix} + \begin{vmatrix} a_{22} & a_{23} \\ a_{32} & a_{33} \end{vmatrix}$$

Suggestion: The coefficient of s is $P'(0)$. Compute this derivative without expanding the determinant $|sI - A|$.

4P. By comparing the result of Problem 3 with the factored form

$$P(s) = (s - \lambda_1)(s - \lambda_2)(s - \lambda_3)$$

deduce that $\lambda_1 + \lambda_2 + \lambda_3 = \text{trace } A$, $\lambda_1\lambda_2\lambda_3 = \det A$.

5P. Following the pattern given in the text for $n = 2$, generalize the equation $AE = E\Lambda$ to the case $n = 3$. If $y = Ex$ this gives $x' = \Lambda x$ as before.

6P. The real eigenvectors E_i and E_j are said to be orthogonal if $E_j^T E_i = 0$. Assuming E_i and E_j orthogonal for all $i \neq j$, verify (a) for $n = 3$ and (b) in general that $E^T E$ is a diagonal matrix with diagonal entries $E_i^T E_i$. This is important in the study of real symmetric matrices.

————————————ANSWERS————————————

1. For economy of layout, the consecutive letters run horizontally rather than vertically as in the problems.

(a) $\begin{pmatrix} e^{-2t} & 3e^{-t} \\ e^{-2t} & 4e^{-t} \end{pmatrix}$ (b) $\begin{pmatrix} e^{-2t} & e^{3t} \\ 4e^{-2t} & -e^{3t} \end{pmatrix}$ (c) $\begin{pmatrix} -e^{-3t} & e^{9t} \\ e^{-3t} & 2e^{9t} \end{pmatrix}$

(d) $\begin{pmatrix} -e^{-t} & 2e^{5t} \\ e^{-t} & e^{5t} \end{pmatrix}$ (e) $\begin{pmatrix} -e^{4t} & 2e^{9t} \\ e^{4t} & 3e^{9t} \end{pmatrix}$ (f) $\begin{pmatrix} -e^{-2t} & -3e^{-t} \\ e^{-2t} & 2e^{-t} \end{pmatrix}$

(g) $\begin{pmatrix} -e^{-7t} & e^{-t} \\ e^{-7t} & 5e^{-t} \end{pmatrix}$ (h) $e^{4t}\begin{pmatrix} \cos 2t + \sin 2t & \cos 2t - \sin 2t \\ 4\cos 2t & -4\sin 2t \end{pmatrix}$

(i) $\begin{pmatrix} e^{4t} & e^{6t} \\ 3e^{4t} & e^{6t} \end{pmatrix}$ (j) $e^{-5t}\begin{pmatrix} -2\cos t + \sin t & \cos t + 2\sin t \\ 5\cos t & -5\sin t \end{pmatrix}$

(k) $\begin{pmatrix} -7e^{-8t} & e^{2t} \\ 3e^{-8t} & e^{2t} \end{pmatrix}$ (l) $e^{-3t}\begin{pmatrix} -\cos 6t + 3\sin 6t & 3\cos 6t + \sin 6t \\ 5\cos 6t & -5\sin 6t \end{pmatrix}$

2. Partial answer: A matrix E is obtained by setting $t = 0$ in the answers to Problem 1. Possible canonical variables are $x = |E|E^{-1}y$ where $|E|E^{-1}$ can be written by inspection. The factor $|E|$ is not necessary but simplifies the result.

2. Three-dimensional systems

If A is an n-by-n matrix, the conditions defining an eigenvalue λ_j and corresponding eigenvector E_j were stated in Section 1. Namely,

$$|\lambda_j I - A| = 0, \qquad AE_j = \lambda_j E_j, \qquad E_j \neq 0.$$

In the case $n = 3$ there are three eigenvalues λ_j, not necessarily distinct, and

$$\Lambda = \begin{pmatrix} \lambda_1 & 0 & 0 \\ 0 & \lambda_2 & 0 \\ 0 & 0 & \lambda_3 \end{pmatrix}.$$

The general theory for $n = 2$ was developed in Section 1 and was extended to $n = 3$ in the problems. Here the case $n = 3$ is illustrated in numerical examples. Since the calculations are decidedly error-prone, we give some suggestions for checking intermediate steps.

EXAMPLE 1. Find the eigenvalues of the matrix A on the left below:

$$\begin{pmatrix} 5 & 6 & 7 \\ 0 & -5 & 3 \\ 0 & -4 & 2 \end{pmatrix}, \qquad \begin{vmatrix} \lambda - 5 & -6 & -7 \\ 0 & \lambda + 5 & -3 \\ 0 & 4 & \lambda - 2 \end{vmatrix}.$$

The determinant on the right is $|\lambda I - A|$. Laplace expansion on the first column yields

$$|\lambda I - A| = (\lambda - 5)(\lambda^2 + 3\lambda + 2) = (\lambda - 5)(\lambda + 1)(\lambda + 2)$$

which is 0 for $\lambda = -2$, -1 or 5. Hence these are the eigenvalues. As a check, note that their sum is 2, which is the trace of A. See Section 1, Problem 4.

EXAMPLE 2. We shall now find the eigenvectors in Example 1. When $\lambda = -2$, -1, or 5 the matrices $A - \lambda I$ in Example 1 are, respectively,

(1) $$\begin{pmatrix} 7 & 6 & 7 \\ 0 & -3 & 3 \\ 0 & -4 & 4 \end{pmatrix}, \quad \begin{pmatrix} 6 & 6 & 7 \\ 0 & -4 & 3 \\ 0 & -4 & 3 \end{pmatrix}, \quad \begin{pmatrix} 0 & 6 & 7 \\ 0 & -10 & 3 \\ 0 & -4 & -3 \end{pmatrix}.$$

In each case the determinant is 0; that is the reason for choosing λ as we did. By definition, the eigenvectors E_j are nontrivial solutions of the linear systems associated with the singular matrices (1).

We denote the eigenvector E_j under consideration by (e_i), so that e_i is an abbreviation for e_{ij}. An efficient procedure in the 3-by-3 case is to set $e_3 = 1$ and thus to get a system of two equations for e_1 and e_2. If this system has no solution, the correct value is $e_3 = 0$, and we are led to a 2-by-2 system for e_1 and e_2.

Let us illustrate in the case of the first matrix (1), which corresponds to $\lambda = -2$. When $e_3 = 1$ the equations are

$$7e_1 + 6e_2 + 7 = 0, \qquad -3e_2 + 3 = 0.$$

The second of these gives $e_2 = 1$ and the first then gives $e_1 = -13/7$. We multiply by 7 to avoid fractions and get E_1 as shown below:

$$E_1 = \begin{pmatrix} -13 \\ 7 \\ 7 \end{pmatrix}, \qquad E_2 = \begin{pmatrix} -23 \\ 9 \\ 12 \end{pmatrix}, \qquad E_3 = \begin{pmatrix} 1 \\ 0 \\ 0 \end{pmatrix}.$$

In Problem 1 you are asked to derive the above vectors E_2 and E_3, using the second and third matrices (1) respectively.

The matrix E so obtained is written with A at its left and Λ at its right, thus:

$$\begin{pmatrix} 5 & 6 & 7 \\ 0 & -5 & 3 \\ 0 & -4 & 2 \end{pmatrix} \begin{pmatrix} -13 & -23 & 1 \\ 7 & 9 & 0 \\ 7 & 12 & 0 \end{pmatrix} \begin{pmatrix} -2 & 0 & 0 \\ 0 & -1 & 0 \\ 0 & 0 & 5 \end{pmatrix}.$$

The product of the first two is AE and of the last two is $E\Lambda$. In Section 1, Problem 5 it was seen that these must be equal, thus giving a check.

The point of the next example is to emphasize that the algebraic results lead to a complete solution of the analytic problems associated with the differential equation.

EXAMPLE 3. In Example 1 the general solution is

$$\alpha_1 E_1 e^{-2t} + \alpha_2 E_2 e^{-t} + \alpha_3 E_3 e^{5t}$$

where the E_j are given by Example 2. A fundamental matrix is

$$U(t) = (E_1 e^{-2t} \quad E_2 e^{-t} \quad E_3 e^{5t})$$

and the canonical variables $x = E^{-1}y$ satisfy

$$x_1' = -2x_1, \qquad x_2' = -x_2, \qquad x_3' = 5x_3.$$

From this the x_j can be found by inspection and the equation $y = Ex$ gives y. In Problem 1 you are asked to compute E^{-1} by reducing (E, I) or by the cofactor transpose, as you prefer, and to show that

$$7x_1 = 4y_2 - 3y_3, \qquad 3x_2 = -y_2 + y_3, \qquad 21x_3 = 21y_1 - 5y_2 + 44y_3.$$

These particular combinations of the unknowns y_j satisfy the uncoupled system above.

The next example shows that there can be a set of linearly independent eigenvectors even if the eigenvalues are not distinct. The characteristic polynomial is less tractable than those considered hitherto, and we preface the example by some remarks on the problem of factorization.

Let p_j be the coefficient of λ^j in $P(\lambda)$ and note that $P(\lambda) = 0$ gives

$$p_0 = -\lambda(p_1 + \lambda p_2 + \cdots + \lambda^{n-1}).$$

If the coefficients and λ are integers this shows that the constant term p_0 is a multiple of λ. In other words, *any integer root must be a divisor of the constant term.* The same argument shows that *any rational root must be an integer.* Once a root λ_1 is found, we know that $\lambda - \lambda_1$ is a factor of $P(\lambda)$. The other factor is obtained by division.

EXAMPLE 4. In Problem 1 you are asked to show that the matrix on the left below has the characteristic equation on the right:

$$\begin{pmatrix} 11 & 12 & 6 \\ -4 & -5 & -2 \\ -4 & -4 & -3 \end{pmatrix}, \qquad \lambda^3 - 3\lambda^2 - 9\lambda - 5 = 0.$$

(It is advisable to expand directly as in Example 1 and to use Problem 3 of Section 1 as a check.) Integral roots, if any, must be divisors of the constant term, which is -5. Hence they can only be ± 1, ± 5. The root -1 is found by trial, and factorization yields

$$P(\lambda) = (\lambda + 1)(\lambda^2 - 4\lambda - 5) = (\lambda + 1)^2(\lambda - 5).$$

The solutions of $(A - \lambda I)E_i = 0$ are not affected when the rows of the coefficient matrix $A - \lambda I$ are divided by nonzero constants. After obvious simplification the matrices $A - \lambda I$ for $\lambda = -1$ and $\lambda = 5$ become, respectively,

$$\begin{pmatrix} 2 & 2 & 1 \\ 2 & 2 & 1 \\ 2 & 2 & 1 \end{pmatrix}, \qquad \begin{pmatrix} 1 & 2 & 1 \\ 2 & 5 & 1 \\ 1 & 1 & 2 \end{pmatrix}.$$

If $E_1 = (e_i)$ the first matrix leads to the single equation

$$2e_1 + 2e_2 + e_3 = 0.$$

An easy way to get linearly independent solutions is to take first $e_3 = 1$, $e_2 = 0$, and then $e_3 = 0, e_2 = 1$. The equations yield $e_1 = -1/2$ and $e_1 = -1$ in these two cases respectively. In Problem 1 you are asked to get a vector E_3 corresponding to $\lambda = 5$. When the E_j are multiplied by suitable constants, the matrix E so obtained is in the center below:

$$\begin{pmatrix} 11 & 12 & 6 \\ -4 & -5 & -2 \\ -4 & -4 & -3 \end{pmatrix} \begin{pmatrix} -1 & -1 & -3 \\ 0 & 1 & 1 \\ 2 & 0 & 1 \end{pmatrix} \begin{pmatrix} -1 & 0 & 0 \\ 0 & -1 & 0 \\ 0 & 0 & 5 \end{pmatrix}.$$

As in previous cases, we have written A, E, Λ side by side to verify that $AE = E\Lambda$.

In the case of a real symmetric matrix, it turns out that the eigenvalues are always real, and eigenvectors corresponding to different eigenvalues are always orthogonal. Proof of these statements is given in Section 3. Here we illustrate in an example.

EXAMPLE 5. Find the eigenvalues for the symmetric matrix

$$A = \begin{pmatrix} 2 & 4 & -6 \\ 4 & 2 & -6 \\ -6 & -6 & -15 \end{pmatrix}.$$

In Problem 1 you are asked to determine the characteristic polynomial $P(\lambda)$ and, in particular, to verify that its constant term is

$$P(0) = -|A| = -324 = -(2^2)(3^4).$$

The rational eigenvalues, if any, must be integers and must be divisors of $P(0)$; but there are so many divisors that it is by no means obvious what to do next.

By sketching a graph of $P(s)$ we are led to surmise that it has a zero near -2, and if 2 is added to each diagonal element of A the resulting matrix has two equal rows. Hence its determinant is 0, and this shows that $\lambda = -2$ is indeed an eigenvalue. In Problem 1 you are asked to obtain the factored form

$$P(\lambda) = (\lambda + 2)(\lambda^2 + 9\lambda - 162)$$

and to get the remaining eigenvalues -18, 9.

EXAMPLE 6. With your cooperation, we shall verify that the eigenvectors in Example 5 are orthogonal. When $\lambda = -18$ the matrix $A - \lambda I$ is obtained by adding 18 to the diagonal elements of A and is displayed on the left below:

$$\begin{pmatrix} 20 & 4 & -6 \\ 4 & 20 & -6 \\ -6 & -6 & 3 \end{pmatrix} \begin{matrix} R_1/2 \\ R_2/2 \\ R_3/3 \end{matrix} \begin{pmatrix} 10 & 2 & -3 \\ 2 & 10 & -3 \\ -2 & -2 & 1 \end{pmatrix} \begin{matrix} R_1 + 3R_3 \\ R_2 + 3R_3 \\ \end{matrix} \begin{pmatrix} 4 & -4 & 0 \\ -4 & 4 & 0 \\ -2 & -2 & 1 \end{pmatrix}.$$

The homogeneous system corresponding to the reduced matrix on the right has an obvious solution $E_1 = (1, 1, 4)^T$.

In Problem 1 you are asked to show that the choice $e_3 = 1$ is not possible when $\lambda = -2$, and that the choice $e_3 = 0$ (which is mandated by the failure of $e_3 = 1$) leads to a solution. You are also asked to solve the equations with $\lambda = 9$ by diagonalization, as was illustrated above for $\lambda = -18$. The final result is the matrix E in the center below:

$$(2) \qquad E^T E = \begin{pmatrix} 1 & 1 & 4 \\ -1 & 1 & 0 \\ -2 & -2 & 1 \end{pmatrix} \begin{pmatrix} 1 & -1 & -2 \\ 1 & 1 & -2 \\ 4 & 0 & 1 \end{pmatrix} = \begin{pmatrix} 18 & 0 & 0 \\ 0 & 2 & 0 \\ 0 & 0 & 9 \end{pmatrix}.$$

The product $E^T E$ is diagonal because the column vectors of E are orthogonal; see Section 1, Problem 6. The equation $Ex = y$ now yields $E^T Ex = E^T y$ and from this, x can be found by inspection. The variables x_j so defined satisfy $x_j' = \lambda_j x_j$ and y is found from $y = Ex$.

PROBLEMS

1P. In Examples 2, 3, 4, 5, and 6 we left a number of gaps, referring in each case to Problem 1. Study these examples and fill in the gaps.

2. The characteristic values are distinct integers. Find them:

(a) $\begin{pmatrix} 1 & -1 & -1 \\ 6 & 9 & 8 \\ -4 & -5 & -4 \end{pmatrix}$ (f) $\begin{pmatrix} 3 & 0 & 4 \\ 0 & 2 & 0 \\ -2 & 0 & -3 \end{pmatrix}$ (k) $\begin{pmatrix} 4 & -4 & 5 \\ 1 & 3 & 1 \\ -2 & 4 & -3 \end{pmatrix}$

(b) $\begin{pmatrix} -2 & -6 & -2 \\ -9 & -3 & -9 \\ -1 & -3 & -1 \end{pmatrix}$ (g) $\begin{pmatrix} 14 & 13 & -14 \\ -4 & -1 & 6 \\ 2 & 3 & 2 \end{pmatrix}$ (l) $\begin{pmatrix} 2 & 4 & 3 \\ 5 & 3 & 5 \\ -4 & -4 & -5 \end{pmatrix}$

(c) $\begin{pmatrix} -1 & 3 & 0 \\ 3 & -1 & 0 \\ -1 & -1 & -2 \end{pmatrix}$ (h) $\begin{pmatrix} 3 & 4 & 1 \\ -3 & -4 & 3 \\ -2 & -2 & 6 \end{pmatrix}$ (m) $\begin{pmatrix} 3 & -6 & -2 \\ 2 & 3 & 2 \\ -2 & 6 & 3 \end{pmatrix}$

(d) $\begin{pmatrix} 5 & 0 & -1 \\ -1 & 3 & -1 \\ -2 & -1 & 4 \end{pmatrix}$ (i) $\begin{pmatrix} -8 & -11 & 16 \\ 2 & 3 & -4 \\ -4 & -5 & 8 \end{pmatrix}$ (n) $\begin{pmatrix} -7 & -3 & -8 \\ 2 & 2 & 2 \\ 7 & 3 & 8 \end{pmatrix}$

(e) $\begin{pmatrix} 0 & 2 & 0 \\ 3 & 0 & 3 \\ 0 & 1 & 0 \end{pmatrix}$ (j) $\begin{pmatrix} 5 & 0 & 6 \\ -1 & 1 & -1 \\ -3 & 0 & -4 \end{pmatrix}$ (o) $\begin{pmatrix} -7 & -12 & -6 \\ 2 & 2 & 2 \\ 7 & 12 & 6 \end{pmatrix}$

3. (abcdefghijklmno) Find eigenvectors for the eigenvalues in Problem 2 and check that $AE = E\Lambda$.

4. (abcdefghijklmno) In Problem 3 find canonical variables $x = E^{-1}y$ and verify at least one equation $dx_j/dt = \lambda_j x_j$.

5. (abcdef) The following matrices have integral eigenvalues not all distinct. Find three linearly independent eigenvectors and check that $AE = E\Lambda$:

(a) $\begin{pmatrix} -3 & -1 & -2 \\ 0 & -2 & 0 \\ 1 & 1 & 0 \end{pmatrix}$ (c) $\begin{pmatrix} 3 & 0 & -1 \\ 0 & 4 & 0 \\ -1 & 0 & 3 \end{pmatrix}$ (e) $\begin{pmatrix} 1 & 3 & 6 \\ 0 & 0 & -2 \\ 0 & 1 & 3 \end{pmatrix}$

(b) $\begin{pmatrix} -1 & -4 & 2 \\ 0 & 1 & 0 \\ -1 & -2 & 2 \end{pmatrix}$ (d) $\begin{pmatrix} 0 & -2 & -2 \\ -3 & -1 & -3 \\ -1 & -1 & 1 \end{pmatrix}$ (f) $\begin{pmatrix} -5 & -4 & 8 \\ 0 & -1 & 0 \\ -4 & -4 & 7 \end{pmatrix}$

6P. Check that the matrix E^T in (2) gives, with very little calculation,

$$18x_1 = y_1 + y_2 + 4y_3, \qquad 2x_2 = -y_1 + y_2, \qquad 9x_3 = -2y_1 - 2y_2 + y_3$$

7. The following symmetric matrices have distinct integral eigenvalues. Find the eigenvalues and the matrix E of eigenvectors. Check that $AE = E\Lambda$, and also that $E^T E$ is diagonal. Thus find canonical variables $x = E^T y$ as in Problem 6:

(a) $\begin{pmatrix} 0 & 0 & -2 \\ 0 & 0 & 0 \\ -2 & 0 & 0 \end{pmatrix}$ (b) $\begin{pmatrix} 3 & 0 & -1 \\ 0 & 3 & 0 \\ -1 & 0 & 3 \end{pmatrix}$ (c) $\begin{pmatrix} -7 & 0 & 2 \\ 0 & -5 & 2 \\ 2 & 2 & -6 \end{pmatrix}$

8. Let $AE = E\Lambda$ where E is nonsingular. Check that the substitution $y = Ex$ in $y' = Ay + f$ gives the following uncoupled system:

$$x' = \Lambda x + g \quad \text{where} \quad g = E^{-1}f$$

If g is continuous deduce for $j = 1, 2, \cdots, n$ that

$$x_j(t) = e^{\lambda_j t} \int g_j(s) e^{-\lambda_j s}\, ds + c_j e^{\lambda_j t}$$

──────────────────────ANSWERS──────────────────────

2. (abcde) (1, 2, 3)   (−12, 0, 6)   (−4, −2, 2)   (2, 4, 6)   (−3, 0, 3)
   (fghij)   (−1, 1, 2)   (4, 5, 6)   (−1, 2, 4)   (0, 1, 2)   (−1, 1, 2)
   (klmno)   (−1, 2, 3)   (−2, −1, 3)   (1, 3, 5)   (0, 1, 2)   (−1, 0, 2)

3. Here are possible matrices E for eigenvalues as above.

(a) $\begin{pmatrix} 0 & 1 & 1 \\ -1 & -2 & -5 \\ 1 & 1 & 3 \end{pmatrix}$ (f) $\begin{pmatrix} -1 & -2 & 0 \\ 0 & 0 & 1 \\ 1 & 1 & 0 \end{pmatrix}$ (k) $\begin{pmatrix} -1 & -3 & -1 \\ 0 & 1 & 1 \\ 1 & 2 & 1 \end{pmatrix}$

(b) $\begin{pmatrix} 2 & -1 & 2 \\ 3 & 0 & -3 \\ 1 & 1 & 1 \end{pmatrix}$ (g) $\begin{pmatrix} 4 & 3 & 5 \\ -2 & -1 & -2 \\ 1 & 1 & 1 \end{pmatrix}$ (l) $\begin{pmatrix} -1 & -1 & -1 \\ 1 & 0 & -1 \\ 0 & 1 & 1 \end{pmatrix}$

(c) $\begin{pmatrix} -1 & 0 & -2 \\ 1 & 0 & -2 \\ 0 & 1 & 1 \end{pmatrix}$ (h) $\begin{pmatrix} -1 & 3 & 1 \\ 1 & -1 & 0 \\ 0 & 1 & 1 \end{pmatrix}$ (m) $\begin{pmatrix} -1 & -3 & -1 \\ -1 & -1 & 0 \\ 2 & 3 & 1 \end{pmatrix}$

(d) $\begin{pmatrix} 1 & 1 & -1 \\ 4 & -2 & 0 \\ 3 & 1 & 1 \end{pmatrix}$ (i) $\begin{pmatrix} 2 & 3 & 7 \\ 0 & -1 & -2 \\ 1 & 1 & 3 \end{pmatrix}$ (n) $\begin{pmatrix} -5 & -1 & -3 \\ 1 & 0 & 1 \\ 4 & 1 & 3 \end{pmatrix}$

(e) $\begin{pmatrix} 2 & -1 & 2 \\ -3 & 0 & 3 \\ 1 & 1 & 1 \end{pmatrix}$ (j) $\begin{pmatrix} -1 & 0 & -2 \\ 0 & 1 & 1 \\ 1 & 0 & 1 \end{pmatrix}$ (o) $\begin{pmatrix} -1 & -6 & -4 \\ 0 & 1 & 1 \\ 1 & 5 & 4 \end{pmatrix}$

4. Before finding E^{-1} check that your matrix E is consistent with the book's answers to Problem 3; it need not be identical. Since verification is part of the problem, no further answer is provided.

5. (abcdef) $(-2, -2, -1)$ $(0, 1, 1)$ $(2, 4, 4)$ $(-4, 2, 2)$ $(1, 1, 2)$ $(-1, -1, 3)$
   Possible matrices of eigenvectors E with eigenvalues as above:

$$\text{(a)} \begin{pmatrix} -2 & -1 & -1 \\ 0 & 1 & 0 \\ 1 & 0 & 1 \end{pmatrix} \quad \text{(c)} \begin{pmatrix} 1 & -1 & 0 \\ 0 & 0 & 1 \\ 1 & 1 & 0 \end{pmatrix} \quad \text{(e)} \begin{pmatrix} 0 & 1 & 3 \\ -2 & 0 & -1 \\ 1 & 0 & 1 \end{pmatrix}$$

$$\text{(b)} \begin{pmatrix} 2 & 1 & -2 \\ 0 & 0 & 1 \\ 1 & 1 & 0 \end{pmatrix} \quad \text{(d)} \begin{pmatrix} 2 & -1 & -1 \\ 3 & 0 & 1 \\ 1 & 1 & 0 \end{pmatrix} \quad \text{(f)} \begin{pmatrix} 2 & -1 & 1 \\ 0 & 1 & 0 \\ 1 & 0 & 1 \end{pmatrix}$$

7. Eigenvalues: (a) $(-2, 0, 2)$ (b) $(2, 3, 4)$ (c) $(-9, -6, -3)$
   Canonical variables x are given in terms of y by $x = E^T y$ where possible values of E^T are:

$$\text{(a)} \begin{pmatrix} 1 & 0 & 1 \\ 0 & 1 & 0 \\ -1 & 0 & 1 \end{pmatrix} \quad \text{(b)} \begin{pmatrix} 1 & 0 & 1 \\ 0 & 1 & 0 \\ -1 & 0 & 1 \end{pmatrix} \quad \text{(c)} \begin{pmatrix} -2 & -1 & 2 \\ 2 & -2 & 1 \\ 1 & 2 & 2 \end{pmatrix}$$

---

3. The general case

Results illustrated for $n = 2$ and $n = 3$ in the foregoing pages are now established for arbitrary n. We begin with a brief review. If A is a square matrix of order n, there are n eigenvalues λ_j, not necessarily distinct, and corresponding eigenvectors E_j. By definition, $E_j \neq 0$ and

(1) $$AE_1 = \lambda_1 E_1, \qquad AE_2 = \lambda_2 E_2, \qquad \cdots, \qquad AE_n = \lambda_n E_n.$$

We set $E = (E_1 \ E_2 \ \cdots \ E_n)$ and $\Lambda = \text{diag} (\lambda_1 \ \lambda_2 \ \cdots \ \lambda_n)$. Equation (1) shows that each column of the matrix

(2) $$U(t) = (E_1 e^{\lambda_1 t} \quad E_2 e^{\lambda_2 t} \quad \cdots \quad E_n e^{\lambda_n t})$$

satisfies $y' = Ay$; that is the way the notion of characteristic values and characteristic vectors arose in the first place. In many cases the vectors E_j can be chosen to be linearly independent, even if the λ_j are not all distinct. Then $U(t)$ gives a fundamental matrix for $y' = Ay$. By Chapter 15, Section 1, Theorem 2, the condition of linear independence is equivalent to $|E| \neq 0$.

(a) Canonical variables

The method of canonical variables depends on the equation $AE = E\Lambda$ which was established for $n = 2$ and $n = 3$ in Section 1. Here is a concise proof for arbitrary n:

$$AE = (AE_1 \ AE_2 \ \cdots \ AE_n) = (\lambda_1 E_1 \ \lambda_2 E_2 \ \cdots \ \lambda_n E_n) = E\Lambda.$$

If $|E| \neq 0$ the substitution $y = Ex$ in $y' = Ay$ yields $Ex' = AEx = E\Lambda x$, hence $x' = \Lambda x$. We summarize as follows:

THEOREM 1. *If $|E| \neq 0$ then $U(t)$ in (2) is a fundamental matrix for $y' = Ay$ and the change of variables $y = Ex$ leads to the uncoupled system $x' = \Lambda x$.*

The equation $AE = E\Lambda$ was found above under the assumption that the columns E_j of E are eigenvectors. Conversely, if $AE = E\Lambda$, where Λ is diagonal, then the columns of E satisfy $AE_j = \lambda_j E_j$. Hence the nonzero columns are eigenvectors and the corresponding entries of Λ are eigenvalues. In most applications, the condition $|E| \neq 0$ of Theorem 1 is ensured by the following:

THEOREM 2. *If the n eigenvalues λ_i are distinct, the corresponding eigenvectors E_i are linearly independent.*

For proof assume (contrary to the desired conclusion) that the E_j are dependent, and let m be the *smallest* number of these vectors that are dependent. Without loss of generality we number the E_j so that the first m vectors E_j are dependent. Thus, there are real or complex constants α_j, not all 0, such that

$$(3) \qquad\qquad \alpha_1 E_1 + \alpha_2 E_2 + \cdots + \alpha_m E_m = 0.$$

We have $m > 1$ since the E_j are nonzero and $m \leq n$ since there are only n vectors in all. Hence $2 \leq m \leq n$. The fact that m is minimal shows that no α_j can be 0, and in particular, $\alpha_2 \neq 0$.

If (3) is multiplied on the left by A the result is

$$(4) \qquad\qquad \alpha_1 \lambda_1 E_1 + \alpha_2 \lambda_2 E_2 + \cdots + \alpha_m \lambda_m E_m = 0.$$

We multiply (3) by λ_1 and then subtract (4) to get

$$\alpha_2(\lambda_1 - \lambda_2)E_2 + \cdots + \alpha_m(\lambda_1 - \lambda_m)E_m = 0.$$

This involves only $m - 1$ vectors E_j. Hence, by the minimal property of m, all the coefficients must be 0 and, in particular, $\alpha_2(\lambda_1 - \lambda_2) = 0$. Since

$\lambda_1 \neq \lambda_2$ we conclude that $\alpha_2 = 0$, while it was seen above that $\alpha_2 \neq 0$. This contradiction shows that our initial assumption of linear dependence was not tenable. Hence the vectors must be linearly independent, as we wanted to show.

(b) Symmetric and Hermitian matrices

Let us recall that the matrix A is symmetric if $A = A^T$, Hermitian if $A = A^*$. Here A^* denotes the conjugate transpose, or adjoint, of A. When A is real $A^* = A^T$ and the terms "Hermitian" and "symmetric" mean the same thing.

In fields such as biomathematics and economics the matrices are usually real. But in electrical engineering and quantum mechanics they are usually complex. To accommodate the latter subjects, our theorems are stated for Hermitian matrices. That case is more general than the real symmetric case, yet the proofs are no more difficult.

If u and v are column vectors of dimension n, the expression v^*u is a matrix of one element, namely,

$$v^*u = (\alpha) \quad \text{where} \quad \alpha = u_1\bar{v}_1 + u_2\bar{v}_2 + \cdots + u_n\bar{v}_n.$$

As stated in Chapter D, the vectors u and v are *orthogonal* if $v^*u = (0)$. The special case $v = u$ gives

$$u^*u = (\beta) \quad \text{where} \quad \beta = |u_1|^2 + |u_2|^2 + \cdots + |u_n|^2.$$

The quantity $\sqrt{\beta}$ so defined is the length of u and is denoted by $\|u\|$. Note that $\|u\|$ is real and is 0 only when $u = 0$.

We shall establish the following:

THEOREM 3.   *Let $A = A^*$. Then all eigenvalues λ of A are real and eigenvectors u, v belonging to different eigenvalues λ, μ are orthogonal.*

For proof, let

(5) $Au = \lambda u, \qquad Av = \mu v, \qquad u \neq 0, \ v \neq 0, \ \lambda \neq \mu.$

Left-multiply the equation $Au = \lambda u$ by u^* and then take the adjoint. Since $(Au)^* = u^*A^*$ the two equations so obtained are:

$$u^*Au = \lambda u^*u, \qquad u^*A^*u = \bar{\lambda}u^*u.$$

By hypothesis $A = A^*$ and $u^*u \neq 0$. Subtraction yields $\lambda - \bar{\lambda} = 0$, which shows that λ is real.

To establish orthogonality of u and v, left-multiply the first equation (5) by v^* and the second by u^*. The results are

(6) $v^*Au = \lambda v^*u, \qquad u^*Av = \mu u^*v.$

Taking the adjoint of the latter equation yields $v^* A^* u = \bar{\mu} v^* u$, hence

(7) $$v^* A u = \mu v^* u.$$

Here we use $A^* = A$ and the fact, proved above, that μ is real. Subtracting (7) from the first equation (6) we get $v^* u = 0$. This shows that u and v are orthogonal and completes the proof.

(c) Diagonalization

If there are n eigenvectors E_j, and if they are mutually orthogonal as in Theorem 3, it is easily checked that

(8) $$E^* E = \operatorname{diag}\left(\|E_i\|^2\right).$$

(See Section 1, Problem 6.) Since the eigenvector E_j is not zero we can divide E_j by $\|E_j\|$ and assume without loss of generality that $\|E_j\| = 1$. In that case equation (8) yields

$$E^* E = I, \quad \text{hence} \quad E^{-1} = E^*.$$

As stated in Chapter C, matrices with this property are called *unitary*.

If E is unitary the equation $AE = E\Lambda$ can be multiplied on the right or left by E^* to yield the equivalent equations

$$A = E\Lambda E^*, \qquad E^* AE = \Lambda.$$

Thus we have established the following when A has n linearly independent eigenvectors. In Chapter 17, Section 3 it is seen that the theorem holds without this restriction:

THEOREM 4. *If $A^* = A$ there is a unitary matrix E such that $E^* AE$ is diagonal.*

Theorem 4 has an interesting geometric interpretation. The column vectors $x = (x_j)$ and $y = (y_j)$ can be written

$$x = x_1 I_1 + x_2 I_2 + \cdots + x_n I_n, \qquad y = y_1 I_1 + y_2 I_2 + \cdots + y_n I_n$$

where I_j are columns of the identity matrix I. It is said that the ordered set $\{I_j\}$ is the *standard basis* and that x_j and y_j are coordinates of x and y relative to the standard basis.

Instead of I_j we could use the columns B_j of any nonsingular matrix B. Since the columns of such a matrix form a basis, x and y can be written uniquely in the form

$$x = \xi_1 B_1 + \xi_2 B_2 + \cdots + \xi_n B_n, \qquad y = \eta_1 B_1 + \eta_2 B_2 + \cdots + \eta_n B_n.$$

Here ξ_j and η_j are the coordinates of x and y relative to the new basis. The above equations are equivalent to

$$x = B\xi, \qquad y = B\eta.$$

In these coordinates the transformation $y = Ax$ becomes $B\eta = AB\xi$, hence

$$\eta = B^{-1}AB\xi.$$

It is said that A and $B^{-1}AB$ are *similar*, because they represent the same transformation to two different bases. Theorem 4 indicates that every Hermitian matrix is similar to a diagonal matrix, and that if the columns of E are used as a basis, the transformation $y = Ax$ takes the simple form $\eta = \Lambda\xi$ in the new coordinates.

PROBLEMS

1P. When the determinant $|sI - A|$ is expanded on the first row, the terms in s^n and s^{n-1} arise from the product of the diagonal terms, namely, from

$$(s - a_{11})(s - a_{22}) \cdots (s - a_{nn})$$

The constant term $P(0)$ in $P(s)$ is $|-A|$. Thus show that $P(s)$ has the form

$$P(s) = s^n - (\text{trace } A)s^{n-1} + \cdots + (-1)^n(\det A)$$

2. As in Section 1, Problem 4, derive the identities

$$\sum_{i=1}^{n} \lambda_i = \sum_{i=1}^{n} a_{ii}, \qquad \prod_{i=1}^{n} \lambda_i = \det A$$

3. Complex eigenvectors. Let A be a real matrix which has a complex eigenvalue $\lambda = \sigma + i\omega$ and a corresponding complex eigenvector $w = u + iv$. By the principle of equating real parts, the following functions both satisfy $y' = Ay$. Show that they are linearly independent if $\omega \neq 0$:

$$e^{\sigma t}(u \cos \omega t - v \sin \omega t), \qquad e^{\sigma t}(u \sin \omega t + v \cos \omega t)$$

Hint: If they are linearly dependent deduce that

$$(\alpha_1 u + \alpha_2 v) \cos \omega t + (-\alpha_1 v + \alpha_2 u) \sin \omega t = 0$$

holds for real constants α_1, α_2 not both 0. Conclude first that

$$\alpha_1 u + \alpha_2 v = 0, \qquad -\alpha_1 v + \alpha_2 u = 0$$

and then, since $\alpha_1^2 + \alpha_2^2 \neq 0$, that $u = v = 0$.

4. This problem and the next show how higher-order equations can be studied by means of eigenvectors. Let A be an n-by-n matrix with distinct negative eigenvalues $\lambda_j = -\omega_j^2$. Check that the function $y = c\cos(\omega t - \theta)$ satisfies $y'' = Ay$ if, and only if, $\omega^2 c = Ac$. Thus get the solution

$$y = E_1\cos(\omega_1 t - \theta_1) + E_2\cos(\omega_2 t - \theta_2) + \cdots + E_n\cos(\omega_n t - \theta_n)$$

where the E_j are characteristic vectors for λ_j and θ_j are arbitrary constants. The values ω_j, or sometimes $\omega_j/2\pi$, are referred to as *characteristic frequencies* and the corresponding solutions $E_j\cos(\omega_j t - \theta_j)$ are the *normal modes* of oscillation.

5. The vibrations of an oscillating mass-spring system are governed by the equations

$$m_1 y_1'' = -k_1 y_1 + k(y_2 - y_1), \qquad m_2 y_2'' = -k_2 y_2 - k(y_2 - y_1)$$

(a) Write as a 2-by-2 system $y'' = Ay$ and, if $m_1 = k = k_1 = 2$, $m_2 = k_2 = 1$ solve in the form

$$\begin{pmatrix} y_1 \\ y_2 \end{pmatrix} = \alpha_1 \begin{pmatrix} 1 \\ 1 \end{pmatrix}\cos(t - \theta_1) + \alpha_2 \begin{pmatrix} 1 \\ -2 \end{pmatrix}\cos(2t - \theta_2)$$

In the first term on the right the two masses oscillate with the same amplitude and phase; in the second the ratio of amplitudes is 2 to 1 and the oscillations are 180° out of phase. These solutions are the normal modes of oscillation referred to in Problem 4.

6. Using the identity $sI - B^{-1}AB = B^{-1}(sI - A)B$ and the product theorem for determinants, show that similar matrices have the same characteristic polynomial. Deduce from Problem 1 that they have the same determinant and the same trace.

Algebra is generous, she often gives more than is asked of her.

—Jean le rond D'Alembert

Chapter 17

THE MATRIX EXPONENTIAL

THE SET of eigenvalues of a square matrix A, with their multiplicities, is called the spectrum of A. If a complex-valued function ψ is defined on the spectrum of A it is possible to compute $\psi(A)$, and a single formula will do for the whole class of matrices with given spectrum. Applying this with $\psi(s) = e^{st}$ we find that $Y = e^{At}$ satisfies $Y' = AY$, $Y(0) = I$. A major advantage of this approach is that multiple eigenvalues require only minor modification and may actually lead to simplification. Once the spectrum is known, the evaluation of e^{At} requires little more than matrix multiplication.

Prerequisite: Chapter 15, Sections 1 and 2, and Chapter 16.

1. Distinct eigenvalues

If a is a number, we know from Chapter 1 that the initial-value problem

$$\frac{dy}{dt} = ay, \qquad y(0) = 1 \quad \text{has the solution } y = e^{at}.$$

Our objective here is to obtain a similar result when a is replaced by a constant n-by-n matrix A. We will define e^{At} in such a way that it is easy to compute, and furthermore,

$$\frac{dY}{dt} = AY, \qquad Y(0) = I \quad \text{has the solution } Y = e^{At}.$$

Throughout Section 1, including the problems, we make the following

SIMPLIFYING ASSUMPTION. The n eigenvalues λ_j of A are distinct.

The numbered theorems and definitions are worded, however, so that they are correct in the case of multiple eigenvalues and they will be applied to that case in the next section.

Some basic relations were derived in Chapter C, Problems 8, 9, and 10, and you may find it helpful to read these problems if you have not already done so. They show that results which are derived here only briefly, or only for $n = 2$, actually hold in full generality.

Before discussing e^{At} we consider polynomials in A. Suppose

$$\phi(s) = q_0 + q_1 s + q_2 s^2 + \cdots + q_m s^m$$

where the coefficients and s are real or complex. Then, by definition,

$$\phi(A) = q_0 I + q_1 A + q_2 A^2 + \cdots + q_m A^m.$$

Note that the constant term q_0 is replaced by $q_0 I$.

Evaluation of a polynomial $\phi(\Lambda)$ is easy when the matrix Λ is diagonal. As an illustration, let

$$\Lambda = \begin{pmatrix} \lambda_1 & 0 \\ 0 & \lambda_2 \end{pmatrix}$$

and let q_j be any real or complex number. Then

$$q_j \begin{pmatrix} \lambda_1 & 0 \\ 0 & \lambda_2 \end{pmatrix}^j = q_j \begin{pmatrix} \lambda_1^j & 0 \\ 0 & \lambda_2^j \end{pmatrix} = \begin{pmatrix} q_j \lambda_1^j & 0 \\ 0 & q_j \lambda_2^j \end{pmatrix}$$

for $j = 0, 1, 2, \cdots$. Summation on j from 0 to m yields

$$\phi(\Lambda) = \begin{pmatrix} \phi(\lambda_1) & 0 \\ 0 & \phi(\lambda_2) \end{pmatrix}$$

with ϕ as above. In condensed notation

(1) $$\phi(\Lambda) = \operatorname{diag} \phi(\lambda_j).$$

Under our simplifying assumption there is an invertible matrix E such that $AE = E\Lambda$, or $A = E\Lambda E^{-1}$. This gives

$$q_j A^j = q_j E \Lambda^j E^{-1} = E q_j \Lambda^j E^{-1}$$

for $j = 0, 1, 2, \cdots$. Summation on j from 1 to m yields

(2) $$\phi(A) = E\phi(\Lambda)E^{-1}.$$

By the results of Chapter C cited above, (1) and (2) hold for arbitrary n. Both equations are basic in the rest of this chapter.

Equations (1) and (2) show that if two polynomials ϕ_1 and ϕ_2 have the same value at each λ_j, then they give the same value for $\phi(A)$. The set of values λ_j is so important that it has a special name:

DEFINITION 1. *The set of eigenvalues of A together with their multiplicities is called the spectrum of A.*

In this section the multiplicities are all 1 and the spectrum is the set of distinct numbers

$$\lambda_1, \lambda_2, \cdots, \lambda_n.$$

The result stated informally above can now be restated as follows:

THEOREM 1. *Let ϕ_1 and ϕ_2 be two polynomials such that $\phi_1 = \phi_2$ on the spectrum of A. Then $\phi_1(A) = \phi_2(A)$.*

(a) Functions of matrices

Theorem 1 allows us to compute $\psi(A)$ for much more general functions than polynomials. Throughout the following discussion, ψ denotes an arbitrary complex-valued function which is defined on the spectrum of A. It need not even be continuous.

DEFINITION 2. *Let ϕ be any polynomial such that $\phi(\lambda) = \psi(\lambda)$ on the spectrum of A. Then $\psi(A) = \phi(A)$.*

By Theorem 1, all such polynomials ϕ yield the same value for $\psi(A)$.

To evaluate $\psi(A)$ we must construct a polynomial that agrees with ψ on the spectrum of A. An easy way of doing this was found by Lagrange. Starting with the characteristic polynomial

$$P(s) = (s - \lambda_1)(s - \lambda_2) \cdots (s - \lambda_n)$$

Lagrange forms additional polynomials

$$L_j(s) = \frac{N_j(s)}{N_j(\lambda_j)}$$

where $N_j(s)$ is obtained by omitting the factor $(s - \lambda_j)$ from $P(s)$. It vanishes at each λ_i for $i \neq j$, but not at λ_j. Hence

(3) $$L_j(\lambda_i) = 0 \quad \text{for } i \neq j, \qquad L_j(\lambda_j) = 1.$$

A moment's thought shows that the *Lagrange polynomial*

$$\phi(s) = L_1(s)\psi(\lambda_1) + L_2(s)\psi(\lambda_2) + \cdots + L_n(s)\psi(\lambda_n)$$

has degree $\leq n - 1$ and satisfies $\phi(\lambda_j) = \psi(\lambda_j)$ for all j. The following examples show how Lagrange polynomials are used to compute e^{At}.

EXAMPLE 1. Let A be a 3-by-3 matrix with eigenvalues 0, 1, 4. To find $\psi(A)$, we start with $P(s) = s(s - 1)(s - 4)$. The three numerators $N_i(s)$ are

$$(s - 1)(s - 4), \qquad s(s - 4), \qquad s(s - 1)$$

and their values at 0, 1, 4 are 4, -3, 12 respectively. Hence the Lagrange polynomial ϕ for ψ is

$$\phi(s) = \frac{(s - 1)(s - 4)}{4}\psi(0) + \frac{s(s - 4)}{-3}\psi(1) + \frac{s(s - 1)}{12}\psi(4).$$

Definition 2 with this choice of ϕ yields

(4) $$\psi(A) = \frac{(A - I)(A - 4I)}{4}\psi(0) + \frac{A(A - 4I)}{-3}\psi(1) + \frac{A(A - I)}{12}\psi(4).$$

EXAMPLE 2. To find e^{At} in Example 1, we set $\psi(s) = e^{st}$ where t is a constant parameter. Thus

$$\psi(0) = 1, \qquad \psi(1) = e^t, \qquad \psi(4) = e^{4t}.$$

Substitution into (4) yields

$$e^{At} = \frac{(A-I)(A-4I)}{4} + \frac{A(A-4I)}{-3}e^t + \frac{A(A-I)}{12}e^{4t}.$$

In Part (b) below, it is seen that this yields a fundamental solution of $Y' = AY$ for the whole class of 3-by-3 matrices A with eigenvalues 0, 1, 4. Also, by Part (c), the nonzero columns of the products

$$(A-I)(A-4I), \qquad A(A-4I), \qquad A(A-I)$$

are characteristic vectors associated with $\lambda = 0, 1, 4$ respectively.

(b) Analytic properties

It is a remarkable fact that functions such as $\sin At$ or e^{At} have analytic properties similar to those of $\sin \lambda t$ or $e^{\lambda t}$. We now explain why this is so. If ϕ is a polynomial, the relations

$$\phi(\Lambda) = \text{diag } \phi(\lambda_j), \qquad \phi(A) = E\phi(\Lambda)E^{-1}$$

were verified by direct calculation. Since we can always find a polynomial ϕ that agrees with ψ on the spectrum of A, Definition 2 has the effect of replacing ϕ by corresponding equations with ψ. In other words,

$$(5) \qquad \psi(\Lambda) = \text{diag } \psi(\lambda_j), \qquad \psi(A) = E\psi(\Lambda)E^{-1}.$$

For example, when $n = 2$

$$(6) \qquad \psi(\Lambda) = \begin{pmatrix} \psi(\lambda_1) & 0 \\ 0 & \psi(\lambda_2) \end{pmatrix}.$$

By definition, the *matrix exponential* e^{At} is the result obtained when $\psi(s) = e^{st}$ where t is a constant parameter. Substitution into (6) yields

$$e^{\Lambda t} = \begin{pmatrix} e^{\lambda_1 t} & 0 \\ 0 & e^{\lambda_2 t} \end{pmatrix}.$$

The right-hand side is differentiable, hence the left-hand side is also. The result of differentiation is

$$\frac{d}{dt}e^{\Lambda t} = \begin{pmatrix} e^{\lambda_1 t}\lambda_1 & 0 \\ 0 & e^{\lambda_2 t}\lambda_2 \end{pmatrix} = \begin{pmatrix} e^{\lambda_1 t} & 0 \\ 0 & e^{\lambda_2 t} \end{pmatrix}\begin{pmatrix} \lambda_1 & 0 \\ 0 & \lambda_2 \end{pmatrix} = e^{\Lambda t}\Lambda.$$

Thus $(d/dt)e^{\Lambda t} = e^{\Lambda t}\Lambda$, as one would expect from the behavior of $e^{\lambda t}$. In Problem 13 you are asked to generalize this equation to diagonal matrices of order n. The calculation is similar to that for $n = 2$.

Setting $\psi(s) = e^{st}$ in (5) we get

$$e^{At} = Ee^{\Lambda t}E^{-1}$$

and, differentiating with the aid of the result found above,

$$\frac{d}{dt}e^{At} = Ee^{\Lambda t}\Lambda E^{-1}.$$

Now comes an important point. If a factor $E^{-1}E$ is inserted between $e^{\Lambda t}$ and Λ, the above product can be written

$$\frac{d}{dt}e^{At} = (Ee^{\Lambda t}E^{-1})(E\Lambda E^{-1}) = e^{At}A.$$

Thus, $(d/dt)e^{At} = e^{At}A$. When $t = 0$ both $e^{\Lambda t}$ and e^{At} reduce to I, and we summarize as follows:

THEOREM 2.   *The matrix exponential satisfies* $(d/dt)e^{At} = e^{At}A$, $e^0 = I$.

In Problem 14 it is seen that the inverse of e^{At} is obtained by simply writing $-t$ for t. For example, the formula for solving $y' = Ay+f$ by variation of parameters is

$$y = e^{At}\int e^{-At}f(t)\,dt.$$

This property of the exponential is very convenient in applications.

(c) A connection with eigenvectors

As in Chapter 16, we use E_j to denote an eigenvector for λ_j. Under our simplifying assumption the functions $e^{\lambda_j t}$ are linearly independent. Let L_j be the coefficient polynomials in the Lagrange polynomial for $\psi(s) = e^{st}$. The equation for e^{At} has the form

(7) $$e^{At} = M_1 e^{\lambda_1 t} + M_2 e^{\lambda_2 t} + \cdots + M_n e^{\lambda_n t}$$

where the M_j are square matrices of order n given by $M_j = L_j(A)$. (Here the subscript does not denote the jth column, as elsewhere in this chapter, but denotes the jth matrix.) The sum satisfies $dU/dt = AU$ by Theorem 2 and hence each term in it satisfies the same equation. That is,

(8) $$\frac{d}{dt}M_j e^{\lambda_j t} = AM_j e^{\lambda_j t}, \qquad j = 1, 2, \cdots, n.$$

This follows from linear independence of the functions $e^{\lambda_j t}$. Equation (8) is equivalent to $\lambda_j M_j = AM_j$, which in turn is equivalent to the following:

(i) The columns of M_j are scalar multiples of E_j.

Thus we can get a characteristic vector E_j by looking at any nonzero column of M_j. Since $e^0 = I$ we have also

(ii) $M_1 + M_2 + \cdots + M_n = I$.

These two conditions show that the sum satisfies $Y' = AY$, $Y(0) = I$ and, by uniqueness, it must agree with e^{At}. Under our simplifying assumption, (i) and (ii) give a definitive check on numerical calculations. When there are multiple eigenvalues, however, it will be seen that Equation (7) requires modification, and this leads to a corresponding modification of (i) and (ii).

(d) The Cayley-Hamilton theorem

If we choose $\phi_1(s) = P(s)$, where P is the characteristic polynomial of A, then $\phi_1(\lambda) = 0$ on the spectrum of A, and Theorem 1 with $\phi_2 = 0$ gives $\phi_1(A) = \phi_2(A) = 0$. This suggests the following interesting result, which is known as the Cayley-Hamilton theorem. The above proof assumes distinct eigenvalues, but the general case is settled in Section 3:

THEOREM 3. *Every square matrix satisfies its own characteristic equation.*

The following example shows how Theorem 3 can be used to verify that e^{At} actually satisfies the expected differential equation.

EXAMPLE 3. In Example 2 we found e^{At} as a sum of three terms. These were constant multiples of

$$U = (A - I)(A - 4I), \qquad V = A(A - 4I)e^t, \qquad W = A(A - I)e^{4t}$$

respectively. We want to check the work by showing that each of these satisfies $dY/dt = AY$. After obvious simplification in the case of V and W, the equations $dY/dt = AY$ for U, V, W reduce to

$$0 = A(A - I)(A - 4I), \quad A(A - 4I) = A^2(A - 4I), \quad 4A(A - I) = A^2(A - I).$$

To check this, recall that the characteristic polynomial in Example 1 was

$$P(s) = s(s - 1)(s - 4) = s^3 - 5s^2 + 4s.$$

Theorem 3 gives $A^3 - 5A^2 + 4A = 0$, which yields each equation above.

PROBLEMS—————————————————————————————————

1P. (a) If α and β are real or complex constants check that

$$(A + \alpha I)(A + \beta I) = A^2 + (\alpha + \beta)A + \alpha\beta I = (A + \beta I)(A + \alpha I)$$

(b) Let ϕ_1 and ϕ_2 be polynomials with real or complex coefficients α_j and β_k, respectively. Sum the identity

$$(\alpha_j A^j)(\beta_k A^k) = (\beta_k A^k)(\alpha_j A^j)$$

on j and k and deduce $\phi_1(A)\phi_2(A) = \phi_2(A)\phi_1(A)$. Definition 2 now yields

$$\psi_1(A)\psi_2(A) = \psi_2(A)\psi_1(A)$$

whenever ψ is defined on the spectrum of A. For example, $e^{At}A = Ae^{At}$.

(c) Show that $(A + B)(A - B) \neq (A - B)(A + B)$ if A and B are square matrices that do not commute. In Part (b) it is important that only a single matrix is involved. The reason why the problem worked out as it did is that A^j and A^k commute.

2P. Let A be a square matrix of order 2 with distinct eigenvalues λ_1 and λ_2.

(a) Find the Lagrange polynomial $\phi(s)$ that agrees with $\psi(s)$ at λ_1 and λ_2 and thus derive the formula

$$\psi(A) = \frac{A - \lambda_2 I}{\lambda_1 - \lambda_2}\psi(\lambda_1) + \frac{A - \lambda_1 I}{\lambda_2 - \lambda_1}\psi(\lambda_2)$$

(b) If A is real and λ is complex, deduce from (a) that

$$e^{At} = \text{Im}\left(\frac{A - \bar{\lambda}I}{\text{Im }\lambda}e^{\lambda t}\right)$$

Hint: Let $\psi(s) = e^{st}$, $\lambda_1 = \lambda$, $\lambda_2 = \bar{\lambda}$.

3. (abcdefghijkl) Using Problem 2 with $\psi(s) = e^{st}$ write down e^{At} for some of the 2-by-2 systems in Chapter 16, Section 1, Problem 1. You can check by verifying that $Y(t) = e^{At}Y(0)$ where $Y(t)$ is the answer from Chapter 16.

4P. A 3-by-3 matrix A has characteristic polynomial $P(s) = s(s^2 - 1)$. Find the Lagrange polynomial ϕ for ψ, then $\psi(A)$, and deduce that

$$2e^{At} = 2(I - A^2) + (A^2 - A)e^{-t} + (A^2 + A)e^t$$

5. Theorem 3 gives $A^3 = A$ for the matrix of Problem 4. Using this verify that the formula there given satisfies $(d/dt)e^{At} = Ae^{At}$.

6P. Check that the following matrix A has $P(s) = s(s^2 - 1)$:

$$A = \begin{pmatrix} -1 & -1 & 1 \\ -2 & -3 & 2 \\ -4 & -5 & 4 \end{pmatrix}$$

Compute A^2 and, by the result of Problem 4, get

$$e^{At} = \begin{pmatrix} 2 & 1 & -1 \\ 0 & 0 & 0 \\ 2 & 1 & -1 \end{pmatrix} + \begin{pmatrix} 0 & 0 & 0 \\ 1 & 2 & -1 \\ 1 & 2 & -1 \end{pmatrix} e^{-t} + \begin{pmatrix} -1 & -1 & 1 \\ -1 & -1 & 1 \\ -3 & -3 & 3 \end{pmatrix} e^{t}$$

7. (abcdefghijklmno) This is best done by means of a hand calculator programmed to compute matrix products. Find e^{At} for some of the 3-by-3 matrices in Chapter 16, Section 2, Problem 2. Check that $e^0 = I$ and that the columns of M_j are multiples of the corresponding eigenvector E_j given in the answers.

8. A real 3-by-3 matrix A has eigenvalues $1, -i, i$. Check that the part of e^{At} associated with $-i$ and i is

$$(A - I)\text{Re} \, \frac{A - iI}{-1 + i} e^{-it}$$

and thus get $2e^{At} = (A^2 + I)e^t - (A^2 - I)\cos t - (A - I)^2 \sin t$.

9. A square matrix of order 4 has $P(s) = s(s^2 - 1)(s - 2)$. Find $\phi(s)$, then $\psi(A)$, and deduce that e^{At} is

$$\frac{(A^2 - I)(A - 2I)}{2} - \frac{(A^2 - A)(A - 2I)}{6} e^{-t} - \frac{(A^2 + A)(A - 2I)}{2} e^t + \frac{A^3 - A}{6} e^{2t}$$

10. Choose $\phi(s) = \sqrt{s}$ in Example 1 to get $6\sqrt{A} = 7A - A^2$ and show by Theorem 3 that it satisfies $(\sqrt{A})^2 = A$. (Actually $\sqrt{A}$ has four values in this case.)

11. Choose $\psi(s) = \cos\sqrt{s}\,t$ in Example 1 to get

$$\cos\sqrt{A}\,t = \frac{(A - I)(A - 4I)}{4} + \frac{A(A - 4I)}{-3} \cos t + \frac{A(A - I)}{12} \cos 2t$$

It can be shown that this satisfies $Y'' + AY = 0$, $Y(0) = I$, $Y'(0) = 0$.

12. Prove that $(d/dt)e^{\Lambda t} = e^{\Lambda t}\Lambda$ for diagonal matrices Λ of arbitrary order.

13P. Prove the first equation below, then the second, then the third:

$$e^{\Lambda s}e^{\Lambda t} = e^{\Lambda(s+t)}, \qquad e^{As}e^{At} = e^{A(s+t)}, \qquad e^{-At}e^{At} = I$$

14. Show (a) that $2 \sin A \cos A = \sin 2A$ and (b) that $\sin^2 A + \cos^2 A = I$.

15. Using (a) a 2-by-2 matrix and (b) a 3-by-3 matrix of your own choice, verify Theorem 3 numerically. Your matrices can lead to easy calculations, yet should not be so special that the verification is trivial.

---

2. Multiple eigenvalues

If the characteristic polynomial $P(s)$ has $(s - \lambda)^m$ as a factor, and no higher power of $s - \lambda$, then λ is said to be an eigenvalue of multiplicity m. There is a general principle in the theory of multiple eigenvalues, which can be stated informally as follows: In addition to $\psi(\lambda)$, use all derivatives up to order $m - 1$.

Let us explain how the above principle applies to Section 1. A function ψ is *defined on the spectrum of A* if the values

$$\psi(\lambda), \qquad \psi'(\lambda), \qquad \cdots, \qquad \psi^{(m-1)}(\lambda)$$

exist at each eigenvalue $\lambda = \lambda_j$ of multiplicity $m = m_j$. Two functions ψ_1 and ψ_2 are such that $\psi_1 = \psi_2$ *on the spectrum of A* if, for each λ as above,

$$\psi_1(\lambda) = \psi_2(\lambda), \qquad \psi_1'(\lambda) = \psi_2'(\lambda), \qquad \cdots, \qquad \psi_1^{(m-1)}(\lambda) = \psi_2^{(m-1)}(\lambda).$$

Definitions 1 and 2 of Section 1 are already worded to allow multiple eigenvalues. With the above conventions, Theorems 1, 2, and 3 of Section 1 retain their validity, though the proofs do not. Instead of discussing proofs we show how problems with multiple eigenvalues are actually solved. Surprising as it may seem, the details of calculation are often easier than when all eigenvalues are distinct.

Our assumption that ψ is defined on the spectrum of A requires existence of at least one derivative at each multiple eigenvalue λ. The notion of derivative is readily extended to allow a complex variable; the defining limit is the same, and for a polynomial or a power series the formula is the same, as in the real case.

If ϕ is any polynomial such that $\phi = \psi$ on the spectrum of A, then $\psi(A) = \phi(A)$ by definition. The only new problem is construction of the polynomial ϕ. The method of Lagrange does not work, because the denominator $N_j(\lambda_j)$ is 0 at a multiple eigenvalue λ_j.

To deal with this problem, recall the equation

$$\psi(A) = L_1(A)\psi(\lambda_1) + L_2(A)\psi(\lambda_2) + \cdots + L_n(A)\psi(\lambda_n)$$

which holds when the λ_j are distinct. The coefficients $L_j(A)$ are polynomials in A, hence each of them is a matrix M_j of order n. Thus the above equation becomes

$$\psi(A) = M_1\psi(\lambda_1) + M_2\psi(\lambda_2) + \cdots + M_n\psi(\lambda_n).$$

It is important that the M_j are independent of ψ.

In the case of multiple eigenvalues the only change is as follows: Instead of the m terms

$$M_{j\,0}\psi(\lambda) + M_{j\,1}\psi(\lambda) + \cdots + M_{j\,m-1}\psi(\lambda)$$

corresponding to an eigenvalue $\lambda = \lambda_j$ of multiplicity $m = m_j$, we must use

$$M_{j\,0}\psi(\lambda) + M_{j\,1}\psi'(\lambda) + \cdots + M_{j\,m-1}\psi^{(m-1)}(\lambda).$$

Note that this agrees with the principle stated informally above. If $\psi(s) = e^{st}$ the nonzero columns of $M_{j\,m-1}$ must be eigenvectors corresponding to λ_m, though no such result holds for the other matrices $M_{j\,k}$.

EXAMPLE 1. A 3-by-3 matrix A has $P(s) = (s-4)^2(s-7)$, so that the spectrum is 4, 4, 7. In this case

$$(1) \qquad \psi(A) = M_1\psi(4) + M_2\psi'(4) + M_3\psi(7)$$

where the M_j are 3-by-3 matrices independent of ψ. To find the M_j we make three independent choices of ψ. The most obvious choices are

$$\psi(s) = 1,\ s,\ s^2$$

for which $\psi(A) = I,\ A,\ A^2$ respectively. But simpler equations are obtained if we choose ψ from among the functions

$$\psi(s) = 1, \qquad s - 4, \qquad (s-4)^2, \qquad s - 7.$$

The choices $\psi(s) = 1,\ s-4,\ (s-4)^2$ in (1) yield, respectively,

$$I = M_1 + M_3, \qquad A - 4I = M_2 + 3M_3, \qquad (A-4I)^2 = 9M_3.$$

The third equation gives M_3, the first then gives M_1, the second gives M_2, and the choice $s-7$ provides a check. It is left for you to verify that

$$\psi(A) = \frac{(A-I)(A-7I)}{-9}\psi(4) - \frac{(A-4I)(A-7I)}{3}\psi'(4) + \frac{(A-4I)^2}{9}\psi(7).$$

EXAMPLE 2. To find e^{At} in Example 1, take

$$\psi(s) = e^{st}, \qquad \psi'(s) = te^{st}.$$

Substitution into the result of Example 1 yields

$$e^{At} = \frac{(A-I)(A-7I)}{-9}e^{4t} - \frac{(A-4I)(A-7I)}{3}te^{4t} + \frac{(A-4I)^2}{9}e^{7t}.$$

EXAMPLE 3. A matrix A has spectrum $\lambda, \lambda, \lambda, \lambda$. To get e^{At} we start from

$$M_1\psi(\lambda) + M_2\psi'(\lambda) + M_3\psi''(\lambda) + M_4\psi'''(\lambda) = \psi(A).$$

The choices $\psi(s) = (s - \lambda)^j$ for $j = 0, 1, 2, 3$ yield

$$M_1 = I, \qquad M_2 = A - \lambda I, \qquad 2!M_3 = (A - \lambda I)^2, \qquad 3!M_4 = (A - \lambda I)^3.$$

From $\psi(s) = e^{st}$ follows $\psi^{(j)}(s) = t^j e^{st}$, so that finally

$$e^{At} = \left(I + (A - \lambda I)t + (A - \lambda I)^2\frac{t^2}{2!} + (A - \lambda I)^3\frac{t^3}{3!} \right)e^{\lambda t}.$$

The result obviously generalizes.

It is a remarkable fact that when A is real and symmetric, or more generally when A is Hermitian, we need to consider only the distinct values of λ_j. There are no derivatives in the equation for $\psi(A)$ and no terms of the form $Mte^{\lambda t}$ in e^{At}. The following theorem is established in Section 3:

THEOREM 1. *If $A = A^*$ there is a unitary matrix E such that $E^*AE = \Lambda$ where Λ is diagonal. The entries of Λ are the eigenvalues of A, repeated according to their multiplicity.*

The unitary matrix of Theorem 1 satisfies $E^*E = I$ by definition, hence the three equations

$$A = E\Lambda E^{-1}, \qquad E^{-1}AE = \Lambda, \qquad E^*AE = \Lambda$$

are equivalent. When any one of them holds, the methods of Section 1 show that construction of $\psi(A)$ can be based upon the distinct eigenvalues λ_j, as if they had multiplicity 1. This fact leads to astonishing simplification, as illustrated in the following example.

EXAMPLE 4. A real symmetric matrix of order n has two distinct eigenvalues, λ and μ, with combined multiplicity $m_1 + m_2 = n$. By the above remarks

$$M_1\psi(\lambda) + M_2\psi(\mu) = \psi(A).$$

Choosing $\psi(s) = s - \lambda$ and $s - \mu$ we get

$$M_2(\mu - \lambda) = A - \lambda I, \qquad M_1(\lambda - \mu) = A - \mu I.$$

From this M_1 and M_2 are found by inspection. The choice $\psi(s) = e^{st}$ yields

$$e^{At} = \frac{A - \mu I}{\lambda - \mu} e^{\lambda t} - \frac{A - \lambda I}{\lambda - \mu} e^{\mu t}.$$

This satisfies $Y' = AY$, $Y(0) = I$. The nonzero columns of $A - \mu I$ are eigenvectors for λ and those of $A - \lambda I$ are eigenvectors for μ. If $n = 1000$, say, the superiority of this method over that of Chapter 16 is obvious.

PROBLEMS

1P. Let A be a 2 by 2 matrix with two equal eigenvalues λ. Show that the methods of this section yield the formula

$$\psi(A) = I\,\psi(\lambda) + (A - \lambda I)\,\psi'(\lambda)$$

2. The following matrices do not have two linearly independent eigenvectors. Taking $\psi(s) = e^{st}$ in Problem 1, find e^{At}:

(a) $\begin{pmatrix} 1 & 1 \\ -1 & 3 \end{pmatrix}$ (b) $\begin{pmatrix} -3 & 2 \\ -2 & 1 \end{pmatrix}$ (c) $\begin{pmatrix} 5 & 1 \\ -1 & 3 \end{pmatrix}$ (d) $\begin{pmatrix} -5 & -2 \\ 2 & -1 \end{pmatrix}$

3. The numbers in parentheses give the spectra for real symmetric matrices A. As in Example 4, find e^{At}:

(a) (0, 0, 1, 1) (b) (1, 1, 1, 2) (c) (3, 3, 4, 4, 4, 4)

4. Verify that the symmetric matrix on the left has spectrum 4, 4, 7 and show that the result of Example 4 gives the formula on the right:

$$A = \begin{pmatrix} 5 & 1 & 1 \\ 1 & 5 & 1 \\ 1 & 1 & 5 \end{pmatrix} \qquad 3e^{At} = \begin{pmatrix} 2 & -1 & -1 \\ -1 & 2 & -1 \\ -1 & -1 & 2 \end{pmatrix} e^{4t} + \begin{pmatrix} 1 & 1 & 1 \\ 1 & 1 & 1 \\ 1 & 1 & 1 \end{pmatrix} e^{7t}$$

5. Check that the matrix of Problem 4 satisfies $(A - 4I)(A - 7I) = 0$ and hence $A^2 = 11A - 28I$. Thus show that the formula of Example 1 simplifies to yield the result of Problem 4.

6. Here are six symmetric matrices each of which has a multiple eigenvalue. Using the result of Example 4, find e^{At}:

(a) $\begin{pmatrix} -2 & 0 & 0 \\ 0 & 0 & 0 \\ 0 & 0 & -2 \end{pmatrix}$ (c) $\begin{pmatrix} 1 & 1 & -2 \\ 1 & 1 & -2 \\ -2 & -2 & 4 \end{pmatrix}$ (e) $\begin{pmatrix} -5 & -2 & -2 \\ -2 & -2 & 4 \\ -2 & 4 & -2 \end{pmatrix}$

(b) $\begin{pmatrix} -1 & 2 & -4 \\ 2 & 2 & 2 \\ -4 & 2 & -1 \end{pmatrix}$ (d) $\begin{pmatrix} 1 & 0 & 0 \\ 0 & 2 & 0 \\ 0 & 0 & 1 \end{pmatrix}$ (f) $\begin{pmatrix} 1 & -1 & -1 \\ -1 & 1 & -1 \\ -1 & -1 & 1 \end{pmatrix}$

7. (abcdef) Review. This takes a good deal of time. In Problem 6 obtain a nonsingular matrix E of eigenvectors by the methods of Chapter 16.

8. (abcdef) Find e^{At} if the spectrum of A is as follows:

$$(0, 1, 1) \quad (3, 3, 3) \quad (2, 2, 5) \quad (0, 0, 0, 1) \quad (1, 1, 0, 0) \quad (0, 1, 1, 2)$$

9. (abcdefghijkl) The following matrices have integral eigenvalues, one of which is of multiplicity 2. Proceeding as in Examples 1 and 2, express e^{At} in terms of A:

(a) $\begin{pmatrix} 0 & 0 & -1 \\ 1 & 0 & -1 \\ 0 & 0 & -1 \end{pmatrix}$ (e) $\begin{pmatrix} 3 & 6 & 3 \\ -1 & 2 & -1 \\ 1 & -3 & 2 \end{pmatrix}$ (i) $\begin{pmatrix} 7 & -3 & -3 \\ 2 & 1 & -2 \\ 3 & -3 & 1 \end{pmatrix}$

(b) $\begin{pmatrix} -1 & 2 & -3 \\ 7 & 4 & 7 \\ -1 & -1 & 2 \end{pmatrix}$ (f) $\begin{pmatrix} 0 & 0 & -3 \\ -1 & 2 & -4 \\ -2 & 0 & -1 \end{pmatrix}$ (j) $\begin{pmatrix} 5 & 2 & 2 \\ 1 & 5 & -1 \\ -2 & 2 & 9 \end{pmatrix}$

(c) $\begin{pmatrix} -3 & -4 & -4 \\ 2 & 3 & 2 \\ 5 & 2 & 7 \end{pmatrix}$ (g) $\begin{pmatrix} 5 & 6 & 5 \\ 1 & 2 & 1 \\ -3 & -3 & -2 \end{pmatrix}$ (k) $\begin{pmatrix} 5 & 9 & -15 \\ 0 & 5 & -2 \\ 1 & 3 & -3 \end{pmatrix}$

(d) $\begin{pmatrix} 4 & 4 & 4 \\ 1 & 2 & 1 \\ -2 & -2 & -1 \end{pmatrix}$ (h) $\begin{pmatrix} -2 & 3 & 3 \\ 2 & -2 & 2 \\ -3 & -3 & -8 \end{pmatrix}$ (l) $\begin{pmatrix} 3 & 1 & 1 \\ -4 & 7 & 2 \\ 1 & -1 & 3 \end{pmatrix}$

10. (abcdefghijkl) Find two independent eigenvectors in Problem 9.

11. (abcdefghijkl) Preferably using a calculator capable of computing matrix products, work out some of the results of Problem 9 explicitly.

12. Let A be a 3-by-3 matrix with eigenvalues λ_1, λ_2, λ_3. Show that the nonzero columns of $(A - \lambda_2 I)(A - \lambda_3 I)$ are eigenvectors for λ_1.

Hint: By the Cayley-Hamilton theorem $(A - \lambda_1 I)(A - \lambda_2 I)(A - \lambda_3 I) = 0$.

13. Let A be a 3-by-3 matrix with spectrum (λ, λ, μ) where $\lambda \neq \mu$. Show that
$$e^{At} = Bte^{\lambda t} + Ce^{\mu t} + (I - C)e^{\lambda t}$$
where $(\lambda - \mu)^2 C = (A - \lambda I)^2$, $(\lambda - \mu)B = (A - \lambda I)(A - \mu I)$. By Problem 12 the nonzero columns of B and C are eigenvectors for λ and μ respectively.

14. Let Λ be a diagonal matrix of order n with only the two distinct eigenvalues λ, μ, and let $A = E^{-1}\Lambda E$. Verify the first of the following by direct computation and deduce the second by use of Section 1, Equation (2):
$$(\Lambda - \lambda I)(\Lambda - \mu I) = 0, \qquad (A - \lambda I)(A - \mu I) = 0$$

————————————ANSWERS————————————

2. $I\,e^{\lambda t} + B\,t e^{\lambda t}$ where $\lambda = 2, -1, 4, -3$. Check that the nonzero columns of B are eigenvectors.

3. (a) $Ae^t - A + I$, (b) $(A - I)e^{2t} - (A - 2I)e^t$, (c) $(A - 3I)e^{4t} - (A - 4I)e^{3t}$

6. (abcdef) Partial answer: Here are the eigenvalues. Example 4 then gives the answer almost by inspection:
$$(-2, -2, 0), \quad (-6, 3, 3), \quad (0, 0, 6), \quad (1, 1, 2), \quad (-6, -6, 3), \quad (-1, 2, 2)$$

7. With eigenvalues as above, possible matrices E are:

(a) $\begin{pmatrix} 0 & 1 & 0 \\ 0 & 0 & 1 \\ 1 & 0 & 0 \end{pmatrix}$ (c) $\begin{pmatrix} 2 & -1 & -1 \\ 0 & 1 & -1 \\ 1 & 0 & 2 \end{pmatrix}$ (e) $\begin{pmatrix} 2 & 2 & -1 \\ 0 & 1 & 2 \\ 1 & 0 & 2 \end{pmatrix}$

(b) $\begin{pmatrix} 2 & -1 & 1 \\ -1 & 0 & 2 \\ 2 & 1 & 0 \end{pmatrix}$ (d) $\begin{pmatrix} 0 & 1 & 0 \\ 0 & 0 & 1 \\ 1 & 0 & 0 \end{pmatrix}$ (f) $\begin{pmatrix} 1 & -1 & -1 \\ 1 & 0 & 1 \\ 1 & 1 & 0 \end{pmatrix}$

8. (a) $e^{At} = (A - I)^2 + (2A - A^2)e^t + (A^2 - A)te^t$
   (b) $e^{At} = Ie^{3t} + (A - 3I)te^{3t} + (A - 3I)^2 t^2 e^{3t}/2$
   (c) $9e^{At} = -(A^2 - 4A - 5I)e^{2t} - 3(A^2 - 7A + 10I)te^{2t} + (A - 2I)^2 e^{5t}$
   (d) $e^{At} = I - A^3 + (A - A^3)t + (A^2 - A^3)t^2/2 + A^3 e^t$
   (e) $e^{At} = (A - I)^2(I + 2A) + (A - I)^2 At + (3A^2 - 2A^3)e^t + (A^3 - A^2)te^t)$
   (f) $2e^{At} = (A - I)^2(2 - A + Ae^{2t}) + 2(2A - A^2)(e^t - te^t + Ate^t)$

9. The spectrum is λ, λ, μ with (λ, μ) as follows. The answers are then given by Problem 13:

(a) $(0, -1)$ (c) $(3, 1)$ (e) $(2, 3)$ (g) $(2, 1)$ (i) $(4, 1)$ (k) $(2, 3)$
(b) $(4, -3)$ (d) $(2, 1)$ (f) $(2, -3)$ (h) $(-5, -2)$ (j) $(7, 5)$ (l) $(4, 5)$

10. Possible eigenvectors with (λ, μ) as above are the columns of the following matrices. These are for (abcdef), those below for (ghijkl):

$\begin{pmatrix} 0 & 1 \\ 1 & 0 \\ 0 & 1 \end{pmatrix}$ $\begin{pmatrix} -1 & -1 \\ -1 & 1 \\ 1 & 0 \end{pmatrix}$ $\begin{pmatrix} -2 & -4 \\ 1 & 1 \\ 2 & 3 \end{pmatrix}$ $\begin{pmatrix} -2 & 0 \\ -1 & -1 \\ 2 & 1 \end{pmatrix}$ $\begin{pmatrix} -3 & -1 \\ -1 & -1 \\ 3 & 2 \end{pmatrix}$ $\begin{pmatrix} 0 & 1 \\ 1 & 1 \\ 0 & 1 \end{pmatrix}$

$\begin{pmatrix} -3 & -1 \\ -1 & -1 \\ 3 & 2 \end{pmatrix}$ $\begin{pmatrix} -1 & -1 \\ 0 & -1 \\ 1 & 1 \end{pmatrix}$ $\begin{pmatrix} 1 & 1 \\ 0 & 1 \\ 1 & 1 \end{pmatrix}$ $\begin{pmatrix} 1 & 1 \\ 0 & -1 \\ 1 & 1 \end{pmatrix}$ $\begin{pmatrix} 9 & 3 \\ 2 & 1 \\ 3 & 1 \end{pmatrix}$ $\begin{pmatrix} -1 & -1 \\ -2 & -3 \\ 1 & 1 \end{pmatrix}$

11. Check that $e^0 = I$ and, by the above answers, that nonzero columns of the coefficients of $te^{\lambda t}$ and of $e^{\mu t}$ are eigenvectors for λ and μ.

3. Theoretical supplement

Although multiple eigenvalues can lead to numerical simplification, as we have seen, they usually complicate the proofs. The following perturbation theorem is useful in this connection:

THEOREM 1. *Let A be a square matrix with multiple eigenvalues. By adding arbitrarily small positive numbers to the diagonal elements of A we can get a new matrix A_ϵ for which all the eigenvalues are distinct.*

For example, let A be of order 100, say, and let the elements be given to 20 decimal places. Then it is possible to change the diagonal elements of A by less than one unit in the 20th decimal and ensure that the resulting matrix has 100 distinct eigenvalues. If A is Hermitian then A_ϵ is also Hermitian and, by results of Chapter 16, the 100 distinct eigenvalues are all real.

First we use Theorem 1 and then we give the proof. Let the diagonal elements of A_ϵ be

$$a_{jj} + \epsilon_j.$$

Theorem 1 means the following: If $\delta > 0$, the vector $\epsilon = (\epsilon_i)$ can be chosen so that

$$0 < \epsilon_j < \delta, \qquad j = 1, 2, \cdots, n$$

and A_ϵ has distinct eigenvalues. We pick a sequence of positive numbers δ tending to 0; for example, $1, 1/2, 1/3, \cdots$. This yields a corresponding sequence $\{\epsilon\}$ tending to 0 for which A_ϵ has distinct eigenvalues. Whenever we write $\epsilon \to 0$, we mean that ϵ is confined to a sequence of that type. The convergence of matrices is interpreted elementwise, so that

$$\lim_{\epsilon \to 0} A_\epsilon = A.$$

An important use of Theorem 1 is to obtain the results of Section 2 by letting $\epsilon \to 0$ in the Lagrange polynomial $\phi_\epsilon(A_\epsilon)$ of Section 1. This technique is illustrated in Problem 4, though it will not be developed in detail here. Additional applications are explained in Parts (abcd) below.

(a) The Cayley-Hamilton theorem revisited

In Section 1 we established the Cayley-Hamilton theorem $P(A) = 0$ for the case of distinct eigenvalues. Hence, if A_ϵ satisfies the conclusion of Theorem 1,

$$P_\epsilon(A_\epsilon) = 0 \text{ where } P_\epsilon(s) = |sI - A_\epsilon|.$$

The coefficients of $P_\epsilon(s)$ are polynomials in the ϵ_j, hence are continuous. The Cayley-Hamilton theorem for A is now obtained as follows:

$$P(A) = \lim_{\epsilon \to 0} P_\epsilon(A_\epsilon) = \lim_{\epsilon \to 0} 0 = 0.$$

The equation $P(A) = 0$ has an interesting application to the computation of polynomials $\phi(A)$. Divide $\phi(s)$ by $P(s)$ to get

$$\phi(s) = P(s)Q(s) + R(s) \tag{1}$$

where the remainder R is 0 or of degree $\leq n - 1$. Upon setting $s = A$ and using the fact that $P(A) = 0$, we get the remarkable formula

$$\phi(A) = R(A). \tag{2}$$

For example, if A is of order 4 and ϕ is of degree 1000, the computation of $\phi(A)$ is reduced to the evaluation of a polynomial $R(A)$ of degree at most 3.

By means of (1) and (2) we shall extend Section 1, Theorem 1, to the case of multiple eigenvalues. The theorem states that $\phi_1(A) = \phi_2(A)$ if ϕ_1 and ϕ_2 agree on the spectrum of A. The hypothesis means that the derivatives up to order $m_j - 1$ agree at each eigenvalue λ_j of multiplicity m_j. Hence the difference $\phi_1 - \phi_2$ has a zero of multiplicity $\geq m_j$ at each λ_j. If R_1 and R_2 are the corresponding remainders, Equation (1) shows that $R_1 - R_2$ also has a zero of multiplicity at least m_j at λ_j. Hence $R_1 - R_2$ has at least n zeros in all, counting multiplicity. Since this polynomial is of degree $n - 1$ at most, unless it is 0, it must be 0. We conclude that $R_1 = R_2$ and (2) then gives $\phi_1(A) = \phi_2(A)$.

(b) The method of Krylov

Once the spectrum is known, the methods of this chapter allow us to solve $Y' = AY$ by elementary operations such as matrix multiplication. But for matrices of high order it is not only difficult to find the spectrum, it is difficult merely to write the characteristic polynomial. There is a general formula for $P(s)$; but if $n = 10$ the formula involves 120, 210, and 252 determinants of respective orders 7, 6, and 5, and many more besides.

Another procedure was developed by the Soviet mathematician A. N. Krylov in 1931. The idea is to start with any convenient vector X_0 and form successive iterates

$$X_1 = AX_0, \quad X_2 = AX_1, \quad X_3 = AX_2, \quad \cdots.$$

By mathematical induction,

$$X_j = A^j X_0, \qquad j = 0, 1, 2, \cdots. \tag{3}$$

Krylov's theory is based upon the following:

THEOREM 2.   *Suppose the vectors X_0, X_1, $\cdots$, X_{n-1} are linearly independent. Then the unique representation*

(4) $$-X_n = q_0 X_0 + q_1 X_1 + \cdots + q_{n-1} X_{n-1}$$

gives the coefficients $p_j = q_j$ in the characteristic polynomial of A.

For example, let A be the 2-by-2 matrix A on the left below. We choose X_0 as in the third column and write $X_1 = AX_0$ and $X_2 = AX_1$ in the fourth and fifth columns:

$$\begin{pmatrix} 1 & 2 & | & 0 & 2 & 10 \\ 3 & 4 & | & 1 & 4 & 22 \end{pmatrix}$$

According to Theorem 2, the coefficients p_j satisfy

$$p_0 \begin{pmatrix} 0 \\ 1 \end{pmatrix} + p_1 \begin{pmatrix} 2 \\ 4 \end{pmatrix} = \begin{pmatrix} -10 \\ -22 \end{pmatrix}.$$

It is left for you to check that this gives $p_1 = -5$, $p_0 = -2$ and that these are indeed the coefficients of $P(s)$.

Theorem 2 follows from the equation $P(A)X_0 = 0$, or, equivalently,

$$A^n X_0 + p_{n-1} A^{n-1} X_0 + \cdots + p_1 A X_0 + p_0 X_0 = 0.$$

We rewrite this using (3) and we transfer the term X_n to the left. The result is

(5) $$-X_n = p_0 X_0 + p_1 X_1 + \cdots + p_{n-1} X_{n-1}.$$

But since the X_j for $j \leq n - 1$ form a basis by hypothesis, there is only one way to write X_n in the form (4) or (5). This gives $q_j = p_j$ and completes the proof. The rest of Krylov's theory is concerned with the efficient determination of the q_j and will not be presented here.

Theorem 2 is not infallible, because the first n vectors X_j may be linearly dependent. In that case there is another conclusion that is just as interesting. Namely, if the first m vectors are linearly independent, but

$$-X_m = q_0 X_0 + q_1 X_1 + \cdots + q_{m-1} X_{m-1},$$

then

$$Q(s) = q_0 + q_1 s + \cdots + q_m s^m$$

is a divisor of $P(s)$. The proof is more difficult than the proof of Theorem 2 and is omitted.

(c) Compactness

Our next application of Theorem 1 depends on a principle of advanced calculus that reads as follows:

COMPACTNESS PRINCIPLE. A bounded sequence $\{s_j\}$ of real numbers has a convergence subsequence.

A *subsequence* is an infinite sequence obtained from an infinite sequence by leaving out some of its terms. For example the subsequence

$$s_1, \ s_4, \ s_9, \ s_{16}, \ s_{25}, \ \cdots$$

is obtained from $\{s_j\}$ by omitting all indices j that are not perfect squares. The principle of compactness is logically equivalent to the completeness axiom of Chapter 12, Section 3, and either principle can be regarded as an axiom.

One of the main uses of compactness is to show that a continuous real-valued function on a bounded closed interval attains its maximum; see Chapter 10, Theorem A. This fact is the basis for Rolle's theorem, which is constantly used in the theory of differential equations.

A subsequence of $\{j\}$ is here denoted by $\{j'\}$, a subsequence of $\{j'\}$ is denoted by $\{j''\}$, and so on. With this notation, the compactness principle is easily extended to a bounded sequence of complex numbers $x + iy$. We first pick a subsequence $\{j'\}$ on which x_j has a limit, and then we pick a subsequence $\{j''\}$ of $\{j'\}$ on which y_j has a limit. It follows that a bounded sequence of complex numbers has a convergent subsequence. This is the main lemma that underlies the fundamental theorem of algebra. The latter was used in Chapter 8 and is indispensable in Chapters 16 and 17. Indeed, without the fundamental theorem of algebra we could not, in general, establish the existence of even a single eigenvalue λ.

The principle of compactness also applies to a sequence of real or complex vectors X_j of the same dimension, d. The sequence is *bounded* if the lengths satisfy $|X_j| \leq M$ for some constant M, or equivalently, if their coordinates satisfy an inequality of this type. Given a bounded sequence of vectors, there is a subsequence $\{j'\}$ on which the first coordinates converge. From that we can select a subsequence $\{j''\}$ on which the second coordinates converge, and so on. The process terminates after d steps and shows that a bounded sequence of vectors has a convergent subsequence.

(d) Diagonalization

If E is a unitary matrix, the equation $E^*E = I$ shows that the columns of E have length 1, hence the elements satisfy $|e_{ij}| \leq 1$. Thus a sequence of unitary matrices of order n can be regarded as a bounded sequence of vectors of dimension $d = n^2$. By the above remarks, it has a convergent subsequence. We shall use this fact to prove that a Hermitian matrix A can be diagonalized by a unitary matrix E.

Let A_ϵ have distinct eigenvalues. Chapter 16, Section 3, Theorem 4 gives a matrix E_ϵ such that

$$(6) \qquad E_\epsilon^* A_\epsilon E_\epsilon = \Lambda_\epsilon, \qquad E_\epsilon^* E_\epsilon = I,$$

where Λ_ϵ is diagonal. Using compactness, we form a subsequence $\{\epsilon'\}$ on which E_ϵ has a limit E. Then $E_\epsilon^* A_\epsilon E_\epsilon$ has the limit $E^* A E$. Equation (6) shows that Λ_ϵ has a limit, which we call Λ. Since each Λ_ϵ is diagonal, so is Λ. Taking the limit as $\epsilon \to 0$ in (6) now yields the desired result,

$$(7) \qquad E^* A E = \Lambda, \qquad E^* E = I.$$

Equation (7) really depends on compactness, or on analytic ideas that are related to compactness. It would be false, for example, if the matrices were over the field of rational numbers. By contrast, the Cayley-Hamilton theorem is equivalent to a collection of polynomial identities in the elements a_{ij}. Since these identities hold for arbitrary complex a_{ij}, corresponding coefficients agree, and the identities remain valid when the a_{ij} belong to any integral domain. A purely algebraic proof is outlined in Problem 6.

(e) Proof of the perturbation theorem

We conclude with a proof of Theorem 1. Since the statement is trivial when $n = 1$, we assume $n \geq 2$ and we use mathematical induction. The induction hypothesis is that Theorem 1 holds for matrices of order $n - 1$.

Expansion of the determinant $P_\epsilon(s) = |sI - A_\epsilon|$ on its top row leads to an equation of form

$$(8) \qquad P_\epsilon(s) = (s - a_{11} - \epsilon_1) D(s) + Q(s).$$

Here $D(s) = s^{n-1} + \cdots$ is the cofactor of the element $s - a_{11} - \epsilon_1$ and $Q(s)$ is a polynomial of degree at most $n - 1$ which does not involve ϵ_1. Equation (8) can be written

$$(9) \qquad P_\epsilon(s) = C(s) - \epsilon_1 D(s)$$

where C and D are independent of ϵ_1 and

$$(10) \qquad C(s) = s^n + \cdots, \qquad D(s) = s^{n-1} + \cdots.$$

Let $\delta > 0$. By the induction hypothesis we can choose ϵ_j satisfying

$$0 < \epsilon_j < \delta, \qquad j = 2, 3, \cdots, n$$

in such a way that $D(s)$ has no multiple zeros. This means that $D(s)$ and $D'(s)$ do not vanish simultaneously. At any multiple zero of $P_\epsilon(s)$ in (9)

(11) $$C(s) = \epsilon_1 D(s), \qquad C'(s) = \epsilon_1 D'(s)$$

and hence $C'(s)D(s) - D'(s)C(s) = 0$. Since the leading term is

$$ns^{n-1}s^{n-1} - (n-1)s^{n-2}s^n = s^{2n-2}$$

this equation has at most $2n-2$ roots s_j. We choose ϵ_1 in such a way that the first equation (11) fails at values s_j where $D(s_j) \neq 0$ and the second fails at values s_j where $D(s_j) = 0$. This is possible because $D(s)$ and $D'(s)$ do not vanish simultaneously. Only $2n-2$ values of ϵ_1 are thus ruled out. Hence the side condition $0 < \epsilon_1 < \delta$ is easily satisfied, and this completes the proof.

PROBLEMS───

1. This pertains to Theorem 2 as applied to the 3-by-3 matrix A on the left:

$$\left(\begin{array}{ccc|cccc} -3 & 2 & -2 & 0 & -2 & 0 & -2 \\ -13 & 9 & -7 & 0 & -7 & -9 & -25 \\ -9 & 6 & -4 & 1 & -4 & -8 & -22 \end{array} \right)$$

Verify the iteration, solve $p_0 x_0 + p_1 x_1 + p_2 x_2 = -x_3$ to get the characteristic polynomial $s^3 - 2s^2 - s + 2$, and check by the usual formula.

2. Proceed as in Problem 1. No answer is provided, but we mention that in (bc) the choice $x_0 = (0, 0, 1)^T$ does not work:

(a) $\begin{pmatrix} 1 & 0 & 2 \\ 0 & 2 & 3 \\ 0 & 0 & 4 \end{pmatrix}$ (b) $\begin{pmatrix} 2 & 0 & 0 \\ 1 & 1 & 0 \\ 3 & 0 & 2 \end{pmatrix}$ (c) $\begin{pmatrix} 1 & 0 & 1 \\ 0 & 1 & 0 \\ -1 & 2 & 0 \end{pmatrix}$

3. (abcd) Taking $x_0 = (0, 0, 0, 1)^T$ find the characteristic polynomial and verify that two of the coefficients are $-\text{trace } A$ and $\det A$:

$$\begin{pmatrix} 1 & 1 & 0 & 0 \\ 1 & 0 & 1 & 0 \\ 1 & 0 & 0 & 1 \\ 1 & 0 & 0 & 0 \end{pmatrix} \begin{pmatrix} 1 & 1 & 0 & 0 \\ 1 & 0 & 1 & 0 \\ 1 & 1 & 0 & 1 \\ 1 & 0 & 0 & 0 \end{pmatrix} \begin{pmatrix} 1 & 1 & 0 & 0 \\ 1 & 0 & 1 & 0 \\ 1 & 1 & 0 & 1 \\ 0 & 1 & 1 & 1 \end{pmatrix} \begin{pmatrix} 1 & -1 & 0 & 2 \\ 0 & 1 & 2 & -1 \\ 3 & -3 & 4 & 0 \\ 1 & 0 & 1 & 1 \end{pmatrix}$$

4. Let a family of 2-by-2 matrices A_ϵ have distinct real eigenvalues λ_1 and λ_2 depending on ϵ. Suppose further that $\lambda_1 \to \lambda$, $\lambda_2 \to \lambda$ as $\epsilon \to 0$. If $\psi(s)$ is a continuously differentiable real-valued function, deduce the result of Section 2, Problem 1 from that of Section 1, Problem 2.

Hint: $\psi(\lambda_1) - \psi(\lambda_2) = \psi'(s)(\lambda_1 - \lambda_2)$ where $s \to \lambda$ as $\epsilon \to 0$.

5. Writing (7) in the form $AE = E\Lambda$, show that each column E_j of E is an eigenvector corresponding to the eigenvalue λ_j.

6. It was established in Chapter E that B adj $B = I|B|$ where adj A is the cofactor transpose. Apply this with $B = sI - A$ to get

$$(sI - A) \text{ adj } (sI - A) = I|sI - A|$$

Check that adj $(sI - A)$ is a polynomial in s of degree $n - 1$ with coefficients C_j, which are matrices of order n. Hence

$$(sI - A)(C_0 + C_1 s + \cdots + C_{n-1}s^{n-1}) = I(p_0 + p_1 s + \cdots + s^n)$$

or, equating coefficients,

$$-AC_0 = p_0 I, \quad C_0 - AC_1 = p_1 I, \quad C_1 - AC_2 = p_2 I, \quad \cdots, \quad C_{n-1} = I$$

Multiply these equations respectively by I, A, A^2, $\cdots$, A^n, and add. The left side telescopes to 0 and the right side gives $P(A)$. Hence $P(A) = 0$, which is the Cayley-Hamilton theorem.

REVIEW PROBLEMS————————————————————————

7. When done in sequence, as intended, problems 7, 8, and 9 are self-checking and no answer is provided. Here are three matrices A containing a constant parameter α. Solve $y' = Ay$ by Chrystal's method (Chapter 15, Section 4) and verify that a term $te^{\lambda t}$ is present unless $\alpha = 1$, in which case there is no such term:

$$\text{(a)} \begin{pmatrix} 4 & -1 & \alpha \\ 0 & 7 & -3 \\ 0 & 0 & 4 \end{pmatrix} \quad \text{(b)} \begin{pmatrix} 1 & 1 & \alpha \\ 0 & 3 & 2 \\ 0 & 0 & 1 \end{pmatrix} \quad \text{(c)} \begin{pmatrix} -1 & \alpha & 3 \\ 0 & -1 & 0 \\ 0 & 1 & 2 \end{pmatrix}$$

8. (abc) Solve Problem 7 by finding e^{At} as in Section 1 or 2.

9. (abc) Try to solve Problem 7 by the method of eigenvectors as in Chapter 16. It will be found that the method succeeds only if $\alpha = 1$.

10. This takes a good deal of time. The following matrices have one real and two complex eigenvalues. Find them, and express a fundamental solution in real form by use of the result of Problem 3. In each case we have taken e_3 to be real:

(a) $\begin{pmatrix} 0 & 7 & 2 \\ 0 & 1 & 0 \\ -1 & 7 & -2 \end{pmatrix}$
(b) $\begin{pmatrix} 1 & 2 & 3 \\ 0 & 6 & -1 \\ 0 & 5 & 4 \end{pmatrix}$
(c) $\begin{pmatrix} 9 & 30 & 2 \\ -1 & -8 & -1 \\ -11 & -20 & -4 \end{pmatrix}$

––––––––––––––––––ANSWERS––––––––––––––––––

3. (a) $s^4 - s^3 - s^2 - s - 1$ (c) $s^4 - 2s^3 - 2s^2 + 2s + 2$
   (b) $s^4 - s^3 - 2s^2 - 1$ (d) $s^4 - 7s^3 + 19s^2 - 19s - 6$

10. Eigenvalues: (a) $1, -1 \pm i$ (b) $1, 5 \pm 2i$ (c) $7, -5 \pm i$.   Solutions:

(a) $e^{-t} \begin{pmatrix} 7e^{2t} & -\cos t + \sin t & -\cos t - \sin t \\ e^{2t} & 0 & 0 \\ 0 & \cos t & \sin t \end{pmatrix}$

(b) $e^{5t} \begin{pmatrix} e^{-4t} & 38\cos 2t + 9\sin 2t & 38\sin 2t - 9\cos 2t \\ 0 & 10\cos 2t - 20\sin 2t & 10\sin 2t + 20\cos 2t \\ 0 & 50\cos 2t & 50\sin 2t \end{pmatrix}$

(c) $e^{-5t} \begin{pmatrix} -e^{12t} & 2\cos t + 2\sin t & -2\cos t + 2\sin t \\ 0 & -\cos t - \sin t & \cos t - \sin t \\ e^{12t} & 2\cos t & 2\sin t \end{pmatrix}$

Mathematics possesses not only truth but supreme beauty, a beauty cold and austere, like that of sculpture, without appeal to any part of our weaker nature, and capable of a stern perfection such as only the greatest art can show.

—Bertrand Russell

Mathematics has the inhuman quality of starlight, brilliant and sharp but cold.

—Hermann Weyl

Chapter 18

STURM-LIOUVILLE THEORY

THIS chapter is concerned with a class of boundary-value problems that was introduced by Charles Sturm and Joseph Liouville in 1836 and has been extensively studied ever since. With historical roots in the application of Fourier series to heat flow, Sturm-Liouville theory is an important aid in solving the partial differential equations of mathematical physics. On the theoretical side, it points the way to a major result of operator analysis known as the spectral theorem. Though the deeper aspects will not be developed here, we go far enough to show that Sturm-Liouville theory is rich in mathematical ideas and powerful in applications.

Prerequisites: Chapter 9, Section 3 and parts of Chapters 10, 12, 13, 16, F.

1. Fundamental principles

Let us begin with an example that sheds light on the main themes of this chapter. If m and n are real constants, the functions $\sin mx$, $\cos mx$ satisfy the first of the following equations and $\sin nx$, $\cos nx$ satisfy the second:

$$u'' = -m^2 u, \qquad v'' = -n^2 v.$$

These equations yield

$$(uv' - vu')' = uv'' - vu'' = (m^2 - n^2)uv,$$

hence

$$uv' - vu' \Big|_{-\pi}^{\pi} = (m^2 - n^2) \int_{-\pi}^{\pi} uv \, dx.$$

Here, and in analogous cases below, $u, u', \cdots$ stand for $u(x), u'(x), \cdots$.

So far, m and n have been unrestricted. We now assume that m and n are integers which, without loss of generality, are nonnegative. In that case $uv' - vu'$ has period 2π, the left side of the above equation is 0, and therefore

$$\int_{-\pi}^{\pi} uv \, dx = 0, \qquad m^2 \neq n^2.$$

When $m = n$ the procedure gives no information, but obviously

$$\int_{-\pi}^{\pi} \sin mx \cos mx \, dx = \frac{1}{2} \int_{-\pi}^{\pi} \sin 2mx \, dx = 0.$$

In general, two functions u and v are *orthogonal* over a given interval if the integral of their product over that interval is 0. What has been shown is that any two different members of the sequence

$$(1) \qquad 1, \ \sin x, \ \cos x, \ \sin 2x, \ \cos 2x, \ \cdots, \ \sin nx, \ \cos nx, \ \cdots$$

are orthogonal over the interval $-\pi \le x \le \pi$.

It turns out that broad classes of functions f can be represented as a uniformly convergent series

$$(2) \qquad f(x) = \frac{a_0}{2} + \sum_1^\infty (a_n \cos nx + b_n \sin nx), \quad -\pi \le x \le \pi.$$

Under this assumption, term-by-term integration is justified by Chapter F, Theorem 1, and we can derive a formula for the coefficients.

To get a_0, integrate from $-\pi$ to π. The only surviving term on the right is πa_0 and division by π gives a_0. To find a_n with $n \ge 1$, multiply by $\cos nx$. The result is

$$f(x) \cos nx = \cdots + a_n \cos^2 nx + \cdots$$

where the terms not written involve a product of $\cos nx$ with some other member of the sequence (1). If we integrate from $-\pi$ to π these products integrate to 0 by orthogonality and the surviving term is

$$\int_{-\pi}^{\pi} a_n \cos^2 nx \, dx = \int_{-\pi}^{\pi} a_n \frac{1 + \cos 2nx}{2} \, dx = \pi a_n.$$

A similar calculation gives πb_n, with the result

$$(3) \qquad a_n = \frac{1}{\pi} \int_{-\pi}^{\pi} f(x) \cos nx \, dx, \quad b_n = \frac{1}{\pi} \int_{-\pi}^{\pi} f(x) \sin nx \, dx.$$

The constant term in (2) is written $a_0/2$ so that the formula for a_n remains valid when $n = 0$.

The constants a_n, b_n in (3) are called *Fourier coefficients* and the equations (3) are the Euler-Fourier formulas; they were stated by Euler as early as 1750. The functions (1) satisfy the boundary-value problem

$$(4) \qquad y'' + \lambda y = 0, \qquad y(-\pi) = y(\pi), \qquad y'(-\pi) = y'(\pi)$$

and, in a sense made precise in Problem 3, Equation (4) has no other solutions. From this point of view, the theory of Fourier series appears as a part of the theory of differential equations.

Our object is to obtain similar results when (4) is replaced by

$$(5) \qquad (py')' + qy + r\lambda y = 0, \qquad a \le x \le b.$$

Here λ is a constant parameter, p, q, r, y are functions of x, and the endpoint derivatives are defined by appropriate one-sided limits. When provided with suitable boundary conditions, (5) is called a *Sturm-Liouville problem*.

At first glance, having y'' and y' in the special combination $(py')'$ seems to restrict generality. However, the identity $p(y'' + p_1 y') = (py')'$ holds if $pp_1 = p'$, hence if p_1 is continuous and

$$p = e^{\int p_1(x)\,dx}.$$

Thus we can transform general linear equations into the form (5). Here is an example.

EXAMPLE 1. The Bessel equation for the Bessel function $J_\nu(x)$ is

$$x^2 y'' + xy' + (x^2 - \nu^2)y = 0, \qquad x > 0.$$

Division by x^2 gives $p_1 = 1/x$, hence $p = e^{\ln x} = x$. After the division, multiplication by x yields

$$(xy')' + \left(x - \frac{\nu^2}{x}\right)y = 0, \qquad x > 0.$$

This is in the form (5) with $p(x) = x$, $q(x) = x$, $r(x) = 1/x$, $\lambda = -\nu^2$.

To facilitate study of the Sturm-Liouville problem, we introduce the following:

BASIC HYPOTHESIS. For $a \le x \le b$ the functions p', q, r are real and continuous. For $a < x < b$ the functions p, r are positive.

The hypothesis of continuity on the closed interval $a \le x \le b$ can be weakened, and the scope of the theory increased, by working on a subinterval (α, β) and then letting $\alpha \to a+$, $\beta \to b-$. If the interval is infinite, we let $\alpha \to -\infty$ or $\beta \to \infty$. These refinements are illustrated in the problems but are not emphasized in the text.

(a) The Lagrange identity

Throughout this chapter we set

(6) $Ty = (py')' + qy, \qquad w = uv' - vu'.$

If u and v are in the domain of T, the *Lagrange identity*

(7) $uTv - vTu = (pw)'$

holds. To see why, note that the terms involving q on the left cancel, so that (7) becomes

$$u(pv')' - v(pu')' = (u(pv') - v(pu'))'.$$

The truth of this is evident upon differentiating the two terms on the right.

An important consequence of (7) is:

THEOREM 1. *Suppose u and v satisfy $Tu + r\lambda u = 0$, $Tv + r\mu v = 0$ where μ, like λ, is a constant parameter. Then the first of the following equations holds in general and the second holds if $p(a)w(a) = p(b)w(b)$:*

$$(8ab) \qquad (\lambda - \mu)ruv = (pw)', \qquad (\lambda - \mu)\int_a^b ruv\,dx = 0.$$

To get the first equation, substitute $vTu = -r\lambda uv$, $uTv = -r\mu uv$ into (7). The second equation follows when the first is integrated from a to b.

The hypothesis $p(a)w(a) = p(b)w(b)$ of Theorem 1 is ensured by suitable boundary conditions. We refer to boundary conditions as *standard* when they have the following three properties:

Reality: If y satisfies them so does $\bar{y}$.

Linearity: If u and v satisfy them, so do cu and $u + v$.

Orthogonality: If u and v satisfy them then $p(a)w(a) = p(b)w(b)$.

For example, the boundary conditions $y(a) = y(b) = 0$ are standard, and so are the conditions $y'(a) = y'(b) = 0$. These are called *Dirichlet conditions* and *Neumann conditions* respectively. Further examples are given later.

Equation (5) can be written in the form $-Ty = \lambda ry$, which resembles the matrix equation $Ay = \lambda y$ of Chapter 16. Guided by this analogy, we introduce the following:

DEFINITION. An eigenfunction is a solution of $Ty + r\lambda y = 0$ that is non-trivial and satisfies standard boundary conditions. The values λ permitting such solutions are eigenvalues.

The statement that y is *nontrivial* means $\|y\| > 0$ where, by definition,

$$\|y\|^2 = \int_a^b r|y|^2\,dx, \qquad \|y\| \geq 0.$$

According to Chapter 16, the eigenvalues for a self-adjoint matrix A are real and eigenvectors for different eigenvalues are orthogonal. Similar results hold for the Sturm-Liouville problem as seen next.

(i) Every eigenvalue is real.

For proof, let λ be a complex eigenvalue with corresponding complex eigenfunction y. Thus

$$Ty + \lambda ry = 0, \qquad T\bar{y} + \bar{\lambda} r\bar{y} = 0$$

where the second equation is obtained by taking the conjugate of the first. The choice $\mu = \bar{\lambda}$, $u = y$, $v = \bar{y}$ in (8b) yields $\lambda = \bar{\lambda}$. Hence λ is real.

By definition, the functions u, v are orthogonal with respect to the weight r if

$$(9) \qquad \int_a^b r(x)u(x)\overline{v(x)}\, dx = 0.$$

When r is not mentioned, it is understood that r is the same as in the equation under discussion. For example, if the equation is $y'' + \lambda y = 0$, "orthogonal" means "orthogonal with respect to the weight $r = 1$."

(ii) Eigenfunctions u, v for different eigenvalues λ, μ are orthogonal.

The proof is simple. Let u and v be eigenfunctions for different eigenvalues λ, μ. By (i) the function $\bar{v}$ is also an eigenfunction for μ, and (ii) follows when (8b) is applied to $u, \bar{v}$.

In Chapter 16 we found that eigenvectors for different eigenvalues are linearly independent. A corresponding result holds here:

(iii) The eigenfunctions in (ii) are linearly independent.

For proof let $c_1 u + c_2 v = 0$, multiply by $r\bar{v}$, and integrate from a to b. By (ii) the result is $\|v\|^2 c_2 = 0$, hence $c_2 = 0$. Similarly, $c_1 = 0$.

(b) Boundary conditions

Additional properties can be established for some standard boundary conditions but not for all. To formulate these, we list the main types of standard boundary conditions:

(P) $y(a) = y(b)$, $\quad p(a)y'(a) = p(b)y'(b)$

(S) $p(a) = p(b) = 0$ and $y'(a)$, $y'(b)$ are defined.

(L) $Ay(a) = By'(a)$, $Cy(b) = Dy'(b)$ where $A^2 + B^2 > 0$, $C^2 + D^2 > 0$.

(M) A condition of type (S) at one end, (L) at the other.

The letter P is used because conditions of this type are often associated with periodic solutions such as those in (1). Verification of the equation $p(a)w(a) = p(b)w(b)$ under conditions (P) is left for you.

In (S) the hypothesis about y' is automatic if the differential equation holds at a and b, but is repeated for emphasis. When $p(a) = 0$ or $p(b) = 0$ the solution usually becomes unbounded near the corresponding endpoint and exists on the closed interval $[a, b]$ only for special choices of λ. Sturm-Liouville problems in which $p(x)$ has a zero on $[a, b]$ are called *singular* and this suggests the letter (S). The condition $p(a)w(a) = p(b)w(b)$ is obviously satisfied.

In (L) it is assumed that A, B, C, D are real constants. The special cases $B = D = 0$ and $A = C = 0$ give Dirichlet and Neumann conditions respectively. To see why (L) makes $p(a)w(a) = p(b)w(b)$, note that the two equations

$$Au(a) = Bu'(a), \qquad Av(a) = Bv'(a)$$

form a linear homogeneous system in the unknowns A, $-B$; the letter L stands for "linear system." Since the unknowns are not both 0 the coefficient determinant is 0, which is to say, $w(a) = 0$. Similarly $w(b) = 0$.

Conditions (M) are mixed; an example is $p(a) = y(b) = 0$. For boundary conditions of type (S), (L), or (M), or briefly of type (SLM), we have $p(a)w(a) = p(b)w(b) = 0$. The following result depends on these equations and fails for boundary conditions (P):

(iv) If the boundary conditions are of type (SLM), any two eigenfunctions for the same eigenvalue λ are linearly dependent.

For proof, let u and v be eigenfunctions for λ. Equation (8a) with $\mu = \lambda$ gives $(pw)' = 0$; hence pw is constant. By the boundary conditions, pw vanishes at a or b. This gives $pw = 0$, hence $w = 0$, and linear dependence follows from the fact that u and v both satisfy the same second-order equation.

Items (i)–(iv) are corollaries of Theorem 1. By contrast, the following useful result depends on different ideas:

THEOREM 2. *Suppose $q \le 0$ and suppose also that the boundary conditions are such that $p(a)y(a)y'(a) \ge p(b)y(b)y'(b)$. Then every eigenvalue for a nonconstant eigenfunction is positive.*

The inequality involving y may be meaningful only if y is real, but by Problem 12 the assumption that y is real involves no loss of generality.

For proof, let y be a real nonconstant eigenfunction for the eigenvalue λ. Multiply the equation $r\lambda y = -(py')' - qy$ by y and integrate to get

$$\lambda \int_a^b ry^2 \, dx = -\int_a^b y(py')' \, dx - \int_a^b qy^2 \, dx.$$

If we write $y(py')'\,dx = y\,d(py')$ and integrate by parts this becomes

$$\lambda \int_a^b ry^2\,dx = \int_a^b \left(p(y')^2 - qy^2\right)\,dx - pyy'\Big|_a^b.$$

Since $q \le 0$ the integral on the right is positive, and the boundary conditions ensure that the integrated term at the far right is nonnegative. Hence $\lambda > 0$.

EXAMPLE 2. Consider the eigenvalue problem

$$(py')' + \lambda ry = 0, \qquad y'(a) = y'(b) = 0.$$

Since $q = 0$, and since $p(a)y(a)y'(a) = p(b)y(b)y'(b) = 0$, Theorem 2 states $\lambda > 0$ unless y is constant. Clearly $y = 1$ is an eigenfunction for $\lambda = 0$, so that constant eigenfunctions must be excluded from Theorem 2.

EXAMPLE 3. Consider the eigenvalue problem

$$y'' + \lambda y = 0; \qquad y(0) = 0, \qquad y(1) + ky'(1) = 0$$

where k is a positive constant. In this case $p = 1$, $q = 0$, and

$$p(0)y(0)y'(0) = 0,$$

$$p(1)y(1)y'(1) = -ky'(1)^2 \le 0.$$

The boundary conditions show that there is no constant eigenfunction and hence, by Theorem 2, every eigenvalue is positive.

Setting $\lambda = \omega^2 > 0$ we get $y = c_0 \sin \omega x + c_1 \cos \omega x$. The boundary condition at 0 yields $c_1 = 0$, and at 1 it gives $\sin \omega + k\omega \cos \omega = 0$. Thus

$y = c_0 \sin \omega x$ where $\tan \omega = -k\omega.$

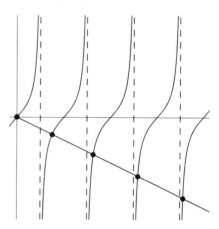

Figure 1

As seen in Figure 1, the equation $\tan \omega = -k\omega$ has one positive root ω_n in each strip

$$n\pi - \frac{\pi}{2} < \omega < n\pi + \frac{\pi}{2}, \qquad n = 1, 2, 3, \cdots.$$

Hence there is an infinite sequence of eigenfunctions $\sin \omega_n x$ with corresponding eigenvalues $\lambda_n = \omega_n^2$. The orthogonality in this case says that

$$\int_0^1 \sin \omega_m x \sin \omega_n x = 0, \qquad m \ne n.$$

PROBLEMS——

1P. Review the boundary-value problems in Chapter 9 Section 3, including Problems 1, 3, and 4 of that section. Also review Problem 1 of Chapter 10, Section 3.

2. This pertains to $y'' + \lambda y = 0$, $0 \le x \le \pi$ with the boundary conditions

$$\text{(a) } y(0) = y(\pi) = 0 \qquad \text{(b) } y'(0) = y'(\pi) = 0$$

Using Theorem 2, show that $\lambda = m^2$ where m is a positive integer in (a) and a nonnegative integer in (b). Obtain eigenfunctions $\sin mx$, $\cos mx$ in the two cases, respectively.

3. (a) Deduce from Theorem 2 that $\lambda \ge 0$ for the eigenvalue problem

$$y'' + \lambda y = 0, \qquad y(-\pi) = y(\pi), \qquad y'(-\pi) = y'(\pi)$$

Thus get $y = c_1 \cos \omega x + c_2 \sin \omega x$ where $\omega = \sqrt{\lambda}$ and verify that the boundary conditions imply $c_2 = c_1 = 0$ unless ω is an integer. This shows that every eigenfunction is a linear combination of those in (1).

(b) If (1) is written $\psi_0, \psi_1, \psi_2, \cdots$ check that ψ_n has exactly n zeros on the open interval $(-\pi, \pi)$. (For example, $\psi_3 = \sin 2x$ has three zeros.) Show also that the zeros for ψ_n lie between those for ψ_{n+1}; in other words, the zeros interlace. These properties are generalized in Section 4.

4. Let y be the eigenfunction in Problem 3 with $\omega = n$, a positive integer, and with real constants c_1, c_2. Let z be another eigenfunction of that form, with constants c_3, c_4, and the same n. Show that the conditions for (y, z) to be linearly independent or orthogonal are, respectively,

$$c_1 c_4 - c_2 c_3 \ne 0 \qquad c_1 c_3 + c_2 c_4 = 0$$

5. Here are four differential equations with their names at the left and certain polynomial solutions at the right:

(a) Legendre: $\left((1-x^2)y'\right)' + \lambda y = 0$, $\quad \lambda = m(m+1)$, $\quad P_m(x)$

(b) Chebychev: $\left((1-x^2)^{1/2}y'\right)' + \lambda(1-x^2)^{-1/2}y = 0$, $\quad \lambda = m^2$, $\quad T_m(x)$

(c) Laguerre: $(xe^{-x}y')' + \lambda e^{-x}y = 0$, $\quad \lambda = m$, $\quad L_m(x)$

(d) Hermite: $(e^{-x^2}y')' + \lambda e^{-x^2}y = 0$, $\quad \lambda = 2m$, $\quad H_m(x)$

The respective intervals are $(-1, 1)$, $(-1, 1)$, $(0, \infty)$, $(-\infty, \infty)$. Verify agreement with equations of the same name in Chapter 10, or proceed as in Example 1 to derive these equations from those in Chapter 10.

6. For nonnegative integers $m \neq n$ check that

(a) $\displaystyle\int_{-1}^{1} P_m(x) P_n(x)\, dx = 0$ (c) $\displaystyle\int_{0}^{\infty} e^{-x} L_m(x) L_n(x)\, dx = 0$

(b) $\displaystyle\int_{-1}^{1} \frac{T_m(x) T_n(x)}{\sqrt{1 - x^2}}\, dx = 0$ (d) $\displaystyle\int_{-\infty}^{\infty} e^{-x^2} H_m(x) H_n(x)\, dx = 0$

If the functions are normalized as in Chapter 12, Section 2 it can be shown that the respective integrals for $m = n \geq 1$ are

(a) $\dfrac{2}{2n + 1}$ (b) $\dfrac{\pi}{2}$ (c) $(n!)^2$ (d) $2^n n! \sqrt{\pi}$

The same holds for $m = n = 0$ except that the integral (b) is π.

Hint : In (cd) integrate Lagrange's identity from 0 to β or from α to β and take limits. The reason for restricting m and n to integer values is to have polynomial solutions.

7. The functions $e^{-x^2/2} H_m(x)$ satisfy $y'' + (2m + 1 - x^2) y = 0$. Using this equation, confirm the result of Problem 6(d).

8. An eigenvalue is *simple* if it has only one linearly independent eigenfunction. Corollary (iv) gives a condition for all eigenvalues to be simple. Show that the eigenvalues in Problem 2 are simple but those in Problem 3 for $\lambda > 0$ are not. Hence boundary conditions (P) must be excluded from (iv).

9. If $\lambda < 0$ in Example 3, show without using Theorem 2 that $y = 0$.

10. Establish the orthogonality in Example 3 without using (ii).

11. It was seen in Chapter 13 that $J_n(x)$ behaves like a constant times x^n as $x \to 0+$. If $m + n > 0$ deduce from the equation of Example 1 that

$$\frac{(m^2 - n^2)}{\beta} \int_{0}^{\beta} J_m(x) J_n(x)\, \frac{dx}{x} = J_n(\beta) J'_m(\beta) - J'_n(\beta) J_m(\beta), \qquad \beta > 0$$

12. (a) Show that $\|u + iv\|^2 = \|u\|^2 + \|v\|^2$. Hence $u + iv$ is nontrivial when, and only when, u or v is nontrivial.

(b) Deduce from (a) that if $y = u + iv$ is a complex eigenfunction then either u or v is a real eigenfunction for the same λ.

(c) If the boundary conditions are of type (SLM) in Part (b), show that $y = cu$ or $y = cv$ where c is a real or complex constant.

2. Fourier series

For $n = 0, 1, 2, \cdots$ let ψ_n be an eigenfunction corresponding to the eigenvalue λ_n. Thus $\|\psi_n\| > 0$ and the ψ_n are continuous. If the λ_j are distinct Corollary (ii) of Section 1 yields the orthogonality condition

$$\int_a^b r\psi_j\overline{\psi_k}\,dx = 0, \qquad j \neq k.$$

Suppose f admits a uniformly convergent expansion

$$\tag{1} f = c_0\psi_0 + c_1\psi_1 + \cdots + c_k\psi_k + \cdots$$

for $a \leq x \leq b$, where the c_k are real or complex constants and the ψ_k are as above. By Chapter F, Theorem 1, uniform convergence justifies the term-by-term integration in the following calculations.

If we multiply (1) by $r\overline{\psi}_k$ and integrate from a to b, all terms $r\psi_j\overline{\psi}_k$ with $j \neq k$ integrate to 0. The surviving term with $j = k$ gives $c_k\|\psi_k\|^2$, so that finally

$$\tag{2} c_k\|\psi_k\|^2 = \int_a^b rf\overline{\psi_k}\,dx.$$

The constants c_k so defined are the Fourier coefficients and the resulting series (1) is the Fourier series of f. Both the coefficients and the series are sometimes referred to as "generalized." A classical Fourier series is the trigonometric series discussed at the beginning of Section 1.

Taking the conjugate of (2) and multiplying by c_k, we get

$$\|\psi_k\|^2 c_k\overline{c_k} = \int_a^b r\overline{f}c_k\psi_k\,dx.$$

If this is summed on k from 0 to ∞, Equation (1) yields

$$\tag{3} \sum_{k=0}^{\infty} \|\psi_k\|^2|c_k|^2 = \int_a^b r|f|^2\,dx = \|f\|^2.$$

Equation (3) is the (generalized) *Parseval equality*. The corresponding result for trigonometric series was stated by the French mathematician Marc-Antoine Parseval in 1805, hence the name.

By definition, the series with coefficients given by (2) is the Fourier series for f even if it does not converge. The Fourier series is denoted by $S(f)$. An equation such as $S(f) = S(g)$ means that $S(f)$ and $S(g)$ have the same coefficients; it says nothing about convergence. To form the Fourier series of f, it is only necessary that each $rf\psi_j$ be integrable, and this condition is always assumed here.

The notation $f \sim S(f)$ expresses the fact that $S(f)$ is the Fourier series of f without saying that the series converges to f. If the Fourier coefficient c_n of f is denoted by $c_n(f)$ it is obvious from the definition that $c_n(f+g) = c_n(f) + c_n(g)$ and $c_n(kf) = kc_n(f)$. This gives

$$S(f+g) = S(f) + S(g), \qquad S(kf) = kS(f), \qquad k \text{ const.}$$

In other words, the Fourier series for $f + g$ can be obtained by termwise addition, and the series for kf is obtained when the series for f is multiplied termwise by k.

Examples of orthogonal polynomials are given in Section 1, Problem 6 and a series involving Bessel functions is used in the next section. Here we illustrate the calculations for trigonometric Fourier series, which represent the earliest, simplest, and in many respects the most important case.

If n is an integer, $\cos nx$ and $\sin nx$ are solutions of $y'' + n^2 y = 0$ on the interval $0 \le x \le \pi$ with Neumann and Dirichlet boundary conditions respectively. Hence each of these sequences is orthogonal with $r = 1$. It is left for you to check that the Fourier series in the first case is

$$f(x) \sim \frac{a_0}{2} + \sum_{k=1}^{\infty} a_k \cos kx, \qquad a_k = \frac{2}{\pi} \int_0^{\pi} f(x) \cos kx \, dx$$

and in the second it is

$$f(x) \sim \sum_{k=1}^{\infty} b_k \sin kx, \qquad b_k = \frac{2}{\pi} \int_0^{\pi} f(x) \sin kx \, dx.$$

These are called the half-range cosine series and the half-range sine series, respectively. The full-range Fourier series of Section 1 has both sines and cosines and is considered on $(-\pi, \pi)$ rather than on $(0, \pi)$.

The three types of Fourier series bear a simple relation to one another. Let us recall that a function F is even if $F(-x) = F(x)$ and odd if $F(-x) = -F(x)$, both equations being required for all x in the domain of F. In an obvious notation,

$$e + e = e, \qquad o + o = o, \qquad ee = e, \qquad eo = o, \qquad oo = e.$$

We shall establish:

(i) If f is even the full-range Fourier series coincides with the half-range cosine series, and if f is odd the full-range Fourier series coincides with the half-range sine series.

This property permits use of the same letters a_n, b_n for coefficients in the full- and half-range series.

For proof, let F be integrable on $[-\pi, \pi]$. It is easily checked that

$$\int_{-\pi}^{\pi} F(x)\,dx = \gamma \int_{0}^{\pi} F(x)\,dx$$

where $\gamma = 2$ if F is even and $\gamma = 0$ if F is odd. Property (i) is obtained by taking $F(x) = f(x)\cos nx$ or $F(x) = f(x)\sin nx$.

In view of (i), the following theorem can be applied to half-range series, provided f is even or odd:

THEOREM 1. *Let $(-\pi, \pi)$ be divided by points x_k into a finite number of open intervals on each of which $f(x)$ has a bounded derivative. Suppose further that f has period 2π. Then the full-range Fourier series for f converges, at each value of x, to*

(4)
$$\frac{f(x-) + f(x+)}{2}.$$

The proof is postponed to Section 5, but we make a few comments. As elsewhere in this book, $f(x-)$ and $f(x+)$ denote left- and right-hand limits, respectively. The fact that these limits exist follows from Chapter 14, Section 1, Theorem 2. At any point where f is continuous the expression (4) reduces to $f(x)$.

In the theory of trigonometric Fourier series, it is customary to assume that the functions have period 2π and that they have the value given by (4) at all points x where (4) is meaningful. For instance, if f is given on $(0, 2\pi)$ then the periodic extension is, initially, undefined at all integral multiples of 2π. The missing definition at these points is now supplied by (4).

If f satisfies the hypothesis of Theorem 1 and is continuous for all x, a classical theorem known as the Dini-Lipschitz test (not proved here) shows that the convergence is uniform.

Unless the Fourier series is absolutely convergent, Theorem 1 applies only if the terms are written in the normal order, that is, in the order of increasing n. The necessity for some such condition is clear from a theorem of Riemann. It states that a real series that converges, but not absolutely, can be rearranged in such a way as to converge to any specified real number.

EXAMPLE 1. Let $f(x) = \pi/2$ for $0 \leq x \leq \pi/2$ and $f(x) = 0$ elsewhere on $(0, \pi)$. The half-range cosine coefficients are

$$a_n = \int_{0}^{\pi/2} \cos nx\,dx = \frac{\sin(n\pi/2)}{n}, \qquad n \geq 1$$

and $a_0 = \pi/2$. The function $\sin(n\pi/2)$ has the values

$$\sin \frac{n\pi}{2} = 1,\ 0,\ -1,\ 0 \quad | \quad \text{repeat}$$

for $n = 1, 2, 3, 4, \cdots$, where the word "repeat" means that the next four values are the same, the next four are the same, and so on. These give the numerators for a_n and the denominators n are $1, 2, 3, \cdots$. Taking account of the term $a_0/2$ we get the half-range cosine series

$$\frac{\pi}{4} + \frac{\cos x}{1} - \frac{\cos 3x}{3} + \frac{\cos 5x}{5} - \frac{\cos 7x}{7} + \cdots.$$

The sum of the series is even because the terms $\cos nx$ are even, and it has period 2π because the terms do. By Theorem 1, the series converges to the function of period 2π shown graphically in Figure 2(a).

EXAMPLE 2. It is easily checked that the sine coefficients for the function in Example 1 satisfy $nb_n = 1 - \cos(n\pi/2)$. Here

$$1 - \cos \frac{n\pi}{2} = 1,\ 2,\ 1,\ 0 \quad | \quad \text{repeat}$$

and the half-range sine series is

$$\frac{\sin x}{1} + 2\frac{\sin 2x}{2} + \frac{\sin 3x}{3} + \frac{\sin 5x}{5} + 2\frac{\sin 6x}{6} + \frac{\sin 7x}{7} + \frac{\sin 9x}{9} + \cdots$$

The function to which this series converges is shown in Figure 2(b). It is odd because the terms $\sin nx$ are odd and it clearly has period 2π.

EXAMPLE 3. Let $f(x) = \pi/2$ for $0 \le x \le \pi/2$ and $f(x) = 0$ elsewhere on $(-\pi, \pi)$. The periodic extension of this function is half the sum of those

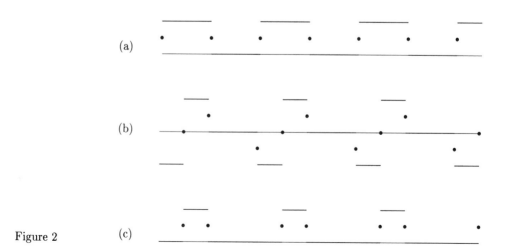

(a)

(b)

Figure 2   (c)

in Figures 2(a) and 2(b), hence the full-range Fourier series is half the sum of the two series obtained in Examples 1 and 2. The function to which this series converges is shown in Figure 2(c).

In many cases the full-range series is best obtained by writing f as a sum of an even and odd function; for example,

$$1 + x + x^2 + x^3 + x^4 + x^5 = (1 + x^2 + x^4) + (x + x^3 + x^5).$$

The two functions on the right are then expanded in half-range cosine and half-range sine series, respectively, and the sum of these series gives the full-range series. For polynomials the calculations are shortened by the following useful result, which is established in Problem 14:

(ii) Let $f(x)$ coincide with a polynomial on $[0, \pi]$ and let

$$\alpha_k = f^{(k)}(0), \qquad \beta_k = f^{(k)}(\pi), \qquad k = 0, 1, 2, \cdots.$$

Then for $n \geq 1$ the coefficient a_n in the half-range cosine series satisfies

$$\frac{\pi}{2} a_n = (-1)^n \left(\frac{\beta_1}{n^2} - \frac{\beta_3}{n^4} + \frac{\beta_5}{n^6} - \cdots \right) - \left(\frac{\alpha_1}{n^2} - \frac{\alpha_3}{n^4} + \frac{\alpha_5}{n^6} - \cdots \right)$$

and the coefficient b_n in the half-range sine series satisfies

$$\frac{\pi}{2} b_n = \left(\frac{\alpha_0}{n} - \frac{\alpha_2}{n^3} + \frac{\alpha_4}{n^5} - \cdots \right) - (-1)^n \left(\frac{\beta_0}{n} - \frac{\beta_2}{n^3} + \frac{\beta_4}{n^5} - \cdots \right).$$

EXAMPLE 4. As an illustration, let us find the half-range cosine series for x^4. The relevant derivatives and their values at 0 and π are

$$4x^3, \qquad 24x; \qquad \alpha_1 = 0, \qquad \alpha_3 = 0; \qquad \beta_1 = 4\pi^3, \qquad \beta_3 = 24\pi.$$

Since $a_0 = 2\pi^4/5$ by direct integration, the first result (ii) gives

$$x^4 \sim \frac{\pi^4}{5} + \sum_{n=1}^{\infty} (-1)^n \left(\frac{8\pi^2}{n^2} - \frac{48}{n^4} \right) \cos nx.$$

By Theorem 1 the series converges to x^4 for $-\pi \leq x \leq \pi$ and to the periodic extension of x^4 outside this interval. See Figure 3. An idea of

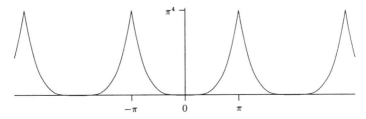

Figure 3

the rate at which the Fourier series converges is given by Figure 4, which shows the function together with the partial sums s_8 and s_{24} of the Fourier cosine series.

In calculations involving Fourier series it is often advantageous to use complex exponentials. If n is an integer, the functions $\psi_n(x) = e^{inx}$ satisfy

$$y'' + n^2 y = 0,$$

$$y(-\pi) = y(\pi), \qquad y'(-\pi) = y'(\pi).$$

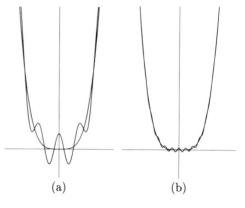

(a)   (b)

Figure 4

By the general theory ψ_m and ψ_n for $m \neq n$ are orthogonal on $(-\pi, \pi)$. An independent verification is given in Problem 2, where it is also seen that $\|e^{inx}\|^2 = 2\pi$. Hence the complex Fourier series is

$$f(x) \sim \sum_{k=-\infty}^{\infty} c_k e^{ikx}, \qquad c_k = \frac{1}{2\pi} \int_{-\pi}^{\pi} f(x) e^{-ikx} \, dx.$$

Since e^{inx} are eigenfunctions for the problem considered at the beginning of Section 1, the complex Fourier series is an alternative form of the real Fourier series obtained there. If f is real, $c_{-k} = \overline{c_k}$ and

$$2c_k = a_k - ib_k, \qquad 2\operatorname{Re} c_k e^{ikx} = a_k \cos kx + b_k \sin kx.$$

Here a_k and b_k are as in Section 1, Equation (3). By the above equations

$$\frac{a_0}{2} + \sum_{k=1}^{n}(a_k \cos kx + b_k \sin kx) = \sum_{k=-n}^{n} c_k e^{ikx} = c_0 + 2\operatorname{Re}\sum_{k=1}^{n} c_k e^{ikx}.$$

Hence either sum can be calculated from the other and the convergence properties of the corresponding infinite series are identical. Note that this equivalence is based on the *symmetric partial sum* of the complex series; that is, the sum from $-n$ to n. Unless the contrary is stated, the symmetric sum is always intended here.

EXAMPLE 5. Let $f(x) = e^{\alpha x}$ on $(-\pi, \pi)$ where α is a real nonzero constant. Then

$$2\pi c_n = \int_{-\pi}^{\pi} e^{\alpha x} e^{-inx} \, dx = \left. \frac{e^{(\alpha - in)x}}{\alpha - in} \right|_{-\pi}^{\pi}.$$

Since $e^{\pm in\pi} = (-1)^n$ division by 2 gives

$$\pi c_n = (\sinh \pi\alpha) \frac{(-1)^n}{\alpha - in} = (\sinh \pi\alpha) \frac{(-1)^n}{\alpha^2 + n^2}(\alpha + in).$$

Hence the complex Fourier series is

$$e^{\alpha x} \sim \frac{\sinh \pi\alpha}{\pi} \sum_{n=-\infty}^{\infty} \frac{(-1)^n}{\alpha^2 + n^2}(\alpha + in)e^{inx}.$$

By Theorem 1 the series converges to $e^{\alpha x}$ for $|x| < \pi$, to $\cosh \pi\alpha$ at $x = \pi$, and to the periodic extension of this function elsewhere.

The complex Fourier series has the following properties:

(iii) If $f(x) = e^{ix}g(x)$ then $c_k(g) = c_{k+1}(f)$.

(iv) If $g(x) = f(x + x_0)$ and f has period 2π, replacing x by $x + x_0$ in the Fourier series for f yields the Fourier series for g.

Property (iii) follows by inspection of the formula for $c_{k+1}(f)$. For (iv), let $c_k = c_k(f)$. Then

$$c_k(g) = \int_0^{2\pi} f(x + x_0)e^{-ikx}\, dx = \int_{x_0}^{2\pi+x_0} f(t)e^{-ik(t-x_0)}\, dt = e^{ikx_0}c_k.$$

The last equation follows from the periodicity and shows that the complex Fourier series for g is

$$\sum_{k=-\infty}^{\infty} c_k e^{ikx_0} e^{ikx} = \sum_{k=-\infty}^{\infty} c_k e^{ik(x_0+x)}.$$

PROBLEMS

1P. If f has period 2π show that the full-range Fourier coefficients can be computed by integration over any interval of length 2π. Hence we can consider the Fourier series on $(-\pi, \pi)$ or on $(0, 2\pi)$ as may be convenient.

2. Verify the orthogonality of $\{e^{inx}\}$ on $(-\pi, \pi)$ by direct evaluation of the relevant integral and also show that $\|e^{inx}\|^2 = 2\pi$.

3. (ab) Check that the Parseval equations for half-range cosine and sine series are, respectively,

$$\frac{a_0^2}{2} + \sum_{k=1}^{\infty} a_k^2 = \frac{2}{\pi} \int_0^{\pi} f(x)^2\, dx \qquad \sum_{k=1}^{\infty} b_k^2 = \frac{2}{\pi} \int_0^{\pi} f(x)^2\, dx$$

4. (cd) Check that the Parseval equations for the full-range Fourier series and the complex Fourier series are respectively

$$\frac{a_0^2}{2} + \sum_{n=1}^{\infty}(a_n^2 + b_n^2) = \frac{1}{\pi}\int_{-\pi}^{\pi} f(x)^2\, dx \qquad \sum_{k=-\infty}^{\infty} |c_k|^2 = \frac{1}{2\pi}\int_0^{2\pi} |f(x)|^2\, dx$$

5. (ab) Obtain the expansions

$$|x| \sim \frac{\pi}{2} - \frac{4}{\pi}\sum_{n=1}^{\infty}\frac{\cos nx}{(2n-1)^2} \qquad |\sin x| \sim \frac{2}{\pi} - \frac{4}{\pi}\sum_{n=1}^{\infty}\frac{\cos 2nx}{(2n)^2 - 1}$$

6. (ab) Obtain the following expansions for x^2 and x^6, respectively:

$$\frac{\pi^2}{3} + \sum_{n=1}^{\infty}(-1)^n\frac{4\cos nx}{n^2} \qquad \frac{\pi^6}{7} + \sum_{n=1}^{\infty}(-1)^n\left(\frac{12\pi^4}{n^2} - \frac{240\pi^2}{n^4} + \frac{1440}{n^6}\right)\cos nx$$

7. (abc) Expand x, x^3, x^5 in half-range sine series and verify that the results agree with those obtained by differentiating the half-range cosine series for $x^2/2$, $x^4/4$, $x^6/6$ as given by Problem 6 and Example 4.

8. Consider the functions $x(\pi - x)$, $(\pi - x)^2$, $\pi(\pi - x)$.

(a) Find the half-range cosine series for each function and check that the third series is the sum of the other two.

(b) Do Part (a) with the word "cosine" replaced by "sine".

9. (ab) Proceed as in Problem 8: $x^3(\pi^2 - x^2)$, $x^2(x^3 - \pi^3)$, $(\pi x)^2(x - \pi)$.

10. Graph the function to which the Fourier series converges (a) in Problems 5,6,7 and (b) in Problem 8. (Seven and six graphs, respectively.)

11. Using Re $(\alpha + in)e^{inx} = \alpha\cos nx - n\sin nx$, write the real form of the series in Example 5. Assuming that β is not an integer, set $\alpha = i\beta$ in your result and deduce the following series for $\cos\beta x$ and $\sin\beta x$, respectively:

$$\frac{\sin\pi\beta}{\pi\beta}\left(1 + 2\sum_{n=1}^{\infty}\frac{(-1)^n\beta^2}{\beta^2 - n^2}\cos nx\right) \qquad \frac{2\sin\pi\beta}{\pi}\sum_{n=1}^{\infty}\frac{(-1)^n n}{n^2 - \beta^2}\sin nx$$

12. (ab) Obtain the two series of Problem 11 by direct use of the Euler-Fourier formulas.

13. (a) Show that the full-range Fourier series for the function $f(x) = 1$ cannot be obtained by differentiating that for x.

(b) If f' is continuous on $[-\pi, \pi]$ and $f(-\pi) = f(\pi)$ show that the complex Fourier series $S(f')$ can be obtained by formal differentiation of $S(f)$. This yields a similar result for real full- and half-range Fourier series.

Hint: Integrate the formula for $c_k(f')$ by parts. This problem has nothing to do with convergence.

14. (a) If f' is continuous, use integration by parts to show that the coefficients for the half-range Fourier series satisfy

$$b_n(f') = -na_n(f) \qquad \pi n b_n(f) = 2f(\pi)(-1)^{n+1} + 2f(0) + \pi a_n(f')$$

(b) Express $b_n(f)$ in terms of $b_n(f'')$ in Part (a) and iterate to establish the equation (iv) for b_n. Applying this to f' yields (iv) for a_n.

15. Open-ended problem. By taking $x = 0$, $\pi/2$, or π in the Fourier series of this section, or by use of the Parseval equations, get as many expressions for powers of π as you can. Here are four examples:

$$\frac{\pi}{4} = \sum_{n=1}^{\infty} \frac{(-1)^{n-1}}{2n-1} \qquad \frac{\pi^2}{6} = \sum_{n=1}^{\infty} \frac{1}{n^2} \qquad \frac{\pi^2}{8} = \sum_{n=0}^{\infty} \frac{1}{(2n+1)^2} \qquad \frac{\pi^4}{90} = \sum_{n=1}^{\infty} \frac{1}{n^4}$$

---

3. Separation of variables

One of the main uses of Sturm-Liouville theory is to solve partial differential equations. Three typical examples are

$$u_t = c^2 u_{xx}, \qquad u_{xx} + u_{yy} = 0, \qquad u_{tt} = c^2 u_{xx}$$

where the subscripts denote partial differentiation. The first of these equations describes heat flow in a uniform insulated bar. The constant c depends on the conductivity, density, and heat capacity of the material and the solution $u(x, t)$ denotes the temperature at point x of the bar at time t. The second equation describes the steady-state temperature $u(x, y)$ at point (x, y) of a uniform sheet and the third describes wave motion of small amplitude on a uniform stretched string. Here the constant c is determined by the tension and density and turns out to be the same as the speed of propagation of the waves.

By a change of scale, constants such as c in the above equations can be reduced to 1, and a length ℓ can be reduced to 1 or π. This use of *dimensionless variables* does not diminish the mathematical content of our investigation, and as seen in Problem 22, it involves no loss of generality.

When partial differential equations are applied to scientific problems the procedure involves the following steps:

(i) Derive the equation from the underlying physical principles.

(ii) Construct a solution satisfying the equation and the boundary conditions.

(iii) Prove continuity in the relevant region including its boundary.

(iv) Establish uniqueness of the solution.

The second item is the only one that depends on Sturm-Liouville theory and is the only one that will be discussed here.

The connection between Sturm-Liouville theory and partial differential equations rests on the following fact. Let F be a function of x alone and G a function of t alone, each with a nonempty domain of definition. If

$$F(x) = G(t)$$

for all values of the variables for which the two sides are defined, then both F and G are constant. To see why, choose any value t_0 in the domain of G. The resulting equation

$$F(x) = G(t_0)$$

shows that F is constant. Similarly, G is constant.

It is said that the variables in $F(x) = G(t)$ are separated, the common constant value of F and G is the separation constant, and the procedure described in this section is called the method of separation of variables. Postponing questions about convergence, we illustrate in four examples.

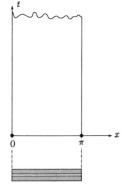

EXAMPLE 1. Heat flow in a bar. A uniform insulated bar extending from $x = 0$ to $x = \pi$ has its ends kept at temperature 0. The initial temperature is $f(x)$ at point x of the bar. We want to find the temperature $u(x,t)$ at x at any subsequent time t. See Figure 5, which shows the bar with the relevant part of the (x, t) plane above it.

This problem can be formulated as follows:

Figure 5

$$u_t = u_{xx}, \qquad u(0,t) = 0, \qquad u(\pi,t) = 0, \qquad u(x,0) = f(x).$$

The differential equation holds for $0 < x < \pi$, $t > 0$ and expresses the law of heat flow. The middle equations describe the boundary conditions and the last equation says that the initial temperature is $f(x)$.

Let us try $u(x, t) = R(x)S(t)$. Substitution into the differential equation gives successively

(1) $$RS' = R''S, \qquad \frac{R''}{R} = \frac{S'}{S}, \qquad \frac{R''}{R} = \frac{S'}{S} = -\lambda.$$

The separation constant is denoted by $-\lambda$ because λ will turn out to be an eigenvalue. In fact,

$$R'' + \lambda R = 0, \qquad R(0) = R(\pi) = 0$$

where the boundary conditions are inherited from those in the original problem. This is a Sturm-Liouville equation with Dirichlet conditions and eigenvalue λ. Aside from a multiplicative constant, the solutions are

$$R(x) = \sin nx, \qquad \lambda = n^2, \qquad n = 1, 2, 3, \cdots.$$

We get $S(t) = e^{-\lambda t}$ from the last equation (1), so that

$$S(t)R(x) = e^{-n^2 t}\sin nx.$$

This function $S(t)R(x)$ is called a *partial solution* because it satisfies part of the problem. Namely, for each n it satisfies both the differential equation and the boundary conditions.

To satisfy the initial condition we try a superposition of partial solutions,

$$u(x, t) = \sum_{n=1}^{\infty} b_n e^{-n^2 t}\sin nx.$$

The series takes the value 0 when $x = 0$ or $x = \pi$, so that the boundary conditions hold for all choices of $\{b_n\}$. Setting $t = 0$ yields

$$f(x) = \sum_{n=1}^{\infty} b_n \sin nx, \quad \text{hence} \quad b_n = \frac{2}{\pi}\int_0^\pi f(\xi)\sin n\xi \, d\xi$$

where we have written the integration variable as ξ to avoid confusion with x. If the above value b_n is substituted into the series for $u(x, t)$ we get

$$u(x, t) = \int_0^\pi K(t, x, \xi)f(\xi) \, d\xi, \qquad K(t, x, \xi) = \frac{2}{\pi}\sum_{n=1}^{\infty} e^{-n^2 t}\sin nx \sin n\xi.$$

The function K is called the *kernel* and the formula is the *kernel representation* of the solution.

EXAMPLE 2.   Heat flow with radiation. We now assume that the bar of Example 1 extends from $x = 0$ to $x = 1$. The temperature at the end $x = 0$ is held to the value 0 as before, but the end $x = 1$ radiates heat into a medium of temperature 0. If the radiation takes place at a sufficiently low temperature, it can be shown that the boundary conditions are

$$u(0, t) = 0, \qquad u_x(1, t) = -ku(1, t)$$

where k is a positive constant. The procedure used in Example 1 leads to the partial solution $S(t)R(x)$ where $S' = -\lambda S$ and

$$R'' + \lambda R = 0, \qquad R(0) = 0, \qquad R'(1) + kR(1) = 0.$$

With $\lambda = \omega_n^2$ and $\tan \omega_n = -k\omega_n$ as in Section 1, Example 3, we get

$$S(t)R(x) = e^{-\omega_n^2 t} \sin \omega_n x.$$

This partial solution satisfies the differential equation and the boundary conditions.

To take care of the initial condition $u(x, 0) = f(x)$ we try

$$(2) \qquad u(x, t) = \sum_{n=1}^{\infty} b_n e^{-\omega_n^2 t} \sin \omega_n x.$$

Since the functions $\sin \omega_n x$ are orthogonal, the initial condition gives

$$f(x) = \sum_{n=1}^{\infty} b_n \sin \omega_n x, \qquad b_n = \frac{1}{\|\sin \omega_n x\|^2} \int_0^1 f(\xi) \sin \omega_n \xi \, d\xi.$$

The value of $\|\sin \omega_n x\|^2$ is found by setting $\omega = \omega_n$ in

$$(3) \qquad \int_0^1 (\sin \omega \xi)^2 \, d\xi = \frac{1}{2}\left(1 - \frac{\sin 2\omega}{2\omega}\right), \qquad \omega \neq 0.$$

Our next example concerns an important problem known as the Dirichlet problem. The *Laplace equation* in two dimensions is $u_{xx} + u_{yy} = 0$. The *Dirichlet problem* is to find a function satisfying Laplace's equation in a given region subject to prescribed values on the boundary. It arises in connection with such topics as heat conduction, ideal fluid flow, diffusion, electrostatics, and gravitation. Aside from physical applications, the problem plays a prominent role in a branch of mathematics known as potential theory.

EXAMPLE 3.   The Dirichlet problem for a disk. We shall solve

$$u_{xx} + u_{yy} = 0, \qquad x^2 + y^2 < 1,$$

with given values on the boundary. As one of several interpretations, this gives the steady-state temperature of a uniform disk when the temperature on the boundary is prescribed.

In polar coordinates $x = r\cos\theta$, $y = r\sin\theta$ the function $U(r, \theta) = u(x, y)$ satisfies

(4) $$r(rU_r)_r + U_{\theta\theta} = 0, \qquad U(1, \theta) = f(\theta).$$

where $f(\theta)$ is a given function with period 2π. The first equation is derived in Problem 16 and the second expresses the boundary condition.

We try $U(r, \theta) = R(r)S(\theta)$ and obtain

$$r(rR')'S + RS'' = 0, \qquad \frac{r(rR')'}{R} = -\frac{S''}{S} = \lambda$$

where λ is the separation constant. This gives

$$S'' + \lambda S = 0, \qquad S(\theta + 2\pi) = S(\theta).$$

The periodicity results from the fact that u is a single-valued function of position. Namely, increasing θ by 2π brings us back to the same point (x, y), hence gives the same value for the solution as before.

From Sections 1 and 2 we know that $\lambda = n^2$ and that, aside from a constant multiplier,

$$S(\theta) = e^{in\theta}, \qquad n = 0, \pm 1, \pm 2, \cdots.$$

The equation $r(rR')' = \lambda R$ is of Euler type and is solved by the trial solution $R(r) = r^m$. When $\lambda = n^2$, it is easily checked that $m = \pm n$. Negative exponents are rejected because they make $R(r)$ infinite at the origin. Thus $m = |n|$ and our partial solution is

$$r^{|n|}e^{in\theta}.$$

This satisfies the partial differential equation and is periodic.

To take account of the boundary condition, we try a superposition

$$U(r, \theta) = \sum_{n=-\infty}^{\infty} c_n r^{|n|} e^{in\theta}.$$

The boundary condition leads to

$$f(\theta) = \sum_{n=-\infty}^{\infty} c_n e^{in\theta}, \qquad c_n = \frac{1}{2\pi} \int_{-\pi}^{\pi} e^{-in\phi} f(\phi)\, d\phi$$

and the kernel representation is obtained by substituting these coefficients into the series. This gives

$$U(r, \theta) = \int_{-\pi}^{\pi} K(r, \theta - \phi) f(\phi)\, d\phi$$

where

$$2\pi K(r,\theta) = \sum_{n=-\infty}^{\infty} r^{|n|} e^{in\theta} = 1 + 2\mathrm{Re} \sum_{n=1}^{\infty} r^n e^{in\theta}.$$

With $\rho = re^{i\theta}$, the formula for the sum of a geometric series yields

$$2\pi K(r,\theta) = 1 + 2\mathrm{Re}\,\frac{\rho}{1-\rho} = \mathrm{Re}\,\frac{1+\rho}{1-\rho} = \mathrm{Re}\,\frac{1+\rho}{1-\rho}\frac{1-\bar{\rho}}{1-\bar{\rho}}.$$

By use of the equations

$$\rho\bar{\rho} = r^2, \qquad \rho - \bar{\rho} = 2ir\sin\theta, \qquad \rho + \bar{\rho} = 2r\cos\theta$$

the value of $K(r,\theta)$ is easily found. The final result is

$$U(r,\theta) = \frac{1}{2\pi} \int_{-\pi}^{\pi} \frac{1-r^2}{1-2r\cos(\theta-\phi)+r^2} f(\phi)\,d\phi.$$

This is *Poisson's formula*, named after the French mathematician Simeon Denis Poisson who derived the formula and investigated some of its uses.

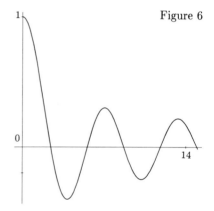

Figure 6

Our next example leads to an important series involving Bessel functions. As a preliminary, we derive some properties of the function $R(x) = J_0(kx)$ where k is a positive constant.

Setting $\nu = 0$ in Section 1, Example 1 gives a differential equation for $J_0(x)$, which becomes

$$(kx)^2 J_0''(kx) + (kx)J_0'(kx) + (kx)^2 J_0(kx) = 0$$

when x is replaced by kx. Since $R'(x) = kJ_0'(kx)$ and $R''(x) = k^2 J_0''(kx)$, this gives

(5) $$(xR')' + \lambda xR = 0 \quad \text{where} \quad \lambda = k^2.$$

It is suggested by Figure 6, and proved in Section 4 Problem 5, that $J_0(x)$ has infinitely many distinct positive zeros k_n. We shall show that

(6) $$\int_0^1 J_0(k_n x)J_0(k_m x)x\,dx = \begin{cases} 0, & \text{if } m \neq n; \\ J_0'(k_n)^2/2, & \text{if } m = n. \end{cases}$$

In view of (5), the result (6) for $m \neq n$ follows from the general theory with $r(x) = x$ and mixed boundary conditions (M). Namely, $p(x) = x$ vanishes at $x = 0$ and $R(x) = J_0(k_n x)$ vanishes at $x = 1$.

For the case $m = n$, multiply the equation $(xR')' = -k^2 xR$ by xR', integrate, and integrate by parts. The result is

$$\frac{1}{2}(xR')^2\Big|_0^1 = -k^2 \int_0^1 x^2 RR' \, dx = -\frac{k^2 x^2 R^2}{2}\Big|_0^1 + k^2 \int_0^1 R^2 x \, dx.$$

Setting $k = k_n$, we get $R(1) = 0$ and $R'(1) = k_n J_0'(k_n)$. This gives (6) for $m = n$. Equation (6) leads to the *Fourier-Bessel series*

$$f(x) = \sum_{n=1}^{\infty} a_n J_0(k_n x), \qquad a_n = \frac{2}{J_0'(k_n)^2} \int_0^1 f(\xi) J_0(k_n \xi)\xi \, d\xi.$$

EXAMPLE 4. The circular drumhead. In polar coordinates let $u(r, \theta, t)$ denote the vertical displacement of a vibrating circular membrane at (r, θ) and at time t. For simplicity we consider the radially symmetric case, $u = u(r, t)$, and we formulate the problem as follows:

$$ru_{tt} = (ru_r)_r, \qquad u(1, t) = 0, \qquad u(r, 0) = f(r), \qquad u_t(r, 0) = 0.$$

The first equation describes the physics of small vibrations and the second is a boundary condition expressing the fact that the rim of the drumhead is fixed. The last two equations state that the initial displacement is given by $f(r)$ and that the initial velocity is 0.

The trial solution $u(r, t) = R(r)S(t)$ leads to

$$rRS'' = (rR')'S, \qquad \frac{S''}{S} = \frac{(rR')'}{rR} = -\lambda.$$

The equation for R is the same as (5) with r replacing x. Hence a solution is $R(r) = J_0(kr)$. In view of the boundary condition $R(1) = 0$ we set $k = k_n$ and get

$$R(r)S(t) = J_0(k_n r)(a_n \cos k_n t + b_n \sin k_n t).$$

This partial solution satisfies the differential equation and the boundary condition. By a brief calculation, which is left for you,

$$u(r, t) = \sum_{n=1}^{\infty} a_n J_0(k_n r) \cos k_n t, \qquad a_n = \frac{2}{J_0'(k_n)^2} \int_0^1 f(\rho) J_0(k_n \rho)\rho \, d\rho.$$

Having explained the gist of the method, we now say a few words about convergence. In problems that lead to a classical Fourier series it suffices to suppose that f or g, as the case may be, is continuous on the relevant interval and satisfies the hypothesis of Section 2, Theorem 1. This ensures that the infinite series given by separation of variables assumes the desired initial or boundary values.

Many applications do not lead to a classical Fourier series, but do lead to a regular Sturm-Liouville problem in the sense of the following definition:

DEFINITION. The Sturm-Liouville problem is regular if p, q, r satisfy the basic hypothesis of Section 1 on the closed interval $a \leq x \leq b$ and the boundary conditions are of type (L).

In such cases, we agree that f or g is so restricted as to come within the scope of the following:

THEOREM 1. *For every regular Sturm-Liouville problem there is a sequence of eigenvalues $\lambda_0 < \lambda_1 < \lambda_2 < \lambda_3 < \cdots$ with $\lambda_n \to \infty$ and a corresponding sequence of eigenfunctions $\{\psi_n\}$. If f satisfies the boundary conditions and has a continuous derivative for $a \leq x \leq b$, the generalized Fourier series of f converges absolutely and uniformly to f on $[a, b]$.*

The proof is not given in this book, though related results can be found in Sections 4 and 5.

Theorem 1 fails for Fourier-Bessel series since the function $p(x) = x$ vanishes at $x = 0$. The problem of the harmonic oscillator in quantum mechanics involves Hermite polynomials on an infinite interval, and again Theorem 1 does not apply. Such cases come under the heading of *singular Sturm-Liouville problems*. They are best considered on their own merits, rather than as part of a comprehensive theory. We state without proof that the Fourier-Bessel series converges to f if f is continuously differentiable for $0 \leq x \leq 1$ and $f(1) = 0$.

In nearly all applications, the initial and boundary conditions are to hold as limits; for example,

$$\lim_{t \to 0+} u(x, t) = f(x) \quad \text{or} \quad \lim_{r \to 1-} U(r, \theta) = f(\theta).$$

Theorems yielding such limit relations from convergence on the boundary are called *Abelian*; the first result of this kind was established by Niels Abel in connection with power series. Although details will not be given here, we mention that, in general, the Abelian limit allows a larger class of functions f than those to which the Fourier series converges.

It is seen in Section 5 that the generalized Fourier coefficients tend to zero. Hence the presence of factors such as

$$e^{-n^2 t}, \qquad e^{-\omega_n^2 t}, \qquad |r|^n, \qquad e^{-ny}$$

in our series solutions allows us to deduce from the theory of uniform convergence (Chapter F) that the partial differential equation is satisfied. If factors of this kind are missing, the problem is dealt with by restricting f to have a rapidly convergent Fourier series or by interpreting the differential equation in a generalized sense. Here too, we are obliged by limitations of space to omit details.

PROBLEMS

If this is your first encounter with the method of separating variables, it is suggested that you ignore questions of convergence and concentrate on getting the correct form of the solution.

1. In Example 1 suppose the initial temperature is $f(x) = x(\pi - x)$; note that this satisfies the boundary conditions. Find $u(x,t)$ and check that the temperature at the midpoint of the bar is

$$u\left(\frac{\pi}{2}, t\right) = \frac{8}{\pi}\left(\frac{e^{-t}}{1^3} - \frac{e^{-3^2 t}}{3^3} + \frac{e^{-5^2 t}}{5^3} - \frac{e^{-7^2 t}}{7^3} + \cdots\right)$$

2. If both ends of the bar in Example 1 are insulated the boundary conditions are $u_x(0,t) = u_x(\pi,t) = 0$. Obtain the solution

$$u(x,t) = \frac{a_0}{2} + \sum_{n=1}^{\infty} a_n e^{-n^2 t} \cos nx, \qquad a_n = \frac{2}{\pi}\int_0^{\pi} f(\xi)\cos n\xi\, d\xi$$

3P. This pertains to $R'' + \lambda R = 0$. For the boundary conditions $R'(0) = R(\pi) = 0$ obtain the solutions $\cos \omega_n x$ with $2\omega_n = 2n - 1$ and for the boundary conditions $R(-\pi) = R(\pi) = 0$ obtain additional solutions $\sin nx$. In both cases n is an arbitrary positive integer.

4. In Example 1 suppose the end $x = 0$ is insulated and the end $x = \pi$ is kept at temperature 0, so that $u_x(0,t) = u(\pi,t) = 0$. With ω_n as in Problem 3, derive the following solution by separation of variables:

$$u(x,t) = \sum_{n=1}^{\infty} a_n e^{-\omega_n^2 t} \cos \omega_n x, \qquad a_n = \frac{2}{\pi}\int_0^{\pi} f(\xi)\cos \omega_n \xi\, d\xi$$

5. The method of images. A bar extending from $-\pi$ to π has both ends at temperature 0, and the initial temperature is an even function $f(x)$. By symmetry, there is no heat flow across the midpoint, hence the right-hand half of this bar satisfies the same boundary-value problem as that in Problem 4. Using the functions in Problem 3, verify this equivalence. See Figure 7.

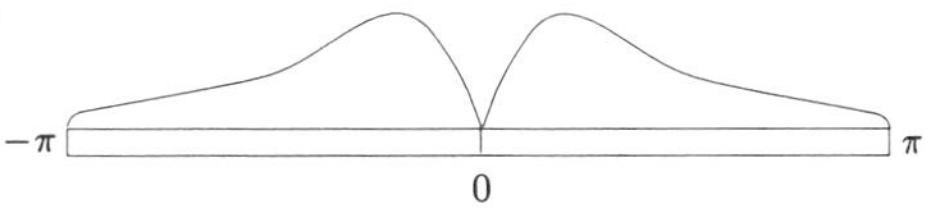

Figure 7

6. Let Example 2 be modified as follows: The bar extends from $x = 0$ to $x = \pi$ and the boundary conditions are $u_x(0,t) = ku(0,t)$, $u(\pi,t) = 0$. (Hence there is a change of scale and the bar is turned end for end.) Solve this version of Example 2. No answer is given, but when $k = 0$ your result should reduce to that of Problem 4.

7. (a) Figure 8 shows a semi-infinite strip of width π in the (x, y) plane. The conditions for steady-state heat flow in the strip are

$$u_{xx} + u_{yy} = 0, \qquad u(0, y) = u(\pi, y) = 0, \qquad u(x, 0) = f(x)$$

The first equation holds for $0 < x < \pi$ and $0 < y < \infty$. It describes the law of steady-state heat flow. The next three equations give boundary conditions on the three edges of the strip. By separation of variables obtain two distinct solutions when $f(x) = \sin x$.

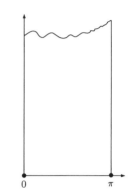

Figure 8

(b) In Part (a) let $u(x, y) \to 0$ uniformly in x as $y \to \infty$. Under this hypothesis it can be shown that there is at most one continuous solution. Reject partial solutions involving e^{ny} with $n > 0$ and get

$$u(x, y) = \sum_{n=1}^{\infty} b_n e^{-ny} \sin nx, \qquad b_n = \frac{2}{\pi} \int_0^{\pi} f(\xi) \sin n\xi \, d\xi$$

8. If $f(x) = \pi/4$ in Problem 7, the boundary conditions at the corners are mutually contradictory. Ignoring this difficulty, obtain the solution

$$u(x, y) = e^{-y} \sin x + e^{-3y} \frac{\sin 3x}{3} + e^{-5y} \frac{\sin 5x}{5} + \cdots$$

9P. The vertical displacement $u(x, t)$ of a vibrating string extending from 0 to π satisfies

$$u_{xx} = u_{tt}, \qquad u(0, t) = 0, \qquad u(\pi, t) = 0$$

The first equation holds for $0 < x < \pi$, $0 < t < \infty$ and describes the physics of small oscillations. The next two equations are boundary conditions stating that the endpoints of the string remain fixed throughout the motion. Obtain the partial solutions $\sin nx \cos nt$, $\sin nx \sin nt$ where n is an integer.

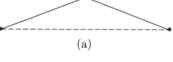

(a)

(b)

Figure 9

10. The plucked string. In Problem 9 let $u(x, 0) = f(x)$, $u_t(x, 0) = 0$. This means that the string is given an initial displacement and is gently released, as in a harp; see Figure 9(a). Obtain the solution

$$u(x, t) = \sum_{n=1}^{\infty} b_n \cos nt \sin nx, \qquad b_n = \frac{2}{\pi} \int_0^{\pi} f(\xi) \sin n\xi \, d\xi$$

11. The struck string. In Problem 9 let $u(x,0) = 0$, $u_t(x,0) = g(x)$. This means that the string has initial displacement zero but is given an initial velocity, as in a piano; see Figure 9(b). Obtain the solution

$$u(x,t) = \sum_{n=1}^{\infty} b_n \sin nt \sin nx, \qquad b_n = \frac{2}{n\pi} \int_0^{\pi} g(\xi) \sin n\xi \, d\xi$$

The sum of this function and that in Problem 10 solves the string problem with arbitrary initial displacement and velocity.

12. A uniform rectangle of width π and height b has its top edge at temperature $f(x)$ and the remaining three edges at temperature 0 as shown in Figure 10. The steady-state temperature u satisfies $u_{xx} + u_{yy} = 0$ for $0 < x < \pi$, $0 < y < b$ with the boundary conditions

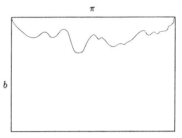

Figure 10

$$u(x,b) = f(x), \qquad u(0,y) = 0, \qquad u(x,0) = 0, \qquad u(\pi,y) = 0$$

Separate variables and express the solution by means of the kernel

$$K(x,y,\xi) = \frac{2}{\pi} \sum_{n=1}^{\infty} \frac{\sinh ny}{\sinh nb} \sin nx \sin n\xi$$

13. Explain how to solve the Dirichlet problem for a square by four uses of the result of Problem 12 with $b = \pi$.

14. Suppose the two lateral edges of the rectangle in Problem 12 are insulated, so that $u_x(0,y) = u_x(\pi,y) = 0$ while the other two boundary conditions are still $u(x,0) = 0$, $u(x,b) = f(x)$. Obtain the solution

$$u(x,y) = \frac{a_0}{2}y + \sum_{n=1}^{\infty} a_n \cos nx \sinh ny$$

Express the coefficients a_n in terms of f, evaluate when $f(x) = c$, a constant, and verify that the resulting solution satisfies both the differential equation and the boundary conditions.

15. By taking $r = 0$ in the result of Example 3 show that the value of U at the origin is the average of its values on the circumference.

16. Verify the symbolic formulas $rD_r = xD_x + yD_y$, $D_\theta = -yD_x + xD_y$. If $U(r,\theta) = u(x,y)$ is twice differentiable deduce that

$$r(rU_r)_r + U_{\theta\theta} = (xD_x + yD_y)^2 u + (-yD_x + xD_y)^2 u = r^2(u_{xx} + u_{yy})$$

17. (a) Fill in the missing details in the discussion of (6) and Example 4.

(b) If the initial conditions in Example 4 are $u(r,0) = 0$, $u_t(r,0) = g(t)$, get the solution

$$u(r,t) = \sum_{n=1}^{\infty} b_n J_0(k_n r) \sin k_n t, \qquad b_n = \frac{2}{k_n J_0'(k_n)^2} \int_0^1 g(\rho) J_0(k_n \rho) \rho \, d\rho$$

18. The displacement v of an elastic membrane subject to uniform gas pressure p satisfies $v_{tt} + p = v_{xx} + v_{yy}$. If the membrane is circular show that the substitution $u = v - p(r^2 - 1)/4$ leads, in the radially symmetric case, to a problem like that in Example 4 or Problem 17.

19. Transient heat flow in a cylinder. The outside of a uniform circular cylinder is kept at the temperature 0 and the initial temperature is a radial function $f(r)$ at distance r from the axis. In polar coordinates the conditions for the temperature $u(r,t)$ are

$$ru_t = (ru_r)_r, \qquad u(1,t) = 0, \qquad u(r,0) = f(r)$$

Set $u(r,t) = R(r)S(t)$, separate variables, and get

$$u(r,t) = \sum_{n=1}^{\infty} a_n e^{-k_n^2 t} J_0(k_n r), \qquad a_n = \frac{2}{J_0'(k_n)^2} \int_0^1 f(\rho) J_0(k_n \rho) \rho \, d\rho$$

20. Steady-state temperature in a cylinder. In cylindrical coordinates (r, θ, z) the steady-state temperature of a uniform semi-infinite circular cylinder depends only on (r, z) and satisfies

$$(ru_r)_r + ru_{zz} = 0, \qquad u(1,z) = 0, \qquad u(r,0) = f(r)$$

Assuming $u \to 0$ as $z \to \infty$ derive the solution

$$u(r,z) = \sum_{n=1}^{\infty} a_n e^{-k_n z} J_0(k_n r), \qquad a_n = \frac{2}{J_0'(k_n)^2} \int_0^1 f(\rho) J_0(k_n \rho) \rho \, d\rho$$

21. The oscillating chain. This problem was proposed and solved in 1732 by the Swiss mathematician Daniel Bernoulli. It is important both in the history of eigenfunction expansions and in the history of Bessel functions.

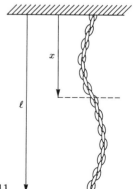

As illustrated in Figure 11, suppose a uniform flexible chain hangs under its own weight and oscillates while remaining in a plane. The lateral displacement u in the case of small oscillations satisfies

$$\rho u_{tt} = (\tau u_x)_x, \qquad \tau = \rho g(\ell - x), \qquad u(0,t) = 0$$

Figure 11

where ℓ is the length of the chain, ρ its density, and g the acceleration of gravity. Check that the function U defined on the left below satisfies the equation on the right:

$$u(x,t) = U(z,t), \qquad \ell - x = z^2, \qquad 4zU_{tt} = g(zU_z)_z$$

Setting $U(z,t) = R(z)S(t)$ obtain partial solutions

$$U(z,t) = J_0(kz)(c_1 \cos\omega t + c_2 \sin\omega t)$$

where $\omega^2 = k^2 g/4$ and $k\sqrt{\ell} = k_n$. If $\ell = 1$, as can be assumed without loss of generality, express $U(z,t)$ in the form

$$\sum_{n=1}^{\infty} a_n J_0(k_n z) \cos\frac{k_n}{2}\sqrt{g}t, \qquad \sum_{n=1}^{\infty} b_n J_0(k_n z) \sin\frac{k_n}{2}\sqrt{g}t$$

when $U(z,0) = f(z)$, $U_t(z,0) = 0$, and $U(z,0) = 0$, $U_t(z,0) = g(z)$, respectively. The coefficients are determined as in Example 4 or Problem 17.

22. A generalization of Example 1 on the interval $0 < x < \ell$ is

$$v_t = c^2 v_{xx}, \qquad v(0,t) = v(\ell,t) = 0, \qquad v(x,0) = g(x)$$

where c is constant. Check that the substitution $v(x,t) = u(\alpha x, \beta t)$ reduces this problem to that in Example 1 if $\alpha = \pi/\ell$, $\beta = (c\alpha)^2$. Deduce from Example 1 with $f(x) = g(\ell x/\pi)$ that

$$v(x,t) = \sum_{n=1}^{\infty} b_n e^{-(n\pi c/\ell)^2 t} \sin\frac{n\pi x}{\ell}, \qquad b_n = \frac{2}{\pi}\int_0^\pi g\left(\frac{\ell\xi}{\pi}\right)\sin n\xi\, d\xi$$

23. Rewrite the formula for b_n in Problem 22 by setting $\eta = \ell\xi/\pi$ and obtain this solution directly by separation of variables; do not use Example 1.

---

4. Comparison

One way to study a Sturm-Liouville problem is by comparison with another Sturm-Liouville problem whose solutions are known. This procedure is based on the following proposition:

(i) If $uTv \geq vTu$ and $p(a)w(a) \geq 0 \geq p(b)w(b)$ then $w = 0$.

For proof, the inequality $uTv \geq vTu$ gives $(pw)' \geq 0$ by Lagrange's identity. Hence pw is (weakly) increasing and the other inequalities (i) yield

$$0 \leq p(a)w(a) \leq p(x)w(x) \leq p(b)w(b) \leq 0, \qquad a < x < b.$$

Thus $pw = 0$ and therefore $w = 0$.

The following consequence of (i) is *Sturm's comparison theorem*:

THEOREM 1. *Let α and β be two consecutive zeros of v and on (α, β) let*

$$Tu + P(x)u = 0, \qquad Tv + Q(x)v = 0, \qquad P(x) \geq Q(x).$$

Then $u(x)$ has a zero on (α, β) or $u(x) = cv(x)$ on (α, β).

If u has no zero on (α, β) we can change the sign of u and v, if necessary, so as to make $u > 0$ and $v > 0$ on (α, β). In that case

$$uTv - vTu = (P - Q)uv \geq 0, \qquad \alpha < x < \beta,$$

and the graphs of u and v are as shown in Figure 12. The figure gives

$$u(\alpha) \geq 0, \ v'(\alpha) \geq 0$$

$$u(\beta) \geq 0, \ v'(\beta) \leq 0.$$

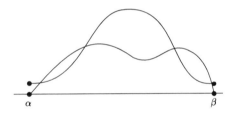

Hence $w(\alpha) \geq 0$, $w(\beta) \leq 0$. From Proposition (i) we get $w = 0$, hence $(u/v)' = 0$, hence $u = cv$. This completes the proof.

Figure 12

EXAMPLE 1. Let u be an eigenfunction for λ and v for μ. If $\lambda > \mu$, there is at least one zero of u between any two consecutive zeros of v. This follows from Theorem 1 with $P = \lambda r$, $Q = \mu r$. The condition $\lambda > \mu$ makes $P > Q$ and rules out the possibility that $u = cv$.

EXAMPLE 2. For $|x| < \infty$ suppose $u'' + P(x)u = 0$ and $P(x) \geq \omega^2$ with $\omega > 0$. Then u has at least one zero in every closed interval of length π/ω. To see why, let (α, β) be a given interval of length π/ω and let

$$v'' + \omega^2 v = 0, \qquad v = \sin \omega(x - \alpha).$$

Theorem 1 shows that u has a zero on (α, β) unless $u = cv$, in which case u has a zero at both α and β.

If we had assumed $P(x) \leq \omega^2$ instead of $P(x) \geq \omega^2$ the conclusion would be that u has at most one zero in each open interval of length π/ω. This follows from Theorem 1 with the roles of u and v interchanged. Strict inequality such as $P > \omega^2$ or $P < \omega^2$ excludes the possibility that $u = cv$ and leads to slightly sharper results. These are used in Example 3.

EXAMPLE 3. Let u be a nontrivial solution of

$$u'' + (q + \lambda)u = 0, \qquad a < x < b$$

and let $m_1 \leq q \leq m_2$ for constants m_j. If the number of zeros of u is $N(\lambda)$, it is to be shown that for large λ

(1)
$$\left| N(\lambda) - \frac{b-a}{\pi}\sqrt{\lambda} \right| < 2.$$

Setting $\omega_j^2 = \lambda + m_j$, we use the comparison equations

$$v'' + \omega_1^2 v = 0, \qquad v'' + \omega_2^2 v = 0.$$

Let the interval (a, b) be divided into N intervals I_k each of length $(b-a)/N$. Then

$$\frac{b-a}{N} > \frac{\pi}{\omega_1} \Rightarrow N(\lambda) \geq N, \qquad \frac{b-a}{N} < \frac{\pi}{\omega_2} \Rightarrow N(\lambda) \leq N.$$

This is so because in the first case there is at least one zero of u in each open interval I_k and in the second case there is at most one zero of u in each closed interval I_k. Choosing N appropriately in the two cases, we are led to

$$\frac{b-a}{\pi}\omega_1 - 1 \leq N(\lambda) \leq \frac{b-a}{\pi}\omega_2 + 1.$$

Since $\omega_j = \sqrt{\lambda + m_j}$ and $\lim_{\lambda \to \infty} (\sqrt{\lambda + m_j} - \sqrt{\lambda}) = 0$ this gives (1).

Theorem 1 depends only on the differential equation. The following more subtle result uses the boundary conditions:

THEOREM 2. *Let u be an eigenfunction for λ and v for μ, where the same boundary conditions of type (SLM) are imposed for u and v. Suppose u and v both have exactly n zeros on (a, b). Then $\lambda = \mu$. Hence u and v are linearly dependent and they have the same zeros.*

In the proof we take $n \geq 2$; if $n = 0$ or $n = 1$ the argument simplifies. The zeros x_j of v divide (a, b) into intervals

$$(a, x_1), \qquad (x_1, x_2), \qquad (x_2, x_3), \qquad \cdots, \qquad (x_n, b).$$

Since the boundary conditions make $p(a)w(a) = p(b)w(b) = 0$, the intervals (a, x_1) and (x_n, b) cause no difficulty in the use of (i). Reasoning as in

Example 2, we find that if $\lambda > \mu$ then u has at least one zero on each interval. Hence u has too many zeros. If $\mu > \lambda$ then u has too few zeros, and therefore $\lambda = \mu$. Section 1, Corollary (iv) now shows that u and v are linearly dependent. This yields the theorem.

The following supplement to Theorem 2 is stated without proof:

THEOREM 3. *When the λ_n are arranged as an increasing sequence, the nth eigenfunction ψ_n of a regular Sturm-Liouville problem has exactly n zeros, $n = 0, 1, 2, \cdots$.*

No matter whether the Sturm-Liouville problem is regular or not, we let λ_n denote the eigenvalue for an eigenfunction with exactly n zeros. The quantity λ_n has various physical interpretations in addition to those mentioned in Chapter 9. For example, λ_0 is connected by way of partial differential equations with the fundamental frequency of a vibrating membrane, with the torsional rigidity of a bar, and with the capacity of a condenser.

The following theorem gives a method of estimating this important parameter numerically. The estimate is theoretically sharp, since the choice $v = \psi_n$ would allow $m_1 = \lambda_n = m_2$:

THEOREM 4. *Let λ_n be the nth eigenvalue for a Sturm-Liouville problem with boundary conditions of type SLM, and let v be any twice-differentiable function that satisfies the boundary conditions and has exactly n zeros on (a, b). Suppose m_j are constants such that*

$$m_1 \leq -\frac{(Tv)(x)}{r(x)v(x)} \leq m_2$$

at all points of (a, b) where $v(x) \neq 0$. Then $m_1 \leq \lambda_n \leq m_2$.

The proof is like the proof of Theorem 2; namely, if $\lambda < m_1$ then ψ_n has too few zeros and if $\lambda > m_2$ it has too many. When $n = 0$ (but only then) a different approach allows all standard boundary conditions. See Problem 7.

EXAMPLE 4. Obtain the estimate $246.49 < \lambda_4 < 246.75$ for the problem

$$y'' + x(1 - x)y + \lambda y = 0, \qquad y(0) = y(1) = 0.$$

Here we choose $v(x) = \sin 5\pi x$, which has four zeros on $(0, 1)$ and satisfies the boundary conditions. The desired estimate follows from

$$-\frac{Tv}{v} = 25\pi^2 - x(1 - x), \qquad 5x \neq 1, 2, 3, 4.$$

The following generalization of a result known as *Sonin's theorem* gives information about the behavior of the maxima and minima between zeros:

THEOREM 5. *Let y be a nontrivial solution of $Py'' + Qy' + y = 0$ on an interval (a, b) on which P is differentiable. Then the successive relative maxima of $|y|$ on (a, b) form an increasing, constant, or decreasing sequence when $P' > 2Q$, $P' = 2Q$, or $P' < 2Q$, respectively.*

The proof is simple. If $z = y^2 + P(y')^2$ the differential equation gives

$$z' = (y')^2(P' - 2Q).$$

Hence z is increasing, constant, or decreasing according as $P' > 2Q$, $P' = 2Q$ or $P' < 2Q$. At a maximum x_k we have $y'(x_k) = 0$, therefore $z(x_k) = y(x_k)^2$, and the result follows.

EXAMPLE 5. The Chebychev equation is

$$(1 - x^2)y'' - xy' + \lambda y = 0$$

where λ is a positive constant. Here $P = (1-x^2)/\lambda$, $Q = -x/\lambda$, and $P' = 2Q$. Theorem 5 says that the maxima of $|y|$ on $(-1, 1)$ all have the same value. This applies to the Chebychev polynomials T_m and also to the infinite-series solutions. A graph of T_5 is shown in Figure 13.

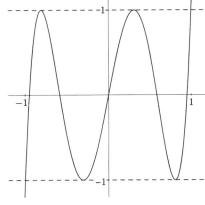

Figure 13

PROBLEMS

1P. A nontrivial solution on (a, ∞) is said to be *oscillatory* if it has a sequence of zeros $\{x_n\}$ with $x_n \to \infty$. Otherwise it is nonoscillatory. Suppose u is a nontrivial solution of $u'' + q(x)u = 0$ on (a, ∞). Show that u is nonoscillatory if $q \le 0$ but is oscillatory if $q \ge \delta$ for a positive constant δ.

2. Let $4x^2 u'' + q(x)u = 0$ on $(1, \infty)$ and let k be constant. If $q(x) \ge k > 1$ show that all nontrivial solutions u are oscillatory, and if $q(x) \le k < 1$ no solution u is oscillatory.

Hint: The comparison equation $4x^2 v'' + kv = 0$ is of Euler type.

3. For the problem $u'' + (2x^2 - 1)u + \lambda u = 0$ on $|x| < 1$ with Neumann boundary conditions $y'(0) = y'(1) = 0$, show that $|4\lambda_n - (n\pi)^2| \le 4$.

Hint: Take $v(x) = \cos(n\pi x/2)$ in Theorem 4.

4. Assuming $x > |\nu|$, deduce from Theorem 5 that the successive maxima of $|J_\nu(x)|$ form a strictly decreasing sequence.

5. By Chapter 10, Section 1, Problem 8, $y = \sqrt{x}J_\nu(x)$ satisfies

$$4x^2 y'' + (4x^2 - 4\nu^2 + 1)y = 0$$

If $N(a)$ denotes the number of zeros of $J_\nu(x)$ on the interval $(0, a)$, show that $N(a)/a \to 1/\pi$ as $a \to \infty$.

6. For $x > |\nu|$ show that the successive maxima of $|\sqrt{x}J_\nu(x)|$ in Problem 5 form an increasing, constant, or decreasing sequence for $|\nu|$ less than, equal to, or greater than $1/2$, respectively.

7. Let $u = \psi_0$ be an eigenfunction for $\lambda = \lambda_0$ that is positive on (a, b). Let v be any twice-differentiable function that is positive on (a, b) and satisfies the same standard boundary conditions as those for u. If a continuous function h is defined by $Tv = -hrv$, use Lagrange's identity to show that

$$\int_a^b (\lambda_0 - h)ruv\, dx = 0$$

Hence $\lambda_0 = h(x_0)$ holds for some $x_0 \in (a, b)$.

8. Let $z = \psi(x)$ be an eigenfunction with eigenvalue λ for a regular Sturm-Liouville problem, and change variables as follows:

$$t = \int_a^x \frac{d\xi}{p(\xi)}\, d\xi, \qquad y(x) = z(t)$$

Verify the symbolic equation $D_t = p(x)D_x$ and thus show that

$$z'' + (Q + R\lambda)z = 0 \quad \text{where} \quad Q(t) = p(x)q(x), \qquad R(t) = p(x)r(x)$$

Hence the zeros can be studied by the same comparison equations as those in Example 3.

9. Show that any zero x_0 of an eigenfunction on (a, b) must be isolated. That is, some open interval containing x_0 has no other zero.

Hint: If $\psi(x_0) = \psi'(x_0) = 0$ at some point $x_0 \in (a, b)$ the results of Chapter 10, Section 4 give $\psi = 0$.

10. If you are familiar with the notion of compactness (Chapter 17, Section 3) deduce from Problem 9 that no eigenfunction for a regular Sturm-Liouville problem can have infinitely many zeros.

5. Convergence

Since an eigenfunction ψ can be multiplied by a nonzero constant, there is no loss of generality in assuming $\|\psi\| = 1$. In that case ψ is said to be *normalized* and is denoted by ϕ rather than ψ. A set of functions that are both orthogonal and normalized is orthonormal with respect to the weight r, or simply *orthonormal*. By definition, the members of an orthonormal set satisfy

$$\int_a^b \phi_m(x)\overline{\phi_n(x)}r(x)\,dx = \begin{cases} 0, & \text{if } m \neq n; \\ 1, & \text{if } m = n \end{cases}$$

and the Fourier coefficients relative to an orthonormal set $\{\phi_n\}$ are

$$c_k = \int_a^b f(x)\overline{\phi_k(x)}\,dx.$$

The theory of orthonormal sets is independent of differential equations and allows a much broader class of functions than used hitherto. The appropriate class of functions is traditionally denoted by L^2; it consists of the complex-valued functions f on $[a, b]$ such that rf is *integrable* and $\|f\| < \infty$. For one of our results the italicized word "integrable" has to be understood in the sense of Lebesgue, but for the others Riemann integrability suffices. The main difference between the Riemann and Lebesgue integral is that the latter allows functions with wildly discontinuous behavior. See Problem 14.

(a) Mean convergence

Consider the problem of approximating $f \in L^2$ by a sum

$$s_n = \sum_{k=1}^n a_k\phi_k, \qquad \phi_k \in L^2$$

where $\{\phi_k\}$ is an orthonormal set and the a_k are constants. A measure of the closeness of the approximation is given by the *mean-square error*

$$(1) \qquad \|f - s_n\|^2 = \int_a^b r|f - s_n|^2\,dx.$$

For simplicity we assume that all functions and constants are real, but the numbered equations are so written that they remain correct in the complex case. By a brief calculation

$$\int_a^b rfs_n\,dx = \sum_{k=1}^n a_kc_k, \qquad \int_a^b rs_n^2\,dx = \sum_{k=1}^n a_k^2$$

where the c_k are the Fourier coefficients as above. From this it follows that

$$\int_a^b r(f - s_n)^2 \, dx = \int_a^b rf^2 \, dx - 2\sum_{k=1}^n a_k c_k + \sum_{k=1}^n a_k^2$$

or equivalently,

$$(2) \qquad \|f - s_n\|^2 = \|f\|^2 - \sum_{k=1}^n |c_k|^2 + \sum_{k=1}^n |a_k - c_k|^2.$$

The expression on the right is least when $a_k = c_k$, and only then. Thus we have obtained a new interpretation of the Fourier coefficients. Namely, they are the unique coefficients that minimize the mean-square error. Note that the above argument uses finite sums, hence does not depend on the theory of convergence.

Setting $a_k = c_k$ in (2) yields the minimum error

$$(3) \qquad \min \|f - s_n\|^2 = \|f\|^2 - \sum_{k=1}^n |c_k|^2.$$

Since $\|f - s_n\| \geq 0$, the minimum is nonnegative. This gives an inequality for the sum in (3), which when $n \to \infty$ becomes

$$(4) \qquad \sum_{k=1}^\infty |c_k|^2 \leq \|f\|^2.$$

Convergence of the series follows from the completeness principle of Chapter 12, Section 3.

Equation (4) is called *Bessel's inequality*. Since the general term of a convergent series tends to 0, Bessel's inequality yields

$$(5) \qquad \lim_{k \to \infty} c_k = 0.$$

By (3) the minimum error $\|f - s_n\|$ tends to 0 as $n \to \infty$ if, and only if, Bessel's inequality becomes the Parseval equality

$$(6) \qquad \sum_{k=1}^\infty |c_k|^2 = \|f\|^2.$$

In this case it is said that s_n *converges in mean* to f. The ingenious and suggestive notation

$$f = \text{l.i.m.} \; s_n \quad \Longleftrightarrow \quad \lim_{n \to \infty} \|f - s_n\| = 0$$

was introduced by the American mathematician Norbert Wiener; the initials l.i.m. stand for "limit in mean." Parseval's equation gives a necessary and sufficient condition for $f = $ l.i.m. s_n where s_n are partial sums of the Fourier series. We use s_n in this sense from now on; that is, $a_k = c_k$.

(b) Closure and completeness

The set $\{\phi_n\}$ is *closed* $f = $ l.i.m. s_n holds for every $f \in L^2$. The set $\{\phi_n\}$ is *complete* if the only functions $f \in L^2$ orthogonal to all of them are the trivial functions; that is, those for which $\|f\| = 0$. If $\{\phi_n\}$ is closed and f is orthogonal to all the ϕ_n, then the Fourier coefficients of f are all 0, and the Parseval equation gives $\|f\| = 0$. Hence *a closed set is complete.* The converse statement, *a complete set is closed,* is also true. However its proof requires the Lebesgue integral and will not be discussed here.

EXAMPLE 1. Let us return to the problem

$$y'' + \lambda y = 0; \qquad y(0) = 0, \qquad y(1) + ky'(1) = 0$$

considered in Section 1, Example 3. There we obtained the eigenfunctions

$$(7) \qquad \psi_n(x) = \sin \omega_n x, \qquad n\pi - \frac{\pi}{2} < \omega_n < n\pi + \frac{\pi}{2}, \qquad n = 1, 2, 3, \cdots$$

where ω_n is given by Figure 1. The figure also applies when $k < 0$, the slope of the line then being positive. But if $k < 0$ it is seen in Problem 2 that there is an additional eigenfunction

$$(8) \qquad \qquad \psi_0(x) = \sinh \omega_0 x, \qquad x, \quad \text{or} \quad \sin \omega_0 x$$

according as $-1 < k < 0$, $k = -1$ or $k < -1$. Since this function is orthogonal to those in (7), the set (7) is not complete.

The concept of completeness has an important bearing on the deeper aspects of Sturm-Liouville theory. An operator equation $Ay = \lambda y$ can be so formulated that it includes the eigenvalue problems of Chapter 16 as well as those considered here. In this abstract setting, completeness of the set of eigenfunctions follows from a general operator-theoretic result known as the *spectral theorem.* When applied to the Sturm-Liouville equation, the spectral theorem gives mean convergence of the Sturm-Liouville series. Uniform convergence is then deduced by using a formula of Green type (Chapter 10, Section 2) together with the Schwartz inequality of Problem 6. The final result is Section 3, Theorem 1.

(c) Pointwise convergence

The theory of trigonometric Fourier series has been a major topic in analysis for about 150 years. Thus it came as a surprise when the contemporary American mathematician Paul Chernoff found a new approach to the problem of convergence in 1980. His method is not only simpler than those used before, but it also gives convergence of the *asymmetric* sums of the complex Fourier series. Most convergence theorems apply only to the symmetric sums.

As in Chapter 14, the function f is said to be *Lipschitzian* at x_0 if there is a constant K such that

$$(9) \qquad\qquad |f(x) - f(x_0)| \le K|x - x_0|$$

for x sufficiently near to x_0. Here is a special case of Chernoff's theorem:

THEOREM 1. *Let f be a function of period 2π that belongs to L^2 on $[-\pi, \pi]$. Suppose further that f is Lipschitzian at x_0. Then*

$$\lim_{m,n \to \infty} \sum_{k=-m}^{n} c_k e^{ikx_0} = f(x_0).$$

Without loss of generality, Item (iv) of Section 2 allows us to assume $x_0 = 0$. The key to the proof is to define $g(x)$ by

$$f(x) - f(0) = g(x)(1 - e^{ix}), \qquad x \ne 0; \qquad g(0) = 0.$$

For $0 < |x| \le \pi$ we have

$$|g(x)| = \left| \frac{f(x) - f(0)}{x} \right| \left| \frac{x}{1 - e^{ix}} \right|.$$

Equation (9) with $x_0 = 0$ shows that the first factor is bounded when x is close to 0. By the definition of the derivative

$$\lim_{x \to 0} \frac{e^{ix} - 1}{x} = \frac{d}{dx} e^{ix} \bigg|_{x=0} = i$$

so that the second factor is also bounded. Away from 0 the functions f and g have the same order of magnitude, hence $g \in L^2$. This is the first step.

For the second step, denote the Fourier coefficients of f by c_k and those of g by d_k. Section 2, Item (iii) gives

$$c_0 = f(0) + d_0 - d_1, \qquad c_k = d_k - d_{k+1}, \qquad k \ne 0.$$

Hence the asymmetric partial sum of the Fourier series for f at $x = 0$ is

$$\sum_{k=-m}^{n} c_k = f(0) + \sum_{k=-m}^{n} (d_k - d_{k+1}).$$

Upon displaying the right-hand sum in expanded form, you will find that it telescopes to $d_{-m} - d_{n+1}$. Since the Fourier coefficients d_k tend to 0 as $|k| \to \infty$ this gives Theorem 1.

If the symmetric partial sums are used, Theorem 1 can be applied to functions that are discontinuous at x_0. Let g satisfy the hypothesis of Theorem 1 except that (9) is replaced by

$$(10) \quad |g(x) - \sigma| \le K|x - x_0|, \ x < x_0; \qquad |g(x) - \tau| \le K|x - x_0|, \ x > x_0$$

for x near x_0, where σ and τ are any real or complex numbers. Then

$$\lim_{n \to \infty} \sum_{k=-n}^{n} c_k e^{ikx_0} = \frac{\sigma + \tau}{2}.$$

To see why, let $x_0 = 0$ as above and define

$$(11) \quad h(x) = \sigma, \ -\pi \le x < 0; \qquad h(0) = \frac{\sigma + \tau}{2}; \qquad h(x) = \tau, \ 0 < x \le \pi.$$

Since $h(x) - h(0)$ is odd, its Fourier series has sine terms only and therefore converges to 0 at $x = 0$. This shows that the series for h converges to $h(0)$ at $x = 0$. It is easily checked that $f = g - h$ satisfies the hypothesis of Theorem 1, including (9) with $x_0 = f(x_0) = 0$. The desired conclusion now follows from

$$S(g) = S(g - h) + S(h).$$

Namely, $S(g - h)$ converges to 0 by Theorem 1 and $S(h)$ converges to $h(0)$.

Equation (10) makes $\sigma = g(x_0-)$ and $\tau = g(x_0+)$. On the other hand if σ and τ are thus defined, and $g'(x)$ is bounded on intervals at the right and left of x_0, the mean-value theorem yields (10). Thus we have proved Section 2, Theorem 2 with the role of f taken by g.

PROBLEMS

1. Verify the first unnumbered equation following (1) by writing out the sums (a) when $n = 3$ and (b) in general.

2. (a) Derive the three solutions in (8).   (b) In Figure 1, show that

$$\lim_{n \to \infty} (\lambda_n - n\pi) = -\frac{\pi}{2}, \qquad \lambda_n = n\pi, \qquad \lim_{n \to \infty} (\lambda_n - n\pi) = \frac{\pi}{2}$$

for $k > 0$, $k = 0$, or $k < 0$, respectively. The study of $\sin \omega_n x$ is a part of the theory of *nonharmonic Fourier series*. In this theory it is shown that the set is complete on $0 \le x \le \pi$ if ω_n behaves like $n\pi - \pi/2$ or like $n\pi$ but not if ω_n behaves like $n\pi + \pi/2$. In the latter case one must adjoin a suitably chosen function $\psi_0 \in L^2$ to get completeness, and the original set is said to have *deficiency* 1. These remarks shed light on Example 1.

3. Let s_n be the partial sums for the Fourier series based on (7) with $k < 0$. Show that any function $f \in L^2$ for which $f = $l.i.m. s_n must be orthogonal to the extra function ψ_0 in (8).   Hint: The orthogonality means $c_0 = 0$.

4. Show that (2) holds for complex functions f, ϕ_n (and hence the consequences of (2) also hold).   Hint: $|\alpha + \beta|^2 = |\alpha|^2 + 2\mathrm{Re}\,\alpha\bar{\beta} + |\beta|^2$.

5. Prove $2|f||g| \le |f|^2 + |g|^2$ and, if $f, g \in L^2$, deduce

$$(*) \qquad 2\int_a^b r|f||g|\,dx \le \int_a^b r|f|^2\,dx + \int_a^b r|g|^2\,dx$$

6. The Schwarz inequality. Under the hypothesis of Problem 5, show that the integral on the left in $(*)$ does not exceed $\|f\|\,\|g\|$.

Hint: Apply Problem 5 to sf and g/s where s is a positive constant. Choose s so as to minimize the right-hand side if $\|f\|\,\|g\| > 0$ and let $s \to \infty$ or $s \to 0$ if $\|f\|\,\|g\| = 0$.

7. The Minkowski inequality. Deduce $\|f + g\| \le \|f\| + \|g\|$ by squaring both sides and using the result of Problem 6.

8. This problem and the next four pertain to the inhomogeneous equation

$$Ty + \lambda ry = rf$$

with standard boundary conditions. The same boundary conditions are imposed on all eigenfunctions ϕ and it is assumed that the various integrals encountered in the analysis exist; see Problems 5, 6, and 7.
    If the inhomogeneous problem has two solutions $y = u$ and $y = v$, with $\|u - v\| > 0$, show that λ is an eigenvalue. (Hence if λ is not an eigenvalue, the inhomogeneous problem has at most one solution.) Hint: Try $\psi = u - v$.

9. Suppose λ is an eigenvalue in Problem 8. Show that the inhomogeneous problem has infinitely many solutions if it has any. Hint: Look at $y + c\psi$.

10. If the inhomogeneous equation of Problem 8 is solvable, show that f is orthogonal to every eigenfunction ψ belonging to λ.

Hint: Apply Lagrange's identity with $u = y$, $v = \bar{\psi}$.

11. In Problem 8 let $\{\phi_n\}$ be the set of all normalized eigenfunctions with corresponding eigenvalues λ_n. Suppose f admits the expansion on the left below, and suppose also that $\lambda \neq \lambda_n$ for every n. In this case, by Problem 8, there is at most one solution. Show that the solution is given by the series on the right, provided that T can be applied to this series term by term:

$$f = \sum_{n=0}^{\infty} f_n \phi_n, \qquad y = \sum_{n=0}^{\infty} \frac{f_n}{\lambda - \lambda_n} \phi_n$$

12. If $\lambda = \lambda_n$ in Problem 11, but f is orthogonal to ϕ_n as required by Problem 10, check that $f_n = 0$. Hence the term with denominator $\lambda - \lambda_n$ does not arise and the series still gives a solution. By Problem 9 there are infinitely many solutions in this case.

13. Open-ended problem. Generalizations of the equations for heat flow and wave motion are respectively

$$ru_t = (pu_x)_x + qy, \qquad ru_{tt} = (pu_x)_x + qu$$

where p, q, r are functions of x as in Section 1. The equations are to hold for $a < x < b$ and $t > 0$, with $u = u(x,t)$. Separate variables and explore the connection with the general eigenvalue problem.

14. The Dirichlet function is defined by $f(x) = 1$ if x is rational, $f(x) = 0$ otherwise. Show that $f(x)$ is discontinuous at every point and is not Riemann integrable on any interval. (The Lebesgue integral of this function exists, however, and is 0 over every interval.)

15. Let $f(x) = 1$ when x is rational, $f(x) = -1$ when x is irrational. Check that $|f|$ is Riemann integrable on every finite interval, though f itself is not. A similar example shows that f need not be Lebesgue integrable if $|f|$ is.

Profound study of nature is the most fertile source of mathematical discoveries.

— Joseph Fourier

Chapter 19

THE PHASE PLANE

MANY PROBLEMS involve two unknown functions, $x(t)$ and $y(t)$, where dx/dt and dy/dt are given in terms of x and y. In such cases a solution can be described very conveniently by plotting the locus of the point $(x(t), y(t))$ in the (x, y) plane. For linear equations with constant coefficients, the resulting curves fall into three main classes, and graphical methods yield detailed information with little effort. A particularly simple case is that in which $x = u(t)$ and $y = u'(t)$ where u satisfies a second-order linear equation with constant coefficients. This case is investigated in detail and more general cases are then reduced to it.

1. A second-order equation

Throughout this chapter p, q, a, b, c, d, k are real constants and primes denote differentiation with respect to t. We begin by considering the linear equation

$$(1) \qquad\qquad u'' + pu' + qu = 0,$$

which has already been studied extensively in this book. If $u(t)$ gives the position at time t of a point moving on the u axis, the function

$$v(t) = \frac{du}{dt} = u'(t)$$

denotes the velocity.

The behavior of the system can be described not only by a graph of $u(t)$ versus t but also by giving the locus of the point $(u(t), v(t))$ in the (u, v) plane. When the (u, v) plane is associated with a differential equation in this way, it is called the *phase plane*. The curves given by $(u(t), v(t))$ as t increases from $-\infty$ to ∞ are called *trajectories* and the set of points occupied by a trajectory is the *orbit* or *trace*. For example, in Chapter 7 it was seen that the phase-plane orbits for $u'' + u = 0$ are circles centered at the origin. The main difference between an orbit and a trajectory is that the latter is associated with the specific parameter t. The parameterization gives the trajectory an orientation, as shown by arrows in the accompanying figures.

Throughout this chapter (u, v) gives the position of a point moving in the phase plane. The velocity of this point is not u' but is the two-dimensional vector

$$(u', v') = \left(\frac{du}{dt}, \frac{dv}{dt} \right) = \mathbf{i}u' + \mathbf{j}v'.$$

The calculation

$$\text{slope of vector} = \frac{dv/dt}{du/dt} = \frac{dv}{du} = \text{slope of curve}$$

shows that the velocity vector is tangent to the orbit; more specifically, the vector gives the direction in which the point (u, v) on the trajectory is moving at any given time. This fact is an aid in sketching the curves.

The analysis of orbits in the phase plane can be simplified by a change of variables known as an affine transformation. An *affine transformation* is defined by

$$\begin{aligned} U &= Au + Bv \\ V &= Cu + Dv \end{aligned} \qquad \begin{vmatrix} A & B \\ C & D \end{vmatrix} \neq 0$$

where A, B, C, D are real constants. Since the determinant of the system is not 0, we can solve for (u, v) in terms of (U, V). As an illustration, each of the three transformations

$$\begin{aligned} U &= 4u - v & U &= 5u + v & U &= u + v \\ V &= u - v & V &= 2u - v & V &= -4u \end{aligned}$$

is affine because the corresponding determinants are $-3, -7, 4$, none of which is 0. These transformations will be used in Examples 1, 2, and 3.

The main features of an affine transformation are best described by saying what happens to a given geometric figure in the (u, v) plane when the figure is mapped into the (U, V) plane. It can be shown that under an affine transformation:

(a) parallel lines go into parallel lines

(b) intersecting lines go into intersecting lines

(c) a square goes into a parallelogram

(d) a circle goes into an ellipse

(e) tangent curves go into tangent curves

(f) any conic goes into a conic of the same type.

The proof of these facts is not essential to your understanding of differential equations and is omitted.

When one geometric figure is obtained from another by an affine transformation, the two figures are said to be *affine equivalent*. The two families of curves in Figure 1 are affine equivalent, so are those in Figure 2, and so are those in Figure 3. The following examples show where these figures came from.

EXAMPLE 1. Let it be required to find the phase-plane orbits of

$$u'' - 5u' + 4u = 0.$$

The characteristic polynomial is

$$s^2 - 5s + 4 = (s - 1)(s - 4)$$

so that, by the theory of Chapter 6 together with $v = u'$,

$$(2) \quad u = c_1 e^t + c_2 e^{4t}, \ v = c_1 e^t + 4c_2 e^{4t}.$$

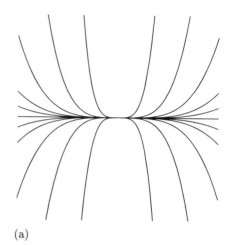

(a)

These are parametric equations of the trajectories. By inspection

$$4u - v = 3c_1 e^t, \qquad u - v = -3c_2 e^{4t}.$$

If we set $U = 4u - v$, $V = u - v$ these equations become

$$(3) \quad U = C_1 e^t, \qquad V = C_2 e^{4t}$$

where $C_1 = 3c_1$ and $C_2 = -3c_2$. From this follows

$$|U|^4 = |C_1|^4 e^{4t}, \qquad |V| = |C_2| e^{4t}.$$

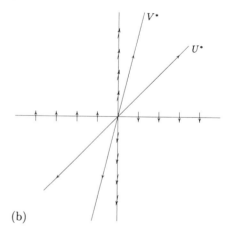

(b)

Assuming $C_1 \neq 0$ we see that $|V|/|U|^4$ is constant and hence

$$(4) \quad |V| = k|U|^4, \qquad k \geq 0 \quad \text{const.}$$

We are interested not just in a single curve (3) but in the whole family of curves obtained for various choices of C_1 and C_2. If (U, V) is a point on one of these curves then each of the points

$$(-U, V), \qquad (U, -V), \qquad (-U, -V)$$

is also on one of these curves. For example, $(U, -V)$ is on the curve (3) obtained when C_2 is changed to $-C_2$.

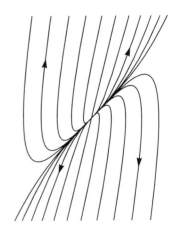

(c)

Figure 1

Thus the family of curves (3), considered as a whole, has the same symmetry properties as the family defined by (4). Either set of curves can be plotted by finding the graphs in the first quadrant and extending to other quadrants by symmetry. In the first quadrant (4) is equivalent to

$$V = kU^4$$

which is plotted with ease. The result after extension by symmetry is shown in Figure 1(a). These are the orbits in the (U, V) plane.

The orbits in the (u, v) plane are obtained by the affine transformation $U = 4u - v$, $V = u - v$ and each of them satisfies the equation

$$|u - v| = k|4u - v|^4 \qquad k = \text{const.}$$

This is not very helpful in sketching the curves, however, and another method is used here.

The U axis in the (U, V) plane is described by the equation $V = 0$, hence its image in the (u, v) plane is the line with equation

$$u - v = 0.$$

This line is labeled U^* in Figure 1(b). In a like manner, the image of the V axis is the line V^* with the equation

$$4u - v = 0.$$

If $v = u'$ the equation $u'' - 5u' + 4u = 0$ yields

$$u' = v, \qquad v' = -4u + 5v$$

and this gives the tangent vector (u', v'). In particular, on the two axes in the (u, v) plane the velocity vector is

$$(0, -4u), \qquad v = 0; \qquad (v, 5v), \qquad u = 0$$

respectively. This information is shown by arrows in Figure 1(b). (The arrows ought to get shorter as they approach the origin, but we are interested only in their direction.) On the lines U^* and V^* the velocity vectors are

$$(u, u), \qquad v = u; \qquad (v, 4v), \qquad v = 4u$$

respectively. They have slopes 1 and 4, the same as the slopes of the lines $v = u$ and $v = 4u$ on which they lie. Hence these lines must be orbits of the differential equation. Indeed, we get

$$u = c_1 e^t, \ v = c_1 e^t; \qquad u = c_2 e^{4t}, \ v = 4c_2 e^{4t}$$

by setting $c_2 = 0$ or $c_1 = 0$, respectively, in (2). These parametric equations give $v = u$ and $v = 4u$, and they also confirm the fact that the point (u, v) moves away from the origin as t increases.

The orbits in the (u, v) plane can now be sketched with ease. Those in the (U, V) plane are tangent to the U axis at the origin and are almost parallel to the V axis at distant points. Hence those in the (u, v) plane are tangent to U^* at the origin and are almost parallel to V^* at distant points. Taking account of the arrows in Figure 1(b), we get the curves in Figure 1(c).

Before giving more examples we introduce some additional terminology. The origin $(0, 0)$ in phase space is called a *critical point* for the equation (1). Critical points are classified into three main types according to the behavior of the characteristic equation

$$s^2 + ps + q = 0.$$

If both roots are real, unequal, and of the same sign the critical point is a *node*. As an illustration, the critical point in Example 1 is a node because the roots are 2, 5. When the roots are real and of opposite signs the critical point is a *saddle*. When they are complex but not purely imaginary the critical point is a *focus*. A node or focus is *stable* if the point $(u(t), v(t))$ moves to the node or focus as a limit for $t \to \infty$ and it is *unstable* if the point moves arbitrarily far away as $t \to \infty$. It will be seen that a node is stable if, and only if, the roots are negative, and a focus is stable if, and only if, the roots have negative real parts. The critical point in Example 1 is an unstable node.

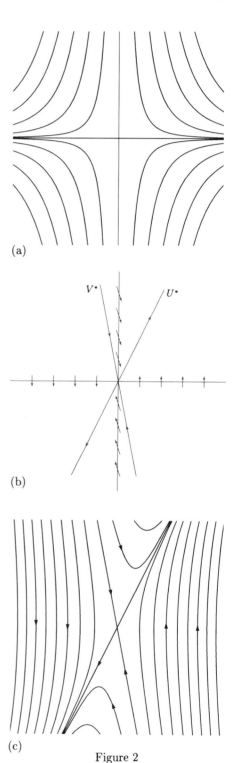

(a)

(b)

(c)

Figure 2

EXAMPLE 2.   For the equation $u'' + 3u' - 10u = 0$ the roots of the characteristic equation are $2, -5$. Since these are real and of opposite signs, the critical point is a saddle. The solution is

$$u = c_1 e^{2t} + c_2 e^{-5t}, \qquad v = 2c_1 e^{2t} - 5c_2 e^{-5t}.$$

By inspection

$$5u + v = 7c_1 e^{2t}, \qquad 2u - v = 7c_2 e^{-5t}.$$

With $U = 5u + v$ and $V = 2u - v$ this gives

$$U = C_1 e^{2t}, \qquad V = C_2 e^{-5t}$$

where $C_1 = 7c_1$, $C_2 = 7c_2$. Hence

$$|U|^5 = |C_1|^5 e^{10t}, \qquad |V|^2 = |C_2|^2 e^{-10t}.$$

These equations yield $|U|^5 |V|^2 = k$ where k is constant. As in the previous case one can plot the locus in the first quadrant of the (U, V) plane and extend to other quadrants by symmetry. The result is shown in Figure 2(a).

     The images of the U and V axes are the lines U^* and V^* with the respective equations

$$2u - v = 0, \qquad 5u + v = 0$$

and the tangent vector is $(u', v') = (v, 10u - 3v)$. It is left for you to check the details of Figure 2(b) and to obtain further verification by setting $c_2 = 0$ or $c_1 = 0$ in the formulas with which we began. The final result is shown in Figure 2(c).

EXAMPLE 3.   As our last example we consider

$$u'' + 2u' + 17u = 0.$$

By the quadratic formula the characteristic equation has roots $-1 \pm 4i$. Hence the solution can be written in the form

$$u(t) = Ae^{-t} \cos(4t - \phi)$$

where A and ϕ are constant; see Chapter 6, Section 3, Problem 7. The derivative is

$$v = -Ae^{-t} \cos(4t - \phi) - 4Ae^{-t} \sin(4t - \phi).$$

By inspection

$$u + v = -4Ae^{-t}\sin(4t - \phi),$$

$$-4u = -4Ae^{-t}\cos(4t - \phi).$$

If we set $U = u + v$, $V = -4u$
the equations become

$$U = -4Ae^{-t}\sin(4t - \phi),$$

$$V = -4Ae^{-t}\cos(4t - \phi).$$

Squaring and adding yield

$$U^2 + V^2 = k^2 e^{-2t}$$

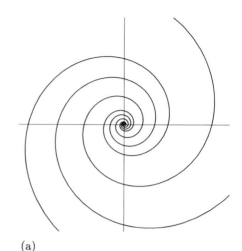

(a)

where $k = -4A$ is constant.
  If the factor e^{-2t} were absent the
locus would be a circle of radius $|k|$
centered at the origin. The effect of
the factor e^{-2t} is to diminish the distance to the origin as t increases, so
that the actual curve is a spiral. See
Figure 3(a). In this case it is not important to get the images U^* and V^*
of the U and V axes, but the tangent
vector

$$(u', v') = (v, -17u - 2v)$$

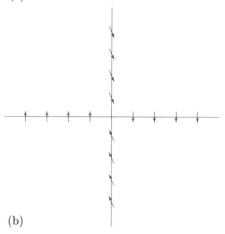

(b)

shows whether the spiral is traversed
in a clockwise or counterclockwise direction. On the axes the vector is

$$(0, -17u), \qquad v = 0$$

$$(v, -2v), \qquad u = 0$$

and the direction is clockwise as in Figure 3(c).
  The factor e^{-2t} in the U, V equations or e^{-t} in the original equations
shows that the point $(u(t), v(t))$ tends
to the origin as $t \to \infty$. The origin
is a *stable focus*. It is stable because

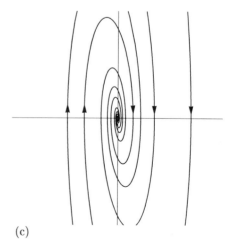

(c)

Figure 3

the point (u, v) tends to $(0,0)$ and it is a focus because the characteristic equation has complex roots not purely imaginary.

PROBLEMS

1P. We shall find the phase-plane orbits of $u'' - 7u' + 10u = 0$.

(a) Check that the roots of the characteristic equation are 2 and 5 (and hence the critical point is an unstable node.) Set $v = u'$ to get

$$u = c_1 e^{2t} + c_2 e^{5t} \qquad v = 2c_1 e^{2t} + 5c_2 e^{5t}$$

(b) Compute $U = 5u - v$, $V = 2u - v$ and check that $|V|^2 = K|U|^5$.

(c) Sketch the curves $V = kU^{5/2}$ and, by symmetry, the curves of Part (b).

(d) The image U^* of the U axis has the equation $v = 2u$. Similarly get the equation of V^* and sketch both images U^* and V^* in the (u, v) plane. Suggestion: Since the v, V^* axes are nearly parallel, it is advisable to use a full $8\frac{1}{2}$-by-11 inch sheet of paper.

(e) Using the tangent vector $(v, -10u + 7v)$, make appropriate arrows on the u, v, U^*, V^* axes and sketch the curves.

(f) Obtain a partial check by the formulas of Part (a).

2. Proceeding as in Examples 1, 2, and 3 analyze the phase-plane orbits:

(a) $u'' + 3u' + 2u = 0$ (d) $u'' + u' - 2u = 0$ (g) $u'' + 4u' + 13u = 0$

(b) $u'' - u = 0$ (e) $u'' + 2u' + 5u = 0$ (h) $u'' + 9u' + 20u = 0$

(c) $u'' - 2u' + 2u = 0$ (f) $u'' - 9u' + 14u = 0$ (i) $u'' + u' - 20u = 0$

─────────────ANSWERS─────────────

Abbreviations: N,S,F = node, saddle, focus; u,s,c,cc = unstable, stable, clockwise, counterclockwise. A saddle is always regarded as unstable, so that the phrase "unstable saddle" is redundant.

2. Partial answers:

  (a) sN, $|2u + v|^2 = k|u + v|$ (f) uN, $|2u - v|^2 = k|7u - v|^7$

  (b) S, $|u^2 - v^2| = k$ (g) scF, $(2u + v)^2 + 9u^2 = k^2 e^{-4t}$

  (c) ucF, $(u - v)^2 + u^2 = k^2 e^{2t}$ (h) sN, $|5u + v|^5 = k|4u + v|^4$

  (d) S, $|u - v||2u + v|^2 = k$ (i) S, $|4u - v|^4|5u + v|^5 = k$

  (e) scF, $(u + v)^2 + 4u^2 = k^2 e^{-2t}$

2. The general case

In the previous examples and problems it was repeatedly observed that the orbits in the case of a node all have the same general character. So do those for a saddle, and those for a focus. We now confirm this behavior in the general case

$$u'' + pu' + qu = 0, \qquad v = u'.$$

If the roots λ and μ of the characteristic equation are real and distinct the solution u and its derivative v are

$$u = c_1 e^{\lambda t} + c_2 e^{\mu t}, \qquad v = \lambda c_1 e^{\lambda t} + \mu c_2 e^{\mu t}.$$

These equations give

$$\mu u - v = (\mu - \lambda)c_1 e^{\lambda t}, \qquad \lambda u - v = (\lambda - \mu)c_2 e^{\mu t}.$$

With $V = \mu u - v$ and $U = \lambda u - v$ we get, as in Section 1,

(1) $$|V|^\mu = K|U|^\lambda, \qquad K \geq 0 \text{ const.}$$

(i) Node. Let λ and μ be unequal real numbers of the same sign, where without loss of generality $\lambda/\mu > 1$. Equation (1) with $k = K^{\lambda/\mu}$ yields

$$|V| = k|U|^{\lambda/\mu}, \qquad k \geq 0 \text{ const.}$$

The curves in the (U, V) plane resemble the parabolas $|V| = k|U|^2$ and the curves in the (u, v) plane are given by an affine transformation. This is the case of a node. The analysis shows that the orbits always have the general character of those shown in Section 1, Figure 1.

(ii) Saddle. If λ and μ have opposite signs let $\mu > 0$ and $\lambda = -\rho < 0$. The curves now take the form

$$|U|^\rho|V|^\mu = K$$

and resemble the hyperbolas $|U||V| = K$. This is the case of a saddle as shown in Section 1, Figure 2.

(iii) Focus. For a focus the roots have the form $\sigma \pm i\omega$ where neither σ nor ω is 0. The solution u and its derivative $v = u'$ are now

$$u = Ae^{\sigma t}\cos(\omega t - \phi), \qquad v = \sigma Ae^{\sigma t}\cos(\omega t - \phi) - \omega Ae^{\sigma t}\sin(\omega t - \phi)$$

where A and ϕ are constant. These equations give

$$\sigma u - v = \omega A e^{\sigma t} \sin(\omega t - \phi), \qquad \omega u = \omega A e^{\sigma t} \cos(\omega t - \phi).$$

The variables $U = \sigma u - v$, $V = \omega u$ satisfy

$$U^2 + V^2 = k^2 e^{2\sigma t}$$

where $k = |\omega A|$. This would represent a circle if $\sigma = 0$; since $\sigma \neq 0$, it represents a spiral. When t increases by $2\pi/\omega$ the point (U, V) returns to the same radial line from which it started, but the exponential factor multiplies its distance to the origin by $e^{2\pi\sigma/\omega}$. The same factor intervenes when we go once around the spiral in the (u, v) plane. The critical point is a focus. The analysis shows that the orbits always have the general character of those shown in Section 1, Figure 3.

In the sequel, the following result is used to reduce more general cases to those considered here:

THEOREM 1. *If two affine transformations are carried out one after the other, the result is again an affine transformation.*

We give the proof in matrix notation, though familiarity with matrices is not assumed elsewhere in this chapter. The matrix form of an affine transformation is

$$\begin{pmatrix} X \\ Y \end{pmatrix} = \begin{pmatrix} A & B \\ C & D \end{pmatrix} \begin{pmatrix} x \\ y \end{pmatrix}, \qquad \begin{vmatrix} A & B \\ C & D \end{vmatrix} \neq 0.$$

If two such transformations are carried out one after the other, the matrix of the resulting transformation is the product of the individual matrices and its determinant is the product of the individual determinants. This shows that the determinant is nonzero and Theorem 1 follows.

PROBLEMS

1P. With $D = d/dt$ the system $x' = ax + by$, $y' = cx + dy$ can be written

$$(D - a)x = by, \qquad (D - d)y = cx$$

(a) Operate on the first equation with $D - d$ and on the second with $D - a$ to get

$$x'' - (a + d)x' + (ad - bc)x = 0, \qquad y'' - (a + d)y' + (ad - bc)y = 0$$

(b) Show that the original system has a nonzero solution satisfying $y = mx$, with m constant, if, and only if, $c + dm = m(a + bm)$.

(c) The characteristic polynomial for the second-order equations in (a) and the quadratic for m in (b) are respectively

$$s^2 - (a+d)s + (ad - bc) = 0, \qquad bm^2 + (a-d)m - c = 0$$

Check that these have the same discriminant.

(d) If m satisfies the second equation of Part (c) show that $a + bm$ satisfies the first.

2P. In the case of a focus let us suppose that you have made a rough sketch of a spiral. According to the text a full turn of the spiral changes the distance to the origin by a factor $e^{2\pi\sigma/\omega}$. If $\lambda = 1 + 8i$ check that this factor is about 2.2, so your rough sketch is probably about right. But if $\lambda = 8 + i$ the factor is

$$e^{16\pi} \doteq 6,761,000,000,000,000,000,000,000$$

and the trajectory goes off the paper so fast that your rough sketch is probably quite wrong. This matter will not be emphasized but you should be aware of it.

3. If the complex number z is written in polar form and ϕ is real the transformation $Z = e^{i\theta}z$ becomes

$$Z = e^{i\theta}re^{i\phi} = re^{i(\theta+\phi)}$$

This represents a *rotation* about the origin through the angle ϕ. Writing

$$X + iY = (\cos\phi + i\sin\phi)(x + iy)$$

and separating into real and imaginary parts, obtain the real form of the rotation in the (x, y) plane:

$$X = x\cos\phi - y\sin\phi, \qquad Y = x\sin\phi + y\cos\phi$$

4. Let S denote the square in the (x, y) plane with vertices at (0,0), (0,1), (1,0), (1,1). Sketch the image of S in the (X, Y) plane under the affine transformations

(a) $\begin{cases} X = 2x \\ Y = 3y \end{cases}$ (b) $\begin{cases} X = -x \\ Y = y \end{cases}$ (c) $\begin{cases} \sqrt{2}X = x - y \\ \sqrt{2}Y = x + y \end{cases}$

The first of these represents a change of the x and y scales, the second a reflection, and the third a rotation.

5. The triangle (P_1, P_2, P_3) formed by three noncollinear points in the (x, y) plane is *positively oriented* if the interior of the triangle lies on your left when you traverse the boundary from P_1 to P_2 to P_3 to P_1, in that order. If the interior lies on your right, the triangle (P_1, P_2, P_3) is *negatively oriented*. See Figure 4.

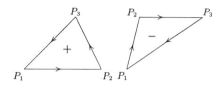

Figure 4

(a) One of the following triangles (P_1, P_2, P_3) is positively oriented and one is negatively oriented. Plot the points and decide which is which:

$$(2, -1), \ (3, 4), \ (-1, 2); \qquad (5, 2), \ (0, 1), \ (-1, 6).$$

(b) It can be shown that an affine transformation preserves orientation if its determinant is positive, and reverses orientation if its determinant is negative. Check this statement by finding the image of each triangle in Part (a) under each of the following affine transformations. No answer is given:

$$X = 2x + 3y, \ Y = x - y; \qquad X = -3x + y, \ Y = x - 2y$$

6. For the system $x' = x$, $y' = -y$ check that the equation $xy = 1$ represents a single hyperbola and two orbits, while $|x||y| = 1$ represents two hyperbolas and four orbits. The fact that a single Cartesian equation may represent the union of two or more orbits causes no practical difficulty, but you should be aware of it.

7. The purpose of this problem is to remind you that parametric equations $x = \xi(t)$, $y = \eta(t)$, and a seemingly equivalent Cartesian equation $\theta(x, y) = 0$, need not describe the same locus. When there is a discrepancy, the parametric equations take precedence.

(a) Given $x' = x$, $y' = y$; $x \, dy = y \, dx$ check that the respective solutions are

$$x = c_1 e^t, \ y = c_2 e^t; \qquad y = cx \text{ or } x = cy$$

The t-system has $x = y = 0$ as sole solution through the origin while the (x, y)-equation has infinitely many.

(b) Given $x' = -y$, $y' = x$; $x \, dx + y \, dy = 0$ check that the respective solutions are

$$x = c \cos(t - t_0), \ y = c \sin(t - t_0); \qquad x^2 + y^2 = c^2$$

Again the t-system has $x = y = 0$ as sole solution through the origin but the (x, y)-equation has none. Further discussion is given in Chapter 2, Section 2 and in Chapter 20.

3. A two-dimensional linear system

The following *equivalence theorem* reduces a general linear system to the simpler cases we have already considered:

THEOREM 1.   *Unless $b = c = 0$ and $a = d$, the phase-plane orbits of*

$$(1) \qquad\qquad x' = ax + by, \qquad y' = cx + dy$$

are affinely equivalent to those of the second-order equation

$$(2) \qquad\qquad u'' + pu' + qu = 0$$

where $p = -(a + d)$ and $q = ad - bc$. Futhermore both x and y satisfy (2).

When the theorem applies, a second affine transformation takes us from (u, u') in (2) to the functions (U, V) of Section 2. Hence the (x, y) trajectories are affinely equivalent to the (U, V) trajectories; see Section 2, Theorem 1.

Readers familiar with 2-by-2 matrices will recognize that (1) can be written in matrix form as

$$\begin{pmatrix} x' \\ y' \end{pmatrix} = \begin{pmatrix} a & b \\ c & d \end{pmatrix} \begin{pmatrix} x \\ y \end{pmatrix}.$$

The characteristic polynomial of the matrix is

$$\begin{vmatrix} a - s & b \\ c & d - s \end{vmatrix} = s^2 - (a + d)s + (ad - bc)$$

which is the same as the characteristic polynomial for (2). In the excluded case $b = c = 0$, $a = d$ the matrix is a scalar multiple of the identity matrix. It is seen in Section 4 that this case is trivial.

We now give the proof. That x and y satisfy (2) was shown in Section 2, Problem 1. To prove the affine equivalence suppose first that $b \neq 0$. Since the phase-plane orbits of (2) are given by

$$u' = v, \qquad v' = -qu - pv$$

we try the affine transformation $u = x$, $v = ax + by$. The first equation (1) is then $u' = v$, as desired. Since $u = x$ satisfies (2), as noted above, the equation $v' = u'' = -qu - pv$ also holds.

If $c \neq 0$ the affine transformation $u = y, v = cx + dy$ reduces the second equation (1) to $u' = v$, and since $u = y$ satisfies (2), the result follows again. In the remaining case (1) is $x' = ax, y' = dy$ with $a \neq d$. The affine transformation

$$u = x + y, \qquad v = ax + dy$$

yields $u' = v$. Since x and y both satisfy (2) the same is true of $u = x + y$ by linearity. This completes the proof.

By definition, the critical point in (1) is a node, saddle, or focus when the critical point in (2) is a node, saddle, or focus, respectively. This equivalence leads to a simple decision procedure as explained next.

(a) Classification of critical points

If λ and μ are the roots of the characteristic equation associated with (2) the identity

$$(s - \lambda)(s - \mu) = s^2 + ps + q$$

shows that

$$\lambda + \mu = -p, \qquad \lambda\mu = q.$$

Furthermore, the discriminant of the quadratic is $p^2 - 4q$. Using these facts, we shall justify the classification of critical points shown in Figure 5.

The roots are complex if $4q > p^2$, and real and unequal if $4q < p^2$. This leads to the classification as "focus" above the parabola of the figure and "node or saddle" below. The positive q axis is excluded in the case of a focus because if $p = 0$ the roots are purely imaginary.

Since $\lambda\mu = q$ the two real roots λ and μ have the same sign if $q > 0$, and they have opposite signs if $q < 0$. This leads to the classification as "node" above the p axis and "saddle" below.

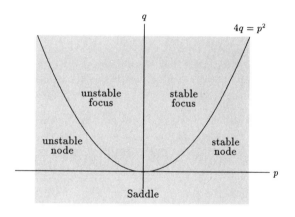

In the case of complex roots the quadratic formula shows that the real part of each root is $-p/2$. Hence the roots both have negative real parts if $p > 0$, positive real parts if $p < 0$. The formula $\lambda + \mu = -p$ gives a similar result for real roots, both of which have the same sign. Namely, the roots are negative if $p > 0$, positive if $p < 0$. These remarks yield the classification "stable" for $p > 0$, "unstable" for $p < 0$, as shown in Figure 5. Our justification of Figure 5 is now complete.

EXAMPLE 1. Classify the critical point for each system:

$$
\begin{array}{lll}
x' = 2x + 4y & \qquad x' = 2x + 3y & \qquad x' = -2x - 3y \\
y' = x + 3y & \qquad y' = x - 4y & \qquad y' = x - 4y
\end{array}
$$

It is convenient to abbreviate each differential equation by giving the matrix of coefficients, thus:

$$\begin{pmatrix} 2 & 4 \\ 1 & 3 \end{pmatrix} \qquad\qquad \begin{pmatrix} 2 & 3 \\ 1 & -4 \end{pmatrix} \qquad\qquad \begin{pmatrix} -2 & -3 \\ 1 & -4 \end{pmatrix}$$

In the first case $p = -5$ and the determinant is $q = 2$. Since

$$4q < p^2, \qquad p < 0, \qquad q > 0$$

Figure 5 shows that the origin is an unstable node. In the second case $p = 2$ and $q = -11$. Since $q < 0$ the critical point is a saddle. In the third case $4q > p^2$, $p > 0$ and the critical point is a stable focus.

EXAMPLE 2.   The classification can be supplemented by plotting the velocity vectors on the x and y axes, namely,

$$(ax, cx) \quad \text{at} \quad (x, 0), \qquad (by, dy) \quad \text{at} \quad (0, y).$$

For example in the third matrix of Example 1 these vectors are $(-2x, x)$ and $(-3y, -4y)$ and the direction of the spiral is counterclockwise. The same method shows that the spiral at a focus is always counterclockwise when c is positive, clockwise when c is negative.

Despite a word of caution in Section 2, Problem 2, the trajectories at a focus can usually be sketched with tolerable accuracy by use of Figure 5 and Example 2. The next example explains what to do at a node or saddle.

EXAMPLE 3.   For the two systems

$$\begin{aligned} x' &= 6x + y \\ y' &= 4x + 3y \end{aligned} \qquad\qquad\qquad \begin{aligned} x' &= -3x + 2y \\ y' &= -3x + 4y \end{aligned}$$

it is left for you to check that the origin is an unstable node and a saddle, respectively, and that the roots (λ, μ) are $(2, 7)$ in the first case, $(-2, 3)$ in the second. We write x corresponding to each root and we get y from $y = x' - 6x$ in the first system, from $2y = x' + 3x$ in the second. The respective results are

$$\begin{aligned} x &= ke^{2t},\ y = -4x \\ x &= ke^{7t},\ y = x \end{aligned} \qquad\qquad \begin{aligned} x &= ke^{-2t},\ 2y = x \\ x &= ke^{3t},\ y = 3x \end{aligned}$$

where k is constant. These are parametric equations for the straight-line orbits.

In the first case $|x|$ and $|y|$ increase with t, since the exponents are positive, and the lines are directed away from the origin. As $t \to -\infty$ the term e^{2t} is much larger than e^{7t}. Hence the orbits are tangent at the origin to

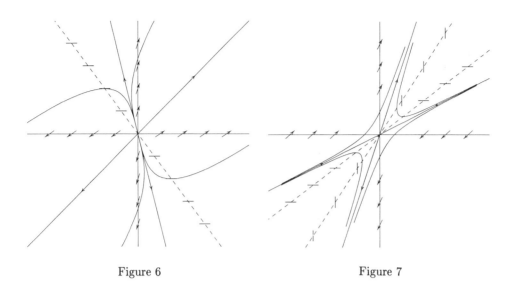

Figure 6   Figure 7

the line $y = -4x$ associated with $\lambda = 2$. As $t \to \infty$ the term e^{7t} dominates and the orbits behave like $y = x$ at distant points. See Figure 6, which contains additional information not considered here.

In the second case the line $2y = x$ is associated with the negative exponent -2 and is directed toward the origin, while the other line is directed away. The trajectories are easily sketched and are shown, with additional information, in Figure 7.

The general character of the trajectories was already given by Theorem 1 and Section 2, and that is the reason why so little calculation was required. For independent confirmation (which is not really needed) one can use a general solution (x, y). In the first system, for example, x and $y = x' - 6x$ yield

$$(3) \qquad \frac{y}{x} = \frac{-4c_1 e^{2t} + c_2 e^{7t}}{c_1 e^{2t} + c_2 e^{7t}} = \frac{-4c_1 + c_2 e^{5t}}{c_1 + c_2 e^{5t}}.$$

If $c_1 c_2 \neq 0$, so that we are not on a straight-line trajectory, it is clear that $x \to 0$, $y \to 0$, $y/x \to -4$ as $t \to -\infty$. This gives an independent proof that the trajectories come in to the origin tangent to the line $y = -4x$. Equation (3) for $t \to \infty$ also confirms the behavior at distant points.

(b) Isoclines and fixed points

Dividing the second equation (1) by the first gives

$$(4) \qquad \frac{dy}{dx} = \frac{cx + dy}{ax + by} = \frac{c + d(y/x)}{a + b(y/x)}.$$

The first equality (4) is valid if $ax + by \neq 0$ and the second if also $x \neq 0$. The expression on the right shows that dy/dx is constant when y/x is constant.

In other words, if m is constant the orbits have the same slope at all points of the line $y = mx$. The line $y = mx$ is an example of an isocline. In general, an *isocline* is a locus on which all solutions of a given differential equation have the same slope. The prefix "iso" means "same".

Among the most helpful isoclines are the x and y axes and the two lines

$$ax + by = 0, \qquad cx + dy = 0$$

on which the tangent to the orbit is vertical or horizontal, respectively. Some of these are shown as dotted lines in Figures 6, 7, and 8.

As a rule, an isocline is *not a solution* of the differential equation. However, if the slope m of the line $y = mx$ agrees with dy/dx in (4) then the isocline is a solution. By inspection of (4) the condition for this to happen is

$$(5) \qquad\qquad m = \frac{c + dm}{a + bm}.$$

With an obvious choice of F the equation has the form $m = F(m)$. The number m is not changed by F and is called a *fixed point*.

In Section 2, Problem 1 it was seen that the quadratic for m given by (5) and the characteristic equation for λ, μ have the same discriminant. Hence if either equation has complex roots so does the other. If either equation has real and distinct roots so does the other. This confirms the fact that there are two straight-line orbits $y = mx$ in the case of a node or saddle and none in the case of a focus.

The line $x = 0$ is an orbit when $b = 0$ and corresponds to $m = \infty$. This case is taken up in Problem 6 and is excluded from the following theorem:

THEOREM 2.   *In the case of a node or saddle with $b \neq 0$:*

(a) *The slopes m of the two orbits $y = mx$ are given by (5).*

(b) *The two values of $a + bm$ satisfy the characteristic equation.*

(c) *The trajectory with orbit $y = mx$ is oriented toward the origin or away from the origin if $a + bm < 0$ or $a + bm > 0$, respectively.*

(d) *For a node, the line of tangency at the origin is the line $y = mx$ for which $|a + bm|$ is smaller.*

Part (a) was established in the above discussion, and both (a) and (b) in Section 2, Problem 1. The substitution $y = mx$ into (1) yields

$$x' = x(a + bm), \quad \text{hence} \quad x = ke^{(a+bm)t}$$

where k is constant. This gives (c) and reconfirms (b). Part (d) is obtained from an equation like (3) with general exponents (λ, μ); the details are easy and are omitted. For examples, see Problems 2 and 3.

PROBLEMS

1. Each matrix stands for a system of differential equations. Thus (a) pertains to the system $x' = -5x + 3y$, $y' = -4x + 2y$. Classify the critical point:

(a) $\begin{pmatrix} -5 & 3 \\ -4 & 2 \end{pmatrix}$ (d) $\begin{pmatrix} 3 & 4 \\ 2 & 1 \end{pmatrix}$ (g) $\begin{pmatrix} -6 & 1 \\ 5 & -2 \end{pmatrix}$ (j) $\begin{pmatrix} -3 & 1 \\ -5 & -7 \end{pmatrix}$

(b) $\begin{pmatrix} 2 & -1 \\ -4 & -1 \end{pmatrix}$ (e) $\begin{pmatrix} 6 & 2 \\ 3 & 7 \end{pmatrix}$ (h) $\begin{pmatrix} 2 & 1 \\ -8 & 6 \end{pmatrix}$ (k) $\begin{pmatrix} -5 & 7 \\ 3 & -1 \end{pmatrix}$

(c) $\begin{pmatrix} 1 & 4 \\ 8 & 5 \end{pmatrix}$ (f) $\begin{pmatrix} 1 & 3 \\ -2 & -4 \end{pmatrix}$ (i) $\begin{pmatrix} 7 & -1 \\ 3 & 3 \end{pmatrix}$ (l) $\begin{pmatrix} -1 & 4 \\ -10 & -5 \end{pmatrix}$

2P. For $x' = 6x + y$, $y' = 4x + 3y$ verify that the origin is an unstable node and draw velocity vectors on the axes as in Example 2. Then:

(a) Check that the fixed-point relation on the left below yields the equations on the right, and plot the corresponding orbits $y = mx$:

$$m = \frac{4 + 3m}{6 + m}, \qquad m^2 + 3m - 4 = 0, \qquad (m - 1)(m + 4) = 0$$

(b) Obviously $y = mx$ yields $x' = (6 + m)x$. Thus get

$$x = ce^{7t}, \qquad m = 1; \qquad x = ce^{2t}, \qquad m = -4$$

This shows that the trajectories are oriented away from the origin. Since $|2|$ is smaller than $|7|$, Theorem 2 says that the other orbits are tangent to the line $y = -4x$ at the origin and they behave like $y = x$ at distant points. Sketch the trajectories.

(c) Verify that the exponents 7 and 2 obtained in Part (b) satisfy

$$\begin{vmatrix} 6 - s & 1 \\ 4 & 3 - s \end{vmatrix} = 0$$

3P. For $x' = -3x + 2y$, $y' = -3x + 4y$, verify that the origin is a saddle and draw velocity vectors on the axes as in Example 2. Then:

(a) From the fixed-point equation, obtain the orbits $2y = x$ and $y = 3x$.

(b) Obviously $y = mx$ yields $x' = (-3 + 2m)x$. Using this, orient the trajectories corresponding to $y = mx$ in (a) and sketch some other trajectories.

(c) Check that the exponents $-2, 3$ obtained in Part (b) satisfy the characteristic equation.

4P. In this problem

$$x' = x - 4y, \quad y' = 4x + y.$$

Check that the origin is an unstable counterclockwise focus and verify the additional features shown in Figure 8.

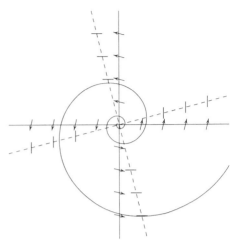

5. (abcdefghijkl) Sketch the trajectories in Problem 1. That is, sketch the orbits, and provide them with arrows showing orientation. For a node or saddle, you will find it instructive to do some as in Example 3, others by use of Theorem 2.

Figure 8

6. For the system $x' = ax$, $y' = cx + dy$ with $a \neq d$ check that

$$x = c_1 e^{at}, \qquad cx + (d - a)y = c_2 e^{dt}$$

Hence $U = x$, $V = cx + (d - a)y$ satisfy $|U|^d = |V|^a k$ and the trajectories can be plotted as in Section 1.

7. Eigenvectors. Problems 7, 8, and 9 depend on Chapter 16, Section 1. We set $X = (x, y)^T$ and we denote eigenvectors for λ, μ by E_1, E_2.

(a) If λ, μ are real and distinct, the lines $y = mx$ of Theorem 2 are obtained by the choices $c_1 = 0$ or $c_2 = 0$ in

$$X = c_1 E_1 e^{\lambda t} + c_2 E_2 e^{\mu t}$$

With this as a clue, deduce Theorem 2.

(b) By means of canonical variables show that the (x, y) orbits in the case of a node or saddle are affinely equivalent to the (U, V) orbits for those cases in Section 2.

(c) Obtain the trajectories in Example 3 by eigenvectors.

8. Using complex eigenvectors, show that the orbits for a focus are affinely equivalent to the (U, V) orbits obtained for that case in Section 2.

9. (abcdefghijkl) Use eigenvectors to sketch the trajectories in Problem 1.

————————————————ANSWERS————————————————

1. (afg) sN　(bcdk) S　(ei) uN　(h) ucF　(jl) scF
2. See Figure 6.　　3. See Figure 7.
5. Partial answer: The equations are the same as those in Chapter 16, Section 1, Problem 1 and answers are given there. Here is a clue to their interpretation:
   (a) $x = c_1 e^{-2t} + 3c_2 e^{-t}$, $y = c_1 e^{-2t} + 4c_2 e^{-t}$

---

4. Degenerate cases

If (p,q) is in any one of the five regions of Figure 5 the point (p_1, q_1) is in the same region provided its distance to (p, q) is sufficiently small. This means that the classification of the critical point as node, saddle, or focus is unchanged by sufficiently small perturbations of the coefficients.

We now consider limiting or degenerate cases that do not exhibit this insensitivity. On the contrary, the classification depends on certain exact equalities, and can change if a single coefficient is altered by an arbitrarily small amount. Since the coefficients are seldom known with infinite precision, the classification as node, saddle, or focus is more fundamental than the cases considered here.

(i) Center. If $4q > p^2$, as for a focus, but $p = 0$, the roots are purely imaginary and the critical point is called a *center*. Instead of being a spiral the orbits are ellipses centered at the origin as shown in Figure 9. To see why, note that the equivalent second-order equation is in this case

$$u'' + qu = 0, \quad q > 0.$$

Setting $q = \omega^2$ and recalling that $v = u'$ we get

$$u = A\cos(\omega t - \phi), \quad v = -\omega A \sin(\omega t - \phi)$$

where A and ϕ are constant. If $A \neq 0$ then

$$\frac{u^2}{A^2} + \frac{v^2}{(\omega A)^2} = 1$$

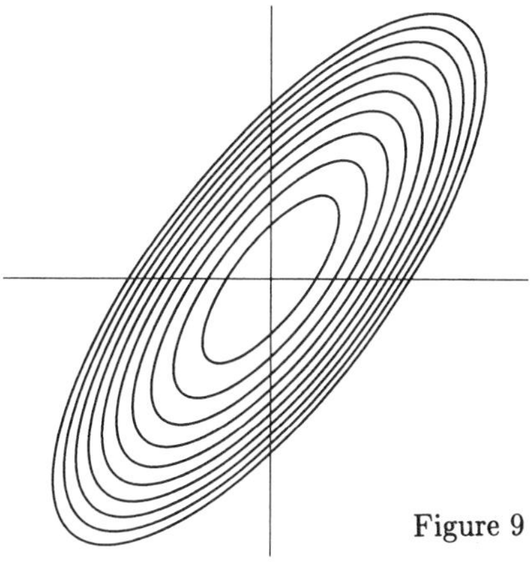

Figure 9

and this is the equation of an ellipse. A different proof was given in Chapter 7 and still another is given in Problem 1. The orbits are plotted by the same method as for a focus.

(ii) Degenerate node. Here we assume $4q = p^2$, $p \neq 0$ and we also assume that we do not have the case $b = c = 0$, $a = d$ that was excluded in the equivalence theorem. The roots of the characteristic equation are both equal to $-p/2$. Since they have the same sign one would expect the orbits to resemble those for a node, and they do. The only significant difference is that the fixed-point equation $m = F(m)$ has two equal roots, and the two straight-line solutions are now coincident. Thus, the orbits are tangent to a line at the origin and almost parallel to the same line at distant points. The proof requires only minor modification of the proof of Section 3, Theorem 2(d). The critical point in this case is a *degenerate node*. It is stable if $p > 0$, unstable if $p < 0$. See Figure 10.

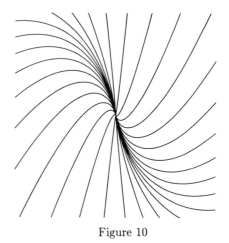

Figure 10

(iii) Singular node. We continue to assume $4q = p^2$ but we consider the special case $b = c = 0$, $a = d \neq 0$ that was excluded above. The linear system reduces to $x' = ax$, $y' = ay$, which can be solved by the methods of Chapter 1. The solution

$$x = c_1 e^{at}, \qquad y = c_2 e^{at}$$

shows that y/x is constant and hence the orbits are radial lines through the origin; see Figure 11. In this case the origin is a *singular node*. It is stable if $a < 0$, unstable if $a > 0$. Plotting the orbits is trivial.

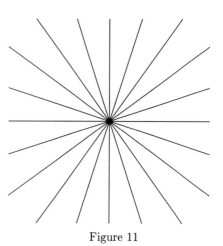

Figure 11

(iv) Vanishing determinant. Next we consider the case $q = 0$, which means that the determinant of the system

$$ax_0 + by_0 = 0, \qquad cx_0 + dy_0 = 0$$

is 0. Hence there is a nontrivial solution (x_0, y_0). Since (kx_0, ky_0) is also a solution for any constant k, there is a whole line of critical points. Thus it makes no sense to talk about "the critical point" and analyze its character.

The linear system is

$$x' = ax + by, \qquad y' = cx + dy, \qquad ad = bc.$$

Clearly $cx' - ay' = 0$, hence $cx - ay$ is constant, and this represents a family of parallel lines unless $a = c = 0$. In a like manner $dx - by$ is constant, and this represents a family of parallel lines unless $b = d = 0$.

(v) Zero coefficients. The only remaining case is $a = b = c = d = 0$. The differential equations $x' = 0$, $y' = 0$ show that every point (x_0, y_0) is a critical point, every solution $(x(t), y(t))$ is constant, and the problem loses interest.

PROBLEMS

1. If $p = 0$, so that $d = -a$, the equation of the orbits can be written in differential form as $(ax + by)\, dy = (cx - ay)\, dx$. Check that

$$cx^2 - 2axy - by^2 = \text{const.}$$

and that in the case of a center you have the additional condition $a^2 + bc < 0$. This shows that the discriminant of the above quadratic is negative and hence the orbits are ellipses centered at the origin.

2P. Assuming that $x \neq 0$ and $a + bu \neq 0$, let $y = ux$ in the left-hand equation to get the separable equation on the right:

$$\frac{dy}{dx} = \frac{cx + dy}{ax + by}, \qquad x\frac{du}{dx} + u = \frac{c + du}{a + bu}$$

This is helpful in some of the problems below.

3. In the following examples the equation of the orbits can be found with ease. Find it, and verify that the orbits have the general character that one would expect from the classification given in the text. Some are degenerate:

(a) $x' = y, y' = 0$ (c) $x' = y, y' = -x$ (e) $x' = x, y' = 2y$

(b) $x' = x, y' = y$ (d) $x' = -x, y' = y$ (f) $x' = x, y' = x + y$

4. For $x' = x + y$, $y' = -x + y$ introduce polar coordinates (r, θ) and show that $r' = r$, $\theta' = -1$, hence $r = ke^{-\theta}$. Check by classifying the critical point.

5. Here are the coefficient matrices for some degenerate cases. Classify the critical point and sketch the orbits:

(a) $\begin{pmatrix} 1 & 2 \\ -3 & -1 \end{pmatrix}$ (d) $\begin{pmatrix} 1 & -1 \\ 2 & -1 \end{pmatrix}$ (g) $\begin{pmatrix} -3 & 2 \\ -2 & 1 \end{pmatrix}$ (j) $\begin{pmatrix} -3 & 0 \\ 0 & -3 \end{pmatrix}$

(b) $\begin{pmatrix} 1 & 1 \\ -1 & 3 \end{pmatrix}$ (e) $\begin{pmatrix} 1 & 2 \\ 2 & 4 \end{pmatrix}$ (h) $\begin{pmatrix} -1 & 1 \\ -2 & 2 \end{pmatrix}$ (k) $\begin{pmatrix} 5 & 1 \\ -1 & 3 \end{pmatrix}$

(c) $\begin{pmatrix} 2 & 0 \\ 0 & 2 \end{pmatrix}$ (f) $\begin{pmatrix} -2 & 3 \\ -3 & 2 \end{pmatrix}$ (i) $\begin{pmatrix} -5 & -4 \\ 7 & 5 \end{pmatrix}$ (l) $\begin{pmatrix} -5 & -2 \\ 2 & -1 \end{pmatrix}$

6. Let $u'' + 2bu' + a^2 u = 0$ where $a > 0$ and $b \geq 0$. In Chapter 7 we considered four cases: undamped, underdamped, critically damped, and overdamped. Show that these correspond respectively to a center, stable focus, stable degenerate node, and stable node in the phase plane.

7. Degenerate node. Let the two equal roots be λ where λ is real and not 0. Thus u and $v = u'$ satisfy

$$u = (c_1 + c_2 t)e^{\lambda t}, \qquad v = \lambda u + c_2 e^{\lambda t}$$

The choice $c_2 = 0$ yields the straight-line orbit. Assuming $c_2 \neq 0$ check that the transformation $U = v - \lambda u$, $V = \lambda u$ gives

$$V = U \left(\frac{\lambda c_1}{c_2} + \lambda t \right) = U \left(\frac{\lambda c_1}{c_2} + \ln \frac{|U|}{|c_2|} \right)$$

If a positive constant k is defined by $\ln k = \lambda(c_1/c_2) - \ln |c_2|$ show that $kV = kU \ln(k|U|)$. Hence all trajectories, except $U = 0$, can be obtained from the single trajectory $V = U \ln |U|$ by expansion about the origin. Sketch this family of curves in the (U, V) plane. An affine transformation of it is shown in Figure 10.

8. Open-ended problem. If (x, y) satisfies a linear system of the type considered in the text, (kx, ky) satisfies it for any positive constant k. The transformation $(x, y) \rightarrow (kx, ky)$ represents an expansion (or contraction) with respect to the origin, hence takes any curve into a curve that is not only similar, but is similarly oriented. Here is a statement which, if true, is interesting and important:

"In a given linear system $x' = ax + by$, $y' = cx + dy$, all nontrivial orbits are similar, and all of them can be obtained from a single one of them by the transformation $(x, y) \rightarrow (kx, ky)$."

How should this statement be interpreted and to what extent is it true? Illustrate by reference to typical orbits and typical types of critical points, including degenerate cases. No answer is provided.

9. Eigenvalues. This problem requires more knowledge of linear algebra than needed elsewhere in this book. Analyze the behavior at a degenerate node by means of the theory of elementary divisors and Jordan canonical form. In particular, show that the orbit $y = mx$ is parallel to the eigenvector and that all orbits, except $y = mx$, are affinely equivalent to the locus $V = U \ln |U|$ obtained in Problem 7.

—————————————ANSWERS—————————————

Further abbreviations: C= center, d= degenerate, ŝ= singular.

3. Partial answer: (a) $y = k$ (c) $x^2 + y^2 = k$ (e) $y = kx^2$
   (b) $y = kx$ (d) $xy = k$ (f) $y = x \ln kx$

4. uF. In this special case the orbits are logarithmic spirals.

5. Partial answer: (ad) cC (bk) udN (c) uŝN (eh) parallel lines
   (f) ccC (gl) sdN (i) ccC (j) sŝN

Mathematical objects are sometimes as peculiar as the most
exotic beast or bird, and the time spent in examining them
may be well employed.

—Hugo Steinhaus

Mathematics and natural philosophy are so useful in the
most familiar occurrences of life and are so peculiarly engaging and delightful as would induce everyone to wish an
acquaintance with them.

—Thomas Jefferson

It may well be doubted whether, in all the range of science,
there is any field so fascinating to the explorer—so rich in
hidden treasures—so fruitful in delightful surprises—as that
of pure mathematics.

—Lewis Carroll

Chapter 20
A SURVEY OF NONLINEAR PROBLEMS

THE THEORY of linear approximation is used here to study competitive and predator-prey relationships in mathematical ecology, and the ecological problems in turn provide a guide to further development of the theory. Thus we are led to the methods of Liapunov and LaSalle, which bear upon the behavior of solutions in the remote future, and to theorems of Poincaré, Bendixson, and Dulac pertaining to existence and nonexistence of periodic solutions. Although the results are stated for two variables, some of them extend easily to problems in any number of variables. Others make essential use of plane topology and cannot be so extended.

Prerequisites: Chapter 19 is essential, Chapters 14 and A desirable.

1. An autonomous system

In this chapter the variables and functions are real and primes denote differentiation with respect to t. Intuitive concepts such as "closed curve" are used freely at first, then examined from a more sophisticated point of view as the exposition progresses.

In a system of the form

$$(1) \qquad x' = f(x, y), \qquad y' = g(x, y),$$

the right-hand side does not involve t explicitly, but only implicitly through the fact that x and y themselves depend on t. Such a system is called *autonomous*. The word "autonomous" means "self-governing" and implies that the slope of a solution-curve at (x, y) is determined solely by (x, y). Indeed, if one equation (1) is divided by the other the result is

$$(2) \qquad \frac{dy}{dx} = \frac{g(x, y)}{f(x, y)} \quad \text{or} \quad \frac{dx}{dy} = \frac{f(x, y)}{g(x, y)}$$

according as $f(x, y) \neq 0$ or $g(x, y) \neq 0$. The left-hand side is the slope of the solution-curve $y = \phi(x)$ or $x = \psi(y)$, as the case may be, and the right-hand side depends only on (x, y). Both formulations (2) are included in

$$(3) \qquad f(x, y)dy = g(x, y)dx.$$

If $x = \xi(t)$, $y = \eta(t)$ is a trajectory of (1), Equation (3) is a differential equation for the orbit or trace. The chain rule shows that

$$\overline{x} = \xi(t - t_0), \qquad \overline{y} = \eta(t - t_0)$$

satisfies (1) on the corresponding interval for $\bar{t} = t - t_0$. In other words, the behavior of an orbit through a point is independent of the time t_0 at which that point is reached.

When $f(x, y)$ and $g(x, y)$ are defined and not both zero, at least one equation (2) is a consequence of (1) and also of (3). But if

$$f(x_0,\, y_0) = g(x_0,\, y_0) = 0$$

neither equation (2) is meaningful at (x_0, y_0), and (x_0, y_0) is called a *critical point.* Since the constant functions defined by $x(t) = x_0$, $y(t) = y_0$ satisfy (1) in this case, the critical point is also called a *stationary point.* The substitution

$$u = x - x_0, \qquad v = y - y_0$$

yields a new system $u' = F(u, v)$, $v' = G(u, v)$ for which the corresponding stationary point is at the origin.

Near a critical point, a system of the form (1) can usually be approximated by a linear system with constant coefficients, and the procedure gives information that would be hard to get in any other way. It is now illustrated in connection with the system

(4) $$x' = x(x + y - 1), \qquad y' = y(x - y + 3).$$

Setting $x' = y' = 0$ in (4) we are led to four cases

$$\begin{cases} x & = 0 \\ y & = 0 \end{cases} \qquad \begin{cases} x & = 0 \\ x - y & = -3 \end{cases} \qquad \begin{cases} x + y & = 1 \\ y & = 0 \end{cases} \qquad \begin{cases} x + y & = 1 \\ x - y & = -3 \end{cases}$$

with corresponding stationary points $(0,0)$, $(0,3)$, $(1,0)$, and $(-1,2)$. A linear approximation near $(0,0)$ is obtained by neglecting quadratic terms in (4) and is $x' = -x$, $y' = 3y$ or, in matrix form,

$$\frac{d}{dt} \begin{pmatrix} x \\ y \end{pmatrix} = \begin{pmatrix} -1 & 0 \\ 0 & 3 \end{pmatrix} \begin{pmatrix} x \\ y \end{pmatrix}.$$

To get linear approximations at $(0,3)$, $(1,0)$, $(-1,2)$ we introduce the respective changes of variables

$$\begin{cases} u & = x \\ v & = y - 3 \end{cases} \qquad \begin{cases} u & = x - 1 \\ v & = y \end{cases} \qquad \begin{cases} u & = x + 1 \\ v & = y - 2 \end{cases}.$$

In terms of (u, v) the equations (4) become

$$\begin{cases} u' & = u(u + v + 2) \\ v' & = (v + 3)(u - v) \end{cases} \qquad \begin{cases} u' & = (1 + u)(u + v) \\ v' & = v(u + v + 4) \end{cases} \qquad \begin{cases} u' & = (u - 1)(u + v) \\ v' & = (v + 2)(u - v) \end{cases}$$

and, omitting quadratic terms, we get the second, third, and fourth matrices below for linear approximations near $(u, v) = (0, 0)$. The first is the matrix found above for (x, y) near $(0, 0)$:

$$(5) \qquad \begin{pmatrix} -1 & 0 \\ 0 & 3 \end{pmatrix}, \quad \begin{pmatrix} 2 & 0 \\ 3 & -3 \end{pmatrix}, \quad \begin{pmatrix} 1 & 1 \\ 0 & 4 \end{pmatrix}, \quad \begin{pmatrix} -1 & -1 \\ 2 & -2 \end{pmatrix}.$$

If (4) is written in the form (1) the functions f and g satisfy

$$\begin{pmatrix} f_x & f_y \\ g_x & g_y \end{pmatrix} = \begin{pmatrix} 2x + y - 1 & x \\ y & x - 2y + 3 \end{pmatrix}.$$

At $(0, 0)$, $(0, 3)$, $(1, 0)$, and $(-1, 2)$ this equation yields the four matrices (5). A corresponding property holds in general as seen below.

For the constant-coefficient linear system

$$x' = ax + by, \qquad y' = cx + dy, \qquad ad \neq bc$$

that was studied in Chapter 19, the behavior of the trajectories near a critical point falls into six classes as illustrated in Figure 1. The classification is determined by the signs of the three quantities

$$p = -(a + d), \qquad q = ad - bc, \qquad 4q - p^2.$$

The error introduced in replacing a nonlinear system by its linear approximation is similar to that introduced by a small change in the coefficients a, b, c, d. Table 1 pertains to cases in which the behavior of the linear system is insensitive to such changes. Cases that are sensitive are presented in Table 2.

Table 1

	stable		unstable		saddle	
p	$+$	$+$	$-$	$-$	$\pm$	0
q	$+$	$+$	$+$	$+$	$-$	$-$
$4q - p^2$	$+$	$-$	$+$	$-$	$-$	$-$
	focus	node	focus	node	saddle	saddle

Table 2

	stable	unstable	ambiguous
p	$+$	$-$	0
$4q - p^2$	0	0	$+$
	degenerate or singular node		center or focus

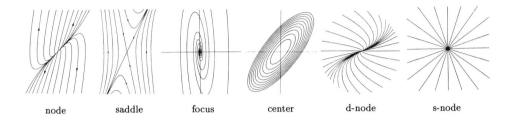

node   saddle   focus   center   d-node   s-node

Figure 1

By results of Chapter 19, a focus or center is counterclockwise if $c > 0$, clockwise if $c < 0$. The degenerate node in Table 2 is obtained when $a \neq d$, the singular node when $a = d$. If $p = 0$ and $q > 0$ the linear system always has a center. However a nonlinear system could have a center or focus, as explained later.

EXAMPLE 1. The pairs (p, q) for the matrices (5) are respectively

$$(-2, -3), \qquad (1, -6), \qquad (-5, 4), \qquad (3, 4).$$

Hence $4q - p^2 = -16$, -25, -9, 7. According to the tables, the corresponding critical points for the linear systems are:

saddle,   saddle,   unstable node,   stable focus.

By inspection of the matrix itself, the focus is counterclockwise. The trajectories are shown in Figure 2. Note that near the critical points

$$(0, 0), \quad (0, 3), \quad (1, 0), \quad (-1, 2)$$

the trajectories have the general behavior illustrated for the corresponding type of critical point in Figure 1. That this is not a coincidence is seen by the following Theorem 1.

Using a standard notation, we denote by $C^{(k)}$ the class of functions with continuous partial derivatives of order k. Thus, the relations

$$f \in C^{(0)}, \quad g \in C^{(1)}, \quad (f, g) \in C^{(2)}$$

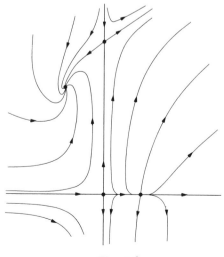

Figure 2

mean respectively that f is continuous, that g has continuous first partial derivatives, and that both f and g have continuous partial derivatives of order 2. If no region is specified, these conditions are to hold in the (x, y) plane.

THEOREM 1.   *Let* $(f, g) \in C^{(2)}$ *in a neighborhood of a critical point* (x_0, y_0). *Then Tables 1 and 2 apply to the system* $x' = f(x, y)$, $y' = g(x, y)$ *with*

$$\begin{pmatrix} a & b \\ c & d \end{pmatrix} = \begin{pmatrix} f_x(x_0, y_0) & f_y(x_0, y_0) \\ g_x(x_0, y_0) & g_y(x_0, y_0) \end{pmatrix}.$$

The proof will not be given here, though a start toward it can be found in items (i)–(vii) below and in the problems. The following comments give useful insight.

Theorem 1 pertains only to the behavior in the immediate neighborhood of (x_0, y_0). It says nothing about what happens at points far away from (x_0, y_0).

A critical point is said to be *isolated* if there is some disk that contains this critical point and no other. The condition $q \neq 0$ holds in all entries of the table and, as seen presently, it ensures that (x_0, y_0) is an isolated critical point. If we allowed $q = 0$, we would have one of the last two degenerate cases in Chapter 19, the critical point would not necessarily be isolated, and Theorem 1 would not apply. In fact, when $a = b = c = d = 0$ the linear system gives no information at all.

The concept of *center* in nonlinear problems requires some care as to its interpretation, and is discussed in the Review Problems at the end of this chapter. Aside from that, the trajectories of the nonlinear system near a critical point are a rather mild distortion of those in the linear case. The nature of the distortion is illustrated in the figures.

We conclude this introductory discussion by giving some general properties of (1) that are used later. If $(f, g) \in C^{(1)}$ it follows from Chapter 14 that (1) has one, and only one, local solution through any given point. Here are some further consequences of the hypothesis $(f, g) \in C^{(1)}$:

(i) If a solution is bounded for $t > 0$ as long as it exists, then it exists for all positive t.

(ii) A critical point cannot lie on any trajectory $x = \xi(t)$, $y = \eta(t)$ with ξ or η nonconstant.

(iii) In a region free of critical points, two orbits cannot cross nor can an orbit cross itself.

(iv) If $\lim_{t \to \infty} (x, y) = (x_1, y_1)$ exists, then (x_1, y_1) is a stationary point.

(v) A periodic solution corresponds to a simple closed orbit.

(vi) Near a critical point (x_0, y_0) the system (1) can be written

(6a)
$$\frac{d}{dt}\begin{pmatrix} x - x_0 \\ y - y_0 \end{pmatrix} = \begin{pmatrix} A & B \\ C & D \end{pmatrix}\begin{pmatrix} x - x_0 \\ y - y_0 \end{pmatrix}$$

where A, B, C, D are functions of (x, y) satisfying

(6b)
$$\lim_{(x,y)\to(x_0,y_0)}\begin{pmatrix} A & B \\ C & D \end{pmatrix} = \begin{pmatrix} f_x(x_0, y_0) & f_y(x_0, y_0) \\ g_x(x_0, y_0) & g_y(x_0, y_0) \end{pmatrix}.$$

(vii) The critical point in Theorem 1 is isolated.

To establish (i), let $0 \le t < t_1$ be the largest interval on which the solution exists for $t \ge 0$, and suppose that (x, y) is bounded on this interval. If $t_1 < \infty$ it follows from Chapter 14 that (x, y) tends to a definite point (x_1, y_1) as $t \to t_1-$. The local solution through (x_1, y_1) gives an extension of the original solution, contradicting the fact that t_1 was maximal.

Properties (ii) and (iii) follow from uniqueness.

To establish (iv), suppose (x_1, y_1) is not a stationary point. This means that at least one of the inequalities $f > 0$, $f < 0$, $g > 0$, or $g < 0$ holds at (x_1, y_1). Without loss of generality let

$$f(x_1, y_1) = -2\alpha < 0;$$

other cases are similar. If $t \ge t_0$, where t_0 is sufficiently large, we have $f(x, y) < -\alpha$ on the trajectory since f is continuous. Hence $x'(t) \le -\alpha$ for $t \ge t_0$. Problem 7 gives $x(t) \to -\infty$, which contradicts the hypothesis $x(t) \to x_1$.

Property (v) pertaining to periodic solutions requires explanation as well as proof. Although an orbit $x = \xi(t)$, $y = \eta(t)$ cannot cross itself, it can return to its starting point and start over. That is, we can have

(7)
$$\xi(T) = \xi(0), \qquad \eta(T) = \eta(0), \qquad T > 0.$$

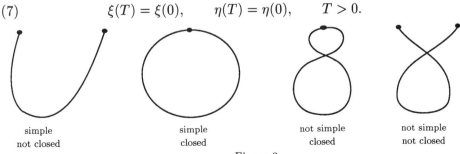

| simple | simple | not simple | not simple |
| not closed | closed | closed | not closed |

Figure 3

A nonconstant orbit with this property is said to be a *closed curve*, or, if it does not intersect itself, a *simple closed curve*. See Figure 3. When (7) holds, uniqueness yields

$$(8) \qquad \xi(T+t) = \xi(t), \qquad \eta(T+t) = \eta(t).$$

This is true because the trajectories defined by

$$x = \xi(T+t), \ y = \eta(T+t); \qquad x = \xi(t), \ y = \eta(t)$$

both satisfy (1) and both go through the same point when $t = 0$. Equation (8) says that the functions ξ and η have period T. Conversely, setting $t = 0$ in (8) yields (7).

Property (vi) is easily established if f and g have convergent Taylor series about (x_0, y_0). For example, since $f(x_0, y_0) = 0$, the Taylor series for f would have the form

$$f(x, y) = f_x(x_0, y_0)(x - x_0) + f_y(x_0, y_0)(y - y_0) + \cdots.$$

A proof of (vi) assuming only $(f, g) \in C^{(1)}$ is outlined in Problem 8.

To establish (vii), write the differential equation in the form (6a). The equations for a critical point (x_1, y_1) are then

$$(9) \qquad \begin{pmatrix} A & B \\ C & D \end{pmatrix} \begin{pmatrix} x_1 - x_0 \\ y_1 - y_0 \end{pmatrix} = \begin{pmatrix} 0 \\ 0 \end{pmatrix}.$$

If $q \neq 0$, that is, if $|ad - bc| > 0$, Equation (6b) gives $|AD - BC| > 0$ in a sufficiently small disk centered at (x_0, y_0). In this disk the sole solution of (9) is $x_1 = x_0$, $y_1 = y_0$, which shows that (x_0, y_0) is isolated. The condition $q \neq 0$ holds in the tables, hence it holds in Theorem 1.

PROBLEMS

1P. Obtain the four matrices (5) by appropriate choice of (x, y) in the matrix of partial derivatives following (5).

2P. Let $k + ax_0 + by_0 = 0$, $m + cx_0 + dy_0 = 0$ and let

$$x' = x(k + ax + by), \qquad y' = y(m + cx + dy)$$

Show that the change of variables $x = u + x_0$, $y = v + y_0$ yields

$$u' = (u + x_0)(au + bv), \qquad v' = (v + y_0)(cu + dv)$$

3. Find the stationary points and classify them by use of Theorem 1:

(a) $\begin{cases} x' &= x(6 - 2x - 3y) \\ y' &= y(4 - 3x - 4y) \end{cases}$

(e) $\begin{cases} x' &= x(-18 - 2x + 2y) \\ y' &= y(3 - x - 3y) \end{cases}$

(b) $\begin{cases} x' &= x(3 - 3x - 4y) \\ y' &= y(6 - 2x - 3y) \end{cases}$

(f) $\begin{cases} x' &= x(6 - 3y) \\ y' &= y(4 - 2x) \end{cases}$

(c) $\begin{cases} x' &= x(6 - 2x + 2y) \\ y' &= y(15 - x - 3y) \end{cases}$

(g) $\begin{cases} x' &= 5x(1 - y) \\ y' &= y(-12 + 3x) \end{cases}$

(d) $\begin{cases} x' &= x(-6 - 2x + 2y) \\ y' &= y(21 - x - 3y) \end{cases}$

(h) $\begin{cases} x' &= x(-6 + 2x - 3y) \\ y' &= y(12 - 4x + 6y) \end{cases}$

4. (abcdefg) Confirm the results of Problem 3 by finding linear approxima-tions near the stationary points. Since Problem 3 gives a check, no further answer is provided.

5P. On a full $8\frac{1}{2}$ by 11 inch sheet of graph paper, plot trajectories for the ma-trices (5) near the corresponding points $(0,0)$, $(0,3)$, $(1,0)$, $(-1,2)$. Then join these local charts by suitable curves to get a *global phase-plane portrait*. See Figure 2.

6. (abcdefg) This problem is time-consuming but instructive. After the manner of Problem 5, obtain global phase-plane portraits for the equations of Problem 3. Since some of the characteristic equations have irrational roots, you may find it helpful to use a hand calculator.

7P. Suppose $x'(t) < -\alpha$ for $t \geq t_0$, where α is a positive constant. If $t \geq t_0$ deduce that that $x + \alpha t$ is decreasing, hence $x(t) + \alpha t \leq x(t_0) + \alpha t_0$, hence $\lim_{t \to \infty} x(t) = -\infty$.

8. Proof of (vi). Assume without loss of generality that $x_0 = y_0 = 0$. Let $f \in C^{(1)}$ in a disk $r < \delta$ and define $F(t) = f(xt, yt)$ where (x, y) is a given point in this disk. The mean-value theorem gives $F(1) - F(0) = F'(\tau)$ for some τ satisfying $0 < \tau < 1$. Compute $F'(\tau)$ by the chain rule and deduce

(*) $$f(x, y) = x f_x(x\tau, y\tau) + y f_y(x\tau, y\tau)$$

The continuity of f_x and f_y yields

(**) $|f_x(x\tau, y\tau) - f_x(0, 0)| < \epsilon(r), \quad |f_y(x\tau, y\tau) - f_y(0, 0)| < \epsilon(r)$

where $\epsilon(r) \to 0$ as $r \to 0$. Thus Property (vi) holds for f with A and B equal respectively to the coefficients of x and y in (*). A similar argument applies to g.

9. If $f \in C^{(2)}$ it follows from results of Chapter 14 that f_x and f_y are locally Lipschitzian. Deduce that $\epsilon(r)$ in (**) can be replaced by Kr where K is constant, and similarly for g. This refinement is needed when Theorem 1 is applied to a degenerate or singular node and is the reason for assuming $(f, g) \in C^{(2)}$. For all other cases, $(f, g) \in C^{(1)}$ suffices.

10P. Make a tracing of Chapter 19, Figure 5. On your tracing put seven small disks to represent the seven possibilities for p, q and $4q - p^2$ in Table 1. Also put three small circles to represent the three possibilities in Table 2. This figure gives insight into Theorem 1.

11. If you are sufficiently familiar with Chapter 14 and with n-variable calculus, extend the main results of this section to $x' = f(x)$ where $x \in R^n$ and $f(x) \in R^n$.

───────────────────────ANSWERS───────────────────────

3. (a) $(0,0)$uN, $(0,1)$S, $(3,0)$sN, $(-12, 10)$sF
   (b) $(0,0)$uN, $(0,2)$sN, $(1,0)$S, $(-15, 12)$S
   (c) $(0,0)$uN, $(0,5)$S, $(3,0)$S, $(6,3)$sF
   (d) $(0,0)$S, $(0,7)$S, $(-3,0)$uN, $(3,6)$(sdN or sŝN)
   (e) $(0,0)$S, $(0,1)$sN, $(-9,0)$uN, $(-6,3)$S
   (f) $(0,0)$uN, $(2,2)$(cC or cF)   (g) $(0,0)$S, $(4,1)$(ccC or ccF)
   (h) The critical points are $(3\alpha, 2\alpha - 2)$ where α is arbitrary. These are not isolated. By (vi), or by direct calculation, $q = 0$ and Theorem 1 does not apply.

6. If it seems impossible to complete the figure, you may have made an error in one of the local approximations.

2. Problems from ecology

Let $x = \xi(t)$ and $y = \eta(t)$ denote the sizes, at time t, of two interacting populations: trees, bacteria, animals, or fish, to name four examples. These functions are nonnegative and a value 0 for either of them means that the corresponding population is extinct.

If $dx/dt = kx$, the constant k measures the rate of growth of x and is called the growth constant. A more general, and often a more realistic model is obtained if k is replaced by a function F whose value depends on one or both populations. Upon making a similar generalization of $dy/dt = ky$, we are led to the autonomous system

(1) $x' = xF(x,y), \quad y' = yG(x,y), \quad t \geq 0; \qquad \xi(0) > 0, \quad \eta(0) > 0.$

The initial conditions mean that both populations are present at the start. Special systems of the form (1) were analyzed by the American biologist Alfred James Lotka and the Italian mathematician Vito Volterra in the decade before 1930, and (1) is a *generalized Lotka-Volterra system*. It has a host of applications not only to ecology but to other disciplines.

Throughout the following analysis, we assume $(F, G) \in C^{(0)}$ in Q and we impose additional hypotheses as needed. The first quadrant is denoted by Q and its interior by Q_0; thus

$$Q = \{(x, y) : x \geq 0, \ y \geq 0\}, \qquad Q_0 = \{(x, y) : x > 0, \ y > 0\}.$$

The origin is a *repeller* for (1) if there is a positive constant δ, independent of (ξ, η), such that the function

$$\rho(t) = \sqrt{\xi(t)^2 + \eta(t)^2}$$

is increasing whenever $\rho < \delta$. This means that the point (x, y) moves away from the origin as t increases.

Here are some properties of (1) that will be used:

(i) The trajectory remains in Q_0 as long as it exists.

(ii) If $F(0, 0) > 0$ and $G(0, 0) > 0$, the origin is a repeller.

(iii) For $(x, y) \in Q$ let $F(x, y) \leq 0$ and $G(x, y) \leq 0$ outside of some disk $r \leq r_0$. Then all solutions are bounded.

(iv) Near a critical point $(x_0, y_0) \in Q_0$ where $(F, G) \in C^{(1)}$ the matrix for the linear approximation is

$$\begin{pmatrix} x_0 F_x(x_0, y_0) & x_0 F_y(x_0, y_0) \\ y_0 G_x(x_0, y_0) & y_0 G_y(x_0, y_0) \end{pmatrix}.$$

To establish (i), write the differential equation for x in the form

$$x' = x\phi(t) \quad \text{where} \quad x = \xi(t), \ y = \eta(t), \ \phi(t) = F(\xi(t), \eta(t)).$$

If $\xi(t_0) = 0$ at some value t_0 this yields $\xi \equiv 0$, contradicting the hypothesis $\xi(0) > 0$. Discussion of y is similar.

For (ii) and (iii) let $r = \rho(t)$, so that $r^2 = x^2 + y^2$ and

$$rr' = xx' + yy' = x^2 F(x, y) + y^2 G(x, y).$$

By continuity the hypothesis (ii) gives $F(x, y) > 0$ and $G(x, y) > 0$ on some disk $r < \delta$. Hence $r' > 0$ for $0 < r < \delta$. This shows that the point (x, y) moves away from the origin whenever it is within the disk $r < \delta$ and (ii) follows. By the same calculation, the hypothesis in (iii) gives a constant r_0

such that $r' \leq 0$ whenever $r > r_0$. Hence r is bounded, and this shows that x and y are bounded also.

If (x_0, y_0) is a critical point in Q_0 then $F(x_0, y_0) = G(x_0, y_0) = 0$. Differentiating xF and yG, we get (iv).

(a) Competitive systems

Let x and y be the numbers of individuals in two competing populations; for example, two varieties of trees competing for sunlight, sheep and goats competing for a limited amount of forage, or two kinds of fish that do not prey on each other but do have a common source of food. Let us suppose that, in the absence of competition, growth would be governed by the logistic equations

$$x' = kx - ax^2, \qquad y' = my - dy^2.$$

The effect of competition is assumed to be negative and jointly proportional to both populations. Adjoining a term $-bxy$ to the first equation and $-cxy$ to the second, we are led to the system

(2) $\qquad\qquad x' = x(k - ax - by), \qquad y' = y(m - cx - dy)$

where all the coefficients are positive constants. This is a mathematical model for competition. Besides the origin, there are two stationary points $(k/a, 0)$, $(0, m/d)$ on the axes, and in the text we assume that there is a stationary point $(x_0, y_0) \in Q_0$. The case in which there is no such point is considered in the problems.

When (2) is written in the notation (1) we have

(3) $\qquad\qquad F(0,0) = k > 0, \qquad G(0,0) = m > 0$

and if $x + y$ is sufficiently large both functions F and G are negative. By Properties (i), (ii), and (iii), the trajectories of (2) exist for $0 \leq t < \infty$, they remain in a bounded region of Q_0, and they stay away from the origin. In Section 5 we prove that no trajectory of (2) is a closed curve. From this it can be deduced that $\xi(t)$ and $\eta(t)$ have limits as $t \to \infty$, and the existence of these limits will be taken for granted here. By Section 1, Item (iv) the limit is a stationary point.

There is a simple graphical procedure for deciding whether the limit is (x_0, y_0) or is one of the stationary points on the axes. The decision is important, because it determines the long-time fate of the two populations. The first step is to plot the lines

$$ax + by = k, \qquad cx + dy = m$$

on which $x' = 0$ and $y' = 0$, respectively. These lines are indicated as xx and yy in Figure 4. Their point of intersection is the stationary point

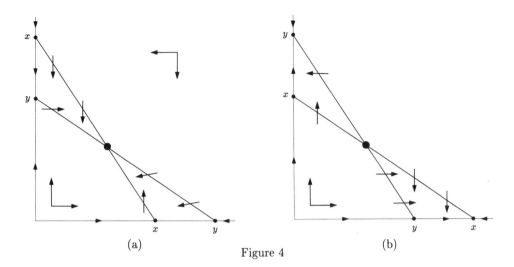

(a)   (b)

Figure 4

$(x_0, y_0) \in Q_0$ whose existence was postulated above. It is isolated if, as we now assume, $ad \neq bc$. On xx the velocity vector is vertical, on yy it is horizontal. To determine the sense of the vector (up or down, right or left) note that at points far from the axes both x' and y' are negative by (2), while at points near the origin they are both positive. This information is summarized by the two pairs of orthogonal vectors in the figure. For want of a better term, such pairs of orthogonal vectors are referred to as *indicators*.

When we move across the line xx from one side to the other, x' in (2) changes sign, going through the value 0 on the line itself. Similarly for y' and the line yy. Hence, starting at either indicator, we reverse the direction of the x component of the velocity vector when crossing line xx and we reverse the direction of the y component when crossing line yy. If one of the two indicators can be moved to a point P without crossing xx or yy, that indicator gives the direction at P directly.

In Figure 4(a) the arrows suggest that the stationary point (x_0, y_0) is stable, and that every trajectory tends to this point for $t \to \infty$ no matter where it starts. Figure 4(b) suggests that the stationary point (x_0, y_0) is unstable and that, except for infinitely rare cases, one or the other population will go to extinction. These conclusions are correct, and are now explored more fully.

(b) The separatrix

Property (iv) above yields the following matrix and values p, q for the linear approximation near (x_0, y_0):

$$\begin{pmatrix} -ax_0 & -bx_0 \\ -cy_0 & -dy_0 \end{pmatrix}, \qquad p = ax_0 + dy_0, \qquad q = x_0 y_0 (ad - bc).$$

The difference between Figure 4(a) and 4(b) is that the xx line slants down more steeply than the yy line in Figure 4(a), less steeply in 4(b). Comparing slopes we see that $ad < bc$ in the first case and $ad > bc$ in the second. The values p and q found above now show that (x_0, y_0) is a stable node in Figure 4(a) and a saddle in Figure 4(b). This confirms the behavior suggested by the direction field. Further confirmation is given by Figure 5.

In Figure 4(b) the only critical point allowing coexistence is unstable, and it is virtually certain that one of the populations will drive the other to extinction. From the standpoint of ecology, the principal problem is to decide from the initial values whether the x-population or the y- population is the one that ultimately survives. The curve separating these two cases is called the *separatrix* and is discussed next.

The fact that (x_0, y_0) is a stationary point is expressed by

$$k = ax_0 + by_0, \qquad m = cx_0 + dy_0.$$

If we use these to eliminate k and m from (2) it is found that

$$u' = -(u + x_0)(au + bv), \qquad v' = -(v + y_0)(cu + dv)$$

where $u = x - x_0$, $v = y - y_0$. Division yields the following equation for the phase-plane orbits:

$$(4) \qquad \frac{dv}{du} = \frac{(v + y_0)(cu + dv)}{(u + x_0)(au + bv)}.$$

The separatrix is determined by the solution (u, v) through the origin with slope α, where α is the positive root of

$$\alpha = \frac{y_0(c + d\alpha)}{x_0(a + b\alpha)}.$$

We shall not give the proof, but the result is made plausible by consideration of the approximating linear system. The separatrix is the emphasized curve in Figure 5(b).

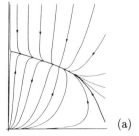

(a)

Figure 5

(b)

The case $k = m$ is of particular interest, because it means that both populations have the same intrinsic growth rates and any difference in survival depends wholly upon the interaction. In this case, by Problem 12, the separatrix is the straight line through $(0,0)$ and (x_0, y_0). In general, however, Equation (4) does not have elementary solutions, and the separatrix must be found by numerical methods. A linear approximation to it is given by $v = \alpha u$ and the method of Problem 13 leads to the quadratic approximation

$$v = \alpha u + \beta u^2 \quad \text{where} \quad \beta = \alpha \frac{(c - a) + (d - b)\alpha}{(2a + 3b\alpha)x_0 - dy_0}.$$

Although we shall not give the proof, it can be shown that

$$\lim_{t \to \infty} (x, y) = \left(\frac{k}{a}, 0\right) \quad \text{or} \quad \lim_{t \to \infty} (x, y) = \left(0, \frac{m}{d}\right)$$

according as the trajectory starts on the same side of the separatrix as the point $(k/a, 0)$ or on the same side as the point $(0, m/d)$. In the first case the x-population and in the second case the y-population is the one that survives. If the initial point is on the separatrix, (x, y) tends to (x_0, y_0) and both populations survive. However, this is a contingency of zero probability and would not be observed in nature.

The right- and left-hand regions bounded by the separatrix are *basins of attraction* for $(k/a, 0)$ and $(0, m/d)$ respectively. In recent years the concept of "basin of attraction" has assumed great importance in differential equations and in the theory of iterative processes.

(c) Predator-prey equations

The equations introduced next are best understood in a specific context. Let x be the number of foxes and y the number of rabbits in two populations. The foxes eat rabbits and the rabbits eat vegetation, of which there is an unlimited supply. In the absence of rabbits the number of foxes would decline exponentially, and in the absence of foxes the number of rabbits would increase exponentially. This situation is described by

$$x' = -kx, \qquad y' = my$$

where k measures the deathrate of foxes and m the birthrate of rabbits.

We assume next that the interaction of rabbits and foxes is jointly proportional to both populations. Growth of the fox population is helped by the rabbits and growth of the rabbit population is inhibited by the foxes. With these thoughts in mind, we add a term bxy to x' and we subtract a term cxy from y'. The result is

$$(5) \qquad x' = x(-k + by), \qquad y' = y(m - cx)$$

where k, m, b, c are positive constants. Equations (5) are the Lotka-Volterra *predator-prey equations*.

The lines $by = k$ and $cx = m$ on which $x' = 0$ and $y' = 0$ intersect at $(m/c, k/b)$ and this yields a stationary point $(x_0, y_0) \in Q_0$. Hence the populations can always coexist in a steady state. It is left for you to check that $p = 0$, $q = bcx_0y_0$ at (x_0, y_0) and hence the stationary point for the linear approximation is a center. This means that the phase-plane orbits are closed curves and the corresponding approximate solutions are periodic. It turns out that the nonlinear system has the same properties, and we now explain why this is so.

Leaving (5) for a moment, consider a system of the form

$$(6) \qquad x' = R(x, y)f(y), \qquad y' = R(x, y)g(x)$$

where f and g are continuous. If F is an integral of f and G of g it is easily checked that

$$F(y) - G(x) = \text{const.}$$

Indeed, the derivative of the expression on the left is $f(y)y' - g(x)x'$ and this reduces to 0 by virtue of the differential equations.

Returning to (5), note that the equations can be written

$$x' = xy\left(\frac{-k}{y} + b\right), \qquad y' = xy\left(\frac{m}{x} - c\right).$$

The functions f, g and their integrals F, G are found by inspection, with the result

$$-k \ln y + by - m \ln x + cx = \text{const.}$$

It is left for you to verify that exponentiation yields

$$(7) \qquad x^m y^k e^{-(cx+by)} = H$$

where H is constant. When interpreted in three dimensions, (7) describes the intersection of the surface

$$(8) \qquad z = x^m y^k e^{-(cx+by)}$$

with the plane $z = H$.

We regard the surface as a mountain in which the locus $z = H$ gives a contour line of constant height H above sea level. The elevation of the surface is positive in Q_0, is zero on the axes, and it tends to 0 on the line $cx + by = L$ as L tends to infinity. By a short calculation the equations

$$\frac{\partial z}{\partial x} = 0, \qquad \frac{\partial z}{\partial y} = 0$$

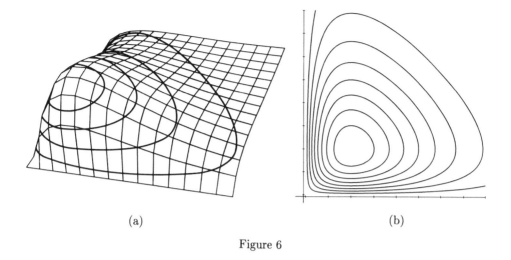

(a)   (b)

Figure 6

yield $x = x_0$ and $y = y_0$, where (x_0, y_0) is the stationary point in Q_0 found above. Hence the maximum of z in Q is unique and is given by

$$H_0 = x_0{}^m y_0{}^k e^{-(cx_0 + by_0)}.$$

This means that the mountain has a single peak and its height is H_0.

A typical example of a surface defined by (8) is shown graphically in Figure 6(a) and contour lines $z = H$ are shown in Figure 6(b). Since these are closed curves, the corresponding solutions are periodic. The interior of the contour line for H is, by definition, the region in Q in which $z > H$. Hence if $H_1 > H_2$ the contour for H_1 is interior to that for H_2. Further information is obtained by considering the behavior as $H \to 0$ and as $H \to H_0$.

From the standpoint of biology, the periodicity is explained as follows: When the fox population is large, the rabbit population declines, and eventually the foxes do not have enough to eat. Then the fox population declines, and when this happens, the rabbit population goes up. The process of successive increase and decline leads to periodic variation in both populations. Such variation has been observed in nature, and formulation of an equation that predicts it was one of the first major successes of mathematical ecology.

(d) Self-limiting terms

When a growth law $y' = my$ is replaced by $y' = my - dy^2$ the extra term $-dy^2$ describes a kind of self-interaction that prevents the population from becoming infinite. It is called the *self-limiting term*. The presence or absence of self-limiting terms can have a profound effect on the behavior of an ecological system.

As an illustration, let us introduce self-limiting terms $-ax^2$ and $-dy^2$ for x and y, respectively, in the predator-prey equations (5). The result is the system

(9) $$x' = x(-k - ax + by), \qquad y' = y(m - cx - dy)$$

where a, b, c, d, k, m are positive constants. The equations for a stationary
point $(x_0, y_0) \in Q_0$ give

$$x_0 = \frac{bm - dk}{ad + bc}, \qquad y_0 = \frac{am + ck}{ad + bc}.$$

Hence $y_0 > 0$ automatically but $x_0 > 0$ only if $bm > dk$. The fact that (9)
need not have a stationary point in Q_0 is an important difference between
(9) and (5).

A second difference is that a stationary point $(x_0, y_0) \in Q_0$ is a center
in (5), but is not a center in (9) even when it exists. To see this, we write
the matrix (iv) for (9) and the corresponding quantities p, q:

$$\begin{pmatrix} -ax_0 & bx_0 \\ -cy_0 & -dy_0 \end{pmatrix}, \qquad p = x_0 a + y_0 d, \qquad q = x_0 y_0 (ad + bc).$$

Since $p > 0$ and $q > 0$ the point (x_0, y_0) is a stable focus or stable node when
$4q - p^2$ is positive or negative, respectively. If $4q = p^2$ the node is singular
or degenerate, but still stable.

These results suggest that the solutions of (9) tend to a unique limit as
$t \to \infty$, and a proof that they do is given in Section 3. If $bm > dk$ the limit
is the stationary point $(x_0, y_0) \in Q_0$, and both populations survive. But if
$bm \le dk$ there is no stationary point in Q_0 and the limit is $(0, m/d)$. Thus
the predators become extinct and only the prey survive. See Figure 7.

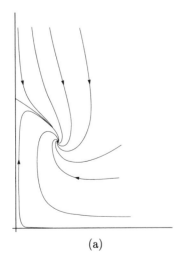

(a)

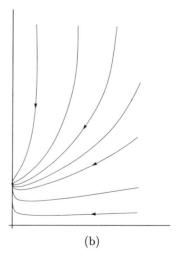

(b)

Figure 7

PROBLEMS————————————————————————————

1. The point of this problem is to emphasize that the precise form of (1) is important for (i). Show that the system $x' = -3x^{2/3}y$, $y' = y$ has solutions

$$x = (c - e^t)^3, \qquad y = e^t$$

some of which extend from the first quadrant to the second.

2P. Near the origin, linear approximations to the competition equations (2) and the general predator-prey equations (9) are respectively

$$x' = kx,\ y' = my; \qquad x' = -kx,\ y' = my$$

Solve, eliminate t, and sketch in four representative cases.

3P. In the competition equations (2) suppose the lines xx and yy do not intersect in Q. Using indicators as in the text, predict the ultimate fate of the populations (a) when the xx line and (b) when the yy line is the farther from the origin.

4. Analyze the following competitive systems by indicators, as in the text:

(a) $\begin{cases} x' &= x(3 - 2x - y) \\ y' &= y(3 - x - 2y) \end{cases}$ (d) $\begin{cases} x' &= x(2 - x - y) \\ y' &= y(5 - x - 4y) \end{cases}$

(b) $\begin{cases} x' &= x(3 - x - 2y) \\ y' &= y(3 - 2x - y) \end{cases}$ (e) $\begin{cases} x' &= x(5 - x - 4y) \\ y' &= y(2 - x - y) \end{cases}$

(c) $\begin{cases} x' &= x(2 - x - 2y) \\ y' &= y(3 - x - y) \end{cases}$ (f) $\begin{cases} x' &= x(6 - x - 2y) \\ y' &= y(3 - 3x - y) \end{cases}$

5. (abcdef) Confirm the results of Problem 4 by finding the character of the critical points.

6. Here are two predator-prey systems. Draw a figure like Figure 4, use it to predict the qualitative behavior, and confirm in detail:

(a) $x' = x(y - 1)$, $y' = y(1 - x)$ (b) $x' = 2x(2 - y)$, $y' = 3y(x - 5)$

7. In the two cases $bm > kd$, $bm < kd$ make figures for (9) analogous to Figure 4.

8. Solve the linear approximation to the predator-prey equations (5) near the stationary point $(m/c, k/b)$ and show that the solutions have period $T = 2\pi/\sqrt{km}$. Show also that the orbits are ellipses with horizontal and vertical axes, and that they are circles if $b^2 m = c^2 k$.

9. Let $x = \xi(t)$, $y = \eta(t)$ be a solution of the predator-prey equations (5) of period T. The mean values of x, y, and xy are defined by the integrals on the left below. Show that they have the values on the right:

$$\frac{1}{T}\int_0^T \xi(t)\,dt = x_0 \qquad \frac{1}{T}\int_0^T \eta(t)\,dt = y_0 \qquad \frac{1}{T}\int_0^T \xi(t)\eta(t)\,dt = x_0 y_0$$

Hint: Write the first equation as $x'/x = -k + by$ and integrate from 0 to T. Since $\ln x$ has period T, the term x'/x integrates to 0 and you get an equation for the mean value of y.

10. Pesticides. In the predator-prey equations (5) suppose the prey is the target population for a pesticide that acts equally on both populations; that is, it subtracts hx from x' and hy from y'. Here h is a positive constant measuring the strength of the pesticide. Using the result of Problem 9, show that the mean value of the target population is increased by the pesticide.

11. Pesticides, continued. Let the basic equations in Problem 10 be the general predator-prey equations (9). Show that the pesticide never drives the prey to extinction, but drives the predator to extinction if

$$h > \frac{bm - dk}{b + d}$$

According to this model, environmental stress affects the predator more than it affects the prey. The same conclusion is suggested by the simpler model of Problem 10.

12. Show that the competition equations (2) admit a solution of the form $y = \alpha x$ if, and only if, $k = m$, and in that case $\alpha = y_0/x_0$. The line $y = \alpha x$ so obtained obviously goes through the stationary point. Show that it has the slope prescribed in the text for a separatrix.

13. Given $x' = x(5 - x - 4y)$, $y' = y(2 - x - y)$ let $u = x - 1$, $v = y - 1$. Check that the (u, v) equation for the orbits is

$$(1 + u)(u + 4v)dv = (1 + v)(u + v)du$$

and that the value α for the slope of the separatrix at $(1, 1)$ is $1/2$. Substitute $v = (u/2) + \beta u^2$ and show that the quadratic terms cancel if, and only if, $\beta = -3/28$.

14. (abcdef) This problem is time-consuming but instructive. Plot the trajectories near each of the four critical points in Problem 4 and thus get a global phase-plane portrait in Q. Some of the characteristic equations have irrational roots.

15. Here are two predator-prey equations with self-limiting terms:

(a) $\begin{cases} x' & = x(-1 - x + y) \\ y' & = y(3 - x - y) \end{cases}$ (b) $\begin{cases} x' & = x(-3 - x + y) \\ y' & = y(1 - x - y) \end{cases}$

Sketch the behavior near the critical points in Q and thus get a global phase-plane portrait in Q.

────────────────────────ANSWERS────────────────────────

2. Partial answer: For (2), $y^k = Cx^m$. When $t \to \infty$, the trajectories approach the origin tangent to the y axis if $k > m$, tangent to the x axis if $k < m$, and in any direction if $k = m$. For (9), $x^m y^k = C$. The trajectories near $(0,0)$ resemble the hyperbolas $xy = C$ with x and y axes as asymptotes.

3. (a) Only the x-population survives. (b) Only the y-population survives.

4. Partial answer: (d) is the basis for Figure 5(a) and (a) is like it. (e) is the basis for Figure 5(b) and (b) is like it. In (c) and (e) just one population survives. See also answers to Problem 5.

5. (a) $(1,1)$sN, $(3/2,0)$S, $(0,3/2)$S   (d) $(1,1)$sN, $(2,0)$S, $(0,5/4)$S

   (b) $(0,3)$sdN, $(3,0)$sdN, $(1,1)$S   (e) $(0,2)$sdN, $(5,0)$sdN, $(1,1)$S

   (c) $(0,3)$sN, $(2,0)$S   (f) $(8,0)$sN, $(0,3)$S

   The character of the critical point $(0,0)$ is given by Problem 3.

6. Partial answer: (a) $(1,1)$cC, $xye^{-(x+y)} = H$. See Figure 6.

   (b) $(5,2)$ccC, $x^{15}y^4 e^{-3x-2y} = H$. Here x denotes prey.

15. Partial answer: (a) $(0,0)$S, $(0,3)$S, $(1,2)$scF

    (b) $(0,0)$S, $(0,1)$sN. See Figure 7.

──

3. General methods

In this section the methods introduced earlier are developed in greater detail and from a more general point of view. We discuss direction fields, closed curves, and existence of limits in the same order in which these topics were taken up in Section 2.

(a) Signs and the direction field

In the system $x' = f(x,y)$, $y' = g(x,y)$ suppose the locus $f(x,y) = 0$ divides the plane into two regions, H^- and H^+, such that

$$f(x,y) < 0 \quad \text{in } H^-, \qquad f(x,y) > 0 \quad \text{in } H^+.$$

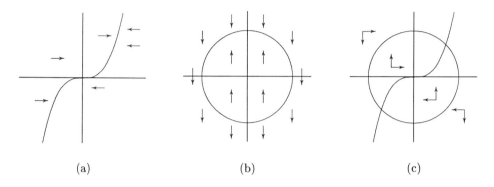

(a)   (b)   (c)

Figure 8

This means that the horizontal component of the velocity vector points to the left in H^-, to the right in H^+. For example, if

(1) $$x' = y - x^3, \qquad y' = 2 - x^2 - y^2$$

then $f(x,y) = y - x^3$, the dividing curve is the cubic $y = x^3$, and the horizontal components are as shown in Figure 8(a).

In a like manner, suppose the locus $g(x, y) = 0$ divides the plane into two regions, V^- and V^+, such that

$$g(x, y) < 0 \quad \text{in } V^-, \qquad g(x, y) > 0 \quad \text{in } V^+.$$

Here the vertical component of the velocity vector points downward in V^-, upward in V^+. For example, in (1) the locus $g(x, y) = 0$ is a circle and the vertical component of the velocity vector is as shown in Figure 8(b).

The indicators in Figure 8(c) summarize Figures 8(ab) and lead to Figure 9(a). The stationary points are the points where the cubic and circle intersect. On the cubic the tangent is vertical, on the circle it is horizontal.

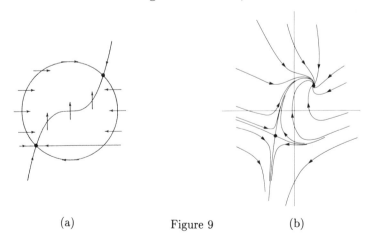

(a)   Figure 9   (b)

The direction is given by the above discussion: to the right above the cubic, to the left below; upward inside the circle, downward outside.

A trajectory can never cross the circle or cubic in a direction contrary to the arrow on it. For example, suppose a trajectory starts *inside the circle* and *above the line* $y = -1$. Since the y component points upward, the curve remains above the line $y = -1$ as long as it is in the circle. It cannot cross the circle at the left of the y axis, because the x component on this part of the circle points into the circle. Thus the trajectory no doubt ends up by spiraling inward toward the point $(1, 1)$. For confirmation see Figure 9(b) and Problem 1.

The same type of analysis can be carried out whenever one has regions H^-, H^+, V^-, V^+ as described here. In Section 2 the separating curves were straight lines. The use of more general curves leads to a significant increase in scope and power.

(b) Starlike and convex curves

One of the simplest ways of representing closed curves is by means of an equation

$$r = \phi(\theta), \qquad 0 \le \theta \le 2\pi$$

in polar coordinates, where ϕ is continuous, positive, and periodic with period 2π. A curve that can be thus described is said to be *starlike with respect to the origin* and its interior is defined by

$$r < \phi(\theta), \qquad 0 \le \theta \le 2\pi.$$

If $\phi \in C^{(1)}$, the curve has no corners and is called *smooth.*

A stronger property than being starlike is expressed by convexity. A simple closed curve C is *strictly convex* if the following holds: Given any two distinct points on C, the line segment joining them is, except for its ends, wholly interior to C. See Figure 10.

If H is a positive constant, the circle $u^2 + v^2 = H$ is obviously a closed curve, starlike with respect to the origin and strictly convex. We shall generalize this remark to a broad class of equations $U(u) + V(v) = H$.

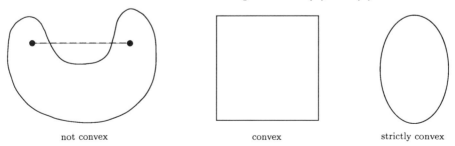

not convex   convex   strictly convex

Figure 10

To see what form the generalization should take, note that u^2 is defined on $(-\infty, \infty)$, it tends to infinity as u approaches either end of the interval, and it vanishes at the origin. Furthermore its derivative is negative when u is negative, positive when u is positive. These properties are incorporated in the following definition:

DEFINITION 1.   *Let $U \in C^{(1)}$ on an open interval I that contains the point 0. Suppose further that $U(0) = 0$, that $U(u) \to \infty$ as u approaches either end of I, and that $uU'(u) > 0$ for $u \in I$, $u \neq 0$. Then U is called an L function.*

In all applications intended here, the functions $U(u)$ and $V(v)$ will turn out to be special cases of the *Liapunov functions* discussed in Section 4. The prefix "L" is for Liapunov.

EXAMPLE 1. Let U be defined on $(-1, \infty)$ by

$$U(u) = \int_0^u \frac{s}{s+1}\, ds = u - \ln(u+1).$$

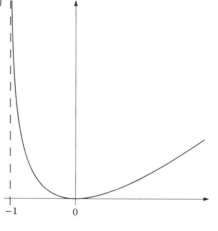

Clearly $U(0) = 0$ and $U(u) \to \infty$ as u approaches $(-1)+$ or ∞. Also

$$U'(u) = \frac{u}{u+1}, \qquad U''(u) = \frac{1}{(u+1)^2}.$$

Hence $uU'(u) > 0$ for $u \neq 0$ and U is an L function. The second equation yields $U'' > 0$, an inequality needed below. See Figure 11.

Figure 11

THEOREM 1.   *Let U and V be L functions. Then the locus defined by $U(u) + U(v) = H > 0$ is a smooth closed curve that is starlike with respect to the origin. If $U'' > 0$ and $V'' > 0$ the curve is strictly convex.*

Instead of requiring that the functions U, V tend to ∞, we could require that they exceed a constant M near the endpoints of their intervals. In that case the conclusion of Theorem 1 holds for $0 < H < M$. This will be clear from the proof, which is given at the end of the section. The following example shows how the result is used.

EXAMPLE 2. The substitution $u = x - x_0$, $v = y - y_0$ in the predator-prey equations of Section 2 yields

$$u' = (u + x_0)bv, \qquad v' = -(v + y_0)cu.$$

Guided by Section 2, Equation (6), we write

$$u' = (u+x_0)(v+y_0)\frac{bv}{v+y_0}, \qquad v' = -(u+x_0)(v+y_0)\frac{cu}{u+x_0}$$

and we introduce the integrals

$$(2) \qquad U(u) = \int_0^u \frac{cs}{s+x_0}\,ds, \qquad V(v) = \int_0^v \frac{bs}{s+y_0}\,ds.$$

Just as in Section 2, it follows that $U(u) + V(v) = H$, where H is constant. By a minor extension of Example 1, the functions U and V are L functions with positive second derivatives, and Theorem 1 now shows that the orbits are strictly convex closed curves. This confirms the main result of Section 2, Part (c), and gives more information than was stated there.

EXAMPLE 3. Let $x'' + g(x) = 0$ with $g \in C^{(1)}$, $g(0) = 0$, and $g'(0) \neq 0$. The substitution $x' = y$ gives

$$\frac{y^2}{2} + G(x) = H, \qquad G(x) = \int_0^x g(\xi)\,d\xi$$

where H is constant. Since $G' = g$ the function $G(x)$ has a local maximum or minimum at 0 when $g'(0) < 0$ or $g'(0) > 0$ respectively. For the corresponding system $x' = y$, $y' = -g(x)$ the matrix in Section 1, Theorem 1 yields $p = 0$, $q = g'(0)$. Hence a local maximum of G at 0 corresponds to a saddle at $(0,0)$. A local minimum of G at 0 gives the condition $uG'(u) > 0$ of Definition 1, and we conclude that the trajectories are simple closed curves. Thus the origin is a center and the solutions near the origin are periodic in t. (This does not follow from Section 1, Theorem 1, because the case $p = 0$, $q > 0$ is ambiguous; it allows either a focus or a center.)

Since x can be replaced by $x - x_0$, these results apply to any critical point $(x_0, 0)$ at which $g'(x_0) \neq 0$. The connection between periodic solutions and a local minimum of G is of central importance in the theory of nonlinear oscillations. An interpretation of H in terms of energy is suggested by Chapter 7, Section 2 and an expression for the period can be found in Problem 8 of that section. Note that (x, y) here corresponds to (u, v) in Theorem 1.

(c) Global asymptotic stability

A major problem in nonlinear theory is to predict the way a system evolves as time goes on. The system might be an ecological community, a set of interconnected electrical components, or planets orbiting around the sun, to name three examples. Sometimes the behavior in the remote future is simple, even though the details of evolution are complicated. The following definition pertains to cases of this kind:

DEFINITION 2.   *The stationary point* (x_0, y_0) *is globally asymptotically stable if all trajectories satisfy*

$$\lim_{t \to \infty} \xi(t) = x_0, \qquad \lim_{t \to \infty} \eta(t) = y_0.$$

As a rule the initial values are confined to a suitable region of the plane, Q_0 for example, and the phrase "all trajectories" is interpreted accordingly. Within this constraint, the limit is independent of the initial conditions.

The following example suggests a connection between Definition 2 and the concept of asymptotic stability as developed in Chapter 8.

EXAMPLE 4.   If $x'' + 3x' + 2x = 0$ the formula $x = c_1 e^{-t} + c_2 e^{-2t}$ shows that $(x, x') \to (0, 0)$ as $t \to \infty$. Hence the origin is globally asymptotically stable for solutions of the corresponding system $x' = y$, $y' = -2x - 3y$.

An application that is interesting both historically and practically is given by the predator-prey equations with self-limiting terms. These are Equations (9) of Section 2, which we repeat for convenience:

$$(3) \qquad x' = x(-k - ax + by), \qquad y' = y(m - cx - dy).$$

The following theorem holds:

THEOREM 2.   *If* (3) *has a stationary point* $(x_0, y_0) \in Q_0$, *this point is globally asymptotically stable for all solutions starting in* Q_0.

For proof, the substitution $u = x - x_0$, $v = y - y_0$ yields

$$(4) \qquad u' = (u + x_0)(-au + bv), \qquad v' = (v + y_0)(-cu - dv).$$

Let us see what happens when we try the same function $U(u) + V(v)$ that was used in Part (b) above. By (2)

$$U'(u) = \frac{cu}{u + x_0}, \qquad V'(u) = \frac{bv}{v + y_0}.$$

Hence, on any trajectory, the function $w(t) = U(u) + V(v)$ satisfies

$$(5) \qquad w'(t) = U'(u)u' + V'(v)v' = -(acu^2 + bdv^2).$$

Since w is decreasing and nonnegative the limit

$$(6) \qquad \lim_{t \to \infty} w(t) = \beta \geq 0$$

exists; see the completeness axiom of Chapter 12, Section 4.

We claim that $\beta = 0$. If not, a moment's thought shows that $u^2 + v^2$ cannot come arbitrarily close to the origin. That is,

$$u^2 + v^2 \geq \alpha$$

where α is a positive constant. But then $w'(t) \leq -\alpha \min(ac, bd)$ by (5), and Section 1, Problem 7 gives $w \to -\infty$. This contradicts the fact that $w \geq 0$. Hence the limit (6) is 0 and we get, in succession,

$$\lim_{t \to \infty} U(u) = 0, \qquad \lim_{t \to \infty} u(t) = 0, \qquad \lim_{t \to \infty} x(t) = x_0.$$

A similar result holds for y and completes the proof.

(d) Proof of Theorem 1

To establish Theorem 1 let $W = U(u) + V(v)$ and set

$$u = r \cos \theta, \qquad v = r \sin \theta.$$

For $r > 0$ the chain rule and the hypothesis of Theorem 1 give

$$\frac{\partial W}{\partial r} = \frac{uU'(u)}{r} + \frac{vV'(v)}{r} > 0.$$

Let $H > 0$ be constant and let θ be regarded, for the moment, as fixed. Clearly $W < H$ when r is small, and $W > H$ when r is sufficiently large. Since W is continuous and strictly increasing as a function of r, the equation $W = H$ has a unique solution $r = \phi(\theta)$. The solution is periodic in θ because W is, and $\phi \in C^{(1)}$ by Chapter A, Theorem E.

To establish the convexity, let (u_0, v_0) and (u_1, v_1) be two points on the curve $W = H$ and let

$$u_s = su_1 + (1-s)u_0, \qquad v_s = sv_1 + (1-s)v_0, \qquad 0 \leq s \leq 1$$

be parametric equations of the line segment joining them. Under the additional hypothesis $U'' > 0$, $V'' > 0$ of Theorem 1, we want to show that (u_s, v_s) is interior to the curve $W = H$ for $0 < s < 1$.

The interior is defined by $W < H$. Hence our problem is to show that the function

$$W(s) = U(u_s) + V(v_s)$$

satisfies $W(s) < H$ for $0 < s < 1$. By the chain rule

$$W''(s) = U''(u_s)(u_1 - u_0)^2 + V''(v_s)((v_1 - v_0)^2 > 0.$$

Since $W(0) = W(1) = H$, the condition $W'' > 0$ yields $W(s) < H$ for $0 < s < 1$ and this completes the proof.

PROBLEMS————————————————————————————————

In these and subsequent problems, H denotes a constant.

1P. For the equations $x' = y - x^3$, $y' = 2 - x^2 - y^2$ in Part (a), classify the stationary points, find linear approximations near them, and thus get a global phase-plane portrait. It should agree with Figure 9.

2. Analyze the trajectories as in Part (a) and Problem 1:

(a) $x' = y - x$, $y' = 2 - x^2 - y^2$ (b) $x' = y - x^2$, $y' = x^2 + y^2 - 2$

3. Tissier's problem is to show that $dy/dx = x - (1/y)$ has exactly one solution on $[0, \infty)$ that is positive and tends to 0 at ∞. As in Part (a) sketch the phase-plane trajectories for $x' = y, y' = xy - 1$ and give reasons for thinking that there is exactly one solution satisfying $0 < y < 1/x$.

4. Sketch the phase-plane trajectories near each critical point, combine in a single figure, and verify agreement with Example 3:

(a) $x' = y$, $y' = x^3 - x$ (b) $x' = y$, $y' = x - x^3$

5. The undamped pendulum. By Chapter 9, Section 2 the motion of an undamped pendulum with $x = \theta$ is described by $x' = y$, $\ell y' = -g \sin x$. Check that

$$\frac{\ell y^2}{2} + g(1 - \cos x) = H$$

We take $|x| \le \pi$; other cases are obtained by periodicity. If $0 < H < 2g$ show that the locus is a smooth closed curve, starlike with respect to the origin, and strictly convex when $H < g$. See Figure 12.

Hint: You have $|x| < \pi$ in the first case and $|x| < \pi/2$ in the second. Use the extension of Theorem 1 given informally in the text.

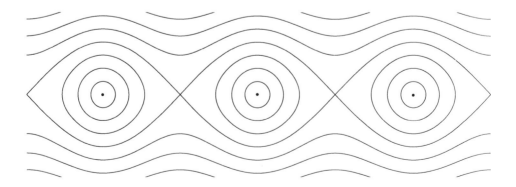

Figure 12

6. If $H > 2g$ in Problem 5 the value of $\ell y^2/2$ is between H and $H - 2g$. Using this as a clue, drop the restriction $|x| < \pi$ and sketch the locus. These curves correspond to an initial velocity so large that the pendulum makes complete revolutions about its pivot. The separatrix is given by $H = 2g$ and is easily sketched. It describes a situation in which the initial conditions are such that the pendulum tends to its unstable equiblirium position (exactly above the pivot) as $t \to \infty$.

7. Convert the equation $x'' = x^3 - x$ into a system by setting $x' = y$ and deduce $2x^2 - x^4 + 2y^2 = H$. Check that $U(x) = 2x^2 - x^4$ satisfies the conditions for an L function if $|x| < 1$ and satisfies $U''(x) > 0$ if $3x^2 < 1$. Deduce that, if $0 < H < 1$, part of the locus is a smooth closed curve starlike with respect to the origin. It is strictly convex if $H < 5/9$.

8. In Theorem 2 suppose $bm \le dk$, so there is no stationary point in Q_0. Repeat the proof with $x_0 = 0$, $y_0 = m/d$ and obtain the same conclusion, $(x, y) \to (x_0, y_0)$.

9. Figure 13 describes an efficient graphical method for sketching the locus defined by $U(u) + V(v) = H$ when the graphs of U and V are known. Explain the figure.

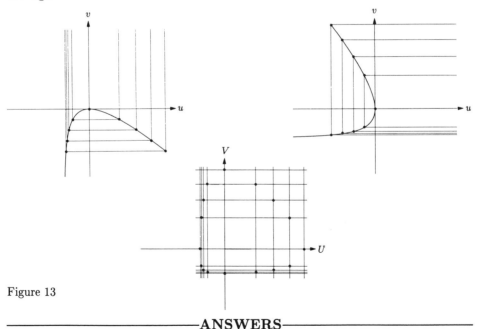

Figure 13

———————————————ANSWERS———————————————

2. Partial answers: (a) $(1,1)$scF, $(-1,-1)$S   (b) $(1,1)$S, $(-1,1)$ucF
3. Partial answer: The desired solution is the separatrix between the solutions that cross the curve $xy = 1$ and those that cross the x-axis. You are not expected to solve Tissier's problem, which is hard.

4. See Chapter 7, Section 2, Figure 4.

8. Since $x > 0$ and $bm \leq dk$ it will be found that (5) holds as an inequality. In other respects the proof is virtually unchanged.

---

4. Liapunov functions

The physical concept of "energy" in mechanical systems leads to a powerful method of studying nonlinear problems. Before taking up the general theory, we illustrate in an example.

EXAMPLE 1. If the displacement in a mass-spring system satisfies

$$x'' + x = 0$$

and the velocity is $y = x'$, the quantity $W(x,y) = (1/2)(x^2 + y^2)$ is constant. This was verified in Chapter 7, where it was seen that W represents the total energy, kinetic plus potential.

Let us now introduce a resistance term that depends on displacement and velocity, so that $x'' + k(x,x')x' + x = 0$. When written as a first-order system, the result is

$$x' = y, \quad y' = -x - k(x,y)y.$$

We assume $k \geq 0$ and $k \in C^{(1)}$. On any trajectory $x = \xi(t)$, $y = \eta(t)$ the function $w = (x^2 + y^2)/2$ satisfies

$$w' = xx' + yy' = -k(x,y)y^2 \leq 0.$$

This shows that w is decreasing. Hence w is bounded for $t \geq 0$; therefore x and y are also bounded, and by Section 1 the solution exists for all $t \geq 0$. Since w is decreasing and nonnegative, w has a limit β as $t \to \infty$. If $\beta > 0$ then (x,y) is almost on the circle $w = \beta$ for all large t, while if $\beta = 0$, both x and y tend to 0. Thus we have obtained a good deal of information without much effort.

In his doctoral dissertation of 1892 the Russian mathematician Alexander Mikhailovich Liapunov introduced more general functions to play the role that in this example is taken by the energy. The main features of Liapunov's theory are explained next.

Throughout our discussion $x = \xi(t)$, $y = \eta(t)$ is a given solution of

(1) $$x' = f(x,y), \qquad y' = g(x,y), \qquad t \geq 0$$

and W is a function of (x,y). For simplicity, we assume $(f,g,W) \in C^{(1)}$. This gives uniqueness and, under the hypotheses of our theorems, it will be seen that all solutions exist for $0 \leq t < \infty$.

If (1) holds, the equation

$$\frac{dW}{dt} = \frac{\partial W}{\partial x}\frac{dx}{dt} + \frac{\partial W}{\partial y}\frac{dy}{dt}$$

reduces to $dW/dt = \dot{W}$ where, by definition,

$$\dot{W}(x,y) = W_x(x,y)f(x,y) + W_y(x,y)g(x,y).$$

It is important that $\dot{W}$ can be computed without knowledge of the trajectories. If

$$w(t) = W(\xi(t), \eta(t))$$

is the value of W on a trajectory, however, then $w'(t) = \dot{W}(\xi(t), \eta(t))$.

By definition, W is a *Liapunov function* for (1) if

(i) $W \in C^{(1)}$, $W(0,0) = 0$, $W \geq 0$, $\dot{W} \leq 0$, and

(ii) $W(x,y)$ bounded implies $|x|$ and $|y|$ bounded.

We say that W is *positive definite* if $W(x,y) > 0$ except when $x = y = 0$, and $\dot{W}$ is *negative definite* if $-\dot{W}$ is positive definite.

As an illustration, consider the system

$$x' = y - x^3, \qquad y' = -x - y^3$$

with $W = x^2 + y^2$. In this case

$$\dot{W}(x,y) = 2(xx' + yy') = -2(x^4 + y^4) \leq 0.$$

Hence W is a Liapunov function, W is positive definite, and $\dot{W}$ is negative definite. The following theorem of Liapunov now shows that the origin is globally asymptotically stable. That is, all solutions tend to $(0,0)$ as $t \to \infty$:

THEOREM 1. *Let $(f,g) \in C^{(1)}$ and suppose the system (1) has a Liapunov function W. Then:*

(a) *All solutions are bounded on $(0, \infty)$ and $\lim_{t \to \infty} w(t)$ exists.*

(b) *If $\dot{W}$ is negative definite then $\lim_{t \to \infty} w(t) = 0$.*

(c) *If $\dot{W}$ is negative definite and W is positive definite, the origin is globally asymptotically stable.*

For proof, the condition $\dot{W} \leq 0$ makes $w'(t) \leq 0$, hence $w(t) \leq w(0)$. This shows w is bounded. Since W is a Liapunov function, $|x|$ and $|y|$ are also bounded, and existence of the solution on $(0, \infty)$ follows from Section 1, (i). The function $w(t)$ has a limit because it is decreasing and bounded below. This gives (a).

The proof of (bc) depends on the following theorem:

THEOREM A. *A continuous real-valued function $\phi(x, y)$ on a bounded closed set attains its maximum and minimum.*

A set is *closed* if it contains all its boundary points. Theorem A will be applied in a ring

$$(2) \qquad\qquad r_0 \leq \sqrt{x^2 + y^2} \leq r_1$$

where r_0 and r_1 are constants satisfying $0 < r_0 < r_1$.

To establish (b), let the limit of $w(t)$ in (a) be β, and note that $w(t) \geq \beta$ because w is decreasing. We have $\beta \geq 0$ since $w \geq 0$, and we want to show $\beta = 0$. If not, choose $r_0 > 0$ so that $W(x, y) < \beta$ for $r \leq r_0$. This can be done because W is continuous and $W(0,0) = 0$. Since $w(t) \geq \beta$ the point (x, y) on the trajectory for $t > t_0$ cannot come into the disk $r \leq r_0$. By (a) the trajectory is bounded, hence it lies in a ring of the type (2). Since $\dot{W} < 0$ in the ring, its maximum is a negative number $-\alpha < 0$. The resulting inequality $w'(t) \leq -\alpha$ makes $w(t) \to -\infty$, contradicting the fact that $w \geq 0$.

For (c) let us suppose that W and $-\dot{W}$ are both positive definite but that nevertheless the function

$$\rho(t) = \sqrt{\xi(t)^2 + \eta(t)^2}$$

does not tend to 0. This means that $\rho(t) \geq r_0$ on a sequence $t = t_n \to \infty$, where r_0 is a positive constant. All values (x, y) on this sequence are in a ring of the type (2), and in such a ring the minimum of $W(x, y)$ is a positive constant, γ. This makes $w(t_n) \geq \gamma$ and contradicts the result of Part (b). Hence $\rho(t) \to 0$, completing the proof.

The condition that $\dot{W}$ be negative definite in Theorem 1 often fails for the Liapunov functions that arise naturally in applications. For example, it fails for the equation considered in Example 1 even if the resistive term $k(x, y)$ is positive. An efficient method of dealing with such cases was discovered by the American mathematician Joseph P. LaSalle in 1960.

As a preliminary to LaSalle's theorem we introduce the following:

DEFINITION. A limit point of a trajectory $x = \xi(t), y = \eta(t)$ is a point (x_1, y_1) such that

$$\lim_{n \to \infty} \xi(t_n) = x_1, \qquad \lim_{n \to \infty} \eta(t_n) = y_1$$

for some sequence $\{t_n\}$, $t_n \to \infty$. The set of all limit points of a given trajectory Λ is called the limit set and is denoted by Λ^+.

For example, if $\lim_{t\to\infty} \xi(t) = x_0$ and $\lim_{t\to\infty} \eta(t) = y_0$ then Λ^+ consists of the single point (x_0, y_0). Conversely, if Λ^+ consists of a single point, then the solution tends to this point as a limit. As another example, if the trajectory has period T then

$$\lim_{n\to\infty} \xi(t + nT) = \lim_{n\to\infty} \xi(t) = \xi(t); \quad \lim_{n\to\infty} \eta(t + nT) = \lim_{n\to\infty} \eta(t) = \eta(t).$$

This holds at each t and shows that, in the case of a periodic solution, Λ^+ coincides with Λ.

Here is a simplified form of LaSalle's theorem:

THEOREM 2. *Let $(f, g) \in C^{(1)}$ and let W be a Liapunov function for (1). Then the limit set Λ^+ for any given trajectory is contained in the set of all orbits whose trajectories satisfy $\dot{W}(x, y) = 0$.*

We give only a partial proof. By Theorem 1 the function

$$w(t) = W(\xi(t), \eta(t))$$

has a limit as $t \to \infty$. If we denote the limit by β this gives

$$\lim_{n\to\infty} w(t_n) = \beta$$

for every sequence $\{t_n\}$, $t_n \to \infty$. Let us choose a sequence that produces a given point $(x_1, y_1) \in \Lambda^+$. Since W is continuous the above remarks yield

$$W(x_1, y_1) = \lim_{n\to\infty} W(\xi(t_n), \eta(t_n)) = \lim_{n\to\infty} w(t_n) = \beta.$$

This is true for every $(x_1, y_1) \in \Lambda^+$, hence W is constant on Λ^+.

It can be shown that any trajectory starting in Λ^+ remains in Λ^+ on its whole interval of definition. In other words, Λ^+ is an *invariant set* for the differential equation. The fact that W is constant in Λ^+ now gives $\dot{W} = 0$ on the trajectory containing (x_1, y_1). Thus every point of Λ^+ is contained in an orbit satisfying $\dot{W} = 0$, and that is what LaSalle's theorem says.

For the $C^{(1)}$ case considered here, LaSalle's theorem contains the main result of Liapunov. To see why, let $\dot{W}$ be negative definite. Then the only trajectory satisfying $\dot{W} = 0$ is $x = y = 0$, hence the limit set consists of the single point $(0, 0)$. This shows that the origin is globally asymptotically stable and yields Theorem 1(bc) even without the hypothesis that W is positive definite.

EXAMPLE 2. Let $x'' + k(x, x')x' + x = 0$ where $k \in C^{(1)}$. With $y = x'$ it was seen in Example 1 that

$$2W = x^2 + y^2 \quad \text{gives} \quad \dot{W} = -k(x, y)y^2.$$

Suppose now that $k(x, y) > 0$ unless $y = 0$. Under this assumption the condition $\dot{W}(x, y) = 0$ gives $y = 0$ and the differential equation then gives $x = 0$. Hence the sole trajectory satisfying $\dot{W} = 0$ is the trivial trajectory $x = y = 0$. By LaSalle's theorem, the origin is globally asymptotically stable. Since $\dot{W}$ is not negative definite, this cannot be deduced from Liapunov's theorem.

The theory developed here and in Section 3 admits a number of variations. For example, a Liapunov function need not be defined in the whole (x, y) plane but may be defined only in some region. As another variation, we may seek to show that a trajectory is bounded, or tends to the origin, only if it starts sufficiently near the origin. Such results are local, in contrast to the global theorems that have been emphasized here.

Instead of formulating a multiplicity of variations, we have singled out a few basic results and have concentrated on the proofs. The thought is that if you master the proofs, you will have no difficulty in applying the theory to cases not explicitly covered by the theorems. A few cases of this kind are given in the problems.

PROBLEMS————————————————————————

1. Taking $W = x^2 + y^2$ in Theorem 1, show that the origin is globally asymptotically stable:

$$x' = y^6 - ye^{xy} - x^3, \qquad y' = xe^{xy} - xy^5 - y^3$$

2P. Show that the conclusion of Example 2 remains valid if $k(x, y)y$ is replaced by any function $K(x, y) \in C^{(1)}$ such that $yK(x, y) > 0$ for $y \neq 0$.

3. The equations $x' = y$, $\ell y' = -g \sin x - ky$ describe motion of a nonlinear damped pendulum with damping constant $k > 0$. Taking $g = k = \ell$ use the energy function from Section 3, Problem 5 to establish asymptotic stability for $|x| < \pi$; see Example 2. Also check that the behavior near the stationary points agrees with Figure 14.

4. Let A, B, C be constant. Show that $W(x, y) = Ax^2 + 2Bxy + Cy^2$ is positive definite if, and only if, $A > 0$, $B > 0$, $B^2 < AC$.

Hint: $AW(x, y) = (Ax + By)^2 + (AC - B^2)y^2$.

5. If W in Problem 4 is positive definite, its minimum on the circle $r = 1$ is a positive constant δ. Deduce that $W(x, y) \geq \delta(x^2 + y^2)$.

Hint: $W(x, y) = r^2 W(x/r, y/r)$ for $r > 0$.

Figure 14

6. Symbiosis. Suppose two populations x and y are described by the logistic equations when they do not interact, and suppose that their interaction adds a positive term bxy to x' and a positive term cxy to y'. The resulting equations

$$x' = x(\pm k - ax + by), \qquad y' = y(\pm m + cx - dy)$$

provide a model for a type of mutual reinforcement known as *symbiosis*. The trouble with this model is that it often predicts that the populations become infinite.

(a) If $k = m = 0$ and $bc < ad$ show that $W = cx + by$ is a Liapunov function and deduce from Theorem 1 that $(x, y) \to (0, 0)$.

(b) If $bc < ad$ and $k = m = 0$ is not assumed show that (x, y) is bounded.

Hint: For (a) use Problem 4, for (b) use Problem 5.

7. Try $W = \alpha x + \beta y$ in Problem 6 and show that an optimum choice of the positive constants α, β is $\alpha = c$, $\beta = b$.

8. Show that the conclusion of Section 3, Theorem 2 remains valid if there is only a single self-limiting term; that is, if $a \geq 0$, $d \geq 0$, $a + d > 0$.

9. Peano's theorem was mentioned without proof in Chapter 14; it gives local existence of solutions of (1) if $(f, g) \in C^{(0)}$. Examine the proof of Theorem 1 and check that $(f, g) \in C^{(0)}$ suffices. By contrast, the hypothesis $(f, g) \in C^{(1)}$ can be weakened only slightly for the theorem of LaSalle. Hence, when stated in full generality, neither theorem contains the other.

10. If you are sufficiently familiar with Chapter 14 and with n-variable calculus, extend the main results of this section to $x' = f(x)$ where $x \in R^n$ and $f(x) \in R^n$.

5. Results depending on plane topology

In general, a *curve* is defined by $x = \xi(t), y = \eta(t)$ for $0 \leq t \leq T$ where $T > 0$ and where ξ and η are continuous. If the conditions

$$\xi(t_1) = \xi(t_2), \qquad \eta(t_1) = \eta(t_2), \qquad 0 \leq t_i < T$$

hold only when $t_1 = t_2$, the curve is *simple*. This means that it does not intersect itself. The curve is *closed* if the endpoint coincides with the initial point, that is, if $\xi(0) = \xi(T)$ and $\eta(0) = \eta(T)$. For reasons given later, a simple closed curve is called a *Jordan curve*. In the case of a Jordan curve, it is helpful to consider that the interval $0 \leq t \leq T$ is wrapped around a circle of circumference T. We can then summarize as follows: A simple unclosed curve is a continuous one-to-one image of a segment. A Jordan curve is a continuous one-to-one image of a circle.

An open set Ω is *connected* if every pair of points in Ω can be joined by a simple curve that lies wholly in Ω. Intuitively, this means that the set does not consist of two or more separate pieces. An open connected set is called a *domain* and is here denoted by D. However, this use of "domain" should be distinguished from its use in such phrases as "the domain of an operator" (Chapter 4) or "an integral domain" (Chapter E).

The *Jordan curve theorem* was first formulated by the French mathematician Marie Ennemond Camille Jordan in 1887. It states that a simple closed curve divides the plane into two domains: an interior domain, which is bounded, and an exterior domain. Despite the fact that the statement seems self-evident, Jordan's proof was inconclusive, and the first correct proof was given by the American mathematician Oswald Veblen in 1905.

A domain D is *simply connected* if, whenever a Jordan curve lies in D, the interior of the curve is also in D. Intuitively, this means that D has no holes. See Figure 15.

We shall use these concepts to study the trajectories of

$$(1) \qquad\qquad x' = f(x, y), \qquad y' = g(x, y).$$

In contrast to most of the results of Sections 1–4, those obtained here depend on the topology of the plane and do not extend to higher dimensions.

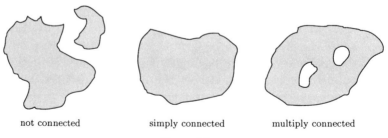

not connected   simply connected   multiply connected

Figure 15

(a) A theorem of Poincaré

For the predator-prey equations, it was seen in Section 2 that the orbits are closed curves surrounding the critical point. The following theorem of Poincaré generalizes this observation:

THEOREM 1.   *Let $(f, g) \in C^{(0)}$ in a simply connected domain D and let C be a simple closed trajectory of (1) lying in D. Then there is at least one critical point inside or on C.*

By "a simple closed trajectory" we mean a trajectory whose orbit is a Jordan curve. If $(f, g) \in C^{(1)}$, uniqueness shows that there can be no critical point on C. The theorem then implies that there is a critical point inside C.

We do not give a proof, but we do explain the principal ideas. These ideas are four in number, and each of them is important in several parts of mathematics.

(i) Let C be the image of the circle $r = 1$. It is possible to fill out C and its interior domain by the sets C_s where C_s is the image of the circle $r = s$, $0 \le s \le 1$. Thus C_s is a curve for $0 < s \le 1$, with $C_1 = C$, and C_0 reduces to a point. The region consisting of C and its interior is filled out in much the same way as the circles $r = s$ for $0 < s \le 1$ and the point $r = 0$ fill out the disk $r \le 1$. See Figure 16.

(ii) If (f, g) does not take the value $(0, 0)$ inside or on C, the angle that this vector makes with the x-axis can be so defined as to be a continuous function of the parameter t on C_s. This function changes by a multiple of 2π when C_s is traversed. If the total change is $\Delta\theta$ the quantity $\Delta\theta/(2\pi)$ is an integer that is denoted by I_s and is called the *index* of C_s with respect to the vector field (f, g). The index is a continuous integer-valued function of s.

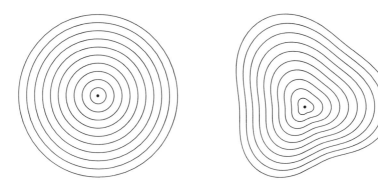

Figure 16

(iii) The differential equation shows that the vector (f, g) is tangent to the closed trajectory C, hence makes exactly one revolution when C is traversed. Thus the index is $+1$ or -1 when C is traversed in the positive or negative direction, respectively. If s is close to 0, however, the curve C_s is almost a point, the vector (f, g) is almost constant on C_s, and the index $I_s = 0$.

(iv) An integer-valued function I_s continuous for $0 \le s \le 1$ must be constant; it cannot change from 1 or -1 to 0 as described above. Hence our assumption that (f, g) does not take the value $(0, 0)$ was false. In other words there must be a point in or on C where $(f, g) = (0, 0)$, and this yields the stationary point whose existence is asserted in the theorem.

(b) The Bendixson-Dulac theorem

The following theorem was established for $R = 1$ by the Swedish mathematician Ivar Bendixson in 1901 and for general R by the French mathematician Henri Dulac in 1933:

THEOREM 2. *Let $(f, g) \in C^{(1)}$ in a simply connected domain D and let R be a $C^{(1)}$ function such that*

$$(2) \qquad (Rf)_x + (Rg)_y \ne 0, \qquad (x, y) \in D.$$

Then the system (1) has no closed-curve trajectory in D.

The proof requires knowledge of line integrals and Green's theorem. If a simple closed trajectory C lies in D we have

$$\int_C R(g\,dx - f\,dy) = \int_0^T R\left(g\frac{dx}{dt} - f\frac{dy}{dt}\right)dt.$$

This reduces to 0 by virtue of the differential equations. Green's theorem now gives

$$\int\int_G ((Rf)_x + (Rg)_y)\,dx\,dy = 0$$

where G is the interior domain of C. Since the integrand does not vanish, it must be either positive or negative throughout D; this follows from the fact that D is connected and from Chapter A, Theorem B. In either case the integral is nonzero, which is a contradiction.

EXAMPLE 1. For the competition equations

$$x' = x(k - ax - by), \qquad y' = y(m - cx - dy)$$

let us assume only that $ad \ge 0$ and that a and d are not both 0. By a little experimenting one is led to the choice $R = 1/(xy)$ in Theorem 2, which gives

$$Rf = \frac{k}{y} - \frac{ax}{y} - b, \qquad Rg = \frac{m}{x} - c - \frac{dy}{x}.$$

It is obvious that $(Rf)_x + (Rg)_y \neq 0$ in Q_0, hence there is no closed-curve orbit in this region. This confirms what was said in Section 2 and makes fewer assumptions about the coefficients than were needed there.

(c) The Poincaré-Bendixson theorem

The following theorem was established by Poincaré in 1880 when the coefficients are polynomials, and also when they have power-series expansions. Poincaré's methods were simplified by Bendixson in 1901 and generalized to the $C^{(1)}$ case:

THEOREM 3. *Let $(f, g) \in C^{(1)}$ in a domain D and let Λ be a trajectory of (1) for $t \geq 0$. Suppose Λ is contained in a bounded closed subset of D that has no critical points. Then Λ^+ is an orbit of the differential equation and is a Jordan curve.*

If Λ itself is a Jordan curve then $\Lambda = \Lambda^+$. In general, however, Λ is not a closed curve, but spirals toward the closed curve Λ^+. When this happens, Λ^+ is called a *limit cycle*. There is a periodic trajectory in either case, and that is the reason why the theorem is important. The Poincaré-Bendixson theorem is one of few known results that assert the existence of a periodic solution for a general class of nonlinear systems.

The ideas of the proof are difficult, and they have only a limited usefulness aside from this one theorem. Hence, not even an outline of the proof will be given here. Instead, we show how the theorem is used.

(d) The van der Pol equation

In 1924 the Dutch applied mathematician and engineer Balthasar van der Pol investigated a series electrical circuit containing a nonlinear resistance. His analysis led to the equation

$$x'' + k(x^2 - 1)x' + x = B \sin \omega t$$

where k, ω, B are constants with $k > 0$. One of his principal results is that the equation with $B = 0$ has a nonconstant periodic solution. This will now be deduced from the Poincaré-Bendixson theorem.

Setting $B = 0$ we get the equivalent system

$$x' = y, \qquad y' = -x - k(x^2 - 1)y$$

which has $(0, 0)$ as the sole critical point. To avoid the trivial solution $(x, y) = (0, 0)$ we consider a trajectory $x = \xi(t)$, $y = \eta(t)$ starting at a distance $r_0 > 0$ from the origin. The radius r^2 satisfies

$$rr' = xx' + yy' = -k(x^2 - 1)y^2.$$

If $|x| \leq 1$ then $rr' \geq 0$. Hence the trajectory does not enter the disk $r < r_1$ where $r_1 = \min(r_0, 1)$.

If $|x| \geq 1$ then $rr' \leq 0$. Hence the trajectory cannot cross the circle $r = r_2 > 1$ from inside to outside on any part of the arc satisfying $|x| > 1$. This information is summarized by saying that, in Figure 17, a trajectory cannot cross either circular arc in a direction contrary to the arrows on it.

In a like manner, on the line $y = y_0$ we have

$$y' = -x - k(x^2 - 1)y_0.$$

This is negative if $|x| > 1$ and

$$ky_0 > \frac{|x|}{x^2 - 1}.$$

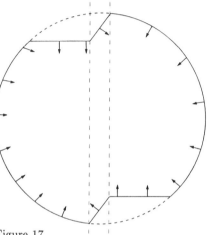

Figure 17

For $|x| \geq 2$, the maximum of the expression on the right is taken on when $|x| = 2$ and is $2/3$. Hence if $ky_0 > 2/3$ the value of y is decreasing on the line segment $y = y_0$, $x \leq -2$. Similarly, if $ky_0 < -2/3$ the value of y is increasing on the segment $y = y_0, x > 2$. This information is summarized by saying that, in Figure 17, the trajectory cannot cross either horizontal segment in a direction contrary to the arrows on it.

To deal with the strip $|x| \leq 2$ let us consider a line $y = y_1 + mx$ with y_1 large and m positive. A downward-pointing normal vector to this line is $\mathbf{N} = (m, -1)$. If the tangent to the trajectory is $\mathbf{T} = (x', y')$ then

$$\mathbf{T} \cdot \mathbf{N} = (y, -x - k(x^2 - 1)y) \cdot (m, -1) = my + x + k(x^2 - 1)y.$$

We take $m = k + 1$. Then for $|x| \leq 2$ we have

$$\mathbf{T} \cdot \mathbf{N} = y + x + kx^2 \geq y - 2 = y_1 + mx - 2 \geq y_1 - 2m - 2$$

This is positive if $y_1 > 2 + 2m$. A similar calculation applies when y_1 is a large negative number. Hence the trajectory cannot cross the two slanting segments of Figure 17 in a direction contrary to the arrows.

Since $r_2, |y_0|, |y_1|$ in the foregoing discussion can be arbitrarily large, there is no difficulty in constructing the figure so that the initial point of the trajectory is in the interior of the simple closed curve forming the boundary. Then the trajectory stays in a ring-shaped region between this curve and the circle $r = r_1$. Since there is no critical point in this region or on its boundary, Theorem 3 shows that a periodic solution exists. For an example, see Figure 18.

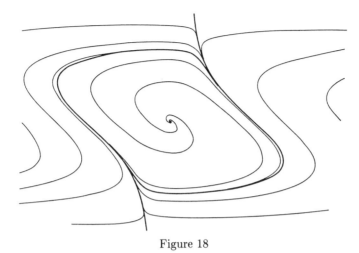

Figure 18

PROBLEMS

1. Let C be a Jordan curve and P a point not on it. In general, P is interior to C if a ray through P crosses C odd number of times, exterior to C if the number of crossings is even. Check this out experimentally.

2. Show that the system $x' = x \sin xy$, $y' = y \cos xy$ has no periodic solution in Q_0, and if the initial factors x, y are omitted, there is no periodic solution in the plane. Hint: Do not use Theorem 2.

3. Taking $R = e^{-ax-by}$ or $R = x^{-a}y^{-b}$ in Theorem 2, show that the following inequalities ensure that there are no closed-curve orbits in the plane or in Q_0, respectively:

$$f_x + g_y < af + bg, \qquad xy(f_x + g_y) < ayf + bxg$$

4P. On a differentiable curve $x = \xi(t)$, $y = \eta(t)$ that does not pass through the origin, the function $\theta = \tan^{-1}(y/x)$ can be defined so as to be differentiable. Assuming this, and taking $r^2 = x^2 + y^2$ as usual, verify the equations

$$rr' = xx' + yy', \qquad r^2\theta' = xy' - yx'$$

5. Example of a limit cycle. If $x' = -y + x(1 - r^2)$, $y' = x + y(1 - r^2)$, check that $r' = r(1 - r^2)$, $\theta' = 1$ for $r > 0$. The separable equation for r is of Bernoulli type and can be solved by taking $r = u^{-1/2}$. Thus verify Figure 19. The orbit $r = 1$ is a limit cycle.

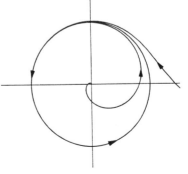

Figure 19

6. Winding number. Here we show how questions of "inside" and "outside" can be settled by analytical computation. Let C be a closed curve of class $C^{(1)}$ not passing through the origin and let θ be as in Problem 4. The quantity $N = \Delta\theta/2\pi$ obtained when C is traversed once is an integer called the *winding number* of C with respect to the origin. Check that

$$N = \frac{1}{2\pi} \int_C \frac{x\,dy - y\,dx}{x^2 + y^2}$$

If C is a Jordan curve, it can be shown that $|N| = 1$ or 0 when the origin is inside of or outside of C, respectively. Similar results hold with "the origin" replaced by any point P.

7. Special case of the Liénard equation. Let the factor $k(x^2 - 1)$ in the van der Pol equation be replaced by $h(x)$ where $h \in C^{(1)}$ is even. Suppose further that $(|x| - 1)h(x) > 0$ for $|x| \neq 1$ and that $x/h(x)$ is bounded for $|x| \geq 2$. Check that the discussion of the text goes through almost without change, so that again there is a periodic solution.

REVIEW PROBLEMS

8. Node. For $x' = -x$, $y' = -2y$, sketch the phase-plane trajectories and note that exactly one of them approaches the origin along the y axis while all others approach the origin tangent to the x axis.

The general definition of a stable node is based upon these two properties, the x and y axes being replaced by two lines as given by the linear approximation. In the case of a degenerate node, however, there is only one line and all orbits approach the critical point tangent to it. For a singular node each line through the critical point has an orbit tangent to it.

9. Focus. In Problems 9, 10, 11, and 12 the notation is as in Problem 4 and $\epsilon(r)$ is a continuous function satisfying $\epsilon(0) = 0$. If

$$x' = -x + y + x\epsilon(r), \qquad y' = -x - y + y\epsilon(r)$$

show that $r' = -r + r\epsilon(r)$. Deduce that $r' < -r/2$ in some disk $r < \delta$, or equivalently, $(\ln r)' < -1/2$. Hence, if the trajectory intersects this disk, it tends to 0 as $t \to \infty$. Show also that $\theta' = -1$ and hence $\theta \to -\infty$ as $t \to \infty$.

The two properties $r \to 0$ and $|\theta| \to \infty$ are used to characterize a stable focus in the general case. For an unstable focus the properties hold with $t \to -\infty$ instead of $t \to \infty$.

10. Singular node. Let $x' = -x - y\epsilon(r)$, $y' = -y + x\epsilon(r)$, and assume $x(0) = 1$, $y(0) = 0$. Check that $r = e^{-t}$ and that $\theta' = \epsilon(r)$. Thus show that

θ has a finite limit as $t \to \infty$ if $\epsilon(r) = \sqrt{r}$ but not if $\epsilon(r) = 1/(\ln r)$, $0 < r < 1$. The first case yields a singular node, the second gives a focus.

In general the limit is finite, and you get a singular node, if the improper integral

$$\int_0^1 \frac{\epsilon(r)}{r} \, dr = \lim_{\delta \to 0+} \int_\delta \frac{\epsilon(r)}{r} \, dr$$

converges. You get a focus if the integral diverges to $-\infty$ or ∞.

11. Center. If $x' = -y + x\epsilon(r)$, $y' = x + y\epsilon(r)$, show that $r' = r\epsilon(r)$, $\theta' = 1$. Conclude that the origin has the character of a stable focus, a center, or an unstable focus, when $\epsilon(r) = -r$, 0, or r, respectively. Similar results hold when $\epsilon(r) < 0$, $= 0$, or > 0, even if $|\epsilon|$ is arbitrarily small. Thus, no matter how well the linear system approximates the nonlinear one, the two systems may behave differently at a center.

12. Center, continued. In Problem 11 let $\epsilon(r)$ have a sequence of positive zeros $r_n \to 0$, and let $\epsilon \in C^{(1)}$. Each zero r_n gives a circular orbit $r = r_n$. Any trajectory starting in the ring-shaped region between two circular orbits remains there by uniqueness, yet $\theta \to \infty$. Sketch the curves when $\epsilon(r) = r^3 \sin \pi/r$.

These properties are used to characterize a center in the nonlinear case. Namely, an isolated critical point (x_0, y_0) is a *center* if every disk centered at (x_0, y_0) contains a trajectory whose orbit is a simple closed curve surrounding the point (x_0, y_0). In the ring-shaped region between two of these closed-curve orbits the trajectory is allowed to have a spiral behavior as in the case of a focus.

Alexander, the king of the Macedonians, began like a wretch to learn geometry. Teach me, saith he, easy things. To whom his master said: These things be the same, and alike difficult unto all.

—Seneca

Chapter 21

THE LAPLACE TRANSFORM I

W E NOW PRESENT a systematic and elegant procedure that is widely
used in circuit analysis and in the study of feedback and control. The theory
takes its form from a symbolic method developed by the English engineer
Oliver Heaviside. It enables one to solve many initial-value problems without
going to the trouble of finding the general solution and then evaluating the
arbitrary constants. The procedure applies to systems of equations, to partial
differential equations, and to integral equations. It often yields results more
readily than other techniques.

1. Basic concepts

The *Laplace transform* of the function f is defined by the equation

$$(1) \qquad F(s) = \int_0^\infty e^{-st} f(t)\, dt = \lim_{r \to \infty} \int_0^r e^{-st} f(t)\, dt$$

provided the limit exists for all sufficiently large s. This means that there is
a number $s_0 = s_0(f)$, depending on f, such that the limit exists whenever
$s > s_0$. If there is no such number s_0 then f has no Laplace transform. The
role of the condition $s > s_0$ is illustrated in the following example.

EXAMPLE 1. If a is a real constant then

$$\int_0^r e^{-st} e^{at}\, dt = \int_0^r e^{t(a-s)}\, dt = \left. \frac{e^{t(a-s)}}{a-s} \right|_0^r .$$

The behavior as $r \to \infty$ depends on the sign of $a - s$. If $s < a$ then $a - s$ is
positive and the expression tends to ∞ as $r \to \infty$. When $s = a$ the value of
the integral is r, which also tends to ∞ as $r \to \infty$. In both cases the integral
defining the Laplace transform is divergent. If $s > a$, however, $e^{r(a-s)}$ tends
to 0 as $r \to \infty$. Hence

$$(2) \qquad \int_0^\infty e^{-st} e^{at}\, dt = \frac{1}{s-a}, \qquad s > a$$

and the integral is convergent. This shows that e^{at} has the Laplace transform
$F(s) = 1/(s-a)$, and that the constant s_0 in the definition can be any
number $\geq a$. If a is a complex constant, the same calculation shows that
$F(s) = 1/(s-a)$ for $s > \operatorname{Re} a$.

Two notations for the Laplace transform will be used here. In the first notation, the connection between a function f and its transform F is indicated by lower case and capital letters. Thus

$$U(s) = \int_0^\infty e^{-st} u(t)\, dt, \qquad V(s) = \int_0^\infty e^{-st} v(t)\, dt$$

and so on. In the second notation the Laplace transform of f is denoted by Lf where L is the Laplace transform operator. Thus

$$Lu = \int_0^\infty e^{-st} u(t)\, dt, \qquad Lv = \int_0^\infty e^{-st} v(t)\, dt$$

and so on. The domain of L is the set of functions f for which Lf exists.

Let us show that L is linear. That is,

$$L(u + v) = Lu + Lv, \qquad L(cu) = cL(u)$$

for all functions u and v in the domain of L and for all constants c. The first of these equations follows by letting $r \to \infty$ in the equation

$$\int_0^r e^{-st}(u(t) + v(t))\, dt = \int_0^r e^{-st} u(t)\, dt + \int_0^r e^{-st} v(t)\, dt.$$

The calculation shows that if $U(s)$ exists for $s > s_0$ and $V(s)$ for $s > s_1$ then $L(u + v)$ exists for $s > \max(s_0, s_1)$. The proof that $L(cu) = cL(u)$ is similar and is left for you.

Since L is linear, all results valid for general linear operators apply to L. In particular, the principle of equating real parts holds provided, as we shall assume, s is real. That is,

$$L(f + ig) = F + iG \quad \Rightarrow \quad Lf = F \quad \text{and} \quad Lg = G.$$

This is true because Lf is real for real f, so that L is a real linear operator. We illustrate in an example.

EXAMPLE 2. If $a = ib$ in Example 1 the result is

$$L(e^{ibt}) = \frac{1}{s - ib} = \frac{s + ib}{s^2 + b^2}.$$

Writing $e^{ibt} = \cos bt + i \sin bt$ and equating real and imaginary parts, we get

(3) $$L(\cos bt) = \frac{s}{s^2 + b^2}, \qquad L(\sin bt) = \frac{b}{s^2 + b^2}.$$

The Laplace transform of a function can often be deduced by calculations that do not involve integration. Sometimes the justification requires a rather careful analysis, which we postpone, and sometimes the justification is trivial. The following examples illustrate both cases.

EXAMPLE 3. Equation (2) can be written in the form

$$\int_0^\infty e^{-st} e^{at}\, dt = (s-a)^{-1}.$$

Proceeding without regard to rigor, let us differentiate both sides with respect to s. Since

$$\frac{d}{ds} e^{-st} = -t e^{-st} \quad \text{and} \quad \frac{d}{ds}(s-a)^{-1} = -(s-a)^{-2}$$

the result is

$$\int_0^\infty e^{-st} t e^{at}\, dt = (s-a)^{-2}.$$

Differentiating again we get a factor $-t^2$ on the left and $-2(s-a)^{-3}$ on the right. Differentiating n times leads to the formula

$$(4) \qquad L(t^n e^{at}) = \frac{n!}{(s-a)^{n+1}}, \qquad n = 0, 1, 2, \cdots$$

where by convention $0! = 1$. Although our argument is suggestive only, both the procedure and the result are in fact correct, as seen later.

EXAMPLE 4. The identity

$$\int_0^\infty e^{-st} e^{-ct} f(t)\, dt = \int_0^\infty e^{-(s+c)t} f(t)\, dt$$

shows that if $Lf(t) = F(s)$ then

$$(5) \qquad L e^{-ct} f(t) = F(s+c).$$

In contrast to (4), Equation (5) is an immediate consequence of the definitions. When applied to the result of Example 2, Equation (5) gives

$$(6) \qquad L(e^{-ct} \cos bt) = \frac{s+c}{(s+c)^2 + b^2}, \qquad L(e^{-ct} \sin bt) = \frac{b}{(s+c)^2 + b^2}.$$

In Chapter G we describe a rather large class of functions that have Laplace transforms. For convenience, the definition is repeated here:

DEFINITION. A real- or complex-valued function $f(t)$ belongs to the class E of functions of exponential type if the following two conditions hold:

(a) *On any interval $[0, r]$ the function is defined and continuous except perhaps at finitely many points, and*

(b) *For constants A and B, depending on f, the inequality $|f(t)| \leq Ae^{Bt}$ holds at all points of $[0, \infty)$ where $f(t)$ is defined.*

If f satisfies these two conditions and $s > B$, a calculation similar to that in Example 1 gives

$$\int_0^\infty e^{-st} |f(t)|\, dt \leq \int_0^\infty e^{-st} Ae^{Bt}\, dt = \frac{A}{s - B}.$$

This suggests that the integral without absolute values exists for $s > B$, and a proof is given in the problems at the end of Chapter 22. Thus if $f \in$ E, then $L(f)$ exists. The class E is used in the following theorem:

THEOREM 1. *Let f be continuous on $[0, \infty)$ and let $f' \in$ E. Then $f \in$ E and $Lf' = sLf - f(0)$.*

If f is defined only on $[0, \infty)$, continuity at $t = 0$ means that

$$f(0) = f(0+) = \lim_{t \to 0+} f(t).$$

This right-hand limit for $f(0)$ must be used in Theorem 1. If $f(t)$ is defined for $t < 0$ as well as for $t \geq 0$, but is discontinuous at $t = 0$, the left-hand limit $f(0-)$ would give an incorrect result.

We now prove Theorem 1. Integration by parts gives

(7)
$$\int_0^r e^{-st} f'(t)\, dt = e^{-st} f(t) \Big|_0^r + s \int_0^r e^{-st} f(t)\, dt.$$

The fact that integration by parts is permissible under the hypothesis of Theorem 1 is established in Chapter G, where it is also shown that $f \in$ E. If

$$|f(t)| \leq Ae^{Bt} \quad \text{and} \quad s \geq B + 1,$$

as we may assume, the integrated term in (7) at the upper limit r satisfies

$$|e^{-sr} f(r)| \leq e^{-(B+1)r} Ae^{Br} = Ae^{-r}.$$

This tends to 0 as $r \to \infty$. Hence letting $r \to \infty$ in (7) gives Theorem 1.

The differential equations considered in this section have solutions y that are sums of functions of the form $ct^j e^{kt}$ with $j = 0, 1, 2, \cdots$. Therefore the solutions as well as their derivatives satisfy the hypothesis of Theorem 1. Using this fact, we take $f = y$ in Theorem 1 to get

$$(8) \qquad\qquad Ly' = sLy - y(0).$$

The choice $f = y'$ in Theorem 1 gives

$$Ly'' = sLy' - y'(0) = s(sLy - y(0)) - y'(0)$$

where the second equality follows from (8). Hence

$$(9) \qquad\qquad Ly'' = s^2 Ly - sy(0) - y'(0).$$

These relations enable us to transform first- or second-order initial-value problems into algebraic equations for Ly. Once Ly is found, the solution y is given by a table of Laplace transforms such as the table on the inside back cover of this book. The procedure is justified by the following theorem, which is established in Chapter G:

UNIQUENESS THEOREM. Let y and $\tilde{y}$ be functions of class E such that their Laplace transforms agree for all large s. Then $y(t) = \tilde{y}(t)$ at all points $t \geq 0$ where both these functions are continuous.

EXAMPLE 5.   Solve the initial-value problem

$$y' - y = e^t, \qquad y(0) = 1.$$

Taking the Laplace transform gives

$$Ly' - Ly = Le^t \quad \text{or} \quad sLy - 1 - Ly = \frac{1}{s-1}$$

where the second equation follows from (8) with $y(0) = 1$ and from (2) with $a = 1$. Hence

$$(s-1)Ly = \frac{1}{s-1} + 1 \quad \text{or} \quad Ly = \frac{1}{(s-1)^2} + \frac{1}{s-1}.$$

By (4) with $a = 1$, $n = 1$, and $n = 0$ we see that

$$Ly = L(te^t + e^t), \quad \text{hence} \quad y = te^t + e^t.$$

EXAMPLE 6.   Solve $y'' + 4y = 0$, $y(0) = 5$, $y'(0) = 6$. Taking the transform gives

$$Ly'' + 4Ly = 0 \quad \text{or} \quad s^2 Ly - 5s - 6 + 4Ly = 0$$

where the second equation follows from (9). Hence

$$Ly = \frac{5s + 6}{s^2 + 4} = 5\frac{s}{s^2 + 4} + 3\frac{2}{s^2 + 4}.$$

Using the two equations (3) with $b = 2$ we get

$$y = 5\cos 2t + 3\sin 2t.$$

Note the manner in which the two constants 5 and 3 were built into the analysis. The correctness of this procedure follows from the fact that L is linear.

The operation by which we recover $f(t)$ from $F(s)$ is called *inverse Laplace transformation* and is denoted by L^{-1}. Linearity of L^{-1} follows from linearity of L. According to the uniqueness theorem, $f(t) = L^{-1}F(s)$ is substantially determined for $t \geq 0$ even if $F(s)$ is known only for large s. The condition that s be large is just what is needed to ensure that the integral defining the transform converges. Hence, whenever we use a Laplace transform, it is assumed that s is real and large enough to justify the calculation.

PROBLEMS—————————————————————————

1P. The purpose of this problem is to help you become familiar with the table of Laplace transforms on the inside back cover of this book. Only in Part (c) are you expected to write anything.

(a) Notice that entries ABCDEF follow from abcdef and (5), and that aA follow from bB with $0! = 1$.

(b) Chapter 7, Section 4, Problem 2 gives

$$\cos ibt = \cosh bt, \qquad \frac{\sin ibt}{ib} = \frac{\sinh bt}{b}$$

   Hence ef and EF follow from cd and CD with b replaced by ib.

(c) Replace b by ib in gG and thus get the two transform pairs

$$t\frac{\sinh bt}{b}, \qquad \frac{2s}{(s^2 - b^2)^2}; \qquad \frac{2b^2}{(s^2 - b^2)^2}, \qquad t\cosh bt - \frac{\sinh bt}{b}$$

2. Find $L1$ and Lt directly from the definition and check by the table.

3. Using the table find the transform $F(s)$ of the following functions $f(t)$:

   (a) $7e^{2t}$ (b) $\cos 2t + \sin 2t$ (c) $e^t(t - \sin t)$ (d) $(t + e^t)^3$

4. Find $L\cos^2 t$ and $L\sin^2 t$. Hint: $2\cos^2 t = 1 + \cos 2t$.

5. Find functions $f(t)$ whose transforms are the following functions $F(s)$:

(a) $\dfrac{6}{s^2 + 1}$ (c) $\dfrac{5s - 6}{s^2 + 4} + \dfrac{2}{s}$ (e) $\dfrac{7(s - 3) + 16}{(s - 3)^2 + 4}$

(b) $\dfrac{s^2 + 4}{s^3}$ (d) $\dfrac{4s + 3}{s^2 - 9} + \dfrac{6}{(s + 1)^3}$ (f) $\dfrac{6s - 1}{(s + 4)^2 - 25}$

6P. Completing the square. The identity $s^2 + 2as = (s + a)^2 - a^2$ will be used here. Check that

$$\frac{2s + 6}{3s^2 - 18s + 15} = \frac{2s + 6}{3(s^2 - 6s + 9) - 12} = \frac{2(s - 3) + 12}{3((s - 3)^2 - 4)}$$

Thus find a function $f(t)$ whose transform is the above function $F(s)$.

7. Find a function $f(t)$ whose transform is the given function $F(s)$:

(a) $\dfrac{s - 2}{s^2 - 4s + 5}$ (b) $\dfrac{2}{s^2 + 2s}$ (c) $\dfrac{s + 8}{2s^2 + 8s + 26}$ (d) $\dfrac{9s + 3}{9s^2 + 6s + 19}$

8P. The *rest solution* of a first-order equation satisfies $y(0) = 0$. If y is the rest solution of the equation on the left below, check that Ly satisfies the other two equations. Thus find y, and verify that it satisfies both the differential equation and the condition $y(0) = 0$:

$$y' + 2y = 1, \qquad sLy + 2Ly = \frac{1}{s}, \qquad Ly = \frac{1}{(s + 1)^2 - 1}$$

9. By means of the Laplace transform, find the rest solution:

(a) $y' + 4y = 8$ (b) $y' - y = e^t$ (c) $y' + 2y = e^{3t}$ (d) $y' = 7y$

10. If $y' + 2y = 4$ and $y(0) = 1$ check that $Ly = (s + 4)/(s^2 + 2s)$. Thus get

$$y = e^{-t}(\cosh t + 3 \sinh t) = 2 - e^{-2t}$$

11. (abcd) By means of the Laplace transform, solve the differential equation subject to the initial condition below it:

$$\begin{array}{llll}
y' + 4y = 0 & y' - y = 3e^{4t} & y' + y = e^t & y' + y = 1 \\
y(0) = 2 & y(0) = 1 & y(0) = 1 & y(0) = 3
\end{array}$$

12. (abcd) Using the Laplace transform, solve subject to the given values of $y_0 = y(0)$ and $y_1 = y'(0)$:

$$\begin{array}{llll}
y'' + y = 0 & y'' - 4y = 0 & y'' + y = 2y' & y'' = y' + 2y \\
y_0 = 1,\, y_1 = 1 & y_0 = 1,\, y_1 = 2 & y_0 = 0,\, y_1 = 1 & y_0 = 3,\, y_1 = 6
\end{array}$$

---------------ANSWERS---------------

3. (a) $7/(s-2)$ (c) $1/((s-1)^4 + (s-1)^2)$
   (b) $(s+2)/(s^2+4)$ (d) $6/s^4 + 6/(s-1)^3 + 3/(s-2)^2 + 1/(s-3)$

4. $(1/2)(1/s \pm s/(s^2+4))$

5. (a) $6\sin t$ (d) $4\cosh 3t + \sinh 3t + 3t^2 e^{-t}$
   (b) $1 + 2t^2$ (e) $e^{3t}(7\cos 2t + 8\sin 2t)$
   (c) $5\cos 2t - 3\sin 2t + 2$ (f) $e^{-4t}(6\cosh 5t - 5\sinh 5t) = (e^t + 11e^{-9t})/2$

6. $3f(t) = e^{3t}(2\cosh 2t + 6\sinh 2t) = 4e^{5t} - 2e^t$

7. (a) $e^{2t}\cos t$ (c) $e^{-2t}((1/2)\cos 3t + \sin 3t)$
   (b) $2e^{-t}\sinh t = 1 - e^{-2t}$ (d) $e^{-t/3}\cos\sqrt{2}t$

8. $y = e^{-t}\sinh t = (1/2)(1 - e^{-2t})$

9. (a) $2(1 - e^{-4t})$ (b) te^t (c) $5y = e^{3t} - e^{-2t}$ (d) 0

11. (a) $2e^{-4t}$ (b) e^{4t} (c) $\cosh t$ (d) $1 + 2e^{-t}$

12. (a) $\cos t + \sin t$ (b) e^{2t} (c) te^t (d) $3e^{2t}$

---

2. The transformed equation

We begin with the equation

(1) $$Ty = y^{(n)} + p_{n-1}y^{(n-1)} + \cdots + p_1 y' + p_0 y = f \in \text{E},$$

where the p_j are constants and the linear operator T is defined by the first equality in (1). The characteristic polynomial of T is

$$P(s) = s^n + p_{n-1}s^{n-1} + \cdots + p_1 s + p_0.$$

When $f(t)$ is discontinuous, as it is in many applications, the word "solution" will be used to mean "generalized solution" in the sense of Chapter G. However, detailed knowledge of Chapter G will not be assumed. All you need to know is that $y^{(n-1)}$ is continuous and that the condition $f \in \text{E}$ ensures that all the derivatives in (1) also belong to E. These two facts make it easy to find Ly.

Calculation of Ly is especially easy for the *rest solution* of (1), which is the solution satisfying null initial conditions

$$y(0) = y'(0) = \cdots = y^{(n-1)}(0) = 0.$$

By Section 1 Theorem 1, these conditions give

$$Ly' = sLy, \qquad Ly'' = sLy' = s^2 Ly, \qquad Ly''' = sLy'' = s^3 Ly$$

and so on. In general, the rest solution satisfies

(2) $$Ly^{(j)} = s^j Ly, \qquad j = 0, 1, 2, \cdots, n.$$

Hence the transform of (1) gives

$$s^n Ly + p_{n-1} s^{n-1} Ly + \cdots + p_1 sLy + p_0 Ly = Lf$$

when we use the fact that L is linear together with (2). Factoring out Ly gives the following:

THEOREM 1. *If y is the rest solution of (1) then $P(s)Ly = Lf$.*

EXAMPLE 1. Find Ly for the rest solution of

$$y^{iv} + 3y''' + 2y'' + 6y' = e^{-t} \sin t + t^2.$$

The Laplace transform of the function $f(t)$ on the right is given by the table and the characteristic polynomial is read off from the left side. Theorem 1 then produces the equation

$$(s^4 + 3s^3 + 2s^2 + 6s)Ly = \frac{1}{(s+1)^2 + 1} + \frac{2}{s^3}$$

and division by the polynomial on the left gives Ly.

If y is not the rest solution but satisfies arbitrary initial conditions

$$y(0) = y_0, \ y'(0) = y_1, \ y''(0) = y_2, \ \cdots, y^{(n-1)}(0) = y_{n-1}$$

the computation of Ly is only slightly more difficult. By results of the previous section

(3) $$Ly' = sLy - y_0, \qquad Ly'' = s^2 Ly - sy_0 - y_1.$$

Since $y''' = (y'')'$ the first equation (3) applied to y'' gives

$$Ly''' = sLy'' - y''(0) = sL''y - y_2.$$

Substituting Ly'' as given by the second equation (3), we get

$$Ly''' = s^3 Ly - s^2 y_0 - sy_1 - y_2.$$

The process continues in an obvious manner and leads to the following:

THEOREM 2. *If $y^{(n)} \in \mathrm{E}$ and $y^{(n-1)}$ is continuous then*

$$Ly^j = s^j Ly - (s^{j-1}y_0 + s^{j-2}y_1 + \cdots + sy_{j-2} + y_{j-1})$$

for $j = 1, 2, \cdots, n$, where $y_j = y^{(j)}(0)$ are the initial values. This holds in particular for the solutions of (1).

EXAMPLE 2. Find Ly if $y(0) = 5$, $y'(0) = 6$, $y''(0) = 7$, and

$$y''' - 3y'' + 2y' = f \in \mathrm{E}.$$

Taking the transform of both sides gives $Ly''' - 3Ly'' + 2Ly' = Lf$. By Theorem 2 with $y_0 = 5$, $y_1 = 6$, $y_2 = 7$ we get

$$(s^3 Ly - 5s^2 - 6s - 7) - 3(s^2 Ly - 5s - 6) + 2(sLy - 5) = Lf.$$

Moving the terms not involving Ly to the right gives

$$(s^3 - 3s^2 + 2s)Ly = Lf + (5s^2 - 9s - 1)$$

and Ly is readily found from this. Note that the coefficient of Ly is the characteristic polynomial $P(s)$, and that the two terms on the right are associated with the forcing function and with the initial conditions respectively.

If the Laplace transform of (1) is taken, the result is

$$(4) \qquad\qquad P(s)Ly = F(s) + P_0(s)$$

where $F = Lf$ and where $P_0(s)$ is a polynomial of degree $n-1$ depending on the initial conditions. A formula for $P_0(s)$ is readily obtained by Theorem 2, but it is better to use Theorem 2 directly as in the foregoing example.

Equation (4) gives

$$(5) \qquad\qquad Ly = \frac{F(s)}{P(s)} + \frac{P_0(s)}{P(s)}.$$

To interpret (5), consider two simpler problems. The first problem is

$$Tu = f(t), \ u(0) = u'(0) = \cdots = u^{(n-1)}(0) = 0$$

which describes a forced response with null initial conditions. The second problem is

$$Tv = 0, \qquad v(0) = y_0, \qquad v'(0) = y_1, \qquad \cdots, \qquad v^{n-1}(0) = y_{n-1}$$

which describes an unforced response with arbitrary initial conditions. By (5) with $P_0 = 0$ or $f = 0$, as the case may be, the two functions u and v satisfy

$$(6) \qquad Lu = \frac{F(s)}{P(s)}, \qquad Lv = \frac{P_0(s)}{P(s)}$$

respectively. Equation (5) indicates that $Ly = Lu + Lv$. Hence

$$Ly = L(u + v), \qquad y = u + v,$$

where the first of these equations follows from linearity and the second from the uniqueness theorem. Thus y is a superposition of the two simpler functions u and v. For dissipative systems u is a representative of the steady state, as discussed in Chapter 8, and v represents a transient that depends on the initial conditions. The separation of steady-state and transient effects embodied in (5) is one of many advantages of the transform method.

The Laplace transform is very convenient for the solution of systems of differential equations. As an illustration, consider the system

$$(7) \qquad \begin{aligned} x' &= ax + by + f, & x(0) &= x_0 \\ y' &= cx + dy + g, & y(0) &= y_0 \end{aligned}$$

where a, b, c, d are constant and where f and g belong to the class E. Assuming that x and y satisfy the hypothesis of Theorem 2 with $n = 1$ we take the Laplace transform of each equation and obtain

$$sX - x_0 = aX + bY + F$$
$$sY - y_0 = cX + dY + G$$

where $X = Lx$, $Y = Ly$, $F = Lf$, and $G = Lg$. Moving the unknowns X, Y to the left and the other terms to the right gives

$$(s - a)X - bY = F + x_0$$
$$-cX + (s - d)Y = G + y_0 \,.$$

For large s the coefficient determinant

$$P(s) = \begin{vmatrix} s - a & -b \\ -c & s - d \end{vmatrix} = s^2 - (a + d)s + (ad - bc)$$

is not 0 and we can use Cramer's rule. The result is

$$P(s)X = \begin{vmatrix} F + x_0 & -b \\ G + y_0 & s - d \end{vmatrix} = \begin{vmatrix} F & -b \\ G & s - d \end{vmatrix} + \begin{vmatrix} x_0 & -b \\ y_0 & s - d \end{vmatrix}$$

where $P(s)$ is the *characteristic polynomial* of the system. It is left for you to get a similar equation for $P(s)Y$. Note that the effects of the forcing function and of the initial condition are separated, much as in the case $Ty = f$ discussed above.

EXAMPLE 3.   Solve the following system by means of the Laplace transform:

$$x' = 6x + y, \qquad x(0) = 2; \qquad y' = 4x + 3y, \qquad y(0) = 7.$$

Taking the transform of each equation, we get

$$sX - 2 = 6X + Y \qquad \text{or} \qquad (s - 6)X - Y = 2$$
$$sY - 7 = 4X + 3Y \qquad \text{or} \qquad -4X + (s - 3)Y = 7.$$

By Cramer's rule applied to the system on the right

$$\begin{vmatrix} s - 6 & -1 \\ -4 & s - 3 \end{vmatrix} X = \begin{vmatrix} 2 & -1 \\ 7 & s - 3 \end{vmatrix} \qquad \text{or} \quad (s^2 - 9s + 14)X = 2s + 1.$$

Thus

$$X = \frac{2s + 1}{s^2 - 9s + 14} = \frac{3}{s - 7} - \frac{1}{s - 2}$$

where the second form is obtained by the same method of partial fractions that is commonly used for integrating rational functions. (A systematic discussion is given in the following section.) By the table of transforms

$$x = 3e^{7t} - e^{2t}.$$

A similar method gives $y = 3e^{7t} + 4e^{2t}$, or y can be found from $y = x' - 6x$.

The transform method applies to linear systems of far greater generality than (7), and the key to such applications is given by Theorems 1 and 2. As an illustration, consider a system of the form

(8)
$$T_{11}x + T_{12}y = f$$
$$T_{21}x + T_{22}y = g$$

where each T_{ij} is an operator like T in (1) with the characteristic polynomial $P_{ij}(s)$. Assuming $f, g \in \mathrm{E}$ the rest solution of (8) satisfies

$$P_{11}(s)X + P_{12}(s)Y = F$$
$$P_{21}(s)X + P_{22}(s)Y = G$$

If the *characteristic polynomial*

$$P(s) = \begin{vmatrix} P_{11}(s) & P_{12}(s) \\ P_{21}(s) & P_{22}(s) \end{vmatrix}$$

does not vanish for large s we can solve for X and Y by Cramer's rule. General initial conditions are introduced by Theorem 2.

Since Cramer's rule applies to n-by-n linear systems, there is no difficulty of principle in extending the foregoing methods to systems

$$\sum_{j=1}^{n} T_{ij}x_j = f_i, \qquad i = 1, 2, \cdots, n$$

with the characteristic polynomial $P(s) = \det P_{ij}(s)$. The Laplace transform thus gives an alternative approach to Chrystal's theory as presented in Chapter 15, Section 4.

PROBLEMS

In all problems, assume that the functions whose transforms you need belong to E and that there is enough continuity for the use of Theorem 2.

1. By inspection, find Ly for the rest solution:

(a) $y''' + y'' - 2y' = f$ (b) $y''' + 2y'' + y' = f$ (c) $y^{iv} + y = f$

2. (abc) Instead of considering the rest solution in Problem 1, assume appropriate initial conditions of the form $y^{(j)}(0) = y_j$, and again find Ly.

3. (abcd) Find $X = Lx$ and $Y = Ly$ for the rest solution:

$$\begin{aligned} x' &= x - y + f & x' &= 2x + 3y + f & x' + y &= x & x' + f &= y \\ y' &= x - y + g & y' &= 3x + 2y + g & y' - x &= g & y' - y &= x \end{aligned}$$

4. (abcd) In the following systems let $x(0) = k$ and $y(0) = 0$. Express $X = Lx$ in a form that shows the separate roles of f and k:

$$\begin{aligned} x' + y &= 0 & x' + x &= y & x' &= 3x + 2y + f & x' + f &= y + x \\ y' + x &= f & y' - x &= f & y' &= 4x + 2y & y' &= x + y \end{aligned}$$

5. Let x and y satisfy the second-order system

$$\begin{aligned} x'' &= ax + by + f, & x(0) &= x_0, & x'(0) &= x_1 \\ y'' &= cx + dy + g, & y(0) &= y_0, & y'(0) &= y_1 \end{aligned}$$

where a, b, c, d are constant. Show that

$$\begin{vmatrix} s^2 - a & -b \\ -c & s^2 - d \end{vmatrix} Lx = \begin{vmatrix} F & -b \\ G & s^2 - d \end{vmatrix} + \begin{vmatrix} sx_0 + x_1 & -b \\ sy_0 + y_1 & s^2 - d \end{vmatrix}$$

and obtain a similar formula for Ly. No answer is given.

6. Find Lx and Ly if $x(0) = 1$, $x'(0) = 2$, $y(0) = 3$, $y'(0) = 4$ and

$$x'' = 3x - 4y, \qquad y'' = 2x - 3y$$

7. Using the Laplace transform, solve $y^{(5)} = 6!t^2$ subject to the initial conditions $y^{(j)}(0) = j!$, $j = 0, 1, 2, 3, 4$.

8P. Partial fractions. It is desired to find constants A, B, C such that the following identity holds for $s \neq 1, 2, 3$:

$$\frac{s^2 - 10s + 13}{(s-1)(s-2)(s-3)} = \frac{A}{s-1} + \frac{B}{s-2} + \frac{C}{s-3}$$

(a) Multiply by $s - 1$ to obtain

$$\frac{s^2 - 10s + 13}{(\not{s}\not{-}\not{1})(s-2)(s-3)} = \frac{A}{\not{s}\not{-}\not{1}} + \frac{B}{s-2}(s-1) + \frac{C}{s-3}(s-1)$$

where the crossed-out factors can be covered by your finger or by a small coin such as a dime. Let $s \to 1$ and get $A = (1 - 10 + 13)/(-1)(-2) = 2$.

(b) By the same technique find $B = 3$ and $C = -4$.

(c) Show that these values of A, B, C lead to an identity when you clear of fractions.

(d) Find a function $f(t)$ whose Laplace transform is the above function $F(s)$.

9P. Partial fractions. It is desired to find constants A, B, C such that the following holds for s real, $s \neq -1$:

$$\frac{5s^2 + 13s + 14}{(s+1)(s^2 + 4s + 5)} = \frac{A}{s+1} + \frac{Bs + C}{s^2 + 4s + 5}$$

(a) By the method of Problem 8, get $A = 3$.

(b) Clearing of fractions now gives

$$5s^2 + 13s + 14 = 3(s^2 + 4s + 5) + (s+1)(Bs + C)$$

Equating the coefficients of s^2 get $5 = 3 + B$, hence $B = 2$. Equating the coefficients of s^0 get $14 = 15 + C$, hence $C = -1$.

(c) Check that with this choice of A, B, C the coefficients of s also agree, thus confirming the calculation.

(d) Find a function $f(t)$ whose Laplace transform is the above function $F(s)$.

―――――――――――――――――――――ANSWERS―――――――――――――――――――――

1. (a) $(s^3 + s^2 - 2s)Ly = F$ (b) $(s^3 + 2s^2 + s)Ly = F$ (c) $(s^4 + 1)Ly = F$
2. Add the following terms to F in the above solutions for Problem 1:
   (a) $y_0(s^2 + s - 2) + y_1(s + 1) + y_2$ (c) $y_0 s^3 + y_1 s^2 + y_2 s + y_3$
   (b) $y_0(s + 1)^2 + y_1(s + 2) + y_2$
3. (a) $s^2 X = (s + 1)F - G,\ s^2 Y = (s - 1)G + F$
   (b) $(s^2 - 4s - 5)X = (s - 2)F + 3G,\ (s^2 - 4s - 5)Y = (s - 2)G + 3F$
   (c) $(s^2 - s + 1)X = -G,\ (s^2 - s + 1)Y = (s - 1)G$
   (d) $(s^2 - s - 1)X = (1 - s)F,\ (s^2 - s - 1)Y = -F$
4. (a) $(s^2 - 1)X = ks - F$ (c) $(s^2 - 5s - 2)X = (ks - 2k) + (Fs - 2F)$
   (b) $(s^2 + s - 1)X = ks + F$ (d) $(s^2 - 2s)X = (ks - k) + (F - Fs)$
6. $(s^4 - 1)X = s^3 + 2s^2 - 9s - 10,\quad (s^4 - 1)Y = 3s^3 + 4s^2 - 7s - 8$
7. $(2/7)t^7 + 1 + t + t^2 + t^3 + t^4$ 8. (d) $f(t) = 2e^t + 3e^{2t} - 4e^{3t}$
9. (d) $f(t) = e^{-2t}(2\cos t - 5\sin t) + 3e^{-t}$

3. Partial fractions

We continue to denote by T the linear operator introduced in the previous section. Suppose the forcing function $f(t)$ is a finite sum of functions of the form

(1) $t^m e^{at}(A\cos bt + B\sin bt)$

where $m \geq 0$ is an integer and a, b, A, B are constant. It follows from Chapter 8, Section 3 that every solution of $Ty = f$ is of the same form, perhaps with a larger value of m. Hence Ly is a rational function, that is, a quotient $Q(s)/P(s)$ of two polynomials. This can be seen from the formula

$$Lt^m e^{kt} = \frac{m!}{(s - k)^{m+1}}$$

which holds when $k = a + ib$ is any complex constant.

Inputs of the type (1) are encountered often and in diverse contexts. The fact that the output has the same form may serve to explain why Laplace transforms $F(s)$ that are rational functions play a dominant role in several areas of technology.

When $F(s)$ is rational, the basic method of recovering $f(t)$ from $F(s)$ is by expansion into partial fractions. You are probably familiar with partial fractions from their use in integrating rational functions. A brief discussion is presented nevertheless, because our approach is based on analysis rather than algebra and may give insights that are new to you. We begin with a review of the following elementary theorems:

THEOREM 1.   *Let $P(s) = s^n + \cdots$ be a polynomial of degree $n \geq 1$ with leading coefficient 1 and let $Q(s) = bs^n + \cdots$ be a polynomial of degree $\leq n$, so that $b = 0$ is allowed. Then*

$$\lim_{s \to \infty} \frac{Q(s)}{P(s)} = \lim_{s \to \infty} \frac{bs^n + \cdots}{s^n + \cdots} = b.$$

THEOREM 2.   *If a polynomial $p(s)$ vanishes when $s = a$ then $p(s)$ is divisible by $s - a$. Hence if two polynomials $p_1(s)$ and $p_2(s)$ of degrees $< n$ agree for n or more distinct values of s, they are identical.*

Theorem 1 is obtained when both numerator and denominator are divided by s^n as illustrated in Problem 1. The first statement in Theorem 2 follows by division, which gives

$$p(s) = (s - a)q(s) + r, \qquad r = p(a)$$

where q is a polynomial of degree lower than that of p. If $p(a) = 0$ the remainder r is also 0, hence p is divisible by $s - a$. For the second statement, if $p_1(s) - p_2(s)$ vanishes at n values a_j then it is divisible by each $s - a_j$. Hence it is divisible by the product and would have degree $\geq n$ unless it is identically 0.

After these preliminaries, let

$$P(s) = (s - a_1)(s - a_2) \cdots (s - a_n)$$

where the a_j are unequal real or complex numbers. If $Q(s)$ is any polynomial of degree less than n, there are constants A_j such that the partial-fraction expansion

$$(2) \qquad \frac{Q(s)}{P(s)} = \frac{A_1}{s - a_1} + \frac{A_2}{s - a_2} + \cdots + \frac{A_n}{s - a_n}$$

holds for $s \neq a_j$; in other words, it holds whenever either side is meaningful. The constants A_j can be determined by clearing of fractions and equating coefficients of corresponding powers of s. But a more efficient method is to multiply by $s - a_j$ and let $s \to a_j$.

In illustration of this last remark consider

$$(3) \qquad \frac{Q(s)}{(s - a_1)(s - a_2)(s - a_3)} = \frac{A_1}{s - a_1} + \frac{A_2}{s - a_2} + \frac{A_3}{s - a_3}$$

where $Q(s)$ is a polynomial of degree 2 or less and where no two a_j are equal. Multiplying by $s - a_1$ and letting $s \to a_1$ gives

$$(4) \qquad A_1 = \frac{Q(a_1)}{(a_1 - a_2)(a_1 - a_3)}.$$

Formulas for A_2 and A_3 are obtained in the same way or by cyclic permutation. In a cyclic permutation of the subscripts 1, 2, $\cdots$, n the subscript 1 goes to 2, 2 goes to 3, $\cdots$, and n goes to 1.

We have shown that, if the expansion (3) is possible, then the A_j are determined by a simple limiting process. However, if the numerator were of degree 3 or more the expansion (3) would not exist. This is true because the left-hand side would not tend to 0 as $s \to \infty$, although the right-hand side obviously tends to 0 for every choice of the constants A_j. Another reason for expecting the expansion to fail when the degree of Q exceeds 2 is that a cubic polynomial involves four arbitrary constants while the right-hand side of (3) has only three.

To show that the expansion is possible when the numerator has degree ≤ 2, we clear of fractions in (3) and get the equivalent form

(5) $Q(s) = A_1(s - a_2)(s - a_3) + A_2(s - a_1)(s - a_3) + A_3(s - a_1)(s - a_2)$.

The above choice of A_1, A_2, and A_3 makes this an equality for $s = a_1, a_2$, and a_3. Hence, the two second-degree polynomials (5) agree for three distinct values of s. By Theorem 2 the polynomials are identical and the desired identity is established. A similar result, with similar proof, holds when there are n values a_j and the degree of Q is less than n.

When applied to the general equation (2) the limit procedure gives

(6) $$A_j = \lim_{s \to a_j} \frac{Q(s)(s - a_j)}{P(s)} = Q(a_j) \lim_{s \to a_j} \frac{s - a_j}{P(s)}.$$

By the definition of derivative

$$\lim_{s \to a_j} \frac{P(s)}{s - a_j} = \lim_{s \to a_j} \frac{P(s) - P(a_j)}{s - a_j} = P'(a_j)$$

where the first equality follows from the fact that $P(a_j) = 0$. Substitution into (6) gives the following theorem:

THEOREM 3.   Let $P(s)$ be a polynomial of degree $n \geq 1$ with n distinct real or complex zeros a_j and let $Q(s)$ be any polynomial of degree $< n$. Then for $s \neq a_j$

$$\frac{Q(s)}{P(s)} = \sum_{j=1}^{n} \frac{A_j}{s - a_j} \quad \text{where} \quad A_j = \frac{Q(a_j)}{P'(a_j)}.$$

So far we have assumed the values a_j to be unequal, so that the equation $P(s) = 0$ has no multiple roots. If a_1 is a root of multiplicity m the term $A_1/(s - a_1)$ in the partial-fraction expansion must be replaced by

(7) $$\frac{A_{11}}{s - a_1} + \frac{A_{12}}{(s - a_1)^2} + \cdots + \frac{A_{1m}}{(s - a_1)^m}$$

where the A_{1j} are constant. Naturally, a similar result applies to the other roots. One way to prove it is by differentiating with respect to the root in question, as illustrated in Problem 12.

Sometimes it is convenient to use quadratic factors in the denominators, instead of linear factors $s - a_j$. The appropriate form is obtained by grouping terms in the basic expansion. For example, corresponding to the quadratic factor $(s - a_1)(s - a_2)$ one gets

$$\frac{A_1}{s - a_1} + \frac{A_2}{s - a_2} = \frac{(A_1 + A_2)s - (A_1 a_2 + A_2 a_1)}{(s - a_1)(s - a_2)}.$$

The numerator is now a general linear function instead of a constant. Repeated factors are treated in a manner analogous to (7). For example, the part of the expansion arising from a triple factor $(s^2 + as + b)^3$ is of the form

$$(8) \qquad \frac{A_{11} + B_{11}s}{s^2 + as + b} + \frac{A_{12} + B_{12}s}{(s^2 + as + b)^2} + \frac{A_{13} + B_{13}s}{(s^2 + as + b)^3}.$$

This can be obtained by twice differentiating the original expansion corresponding to the single factor $s^2 + as + b$. The differentiation is with respect to b.

The following theorem is an aid in checking whether a partial-fraction expansion is in the correct form:

THEOREM 4. *Let P be a polynomial of degree $n \geq 1$ and Q a polynomial of degree $< n$. Then the number of arbitrary constants in the partial-fraction expansion of $Q(s)/P(s)$ equals the degree n of the denominator. This is the same as the number of constants needed to specify the numerator Q.*

The proof follows by inspection of the three forms (2), (7), (8). Thus in (2) the degree of the denominator is n and there are n constants A_j. In (7) the degree of the denominator-factor $(s - a_1)^m$ is m and there are m constants A_{1j}. When (8) is generalized to m factors the number of constants is $2m$ and this is the degree of $(s^2 + as + b)^m$.

As an illustration, consider

$$\frac{Q(s)}{s^3(s^2 + 4)^2(s - 3)(s + 5)}$$

where $Q(s)$ is a polynomial of degree ≤ 8. The appropriate form for the partial fraction expansion is

$$\frac{A}{s^3} + \frac{B}{s^2} + \frac{C}{s} + \frac{Ds + E}{(s^2 + 4)^2} + \frac{Fs + G}{s^2 + 4} + \frac{H}{s - 3} + \frac{I}{s + 5}$$

which has nine constants $A, B, \cdots, I$.

The best way to learn efficient methods of partial-fraction expansion is to use them yourself, with guidance. Hence we have presented a series of partially solved problems instead of text examples. Efficiency is important. One of the main advantages of the transform method is that it gives the solution of the initial-value problem directly; you need not write the general solution

$$y = c_1 e^{a_1 t} + c_2 e^{a_2 t} + \cdots + c_n e^{a_n t} + u$$

and then find the constants c_j. However, determination of the c_j involves n linear equations in n unknowns c_j, and equating coefficients leads to n linear equations in n unknowns A_j. Unless efficient methods are used for the A_j, there is little reason to prefer that procedure over the other.

PROBLEMS

In all problems, $Q(s)$ introduced without explanation denotes a polynomial of degree lower than that of the denominator.

1P. Using the identity $(6s^4 + 5)/(2s^4 + s^2) = (6 + 5/s^4)/(2 + 1/s^2)$ check that the limit of the fraction on the left as $s \to \infty$ is 3.

2P. This problem pertains to the expansion

$$\frac{Q(s)}{(s-1)(s-2)(s+1)(s-3)} = \frac{A}{s-1} + \frac{B}{s-2} + \frac{C}{s+1} + \frac{D}{s-3}$$

Preferably without writing much, multiply by $s - 1$ and let $s \to 1$ to get $A = Q(1)/(-1)(2)(-2) = Q(1)/4$. Similarly find B, C, D.

3P. This problem pertains to the expansion

$$\frac{Q(s)}{(s-1)^3(s+2)(s-3)} = \frac{A}{(s-1)^3} + \frac{B}{(s-1)^2} + \frac{C}{s-1} + \frac{D}{s+2} + \frac{E}{s-3}$$

(a) Multiply by $(s-1)^3$ and let $s \to 1$ to get $A = -Q(1)/6$.

(b) Find D and E.

(c) Multiply by s and let $s \to \infty$ to get $C + D + E = b$ where b is the leading coefficient of $Q(s) = bs^4 + \cdots$. This gives C.

(d) Set $s = 0$ to get $Q(0)/6 = -A + B - C + D/2 - E/3$, which gives B.

The gist of this important method is that the values $s = 1, -2, 3$ were used to find A, D, E and any other two values can be used to find the remaining coefficients B, C. The best to pick are, in order, $\infty, 0, 1, -1, 2, -2, \cdots$. Letting $s \to \infty$ in the equation as it stands produces the equation $0 = 0$, which gives no information. That is why we multiplied by s before letting $s \to \infty$.

4P. This problem pertains to the expansion

$$\frac{Q(s)}{(s+1)(s+3)(s-2)(s^2+2s+2)} = \frac{A}{s+1} + \frac{B}{s+3} + \frac{C}{s-2} + \frac{Ds+E}{s^2+2s+2}$$

(a) Find A, B, C more or less by inspection.

(b) Multiply by s and let $s \to \infty$ to get $A + B + C + D = b$ where b is the leading coefficient of $Q(s) = bs^4 + \cdots$. This gives D.

(c) Check that setting $s = 0$ gives an equation from which E can be found.

5P. Although Q need not be real in Problems 2, 3, 4, we assume Q is real here because the calculation simplifies in that case. This problem pertains to the expansion

$$\frac{Q(s)}{(s^2+1)^2(s^2+4)} = \frac{As+B}{(s^2+1)^2} + \frac{Cs+D}{s^2+1} + \frac{Es+F}{s^2+4}$$

(a) Multiply by $(s^2+1)^2$ and let $s \to i$ to get $Ai + B = Q(i)/3$. Equating real and imaginary parts gives both A and B.

(b) Similarly get E and F.

(c) Check that multiplying by s and letting $s \to \infty$ gives C, and the choice $s = 0$ gives D.

6. Expand in partial fractions:

(a) $\dfrac{4s^2+5s-12}{(s-1)(s-2)(s^2+s+1)}$ 　　　　(b) $\dfrac{4(s^4-s^3+1)}{(s-2)(s+1)^2(s-1)^2}$

7. Without finding the coefficients, write the correct form for the partial-fraction expansion. No answer is provided, but Theorem 4 tells you how many constants you should have:

$$\frac{2s^{15}+6s^7-9s^2+4s+17}{(s^2+1)(s^2+s+1)^3(s+4)(s+2)^2s^5}$$

8. Find $f = L^{-1}F$ by partial fractions and also by completing the square, and verify the equivalence of the results. Since the problem is self-checking, no answer is provided:

(a) $\dfrac{s+1}{s^2-4s+3}$ 　　(b) $\dfrac{2s+3}{s^2+s}$ 　　(c) $\dfrac{s-6}{s^2+2s-35}$ 　　(d) $\dfrac{1}{s^2-3s+2}$

9. Find the inverse transform $f = L^{-1}F$:

(a) $\dfrac{2s^3+6s^2+21s+52}{s(s+2)(s^2+4s+13)}$ 　　(b) $\dfrac{s^2-18s+5}{(s-1)^3(s-2)(s-3)}$

10. Using the Laplace transform, solve $y' + y = 2\sin t$, $y(0) = 1$, and check that your solution satisfies both the differential equation and the initial condition.

11. Using the answer to Section 2, Problem 6, solve that problem completely.

12. If $Q(s)$ is an arbitrary quadratic and $a \neq b$, long division followed by partial-fraction expansion gives

$$\frac{Q(s)}{(s-a)(s-b)} = A_0 + \frac{A}{s-a} + \frac{B}{s-b}$$

(a) Show that A_0 is the limit of the fraction on the left as $s \to \infty$, hence is independent of a and b.

(b) With $A' = dA/da$, $B' = dB/da$, differentiate with respect to a to get

$$\frac{Q(s)}{(s-a)^2(s-b)} = \frac{A}{(s-a)^2} + \frac{A'}{s-a} + \frac{B'}{s-b}$$

───────────────────────**ANSWERS**───────────────────────

2. $B = -Q(2)/3$, $C = -Q(-1)/24$, $D = Q(3)/8$

3. $D = Q(-2)/135$, $E = Q(3)/40$

4. $A = -Q(-1)/6$, $B = Q(-3)/50$, $C = Q(2)/150$

5. Partial answer: $2Ei + F = Q(2i)/9$

6. Partial answer: (a) The sum of the coefficients is -4; their product is 24.

   (b) The sum of the coefficients is 2; their product is -4.

9. (a) $2 - e^{-2t} + e^{-2t}(\cos 3t - 2\sin 3t)$ 10. $2e^{-t} - \cos t + \sin t$

   (b) $e^t(-3t^2 - 17t - 22) + 27e^{2t} - 5e^{3t}$

11. $x = -4e^t + 5\cos t + 6\sin t$, $y = -2e^t + 5\cos t + 6\sin t$

4. Initial-value problems

    The transform method has two different aspects: to find the transform of the solution, and to recover the solution when its transform is known. These problems were considered separately in Sections 2 and 3 and are now considered together. We begin by giving some properties of the Laplace transform that allow a partial check on the computations. For example, if you are led to believe that a function $y \in E$ has $Ly = 3s + 1$, the following theorem shows that this must be wrong:

THEOREM 1. *If $f \in$ E then $\lim\limits_{s \to \infty} F(s) = 0$.*

For proof, the hypothesis $f \in$ E gives

$$|F(s)| \leq \int_0^\infty e^{-st} A e^{Bt}\, dt = \frac{A}{s - B}, \qquad s > B$$

for constants A and B. This shows that $|sF(s)|$ is bounded for large s, which is a stronger conclusion than that of the theorem.

If we know only that $F(s)$ exists, rather than that $f \in$ E, it is no longer true that $|sF(s)|$ has to be bounded. For example, $Lt^{-1/2} = \sqrt{\pi}s^{-1/2}$ as seen in Chapter 22. However, the weaker conclusion $F(s) \to 0$ of Theorem 1 remains true if $F(s)$ exists for a single value $s = s_0$; the hypothesis $f \in$ E is not necessary. We omit the proof, which is rather long.

It follows that, if $F(s)$ does not tend to 0 as $s \to \infty$, then $F(s)$ cannot be the transform of any function f; for example, there is no function f whose transform is the function $F(s) = 1$. This may serve to explain why the rational functions that we encountered as transforms $F(s)$ always had a numerator of lower degree than the denominator, and did not require a preliminary division for the partial-fraction expansion.

Applying Theorem 1 to $Lf' = sLf - f(0)$, we get:

THEOREM 2. *If $f' \in$ E and f is continuous then $\lim\limits_{s \to \infty} sF(s) = f(0)$.*

The following theorem has a narrower range of applicability than the results above but is stated for completeness.

THEOREM 3. *If $f \in$ E and $\lim\limits_{t \to \infty} f(t) = k$ then $\lim\limits_{s \to 0+} sF(s) = k$.*

The converse is false. For example, $sF(s)$ tends to 0 as $s \to 0+$ when $f(t) = \sin t$, but $\sin t$ has no limit as $t \to \infty$.

A formal proof can be given as suggested in Problem 12. Here we explain why the theorem is true. The hypothesis $\lim f(t) = k$ means that if t is large then $f(t)$ is arbitrarily close to k. If we had $f(t) = k$ for $t \geq t_0$, instead of $f(t) \approx k$, we could write

$$\int_0^\infty e^{-st} f(t)\, dt = \int_0^{t_0} e^{-st} f(t)\, dt + \int_{t_0}^\infty e^{-st} k\, dt.$$

When $s > 0$ the first term on the right admits a bound independent of s and the second term is ke^{-st_0}/s. These two remarks give

$$\lim_{s \to 0+} sF(s) = 0 + \lim_{s \to 0+} ke^{-st_0} = k.$$

PROBLEMS————————————————————————————

Many of these problems are more time-consuming than the average problem
in this text.

1. Assuming $y^{iv} \in E$ and y''' continuous, deduce the following equation
from the corresponding equation for y''':

$$Ly^{iv} = s^4 Ly - (s^3 y(0) + s^2 y'(0) + sy''(0) + y'''(0))$$

2P. We are going to solve the initial-value problem

$$y''' - y' = 4\sin t, \qquad y(0) = 8, \qquad y'(0) = 0, \qquad y''(0) = 4$$

(a) Check that

$$Ly = \frac{8s^3 + 4s}{(s-1)(s+1)(s^2+1)}$$

(b) Find two coefficients in the partial-fraction expansion by inspection and
the other two by letting $s \to \infty$ and by taking $s = 0$.

(c) Deduce from (b) that $y = 6\cosh t + 2\cos t$ and check that this function
satisfies both the initial conditions and the differential equation.

3. Solve by the Laplace transform:

(a) $y'' - 3y' + 2y = 4$, $\qquad\qquad$ $y(0) = 2$, $\quad y'(0) = 3$
(b) $y'' + 4y = 6\sin t$, $\qquad\qquad$ $y(0) = 6$, $\quad\; y'(0) = 0$
(c) $y''' + y'' = 6(e^t + t + 1)$, $\quad y(0) = y'(0) = y''(0) = 0$
(d) $y''' - y' = 2$, $\qquad\qquad\qquad\;$ $y(0) = y'(0) = y''(0) = 4$
(e) $y''' - y' = 6 - 3t^2$, $\qquad\qquad$ $y(0) = y'(0) = y''(0) = 1$

4P. We are going to find the rest solution of $(D+1)^3(D+2)y = 6$.

(a) Noting that $P(s) = (s+1)^3(s+2)$, find Ly by inspection.

(b) By using $s \to -1$, $s \to -2$, $s \to 0$ and $s \to \infty$, show that four of the
five coefficients in the partial-fraction expansion have the values $-6, 3, 3$,
and -6, respectively.

(c) To get the remaining coefficient you could (as always) clear of fractions
and equate coefficients. But it is more efficient to use any value other
than the values $s = -1, -2, 0, \infty$ that were already used. By taking
$s = 1$ show that the coefficient not found in Part (b) is 0, and deduce
that

$$y = -3t^2 e^{-t} - 6e^{-t} + 3e^{-2t} + 3$$

5. Using the Laplace transform, find the rest solution:

(a) $y'' - y = 2\sin t$ (d) $y'' + y' = 1 + 2t$ (g) $y'' - 2y' = 20e^{-t}\cos t$

(b) $y'' + 2y' = 5y$ (e) $y'' + 4y' + 3y = 6$ (h) $y'' + y = 2 + 2\cos t$

(c) $y'' + y = 4t\sin t$ (f) $y'' - 2y' = 8(t + e^{2t})$ (i) $y'' - y' = 30\cos 3t$

6. Solve by the Laplace transform:

(a) $y''' + 2y'' - y' - 2y = 24e^{-3t} + 48t^2$, $y(0) = y'(0) = y''(0) = 0$

(b) $(D^2 + 6D + 7)^2 y = 0$, $y(0) = y'(0) = y''(0) = 0$, $y'''(0) = 4\sqrt{2}$.

7P. We are going to solve the system

$$x' + 2y' + x - y = 25, \qquad 2x' + y = 25e^t, \qquad x(0) = 0, \qquad y(0) = 25$$

(a) Transform both equations and solve for Lx to get

$$Lx = \frac{25}{s(s-1)^2(4s+1)}$$

(b) By partial fractions get $x = 25 - 9e^t + 5te^t - 16e^{-t/4}$.

(c) Find y from $y = 25e^t - 2x'$ or by computation of Ly, as you prefer.

8. (abcd) Solve subject to the initial conditions $x(0) = 1$, $y(0) = 0$:

$$x' + 5x = 3y \quad x' + 5x = y \quad x' + 6x = 2y \quad x' = x + 2y$$
$$y' + 4x = 2y \quad y' + 3y = -x \quad y' + 7y = 3x \quad y' + 2x = -3y$$

9. (abcd) Solve Problem 8 with the initial conditions $x(0) = 0$, $y(0) = 1$.

10. (abcd) Find Lx and Ly in Problem 8 when x' and y' are replaced respectively by x'' and y''. Use the initial conditions

$$x(0) = 1, \qquad x'(0) = 0, \qquad y(0) = 0, \qquad y'(0) = 1$$

11. Using the Laplace transform, solve the given system:

(a) $y' + 3y + z' + 2z = e^{-2t}$, $2y' + 2y + z' + z = 1$, $y(0) = z(0) = 0$

(b) $y' + z' = z' + w' = w' + y' = y$, $y(0) = z(0) = w(0) = 1$

12. Prove Theorem 3 by writing $f(t) = k + g(t)$ where $g(t) \to 0$ as $t \to \infty$.

ANSWERS
---

3. (a) $2 - 3e^t + 3e^{2t}$ (c) $t^3 - 6t + 6\sinh t$ (e) $t^3 + e^t$

    (b) $2\sin t + 6\cos 2t - \sin 2t$ (d) $-e^{-t} + 5e^t - 2t$

5. (a) $\sinh t - \sin t$ (e) $e^{-3t} - 3e^{-t} + 2$ (i) $3e^t - 3\cos 3t - \sin 3t$

    (b) 0 (f) $1 - 2t - 2t^2 - e^{2t} + 4te^{2t}$

    (c) $t\sin t - t^2\cos t$ (g) $3e^{2t} - 5 + 2e^{-t}(\cos t - 2\sin t)$

    (d) $1 - e^{-t} + t^2 - t$ (h) $2 - 2\cos t + t\sin t$

6. (a) $4e^{-2t} + 17e^t + 42e^{-t} - 3e^{-3t} - 24t^2 + 24t - 60$

    (b) $e^{-3t}(\sqrt{2}t\cosh\sqrt{2}t - \sinh\sqrt{2}t)$

8. Partial answers: (a) $y = 4e^{-2t} - 4e^{-t}$ (c) $5y = 3(e^{-4t} - e^{-9t})$

                    (b) $y = -te^{-4t}$ (d) $y = -2te^{-t}$

9. Partial answers: (a) $y = 4e^{-t} - 3e^{-2t}$ (c) $5y = 2e^{-4t} + 3e^{-9t}$

                    (b) $y = e^{-4t}(1 + t)$ (d) $y = e^{-t}(1 - 2t)$

10. (a) $X = (s^3 - 2s + 3)/(s^4 + 3s^2 + 2)$, $Y = (s^2 - 4s + 5)/(s^4 + 3s^2 + 2)$

    (b) $X = (s^3 + 3s + 1)/(s^2 + 4)^2$, $Y = (s^2 - s + 5)/(s^2 + 4)^2$

    (c) $X = (s^3 + 7s + 2)/(s^4 + 13s^2 + 36)$, $Y = (s^3 + 3s + 6)/(s^4 + 13s^2 + 36)$

    (d) $X = (s^3 + 3s + 2)/(s^2 + 1)^2$, $Y = (s^2 - 2s - 1)/(s^2 + 1)^2$

11. (a) $y = 2 + e^{-2t} - (3 + t)e^{-t}$, $z = -3 - 2e^{-2t} + (5 + 2t)e^{-t}$

    (b) $y = z = w = e^{t/2}$

---

5. Additional properties

The scope of the transform method is increased by a number of general properties, some of which were already mentioned in the preceding discussion. For example we showed in Section 1 that L is linear and we derived the *first shift theorem*

(1) $$Le^{-ct}f(t) = F(s + c).$$

Equation (1) is true under the sole hypothesis that Lf exists, and it is very important that the constant c can be complex. Further properties of this sort are obtained now. However, some of these have a restriction on f, and all require that c be real. The reason is that the statements involve $f(t - c)$ or $f(ct)$; and a function $f(t)$ of the real variable t cannot ordinarily be extended, in any reasonable way, to allow complex values of the variable. Functions like t^2, $\sin t$, or e^t, which do allow such an extension, illustrate the exception rather than the rule.

Equation (1) has the counterpart

(2) $Lf(t - c) = e^{-sc}F(s)$ if $c \geq 0$ and $f(t) = 0$ for $t < 0$

which is called the *second shift theorem*. For proof, let $u = t - c$ and $t = c + u$. The result is

$$\int_0^\infty e^{-st} f(t - c)\, dt = \int_{-c}^\infty e^{-s(c+u)} f(u)\, du = e^{-sc} \int_{-c}^\infty e^{-su} f(u)\, du.$$

The limits $(-c, \infty)$ can be changed to $(0, \infty)$ since $f(u) = 0$ for $u < 0$, and (2) follows. To emphasize the side condition $f(t) = 0$ for $t < 0$, the function $f(t)$ is often replaced by $h(t)f(t)$ where $h(t) = 0$ for $t < 0$ and $h(t) = 1$ for $t \geq 0$. See entry h of the table and see also Chapter 22, Section 1.

If $c > 0$ the change of variable $u = ct$, $t = u/c$ gives

(3) $$Lf(ct) = \frac{1}{c} F\left(\frac{s}{c}\right), \qquad c > 0;$$

the easy proof is left for you. Equation (3) has the counterpart

(4) $$F(cs) = L\frac{1}{c} f\left(\frac{t}{c}\right), \qquad c > 0.$$

This does not require a new derivation, but follows when (3) is written with $1/c$ instead of c. Although there is no official terminology, either result (3) or (4) will be called the *scaling theorem*.

In Section 1 we obtained the *first differentiation theorem*

(5) $$Lf'(t) = sF(s) - f(0)$$

under the hypothesis that $f' \in E$ and f is continuous. The *first integration theorem*

(6) $$L \int_0^t f(u)\, du = \frac{F(s)}{s}$$

holds if $f \in E$. It is obtained by applying (5) to the integral in (6).

The *second differentiation theorem* and the *second integration theorem* are, respectively,

(7) $$F'(s) = -Ltf(t), \qquad \int_s^\infty F(s)\, ds = L\frac{f(t)}{t}.$$

The equation on the left holds if $f \in E$, that on the right if $f(t)/t \in E$. Since

$$\frac{d}{ds} e^{-st} = -te^{-st}, \qquad \int_s^\infty e^{-st}\, ds = \frac{1}{t}$$

these results agree with those obtained by formal differentiation or integration of the Laplace transform.

For proof that $F'(s) = -Ltf(t)$ let $f \in E$ and let

$$F_n(s) = \int_0^n e^{-st} f(t)\, dt, \qquad G_n(s) = -\int_0^n e^{-st} t f(t)\, dt$$

for $n = 1, 2, 3, \cdots$. Since the range of integration is finite it is easily checked that

$$F_n'(s) = G_n(s).$$

(The proof involves little more than the definition of the derivative and is omitted.) For large s we have

$$\lim_{n\to\infty} F_n(s) = F(s), \qquad \lim_{n\to\infty} G_n(s) = G(s)$$

where $F = Lf(t)$ and $G = -L(tf(t))$. By results of Chapter G the convergence is uniform, and Chapter F, Theorem 2 gives $F'(s) = G(s)$. This completes the proof of the first equation (7). In Problem 8 it is seen that the second follows from the first.

EXAMPLE 1. If $f(t) = \sin t$, the second integration theorem gives

$$L\frac{\sin t}{t} = \int_s^\infty \frac{du}{1+u^2} = \tan^{-1} u \Big|_s^\infty = \frac{\pi}{2} - \tan^{-1} s.$$

It is surprising that such a result can be obtained so easily, because the indefinite integral associated with the transform of $(\sin t)/t$ is not elementary even when $s = 0$.

EXAMPLE 2. Applying (6) to the result of Example 1 gives the transform of the sine integral,

$$L\left(\int_0^t \frac{\sin u}{u}\, du\right) = \frac{1}{s}\left(\frac{\pi}{2} - \tan^{-1} s\right).$$

PROBLEMS

1P. Obtain the inverse transform $(1/2)\sin(t/2)$ by direct use of the first expression and also by applying the scaling theorem with $c = 2$ to the second expression:

$$\frac{1}{1+4s^2} = \frac{1}{4}\frac{1}{(1/4)+s^2} \qquad \frac{1}{1+4s^2} = \frac{1}{1+(2s)^2}$$

2. Find the inverse transform $f(t)$ in two ways as in Problem 1. Since the procedure is self checking, no answer is provided:

(a) $\dfrac{1}{1-9s^2}$ (b) $\dfrac{s-1}{9(s-1)^2+4}$ (c) $\dfrac{1}{4s^2+s}$ (d) $\dfrac{s}{(4s^2+9)^2}$

3. If a and b are real constants, with $b > 0$, deduce the equation

$$Le^{-at}\frac{\sin bt}{t} = \frac{\pi}{2} - \tan^{-1}\frac{s+a}{b}$$

from the result of Example 1. This method of extension applies to several problems below. Hint: Use the first scaling theorem with $c = b$ followed by the first shift theorem with $c = a$.

4. (abc) Show by the second integration theorem that

$$L\frac{1-e^{-t}}{t} = \ln\frac{s+1}{s} \qquad L\frac{1-\cos t}{t} = \frac{1}{2}\ln\frac{s^2+1}{s^2} \qquad L\frac{\sinh t}{t} = \frac{1}{2}\ln\frac{s+1}{s-1}$$

5. (abc) What results do the first integration theorem give when applied to the equations in Problem 4? Since the answers can be written by inspection, no answer is provided.

6. Let $u(t) = 1$ for $0 \le t < 1$ and $u(t) = 0$ elswhere. Let $v(t)$ be obtained by integrating $u(t)$ from 0 to t, and $w(t)$ by integrating $v(t)$ from 0 to t. Sketch the graphs of u, v, and w for $t \ge 0$ and obtain their transforms

$$s^{-j}(1 - e^{-s}), \qquad j = 1, 2, 3$$

7. (abcd) For $\sinh t \sin t$, $\sinh t \cos t$, $\cosh t \sin t$, $\cosh t \cos t$, respectively, obtain the following Laplace transforms:

$$\frac{2s}{s^4+4} \qquad \frac{s^2-2}{s^4+4} \qquad \frac{s^2+2}{s^4+4} \qquad \frac{s^3}{s^4+4}$$

Suggestion: Express the hyperbolic functions in terms of exponentials and apply the first shift theorem.

8. If $H(s) = L(f(t)/t)$ where $f(t)/t \in$ E, deduce $H'(s) = -F(s)$ from the second differentiation theorem. Section 4, Theorem 1 shows that $H(s) \to 0$ as $s \to \infty$. Thus determine the constant of integration and obtain the second integration theorem.

> I do not refuse my dinner, just because I don't understand the process of digestion.
>
> —Oliver Heaviside

Chapter 22
THE LAPLACE TRANSFORM II

THE LAPLACE transform is useful in dealing with discontinuous inputs (closing of a switch) and with periodic functions (sawtooth waves, rectified sine waves). Analysis of the effect of such inputs proceeds most smoothly in the frequency domain, that is, in the domain of the transform-variable s. However, an integral expression known as the convolution allows an easy passage from the frequency domain to the time domain and leads to explicit solutions even in the case of a general forcing function $f(t)$. The convolution is an aid in solving integral equations and it contributes to the theory of the gamma function, which generalizes $n!$ to nonintegral values of n.

1. The frequency domain

For suitable functions $f(t)$ we have defined the Laplace transform by

$$Lf = \int_0^\infty e^{-st} f(t) \, dt = F(s).$$

This formula represents a superposition of exponential functions, e^{-st}, where the superposition is over t and s is a parameter. By contrast, an expression such as

$$c_1 e^{s_1 t} + c_2 e^{s_2 t} + \cdots + c_n e^{s_n t}$$

describes a superposition over s with t as a parameter.

In most applications t is the time and the domain of the real variable t is the *time domain*. But the product st in e^{-st} is necessarily a pure number. One way to see this is by the Taylor series

$$e^{-st} = 1 - st + \frac{(st)^2}{2!} - \frac{(st)^3}{3!} + \cdots.$$

If st had any units, such as centimeters or seconds, the terms in the series would all have different units and the summation would make no sense. Since st is a pure number, s must have the units of $1/t$. Hence it represents the number of something dimensionless (such as cycles) per second, and has the units of frequency. This interpretation agrees with the familiar fact that ω in $e^{i\omega t}$ represents the angular frequency of the corresponding harmonic oscillations, $\cos \omega t$ and $\sin \omega t$.

It is often advantageous to let s be complex. In that case s is referred to as the *complex frequency* and the domain of the complex variable s is the *frequency domain*. Unless the contrary is indicated by context, however, we shall continue to regard s as real.

In those scientific fields that make robust uses of the Laplace transform there is a growing tendency to work in the frequency domain rather than in the time domain. It is an important feature of the frequency domain that many transcendental operations on $f(t)$ correspond to algebraic operations on $F(s)$. For example, the formula

$$L \int_0^t f(t)\, dt = \frac{F(s)}{s}$$

shows that integration corresponds to division by s.

(a) Discontinuous inputs

According to Chapter G, the function $f(t)$ is admissible on a finite interval $[a, b]$ if it is defined and continuous except at finitely many points of this interval, and is bounded on the set of points where it is defined. For convenience, we write $f \in A$ to indicate that f is admissible on $[0, a]$ for every a, and we refer to such functions as *admissible*. The class E of functions introduced in Chapter G is then the class of admissible functions that are at most of exponential growth. The condition $f \in A$ ensures that f is Riemann integrable on every finite interval $[0, a]$, and $f \in E$ ensures existence of the Laplace transform. That is why these two classes are appropriate.

A *simple discontinuity* c is a discontinuity at which the left- and right-hand limits

$$\lim_{t \to c-} f(t) = f(c-), \qquad \lim_{t \to c+} f(t) = f(c+)$$

exist and are not equal. A function f is said to be *piecewise continuous* if it is admissible and all of its discontinuities are simple. In the theory of Chapter 21 it did not matter whether the discontinuities were simple or not. We want the discontinuities to be simple here, however, for the following theorem:

THEOREM 1. *Let $f'(t) \in E$ and let $f(t)$ be continuous except at a single value $t = c > 0$ where $f(t)$ has a simple discontinuity. Then*

$$Lf'(t) = sLf(t) - f(0+) - J(c)e^{-sc}$$

where $J(c) = f(c+) - f(c-)$ is the jump of $f(t)$ at c.

This follows by addition of the two equations

$$\int_0^c e^{-st} f'(t)\, dt = e^{-st} f(t) \Big|_0^c + s \int_0^c f(t)e^{-st}\, dt$$

$$\int_c^\infty e^{-st} f'(t)\, dt = e^{-st} f(t) \Big|_c^\infty + s \int_c^\infty f(t)e^{-st}\, dt$$

which are obtained by partial integration. As in the more detailed discussion given in Chapter 21, Section 1, the integrated term at ∞ disappears when s is large. To see why one gets one-sided limits, replace c by $c - \delta$ in the first equation, by $c + \delta$ in the second, and let $\delta \to 0+$.

If $f(t)$ has finitely many discontinuities, all simple, there is a term $J(c)e^{-sc}$ for each of them. Applying Theorem 1 to $f^{(n)}$ instead of f and solving for $Lf^{(n)}$, we get

$$(1) \qquad Lf^{(n)} = \frac{Lf^{(n+1)}}{s} + \frac{J_n(0) + e^{-sc_1}J_n(c_1) + e^{-sc_2}J_n(c_2) + \cdots}{s}$$

where $J_n(c_j)$ is the jump of $f^{(n)}$ at c_j. For uniformity of notation we have considered that $f(t) = 0$ for $t < 0$ and hence that

$$J_n(0) = f^{(n)}(0+).$$

The form of Equation (1) alerts you to the fact that a term $J_n(c_j)$ is needed at each discontinuity.

(b) Piecewise polynomial inputs

We shall find Lf if $f(t) = 0$ when $t > 2$ and

$$f(t) = t^3, \qquad 0 \le t \le 1; \qquad f(t) = (2 - t)^3, \qquad 1 \le t \le 2.$$

The graph of $f(t)$ is shown in Figure 1(a). The function f is defined by a polynomial on each of finitely many intervals and is said to be *piecewise polynomial*; the word "polynomial" here is an adjective not a noun. Equation (1) is especially useful for such functions, because sufficiently high derivatives of a polynomial are 0.

We begin by differentiating $f(t)$ on the intervals $(0, 1)$, $(1, 2)$, and $(2, \infty)$. The first row in the following table is obtained by copying the definition of $f(t)$ and the other rows follow by inspection:

Table 1

interval	$(0, 1)$	$(1, 2)$	$(2, \infty)$
$f^{(0)}(t)$	t^3	$(2 - t)^3$	0
$f^{(1)}(t)$	$3t^2$	$-3(2 - t)^2$	0
$f^{(2)}(t)$	$6t$	$6(2 - t)$	0
$f^{(3)}(t)$	6	-6	0

The next table is found by evaluating these derivatives at $0, 1, 2$. The limits at $0+$ and $1-$ are obtained from values on $(0, 1)$, hence from the first column of Table 1. The limits at 1+ and 2− are obtained from the second column, and the limit 0 at 2+ is obtained from the last column:

Table 2

n	$f^{(n)}(0+)$	$f^{(n)}(1-)$	$f^{(n)}(1+)$	$f^{(n)}(2-)$	$f^{(n)}(2+)$
0	0	1	1	0	0
1	0	3	−3	0	0
2	0	6	6	0	0
3	6	6	−6	−6	0

The values of $J_n(0)$ are given by the first column of Table 2, $J_n(1)$ is given by the third column minus the second, and $J_n(2)$ is given by the last column minus the fourth. Thus we get a third table:

Table 3

n	$J_n(0)$	$J_n(1)$	$J_n(2)$
0	0	0	0
1	0	−6	0
2	0	0	0
3	6	−12	6

The first row has only zeros because $f(t)$ is continuous as seen by Figure 1(a). For continuous functions f, the jump is 0 at every value c. We have included this third table only for clarity of exposition. In practice, one could insert the three columns of Table 3 between appropriate columns of Table 2.

Since $f^{(4)}(t) = 0$ except at 0,1,2 where it is undefined, Equation (1) with $n = 3$ gives

$$Lf^{(3)} = \frac{6 - 12e^{-s} + 6e^{-2s}}{s} = 6\frac{(1 - e^{-s})^2}{s}.$$

Here we used the last line of Table 3. The third line of the table and (1) show that $Lf^{(2)}$ is obtained when the above formula is divided by s. The second line of the table then gives

$$Lf' = 6\frac{(1 - e^{-s})^2}{s^3} - 6\frac{e^{-s}}{s}$$

and, referring to the first line of the table, we see that division by s gives Lf. Direct calculation of Lf from its definition involves elementary integrals, but the above method is easier. It applies to all piecewise polynomial functions.

In the course of the computation we obtained the Laplace transforms of f', f'', f''' automatically. The graphs of these functions are shown in Figure 1(bcd). It does not matter how the functions are defined, or even whether they are defined, at points of discontinuity. This fact is indicated by vertical lines at the jumps in Figures 1 and 3. Graphs with the jumps filled in are called *maximal*. They are used for mathematical reasons in some parts of the theory of differential equations, but are introduced for visual reasons here.

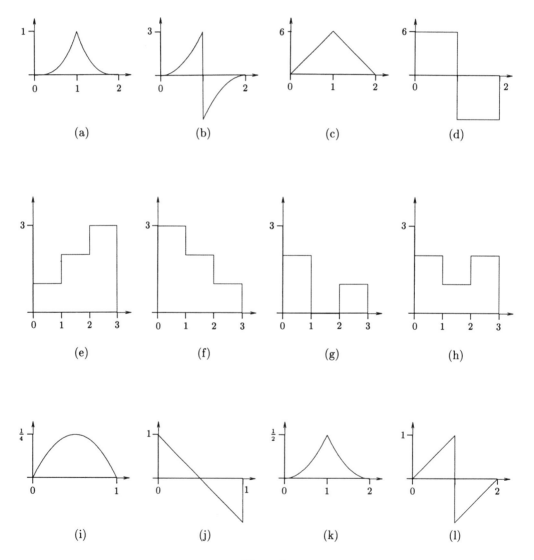

Figure 1

(c) The Heaviside unit function

The *Heaviside unit function* is defined by the equations

$$h(t) = 0, \qquad t < 0; \qquad h(t) = 1, \qquad t \geq 0.$$

Clearly $Lh = L1 = 1/s$, a fact used below.

One of the main uses of the Heaviside function is in connection with the second shift theorem,

$$Lf(t - c) = e^{-sc} Lf(t), \qquad c \geq 0,$$

which requires the side condition $f(t) = 0$ for $t < 0$. If $f(t)$ is defined for $t < 0$ (this is essential) then $h(t)f(t)$ agrees with $f(t)$ for $t \geq 0$ and is 0 for $t < 0$. Hence, it satisfies the side condition. If $0 \leq a < b$ the first of the functions

$$h(t - a)f(t), \qquad h(t - b)f(t)$$

agrees with $f(t)$ for $t \geq a$ and is 0 for $t < a$. The second of these functions agrees with $f(t)$ for $t \geq b$ and is 0 for $t < b$. A moment's thought shows then that the difference

$$h(t - a)f(t) - h(t - b)f(t)$$

agrees with $f(t)$ on the interval $[a, b)$ and is 0 elsewhere. This function describes the effect of closing a switch at time $t = a$ and opening it at $t = b$. Since $Lh(t) = 1/s$ the second shift theorem gives

(2) $$Lh(t - c) = \frac{e^{-cs}}{s}, \qquad c \geq 0.$$

This equation is used in Example 1.

EXAMPLE 1.   The response y to an input f satisfies $Ty = f$ where T is a linear differential operator with constant coefficients. Find the frequency-domain response corresponding to the rest solution, if a switch to a unit input is closed at time $t = a \geq 0$ and opened at time $t = b > a$.

The desired frequency-domain response is $Y = Ly$ where y is the rest solution of

$$Ty = h(t - a) - h(t - b).$$

There is no factor $f(t)$ because the term *unit input* means $f = 1$. Equation (2) with $c = a$ and $c = b$ yields

$$P(s)Y = \frac{e^{-as} - e^{-bs}}{s}$$

where $P(s)$ is the characteristic polynomial of T. Division by $P(s)$ gives Y.

EXAMPLE 2. If $f_0(t)$ represents one arch of a sine curve, so that $f_0(t) = \sin t$ for $0 \le t < \pi$ and $f_0(t) = 0$ elsewhere, find $F_0 = Lf_0$. Since

$$f_0(t) = h(t)\sin t - h(t - \pi)\sin t = h(t)\sin t + h(t - \pi)\sin(t - \pi)$$

the second shift theorem with $c = 0$ and with $c = \pi$ gives

$$F_0(s) = (1 + e^{-s\pi})L\sin t = \frac{1 + e^{-s\pi}}{1 + s^2}.$$

As a check, you can equate imaginary parts in

$$\int_0^\pi e^{-st}e^{it}dt = \frac{e^{(i-s)t}}{i - s}\Big|_0^\pi = \frac{1 + e^{-s\pi}}{s - i}.$$

(d) Periodic functions

Let f be a periodic function of period p, so that $p > 0$ and

$$f(t + p) = f(t)$$

for all t. We assume $f \in E$, which holds if $f(t)$ is admissible on an interval covering a period, such as $0 \le t \le p$.

The function

(3) $$f_0(t) = h(t)f(t) - h(t - p)f(t)$$

agrees with $f(t)$ on $0 \le t < p$ and is 0 elsewhere. Hence f_0 represents f on a single period, and f will be called the *periodic extension* of f_0. Since f is periodic we can replace $f(t)$ by $f(t - p)$ in the second term of (3). Taking the Laplace transform then gives

$$Lf_0 = Lf - e^{-ps}Lf$$

and, solving for Lf,

(4) $$Lf = \frac{Lf_0}{1 - e^{-ps}} \quad \text{where} \quad Lf_0 = \int_0^p e^{-st}f(t)dt.$$

This is the basic formula for the Laplace transform of a periodic function.

EXAMPLE 3. A *full rectified wave* is defined by

$$f(t) = |\sin t|$$

and has the graph shown in Figure 2(a), while for a half-rectified wave every other arch is omitted as shown in Figure 2(b). For the full wave $f_0(t) = \sin t$ on the interval $0 \le t < \pi$ whereas the half wave has

$$f_0(t) = \sin t, \quad 0 \le t < \pi; \quad f_0(t) = 0, \quad \pi \le t < 2\pi.$$

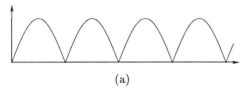

(a)

Figure 2

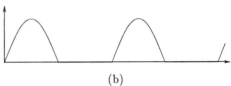

(b)

In both cases Lf_0 is given by $F_0(s)$ from Example 2. Hence by (4) the respective Laplace transforms are

$$\frac{F_0(s)}{1 - e^{-\pi s}}, \qquad \frac{F_0(s)}{1 - e^{-2\pi s}}.$$

We have taken $p = \pi$ in the first case and $p = 2\pi$ in the second. Substituting $F_0(s)$ as given by Example 2 yields

$$\frac{1}{1 + s^2} \frac{1 + e^{-\pi s}}{1 - e^{-\pi s}}, \qquad \frac{1}{1 + s^2} \frac{1}{1 - e^{-\pi s}}$$

where the second expression was simplified by cancellation of a common factor. These two functions give the Laplace transform of the full-rectified and half-rectified waves, respectively.

PROBLEMS

1. (efgh) A *step function* is a piecewise polynomial function in which each polynomial is constant. Since the derivative is 0 except at points of discontinuity, the choice $n = 0$ in (1) gives

$$sF(s) = J(0) + e^{-sc_1} J(c_1) + e^{-sc_2} J(c_2) + \cdots$$

for step functions $f(t)$, where $J(c) = J_0(c)$ is the jump of $f(t)$ at $t = c$. Use this to find $sF(s)$ for step functions that are defined on (0,3) by the graphs in Figure 1(efgh) and are 0 elsewhere.

Hint for (e): By inspection, $J(0) = J(1) = J(2) = 1$, $J(3) = -3$.

2. (efgh) Solve Problem 1 by use of $sLh(t - c) = e^{-sc}$, $c \geq 0$

Hint for (e): $f(t) = h(t) + h(t - 1) + h(t - 2) - 3h(t - 3)$.

3. What function must the Laplace transforms of f, f', f'', f''' in Part (b) be divided by to get the transforms of the periodic functions shown in Figure 3(abcd)? The period is $p = 2$ in each case.

4. What function must the transforms $F(s)$ in Problems 1 and 2 be divided by to get the transforms of the periodic functions shown in Figure 3(efgh)? The period is $p = 4$ in each case.

5. Let $f(t) = t(1 - t)$ for $0 \le t \le 1$ and $f(t) = 0$ elsewhere. The graphs of f and f' are shown in Figures 1(i) and 1(j).

(a) Starting at the left-hand edge of your paper, make a two-column table showing f, f', f'' on (0,1) versus $n = 0, 1, 2$.

(b) To the above two columns add two more giving the values of $f^{(n)}$ at 0+ and 1−. The values at 1+ are all 0.

(c) To your four columns add two more giving $J_n(0)$ and $J_n(1)$.

(d) Thus find the Laplace transforms of f, f' and of the periodic functions whose graphs are shown in Figure 3(ij).

6. Let $f(t) = 2t$ on (0,1), $f(t) = 3 - t$ on (1,3), and $f(t) = 0$ elsewhere.

(a) Sketch the graph of $f(t)$ and note that, by continuity, $J_0(c) = 0$ at each c.

(b) Find $J_1(c)$ at $c = 0, 1, 3$ by inspection of $f'(t)$.

(c) Thus find the transform of f and of its periodic extension of period 3.

(d) Sketch the graph of the periodic extension.

7. Let $f(t) = t^2/2$ for $0 \le t \le 1$, $f(t) = (t - 2)^2/2$ for $1 \le t \le 2$, and $f(t) = 0$ elsewhere. The graphs of f and f' are shown in Figures 1(k) and 1(ℓ). Proceeding as in Example 1 or as in the foregoing problems, show that the Laplace transforms of f'', f', and f are respectively

$$\frac{1 - e^{-2s}}{s}, \qquad \frac{1 - e^{-2s}}{s^2} - \frac{2e^{-s}}{s}, \qquad \frac{1 - e^{-2s}}{s^3} - \frac{2e^{-s}}{s^2}$$

8. If the three functions $F(s)$ of Problem 7 are divided by $1 - e^{-2s}$ you get the Laplace transforms of three periodic functions $f(t)$. Sketch the graphs of these periodic functions.

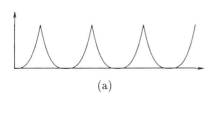

(a)

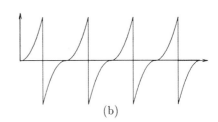

(b)

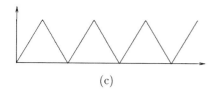

(c)

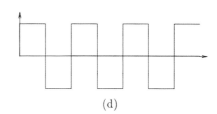

(d)

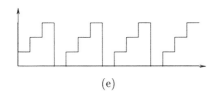

(e)

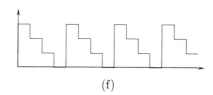

(f)

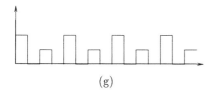

(g)

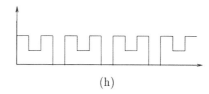

(h)

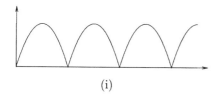

(i)

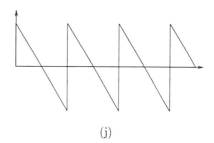

(j)

Figure 3

9. Let $f(t)$ with Laplace transform $F(s)$ be admissible on $0 \le t \le 1$ and 0 outside this interval. We can modify f in three ways: Multiply f by a constant a; change the t scale to get $f(t/b)$, which may be nonzero for $0 \le t \le b$; and shift the graph of f to the right by an amount c. If we do all three of these operations we get

$$g(t) = af\left(\frac{t-c}{b}\right), \qquad b > 0,\ c \ge 0$$

Check that $G(s) = abe^{-cs}F(bs)$. The existence of such a formula explains why we did not introduce parameters a, b, c in the examples and problems of this section.

10P. The Heaviside function. Let $c > 0$ and let $f(t)$ be defined on $(-\infty, \infty)$. For $t \ge 0$ the two functions $f(t)$ and $h(t)f(t)$ agree, but the functions

$$h(t)f(t) \qquad h(t)f(t-c) \qquad h(t-c)f(t-c) \qquad h(t-c)f(t)$$

are, as a rule, entirely different from one another. This difference is explored here when $c = 1$ and $f(t) = 1 + t^2$.

(a) Graph the above four functions for $t \ge 0$.

(b) Find $Lh(t)(1 + t^2) = L(1 + t^2)$.

(c) Find $Lh(t)(1 + (t-1)^2) = L(t^2 - 2t + 2)$.

(d) Find $Lh(t-1)(1 + (t-1)^2)$ from Part (b).

(e) Find $Lh(t-1)(1 + t^2)$. Hint: $1 + t^2 = 1 + (1 + (t-1))^2 = ?$

11. Proceed as in Problem 10. Each item (a–f) has eight things to do:

(a) $f(t) = 2t + 2$, $c = 2$ (c) $f(t) = t^3$, $c = 1$ (e) $f(t) = \sin t$, $c = \pi/2$

(b) $f(t) = e^t$, $c = 4$ (d) $f(t) = te^t$, $c = 1$ (f) $f(t) = \cos t$, $c = \pi/2$

12. This problem requires knowledge of the geometric series for $1/(1-r)$. Setting $r = e^{-cs}, c > 0$ and proceeding without regard to rigor, give reasons for believing that, in general,

$$\frac{F(s)}{1 - e^{-sc}} = L((h(t)f(t) + h(t-c)f(t-c) + h(t-2c)f(t-2c) + \cdots)$$

If $c = p$ and if $f(t) = f_0(t)$ is 0 outside the interval $(0, c)$, this result is equivalent to the formula for the transform of a periodic function. The method in the text avoids questions about whether the series can be integrated term by term.

———————————————ANSWERS———————————————

1. (e) $1 + e^{-s} + e^{-2s} - 3e^{-3s}$ (g) $2 - 2e^{-s} + e^{-2s} - e^{-3s}$

    (f) $3 - e^{-s} - e^{-2s} - e^{-3s}$ (h) $2 - e^{-s} + e^{-2s} - 2e^{-3s}$

3. $1 - e^{-2s}$ 4. $1 - e^{-4s}$

5. Partial answer, Figure 3(i): $(1 - e^{-s})Lf = 2s^{-3}(e^{-s} - 1) + s^{-2}(e^{-s} + 1)$

6. Partial answer (c): $s^2(1 - e^{-3s})Lf = 2 - 3e^{-s} + e^{-3s}$ for periodic extension

10. (b) $(s^2 + 2)/s^3$ (d) $e^{-s}(s^2 + 2)/s^3$

    (c) $2(1 - s + s^2)/s^3$ (e) $2e^{-s}(s^2 + s + 1)/s^3$

11. Answers aside from graphs:

    (a) $(2 + 2s)/s^2$, $(2 - 2s)/s^2$, $e^{-2s}(2 + 2s)/s^2$, $e^{-2s}(2 + 6s)/s^2$

    (b) $1/(s - 1)$, $e^{-4}/(s - 1)$, $e^{-4s}/(s - 1)$, $e^{4-4s}/(s - 1)$

    (c) $6/s^4$, $(6 - 6s + 3s^2 - s^3)/s^4$, $6e^{-s}/s^4$, $e^{-s}(s^3 + 3s^2 + 6s + 6)/s^4$

    (d) $1/(s - 1)^2$, $e^{-1}(2 - s)/(s - 1)^2$, $e^{-s}/(s - 1)^2$, $se^{1-s}/(s - 1)^2$

    (e) $1/(s^2 + 1)$, $-s/(s^2 + 1)$, $e^{-\pi s/2}/(s^2 + 1)$, $se^{-\pi s/2}/(s^2 + 1)$

    (f) $s/(s^2 + 1)$, $1/(s^2 + 1)$, $se^{-\pi s/2}/(s^2 + 1)$, $-e^{-\pi s/2}/(s^2 + 1)$

2. Convolution

Consider a linear system in which the effect at the present time t of a stimulus $f(t_1)\,dt_1$ at the past time t_1 is proportional to the stimulus. On physical grounds we assume that the proportionality constant depends only on the elapsed time $t - t_1$ and hence has the form $g(t - t_1)$. The effect at the present time t is therefore

$$f(t_1)g(t - t_1)dt_1.$$

Since the system is linear, the response to the whole past history can be obtained by adding these separate effects and we are led to the integral

(1) $$w(t) = \int_0^t f(t_1)g(t - t_1)\,dt_1.$$

The lower limit is 0 because the process is assumed to have started at time $t = 0$; in other words, $f(t_1) = 0$ for $t_1 < 0$.

The expression (1) is called the *convolution* of f and g. It gives the response at the present time t as a weighted superposition over the inputs at times $t_1 \le t$. The weighting factor $g(t - t_1)$ characterizes the system and $f(t_1)$ characterizes the past history of the input. Because of this physical interpretation, the convolution is sometimes called the *superposition integral*.

To ensure existence of the integral, we assume from now on that $f \in A$ and $g \in A$, where A is the class of admissible functions introduced in Section 1. The calculation of a convolution is illustrated in the following example.

EXAMPLE 1.   Let $f(t) = e^{B_1 t}$ and $g(t) = e^{B_2 t}$ where B_1 and B_2 are unequal constants. The convolution $f * g$ is

$$\int_0^t e^{B_1 t_1} e^{B_2(t-t_1)} dt_1 = e^{B_2 t} \int_0^t e^{(B_1 - B_2)t_1} dt_1 = \frac{e^{B_1 t} - e^{B_2 t}}{B_1 - B_2}.$$

Since $Le^{ct} = 1/(s-c)$ for any constant c we have the remarkable equation

$$L(f * g) = \frac{1}{B_1 - B_2} \left(\frac{1}{s - B_1} - \frac{1}{s - B_2} \right) = \frac{1}{s - B_1} \frac{1}{s - B_2} = (Lf)(Lg).$$

This is a special case of the convolution theorem as discussed next.

The function w in (1) is denoted by $f * g$, so that $(f * g)(t) = w(t)$ is the value of $f * g$ at the argument t. Naturally the meaning of $*$ does not depend on the letters f and g. For example if $a \in A$ and $b \in A$ then

$$(a * b)(t) = \int_0^t a(t_1) b(t - t_1) dt_1.$$

In Problems 2, 3, and 4 it is seen that the $*$ operation acts like ordinary multiplication in that, if a, b, c are admissible,

$$a * (b + c) = a * b + a * c, \qquad a * b = b * a, \qquad a * (b * c) = (a * b) * c.$$

The following theorem is established at the end of this section. It shows that convolution in the time domain really does correspond to multiplication in the frequency domain:

CONVOLUTION THEOREM. If $f \in E$ and $g \in E$ then $f * g \in E$ and $L(f * g) = (Lf)(Lg)$.

The importance of the convolution could hardly be overestimated. It plays a prominent role in the study of heat conduction, wave motion, plastic flow and creep, and in other branches of mathematical physics. It is also encountered in those areas of sociology, economics, ecology, and genetics in which an effect at a given time t_1 induces a delayed response at a later time t. The convolution theorem gives a bridge between the time and frequency domains, as illustrated in the following example.

EXAMPLE 2. Consider the problem

$$y'' + \omega^2 y = \omega^2 f(t), \qquad y(0) = y'(0) = 0$$

where ω is constant and $f \in$ E. Taking the transform yields

(2) $(s^2 + \omega^2)Ly = \omega^2 Lf$ or $Ly = (Lf)\dfrac{\omega^2}{s^2 + \omega^2}.$

By the table of transforms

$$\frac{\omega^2}{s^2 + \omega^2} = Lg \quad \text{where} \quad g(t) = \omega \sin \omega t.$$

Hence Equation (2) can be written

$$Ly = (Lf)(Lg) = L(f * g)$$

where the second equality follows from the convolution theorem. By uniqueness $y = f * g$ or, more explicitly,

$$y(t) = \omega \int_0^t f(t_1) \sin \omega(t - t_1) \, dt_1.$$

Here we have a formula for the rest solution corresponding to the arbitrary forcing function f.

The method of this example applies to the general equation $Ty = f(t)$ where $f \in$ E and where T is any linear operator with constant coefficients. Namely, let $P(s)$ be the characteristic polynomial of T and let g be a function such that

(3) $Lg = \dfrac{1}{P(s)}.$

Then the rest solution of $Ty = f(t)$ satisfies $P(s)Ly = Lf$ or

(4) $Ly = (Lf)\dfrac{1}{P(s)} = (Lf)(Lg) = L(f * g)$

and $y = f * g$ by uniqueness as in Example 1. The two equations

$$Y(s) = \frac{F(s)}{P(s)}, \qquad y = f * g$$

describe the solution in the frequency and time domains, respectively.

(a) A more general class of functions

Although the above argument requires $f \in$ E the final result $y = f * g$ is valid for $f \in$ A, which is a much larger class. For example, it holds for functions like

$$f(t) = e^{e^t} \quad \text{or} \quad f(t) = e^{t^2} \sin e^t$$

neither of which is in E. To see why $f \in A$ suffices, let $f \in A$ and let

$$f_a(t) = f(t) - h(t-a)f(t), \qquad a > 0.$$

Since $f_a(t) = 0$ for $t \geq a$ where it is defined, we see that $f_a \in E$. Equation (4) with f_a instead of f gives

$$y = f_a * g$$

for the rest solution of $Ty = f_a$. If $t < a$, however, we have $f_a(t) = f(t)$ and the equation and its solution become

$$Ty = f(t) = f_a(t), \qquad y = (f_a * g)(t) = (f * g)(t).$$

This shows that the formula $y = (f * g)(t)$ is valid for $0 < t < a$. Since a is arbitrary, the formula is actually valid for all positive t; and that is what we wanted to establish. Many other results can be extended in the same way, so that the method of Laplace transformation has greater scope than appears at first sight.

(b) Proof of the convolution theorem

If $f \in A$ and $g \in A$, it can be shown that $f * g$ is continuous, hence $f * g \in A$. (The proof is omitted.) In order to take the Laplace transform, we need the fact that, if $f \in E$ and $g \in E$ then $f * g \in E$. This will be established now under the assumption that $f * g \in A$ is already known.

The hypothesis $f \in E$ and $g \in E$ gives

$$|f(t)| \leq A_1 e^{B_1 t}, \qquad |g(t)| \leq A_2 e^{B_2 t}$$

where A_j and B_j are constant and where, without loss of generality, $B_1 \neq B_2$. By Example 1

$$|(f * g)(t)| \leq \int_0^t A_1 e^{B_1 t_1} A_2 e^{B_2(t-t_1)} dt_1 = A_1 A_2 \frac{e^{B_1 t} - e^{B_2 t}}{B_1 - B_2}.$$

This gives $f * g \in E$ and even the stronger result $|f| * |g| \in E$. Hence for large s the iterated integral defining $L(f * g)$ is absolutely convergent, a fact needed below.

To simplify the proof we define $f(t) = 0$ and $g(t) = 0$ for all negative t. The Laplace transforms can then be written

$$Lf = \int_{-\infty}^{\infty} e^{-st} f(t)\, dt, \qquad Lg = \int_{-\infty}^{\infty} e^{-st} g(t)\, dt$$

and the convolution $w = f * g$ is

$$w(t) = \int_{-\infty}^{\infty} f(t_1) g(t - t_1) \, dt_1.$$

Indeed, the lower limit $-\infty$ can be replaced by 0 because $f(t_1) = 0$ when $t_1 < 0$ and the upper limit ∞ can be replaced by t because $g(t - t_1) = 0$ when $t - t_1 < 0$. This latter remark shows also that $w(t) = 0$ for $t < 0$, and hence we can integrate from $-\infty$ to ∞ in computing Lw.

After these preliminaries the proof of the convolution theorem is easily completed. The value of Lw is

$$\int_{-\infty}^{\infty} \left(\int_{-\infty}^{\infty} f(t_1) g(t - t_1) \, dt_1 \right) e^{-st} \, dt$$

or also

$$\int_{-\infty}^{\infty} \left(\int_{-\infty}^{\infty} g(t - t_1) e^{-st} \, dt \right) f(t_1) \, dt_1$$

where the inversion of order is justified by absolute convergence. If we set $t - t_1 = t_2$, $t = t_1 + t_2$ we see that the inner integral on the right is

$$\int_{-\infty}^{\infty} g(t_2) e^{-s(t_1 + t_2)} \, dt_2 = e^{-st_1} Lg.$$

Here we used

$$e^{-s(t_1 + t_2)} = e^{-st_1} e^{-st_2}$$

and we moved the factor e^{-st_1} outside the integral sign. Thus our formula for Lw yields the desired result,

$$Lw = \int_{-\infty}^{\infty} e^{-st_1} (Lg) f(t_1) \, dt_1 = (Lf)(Lg).$$

PROBLEMS

Here and in Sections 3 and 4 below, letters f, g, a, b, c introduced without explanation denote admissible functions.

1P. We have occasionally used the same letter for the upper limit in an integral as for the variable of integration. This notation was used only when there was no danger of its causing confusion. It would cause great confusion in the convolution integral, however, and is discussed now.

(a) Show that if the integration variable t_1 is replaced by the upper limit t in the formula for $f * g$ the resulting expression makes mathematical sense, but is hopelessly wrong as a definition of the convolution. Whenever the upper limit occurs as an extra variable in the integrand, as in this case, you must use another letter to indicate the variable of integration.

(b) The objection to an equation such as

$$\int_0^x \cos x \, dx = \sin x$$

stems from the rule of substitution, which requires that if you replace x by a numerical value such as $x = 2$ on one side of an equality you must do the same on the other side. Check that if you replace x by 2 in all the places where it occurs, the above equation becomes nonsensical. Does the same objection apply to the following?

$$\int \cos x \, dx = \sin x + c$$

2. Verify the formula $a * (b + c) = a * b + a * c$.

3. By the change of variable $t - t_1 = t_2$ show that $a * b = b * a$.

4. Assuming $a, b, c \in$ E, as we may by Part (b), it was shown in the text that $b * c \in$ E, and by a second use of the same result, $a * (b * c) \in$ E. Similarly, $(a * b) * c \in$ E. Hence by the convolution theorem

$$La * (b * c) = (La)L(b * c) = (La)(Lb)(Lc)$$

Obtain the same result for $L(a * b) * c$ so that, by the uniqueness theorem for the transform, $a * (b * c) = (a * b) * c$.

5P. We are going to find the rest solution of $(D + 1)(D + 2)y = f(t)$. Check that the characteristic polynomial satisfies

$$\frac{1}{P(s)} = \frac{1}{s+1} - \frac{1}{s+2} = Lg, \quad \text{where} \quad g(t) = e^{-t} - e^{-2t}.$$

Deduce that $Ly = (Lf)(Lg)$ and, from this, that

$$y = f * g = e^{-t} \int_0^t f(t_1)e^{t_1} \, dt_1 - e^{-2t} \int_0^t f(t_1)e^{2t_1} \, dt_1$$

6. Find the rest solution:

(a) $y' + y = f(t)$ (b) $y'' = f(t)$ (c) $y'' - y = f(t)$ (d) $y'' + 2y' + 2y = f(t)$

7. (abcd) Evaluate your general formulas in Problem 6 when $f(t) = 1$ and verify that the solutions so obtained satisfy both the initial conditions and the differential equation. Although we shall not insist upon it, a partial check of this kind is advisable whenever you have a general formula involving an arbitrary forcing function f.

8. If a and b are constant, find the rest solution (a) when $a \neq b$ and (b) when $a = b$:

$$y'' + (a + b)y' + aby = f(t)$$

9. This problem is rather long. If w_1 is constant find the rest solution of

$$y'' + w^2 y = w^2 \sin w_1 t$$

by the general formula obtained in Example 2, and also by the method of undetermined coefficients. Carry out the calculation (a) when $w_1 \neq w$ and (b) when $w_1 = w$. Since you are asked to check your work, no answer is given. The point being made here is that, for problems to which they apply, the elementary techniques of Chapters 6 and 8 are usually more efficient than general methods, and are not made obsolete by the latter.

10P. If $F(s) = 1/s + e^{-s}/s^2 - 2e^{-2s}/s^3$ the second shift theorem gives

$$f(t) = 1 + h(t - 1)(t - 1) - h(t - 2)(t - 2)^2$$

Find $f(t)$ for $0 \leq t < 1$, $1 \leq t < 2$, $2 \leq t$ and sketch its graph. Is $f(t)$ continuous for $t \geq 0$?

11. If $F(s)$ is as follows, find $f(t)$ and sketch its graph for $t \geq 0$:

(a) $\dfrac{1 - e^{-2s}}{s}$ (c) $\dfrac{3 + e^{-s} + e^{-4s}}{s}$ (e) $\dfrac{2 - e^{-s}}{s^2}$ (g) $\dfrac{e^{-s} - e^{-2s}}{s^2}$

(b) $\dfrac{6 - 6e^{-3s}}{s^4}$ (d) $\dfrac{2 - 2^{1-s}}{1 - s}$ (f) $\dfrac{s + se^{-\pi s}}{1 + s^2}$ (h) $2\dfrac{1 - e^{-\pi s}}{1 + 4s^2}$

12. Let $f(t)$ be continuous at $t = c$ where $c > 0$. Show that

$$\lim_{t \to c-} h(t - c)f(t) = 0, \qquad \lim_{t \to c+} h(t - c)f(t) = f(c)$$

Hence $h(t - c)f(t)$ is continuous at c if, and only if, $f(c) = 0$.

Hint: Continuity at c means $\lim_{t \to c} f(t) = f(c)$. This problem does not need an (ϵ, δ) argument.

────────────────────────────ANSWERS────────────────────────────

1. (b) yes

6. (a) $e^{-t} \int_0^t f(t_1)e^{t_1} \, dt_1$ (c) $\int_0^t f(t_1) \sinh(t - t_1) \, dt_1$

   (b) $\int_0^t f(t_1)(t - t_1) \, dt_1$ (d) $e^{-t} \int_0^t f(t_1)e^{t_1} \sin(t - t_1) \, dt_1$

8. (a) $(b-a)y = \int_0^t f(t_1)(e^{-a(t-t_1)} - e^{-b(t-t_1)})\,dt_1$

(b) $\int_0^t f(t_1)(t-t_1)e^{-a(t-t_1)}\,dt_1$

10. Partial answer: $1,\quad t,\quad 5t - t^2 - 4.$ Yes.

11. Partial answers:

(a) $1 - h(t-2)$ (e) $2t - h(t-1)(t-1)$

(b) $t^3 - h(t-3)(t-3)^3$ (f) $(\cos t)(1 - h(t-\pi))$

(c) $3 + h(t-1) + h(t-4)$ (g) $h(t-1)(t-1) - h(t-2)(t-2)$

(d) $e^t(h(t-\ln 2) - 2)$ (h) $\sin(t/2) + h(t-\pi)\cos(t/2)$

---

3. Some uses of the convolution

The scope and versatility of the convolution theorem will be illustrated in a variety of problems taken from different fields.

(a) The Heaviside expansion theorem

Let T be the usual linear operator with constant coefficients. We assume that its characteristic polynomial $P(s)$ has only simple zeros s_j. By Chapter 21, Section 3 the partial-fraction expansion of $1/P$ is

$$\frac{1}{P(s)} = \sum_{j=1}^n \frac{A_j}{s - s_j} \quad \text{where} \quad A_j = \frac{1}{P'(s_j)}.$$

The equation $Le^{s_j t} = 1/(s - s_j)$ yields

$$\frac{1}{P(s)} = Lg \quad \text{where} \quad g(t) = \sum_{j=1}^n A_j e^{s_j t}.$$

If $f \in A$ the results of the previous section now give $y = f * g$ for the rest solution of $Ty = f$ or, substituting the above expression for g,

(1) $$y(t) = \int_0^t f(t_1) \sum_{j=1}^n A_j e^{s_j(t-t_1)}\,dt_1.$$

This is the *Heaviside expansion theorem.* It gives the rest solution of $Ty = f$ with an arbitrary function $f \in A$. The structure of the formula becomes clearer if we write

$$e^{s_j(t-t_1)} = e^{s_j t}e^{-s_j t_1}$$

and move the factors $e^{s_j t}$ outside of the integral sign. This is done in the following example.

EXAMPLE 1. For the equation $y'' + 9y' + 20y = f(t) \in A$ the characteristic polynomial satisfies

$$P(s) = s^2 + 9s + 20, \qquad P'(s) = 2s + 9, \qquad P(s) = (s+4)(s+5).$$

Hence $s_1 = -4$, $s_2 = -5$, $A_1 = 1/P'(-4) = 1$, $A_2 = 1/P'(-5) = -1$ and the rest solution is

$$y(t) = e^{-4t} \int_0^t f(t_1)e^{4t_1}\, dt_1 - e^{-5t} \int_0^t f(t_1)e^{5t_1}\, dt_1.$$

As a check, it is readily verified that the formula gives the correct result when $f(t) = 1$, namely,

$$y = \frac{1}{20} - \frac{1}{4}e^{-4t} + \frac{1}{5}e^{-5t}.$$

Comparing the Heaviside formula with the solution obtained in Chapter 8, we note that (1) is a sum of expressions

$$A_j e^{s_j t} \int_0^t f(t_1)e^{-s_j t_1}\, dt_1.$$

By contrast, the solution in Chapter 8 required iteration of equations of the form

$$T(s_j, 0)u = e^{s_j t} \int_0^t u(t_1)e^{-s_j t_1}\, dt_1.$$

This method leads to an n-fold integration while (1) requires only a single integration.

(b) The Heaviside superposition principle

It is a remarkable fact that the Heaviside unit function $h(t)$ contains within it the essence of the general case. Namely, if $f'(t) \in A$ and $f(t)$ is continuous, the rest solution of $Ty = f$ can be obtained from the rest solution of $Ty = h$. To see how this is done, let $Ty = f$ and $Tk = h$, both being rest solutions, and let $f_0 = f(0)$. We use the three equations

$$P(s)Ly = Lf, \qquad P(s)Lk = \frac{1}{s}, \qquad sLf = Lf' + f_0.$$

These give

$$Ly = (sLf)\frac{1}{sP(s)} = (Lf' + f_0)Lk = (Lf')(Lk) + L(f_0 k).$$

Hence

$$L(y - f_0 k) = (Lf')(Lk) = L(f' * k)$$

where the last equation follows from the convolution theorem. By uniqueness $y - f_0 k = f' * k$ or, writing in full,

$$(2) \qquad y(t) = \int_0^t f'(t_1) k(t - t_1) dt_1 + f(0) k(t).$$

Equation (2) is the *Heaviside superposition principle*.

EXAMPLE 2. Assuming that f satisfies the needed hypotheses, use (2) to find the rest solution of

$$y'' - y = f(t).$$

First we find the rest solution k of

$$k'' - k = h(t).$$

By a short calculation

$$Lk = \frac{1}{s(s^2 - 1)} = \frac{1/2}{s - 1} + \frac{1/2}{s + 1} - \frac{1}{s}$$

and by another short calculation

$$k(t) = \cosh t - 1.$$

The superposition principle in this case is therefore

$$y(t) = \int_0^t f'(t_1)(\cosh(t - t_1) - 1) \, dt_1 + f(0)(\cosh t - 1).$$

As a check, note that when $f(t) = 1$ the formula produces the correct solution $y = \cosh t - 1$ by inspection.

(c) The tautochrone

An integral equation of Volterra type is an equation of the form

$$y(t) = f(t) + \int_0^t g(t_1) y(t - t_1) \, dt_1$$

where f and g are known and y is to be found. If the various functions belong to E the convolution theorem gives

$$(3) \qquad Ly = Lf + (Lg)(Ly) \quad \text{or} \quad Ly = \frac{Lf}{1 - Lg}.$$

From Ly we can, in principle, find y. By Chapter 21, Section 4, Theorem 1 the expression $Lg = G(s)$ tends to 0 as $s \to \infty$ and hence the denominator $1 - Lg$ in (3) does not vanish when s is large. An integral equation of Volterra type that arose in a problem of historical importance is discussed next.

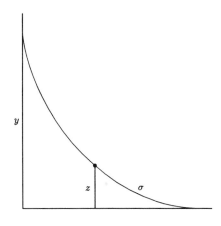

Figure 4

Starting from rest, a particle slides down a frictionless curve under gravity as shown in Figure 4. It is required to determine the shape of the curve so that the time of descent will be independent of the starting point. A curve of this kind is called a *tautochrone* from two Greek words meaning "same" and "time". The problem was solved by the Dutch mathematician Christian Huygens in 1673 as part of a profound study of the mathematics of pendulum clocks.

If the particle starts at height y its speed v when its height is z can be found by equating kinetic and potential energies. The result is

$$(4) \qquad \frac{1}{2}mv^2 = mg(y - z) \quad \text{or} \quad v = \sqrt{2g}\sqrt{y - z}$$

where m is the mass of the particle and g is the acceleration of gravity. Since the letter s is needed for the Laplace transform, we denote arc along the curve from its lowest point to the particle by σ. The time for descent is

$$\int_0^{\sigma(y)} \frac{d\sigma}{v} = \int_0^y \frac{1}{v}\frac{d\sigma}{dz}\,dz = \int_0^y \frac{1}{v}\phi(z)\,dz$$

where $\phi(y) = d\sigma/dy$ so that $\phi(z)$ is $d\sigma/dy$ at $y = z$. Since the time is constant and since v is given by (4), the problem reduces to

$$\int_0^y \phi(z)(y - z)^{-1/2}\,dz = c_0$$

where c_0 is constant. The left side is a convolution, the variables being (y, z) instead of (t, t_1). To take the Laplace transform, we multiply by $e^{-sy}dy$ and integrate. The result is

$$L(\phi * y^{-1/2}) = Lc_0 \quad \text{or} \quad (L\phi)(Ly^{-1/2}) = \frac{c_0}{s}.$$

The second equation follows from the convolution theorem. By entry b of the table of transforms with $n = -1/2$, or also by Problem 9,

$$(5) \qquad L(t^{-1/2}) = ks^{-1/2}, \qquad L^{-1}s^{-1/2} = k^{-1}t^{-1/2}$$

where k is a positive constant whose value need not concern us. (Actually $k = \sqrt{\pi}$ as explained in Section 4 below.) The above calculation gives

$$L\phi(y) = c_1 s^{-1/2}, \qquad \phi(y) = c y^{-1/2}$$

where c_1 and c are constant. Since $\phi(y) = d\sigma/dy$, where σ is arc along the curve, the equation reduces to

(6) $$1 + \left(\frac{dx}{dy}\right)^2 = \frac{c^2}{y} \quad \text{or} \quad dx = \sqrt{\frac{c^2}{y} - 1}\, dy.$$

It is seen in Problem 8 that the curve is a cycloid.

If a particle starts at a point P and slides down a frictionless curve to the point Q, one may ask what curved path minimizes the time. The minimizing curve is called a "brachistochrone" from Greek words meaning "shortest time." It is remarkable that the solution is again a cycloid, rather than a straight line as one might at first expect. This famous problem led to the founding of a branch of mathematics known as the calculus of variations. As in the case of the tautochrone, by far the main part of the problem is the *formulation* of the differential equation, not its solution. Once the equation is correctly formulated, its solution was almost as routine then as it is today.

PROBLEMS

1. (a) Find the rest solution of $y'' + 3y' + 2y = f(t)$ by the Heaviside expansion theorem and check that the result is the same as in Problem 5 of the preceding section.

    (b) Evaluate the solution (a) when $f(t) = h(t)$ and verify that it satisfies both the initial conditions and the differential equation.

2. Taking $f(t) = 20e^{3t}$ in Problem 1, find the rest solution (a) by the Heaviside expansion theorem and (b) by the Heaviside superposition principle.

3. (ab) Taking $f(t) = 4t$, proceed as in Problem 2.

4. We consider the following integral equation:

$$y(t) = f(t) + e^{-t} \int_0^t y(t_1) e^{t_1}\, dt_1$$

(a) Move the factor e^{-t} under the integral sign to get a recognizable convolution, and deduce successively that

$$y = f + y * e^{-t}, \qquad Ly = Lf + \frac{Ly}{s+1}, \qquad Ly = \frac{F(s)(1+s)}{s}$$

(b) When $f(t) = \cos t$ get $y = \sin t + \cos t$ by use of the result (a).

(c) Assuming f differentiable, multiply the original integral equation by e^t, differentiate, and simplify to get

$$y' = f + f', \quad y(0) = f(0), \quad \text{hence} \quad y = \int_0^t f(t)\, dt + f(t)$$

The initial condition comes from the integral equation. Although you assumed f differentiable in the course of the derivation, the final result is valid without this hypothesis.

(d) Obtain the result (b) by the formula (c).

5. We consider the integral equation

$$y(t) = f(t) + \int_0^t y(t_1) \sin(t - t_1)\, dt_1$$

(a) Show that $Ly = F(s)(1 + s^2)/s^2$.

(b) When $f(t) = 12t^2$ show that $y = t^4 + 12t^2$.

6. Let T be a linear differential operator with constant coefficients and with the characteristic polynomial $P(s)$. For the integro-differential equation on the left below show that formal use of the Laplace transform gives the equation on the right, where $P_0(s)$ is a polynomial depending on the initial conditions:

$$Ty + y * g = f, \qquad Ly = \frac{F(s) + P_0(s)}{G(s) + P(s)}$$

7. Find the rest solution of $y^{(n)} = f(t)$ by repeated integration and also by use of the convolution theorem. Thus get the formula

$$\int_0^t \int_0^t \cdots \int_0^t f(t)\, (dt)^n = \int_0^t \frac{(t - t_1)^{n-1}}{(n-1)!} f(t_1)\, dt_1$$

On the left, the upper limit of each integral becomes the variable of integration for the next.

8. Since $\csc^2 \theta - 1 = \cot^2 \theta$ the second equation (6) leads to an integral that you could evaluate by setting $y = c^2 \sin^2 \theta$. Make this substitution in the equation as it stands and thus show, after simplification, that

$$dx = 2c^2 \cos^2 \theta\, d\theta$$

Using $2\cos^2 \theta = 1 + \cos 2\theta$ and a similar formula for the sine, deduce that

$$x = \frac{c^2}{2}(2\theta + \sin 2\theta), \qquad y = \frac{c^2}{2}(1 - \cos 2\theta)$$

With $2\theta = \phi$ these are recognized as the parametric equations of a cycloid.

9. In the definition of $L(t^{-1/2})$ set $st = t_1, t = t_1/s$ to get

$$L(t^{-1/2}) = s^{-1/2} \int_0^\infty e^{-t_1} t_1^{-1/2} \, dt_1$$

If the integral is denoted by k then k is a positive constant and we get the formulas (5) that were used in the discussion of the tautochrone.

─────────────────────ANSWERS─────────────────────

1. (b) $2y = 1 - 2e^{-t} + e^{-2t}$ 2. $e^{3t} + 4e^{-2t} - 5e^{-t}$ 3. $-3 + 2t + 4e^{-t} - e^{-2t}$

---

4. The gamma function

If n is a positive integer the factorial $n!$ is defined by

$$1! = 1, \qquad 2! = 1 \cdot 2, \qquad 3! = 1 \cdot 2 \cdot 3, \qquad 4! = 1 \cdot 2 \cdot 3 \cdot 4$$

and so on. Thus $5! = 120$. It is natural to inquire whether one can give a sensible meaning to $n!$ when n is not an integer. As seen presently, this can be done. For example, we shall derive the astonishing formula

$$\left(-\frac{1}{2}\right)! = \sqrt{\pi}.$$

The main step in the proof involves the convolution theorem.

Because the habit of regarding n as an integer is so strong, we use p instead of n here. After the development is completed, we revert to n again, with the assurance that n need not be an integer. This applies, for example, to n in the table of transforms wherever it occurs.

Let us start from the equation

$$L(t^p) = \frac{p!}{s^{p+1}}, \qquad p \geq 0$$

which reduces to $p!$ when p is a positive integer and $s = 1$. That fact is now used to define $p!$ even if p is not an integer. In other words, by definition,

(1) $$p! = \int_0^\infty e^{-t} t^p \, dt, \qquad p \geq 0.$$

(a) The recurrence formula

Let us apply the first differentiation theorem

$$Lf'(t) = sLf(t) - f(0)$$

to $f(t) = t^p$ with $p \geq 1$. Since $f(0) = 0$ we get

$$pLt^{p-1} = sLt^p$$

and the choice $s = 1$ gives $p(p-1)! = p!$. Note that $(t^p)' = pt^{p-1}$ for $t \geq 0$ and $p \geq 1$, without any assumption that p is an integer.

The previous result can be written

$$(2) \qquad\qquad (p-1)! = \frac{p!}{p}$$

for $p \geq 1$. As p ranges over the intervals

$$(1,2), \quad (0,1), \quad (-1,0), \quad (-2,-1), \quad (-3,-2), \quad \cdots$$

the variable $p - 1$ ranges over the intervals

$$(0,1), \quad (-1,0), \quad (-2,-1), \quad (-3,-2), \quad (-4,-3), \quad \cdots$$

respectively. Using (2) with $1 < p < 2$ we get the value of $p!$ for $0 < p < 1$. Then we use (2) again to get the value when $-1 < p < 0$, and so on. Thus $p!$ is defined whenever p is real and not a negative integer.

Since $0! = 1$ the process gives

$$\lim_{p \to 0-} (p-1)! = \lim_{p \to 0-} \frac{p!}{p} = -\infty$$

and similarly, the limit as $p \to 0+$ is $+\infty$. This describes the behavior of $p!$ as $p \to -1$. By the recurrence formula (2) similar behavior is found whenever p is a negative integer, with an alternation of sign. The graph of $p!$ is shown in Figure 5 and the graph of $1/p!$ in Figure 6.

If we define $1/p! = 0$ when p is a negative integer, it turns out that the function $1/p!$ has derivatives of all orders at every value of p. Furthermore the formula

$$p\frac{1}{p!} = \frac{1}{(p-1)!}$$

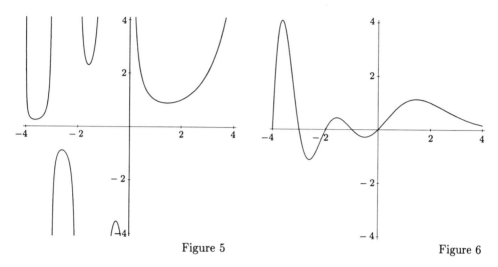

Figure 5   Figure 6

holds at every p without exception. The properties of $p!$ and $1/p!$ are analogous to those of $\csc \pi p$ and $\sin \pi p$, respectively.

The theory sketched above is due to Leonhard Euler, who started from the equation

$$\Gamma(p) = \int_0^\infty e^{-t} t^{p-1} \, dt.$$

The symbol Γ is the Greek letter "gamma" and the function $\Gamma(p)$ is called the *gamma function*. In terms of the factorial function $p!$ we have

$$p! = \Gamma(p+1), \qquad \Gamma(p) = (p-1)!.$$

The equation $p\Gamma(p) = \Gamma(p+1)$ is the recurrence formula $p(p-1)! = p!$ in Euler's notation.

(b) The beta function

One of the main uses of the gamma function is in the evaluation of integrals. As an illustration we consider the function

(3)
$$B(a, b) = \int_0^1 x^{a-1} (1 - x)^{b-1} \, dx$$

where a and b are positive constants. The letter B is a capital Greek beta and $B(a, b)$ is called the *beta function*.

The conditions $a > 0$ and $b > 0$ allow the integrand to become infinite at $x = 0$ and $x = 1$, but this does no harm since the integrals

$$\int_0^1 x^{a-1} \, dx = \frac{1}{a}, \qquad \int_0^1 (1 - x)^{b-1} \, dx = \frac{1}{b}$$

are both convergent; see Problem 16. As an illustration, the substitution $x = \sin^2 \theta$ gives the following value for $B(1/2, 1/2)$:

$$\int_0^1 x^{-1/2}(1-x)^{-1/2}\, dx = \int_0^{\pi/2} (\sin\theta \cos\theta)^{-1}(2\sin\theta\cos\theta)\, d\theta = \pi.$$

The beta function has a close relation to the convolution

$$w(t) = t^{a-1} * t^{b-1} = \int_0^t t_1^{a-1}(t-t_1)^{b-1}\, dt_1.$$

Indeed, the change of variable $t_1 = xt$ for $t > 0$ gives

(4) $$w(t) = \int_0^1 (xt)^{a-1}(t-xt)^{b-1}t\, dx = t^{a+b-1}B(a,b).$$

By the convolution theorem

$$Lw(t) = (Lt^{a-1})(Lt^{b-1}) = B(a,b)Lt^{a+b-1}$$

where the second equality follows from (4). Evaluating the transforms by the table we get

$$\frac{(a-1)!}{s^a}\frac{(b-1)!}{s^b} = \frac{(a+b-1)!}{s^{a+b}}B(a,b).$$

Hence, solving for $B(a,b)$,

(5) $$B(a,b) = \frac{(a-1)!(b-1)!}{(a+b-1)!} = \frac{\Gamma(a)\Gamma(b)}{\Gamma(a+b)}.$$

This is the Euler formula for the beta function. Since $\Gamma(1) = 0! = 1$ the choice $a = b = 1/2$ gives

$$\pi = B(1/2, 1/2) = (\Gamma(1/2))^2, \quad \text{hence} \quad \Gamma(1/2) = \sqrt{\pi}.$$

The value π on the left was obtained above. Thus we have derived the formula for $(-1/2)!$ that was mentioned at the beginning of this section.

REVIEW PROBLEMS————————————————————

1. The definition of $p!$ by recursion depends on the formula $p! = p(p-1)!$ or, equivalently, $\Gamma(p+1) = p\Gamma(p)$. Check that if we take $1! = 1$, as is natural, the formula gives $0! = 1$. Check also that $0! = 1$ agrees with the result of setting $p = 0$ in the integral definition of $p!$.

2. By substituting $t = x^2$ in the integral that defines $\Gamma(1/2)$ show that the equation $\Gamma(1/2) = \sqrt{\pi}$, which was derived in the text, is equivalent to

$$\sqrt{\pi} = 2 \int_0^\infty e^{-x^2}\, dx = \int_{-\infty}^\infty e^{-x^2}\, dx$$

3. In three ways, we will find a function f such that

$$Lf = \frac{1}{s^2(s^2+1)}$$

(a) Get the answer $t - \sin t$ by applying the first integration theorem twice to the formula for $L(\sin t)$.

(b) Get the answer by partial fractions.

(c) Note that $1/s^2 = Lt$, $1/(s^2+1) = L\sin t$ and get the answer by the convolution theorem. Suggestion: $u = t - t_1$, $dv = \sin t_1\, dt_1$.

4. In three ways, obtain functions whose transforms are as follows. Since the problem is self-checking, no answer is provided:

(a) $\dfrac{1}{s^2(s^2-1)}$ (b) $\dfrac{1}{s^3(s^2+1)}$ (c) $\dfrac{1}{s^3(s^2-1)}$ (d) $\dfrac{8}{s^2(s-1)(s-2)}$

5. The notation $h(t-c) = h_c(t)$ is often used, and will be used here. If $0 \le a \le b$ show by the convolution theorem that

$$\int_0^t h_a(t_1)h_b(t-t_1)\, dt_1 = \int_0^t h_{a+b}(t_1)\, dt_1$$

6. Let $I(s) = L(e^{-t^2/2})$. By the second differentiation theorem express $I'(s)$ as an integral, integrate by parts with $u = e^{-st}$, $dv = -te^{-t^2/2}$, and thus show that $I'(s) = -1 + sI(s)$. Solve this first-order linear equation and deduce that

$$L(e^{-t^2/2}) = e^{s^2/2} \int_s^\infty e^{-t^2/2}\, dt$$

7. In using the second shift theorem it is helpful to have formulas that express $\sin t$ and $\cos t$ in terms of $\sin(t-c)$ and $\cos(t-c)$. Consider the equations

$$\sin t \cos c - \cos t \sin c = \sin(t-c)$$

$$\sin t \sin c + \cos t \cos c = \cos(t-c)$$

to be two simultaneous linear equations in the unknowns $\sin t$, $\cos t$. Check that the coefficient determinant is 1 and solve by Cramer's rule to get

$$\sin t = \cos c \sin(t-c) + \sin c \cos(t-c)$$

$$\cos t = \cos c \cos(t-c) - \sin c \sin(t-c)$$

8. Using the result of Problem 7 show that

$$Lh(t-c)\sin t = \frac{e^{-sc}}{s^2+1}(\cos c + s\sin c)$$

$$Lh(t-c)\cos t = \frac{e^{-sc}}{s^2+1}(s\cos c - \sin c)$$

9. Confirm the result of Problem 8 by equating real and imaginary parts in the equation

$$\int_c^\infty e^{(-s+i)t}\,dt = \frac{e^{(-s+i)c}}{s-i}$$

10. An oscillatory system is described by $y'' + y = f(t)$. If a switch to a unit input is closed at time $t = 0$ and opened at time $t = c > 0$ the differential equation is

$$y'' + y = h(t) - h(t-c)$$

(a) Show that for the rest solution

$$Ly = (1 - e^{-sc})\left(\frac{1}{s} - \frac{s}{s^2+1}\right)$$

(b) Deduce from the second shift theorem that

$$y = h(t)(1 - \cos t) - h(t-c)(1 - \cos(t-c))$$

and hence that $y = 1 - \cos t,\ 0 \le t \le c;\ y = \cos(t-c) - \cos t,\ c \le t$.

(c) Show that y and y' are continuous at c but y'' is not, so that the differential equation fails. More specifically, show that the jumps in the second derivative are $J_2(0) = 1,\ J_2(c) = -1$. These reflect the discontinuity of the forcing function.

11. This is difficult. In an oscillatory system described by $y'' + \omega^2 y = f(t)$ with $\omega^2 \ne 1$ a switch to a sinusoidal input is closed at time $t = 0$ and opened at time $t = c > 0$. Thus the differential equation is

$$y'' + \omega^2 y = h(t)\sin t - h(t-c)\sin t$$

(a) Using the result of Problem 8, then 7, show that $(\omega^2 - 1)y$ is

$$\sin t - \frac{\sin \omega t}{\omega} + h(t-c)\left(\frac{\sin \omega(t-c)}{\omega}\cos c + \cos\omega(t-c)\sin c - \sin t\right)$$

(b) Setting $h(t-c) = 0$ for $t < c$ and $h(t-c) = 1$ for $t \ge c$ obtain specific formulas on these two intervals. Thus show that $y(0) = y'(0) = y''(0) = 0$ and that y and y' are continuous at c. That is, $J_0(c) = J_1(c) = 0$.

(c) Show however that $J_2(c) = -\sin c$, which is 0 if, and only if, $c = n\pi$ for an integer n. This behavior is explained by the fact that the function $h(t-c)\sin t$ is continuous at c if, and only if, $\sin c = 0$.

12. (abc) By setting $e^{-t} = u$, $t = u^2$, or $t^p = u$ and $q = 1/p$ in the three cases respectively, show that for $p > 0$

$$\int_0^1 \left(\ln \frac{1}{u}\right)^p du = p! \qquad 2\int_0^\infty u^{2p+1}e^{-u^2}\, du = p! \qquad \int_0^\infty e^{-u^q}\, du = p!$$

13. (a) If k is constant, prove the following by mathematical induction:

$$Lt^n e^{kt} = \frac{n!}{(s-k)^{n+1}}, \qquad n = 0,\, 1,\, 2,\, 3,\, \cdots$$

(b) Take $k = a + ib$ and equate real parts to get the formula

$$Lt^n e^{at}\cos bt = \frac{n!}{((s-a)^2 + b^2)^{n+1}}\ \mathrm{Re}(s - a + ib)^{n+1}$$

(c) When $a = 0$ and $b = 1$ the result (b) and its counterpart for the sine are

$$Lt^n \cos t = \frac{n!}{(s^2+1)^{n+1}}\ \mathrm{Re}(s+i)^{n+1}, \quad Lt^n \sin t = \frac{n!}{(s^2+1)^{n+1}}\mathrm{Im}(s+i)^{n+1}$$

Using this, make a table of $Lt^n \cos t$ and $Lt^n \sin t$ for $n = 0, 1, 2, 3, 4$. No answer is provided. However, you may want to keep a record of this table for future use.

14. In this problem we give a sharpened and improved form of Chapter 21, Section 4, Theorems 1 and 2. Suppose $f(t)$ has a Taylor series of the form

$$f(t) = f(0) + f'(0)t + f''(0)\frac{t^2}{2!} + \cdots + f^{m-1}(0)\frac{t^{m-1}}{(m-1)!} + g(t)t^m$$

where $g(t) \in E$. Deduce that

$$F(s) = \frac{f(0)}{s} + \frac{f'(0)}{s^2} + \cdots + \frac{f^{m-1}(0)}{s^m} + \frac{M(s)}{s^{m+1}}$$

where $M(s)$ is bounded for large s.

15. Here we develop the theory of improper integrals to the extent needed for the Laplace transform. Let $f = u + iv$ be an admissible function defined on $[0, \infty)$ and let $F(R)$ denote the integral of f from 0 to R. If $F(R)$ has a limit as $R \to \infty$ it is said that the integral *converges*, or that f is *integrable* on $[0, \infty)$, and by definition

$$\int_0^\infty f(t)\, dt = \lim_{R \to \infty} \int_0^R f(t)\, dt$$

Otherwise the integral *diverges* and no value is assigned to it.

(a) If $|f(t)|$ is integrable on $[0, \infty)$ prove that $f(t)$ is too. In other words, an absolutely convergent integral is convergent.

Hint: Since $|u| \leq |f|$ we have $0 \leq |f| - u \leq 2|f|$. Deduce from the completeness axiom (Chapter 12, Section 3) that $|f| - u$ is integrable on $[0, \infty)$. The fact that u is integrable now follows from $u = |f| - (|f| - u)$.

(b) Let $|f(t)| \leq Ae^{Bt}$ where A and B are constant and let $s \geq B + 1$. Using $|e^{-st} f(t)| \leq Ae^{-t}$, show that the Laplace transform of f exists and that

$$\left| \int_0^\infty e^{-st} f(t)\, dt - \int_0^R e^{-st} f(t)\, dt \right| \leq \frac{A}{e^R}$$

16. In this problem we explore the possibility of defining Lf for a class E^* of functions that is larger than the class E. The statement $f \in E^*$ means

$$|f(t)| \leq Mt^b,\ 0 < t < 1; \qquad |f(t)| \leq Ae^{Bt},\ 1 \leq t < \infty$$

where M, b, A, B are constants, depending on f, with $b > -1$. It is also assumed that $f(t)$ is defined and continuous except at finitely many points on any bounded closed subinterval of $(0, \infty)$, hence is Riemann integrable on such a subinterval. The Laplace transform is defined by integrating from ϵ to R and then letting $\epsilon \to 0+$ as well as $R \to \infty$.

(a) If $f \in E^*$ check that the integral defining Lf is absolutely convergent and deduce, from an extension of Problem 15, that Lf exists.

(b) Taking $b = -1/2$, show that Lf can exist even if f is not in E.

(c) Taking $f(t) = g(t) = t^{-1/2}$, show that if two functions belong to E^* their product need not belong to E^*.

(d) Show that $\sin(e^{t^2})$ is in E but its derivative is not even in E^*.

Quiet water becomes stagnant. Iron rusts from disuse. So doth inactivity sap the vigor of the mind.

—Leonardo da Vinci

Chapter 23
THE LAPLACE TRANSFORM III

THE PURPOSE of this chapter is to give an idea of recent developments, some of which are the subject of current research in engineering and biomathematics. We begin by discussing a singular input that is sometimes called the Dirac delta function, though it is not a function. This concept has its origin in mathematical physics, yet has been the chief inspiration for a branch of pure mathematics known as the theory of distributions. The notions of transfer function and feedback are of continuing interest in engineering, as are the concepts of compartmentalization and parameter identification in biomathematics. All of these topics were developed chiefly in the twentieth century.

1. The Dirac distribution

Here we consider the response of a system to an impulse that acts for a very short time but produces a large effect. The physical situation is exemplified by a lightning stroke on a transmission line or a hammer blow on a mechanical system.

To formulate the idea of an impulse, let a be a small positive constant and let $\delta_a(t)$ be the function whose graph is shown in Figure 1. That is,

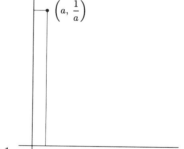

Figure 1

$$\delta_a(t) = \frac{1}{a}, \qquad 0 \le t < a; \qquad \delta_a(t) = 0 \quad \text{elsewhere.}$$

Although the function is nonzero only on a small interval, the area under the curve is the area of the rectangle, which is 1. The Laplace transform is

(1) $$L\delta_a(t) = \int_0^a \frac{1}{a} e^{-st} dt = \frac{1 - e^{-sa}}{sa}.$$

By l'Hospital's rule with $h = sa$

$$\lim_{a \to 0} \frac{1 - e^{-sa}}{sa} = \lim_{h \to 0} \frac{1 - e^{-h}}{h} = 1.$$

It is a conceptual aid to introduce an expression $\delta(t)$ that describes the effect of $\delta_a(t)$ as $a \to 0$ and to say that

$$L\delta(t) = 1.$$

The symbol $\delta(t)$ is called the Dirac delta function or the unit impulse. It acts as if it were a function with the two properties

(2) $$\delta(t) = 0 \quad \text{for} \quad t \neq 0, \quad \int_0^1 \delta(t)dt = 1.$$

First the usefulness and then the mathematical meaning of $\delta(t)$ will be explored here. Let us state at the outset, however, that *no function with the properties* (2) *exists* and hence, whatever else $\delta(t)$ may be, it is not a function of t.

Actually, $\delta(t)$ is an example of a *distribution*. A distribution is characterized, not by giving its value $\delta(t)$ at each t, but by giving its value $\delta\{\phi\}$ on a suitable class of functions ϕ. The theory of distributions has its roots in both physics and mathematics. Paul Dirac, originator of the Dirac delta function, received the Nobel prize in 1933 for research in quantum mechanics. Laurent Schwartz was awarded a Fields medal (a mathematical equivalent of the Nobel prize) in 1950 for his creation of the theory of distributions.

(a) An example

Consider a mass-spring system in which the displacement y satisfies

$$y'' + y = f(t), \qquad y(0) = y'(0) = 0.$$

To find the response to a unit impulse at time $t = 0$, replace the forcing function $f(t)$ by $\delta(t)$:

$$y'' + y = \delta(t), \qquad y(0) = y'(0) = 0.$$

Taking the Laplace transform and recalling that $L\delta = 1$ we get in succession

$$s^2 Ly + Ly = 1, \qquad Ly = \frac{1}{s^2+1}, \qquad y = h(t)\sin t.$$

The factor $h(t)$ is inserted to emphasize that the rest solution is 0 for $t < 0$ as shown in Figure 2.

The solution y is continuous for all t and it satisfies the differential equation for $t \neq 0$. At $t = 0$, however, it satisfies neither the differential equation nor the initial conditions, and is not even differentiable. Indeed, $y'(0-) = 0$ and $y'(0+) = 1$, as seen from the fact that

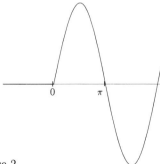

Figure 2

$$y'(t) = 0' = 0, \qquad t < 0; \qquad y'(t) = (\sin t)' = \cos t, \qquad t > 0.$$

The unit impulse produces a jump of magnitude 1 in $y'(t)$ at $t = 0$, and the graph of y versus t has a corner there.

To see what is going on, let us solve

$$(3) \qquad\qquad y'' + y = \delta_a(t), \qquad y(0) = y'(0) = 0$$

and find the behavior as $a \to 0$. By (1) the Laplace transform of (3) gives

$$aLy = \frac{1}{s^2 + 1} \frac{1 - e^{-as}}{s} = (1 - e^{-as}) \left(\frac{1}{s} - \frac{s}{s^2 + 1} \right).$$

The second shift theorem now yields

$$ay = h(t)(1 - \cos t) - h(t - a)(1 - \cos(t - a)).$$

It is easily checked that $y(t)$ is

$$(4) \qquad\qquad 0, \qquad \frac{1 - \cos t}{a}, \qquad \frac{\cos(t - a) - \cos t}{a}$$

on the three intervals $t \leq 0$, $0 \leq t \leq a$, $a \leq t$ respectively.

To study the behavior as $a \to 0$, note that

$$0 \leq \frac{1 - \cos t}{a} \leq \frac{1 - \cos a}{a} \to 0, \qquad a \to 0$$

where the inequality holds for $0 \leq t \leq a$ and where the limit as $a \to 0$ is obtained by l'Hospital's rule. L'Hospital's rule also shows that the limit of the third expression (4) is $\sin t$ as $a \to 0$. Hence letting $a \to 0$ in (4) gives the same solution as was obtained by use of $\delta(t)$.

Two features of this example are worthy of note. First, an uncritical use of $\delta(t)$ gave the correct result as obtained by a conventional passage to the limit; and second, the conventional method was much more difficult. These two features are typical. The increased difficulty of conventional methods is very pronounced when the basic problem is complicated, and may serve to show why study of $\delta(t)$ is worthwhile.

(b) Test functions

In the foregoing discussion, we did not try to define the unit impulse. Instead, we found *the response of a system to the unit impulse*. The distinction might seem like mere hair-splitting, but it is a vital distinction in the present context. The basis for the theory of $\delta(t)$ rests on the fact that $\delta(t)$ is never used alone, but only in combination with other functions. Instead of engaging in a futile attempt to define $\delta(t)$ for each value of t, we define the action of $\delta(t)$ on other functions $\phi(t)$. The functions ϕ used in this development are called *test functions*. In the general theory of distributions, test functions are assumed to have derivatives of all orders and to vanish outside a finite interval. Here we require only that $\phi(t)$ be continuous for all t and have as many continuous derivatives as are needed to make the equations meaningful. We are not trying to prove anything, but merely to show where certain formulas come from.

Let us agree that the result of applying $\delta(t)$ to $\phi(t)$ is $\phi(0)$ and let us express this fact in the symbolic form

$$(5) \qquad \int_{-\infty}^{\infty} \delta(t)\phi(t)\,dt = \phi(0).$$

It is not $\delta(t)$ but the whole formula on the left that is being defined here.

If $\delta(t)$ were an ordinary function, and if the integral in (5) were an ordinary integral, a change of variable would give

$$\int_{-\infty}^{\infty} \delta(t-c)\phi(t)\,dt = \int_{-\infty}^{\infty} \delta(t)\phi(t+c)\,dt.$$

By (5) the right-hand side gives the value of $\phi(t+c)$ at $t = 0$, that is, it gives $\phi(c)$. This is now taken as the definition of the left-hand side. Thus, by definition,

$$(6) \qquad \int_{-\infty}^{\infty} \delta(t-c)\phi(t)\,dt = \phi(c).$$

If $\delta(t)$ were a continuously differentiable function vanishing for large $|t|$, an integration by parts would give

$$\int_{-\infty}^{\infty} \delta'(t)\phi(t)\,dt = -\int_{-\infty}^{\infty} \delta(t)\phi'(t)\,dt.$$

By (5) this is $-\phi'(0)$, and $-\phi'(0)$ is taken, accordingly, as the definition of the integral on the left. The same line of thought is readily extended to higher derivatives and to the argument $t - c$ as in (6). The result is the definition

$$(7) \qquad \int_{-\infty}^{\infty} \delta^{(n)}(t-c)\phi(t)\,dt = (-1)^n \phi^{(n)}(c).$$

(c) The transform of Dirac's distribution

The theory of $\delta(t)$ goes most smoothly when the interval of integration is $(-\infty, \infty)$. However, suitable conventions allow integration over an interval

$[a, b]$, as explained next. Let $H(t) = 1$ for $a \leq t \leq b$ and let $H(t) = 0$ elsewhere. If the integral were an ordinary integral we would have

$$\int_a^b \phi(t)\delta(t - c)\, dt = \int_{-\infty}^{\infty} H(t)\phi(t)\delta(t - c)\, dt.$$

According to the above discussion, the integral on the right ought to be $H(c)\phi(c)$. This is $\phi(c)$ if c is on the interval $[a, b]$ and 0 if not. That is now taken as the definition of the integral on the left. In particular, if the interval is $[0, \infty)$, as it is for the Laplace transform, we get $\phi(c)$ for $c \geq 0$ and 0 for $c < 0$. The other formulas obtained above are treated similarly.

If we make the special choice

$$\phi(t) = e^{-st}, \qquad s = \text{const.},$$

then $\phi(c) = e^{-sc}$ and the foregoing remarks yield

$$L\delta(t - c) = \int_0^{\infty} \delta(t - c)e^{-st}\, dt = e^{-sc}, \qquad c \geq 0.$$

When $c < 0$ the result is 0, since c is not on the interval of integration. Note that the choice $c = 0$ gives the formula $L\delta(t) = 1$ with which this discussion began.

In a like manner, when the above considerations are applied to

$$\phi(t) = e^{-st}, \qquad \phi^{(n)}(t) = (-s)^n e^{-st},$$

Equation (7) gives

$$L\delta^{(n)}(t - c) = s^n e^{-sc}, \qquad c \geq 0.$$

The result is 0 when $c < 0$, since in that case c is not on the interval of integration. These are actually definitions, but the discussion shows that the definitions are consistent with the ordinary rules of calculus. This is the reason why $\delta(t)$ can be treated as a function of the real variable t even though it is not such a function.

(d) Further examples

Let us consider the problem $y'' + y = \delta(t - c)$, $y(0) = y'(0) = 0$ where c is a positive constant. This is the same problem as in Part (a), except that the impulse occurs at time $t = c$ rather than at time $t = 0$. The Laplace transform gives

$$(s^2 + 1)Ly = e^{-sc}, \qquad Ly = e^{-sc}\frac{1}{s^2 + 1}, \qquad y = h(t - c)\sin(t - c).$$

Thus the response is the same as before, shifted by the amount c. This is what one would expect on physical grounds.

As another illustration, let T be a linear operator with constant coefficients and with the characteristic polynomial $P(s)$. We consider the rest solutions y and k of the two equations

$$Ty = f(t), \qquad Tk = \delta(t)$$

where $f \in$ E. Clearly $P(s)Ly = Lf$, $P(s)Lk = 1$. Hence

$$Ly = \frac{1}{P(s)} Lf = (Lk)(Lf) = L(k * f)$$

and by uniqueness of the transform, $y = k * f$. Thus the formula

$$y(t) = \int_0^t k(t_1) f(t - t_1) \, dt_1$$

gives the rest solution of $Ty = f$ in terms of the rest solution of $Tk = \delta$. If we determine k from

$$Lk = \frac{1}{P(s)}$$

by partial fractions, we get the Heaviside expansion theorem. The above formula does not assume that the zeros of $P(s)$ are simple.

PROBLEMS

1. If $y = 0$ for $t < 0$ and $y' = \delta(t)$ for $t > 0$ the Laplace transform suggests that $sLy = 1$. Assuming this, conclude that y agrees with the Heaviside function $h(t)$ except perhaps at $t = 0$ where neither the physics nor the mathematics of the problem is clear. In that sense $h'(t) = \delta(t)$.

2. Let $y = 0$ for $t < 0$ and let as many derivatives as possible be continuous at the origin. If $n \geq 1$ and $y^{(n)} = \delta(t)$ show that $y^{(n-1)}$ has a jump of magnitude 1 at $t = 0$. Problem 1 pertains to the case $n = 1$.

3. Explore continuity of solutions of $y'''(t) = f(t)$ with $f(t)$ as follows. It is assumed that the equation holds for $t > 0$ and that $y(t) = 0$ for $t \leq 0$:

$$t \qquad 1 \qquad \delta(t) \qquad \delta'(t) \qquad \delta''(t)$$

4. Given $y'' + ay' + by = f(t)$, where a and b are constant and $f \in$ E, show that the effect of replacing $f(t)$ by $f(t) + c\delta(t)$ is the same as the effect of increasing the initial value $y'(0)$ by the constant c. A similar result holds for equations of order n.

5. With A, a, b, c constant and $c > 0$, consider the problem

$$y'' + y = A\delta(t - c), \qquad y(0) = a, \qquad y'(0) = b$$

The question is: Under what conditions, if any, is $y(t) = 0$ for $t \geq c$? Otherwise expressed: Can you choose the strength and location of the impulse in such a way as to cancel the oscillation completely?

(a) By the second shift theorem find $Lh(t - c)\sin(t - c)$ from the known value of $Lh(t)\sin t$.

(b) Find Ly and, using the result (a), show that

$$y(t) = a\cos t + b\sin t + Ah(t - c)\sin(t - c)$$

(c) Show that $y(t) = 0$ for $t \geq c$ if, and only if, $y(c) = 0$ and

$$A = \sqrt{a^2 + b^2}, \qquad a\sin c \geq b\cos c; \qquad A = -\sqrt{a^2 + b^2}, \qquad a\sin c \leq b\cos c$$

Why is it necessary to choose c so that $y(c) = 0$?

6P. Let c be a positive constant. Express the action of $\delta(ct)$ on a function $\phi(t)$ by an integral like (6), take $u = ct$ as new variable, and thus motivate the definition $\delta(ct) = (1/c)\delta(t)$.

7. A uniform weightless beam of length L carries a concentrated load P_0 at its center as shown in Figure 3. By the Bernoulli-Euler law the equation of the central line for small static deflection is

$$EIy^{iv}(x) = P_0\delta\left(x - \frac{L}{2}\right)$$

where EI is a positive parameter depending on the material and cross-section shape of the beam. The following reduction is used in Problems 8–11 below.

(a) Set $\overline{x} = bx$, $y(x) = az(\overline{x})$ to get

$$EIab^4 z^{iv}(\overline{x}) = P_0\delta\left(x - \frac{L}{2}\right)$$

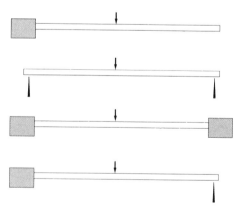

Figure 3

(b) Choose $b = 2/L$ so that the range of $\bar{x}$ is $0 \le \bar{x} \le 2$ and note that with this choice Problem 6 gives

$$\delta\left(x - \frac{L}{2}\right) = \delta\left(\frac{L}{2}\left(\bar{x} - 1\right)\right) = \frac{2}{L}\delta\left(\bar{x} - 1\right)$$

(c) Choose $a = P_0 L^3/(48EI)$ and thus reduce the differential equation to

$$z^{iv}(\bar{x}) = 6\delta(\bar{x} - 1), \qquad 0 \le \bar{x} \le 2$$

8. A weightless beam clamped at one end, free at the other, bears a concentrated load at its center as shown in Figure 3(a). The equation and boundary conditions are

$$y^{iv}(x) = 6\delta(x - 1), \qquad y(0) = y'(0) = y''(2) = y'''(2) = 0$$

(a) With $y''(0) = 2a$, $y'''(0) = 6b$ find Ly and

$$y = ax^2 + bx^3 + (x - 1)^3 h(x - 1)$$

(b) Determine a and b from the boundary conditions at $x = 2$. Thus find y and the maximum deflection.

(c) Your function y should be linear for $x \ge 1$. Why is this to be expected?

9. If the beam of Problem 8 is supported by a pivot at both ends as shown in Figure 3(b) the boundary conditions become

$$y(0) = y''(0) = y(2) = y''(2) = 0$$

(a) By the Laplace transform find a two-parameter family of solutions y of the differential equation, depending on $y'(0) = a$, $y'''(0) = 6b$.

(b) Determine a and b to satisfy the boundary conditions at $x = 2$. It is to be expected that $y(x) = y(2 - x)$, $0 \le x \le 1$. Why? Check that the maximum deflection is $y(1) = 1$.

10. If the beam of Problem 8 is clamped at both ends as shown in Figure 3(c) the boundary conditions are

$$y(0) = y'(0) = y(2) = y'(2) = 0$$

Find y and check that the maximum deflection is $1/4$. Hence the maximum deflections in Figures 3(abc) are in the ratio 20:4:1, respectively. By Problem 7 the same ratios would be found in the general case.

11. Discuss Problem 8 when one end of the beam is clamped and the other pivoted as shown in Figure 3(d).

12. Instead of $\delta_a(t)$ one could consider any positive integrable function $\eta_a(t)$ that vanishes outside the interval $0 < t < a$ and has unit area.

(a) Sketch such a function, different from $\delta_a(t)$, for a rather small.

(b) Using the inequality $e^{-sa}\eta_a(t) \le e^{-st}\eta_a(t) \le \eta_a(t)$, which holds for $t \ge 0$ and $s \ge 0$, show that $L\eta_a$ lies between e^{-sa} and 1, and hence $L\eta_a(t) \to 1$ as $a \to 0$.

──────────────────────────ANSWERS──────────────────────────

3. y''', y'', y', y continuous, respectively, and higher derivatives discontinuous. In the last case y is discontinuous. Compare Problem 2.

5. (c) At a point where $y(c) = 0$ the impulse can control the derivative and make $y'(c) = 0$ too. Then $y(t) = 0$ for $t > c$ by uniqueness. But at a point where $y(c) \ne 0$ no choice of the impulse can reduce $y(c)$ to 0. Such a reduction would violate the continuity that was observed in the previous problems.

8. (b) $a = 3$, $b = -1$, $y(2) = 5$ (c) Because there is no force acting on the weightless beam for $x > 1$, hence no reason for the beam to bend.

9. (a) $y = ax + bx^3 + (x-1)^3 h(x-1)$ (b) $a = 3/2$, $b = -1/2$. By symmetry.

10. $4y(x) = 3x^2 - 2x^3 = 4y(2-x)$, $0 \le x \le 1$

11. Partial answer: $16y(x) = 18x^2 - 11x^3 + 16h(x-1)(x-1)^3$.

──

2. The transfer function

Throughout this section T is a linear differential operator with constant coefficients and with the characteristic polynomial $P(s)$. Before taking up the Laplace transform, we found that the equation

$$Tx = Ae^{st}$$

could be solved, in general, by the trial solution $x = Be^{st}$. Under the assumption that $P(s) \ne 0$ the result was $P(s)Be^{st} = Ae^{st}$ or

(1) $B = G(s)A$ where $G(s) = \dfrac{1}{P(s)}$.

The function $G(s) = 1/P(s)$ that transforms the input amplitude A into the output amplitude B is called the *transfer function*. The main property of a transfer function is given by the equation

(2)   (output) = (transfer function)(input).

Depending on the nature of the input and output, other words are used to describe essentially the same thing. In circuit theory, for example, a transfer function can appear under the guise of an impedance, an admittance, a transmission coefficient, or a reflection coefficient. The word "transfer function" is used here because it is noncommital as to the specific interpretation.

Suppose now that we characterize the input f and output x by their transforms, F and X. If $f \in E$ the rest solution of the equation $Tx = f$ satisfies

$$(3) \qquad P(s)Lx = Lf \quad \text{or} \quad X = G(s)F$$

where $G(s) = 1/P(s)$ is the transfer function as defined above. The difference between this equation and (1) is that (1) applies to exponential inputs only, while (3) allows the arbitrary input f. This increased scope of the transfer function is one of the advantages of working in the frequency domain. Note also that the passage from F to X involves merely a multiplication, while the passage from f to x requires the solution of a differential equation.

The theory of transfer functions embraces a vast field and only a few aspects of this theory will be touched upon here. It is convenient to introduce the following conventions: Whenever we use (2) (or an analog of (2) in different notation) we assume that the input is in E and that the initial conditions are 0. In the interests of brevity, the same letter G is used to identify both a system and its transfer function.

(a) Feedback

The transfer function is especially useful when two or more systems are connected together, so that the output of one system becomes the input of another. A simple case of this kind is shown in Figure 4. For uniformity of notation we denote the input by X_0 rather than F. The coupled system is then described by

$$X_1 = G_1(s)X_0, \qquad X_2 = G_2(s)X_1$$

and hence

$$X_2 = G_2(s)G_1(s)X_0.$$

Thus the lower system in the figure is equivalent to the upper one. By repetition a similar result is obtained for any number of systems connected in this fashion. The transfer function for the whole system is the product of the individual transfer functions.

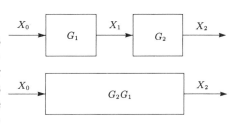

Figure 4

A more elaborate interconnection is shown in Figure 5. Here the basic system is G_1 and the other system G_2 provides a *feedback*. What is meant by this term is that the output is fed back through G_2 and used to modify the input. The encircled plus sign means that the feedback output is added to the input. Hence the total input to G_1 is

$$X_1 = X_0 + G_2X_2.$$

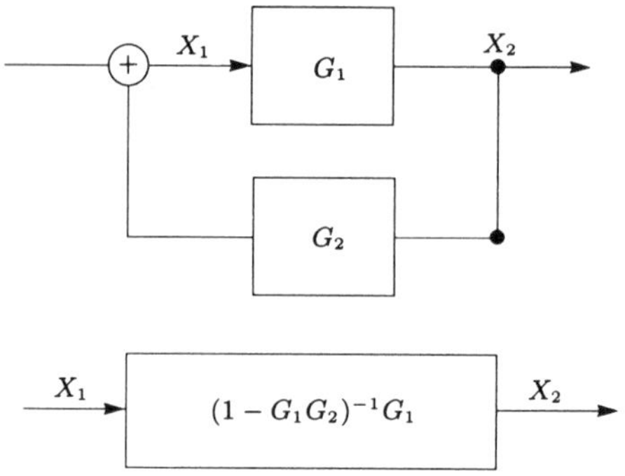

Figure 5

Using the input-output relation (2) as applied to G_1, we get

$$X_2 = G_1(X_0 + G_2X_2) \quad \text{or} \quad (1 - G_1G_2)X_2 = G_1X_0.$$

Therefore the transfer function for the feedback system is

$$(4) \qquad\qquad G = (1 - G_1G_2)^{-1}G_1$$

as indicated in the lower figure.

If the encircled plus is changed to minus, the output of G_2 is subtracted from the input to G_1 and the transfer function for the system is given by

$$(5) \qquad\qquad G = (1 + G_1G_2)^{-1}G_1.$$

Equations (4) and (5) describe positive feedback and negative feedback, respectively.

In many studies of feedback G_2 is replaced by kG_2, where k is a real constant that determines the strength of the feedback. Equation (4) then becomes

$$G = (1 - kG_1G_2)^{-1}G_1$$

and the choices $k = 1, k = -1$ yield both the foregoing results. By changing the constant k one can change the transfer function of the system. This possibility of control is the reason for introducing feedback in the first place.

(b) The problem of identification

So far, we have taken the view that the differential equation and the input (or forcing function) are known, and that the output (or solution) is to be found. In the problem of identification, the situation is different. Here you put in a known input, you measure the output, and you try to find the differential equation. This is a common point of view in biomathematics and in mathematical medicine. After measuring the response to a known regime of drug dosage, for example, one can try to discover the mechanism by which that response was produced. The word *identification* is used because the objective is to identify the main parameters of the system and to determine their values.

It should be said at the outset that the most one can hope for is to find the coefficients in the operator T, and if many choices of system parameters give the same T, the situation cannot be further disentangled by mere study of the input-output relation. As an illustration, the vibrations of a lossless mass-spring system are characterized by the two parameters m and k, the mass and the stiffness constant. However, even complete knowledge of the differential equation

$$my'' + ky = kf(t),$$

yields only the ratio k/m, not k and m separately. Considerations of this kind pervade the subject of parameter identification and are illustrated in the accompanying problems.

The differential operator T is determined by its characteristic polynomial $P(s)$, and $P(s)$ is determined if we know the system function $G(s) = 1/P(s)$. Thus, the identification problem reduces to the problem of finding the transfer function when the input and output are known.

The input $f(t) \in E$ will be termed *nontrival* if its transform $F(s)$ is not identically 0. It is a remarkable fact that the output associated with any nontrivial input suffices for the complete determination of $G(s)$. Though details will not be given here, this can be deduced from the equation $X(s) = G(s)F(s)$. For example the inputs $f(t) = h(t)$ and $f(t) = \delta(t)$ yield, respectively,

$$G(s) = sX(s), \qquad f = h; \qquad G(s) = X(s), \qquad f = \delta.$$

The second equation shows that *the transfer function is the output when the input is* $\delta(t)$. That is, in part, the reason why the unit impulse is often used as input in biomathematics. It can be approximated in various ways depending on the experimental situation; in some cases, for example, by sudden injection of a drug into the bloodstream.

The fact that any nontrivial input-output relationship determines $G(s)$ means that no further information, beyond a reduction of experimental error, can be obtained by use of a variety of different inputs. This is a basic limitation in the problem of parameter identification that could hardly have been foreseen without mathematical analysis.

(c) Compartments

Many problems in biomathematics lead to partial differential equations. One way of dealing with such problems is to divide the system into compartments each of which can be characterized, approximately, by a single

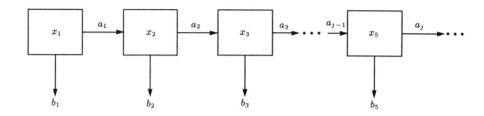

Figure 6

variable. Interaction of the compartments leads to a system of ordinary differential equations.

A compartmental system that has an interesting mathematical structure is shown in Figure 6. The simplest interpretation is in terms of chemical physics rather than biomathematics. Let us suppose that the variable x_j attached to the jth compartment represents the number of grams of the jth decay product in a radioactive chain. The rate of increase of x_j is given by the rate at which x_{j-1} decays into x_j, minus the rate at which x_j itself decays to form new products. Thus we are led to the system

$$(6) \qquad \frac{dx_j}{dt} = a_{j-1}x_{j-1} - a_j x_j - b_j x_j$$

where the a_j and b_j are constant. The term $-a_j x_j$ on the right describes the contribution of x_j to x_{j+1} and the term $-b_j x_j$ describes its contribution to other decay products that are not being followed in detail.

For the first compartment there is no input from previous variables x_j, but we assume that an amount I of the material x_1 was present initially. Thus the first equation is

$$(7) \qquad \frac{dx_1}{dt} = -a_1 x_1 - b_1 x_1, \qquad x_1(0) = I.$$

If the initial values of the other variables are 0, as now assumed, the transformed equations are

$$sX_1 - I = -(a_1 + b_1)X_1, \qquad sX_j = a_{j-1}X_{j-1} - (a_j + b_j)X_j.$$

With $c_j = a_j + b_j$ the equations for $j = 1, 2, 3, \ldots$ give

$$X_1 = \frac{I}{s + c_1}, \qquad X_2 = \frac{a_1}{s + c_2}X_1, \qquad X_3 = \frac{a_2}{s + c_3}X_2$$

and so on. Calculation of X_j for small values of j suggests that

$$X_j = \frac{a_1 a_2 \cdots a_{j-1}}{(s+c_1)(s+c_2)\cdots(s+c_j)} I, \qquad j \geq 1,$$

and a proof by mathematical induction is not difficult.

The fraction multiplying I in this formula is the transfer function $G_j(s)$ for the jth compartment. It is important that the denominator of $G_j(s)$ appears in factored form. This means that computation of the partial-fraction expansion is easy even if j is large.

To see how such a compartmentalized system might arise in biomathematics, let us regard the x_j as being concentrations of a drug at different places in the bloodstream. The main chain involving the a_j describes the progress of the drug to the kidney and the terms involving b_j describe losses to other tissues on the way. If the b_j are all 0 the system is said to be *closed*.

In some applications the compartments are genuine, and are not just a mathematical fiction introduced to simplify the analysis. For example, a system much like the one considered above arises when a polluting agent is transported in a stream joining a series of lakes. The b_j terms now describe losses such as leaching into the ground.

(d) Lumped-parameter circuits

In circuit theory it is customary to write $\sqrt{-1} = j$ since i denotes the current, and to write $\mathcal{L}$ for the Laplace transform operator, since L denotes the inductance. Let us recall that the voltage v across a capacitor is related to the charge Q by $v = SQ$ where $S = 1/C$ is the elastance. Since the current is $i = dQ/dt$ an integration gives

$$v = SQ = S \int_0^t i \, dt + SQ(0).$$

In most cases the initial charge $Q(0)$ is 0 and this condition is assumed here.

If the capacitor is connected in series with a resistor of resistance R, an inductor of inductance L and a source of voltage $v(t)$ the voltage equation is

$$L\frac{di}{dt} + Ri + S \int_0^t i \, dt = v(t).$$

The transform of this with $i(0) = 0$ gives

(8) $$(sL + R + s^{-1}S)I = V.$$

The conditions $Q(0) = 0$, $i(0) = 0$ are already built into the analysis.

For a circuit with n independent loops a calculation of the same sort can be carried out for each loop. The result after transformation is a system of the form

$$(9) \qquad \sum_{j=1}^{n}(sL_{ij} + R_{ij} + s^{-1}S_{ij})I_j = V_i$$

where i_j is the current and v_j the impressed voltage associated with the jth loop. The resistances R_{ij} and elastances S_{ij} are 0 for $i \neq j$ but the inductances L_{ij} need not have this property because of the possibility of mutual inductance between branches.

If you are familiar with matrices, you will recognize that (9) can be written in the form (8) where L, R, S are matrices with R and S diagonal, and where I and V denote the current and voltage vectors, respectively. The matrix equation is *the equilibrium equation for the circuit on a loop basis.*

Equation (8) or its analog (9) in matrix form can be written

$$V = ZI, \qquad I = YV$$

where
$$Z = sL + R + s^{-1}S, \qquad Y = (sL + R + s^{-1}S)^{-1}.$$

In the first case I is considered to be input, V is output, and the transfer function Z is an impedance. In the second case V is considered to be input, I is output, and Y is an admittance.

Circuits that are characterised by numbers R_{ij}, L_{ij}, S_{ij} as above are said to be *lumped parameter*, to distinguish them from *distributed parameter* systems such as a waveguide or a submarine cable. When these latter are approximated by lumped-parameter systems, as is sometimes done in electrical engineering, the procedure is conceptually similar to the use of compartments to analyze the distributed-parameter systems of biomathematics.

PROBLEMS

1. The stabilizing effect of negative feedback. One of the reasons for introducing negative feedback is to keep the system characteristics more or less constant when some of the components change. For a numerical illustration consider
$$y = \frac{G_1}{1 + G_1G_2}x$$

where the G_j instead of being general transfer functions are merely constants. The characteristics of the overall system are measured by the ratio y/x of output to input. Let us suppose that G_1 jumps from its design value of 100 to a new value 10,000. Without feedback this would multiply y/x by 100.

(a) If $G_2 = 0.1$, so there is a small amount of negative feedback, check that the above malfunction in G_1 changes y/x from about 9.1 to about 10. In other words, it produces a 10% change is the system characteristics.

(b) When G_1 malfunctions as above, how much feedback is needed to keep the characteristics constant within 1%? Within 0.01%?

2. In the cases we have been considering the transfer functions are reciprocals of polynomials, $G_j = 1/P_j$. Show that the formula for negative feedback then becomes

$$\frac{G_1}{1 + G_1 G_2} = \frac{P_2}{1 + P_1 P_2}$$

3. In the preceding problem let $P_1(s) = s^2 + as + b$ and $P_2(s) = k$ where $a > 0$, b, and $k > 0$ are constant. Show that the denominator $1 + P_1 P_2$ has all its zeros in the left half plane if, and only if, $G_2 > -b$. This is the condition for the output to be stable in the sense that its long-time behavior is independent of the initial conditions.

Hint: The statement "s is in the left half plane" means Re $s < 0$. As seen in Chapter 8, Section 2, Problem 7, the roots s_j of a real quadratic satisfy Re $s_j < 0$ if, and only if, all the coefficents have the same sign.

4. In many applications the input x and output y are related by $T_2 y = T_1 x$ where T_j are linear differential operators like the operator T considered hitherto. The equation $Ty = x$ is the special case $T_1 = 1$.

(a) If P_j is the characteristic polynomial of T_j check that

$$P_2(s)Y(s) = P_1(s)X(s) + P_0(s)$$

where $P_0(s)$ is a polynomial depending on the initial conditions of both x and y.

(b) Setting the initial conditions equal to 0 gives the transfer function $G(s) = P_1(s)/P_2(s)$ for the above problem. If $G(s) = Lg$ express y by the convolution theorem.

(c) We can have a function g as in Part (b) only if the degree of P_1 is less than that of P_2. Why? What happens to the formula of Part (b) if $P_1 = P_2$?

5. With k constant, let $Ty = x + kg * y$ where T is a linear operator with characteristic polynomial $P_1(s) = 1/G_1(s)$ and where $Lg = G_2(s)$. Assuming null initial conditions, solve this integro-differential equation for y and check that the result agrees with the general feedback formula obtained in the text.

6. In the problem of identification, consider $Ty = 0$ instead of $Ty = f(t)$ or $Ty = \delta(t)$ as in the text, and impose the initial conditions

$$y(0) = 1, \quad y'(0) = 0, \cdots, y^{(n-1)}(0) = 0$$

Here n is the degree of the characteristic polynomial $P(s)$ for T.

(a) Check that $Ly^{(j)} = s^j Ly - s^{j-1}$ for $j = 1, 2, \cdots, n$.

(b) Thus show that knowledge of the response $Y(s)$ gives no information if $P(0) = 0$ (aside from the information that, in fact, $P(0) = 0$).

(c) If $P(0) \neq 0$, however, show that $P(s)$ is fully determined by $Y(s)$.

7. In the compartment problem of Part (c) in the text suppose $n = 3$ and suppose the frequency-domain output

$$X_3(s) = \frac{a_1 a_2}{(s + c_1)(s + c_2)(s + c_3)}$$

is known on some small interval; actually knowledge of $X_3(s)$ for four distinct values of s suffices. To what extent are the system parameters a_j and b_j determined by this information?

8. Solve Problem 7 if you know $X_1(s)$, $X_2(s)$, and $X_3(s)$ on a small interval.

9. Generalize (a) Problem 7 and (b) Problem 8 to the case of arbitrary n. No answer is provided.

10. One method of determining the ratio e/m of charge to mass for an electron leads to the system

$$mx'' + Hey' = eE, \qquad my'' - Hex' = 0$$

The initial values are $x(0) = x'(0) = y(0) = y'(0) = 0$ and m, e, H, E are positive constants.

(a) Solve by the Laplace transform and show that the path of a particle with coordinates (x, y) is given by

$$x = a(1 - \cos \omega t), \qquad y = a(\omega t - \sin \omega t)$$

where $\omega = He/m$, $a = E/\omega H$. This is a cycloid generated by a circle of radius a.

(b) How can you determine the system-parameter e/m from knowledge of the path and of the values E, H set by the experimenter?

11. This problem requires knowledge of matrix algebra. In the discussion of feedback given in the text, assume that k is a scalar but that $G_1(s)$ and $G_2(s)$ are square matrices of the same size. Do the feedback formulas remain valid?

——————————————————ANSWERS——————————————————

1. (b) $G_2 = 1$, 100, approximately.

4. (b) $y = g * x$ (c) Because $G(s) \to 0$ as $s \to \infty$ for every function g. If $G = 1$ then $g = \delta$ and the formula $y = \delta * x$ gives $y = x$.

6. (b) If $P(0) = 0$ it will be found that $Y(s) = 1/s$ and that $P(s)$ cancels out of the equation. This agrees with the fact that $y = 1$ is the unique solution of the initial-value problem, no matter what the other coefficients of P may be. The fact that $y = 1$ is a solution also shows that $P(0) = 0$. (c) $P(s)/P(0) = 1/(1 - sY(s))$. The denominator is not identically 0, because we do not have the solution $y = 1$ if $P(0) \neq 0$. Since the leading term of $P(s)$ is s^n, knowledge of $P(s)/P(0)$ determines first $1/P(0) = \lim_{s \to \infty} s^{-n} P(s)/P(0)$ and then it determines $P(s)$.

7. You know $a_1 + b_1$, $a_2 + b_2$, $a_3 + b_3$, $a_1 a_2$ but you can't find the individual values of any of the parameters a_j or b_j.

8. From X_1 you get c_1, from X_2 you then get a_1 and c_2, and from X_3 you then get a_2 and c_3. Hence you know a_1, b_1, a_2, b_2, and $a_3 + b_3$. Only a_3 and b_3 resist separate evaluation.

10. If the electron beam is in a partially evacuated chamber, the cycloid is visible. The value of a can be found by the depth $2a$ of one arch or by the distance $2\pi a$ between cusps. The ratio of charge to mass is given by $e/m = E/(Ha^2)$.

11. Yes, and that is the reason the formulas were written with a particular order of the factors and with a negative exponent. For matrices, multiplication is noncommutative and B/A is meaningless.

———

The book of Nature is written in mathematical characters.

—Galileo

Each natural science is real science only in so far as it is mathematical.

—Emmanuel Kant

Chapter 24

NUMERICAL METHODS

W HETHER for qualitative results such as a graph or for quantitative results such as the window of safety of a returning spacecraft, numerical analysis of differential equations has become an indispensable tool of great power. It does not consist merely of an enumeration of methods, but includes a study of the reasons why they fail. In this chapter the concepts of accuracy, stability, stiffness, and time stability are illustrated both by examples and by general theorems. To give a comparison with the exact solution, specific algorithms are tested against differential equations that can be solved with ease. But this aspect of the exposition should not obscure the ultimate objective, which is to get accurate numerical answers when no elementary solution exists.

Prerequisites: Facility in the use of Taylor series is assumed; see Chapter B. Some of the problems require knowledge of computer programming.

1. Introduction

The goal of this chapter is to obtain numerical solutions to differential equations by use of a computer. A program of computation usually involves a large number m of separate steps. As m increases the theoretical error goes down but the computational errors go up. If the latter overwhelm the results before sufficient accuracy is reached, the problem is intractable with the method being used. Thus the science of computation involves a compromise between the demands of theoretical and numerical precision. It has subtle features that are easily overlooked on first acquaintance and it leads us to think about familiar topics in a new way.

(a) Algorithms

An algorithm is a definite, fully-defined procedure for solving a given class of problems. Even if two algorithms are mathematically equivalent, in the sense that they give the same theoretical result, they may be entirely different from the point of view of numerical computation. For example, when evaluating a polynomial

$$P(x) = a_0 x^m + a_1 x^{m-1} + \cdots + a_{m-1} x + a_m$$

one should arrange the calculations as follows:

$$u_1 = a_0 x, \qquad u_2 = (u_1 + a_1)x, \qquad \cdots, \qquad u_{n+1} = (u_n + a_n)x, \qquad \cdots.$$

Then $u_m + a_m = P(x)$. As an illustration, if $m = 5$

$$P(x) = (((((a_0x + a_1)x + a_2)x + a_3)x + a_4)x + a_5.$$

This method is commonly named after Horner (1819) but is actually due to Newton. It requires only m multiplications for a polynomial of degree m and is easily programmed on a computer. The number of multiplications in a computation gives a rough measure of its cost.

The above algorithm is iterative; that is, it involves a passage from n to $n + 1$ in a single formula. Nearly all algorithms for the numerical solution of differential equations are iterative. In the early days of computation it was a common practice to make a list of intermediate results for use in subsequent steps. But this pencil-and-paper technique has long been obsolete. The procedure currently used, which is much faster, is to substitute $n + 1$ for n repeatedly in a single basic formula. Values computed at step n are automatically stored for use in step $n + 1$.

The choice of algorithm can have a profound effect on the accuracy of a calculation. For example, let us use the quadratic formula to get the root $x = 1$ of

$$1.0000x^2 - 88,888x + 88,887 = 0.$$

Assuming that the coefficients are correct to 5 significant digits and rounding each step of the calculation to that accuracy, we find the approximate root

$$44,444 - \sqrt{1,975,200,000} \approx 0.8.$$

This is barely correct to one digit. To be sure, rounding off the answer 0.8 yields 1, but all digits after the first remain in doubt.

It might be thought that the root $x = 1$ depends critically on the coefficients, and that if the latter are given only to 5 digits then no more than one correct digit can be expected in the answer. But this is not the case. For example, if the equation is changed to

$$1.0000x^2 - 88,888x + 90,000 = 0$$

the smaller root, correct to 5 digits, is 1.0125.

The trouble does not come from any inherent instability in the problem. It comes, rather, from the use of an inappropriate algorithm. A better way to compute the smaller root is to get the larger root by the quadratic formula and then use the fact that the product of the roots is the constant term. You can easily convince yourself that this procedure does not lead to the degradation of accuracy seen above.

In conclusion, we mention that such words as *method, procedure* and *formula* are sometimes used instead of *algorithm*. The object is to avoid excessive repetition and no mathematical distinction is intended.

(b) Roundoff error

Since a computer can store only a finite number of decimal digits at any given time, it is necessary to round off the results of decimal calculation. This operation introduces an error, called roundoff error, that was already illustrated above. Namely, we found that roundoff in a single use of the quadratic formula led to a loss of four significant digits. It is not surprising that similar difficulty arises, in aggravated form, when a problem involves hundreds or thousands of steps.

There is, however, one favorable aspect. If at each step the average error introduced by roundoff is ϵ, in n steps the probable roundoff error is of the order $\epsilon\sqrt{n}$, and not ϵn as one might expect. This is so because the individual errors act like independent random variables with mean value zero, and positive and negative errors tend to cancel. It is as if the errors perform a random walk, with positive errors representing a step to the right, negative errors a step to the left. (Such desirable behavior is not found, however, if the computer merely drops decimals beyond a certain point, instead of rounding.)

Roundoff error is sometimes estimated by use of a quantity called the *machine epsilon*. This is the smallest $\epsilon > 0$ such that the computer can distinguish $1 + \epsilon$ from 1. For example if a computer is programmed to keep 8 significant digits it will act as if

$$1.00000005 = 1.0000001 > 1, \qquad 1.00000004 = 1.0000000 = 1.$$

Two conventions that are often used in numerical work will be followed here. First, if the result of a numerical calculation is stated with a given number of significant digits, its truth is asserted only to within that accuracy. For example, "$\pi = 3.14$" means "To three significant digits, $\pi = 3.14$." Second, we ignore roundoff errors in most of the theoretical development and assume that our various approximate formulas are computed exactly.

(c) Truncation error

A formula that is valid in the limit as $h \to 0$ is often used as if it were true for a small positive value of the constant h. The resulting error is an example of truncation error. In general, *truncation error* is introduced whenever a formula in a numerical calculation is only approximate (and would therefore give a wrong answer even if computed exactly.) For example, it is a familiar fact that

$$e = \lim_{h \to 0} (1 + h)^{1/h}.$$

If we use this with $h = 0.01$ instead of passing to the limit, the error in the resulting equation

$$e \approx (1.01)^{100} = 2.705$$

is due to truncation.

A convenient notation for describing truncation errors was introduced by the German mathematician Edmund Landau in connection with the analytic theory of numbers. If f and g are functions defined for small positive values of a parameter h, Landau's notation $f = O(g)$ means

$$|f(h)| \leq (\text{const.})|g(h)| \quad \text{as} \quad h \to 0 + .$$

The equation $f = O(g)$ is read aloud thus: "f is oh of g." Both f and g can be real, complex, or vector valued. Here, however, we shall have $g(h) = h^p$ where p is a nonnegative integer. The symbol $O(h^p)$ by itself is thought to mean "terms of order h^p" and an error of magnitude $O(h^p)$ is said to be of order p. Here are some illustrations:

$$9h^2 + h^3 = O(h^2), \quad h^2 = O(h), \quad \sin h = h + O(h^3), \quad e^h = 1 + h + O(h^2).$$

The last two follow from the Taylor series for $\sin h$ and e^h, respectively. More generally, we get a truncation error of order p whenever a convergent Taylor series is cut off, or *truncated*, just after its pth term. This fact is used below.

An equation such as $f = O(h)$ does not denote equality in the sense of logical identity, but signifies merely that f belongs to a certain class of functions. Algebraic properties suggested by the notation are interpreted by going back to the definition. For example, the implication

$$|\phi(h)| \leq Jh^p, \qquad |\psi(h)| \leq Kh^q \quad \Rightarrow \quad |\phi(h)\psi(h)| \leq JKh^{p+q}$$

yields first an interpretation, then a proof, of the equation

$$O(h^p)O(h^q) = O(h^{p+q}).$$

As another example, let us recall that the function f is *Lipschitzian* if there is a constant K such that, for all relevant values of the variables,

$$|f(t, y + z) - f(t, y)| \leq K|z|.$$

Taking $z = O(h^p)$ we get

$$f(t, y + O(h^p)) = f(t, y) + O(h^p).$$

This is a concise way of saying that a change in the second argument of a Lipschitzian function f produces a change of at most the same order of magnitude in f itself.

(d) Richardson extrapolation

Suppose a numerical algorithm produces an approximation $y(x, h)$ to the desired exact answer, $y(x)$. In the applications we have in mind y will be the solution to a differential equation and h will be a parameter called the step size. The smaller the step size, the more steps are needed to calculate $y(x, h)$. By definition, an algorithm of this kind is *convergent* if

$$\lim_{h \to 0+} y(x, h) = y(x).$$

The rate of convergence is generally measured by an estimate of the form

$$(1) \qquad\qquad y(x) - y(x, h) = O(h^p), \qquad p = \text{const.}$$

If $y(x) - y(x, h)$ admits a Taylor expansion

$$y(x) - y(x, h) = b_0(x) + b_1(x)h + b_2(x)h^2 + \cdots + b_p(x)h^p + O(h^{p+1})$$

it is easily checked that (1) implies $b_j(x) = 0$ for $j < p$. Hence

$$(2) \qquad\qquad y(x) - y(x, h) = c(x)h^p + O(h^{p+1})$$

where $c(x) = b_p(x)$. This is a sharpened form of (1).

Equation (2) forms the basis for a simple but powerful procedure known as Richardson extrapolation. Let us write (2) with step size h and step size $2h$, thus:

$$y(x) = y(x, h) + c(x)h^p + O(h^{p+1}), \quad y(x) = y(x, 2h) + c(x)(2h)^p + O(h^{p+1}).$$

If the first of these equations is multiplied by 2^p and the second is then subtracted, the result is

$$(3) \qquad\qquad (2^p - 1)y(x) = 2^p y(x, h) - y(x, 2h) + O(h^{p+1}).$$

This gives $y(x)$ in terms of computed quantities within $O(h^{p+1})$ even though, by (1), the basic method was accurate only within $O(h^p)$. Since one should use at least two different step sizes anyway, as a check, Richardson's procedure is very effective in applications.

To make free use of results such as these, we assume whenever convenient (and without further comment) that functions such as $y(t)$ or $y(t, h)$ have convergent Taylor series expansions. This assumption simplifies the exposition but does not follow automatically from the hypotheses in Part (e) below. See Section 2, Problem 12.

(e) The basic differential equation

Having given a brief discussion of numerical methods in general, we turn to the specific problem

$$(4) \qquad\qquad y' = f(t, y), \qquad t \geq 0; \qquad y(0) = y_0$$

with which this chapter is chiefly concerned. It is assumed that f is continuous and Lipschitzian in a rectangle sufficiently large to contain all values (t, y) that arise in the analysis. By Chapter 14, the solution $y(t)$ exists and is unique. Until further notice the functions and variables are real. However, the main results extend to complex vector-valued functions; this gives a corresponding extension to higher-order equations and to systems. Since $t - t_0$ could be taken as a new variable, prescribing $y(0)$ rather than $y(t_0)$ involves no loss of generality.

Our problem is to approximate the solution of (4) numerically at any given point $t = x$. To this end the interval $[0, x]$ is divided into m equal parts by points

$$t_0 = 0, \qquad t_1 = h, \qquad t_2 = 2h, \qquad t_3 = 3h, \qquad \cdots, \qquad t_m = mh = x.$$

The positive constant $h = x/m$ is the step size. A suitable function $y_n = y(t_n, h)$ is computed successively at each point t_n, starting with y_0 at $t_0 = 0$, and $y(t_m, h) = y(x, h)$ yields an approximation to $y(x)$.

Most algorithms of this kind are based upon approximate identities that are obtained by neglecting certain error terms. If the term neglected is $O(h^{p+1})$ the identity is said to be accurate to order p. For example, the approximate identities obtained by dropping error terms in Problem 1 are accurate to orders 1, 2, 1, 2 and 4 respectively. That they are not accurate to a higher order is seen in Problem 2.

In addition to the identity error mentioned above, another quantity of importance is the local error. This measures the difference between what the algorithm actually produces in a single step and what would be obtained by following a trajectory of the differential equation. It turns out that the local error usually has the same order of magnitude as the identity error in the approximate identity that gave rise to the algorithm.

Most important is the global error $y(x) - y(x, h)$. If the local error is $O(h^{p+1})$ at each step, one would naturally expect the error after $m = x/h$ steps to have the order of magnitude $(x/h)O(h^{p+1}) = O(h^p)$. In that case it would be correct to say, "A local error of order $p + 1$ yields a global error of order p." Despite its plausibility, this statement is not always true. A principal goal of the theory is to characterize those algorithms for which it holds.

PROBLEMS————————————————————————————————————

The letter c or C accompanying a problem means that it requires a calculator or a programmable computer, respectively.

1P. Given the convergent Taylor series

$$y(t + h) = y(t) + hy'(t) + \frac{h^2}{2}y''(t) + \frac{h^3}{6}y'''(t) + \cdots$$

and the corresponding series with y' in place of y, check that:

(a) $y(t + h) = y(t) + hy'(t) + O(h^2)$

(b) $y(t + h) - y(t - h) = 2hy'(t) + O(h^3)$

(c) $y(t + h) - y(t) = hy'(t + h) + O(h^2)$

(d) $y(t + h) - y(t) = \frac{h}{2}(y'(t) + y'(t + h)) + O(h^3)$

(e) $y(t + h) - y(t - h) = \frac{h}{3}(y'(t - h) + 4y'(t) + y'(t + h)) + O(h^5)$

Hint for (e): Write the series for $y(t + h) - y(t - h)$ and $y'(t + h) + y'(t - h)$ through terms in h^4.

2P. (abcde) In most cases of interest here, an approximate identity is accurate to order p if, and only if, it is an exact identity for all polynomials of degree $\leq p$. By trying $y = t^p$ for suitable p verify this behavior in Problem 1, and at the same time show that the exponent in the error terms cannot be increased. For instance, in Problem 1(a) try $y = 1$, t, t^2.

3P. (abcde) In Problem 1 replace $y(t - h)$, $y(t)$, $y(t + h)$ by y_{n-1}, y_n, y_{n+1}, and $y'(t - h)$, $y'(t)$, $y'(t + h)$ by f_{n-1}, f_n, f_{n+1}, respectively, and drop the O term. This produces all but one of the algorithms discussed in Section 2.

4P. Let q be a positive integer and let each ϕ_n be a real-valued function defined for all real values of the variables. Prove by mathematical induction that the recurrence relation

$$y_{n+1} = \phi_n(y_n, y_{n-1} \cdots, y_{n-q+1}), \qquad n \geq q - 1$$

has one and only one solution $\{y_n\}$ satisfying prescribed initial conditions

$$y_0 = Y_0, \qquad y_1 = Y_1, \qquad \cdots, \qquad y_{q-1} = Y_{q-1}$$

5P. The *Fibonacci sequence* is defined by $y_{n+1} = y_n + y_{n-1}$ with the initial conditions $y_0 = y_1 = 1$.

(a) Find the first seven y_n and verify that $y_7 = 21$.

(b) Check that $\{y_n\} = \{s^n\}$ satisfies $y_{n+1} = y_n + y_{n-1}$ if, and only if, s is one of the roots s_j of $s^2 = s + 1$.

(c) With s_j as in (b) verify that $z_n = c_1 s_1^n + c_2 s_2^n$ satisfies $z_{n+1} = z_n + z_{n-1}$ for every choice of the constants c_j.

(d) Choose c_j in (c) so that $\{z_n\}$ has the same inital values as $\{y_n\}$, in which case Problem 4 gives $y_n = z_n$. Thus get the formula

$$y_n = \frac{1}{\sqrt{5}}\left(\frac{1+\sqrt{5}}{2}\right)^{n+1} - \frac{1}{\sqrt{5}}\left(\frac{1-\sqrt{5}}{2}\right)^{n+1}$$

6C. Working to as many digits as you can, find y_{100} in Problem 5 by use of the recurrence relation and compare with the formula of Part (d).

7P. (a) Solve Richardson's formula (3) for $y(x)$ when $p = 1, 2, 3, 4$.

(b) If the step size is changed from κh to h, where $\kappa > 1$ is constant, show that (3) holds with κ in place of 2.

8. (a) Show that $h \sin(1/h) = O(h)$ and that nevertheless $h \sin(1/h)$ does not have the form $ch + O(h^2)$ where c is constant.

(b) Interpret and prove: $O(h^p) + O(h^p) = O(h^p)$.

(c) Briefly discuss $O(h^4) + O(h^5) = O(h^4)$ and state a generalization that contains both this result and the result (b). No answer is provided.

9. If we required $f(t, y)$ to be globally Lipschitzian, that is, Lipschitzian for $-\infty < y < \infty$, show that the equation $y' = y^2$ would be excluded on every interval $0 \le t \le b$ no matter how short.

10C. If $P(x)$ denotes the sum of the first 101 terms of the series for e^x, set up a program for computing $P(x)$ by Horner's algorithm or by direct summation of terms as you prefer. Using this, evaluate $e^x/P(x) \approx 1$ for $x = 20, 100, -20$ and explain.

11c. Apply the method of Richardson extrapolation to the Taylor series for $\sin h$ and thus show that

$$3^5 \sin h = \sin 3h + 240h - 36h^3 + O(h^7)$$

Taking $h = \pi/18$ find $\sin 10°$ and compare with the exact value.

12. The notation $f = o(g)$ means that $f(h)/g(h) \to 0$ as $h \to 0+$. It is said aloud thus: "f is little oh of g." With such choices as $h = |\Delta x|$ or $|\Delta t|$ this notation gives an efficient means of deriving differential equations, both ordinary and partial.

(a) Show that the real-valued function ϕ is differentiable at x if, and only if,

$$\phi(x + \Delta x) - \phi(x) = c(x)\Delta x + o(|\Delta x|)$$

for some function $c(x)$, and in that case $c(x) = \phi'(x)$.

(b) The equation $dT/d\theta = \mu T$ for the tension T of a slipping belt is derived in Chapter 3, Section 2, Part (e). Write out this derivation in o-notation.

13. If f has derivatives of all orders the chain rule shows that all solutions of $y' = f(t, y)$ also have derivatives of all orders. Conversely, suppose $y' = f(t, y)$ has a nontrivial, infinitely differentiable solution. Does it follow that at least one of the partial derivatives f_t or f_y must exist?

————————————ANSWERS————————————

The c problems were done on an HP-32S scientific calculator, the others on a NEXT with machine epsilon $2^{-52} \approx 2 \cdot 10^{-16}$. Different equipment may give somewhat different results.

6. $y_{100} = 573147844013817084101$

7. (a) Partial answer: $p = 3$ gives $y(x) \approx (8/7)y(x, h) - (1/7)y(x, 2h)$

10. 1.0000000000000, 1.89911087960. When $x = -20$ we got about .01 for direct sum, about 0.3 for Horner. The main source of error is truncation when $x = 100$ and roundoff when $x = -20$. It is for you to explain why.

11. $\sin 10° \approx 0.17364816987$

13. No. Look at $y' = (t - y)\phi(t, y) + 1$.

2. Five algorithms

Here we discuss five specific algorithms, each of which has both historical and practical importance. As in the last section $y(t)$ denotes a solution of the initial-value problem

$$y' = f(t, y), \qquad y(0) = y_0.$$

The successive values of the approximation $y(t, h)$ are denoted by y_n and corresponding values of f by f_n. Thus

$$y_n = y(t_n, h), \qquad f_n = f(t_n, y_n), \qquad t_n = nh.$$

The passage from approximate identities for $y(t)$ to numerical algorithms for $y(t, h)$ is facilitated by the equations

$$y(t - h, h) = y_{n-1}, \qquad y(t, h) = y_n, \qquad y(t + h, h) = y_{n+1}, \qquad t = t_n.$$

It turns out that each of the five algorithms has global accuracy of order p, where $p \geq 1$ is the order of accuracy of the approximate identity giving rise to the algorithm. By results to be established in Section 3, this shows that the algorithm is convergent and it gives the correct value p for Richardson extrapolation.

(a) Euler's method

Euler's method was introduced in Chapter 14, Section 1 and is presented again now. The definition of derivative gives

$$\frac{y(t + h) - y(t)}{h} \approx y'(t).$$

Ordinarily this is used as a means of approximating $y'(t)$. Here, however, we use it to approximate $y(t + h)$:

(1) $$y(t + h) \approx y(t) + hy'(t) = y(t) + hf(t, y(t)).$$

Replacing $\approx$ by $=$ and $y(t)$ by $y(t, h)$ we are led to the formula (Euler 1768)

(2) $\qquad y_{n+1} = y_n + hf_n.$

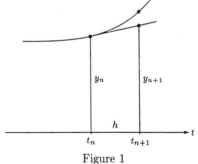

A geometric interpretation is given in Figure 1. By Section 1, Problem 1(a) the identity (1) is accurate to order 1.

Figure 1

EXAMPLE 1. Let us apply this method to the *test problem*

(3) $$y' = 2ty, \qquad y(0) = 1, \qquad y(1) = ?$$

Here $f(t, y) = 2ty$, $f_n = 2nhy_n$ and the exact solution is

$$y(t) = e^{t^2}, \qquad y(1) = 2.718281828459045 \cdots.$$

Euler's formula gives $y_{n+1} = y_n(1 + 2nh^2)$. The sequence

$$y_0 = y_1 = 1, \quad y_2 = 1 + 2h^2, \quad y_3 = (1 + 4h^2)y_2, \quad y_4 = (1 + 6h^2)y_3, \quad \cdots$$

is easily computed, with the stopping-value $n = 4$ or $n = 9$ for $h = .2$ or $h = .1$. This makes the last computed value y_5 or y_{10} in the two cases, respectively, and yields

$$y(1, .2) = 2.0505830, \qquad y(1, .1) = 2.3346334.$$

By Richardson extrapolation with $p = 1$ we get

$$y(1) \approx 2y(1, .1) - y(1, .2) = 2(2.3346334) - 2.0505830 = 2.62.$$

(b) The midpoint formula

A modification of Euler's method is illustrated in Figure 2. Here we go from (t_{n-1}, y_{n-1}) to (t_{n+1}, y_{n+1}) by following a line whose slope agrees with the slope f_n at the midpoint. The figure leads to the *midpoint formula*

$$y_{n+1} = y_{n-1} + 2hf_n$$

(Nyström 1925). This is a two-step method because we have to know both y_0 and y_1 to get started. In general, a q-step method gives y_{n+1} in terms of q consecutive earlier values y_j. Since the midpoint formula is based on the identity of Section 1, Problem 1(b), it is accurate to order $p = 2$.

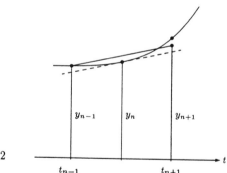

Figure 2

EXAMPLE 2. For the test problem (3) the midpoint formula becomes

$$y_{n+1} = y_{n-1} + 4nh^2 y_n.$$

If we did not know the exact solution, y_1 would be computed by a one-step method based on a subdivision of the interval $[0, h]$. Here, however, we use our knowledge of the exact solution to get $y_1 = e^{h^2}$, hence

$$y_1 = 1.04081108, \qquad h = .2; \qquad y_1 = 1.0100502, \qquad h = .1.$$

The final results are recorded in the table at the end of this section.

(c) The trapezoid rule

A general method for constructing algorithms can be based on integration. Let $y' = f(t, y)$ and suppose for the moment that

$$y_n = y(t_n), \qquad y_{n+1} = y(t_{n+1}).$$

Integrating $y'(t)$ from t_n to t_{n+1} yields

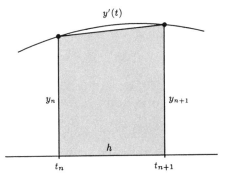

Figure 3

$$(4) \quad y_{n+1} = y_n + \int_{t_n}^{t_{n+1}} f(t, y(t)) \, dt.$$

Since we are following the integral curve from t_n to t_{n+1}, the local error is zero. Such an algorithm at each step, starting from $t = 0$, would yield $y(t)$ exactly. It cannot be implemented because the integrand $f(t, y(t))$ depends on the unknown function y. But if the integral is approximated by a suitable quadrature formula, Equation (4) provides a practical means of computation.

For example, the approximation illustrated in Figure 3 yields the *trapezoid rule*

$$(5) \qquad\qquad y_{n+1} = y_n + \frac{h}{2} \left(f(t_n, y_n) + f(t_{n+1}, y_{n+1}) \right).$$

This is based on the approximate identity

$$(6) \qquad\qquad y(t + h) \approx y(t) + \frac{h}{2} \left(y'(t) + y'(t + h) \right)$$

for which Section 1, Problem 1(d) gives $p = 2$.

Since y_{n+1} occurs in f on the right of (5), the equation (5) determines y_{n+1} only implicitly. In general, a method is *implicit* if it does not give y_{n+1} directly in terms of preceding values but only gives an equation from which, in principle, y_{n+1} can be found. It is a universal characteristic of implicit methods that h must be sufficiently small to ensure that the implicit equations have a solution. For the test problem Equation (5) is easily solved for small h, as seen in Problem 1(c). Numerical results are given in the table.

(d) Heun's formula

One way to go from the implicit equation (5) to an explicit equation is to replace y_{n+1} on the right by an approximate value. For example the choice

$$y_{n+1} \approx y_n + h f_n$$

leads to the formula (Heun 1900)

$$y_{n+1} = y_n + \frac{h}{2}\left(f(t_n, y_n) + f(t_{n+1}, y_n + hf_n)\right).$$

Results for the test problem are given in Problem 1(d) and in the table.

The equation $y_{n+1} = y_n + hf_n$ that was used to predict the value of y_{n+1} is called the predictor and the final formula for y_{n+1} so obtained, in this case Heun's formula, is called the corrector. In general, a *predictor-corrector* method is a method in which a predicted value of y_{n+1} is used to simplify a more accurate, but usually implicit, formula for y_{n+1}.

If $y'(t) = f(t, y(t))$ then $y'(t + h) = f(t + h, y(t + h))$. Using this in (6) and then replacing $y(t + h)$ by $y(t) + y'(t)h + O(h^2)$, we get

$$(7) \qquad y(t + h) = y(t) + \frac{h}{2}\left(y'(t) + f(t + h,\ y(t) + hy'(t)) + O(h^3).\right.$$

(The error is still $O(h^3)$ because the Lipschitzian function f in which we made a change of magnitude $O(h^2)$ is preceded by the factor h.) Equation (7) without the O term is the approximate identity that underlies Heun's method. It is accurate to order 2, hence $p = 2$.

In contrast to other cases above, Equation (7) holds only for solutions of $y' = f(t, y)$ and not for all sufficiently differentiable functions y. Also it involves y in a decidedly nonlinear fashion. We shall see later that Heun's formula is typical of a broad class of algorithms called Runge-Kutta methods while the others are linear multistep methods as discussed in Section 3.

(e) The Milne-Simpson formula

The equation corresponding to (4) for passage from t_{n-1} to t_{n+1} is

$$y_{n+1} = y_{n-1} + \int_{t_{n-1}}^{t_{n+1}} f(t, y(t))\, dt.$$

If the integral is approximated by Simpson's rule we get the formula

$$y_{n+1} = y_{n-1} + \frac{h}{3}(f_{n-1} + 4f_n + f_{n+1})$$

(Milne 1926). This is both two-step and implicit. Since it is based on the approximate identity

$$y(t + h) - y(t - h) \approx \frac{h}{3}(y'(t - h) + 4y'(t) + y'(t + h))$$

Section 1, Problem 2(e) shows that $p = 4$. For the test problem, see Problem 1(e) and the table.

	algorithm	$h = 0.2$	$h = 0.1$	$h = 0.05$	p
a	Euler	2.05058	2.334633	2.5106623	1
b	midpoint	2.59509	2.683557	2.7093196	2
c	trapezoid	2.79397	2.736598	2.7228243	2
d	Heun	2.68138	2.709057	2.7159898	2
e	Milne-Simpson	2.72060	2.718437	2.7182917	4
f	Adams-Bashforth	2.68588	2.713345	2.7178396	4
g	Moulton	2.72415	2.718814	2.7183215	4
h	Adams-Moulton	2.71827	2.718486	2.7183079	4
i	Milne predictor	2.69481	2.716304	2.7181465	4
j	Runge-Kutta	2.71811	2.718270	2.7182811	4

PROBLEMS

The letters (abcde) in Problems 1-9 correspond to those in the table. Implicit algorithms are applied only to equations that allow explicit solution.

1. The following recursion formulas for the test problem were not given in the text. Verify them:

(c) $y_{n+1} = \dfrac{1 + h^2 n}{1 - h^2(n+1)} y_n, \qquad hx < 1$

(d) $y_{n+1} = (1 + (2n+1)h^2 + 2n(n+1)h^4)y_n$

(e) $y_{n+1} = \dfrac{8nh^2 y_n + (3 + 2h^2(n-1))y_{n-1}}{3 - 2h^2(n+1)}, \qquad hx < \dfrac{3}{2}$

2c. (abcde) For the test problem find $y(.4, .2)$.

3C. (abcde) Verify the numerical answers in the table. Here and in analogous cases elsewhere, the preceding c problem is an aid in checking your program.

4c. (abcde) Repeated Richardson extrapolation. Get an improved value X from the entries for $h = .2$ and $h = .1$ in the table. Also get an improved value Y from the entries for $h = .1$ and $h = .05$. Finally, using $p + 1$ instead of p, get an improved value Z from your X and Y.

5c. (abcde) Suppose the quantity $E(h) = y(x) - y(x, h)$ is of order $O(h^{p+1})$ and has a Taylor series. Then $E(h)/h^p$ tends to a finite limit as $h \to 0+$. For the test problem find three values of $E(h)/h^p$ from the table.

6c. (abcde) Check that $(1 + t^2)y' = 4$, $y(0) = 0$ yields $y(1) = \pi$. Then approximate π by using $h = 1/4$ in the algorithms of the text.

7C. (abde) In Problem 6 take $h = .1$, $h = .05$, and extrapolate. We omit (c) because (c)=(d) when f is independent of y. See answer to Problem 6.

8c. (acd) Given $y' = -20y$, $y(0) = 1$ approximate $y(1)$ by the algorithms of the text, taking $h = .2$, $.1$, $.05$, $.01$, $.0025$. This can be done as a c problem because y_n has the form α^n where $\alpha = \alpha(h)$.

9C. (be) Do Problem 8 for (be) by actually programming the algorithms. The strange behavior you will encounter is explained by the discussion of *stiff problems* in Section 5.

10P. Backward Euler. Estimating the integral (4) by the area of a suitable rectangle, get the respective equations

$$y_{n+1} = y_n + h f_n, \qquad y_{n+1} = y_n + h f_{n+1}$$

The first of these is already familiar and the second is the implicit or *backward* Euler formula. Check that the average of these two equations yields the trapezoid formula. Referring to Section 1, Problem 1(c), show that $p = 1$ in the second case.

11c. Apply the backward Euler formula of Problem 10 to Problem 8, using h as prescribed there. Partial answer: $y(1, .0025) = 3.3441086 \cdot 10^{-9}$.

12. This is difficult. For $0 < c < 1$ let $f(t)$ be the function shown in Figure 4. Namely

$$f(t) = 0 \quad \text{or} \quad f(t) = t - c$$

for $t \le c$ or $t > c$ respectively. If an approximate solution to $y' = f(t)$, $y(0) = 0$ is found by Euler's method show that

$$2y(1, h) = (1 - c)^2 - (1 - c)h + \delta(1 - \delta)h^2$$

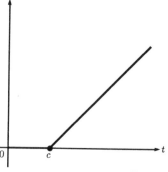

Figure 4

where $\delta = nc - [nc]$ and $hn = 1$. Here $[x]$ is the greatest-integer function. Deduce that $y(1, h)$ does not have a Taylor series for any positive value of h.

-----------------------ANSWERS-----------------------

2. 1.0800   1.1665   1.1775   1.1714   1.1736
4. $X = 2.618684$ 2.713046   2.717474   2.718283   2.718293
   $Z = 2.7093604$ 2.7186016   2.7183415   2.7183033   2.7182817
5. Partial answer: $E(.05)/(.05)^P = 4.152$ 3.585 -1.817 0.917 -1.579
6. 3.3811765   3.1623529   3.1311765   3.1311765   3.1415686
7. Last answer: 3.1424259850   3.1415926920   3.1415926530   3.1415926556
8. Partial answer: $10^9 y(1, .0025) = 1.2286894$ 2.0525801   2.079064
9. (b) -449 41907   741088   223932   4756
   (e)   2.07 -2.28 $-.548$ $-.00059$ $6.56 \cdot 10^{-7}$

---

3. Linear multistep formulas

In scientific work the algorithms of Section 2 are often replaced by others that are more elaborate but have better convergence properties. Some of the most popular are described next. As in Sections 1 and 2, we set

$$t_j = jh, \qquad y(t_j, h) = y_j, \qquad f(t_j, y_j) = f_j.$$

The explicit four-step algorithm

$$(1) \qquad y_{n+1} = y_n + \frac{h}{24}(55f_n - 59f_{n-1} + 37f_{n-2} - 9f_{n-3})$$

was proposed in a book by the British mathematical scientists Adams and Bashforth in 1883, and in 1926 the American astronomer Moulton introduced the implicit three-step formula

$$(2) \qquad y_{n+1} = y_n + \frac{h}{24}(9f_{n+1} + 19f_n - 5f_{n-1} + f_{n-2}).$$

To get an explicit algorithm he replaced the term y_{n+1} in f_{n+1} on the right by (1). Thus the *Adams-Bashforth formula* (1) is used as a predictor and the *Moulton formula* (2) as a corrector. The final result, known as the *Adams-Moulton formula*, is thought by many to be the best of the three.

The use of such formulas in specific problems is much the same as for the simpler methods of Section 2 and is taken up in the problems. Here we discuss a large class of algorithms from a general point of view. This unified approach was developed by the Swedish mathematician Germund Dahlquist in 1956. It applies to both Euler formulas, to the midpoint, trapezoid, Milne-Simpson, Adams-Bashford and Moulton formulas, and to another formula of Milne used as an example below.

A *linear multistep formula*, abbreviated LMF, is a relation of the form

(3) $y_n + \alpha_1 y_{n-1} + \cdots + \alpha_q y_{n-q} = h(\beta_0 f_n + \beta_1 f_{n-1} + \cdots + \beta_q f_{n-q})$

where α_j and β_j are constants. We assume that α_q and β_q are not both 0, since otherwise we could reduce q. The word "linear" is used because both y_j and f_j occur linearly in the equation. If $y_{n-1}, \cdots, y_{n-q}$ are known then (3) gives

$$y_n - \beta_0 h f(t_n, y_n) = \text{known quantity.}$$

Thus (3) is a q-step method. It is explicit if $\beta_0 = 0$, implicit if $\beta_0 \neq 0$.

EXAMPLE 1.   A formula called the *Milne predictor* is

$$y_{n+1} = y_{n-3} + \frac{4h}{3}(2f_n - f_{n-1} + 2f_{n-2}).$$

To compare with (3), move y_{n-3} to the left and replace n by $n-1$:

$$y_n - y_{n-4} = \frac{4h}{3}(2f_{n-1} - f_{n-2} + 2f_{n-3}).$$

Thus $\alpha_4 = -1$, $\beta_1 = 8/3$, $\beta_2 = -4/3$, $\beta_3 = 8/3$. The remaining coefficients are zero.

(a) The Dahlquist relations

Dahlquist's theory is based upon a class of identities of the form

(4) $\displaystyle\sum_{j=0}^{q} \alpha_j y(t - jh) = h \sum_{j=0}^{q} \beta_j y'(t - jh) + \mathrm{O}(h^{p+1})$, $\alpha_0 = 1$.

The constants α_j and β_j are determined so that (4) holds for all functions y having a convergent Taylor series. Dropping the O term leads to an approximate identity with order of accuracy p, and this in turn leads to the algorithm (3). The largest p for which (4) holds is the *order* of the multistep method.

To find the conditions on α_j and β_j, note that the terms involving h^k in the Taylor series for $y(t - hj)$ and $hy'(t - hj)$ are, respectively,

$$\frac{(-jh)^k}{k!} y^{(k)}(t), \qquad \frac{h(-jh)^{k-1}}{(k-1)!} y^{(k)}(t).$$

Equating coefficients of h^k on the left and right of (4) we get

$$\sum_{j=0}^{q} \alpha_j \frac{(-jh)^k}{k!} y^{(k)}(t) = \sum_{j=0}^{q} \beta_j \frac{h(-jh)^{k-1}}{(k-1)!} y^{(k)}(t).$$

We now divide out the common factor $h^k y^{(k)}(t)$ and we multiply by $(-1)^k k!$. The result is

(5a)
$$\sum_{j=0}^{q} \alpha_j j^k = -k \sum_{j=0}^{q} \beta_j j^{k-1}.$$

These are the Dahlquist relations. If they hold for

(5b)
$$k = 0, 1, \cdots, p$$

then all terms of the Taylor series up to and including those in h^p on both sides agree, and there remains an error term $O(h^{p+1})$. Thus (5) is necessary and sufficient for (4).

Since $\alpha_0 = 1$ we have $2q+1$ unknowns to satisfy the $p+1$ equations (5). If $p \le 2q$ there are at least as many unknowns as equations and it can be shown that a solution exists. For example, the Milne formula of Section 2, Part (e) is a linear two-step formula with $p = 4$. What should be emphasized is that the Dahlquist conditions embrace an enormous variety of algorithms.

If $p = 0$ the error in (4) is of the same order as the right-hand terms in (3), and only when $p \ge 1$ are useful results to be expected. A necessary and sufficient condition for $p \ge 1$ is given by (5) with $k = 0$ and $k = 1$:

(6a) $\alpha_0 + \alpha_1 + \cdots + \alpha_q = 0$

(6b) $\alpha_1 + 2\alpha_2 + \cdots + q\alpha_q = -(\beta_0 + \beta_1 + \cdots + \beta_q).$

In terms of the polynomials

$$\rho(s) = \alpha_0 + \alpha_1 s + \cdots + \alpha_q s^q, \qquad \sigma(s) = \beta_0 + \beta_1 s + \cdots + \beta_q s^q$$

these equations state that $\rho(1) = 0$ and $\rho'(1) = -\sigma(1)$.

The polynomials ρ and σ are the *characteristic polynomials* and the LMF (3) is *consistent* if $p \ge 1$. Using these definitions, we summarize as follows:

THEOREM 1. *A linear multistep formula is consistent if, and only if, its characteristic polynomials satisfy $\rho(1) = 0$ and $\rho'(1) = -\sigma(1)$.*

EXAMPLE 2. For the Milne predictor Example 1 gives

$$\rho(s) = 1 - s^4, \qquad \sigma(s) = \frac{4}{3}(2s - s^2 + 2s^3).$$

Since $\rho(1) = 0$ and $\rho'(1) = -\sigma(1)$ Theorem 1 indicates that (5) holds for $k = 0$ and $k = 1$. If $k \ge 2$ the relations (5) simplify to

$$4^k = \frac{4}{3}k(2 - 2^{k-1} + 2 \cdot 3^{k-1}).$$

This holds for $k = 3$ and $k = 4$ but not for $k = 5$. Hence the order of the Milne predictor is 4.

EXAMPLE 3. By use of undetermined coefficients together with Theorem 1 we can often construct an LMF with prescribed properties. For example, let us find the 2-step explicit LMF of the highest possible order. Since $\beta_0 = 0$ for an explicit algorithm, our starting point is

$$y_n + \alpha_1 y_{n-1} + \alpha_2 y_{n-2} = h(\beta_1 f_{n-1} + \beta_2 f_{n-2}).$$

The Dahlquist relations (5) are $\alpha_1 + \alpha_2 = -1$ for $k = 0$ and

$$\alpha_1 + \alpha_2 2^k = -k(\beta_1 + \beta_2 2^{k-1}), \qquad k \geq 1.$$

The equations with $k = 0$ and $k = 1$ are easily solved for α_1 and α_2, with the result

$$\alpha_1 = \beta_1 + \beta_2 - 2, \qquad \alpha_2 = 1 - \beta_1 - \beta_2.$$

If we use these to eliminate α_1 and α_2 from the equations

$$\alpha_1 + 4\alpha_2 = -2\beta_1 - 4\beta_2, \qquad \alpha_1 + 8\alpha_2 = -3\beta_1 - 12\beta_2$$

corresponding to $k = 2$ and $k = 3$, we get a linear system with the unique solution $\beta_1 = 4$, $\beta_2 = 2$. Hence $\alpha_1 = 4$, $\alpha_2 = -5$, and the required LMF is

$$y_n + 4y_{n-1} - 5y_{n-2} = h(4f_{n-1} + 2f_{n-2}).$$

(b) The local error

Suppose the values y_j in the LMF happen to agree with the values $y(t_j)$ of the exact solution for $n - q \leq j < n$. In this case, by definition, the *local error* is $y(t_n) - y_n$. We shall prove the following:

THEOREM 2. *The local error in a linear multistep formula of order p is of order $p + 1$.*

For proof, since $t_n - jh = t_{n-j}$ the choice $t = t_n$ in (4) gives

$$(7) \quad y(t_n) + \sum_{j=1}^{q} \alpha_j y(t_{n-j}) = h\beta_0 y'(t_n) + h\sum_{j=1}^{q} \beta_j y'(t_{n-j}) + O(h^{p+1}).$$

On the other hand Equation (3) is equivalent to

$$(8) \qquad y_n + \sum_{j=1}^{q} \alpha_j y_{n-j} = h\beta_0 f_n + h\sum_{j=1}^{q} \beta_j f_{n-j}.$$

We have $y(t_{n-j}) = y_{n-j}$ for $1 \leq j \leq q$ by hypothesis, and for such j the differential equation gives

$$y'(t_{n-j}) = f(t_{n-j}, y(t_{n-j})) = f(t_{n-j}, y_{n-j}) = f_{n-j}.$$

Thus the terms under the summation signs in Equations (7) and (8) agree. By subtraction

$$y(t_n) - y_n = h\beta_0 f(t_n, y(t_n)) - h\beta_0 f(t_n, y_n) + \mathrm{O}(h^{p+1}).$$

If we set $z_n = y(t_n)$ and use the Lipschitz condition for f the result is

$$|z_n - y_n| \leq K\beta_0 h |z_n - y_n| + \mathrm{O}(h^{p+1}).$$

Since our estimates are wanted only for $h \to 0+$ we assume without loss of generality that $|K\beta_0 h| \leq 1/2$. Multiplying the above equation by 2 gives first

$$2|z_n - y_n| \leq |z_n - y_n| + \mathrm{O}(h^{p+1})$$

and then $|z_n - y_n| \leq \mathrm{O}(h^{p+1})$. This completes the proof.

The somewhat artificial assumption that $y_j = y(t_j)$ for $n - q \leq j < n$ can be weakened. First, we need not require that y be the solution with $y(t_0) = y_0$ but y can be any solution. In the case of a one-step method, for example, we can impose the initial condition $y(t_{n-1}) = y_{n-1}$ without making any special assumptions. The proof with the new definition of $y(t)$ is unchanged.

For multistep methods some artificiality remains, because we cannot reasonably suppose that both (t_{n-2}, y_{n-2}) and (t_{n-1}, y_{n-1}) lie on the same trajectory. But it would suffice to assume the weaker condition

(9) $$y(t_j) - y_j = \mathrm{O}(h^p), \qquad n - q \leq j < n.$$

This produces a change of order $\mathrm{O}(h^p)$ in the arguments of f. Since f is preceded by a factor h the additional error introduced into the final equation is $\mathrm{O}(h^{p+1})$ and the proof is completed as before. We shall presently see that, for linear multistep formulas of the kind that are useful in applications, the global error satisfies

$$y(t) - y(t, h) = \mathrm{O}(h^p).$$

Hence (9) holds with $y(t)$ as in our original definition of local error.

(c) Stability

For a consistent LMF the local error is $\mathrm{O}(h^2)$. Hence the global error after x/h steps ought to be $\mathrm{O}(h)$, which tends to 0 with h. This line of thought may seem reasonable, but it is shown by the following example to be incorrect.

EXAMPLE 4.   For the linear two-step method defined by

$$y_n = 2y_{n-1} - y_{n-2}$$

the characteristic polynomials are $\rho(s) = (1-s)^2$ and $\sigma(s) = 0$. By Theorem 2 the method is consistent. But, since f is not involved in the formula, there is no connection with the differential equation. Hence there is no reason for $y(x, h)$ to converge to $y(x)$. Still worse, the fact that $y_n = n$ satisfies the recurrence relation shows that $\{y_n\}$ need not converge at all.

The behavior exhibited in this example is not exceptional but typical. In general, a consistent LMF does not converge unless it has a special property, known as stability, which is discussed next.

Whenever $\{y_n\}$ converges, the terms f_j on the right of (3) remain bounded as $h \to 0+$ and the limiting form of the equation is

$$(10) \qquad\qquad y_n + \alpha_1 y_{n-1} + \cdots + \alpha_q y_{n-q} = 0.$$

An equation of this form is called a *difference equation*. If any solution $\{y_n\}$ is unbounded as $n \to \infty$ one might expect, and it can be proved, that many choices of f and of the initial condition lead to divergence. Thus we are led to the following definition and theorem:

THEOREM 3.   *By definition, the linear multistep method* (3) *is stable if all solutions of* (10) *are bounded. A necessary and sufficient condition for stability is that all zeros s of ρ satisfy $|s| \leq 1$ and that all zeros with $|s| = 1$ be simple.*

The proof is outlined in Part (d). As an illustration, Example 2 gives $\rho(s) = s^4 - 1$. The zeros ± 1 and $\pm i$ are simple, hence Milne's predictor is stable. But in Example 3 we have $\rho(s) = (s - 1)^2$ and there is a double root $s = 1$. Hence the corresponding LMF is unstable.

The connection between stability and convergence is far from obvious, and the discovery that there is such a connection represents a major advance in numerical analysis. A fundamental theorem of Dahlquist says that an LMF is convergent if, and only if, it is both consistent and stable. Here "convergent" means "uniformly convergent on an interval for each Lipschitzian function f." Naturally, nothing can be concluded from the fact that the algorithm converges for a particular function like $f = 0$.

When the functions are sufficiently differentiable, the condition of stability is also just what is needed to guarantee that a local error of order $p+1$ yields a global error of order p. Further development along these lines leads to the *Dahlquist stability barriers*. These state that the order of accuracy p of a stable q-step LMF can be at most $q + 2$ if q is even, at most $q + 1$ if q is odd, and at most q if the formula is explicit. These results are deep and the proof will not be discussed here.

(d) The difference equation

It has been seen that the question of convergence hinges on the question whether the solutions of (10) are bounded. In discussing this question we assume $\alpha_q \neq 0$; otherwise q could be diminished. The function $y_j = s^j$ satisfies (10) if

$$s^n + \alpha_1 s^{n-1} + \cdots + \alpha_q s^{n-q} = 0$$

or, multiplying by s^{q-n},

$$s^q + \alpha_1 s^{q-1} + \cdots + \alpha_q = 0.$$

This is the same as $\rho(s) = 0$. If any $|s_j| > 1$ the corresponding function s_j^n is unbounded. Thus $|s_j| \leq 1$ is necessary for boundedness of all solutions.

We now inquire whether the condition $|s_j| \leq 1$ is also sufficient. Let us suppose first that ρ has q distinct zeros $s_1, s_2, \cdots, s_q$. Each s_k^j satisfies (10) and by linearity the expression

$$z_n = c_1 s_1^n + c_2 s_2^n + \cdots + c_q s_q^n$$

also satisfies (10) for all choices of the constants c_j. We want to show that every solution can be represented in this form.

To this end, let y_n be a given solution and choose the constants c_j in such a way that the initial conditions for z_n agree with those for y_n:

$$z_0 = y_0, \qquad z_1 = y_1, \qquad \cdots, \qquad z_{q-1} = y_{q-1}.$$

For example, when $q = 3$ these equations are

$$c_1 + c_2 + c_3 = y_0$$
$$c_1 s_1 + c_2 s_2 + c_3 s_3 = y_1$$
$$c_1 s_1^2 + c_2 s_2^2 + c_3 s_3^2 = y_2.$$

The determinant D of this system is shown on the left below together with a certain Wronskian W on the right:

$$D = \begin{vmatrix} 1 & 1 & 1 \\ s_1 & s_2 & s_3 \\ s_1^2 & s_2^2 & s_3^2 \end{vmatrix} \qquad W = \begin{vmatrix} e^{s_1 t} & e^{s_2 t} & e^{s_3 t} \\ s_1 e^{s_1 t} & s_2 e^{s_2 t} & s_3 e^{s_3 t} \\ s_1^2 e^{s_1 t} & s_2^2 e^{s_2 t} & s_3^2 e^{s_3 t} \end{vmatrix}.$$

By results of Chapter 11 the Wronskian never vanishes, and on the other hand it reduces to D when $t = 0$. Hence $D \neq 0$. A similar argument gives $D \neq 0$ for arbitrary q, hence the constants c_j can actually be found.

When the c_j are thus chosen, we have $y_n = z_n$ by Section 1, Problem 4. In other words

$$(11) \qquad\qquad y_n = c_1 s_1^n + c_2 s_2^n + \cdots + c_q s_q^n.$$

Hence the condition $|s_j| \leq 1$ for all j is sufficient for boundedness of the arbitrary solution y_n, provided the roots s_j of the equation $\rho(s) = 0$ are distinct.

When s is a root of multiplicity k it can be shown that the corresponding family of solutions is

$$s^n, \qquad ns^n, \qquad \cdots, \qquad n^{k-1}s^n.$$

If $|s| < 1$ each of these expressions tends to 0 as $n \to \infty$, hence is bounded. But if $|s| = 1$ every term except the first is unbounded. Thus we are led to the requirement that all roots of magnitude 1 be simple, as stated in Theorem 3.

PROBLEMS

In multistep methods use the exact solution to get started. In C problems it will be found that the relevant implicit algorithms can be solved explicitly.

1. (abcefg) As in Examples 1 and 2 discuss methods (abce) of Section 2. In particular, show that the orders are 1, 2, 2, 4 in agreement with the results given there. Also show that the Adams-Bashforth formula and the Moulton formula are stable and of order 4. We omit Heun's method (d) because it is not an LMF.

2c. (fghi) Verify the entries (fghi) of the table in Section 2 for $h = .2$. This can be a c problem because only two or three steps are needed.

3C. (fghi) Do Problem 2 with $h = .1$ and $h = .05$.

4. Fill in the missing details in the discussion of Example 3. Show also that the order is 3 and that the algorithm is unstable.

5. (ab) In general, an *Adams-Bashforth formula* is an LMF in which $\beta_0 = 0$ and $\rho(s) = s - 1$. A *backwards differentiation formula* is an LMF in which $\sigma(s) = \beta_0$. In the following examples of each type, show that the order is 2 for just one choice of the coefficients and that the resulting algorithms of order 2 are stable. To check your work, see Section 5, Equations (3):

$$y_n - y_{n-1} = \beta_1 h f_{n-1} + \beta_2 h f_{n-2} \qquad y_n + \alpha_1 y_{n-1} + \alpha_2 y_{n-2} = h\beta_0 f_n$$

6. If s is a root of multiplicity 2, so that $\rho(s) = \rho'(s) = 0$, show that the difference equation (10) has the solution ns^n as well as s^n.

---

4. Runge-Kutta formulas

An important method was introduced by the German mathematician Carl David Runge in 1895 and was developed further by another German, Wilhelm Kutta, in 1901. Though Runge is much better known, most of the early Runge-Kutta methods are due to Kutta.

The basic idea is to introduce intermediate values

$$t_{n+c}, \qquad y_{n+c}, \qquad f_{n+c} = f(t_{n+c}, y_{n+c})$$

as an aid in getting from y_n to y_{n+1}. The process involves just one step but several stages. This means that a single use of the algorithm takes us from y_n to y_{n+1} through a number of intermediate values y_{n+c}. In the discussion of local error we drop the condition $y(t_0) = y_0$ and the correspondence with solutions of $y' = f(t, y)$ is as follows:

$$t_{n+c} = t + hc, \qquad y_n = y(t), \qquad f_n = y'(t), \qquad t = t_n.$$

It will be seen that the local error $y_{n+1} - y(t + h)$ depends on the error in the approximate equations

$$y_{n+c} \approx y(t + ch), \qquad f_{n+c} \approx y'(t + ch), \qquad t = t_n.$$

Runge-Kutta algorithms are described by an array of coefficients called a *tableau*, plural tableaux. Here is the tableau for one of Kutta's formulas of 1901, with equations involving intermediate values f_{n+c_i} on the right:

0	0	0	0	0		$k_1 = f(t_n, y_n)$
$\frac{1}{2}$	$\frac{1}{2}$	0	0	0		$k_2 = f\left(t_n + \dfrac{h}{2},\, y_n + \dfrac{h}{2}k_1\right)$
$\frac{1}{2}$	0	$\frac{1}{2}$	0	0		
1	0	0	1	0		$k_3 = f\left(t_n + \dfrac{h}{2},\, y_n + \dfrac{h}{2}k_2\right)$
	$\frac{1}{6}$	$\frac{1}{3}$	$\frac{1}{3}$	$\frac{1}{6}$		$k_4 = f\left(t_n + h,\, y_n + hk_3\right)$

After the k_i are thus determined, the step from n to $n + 1$ is made by the formula

$$(1) \qquad\qquad y_{n+1} = y_n + \frac{h}{6}(k_1 + 2k_2 + 2k_3 + k_4).$$

This is a four-stage method because there are four intermediate values. The connection of the tableau with the equations is explained in Part (b). Suffice it to say here that the bottom row of the tableau gives the coefficients in (1) and the remaining entries describe the equations at the right of the tableau. The above method is designated by the words "Runge-Kutta" in the table of Section 2, but in the rest of this chapter it is preceded by the word "classical" to distinguish it from other examples.

EXAMPLE 1.   In the test equation of Section 2 the step from $t = 0$ to $t = h$ is accomplished as follows. Using $f(t, y) = 2ty$, $t_n = 0$ and $y(0) = 1$, compute

$$k_1 = 0, \qquad k_2 = h, \qquad 2k_3 = 2h + h^3, \qquad k_4 = 2h + 2h^3 + h^5.$$

Then use (1) to get

$$y_1 = 1 + h^2 + \frac{1}{2}h^4 + \frac{1}{6}h^6.$$

Since the exact value is $y(h) = e^{h^2}$, the error is $O(h^8)$.

We have displayed these equations to help you understand the algorithm. It should be mentioned, however, that persons making extensive use of Runge-Kutta methods usually program the computer by looking at the tableau. Except for the definition of f, no formulas are needed.

Our next example illustrates a simple but powerful method of determining local accuracy. It depends on the fact that if $y' = f(t, y)$ then

(2) $$f(t + hc, \, y(t + hc)) = y'(t + hc).$$

The second argument $y(t + hc)$ will generally have an error term Eh or Eh^2 where E is bounded as $h \to 0+$. To take account of this, we introduce the following condition on f:

DEFINITION. The function f is admissible if f_{yx} and f_{yy} are bounded.

The boundedness is to hold in the rectangle mentioned in Section 1. When f_{yy} is bounded we have

(3) $$f(t + hc, \, y(t + hc) + Eh) = y'(t + hc) + H(hc)Eh + O(h^2)$$

where $H(hc) = f_y(t + hc, \, y(t + hc))$. This follows from (2) and from Taylor's formula with remainder applied to the function $\phi(z) = f(t + hc, z)$ at the point $z = y(t + hc)$. If f_{yx} is also bounded, results of Chapter 14 show that f_y is Lipschitzian with respect to both variables (t, y). This property is not used in Example 2 but is needed later.

EXAMPLE 2.   It is seen in Part (b) that the most general two-stage, explicit Runge-Kutta algorithm has the form

$$y_{n+1} = y_n + hb_1 f_n + hb_2 f(t + hc, \, y_n + ha f_n).$$

We want conditions on the constants a, b_1, b_2, c for the algorithm to have local order $p = 2$. This amounts to the statement that if y is an arbitrary solution of $y' = f(t, y)$, with arbitrary f then

$$(4) \qquad y(t + h) - y(t) = hb_1 y'(t) + hb_2 f(t + hc,\ y(t) + ha\, y'(t)) + O(h^3).$$

The Taylor series for $y(t + hc)$ gives

$$y(t) + ha\, y'(t) = y(t + hc) + Eh + O(h^2)$$

where $E = (a - c)y'(t)$. Substitution into (3) yields

$$f(t + hc,\ y(t) + ha\, y'(t)) = y'(t + hc) + H(hc)Eh + O(h^2)$$

and (4) becomes

$$(5) \qquad y(t + h) - y(t) = hb_1 y'(t) + hb_2 y'(t + ch) + b_2 H(hc)Eh^2 + O(h^3).$$

To get rid of the term containing E we set $E = 0$, hence $a = c$. It is easily checked that the resulting equation

$$y(t + h) - y(t) = hb_1 y'(t) + hb_2 y'(t + ch) + O(h^3)$$

holds if $b_1 + b_2 = 1$ and $b_2 c = 1/2$. These conditions together with $a = c$ are sufficient for (4). Their necessity is established in Problem 3.

(a) The general case

A Runge-Kutta formula of r stages is defined by the following conditions:

AB: $\displaystyle y_{n+c_i} = y_n + h \sum_{j=1}^{r} a_{ij} f_{n+c_j}$ $\displaystyle \sum_{j=1}^{r} a_{ij} = c_i$

CD: $\displaystyle y_{n+1} = y_n + h \sum_{j=1}^{r} b_j f_{n+c_j}$ $\displaystyle \sum_{j=1}^{r} b_j = 1$

It is understood that $i = 1, 2, \cdots, r$ in AB. Thus we start with the system A of r equations in r unknowns y_{n+c_i}. After these are determined, the step from y_n to y_{n+1} is given by C. To ensure that the intermediate values t_{n+c} remain in the interval where f is defined, some authors impose the further condition

$$0 \leq c_i \leq 1, \qquad i = 1, 2, \cdots, r.$$

However these inequalities play no other role and will not be emphasized.

If $y' = f(t, y)$, $t = t_n$, and $y(t_n) = y_n$, it is seen in Problem 6 that

(6) $$f_{n+c_j} = y'(t) + O(h).$$

Hence C becomes

$$y_{n+1} = y(t) + h \sum_{j=1}^{r} b_j y'(t) + O(h^2).$$

The Taylor series for $y(t + h)$ now shows that D is necessary and sufficient for

$$y_{n+1} = y(t + h) + O(h^2).$$

This says that the method is of order $p \geq 1$, hence is no worse than Euler's method. But a more profound reason for imposing it is that, if the error is as large as $O(h)$, all terms having h as a factor in AC can be neglected and the whole process becomes nonsensical. Thus Condition D for Runge-Kutta methods is much like the requirement of consistency for Dahlquist methods.

We now explain the reason for B. Substitution of (6) into A yields

$$y_{n+c_i} = y(t) + h \sum_{j=1}^{r} a_{ij} y'(t) + O(h^2).$$

Hence B is necessary and sufficient for

(7) $$y_{n+c_j} = y(t + c_j h) + O(h^2).$$

Problem 12 shows that this cannot be deduced from the hypothesis $p \geq 1$, or even from $p \geq 2$, but it is an indispensable aid in developing the theory and is postulated for that reason.

These remarks are now carried further. Equation (7) yields

(8) $$f_{n+c_j} = y'(t + c_j h) + O(h^2)$$

and substitution into C then gives

$$y_{n+1} = y(t) + h \sum_{j=1}^{r} b_j y'(t + c_j h) + O(h^3).$$

Comparing with the Taylor series for $y(t + h)$ we see that

(9) $$\sum_{j=1}^{r} b_j = 1, \qquad \sum_{j=1}^{r} b_j c_j = \frac{1}{2}$$

are necessary and sufficient for the formula to have order ≥ 2. Note that the first condition is D again.

If f is admissible the procedure can be carried still further. Equation (8) together with A yields

$$y_{n+c_i} = y(t) + h \sum_{j=1}^{r} a_{ij} y'(t + c_j h) + O(h^3).$$

By a short calculation, using this and Equation B, it will be found that

$$y_{n+c_i} = y(t + hc_i) + E_i h^2 + O(h^3)$$

where

$$E_i = \frac{1}{2} c_i^2 y''(t) - \sum_{j=1}^{r} a_{ij} c_j y''(t).$$

Equation (3) with $E_i h^2$ instead of Eh then gives

$$f_{n+c_i} = y'(t + hc_i) + H(hc_i) E_i h^2 + O(h^4).$$

For admissible functions, $H(hc_i) = K + O(h)$ where $K = H(0)$. This substitution introduces an error $O(h^3)$ and Equation C then gives

$$y_{n+1} = y(t) + h \sum_{i=1}^{r} b_i y'(t + hc_i) + Kh \sum_{i=1}^{r} b_i E_i + O(h^4).$$

Setting the coefficient of K equal to 0 and comparing with the Taylor series for $y(t + h)$, we find that (9) and the two additional conditions

$$(10) \qquad \sum_{i=1}^{r} b_i c_i^2 = \frac{1}{3}, \qquad \sum_{i,j=1}^{r} a_{ij} b_i c_j = \frac{1}{6}$$

are sufficient for the algorithm to have order ≥ 3. That they are necessary is seen in Problem 9.

Order conditions for Runge-Kutta methods are analogous to the Dahlquist relations for an LMF, but they are nonlinear and their complexity increases with the order. For example, four additional nonlinear equations are needed to ensure that $p = 4$ and nine more equations are needed to bring the order up to $p = 5$.

(b) The tableau

As was already illustrated in the introduction to this section, Runge-Kutta formulas are described by a tableau consisting of the matrix (a_{ij})

bordered on the left by the column vector (c_j) and on the bottom by the row vector (b_j). Although it is not customary, we shall often put a 1 in the bottom left corner.

EXAMPLE 3. Here are two tableaux with $r = 2$:

c_1	a_{11}	a_{12}
c_2	a_{21}	a_{22}
1	b_1	b_2

0	0	0
c	a	0
1	b_1	b_2

The first of these specifies the general case. Equation B of the definition says that the element c_j at the left is the sum of the other elements in its row, and D says that the 1 at the left is the sum of the other elements in its row. The extension to arbitrary r is immediate.

Equations A for this case can be written in matrix form as follows:

$$\begin{pmatrix} y_{n+c_1} \\ y_{n+c_2} \end{pmatrix} = \begin{pmatrix} y_n \\ y_n \end{pmatrix} + h \begin{pmatrix} a_{11} & a_{12} \\ a_{21} & a_{22} \end{pmatrix} \begin{pmatrix} f_{n+c_1} \\ f_{n+c_2} \end{pmatrix}$$

Since $f_{n+c_1} = f(t_n + c_1 h, y_{n+c_1})$ and $f_{n+c_2} = f(t_n + c_2 h, y_{n+c_2})$ by definition, the formula is implicit if any one of a_{11}, a_{12}, a_{22} is nonzero. In that case one of the equations contains some y_{n+c} both by itself and in the argument of f_{n+c}. In general, a Runge-Kutta method is explicit if, and only if, all elements on or above the diagonal of the matrix (a_{ij}) in the tableau are 0.

Hence an explicit algorithm with $r = 2$ must have the form given on the right above. It is left for you to verify that the latter describes the algorithm that was analyzed in Example 2. The two conditions

$$a = c, \qquad b_1 + b_2 = 1$$

found there are just what is needed to define a Runge-Kutta formula, and $b_2 c = 1/2$ ensures that the order is 2. This agrees with the general criterion for order 2 from Part (a).

EXAMPLE 4. Here are two tableaux with $r = 3$:

c_1	a_{11}	a_{12}	a_{13}
c_2	a_{21}	a_{22}	a_{23}
c_3	a_{31}	a_{32}	a_{33}
1	b_1	b_2	b_3

0	0	0	0
1	1	0	0
$\frac{1}{2}$	$\frac{1}{4}$	$\frac{1}{4}$	0
1	$\frac{1}{6}$	$\frac{1}{6}$	$\frac{2}{3}$

The interpretation of Equations ABCD in terms of the first tableau is similar to that in Example 3. Here we show that the second tableau represents a

Runge-Kutta method of order 3. Each element in the left-hand column is the sum of the other three elements in its row, so that conditions BD defining a Runge-Kutta method are fulfilled. In particular, the first condition (9) holds. Since several coefficients are 0, the remaining conditions (9), (10) reduce to

$$b_2c_2 + b_3c_3 = \frac{1}{2}, \qquad b_2c_2^2 + b_3c_3^2 = \frac{1}{3}, \qquad a_{32}b_3c_2 = \frac{1}{6}.$$

These are easily verified, hence $p \geq 3$. That $p \leq 3$ follows from Problem 7.

(c) The order of the classical algorithm

It is seen in Problem 8 that the order p of the classical Runge-Kutta algorithm satisfies $p \leq 4$. Assuming f admissible, we now prove $p \geq 4$, hence $p = 4$. The classical formulas correspond to the following substitutions for functions $y(t)$:

$$k_1 = f(t, y(t)) = y'(t) \qquad\qquad k_3 = f(t + \frac{h}{2}, y(t) + \frac{h}{2}k_2)$$

$$k_2 = f(t + \frac{h}{2}, y(t) + \frac{h}{2}k_1) \qquad\qquad k_4 = f(t + h, y + hk_3).$$

Given $y' = f(t, y)$, it is to be shown that

$$(11) \qquad y(t + h) - y(t) = \frac{h}{6}(k_1 + 2k_2 + 2k_3 + k_4) + O(h^5).$$

We begin by reformulating two results obtained earlier. If t is replaced by $t + h$ and h by $h/2$ the result of Section 1, Problem 1(e), is

$$(12) \qquad y(t + h) - y(t) = \frac{h}{6}\left(y'(t) + 4y'(t + \frac{h}{2}) + y'(t + h)\right) + O(h^5).$$

Equation (3) with $c = 1/2$, $J = H(h/2)$, and Eh^2 in place of Eh yields

$$(13) \qquad f(t + \frac{h}{2}, y(t + \frac{h}{2}) + Eh^2) = y'(t + \frac{h}{2}) + JEh^2 + O(h^4).$$

With your cooperation, we now give the proof. Let us write $x \doteq y$ to mean $x = y + O(h^4)$. Using the values E_j given in Problem (10), you are asked in that problem to verify the equations on the left:

$$y(t) + \frac{h}{2}k_1 \doteq y(t + \frac{h}{2}) + E_2h^2 \qquad\qquad k_2 \doteq y'(t + \frac{h}{2}) + JE_2h^2$$

$$y(t) + \frac{h}{2}k_2 \doteq y(t + \frac{h}{2}) + E_3h^2 \qquad\qquad k_3 \doteq y'(t + \frac{h}{2}) + JE_3h^2$$

$$y(t) + hk_3 \doteq y(t + h) + E_4h^2 \qquad\qquad k_4 \doteq y'(t + h) + JE_4h^2.$$

The first two equations on the right follow from those on the left and (13). For the third use (13) with h instead of $h/2$, noting that $H(h) = H(h/2) + O(h)$. Since $E_4 = O(h)$, the error so introduced is $O(h^4)$.

If we form the expression on the right of (11), the principal terms give $y(t + h) - y(t)$ within $O(h^5)$ by (12) and it is seen in Problem 10 that the principal error terms give

$$\frac{Jh}{6}\left(2E_2 + 2E_3 + E_4\right)h^2 = \frac{J^2}{6}\left(E_2 + E_3\right)h^4 = O(h^5).$$

When multiplied by $h/6$ the $O(h^4)$ error terms introduced by $\doteq$ also give $O(h^5)$. This completes the proof.

From this result together with others established here, it follows that the order $p = 1, 2, 3, 4$ can be achieved by an explicit Runge-Kutta method with $r = 1, 2, 3, 4$ stages, respectively. For $p = 5, 6, 7, 8$ the corresponding minimum number of stages can be shown to be $6, 7, 9, 11$. But the minimum number of stages needed for an explicit method with prescribed order $p \geq 9$ is still unknown.

PROBLEMS

1c. In the test problem of Example 1 approximate $y(1/2)$ by one use of the classical Runge-Kutta formula and verify that it agrees with the formula given there. Then repeat to get $y_2 = 2.7131450$.

2C. Confirm the entries (j) in the table at the end of Section 2.

3P. This solution of Example 1 is easier than that in the text and does not require use of f_y. It also gives necessity. Show that (4) in the two cases

$$f(t, y) = t, \quad y(t) = \frac{t^2}{2}; \qquad f(t, y) = y, \quad y(t) = e^t$$

yields $b_1 + b_2 = 1$, $b_2c = 1/2$ and $b_1 + b_2 = 1$, $ab_2 = 1/2$, respectively. These results give $a = c$ so that (4) becomes

$$y(t + h) - y(t) = hb_1 f(t, y(t)) + hb_2 f(t + hc, y(t) + hcy'(t)) + O(h^3)$$

Using $y(t) + hcy'(t) = y(t + h) + O(h^2)$ show that $b_1 + b_2 = 1$ and $b_2c = 1/2$ are necessary and sufficient for this. The resulting algorithm is expressible in terms of the single parameter $b = b_2$ and it reduces to the Heun formula when $b = 1/2$.

4. (ab) As in Example 4, show that the algorithms corresponding to the following tableaux are both Runge-Kutta algorithms of order 3:

0	0	0	0
$\frac{2}{3}$	$\frac{2}{3}$	0	0
$\frac{2}{3}$	$\frac{1}{3}$	$\frac{1}{3}$	0
1	$\frac{1}{4}$	0	$\frac{3}{4}$

0	0	0	0
$\frac{1}{2}$	$\frac{1}{2}$	0	0
1	-1	2	0
1	$\frac{1}{6}$	$\frac{2}{3}$	$\frac{1}{6}$

5. In the tableau at the left we have already built in the conditions B by setting $a_{21} = c_2$ and $a_{32} = c_3$:

0	0	0	0
c_2	c_2	0	0
c_3	0	c_3	0
1	b_1	b_2	b_3

0	0	0	0
$\frac{1}{3}$	$\frac{1}{3}$	0	0
$\frac{2}{3}$	0	$\frac{2}{3}$	0
1	$\frac{1}{4}$	0	$\frac{3}{4}$

Write down the equations needed for the left-hand algorithm to be of order 3. Then show that $b_2 = 0$ if, and only if, $c_2 = 1/3$, in which case the algorithm reduces to that on the right.

6. If $c = c_j$, $y' = f(t, y)$, $y(t_n) = y_n$, and $\phi \doteq \psi$ means $\phi = \psi + O(h)$, Equation A gives $y_{n+c} = y(t) + O(h)$. Thus explain the following:

$$f_{n+c} \doteq f(t + hc, y(t)) \doteq f(t + hc, y(t + hc)) = y'(t + hc) \doteq y'(t)$$

7. Use the special case $f(t, y) = t^m$, $y(t) = t^{m+1}/(m + 1)$ in C to obtain the following necessary conditions for the general Runge-Kutta algorithm to have order p. Namely, the first is necessary for $p = 1$, the first and second for $p = 2$, the first, second and third for $p = 3$, and so on:

$$\sum_{i=1}^{r} b_i = 1, \qquad \sum_{i=1}^{r} b_i c_i = \frac{1}{2}, \qquad \sum_{i=1}^{r} b_i c_i^2 = \frac{1}{3}, \qquad \sum_{i=1}^{r} b_i c_i^3 = \frac{1}{4}, \quad \cdots$$

8. Show that the classical Runge-Kutta algorithm satisfies the conditions of the text for $p = 1, 2, 3$, and also the condition of Problem 7 for $p = 4$, but it does not satisfy the condition of Problem 7 for $p = 5$.

9. Using the result of Problem 7 for $p = 3$, simplify the equation leading to (10) and thus show that both conditions (10) are necessary.

10. In Part (c) show that E_2, E_3, E_4 are respectively as follows, and confirm the calculations at the end of the proof:

$$-\frac{y''(t)}{8} - \frac{y'''(t)}{48}h, \qquad \frac{y''(t)}{8} + \frac{y'''(t)}{24}h + JE_2\frac{h}{2}, \qquad -\frac{y'''(t)}{24}h + JE_3h$$

11. Show that the classical Runge-Kutta method and the Milne-Simpson method both reduce to Simpson's rule when $f(t,y) = g(t)$.

12. If $a_{ij} = 0$ except for $a_{21} = a$, and if conditions BD are dropped, check that the general two-stage algorithm yields

$$f_{n+c_1} = f(t + c_1h, y_n), \qquad f_{n+c_2} = f(t + c_2h, \ y_n + ha\, f_{n+c_1})$$

Assuming f admissible, show that the resulting algorithm has order ≥ 2 if, and only if,

$$b_1 + b_2 = 1, \qquad b_1c_1 + b_2c_2 = \frac{1}{2}, \qquad b_2a = \frac{1}{2}$$

13. Implicit differentiation. This problem requires knowledge of Taylor series in two variables. We write $f_y = f_y(t, y(t))$ and similarly for other derivatives such as f_t or f_{yy}. It is assumed that f has as many derivatives as needed. From $y' = f(t,y)$ follows $y'' = f_t + f_y f$ and similar results for higher derivatives. Hence the Taylor series for $y(t+h)$ can be written in terms of f and its derivatives. The Taylor series for $f(t + \alpha h, y + \beta h)$ can also be expressed in terms of f and its derivatives. Thus obtain the result of Problem 12.

14. (abcd) This ranges from easy to very difficult. Use the technique of Problem 13 to derive the conditions given in the text for $p = 1, 2, 3, 4$ where the last item $p = 4$ refers to the result of Part (c).

---

5. Additional topics

Some interesting conclusions are suggested by the equation

(1) $$y' = -100(y - \cos t) - \sin t.$$

This has an obvious solution $y = \cos t$ and the general solution is

(2) $$y(t) = \cos t + ce^{-100t} \quad \text{where} \quad c = y(0) - 1.$$

For moderately large t, say $t = 1$ or 2, Equation (2) shows that $y(t)$ is very insensitive to errors in the initial value. Since a roundoff or truncation error at the nth step is much like an initial-value error at t_n, one might naturally expect a high degree of numerical stability in the calculation of y.

To investigate this point let us try the following algorithms:

$$(3) \qquad 2y_{n+1} = 2y_n + 3hf_n - hf_{n-1}, \qquad 3y_{n+1} = 4y_n - y_{n-1} + 2hf_{n+1}.$$

By Section 3, Problem 5 these are stable algorithms of order 2. They are designated as AB2 and BD2 respectively. The results for (1) with $y(0) = 1$ are shown in the accompanying table. Note that the first method, which is explicit, yields hopelessly bad approximations until $h \leq .01$ while the second, implicit method is much better:

	$h = 0.2$	$h = 0.1$	$h = 0.05$	$h = 0.02$	$h = .01$
AB2	14.4	$-6 \cdot 10^4$	$-2 \cdot 10^7$	$-6 \cdot 10^{10}$	.540302
BD2	.5404	.540330	.540309	.5403034	.5403026

Equation (1), and others that exhibit similar behavior, are said to be *stiff*. A formal definition is not easily given, but here are some of the main features:

(i) For stiff problems, many popular algorithms exhibit an extreme numerical instability that is not connected with any actual instability of the problem.

(ii) Because of (i), it is necessary to use a much smaller step-size h than one would expect from the local error.

Item (i) is worded so as to distinguish instability of numerical algorithms from true instability of the underlying problem. If the coefficient -100 on the right of (1) is changed to 100 the exponent $-100t$ in (2) becomes $100t$ and the problem itself is unstable. A corresponding numerical instability then reflects a genuine aspect of the situation.

Stiff problems are sometimes associated with simultaneous processes that take place on widely different time scales. This characterization applies to the first of the following stiff equations though not to the others:

$$y' = -99y + t, \qquad y' = -99y, \qquad y' = -99y + e^{-98t}.$$

Stiff problems also arise in several fields of science, for example in the theory of the kinetics of chemical reactions.

To see why (1) causes difficulty, let z denote the difference of two solutions. One of these is thought to be the desired solution while the other has an incorrect initial value. Linearity gives $z' = -100z$ and the first algorithm (3) yields

$$z_{n+1} = z_n - 50h(3z_n - z_{n-1}).$$

The complete solution of this difference equation is on the left below where s_j are the two roots, supposed unequal, of the quadratic on the right:

$$(4) \qquad z_n = c_1 s_1^n + c_2 s_2^n, \qquad s^2 = s(1 - 150h) + 50h.$$

We number the roots so that $|s_2| \geq |s_1|$. Since their product is $-50h$, clearly $|s_2| > 1$ if $h > .02$. In this case $\{z_n\}$ diverges as $n \to \infty$, (unless c_2 happens to be 0), even though $z(t)$ tends rapidly to 0 as $t \to \infty$. Closer analysis shows that the dividing line is $h = .01$. That is, if $h > .01$ then $|s_2| > 1$ and if $h \leq .01$ both roots satisfy $|s_j| \leq 1$. This behavior is reflected in the first row of the table.

When $|s_2| > 1$ the term $c_2 s_2^n$ is a *parasitic term* that comes from the algorithm and has nothing to do with the differential equation. For the second algorithm, on the contrary, Problem 2 shows that both roots satisfy $|s_j| < 1$. This explains the superior numerical results and points the way to a general theory.

(a) Time stability

If λ is a complex constant, the problem

$$y' = \lambda y, \qquad y(0) = 1, \qquad \text{Re } \lambda \leq 0$$

is often used to test algorithms. It is called the *model problem*. The condition Re $\lambda \leq 0$ ensures that y is bounded and motivates the following definition:

DEFINITION. An algorithm is time stable for specified values (λ, h) if it produces a bounded sequence $\{y_n\}$ when applied to the model problem.

It is seen later that (λ, h) occur only in the combination λh. The set of complex values λh for which $\{y_n\}$ is bounded is called the *stability region* of the algorithm. The algorithm is said to be *absolutely stable*, or *A-stable*, if it always produces a bounded sequence when applied to the model problem. This means that the stability region contains the half plane Re $\lambda \leq 0$.

Since $n = x/h$ the condition of convergence introduced earlier resembles the condition of time stability discussed now. But there is an important difference, in that $n \to \infty$ with $h \to 0+$ in the one case while $n \to \infty$ with fixed h in the other. For example, when the Euler algorithm is applied to the model problem the first of the following equations expresses the fact that the algorithm converges while the second must be investigated in the study of time stability:

$$\lim_{h \to 0+} (1 + \lambda h)^{x/h} = e^{\lambda x}, \qquad \lim_{n \to \infty} (1 + \lambda h)^n = ?$$

EXAMPLE 1. For the Euler and backward Euler methods the choice $f(t, y) = \lambda y$ given by the model problem yields

$$y_{n+1} = y_n + \lambda h y_n, \qquad\qquad y_{n+1} = y_n + \lambda h y_{n+1}$$

respectively. The substitution $y_n = s^n$ leads to

$$s = 1 + \lambda h, \qquad s = \frac{1}{1 - \lambda h}$$

with the corresponding stability regions $|1 + \lambda h| \leq 1$, $|1 - \lambda h| \geq 1$.

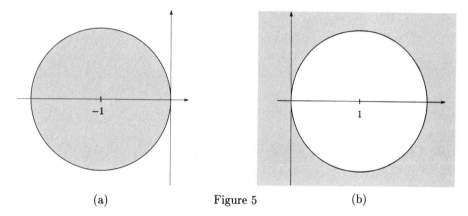

(a)   Figure 5   (b)

These denote respectively the interior and the exterior of a circle as shown by the shaded regions in Figure 5. The region in Figure 5(b) contains the left half plane and that in Figure 5(a) does not. Hence the backward Euler algorithm is A-stable and the explicit Euler algorithm is not. Only now, at this late point in our exposition, can we explain the reason for using implicit algorithms: They are used to ensure time stability.

(b) Two theorems

If the general linear multistep formula of Section 3 is applied to the model problem, the result is

$$y_n + \alpha_1 y_{n-1} + \cdots + \alpha_q y_{n-q} = \lambda h (\beta_0 y_n + \beta_1 y_{n-1} + \cdots + \beta_q y_{n-q}).$$

It is easily checked that $y_n = s^n$ is a solution if, and only if, $\rho(s) = \lambda h \sigma(s)$. The discussion leading to Section 3, Theorem 3 now yields:

THEOREM 1. *A linear multistep formula is time stable for a particular λh if, and only if, the equation $\rho(s) = \lambda h \sigma(s)$ has the following properties: All roots satisfy $|s| \leq 1$, and all roots with $|s| = 1$ are simple.*

Though the proof will not be given here, it follows from Theorem 1 that every A-stable LMF must be implicit and of order at most 2. This result is due to Dahlquist.

To discuss the Runge-Kutta algorithm, we introduce vector-matrix notation as suggested by Section 4, Example 3. Namely,

$$Y_{n+c} = \text{col}\,(y_{n+c_j}), \qquad F_{n+c} = \text{col}\,(f_{n+c_j}), \qquad B = \text{row}\,(b_j)$$

where the prefix "col" or "row" denotes a column vector or row vector, respectively. By Y_n we mean the r-dimensional column vector whose elements are y_n; this agrees with Y_{n+c} when $c = 0$. Note also that $Y_n = Uy_n$ where U denotes the r-dimensional column vector

$$U = \text{col}\,(1, 1, \cdots, 1)$$

With $A = \text{matrix}(a_{ij})$ the main Runge-Kutta formulas AC are

$$Y_{n+c} = Y_n + hAF_{n+c}, \qquad y_{n+1} = y_n + hBF_{n+c}.$$

The last term is a 1-by-1 matrix that has been identified with its single element; $(\alpha) = \alpha$ when α is a scalar. For the case $f(t, y) = \lambda y$ these results take the form

$$Y_{n+c} = Y_n + h\lambda AY_{n+c}, \qquad y_{n+1} = y_n + h\lambda BY_{n+c}.$$

If $I - h\lambda A$ is nonsingular the first equation yields

$$Y_{n+c} = (I - h\lambda A)^{-1}Y_n.$$

Substituting into the second equation and using $Y_n = Uy_n$, we get

$$y_{n+1} = y_n + h\lambda B(I - h\lambda A)^{-1}Uy_n.$$

With $\tau(s) = 1 + sB(I - sA)^{-1}U$ the result becomes

$$y_{n+1} = y_n\tau(h\lambda), \quad \text{hence} \quad y_n = \tau(h\lambda)^n.$$

Clearly $\{y_n\}$ is bounded if, and only if, $|\tau(h\lambda)| \le 1$. Hence we have:

THEOREM 2.   *For a Runge-Kutta algorithm with parameters A, B as above, the stability region is defined by $|\tau(h\lambda)| \le 1$ where the function τ is given by $\tau(s) = 1 + sB(I - sA)^{-1}U$.*

The application of Theorem 2 to specific cases is best carried out by use of the theory of functions of a complex variable and is not discussed here. We

have presented the theorem nevertheless, because it has a central importance in the deeper study of Runge-
Kutta algorithms and it provides the
principal reason for using them. Figure 6 shows the stability region for
the classical Runge-Kutta algorithm
of 1901. It is known that the order p of
a Runge-Kutta method with r stages
always satisfies $p \leq 2r$ and that A-
stable methods with $p = 2r$ exist for
each r. However, the proof is difficult
and is not discussed here.

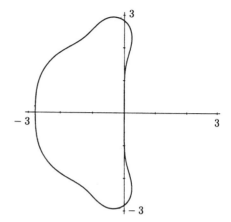

Figure 6

(c) The implementation of algorithms

Here we describe what one actually does when confronted with a differential equation that seems to require a computational approach. We say a
few words about qualitative methods, the choice of algorithm, the determination of step size, the use of vector notation, and the availability of standard
programs. It should be mentioned at the outset that if the starting point
is t_0 rather than 0, you can set $t_n = t_0 + nh$ instead of $t_n = nh$. You can
also progress in the backwards direction by letting h be negative and using
$|h|$ in estimates of error. The same effect is achieved by a change of variable
$t \to t_0 + t$ or $t \to t_0 - t$ in the initial statement of the problem.

(i) Preliminary study. Before undertaking an elaborate calculation, or even
choosing an algorithm, one should get whatever qualitative information can
be easily obtained. This preliminary step may reveal unsuspected sources
of difficulty and it also gives an idea of what to expect when the solution is
computed numerically.

In the case of a first-order scalar equation, the method of isoclines discussed in Chapter 19 is a useful source of qualitative information. It yields
short segments, called *line elements*, that are tangent to the solution at
points of the isocline. For example if

$$\frac{dy}{dx} = (y - x)^{1/3}$$

the line $y - x = 8$ gives an isocline at all points of which the slope of the
solutions is 2. By use of a parallel ruler one can draw many line elements
without the aid of a computer.

A computer allows us to construct a line element at every point of a
grid; for example at the points $(.05m, .05n)$ where m and n range over a set
of integers. One advantage of this approach is that the density of the line
elements is uniform. Such a field for the above equation is shown in Figure
7(a) and some of the trajectories are shown in Figure 7(b).

A corresponding technique for certain types of second-order equations and 2-by-2 systems leads to the concept of the phase plane. It is introduced in Chapter 7, carried further in Chapter 19, and applied to nonlinear problems in Chapter 20. The main idea is to linearize by expansion in Taylor series near the stationary points, thus obtaining a local approximation. Further examples of local approximation can be found in the problems of Chapter 4, Section 2 and in Chapters 12 and 13.

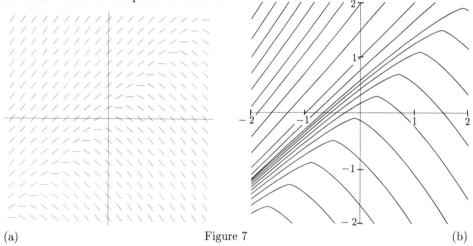

(a)   Figure 7   (b)

(ii) The choice of algorithm. As a general rule it is best to use A-stable algorithms for stiff problems, even though A-stable algorithms are usually implicit. The extent to which a problem is stiff can sometimes be explored by trying two methods, one time stable and the other not. For example, if the backward Euler method is much better than the standard Euler method with the same h, the equation is probably stiff. Implicit algorithms require a subroutine, often based on iteration, to solve the implicit equations. Since this requirement increases the complexity of the program and also increases the time of calculation, the use of implicit algorithms is more a matter of necessity than convenience.

One must also choose between linear multistep formulas and formulas of Runge-Kutta type. The former are usually better if f is costly to evaluate. The reason is that, after we get started, each step of an LMF requires only a single new evaluation. All other evaluations needed were already found in the previous step. By contrast, the evaluations of f in one step of a Runge-Kutta method cannot, in general, be used in the next.

A disadvantage of multistep formulas is that a separate one-step formula, called a *driver*, is needed to get started. Either the Euler method with a small step size or a Runge-Kutta method is commonly used as a driver. Summing up, we can say that LMFs are often faster than Runge-Kutta methods, but they are harder to start and are not as time stable.

(iii) Adaptive step size. Once an algorithm has been chosen, the step size for desired accuracy can be estimated by the order of the algorithm and determined by a stopping procedure. For example, keep dividing h by 2 until the last two computed results agree within a prescribed tolerance. Richardson extrapolation on these values then yields the final result. It should be emphasized, however, that you might have to reduce h so much that roundoff errors overwhelm the solution. In that case the problem may be intractable with the approach you are using.

In many problems it is best to let the step size vary with t. This leads to the important concept of *adaptive step size*. A popular method is to use a smaller step size near values t where $y'(t)$ changes rapidly, and to estimate the rate of change of $y'(t)$ by

$$y''(t) \approx \frac{y'(t_{n+1}) - y'(t_n)}{t_{n+1} - t_n} \approx \frac{f_{n+1} - f_n}{t_{n+1} - t_n}, \qquad t = t_n.$$

The program chooses h proportional to the estimated values of $|1/y''(t)|$ or $|y(t)/y''(t)|$, with an upper and lower threshold to preclude values that are too large or too small. Richardson extrapolation can be used when the step size varies in known, controlled amounts; but when the variation is erratic and unpredictable, extrapolation is best abandoned.

(iv) Truncation error. Another type of adaptive step size depends on an estimation of truncation error. The general technique is the same as in Richardson extrapolation but the goal is different. We illustrate in connection with the classical Runge-Kutta formula, which is one of the most popular and most important examples.

By Section 4, Part (c) the local truncation error $T(h)$ for the classical Runge-Kutta algorithm has the form

$$T(h) = ch^5 + O(h^6).$$

In view of our assumptions regarding existence of Taylor series, the quantity $c = c(h)$ is Lipschitzian. Any change in c on the interval $(t, t + 2h)$ is $O(h)$, and this is absorbed by the term $O(h^6)$. In the following discussion, therefore, c is treated as constant.

Let us apply the classical Runge-Kutta formula in one step, and also in two steps, to go from t to $t + 2h$. If $x \doteq y$ means $x = y + O(h^6)$ the two procedures yield

$$y(t + 2h) \doteq y(t, 2h) + (2h)^5 c, \qquad y(t + 2h) \doteq y(t, h) + 2h^5 c.$$

The second equation expresses the fact that the truncation errors ch^5 in going from t to $t + h$ and from $t + h$ to $t + 2h$ are the same within $O(h^6)$. Subtraction yields the following formula for the truncation error ch^5 in terms of computed quantities:

$$T(h) = ch^5 \doteq \frac{y(t, h) - y(t, 2h)}{30}.$$

The new step size h_0 is so determined that $T(h_0)$ has a prescribed value δ_0. If $\delta = T(h) = ch^5$ then

$$\frac{\delta_0}{\delta} = \left(\frac{h_0}{h}\right)^5, \quad \text{hence} \quad h_0 = h\left(\frac{\delta_0}{\delta}\right)^{1/5}.$$

The quantity $\delta = T(h)$ is known from the previous calculation and δ_0 is known because it was prescribed at the start.

In practice one would multiply h_0 by a factor like .8, for safety, and one would introduce an upper and lower threshold to keep the algorithm from choosing values that are too large or too small. Also h might be increased instead of being decreased. Similar remarks apply to other cases discussed below.

The equations could be used, instead, to carry out a Richardson extrapolation at each step. Such a procedure is called *active extrapolation* to distinguish it from extrapolation done only at the end of the whole calculation. As a rule, however, an estimate of the truncation error is of more value than the extra power of h given by extrapolation. The use of an adaptive step size as described above can easily speed up a calculation by a factor of 100 and can change an intractable problem into a problem that can be handled with ease.

(v) The Fehlberg procedure. An interesting method of estimating truncation error was proposed by Fehlberg in a series of papers starting around 1960. The basic idea is to use two Runge-Kutta methods of orders p and $p+1$, the second method being based, as far as possible, on the same evaluations of f as the first. The greater accuracy of the second method gives an estimate for the truncation error in the first, and the step size is reduced until this meets a prescribed tolerance.

We illustrate by developing the Fehlberg R_{23} method, which uses Runge-Kutta algorithms of orders 2 and 3. With $t = t_n$ and $y = y_n$ let

$$k_1 = f(t, y), \qquad k_2 = f(t + h, y + hk_1), \qquad k_3 = f(t + \frac{h}{2}, y + \frac{h}{4}k_1 + \frac{h}{4}k_2).$$

The evaluations of f needed for k_1 and k_2 are common to both of the algorithms

$$y_{n+1} = y + \frac{h}{2}k_1 + \frac{h}{2}k_2 + T(h) + \mathrm{O}(h^4)$$

$$y_{n+1} = y + \frac{h}{6}k_1 + \frac{h}{6}k_2 + \frac{2h}{3}k_3 + \mathrm{O}(h^4).$$

The first of these is Heun's method, which is of order 2. In Section 4, Example 4 it was seen that the second algorithm has order 3. By subtraction we get the following estimate for the truncation error $T(h) = ch^3$ of the first, less accurate, algorithm:

$$-T(h) = \frac{1}{3}hk_1 + \frac{1}{3}hk_2 - \frac{2}{3}hk_3 + \mathrm{O}(h^4).$$

We reduce h until this meets a prescribed tolerance, and the final result for that step is then based on the second, more accurate algorithm. A program based on this procedure is shown below:

input t_0, t_f, y_0, $f(t,y)$, h_{max}, h_{min}, h_0, Δ
$t = t_0$, $\quad y = y_0$
while $(t < t_f \text{ and } h \geq h_{min})$ {
$\quad$ **if** $t + h > t_f$ **then** $h = t_f - t$
$\quad k_1 = f(t,h)$, $\quad k_2 = f(t+h, y+hk_1)$, $\quad k_3 = f(t+h/2 \, y+h*(k_1+k_2)/4)$
$\quad \delta = (h/3) * |k_1 + k_2 - 2k_3|$
$\quad \tau = \Delta * \max(|y|, 1)$
$\quad$ **if** $\delta \leq \tau$ **then** { $t = t + h$, $\quad y = y + h * (k_1 + k_2 + 4k_3)/6$ }
$\quad$ **if** $\delta \neq 0$ **then** $h = \min(h_{max}, 0.9 * h * (\tau/\delta)^{1/3})$ $\quad\quad$ }
if $(t < t_f)$ **then print** "$h < h_{min}$, cannot converge to desired accuracy"

Here h_0 = initial step size, Δ = desired accuracy (positive), and $|\cdot|$ may be a vector norm. If all goes well the program will terminate with the final value of the variable y equal to $y(t_f)$ to an accuracy of Δ.

(vi) Vector notation. So far, we have assumed that all functions are real or complex valued. In Chapter 14 results originally established for scalar equations are extended to vector equations with no change in notation and with only minor change in the proofs. These methods apply to this chapter. Hence the algorithms illustrated here apply virtually without change to higher-order equations and to systems. All that is needed is a computer program capable of handling vector algebra, and such programs are commonplace.

(vii) Standard programs. So much reliable software for solving differential equations has been developed over the years that it seems pointless to write your own, except for personal edification or for specialized problems. In general, most of this standard software uses adaptive step sizes and a combination of algorithms to handle both stiff and nonstiff problems. Popular programs can be found in the NAG, SLATEC, and IMSL libraries. For more information consult:

IMSL, Inc., 2500 Park West Tower One, 2500 City West Blvd., Houston, Texas 77042–3020. Tel 800-222-IMSL

Jack J. Dongarra and Eric Grosse, *Distribution of mathematical software via electronic mail*, Communications of the ACM, **30**, 5, May 1987, pp. 403–407.

Good programs can also be obtained free of charge by electronic-mail request to the NETLIB operated by Bell Labs.

It is, in part, because of the existence of these services that the design of programs has not been emphasized here. We have tried instead to impart a solid theoretical background, so that you can understand what the designer of a program is trying to do.

PROBLEMS————————————————————————————————

Except in the case of Problem 5, the remarks introducing the problem set in Section 3 apply here. Problems 6–17 show how results from other parts of the text are used in numerical analysis.

1C (fghij) Apply the last five methods of the table (Section 2) to find $y(1)$ if $y' = -20y$, $y(0) = 1$. Take $h = .2, .1, .05, .01$. The answer is given for $h = .01$ only.

2. For the second algorithm (3), write the difference equation corresponding to $f(t, y) = -100y$ and show that the roots s of the associated quadratic satisfy $(s - 1)(3s - 1) = -200hs^2 < 0$. It follows that $1/3 < s < 1$.

3. (bcd) Referring to Section 2, apply the midpoint, trapezoid and Heun formulas to the model problem. Then set $y_n = s^n$ and get the following equations for s in cases (bcd) respectively:

$$s^2 = 1 + 2\lambda hs \qquad s(2 - \lambda h) = (2 + \lambda h) \qquad s = 1 + \lambda h + \frac{1}{2}(\lambda h)^2$$

In (b) the product of the roots is 1, hence for stability $|s_1| = |s_2| = 1$. Let $s = e^{i\phi}$ and deduce that the stability region consists of the segment joining $-i$ to i. This shows that the midpoint method is far from being absolutely stable.

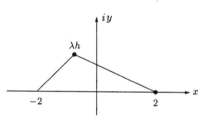

Figure 8

In (c) deduce from Figure 8 that the stability region is the half plane Re $\lambda \leq 0$. It follows that the trapezoid formula is absolutely stable, but only barely.

In (d) show that the choice $\lambda = i\alpha$ with $\alpha \neq 0$ gives $|s| > 1$ for all $h > 0$, hence the Heun method is not absolutely stable. The stability regions for these three methods are shown in Figure 9.

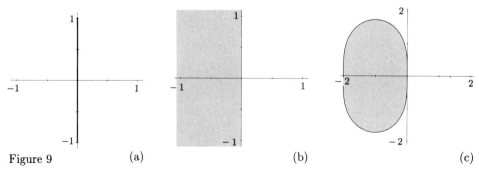

Figure 9   (a)   (b)   (c)

4C. Verify the results in the table (a) for AB2 and (b) for BD2.

5C. Open-ended problem; no answer is given. Apply selected algorithms to the nonlinear equation considered in Problem 11 below and compare your results with the polynomial approximation given there. The exact solution can be expressed in terms of Bessel functions but is not elementary.

6. Monotone operators. This uses the notation and the method of proof of Chapter 10, Section 3, Theorem 2. Given $f(x, p, q \uparrow)$ and $g(x, q \uparrow)$, let F and G be strictly increasing and define

$$Tu = F(u) - f(x, u', u''), \qquad Ru = G(u) - g(x, u_\nu)$$

Show that $Tu \leq Tv$, $Ru \leq Rv \Rightarrow u \leq v$. Operator pairs (T, R) with this property are said to be monotone in the sense of Collatz.

7. For the solution y on the left below obtain the estimate on the right:

$$y^5 - y'' = 2, \quad y(0) = y(1) = 0; \qquad 1.9995x(1 - x) \leq y \leq 2.0000x(1 - x)$$

Hint: Apply Problem 6 with $u = cx(1 - x)$, $v = y$ where c is a suitable constant. Then apply Problem 6 again with $u = y$.

8. The point of this problem and the next is to show that the analytic stability often observed for stiff problems does not depend on linearity. Assuming $t \geq 0$, show that the solution of the equation on the left satisfies the inequality on the right:

$$y' + e^y = -100(y - t) + e^t + 1, \qquad |y(t) - t| \leq |y(0)|e^{-100t}$$

Hint: In the notation of Chapter A, Problem 23, verify that

$$\frac{f(t, u) - f(t, v)}{u - v} \leq -100, \qquad u \neq v$$

9. Given $y(0) = 1 + \delta$ proceed as in Problem 8:

$$y' + y^3 = -100(y - \cos t) + y \cos^2 t - \sin t, \qquad |y(t) - \cos t| \leq \delta e^{-99t}$$

10. Emden's problem. Find the first positive zero of y if

$$xy'' + 2y' + xy = 0, \qquad y(0) = 1, \qquad y'(0) = 0$$

Hint: The indicial equation (Chapter 13) has a root $\rho = 0$, hence there is a power-series solution. This problem arises in astrophysics. The answer is π.

11. In this problem Picard's method from Chapter 14 is applied to

$$y' = t^2 + y^2, \qquad y(0) = 0, \qquad 0 \le t < 1$$

(a) Show that the third Picard approximation is

$$y_3(t) = \frac{t^3}{3} + \frac{t^7}{63} + \frac{2t^{11}}{2,079} + \frac{t^{15}}{59,535}$$

(b) By solving $t^2 + y^2 = M^2$ and $y = Mt$ simultaneously as suggested by Figure 10, check that the following values for M, M_0, and K can be used in the estimates of Chapter 14, Section 3, Problem 2:

$$M = \frac{t}{\sqrt{1 - t^2}}, \qquad M_0 = t^2 \qquad K = 2Mt$$

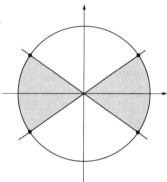

Figure 10

For $0 \le t \le .5$ and $0 \le t \le .1$, respectively, deduce that

$$\left| y(t) - \frac{t^3}{3} \right| < 1.5 \cdot 10^{-4}, \qquad \left| y(t) - \frac{t^3}{3} - \frac{t^7}{63} \right| < 4 \cdot 10^{-13}$$

12. (a) As $t \to 0$, Problem 11 shows that $|y(t) - y_n(t)|$ has the order of magnitude of $|t|^{3n+3}$. Furthermore $y_n(t)$ is a polynomial. Using these two facts show that $y_n(t)$ must agree with the Taylor expansion of $y(t)$ about $t = 0$ up to and including the term in t^{3n+2}.

(b) The Taylor expansion can be obtained by setting $t = 0$ in

$$y' = t^2 + y^2, \qquad y'' = 2t + 2yy', \qquad y''' = 2 + 2yy'' + 2(y')^2$$

and so on. Continue this process and show that the Taylor expansion agrees with $y_3(t)$ from Problem 11 through the term $t^7/63$.

13. The Taylor expansion of Problem 12 can be obtained by substituting $y = a_0 + a_1 t + a_2 t^2 + \cdots$ and equating coefficients. Thus obtain the four values $a_0 = a_1 = a_2 = 0$, $a_3 = 1/3$. Using this result to simplify the calculations, show that $a_4 = a_5 = a_6 = 0$ and find a_7. Using this to simplify the calculations, verify that your series agrees with $y_3(t)$ from Problem 11 through the term in t^{11}, as it should by Problem (12a).

14. (abcdefghij) This requires familiarity with Chapter 14, Section 4 and uses the same notation. With $y^1 = u$, $y^2 = u'$ check that the second-order problem on the left becomes the first-order problem on the right:

$$u'' + u = t$$
$$u(0) = 0, \ u'(0) = 1 \qquad \begin{pmatrix} y^1 \\ y^2 \end{pmatrix}' = \begin{pmatrix} y^2 \\ t - y^1 \end{pmatrix}, \quad \begin{pmatrix} y^1 \\ y^2 \end{pmatrix}_0 = \begin{pmatrix} 0 \\ 1 \end{pmatrix}$$

The first two steps of Euler's algorithm in this case are:

$$\begin{pmatrix} y^1 \\ y^2 \end{pmatrix}_1 = \begin{pmatrix} 0 \\ 1 \end{pmatrix} + h \begin{pmatrix} 1 \\ 0 \end{pmatrix}, \qquad \begin{pmatrix} y^1 \\ y^2 \end{pmatrix}_2 = \begin{pmatrix} h \\ 1 \end{pmatrix} + h \begin{pmatrix} 1 \\ 0 \end{pmatrix}$$

Study the above equations until you understand them. Then, using the same notation, show that the ten algorithms in the table of Section 2 are all satisfied by $y_n = (nh, 1)^T$, $f_n = (1, 0)$.

15C. (abcdefghij) Program the algorithms in the table of Section 2 to solve the equation of Problem 14 with $y(0) = y_0$ and check by Problem 14.

16C. (abcdefghij) Use your program from Problem 15 with $h = .01$ to estimate $y(2)$ in Problem 14 when $y(0) = (1, 1)^T$.

17. The concept of "solution" is generalized in Chapter 2, Section 2 and also in Chapter G. With this as a clue, show by any method that the equation $|x|(y - |x|)\, dy = x(y - 2|x|)\, dx$ leads to the figure at the bottom of this page.

───────────────────────**ANSWERS**───────────────────────

1. $10^9\, y(1, .01) = 2.09186$ 2.05915 2.05654 $5.89791 \cdot 10^9$ 2.06180

16. (abcd) $\begin{pmatrix} 1.593935 \\ .09072106 \end{pmatrix}$ $\begin{pmatrix} 1.583823 \\ .09071644 \end{pmatrix}$ $\begin{pmatrix} 1.583868 \\ .09079564 \end{pmatrix}$ $\begin{pmatrix} 1.583823 \\ .09071622 \end{pmatrix}$

(efg) $\begin{pmatrix} 1.58385316335 \\ .090702573220 \end{pmatrix}$ $\begin{pmatrix} 1.58385316974 \\ .090702570414 \end{pmatrix}$ $\begin{pmatrix} 1.58385316298 \\ .090702573388 \end{pmatrix}$

(hij) $\begin{pmatrix} 1.58385316297 \\ .090702573362 \end{pmatrix}$ $\begin{pmatrix} 1.58385316486 \\ .090702571534 \end{pmatrix}$ $\begin{pmatrix} 1.58385316361 \\ .090702573106 \end{pmatrix}$

Chapter A
SOLUTION BY QUADRATURE

When the solution of a differential equation is expressed by a formula involving one or more integrations, it is said that the equation is solvable by quadrature, and the formula is called a closed-form solution. The term "quadrature" has its historical origin in the connection of integration with area; in plane geometry, a problem of quadrature, such as *the quadrature of the circle*, is a problem about the area of a plane figure.

Not all differential equations can be solved by quadrature. For example the second-order equation

$$y'' + p(x)y' + q(x)y = f(x)$$

cannot be so solved, in general, even if $p = f = 0$. Except in special cases, the *Riccati equation*

$$y' = A(x) + B(x)y + C(x)y^2$$

also cannot be solved by quadrature.

When we do have a closed-form solution, the problems of existence and uniqueness are greatly simplified. If the formula produces a solution, it provides an existence proof; and if every solution is given by the formula, that fact leads to uniqueness. Several methods of solution by quadrature were explained and illustrated in Chapters 1–6. These methods are now discussed from a more general point of view.

(a) The first-order linear equation

We begin with an example that is already familiar.

EXAMPLE 1. Let $g(x)$ be continuous on an open interval I and let $x_0 \in I$. Then the initial-value problem

$$\frac{du}{dx} = g(x),\ x \in I; \qquad u(x_0) = y_0$$

has one, and only one solution. The solution exists on the whole interval I and is given by the formula

(1) $$u = \int_{x_0}^{x} g(\xi)\, d\xi + y_0.$$

That u in (1) satisfies $u(x_0) = y_0$ is obvious, and $du/dx = g(x)$ follows from the fundamental theorem of calculus. Hence (1) gives existence. To show uniqueness, let v be another solution. The differential equation gives $(u - v)' = 0$ on I, hence $u - v$ is constant. Since $u(x_0) = v(x_0)$, the constant value of $u - v$ is 0. Thus $u = v$.

Example 1 is the special case $p = 0$ of the following:

THEOREM 1.   *Let $p(x)$ and $f(x)$ be continuous on an open interval I containing the point x_0. Then the initial-value problem*

$$y' + p(x)y = f(x), \qquad y(x_0) = y_0$$

has one, and only one, solution. The solution exists on the whole interval I and is given by

$$y = e^{-P(x)} \left(\int_{x_0}^{x} e^{P(t)} f(t) \, dt + y_0 \right) \quad \text{where} \quad P(x) = \int_{x_0}^{x} p(t) \, dt.$$

For proof, we use the method of variation of parameters. The function

$$v(x) = e^{-P(x)}$$

satisfies $v' + pv = 0$ and $v(x_0) = 1$. Also v does not vanish anywhere on I. If we set $y = uv$ the differential equation $y' + py = f$ becomes

$$u'v + uv' + puv = f, \qquad u'v = f, \qquad u' = \frac{f}{v}.$$

Here the second equality follows from $v' + pv = 0$. Thus

$$\frac{du}{dx} = e^{P(x)} f(x), \qquad u(x_0) = \frac{y(x_0)}{v(x_0)} = y_0.$$

Example 1 gives u and $y = uv$ then yields the formula of Theorem 1.

(b) The separated equation

Here and below, R is the rectangular region defined by

$$a < x < b, \qquad c < y < d$$

with constants a, b, c, d satisfying $a < b$, $c < d$. A function $f(x)$ given initially for $a < x < b$ can be extended in a natural way to be defined on R, and similarly for $g(y)$. Namely,

$$f(x, y) = f(x), \qquad g(x, y) = g(y), \qquad (x, y) \in R.$$

This extension simplifies the statement of Theorem 2.

THEOREM 2. *Let $f(x)$ and $g(y)$ be continuous for $(x,y) \in R$ and suppose further that f and g are not simultaneously zero at any point of R. Then the equation $f(x)\,dx = g(y)\,dy$ has one, and only one, solution through each point $(x_0, y_0) \in R$. The solution is given implicitly by*

$$(2) \qquad\qquad \int_{x_0}^{x} f(\xi)\,d\xi = \int_{y_0}^{y} g(\eta)\,d\eta$$

and can be expressed on its entire interval of definition in the form $y = \phi(x)$ or in the form $x = \psi(y)$.

It is essential that f and g do not vanish simultaneously. For example, $x\,dx = -y\,dy$ has no solution through the origin and $x\,dx = y\,dy$ has two.

The proof depends on the following theorems from calculus:

Theorem A. *If $x = G(y)$ has a positive derivative on an interval J then the inverse function $y = G^{-1}(x)$ exists and has a positive derivative for x on the corresponding interval I.*

THEOREM B. *If a continuous real-valued function changes sign on an interval then it has a zero on the interval.*

To establish Theorem 2, note that either $f(x) \neq 0$ for $a < x < b$, or $g(y) \neq 0$ for $c < y < d$. Namely, if $f(x) = 0$ at a point x_1 of the first interval and $g(y) = 0$ at a point y_1 of the second, then f and g are both 0 at (x_1, y_1). This contradicts the hypothesis that f and g are not 0 simultaneously at any point of R. Theorem B shows that a continuous function that does not vanish on an interval must be positive or negative on the whole interval. Since the roles of x and y can be interchanged, and since the equation can be multiplied by -1, we assume without loss of generality that

$$(3) \qquad\qquad g(y) > 0 \quad \text{for} \quad c < y < d.$$

Let the integrals in (2) be denoted by $F(x)$ and $G(y)$ respectively, so that (2) becomes

$$(4) \qquad\qquad F(x) = G(y).$$

The fundamental theorem of calculus gives $G' = g$ and $g > 0$ by (3). Theorem A shows that G has a differentiable inverse G^{-1}; hence (4) can be written in the form

$$y = G^{-1}(F(x))$$

which shows that dy/dx exists. Differentiating (4) we get

$$F'(x) = G'(y)\frac{dy}{dx}, \quad \text{or} \quad f(x) = g(y)\frac{dy}{dx}.$$

Thus $f(x)dx = g(y)dy$. Since (2) gives $y = y_0$ when $x = x_0$ we have established existence of a solution in the explicit form $y = \phi(x)$.

To establish uniqueness let y be one solution and z another. Under the same hypothesis on g as above, the equation

$$\frac{dz}{dx} = \frac{f(x)}{g(z)}$$

shows that dz/dx exists (and is in fact continuous) for $(x, z) \in R$. Let

(5) $$u = G(y), \qquad v = G(z).$$

The differential equation yields

$$\frac{du}{dx} = G'(y)\frac{dy}{dx} = g(y)\frac{dy}{dx} = f(x)$$

and similarly, $dv/dx = f(x)$. Since u and v have the same derivative, their difference $u - v$ is constant on any interval on which both functions exist. By the initial conditions $u = v$ at x_0, hence the constant is 0. The equation $y = z$ now follows from (5).

If the hypothesis concerning continuity and simultaneous zeros of f and g holds in the closed rectangle $a \leq x \leq b$, $c \leq y \leq d$ the solution can be extended in both directions from (x_0, y_0) until it meets a definite point of the boundary as illustrated in Figure 1. This follows from the integral formulas and also from results of Chapter 14.

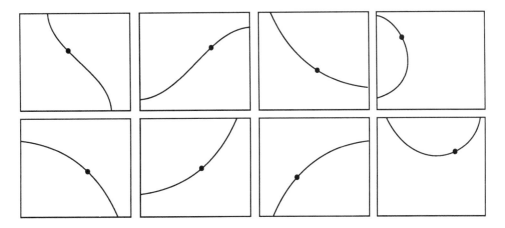

Figure 1

(c) A limit theorem

Let $m \geq 1$ be constant, let (α, β) be an interval, and consider the equation

$$\frac{dy}{dx} = h(y)(\beta - y)^m, \qquad x > 0; \qquad \alpha < y(0) < \beta.$$

We assume that $h(y)$ is continuous and positive for $\alpha < y \leq \beta$. As long as y remains on the interval (α, β) the positivity of $h(y)$ ensures that y is increasing. Thus (α, β) is an *interval of attraction* for β in the sense of Chapter 4, Section 3. The following theorem completes and generalizes the discussion given there:

THEOREM 3. *Under the above conditions,* $\lim\limits_{x \to \infty} y(x) = \beta.$

According to Chapter 10, Section 3, Theorem A the function $h(y)$ has a positive minimum M_1 and a positive maximum M_2 on the interval $y(0) \leq y \leq \beta$. Thus, as long as $y(0) \leq y < \beta$, we have

$$(6) \qquad\qquad M_1(\beta - y)^m \leq \frac{dy}{dx} \leq M_2(\beta - y)^m.$$

Dividing by $(y - \beta)^m$ and integrating between corresponding limits yields

$$(7) \qquad\qquad M_1 x \leq \int_{y(0)}^{y(x)} \frac{d\eta}{(\beta - \eta)^m} \leq M_2 x.$$

The left-hand inequality (7) shows that the upper limit $y(x)$ of the integral cannot stay less than some constant $\beta_0 < \beta$ as $x \to \infty$. Since the integral with $y = \beta$ is divergent, the right-hand inequality shows that $y(x)$ cannot reach the value β at any finite value of x. These remarks give Theorem 3.

The foregoing discussion shows that differential inequalities, as well as equations, can be "solved" by quadrature. However, the process requires caution. It is easily proved, by differentiation of (7), that (6) implies (7); but (7) does not even imply that y is a differentiable function of x.

(d) Homogeneous equations

A function $f(x, y)$ is *homogeneous of degree* m if the following holds whenever (x, y) is in the domain of f:

$$f(tx, ty) = t^m f(x, y), \qquad t > 0.$$

If P and Q are homogeneous functions of the same degree, and P/Q has a nonempty domain of definition, the calculation

$$\frac{P(tx, ty)}{Q(tx, ty)} = \frac{t^m P(x, y)}{t^m Q(x, y)} = \frac{P(x, y)}{Q(x, y)}$$

shows that $g = P/Q$ is homogeneous of degree 0. Setting $t = 1/x$ then gives

$$g(x, y) = g(tx, ty) = g\left(1, \frac{y}{x}\right) = f\left(\frac{y}{x}\right)$$

for $x > 0$, where $f(s) = g(1, s)$. Hence, under the above conditions, the following are essentially equivalent:

$$P(x, y)\, dx = Q(x, y)\, dy, \qquad \frac{dy}{dx} = f\left(\frac{y}{x}\right).$$

Both equations are called *homogeneous*. Note, however, that this use of the term is different from its use in connection with a linear equation $Ty = 0$.

THEOREM 4. *Let $y = \phi(x)$ satisfy $y' = f(y/x)$ on an interval I. Then:*

(i) *If $c \neq 0$ the function z given by $cz = \phi(cx)$ satisfies $z' = f(z/x)$ on a corresponding interval I_c.*

(ii) *If $x \neq 0$ and $f(u) \neq u$ the substitution $u = y/x$ gives*

$$\frac{du}{f(u) - u} = \frac{dx}{x}.$$

Proposition (i) follows from $z' = \phi'(cx) = f(\phi(cx)/cx) = f(z/x)$. To get (ii) note that $y = xu$ yields $y' = xu' + u = f(u)$. Solve for $u' = du/dx$ and separate variables. The procedure fails at a point u_0 where $u_0 = f(u_0)$, but there is the obvious solution $y = u_0 x$ in that case.

EXAMPLE 2. If $xy \neq x^2$ the equation $y'(xy - x^2) = y^2$ can be written

$$\frac{dy}{dx} = \frac{y^2}{xy - x^2} = \frac{(y/x)^2}{(y/x) - 1}.$$

Thus $f(u) = u^2/(u - 1)$. The equation $u = f(u)$ has $u = 0$ as sole solution, and this gives $y = 0$. When $xu \neq 0$ the result (ii) yields

$$\frac{u - 1}{u}\, du = \frac{dx}{x}, \qquad u - \ln|u| = \ln|x| + c_1, \qquad u = \ln|cxu|$$

where $\ln|c| = c_1$. Choosing the sign of c to agree with the sign of xu, we drop the absolute value. Since $u = y/x$ the final result is

$$y = x \ln(cy) \quad \text{or} \quad cy = cx \ln(cy).$$

The second form agrees with Theorem 1(i) and indicates that all these implicitly defined curves are similar to the single curve obtained when $c = 1$.

(e) Exact differentials

In Parts (efg) we supplement the theory of exact equations, implicit solutions, and integrating factors as outlined in Chapter 2, Section 2. It is assumed that the functions P, Q, F are defined on the rectangle R of Part (b) and that

$$(x_0, y_0) \in R, \qquad (x, y) \in R.$$

Partial derivatives are denoted by subscripts; thus,

$$P_x = \frac{\partial P}{\partial x}, \qquad Q_y = \frac{\partial Q}{\partial y}, \qquad F_{xy} = \frac{\partial^2 F}{\partial y \partial x}.$$

The last equation agrees with the convention that $F_{xy} = (F_x)_y$.

An expression $\omega = P\,dx + Q\,dy$ is said to be *exact* in R if there exists a function F with continuous partial derivatives such that $\omega = dF$. In other words

(8) $$P(x, y)\,dx + Q(x, y)\,dy = F_x(x, y)\,dx + F_y(x, y)\,dy$$

for $(x, y) \in R$, where x, y, dx, and dy are independent variables. If ω is exact, the equation $\omega = 0$ is also said to be exact. Taking first $dx = 1$, $dy = 0$ and then $dx = 0$, $dy = 1$ we see that (8) holds if, and only if,

(9) $$P(x, y) = F_x(x, y), \qquad Q(x, y) = F_y(x, y), \qquad (x, y) \in R.$$

This leads to the following:

THEOREM 5. *Suppose P and Q have continuous first partial derivatives in R. Then $P\,dx + Q\,dy$ is exact in R if, and only if, $P_y = Q_x$. In that case $P\,dx + Q\,dy = dF$ where*

(10a) $$F(x, y) = \int_{x_0}^{x} P(\xi, y)\,d\xi + \int_{y_0}^{y} Q(x_0, \eta)\,d\eta.$$

The formula (10a) is an example of a *line integral;* the path of integration is shown in Figure 2(a). The path in Figure 2(b) leads to

(10b) $$F(x, y) = \int_{x_0}^{x} P(\xi, y_0)\,d\xi + \int_{y_0}^{y} Q(x, \eta)\,d\eta$$

and if $(x_0, y_0) = (0, 0)$ the path in Figure 2(c) gives

(10c) $$F(x, y) = x \int_{0}^{1} P(tx, ty)\,dt + y \int_{0}^{1} Q(tx, ty)\,dt.$$

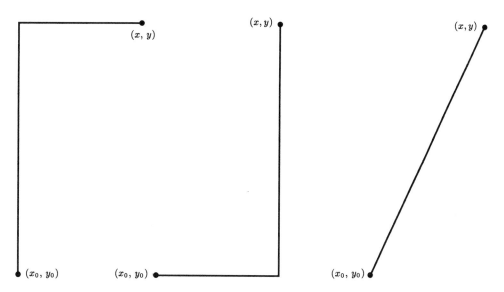

Use of these equations does not presuppose familiarity with the theory of line integrals. However, for readers who have such familiarity, we mention that the general formula is

$$F(x,y) = \int_{(x_0,y_0)}^{(x,y)} (P\,dx + Q\,dy)$$

and instead of being a rectangle, R can be any simply-connected region. The concept of simple connectivity is discussed in Chapter 20, Section 5.

Theorem 5 depends on the following results from calculus:

THEOREM C. *The equation $F_{xy} = F_{yx}$ holds at any point where both these derivatives are continuous.*

THEOREM D. *If P and P_y are continuous for $(x,y) \in R$ then*

$$\frac{\partial}{\partial y} \int_{x_0}^{x} P(\xi, y)\,d\xi = \int_{x_0}^{x} P_y(\xi, y)\,d\xi.$$

To prove Theorem 5, suppose $P\,dx + Q\,dy$ is exact. Then (9) gives

$$P_y = F_{xy}, \qquad Q_x = F_{yx}$$

and these are equal by Theorem C. Conversely, if $P_y = Q_x$ let F be defined by (10a). Then the fundamental theorem of calculus gives $F_x = P$. Using both Theorem D and the fundamental theorem of calculus, we get

$$F_y(x,y) = \int_{x_0}^{x} P_y(\xi, y)\,d\xi + Q(x_0, y).$$

Since $P_y = Q_x$ this yields the desired relation $F_y = Q$, as follows:

$$F_y(x, y) = \int_{x_0}^{x} Q_x(\xi, y)\, d\xi + Q(x_0, y) = Q(x, y) - Q(x_0, y) + Q(x_0, y).$$

EXAMPLE 3. The expression $(2xy + 1)\, dx + (x^2 + 4y)\, dy$ is exact since both P_y and Q_x are equal to $2x$. Equation (10a) with $x_0 = y_0 = 0$ gives

$$F(x, y) = \int_0^x (2\xi y + 1)\, d\xi + \int_0^y 4\eta\, d\eta = x^2 y + x + 2y^2.$$

EXAMPLE 4. Solve Example 3 by indefinite integrals. Since $F_x = P$ we try

$$F(x, y) = \int (2xy + 1)\, dx = x^2 y + x + C(y)$$

where the constant of integration C can depend on y. The equation $F_y = Q$ leads to $x^2 + C'(y) = x^2 + 4y$. Hence $C(y) = 2y^2 + c$, which gives essentially the same result as before. This procedure has the advantage that it is easy to understand and easy to remember. But for theoretical discussion indefinite integrals are less convenient than line integrals, and they are very unsuitable when R is replaced, as suggested above, by a more general region.

(f) Parametric and implicit solutions

Let $x = \xi(t)$ and $y = \eta(t)$ be differentiable on an open interval I. In the notation of Theorem 5 suppose also that $P = F_x$ and $Q = F_y$, where F has continuous partial derivatives. The chain rule gives

$$\frac{d}{dt} F(x, y) = F_x \frac{dx}{dt} + F_y \frac{dy}{dt} = P \frac{dx}{dt} + Q \frac{dy}{dt}$$

and hence, for $t \in I$,

$$F(\xi, \eta) = \text{const.} \iff P(\xi, \eta)\xi' + Q(\xi, \eta)\eta' = 0.$$

When the second of the above equalities holds, it is said that (ξ, η) is a *parametric solution* of $P\, dx + Q\, dy = 0$. Parametric solutions can be obtained from an equation $F(x, y) = c$ by use of the following theorem:

THEOREM E. *Let $F(x, y)$ have continuous partial derivatives in some disk centered at (x_0, y_0) and suppose further that*

$$F(x_0, y_0) = c, \qquad F_y(x_0, y_0) \neq 0.$$

Then the equation $F(x, y) = c$ has a unique solution $y = \phi(x)$ on some open interval I containing x_0, and $\phi'(x)$ is continuous on I. A corresponding statement holds with the roles of x and y interchanged.

By Theorem E, we can solve the equation $F(x, y) = 0$ for $y = \phi(x)$ or for $x = \psi(y)$, according as $F_y \neq 0$ or $F_x \neq 0$ respectively. One or the other of these conditions is always available at any given point (x_0, y_0) if

$$(11a) \qquad\qquad (F_x)^2 + (F_y)^2 > 0.$$

Furthermore, the chain rule shows that $y = \phi(x)$ or $x = \psi(y)$, as the case may be, actually satisfies the differential equation. Thus we get parametric solutions with $t = x$ at some parts of the solution-curve, $t = y$ at others.

Geometrically, (11a) means that the surface defined by $z = F(x, y)$ is smooth. In a like manner,

$$(11b) \qquad\qquad (\xi')^2 + (\eta')^2 > 0$$

means that the curve $x = \xi(t)$, $y = \eta(t)$ is smooth. Equation (11b) is usually imposed on parametric solutions to rule out the trivial case in which both ξ and η are constant.

(g) Integrating factors

Let us recall that a *general solution* of

$$(12) \qquad\qquad P(x, y)\, dx + Q(x, y)\, dy = 0$$

in R is a family of solutions such that at least one member of the family passes through any given point $(x_1, y_1) \in R$. The gist of the result of Part (f) is that if an exact equation has a general solution, the solution can be written in the form $F(x, y) = c$.

Under mild hypotheses a partial converse holds. Namely, if (12) has a general solution given implicitly by $F(x, y) = c$, then (12) becomes exact when multiplied by a suitable nonvanishing function r. Such a function is called an *integrating factor*. The condition $r \neq 0$ is required because otherwise an alleged "solution" might satisfy $r(P\, dx + Q\, dy) = 0$ without satisfying (12).

THEOREM 6. *Let $F(x, y) = c$ give a general solution of (12) in R. Suppose that P, Q, F have continuous partial derivatives and that neither Q nor F_y is zero at any point of R. Then F_y/Q is an integrating factor for $P\, dx + Q\, dy$.*

For proof, let (x_1, y_1) be any given point in R. By hypothesis the equation $F(x, y) = c$ defines an implicit solution through this point, and by Theorem E that solution can be written in the explicit form $y = \phi(x)$. Equation (12) and $F(x, y) = c$ give, respectively,

$$P + Qy' = 0, \qquad F_x + F_y y' = 0.$$

Multiply the first of these by F_y, the second by Q, and subtract. The result is $PF_y = QF_x$. This is equivalent to the left-hand equation below; the other is obvious:

$$\frac{F_y}{Q}P = F_x, \qquad \frac{F_y}{Q}Q = F_y.$$

These hold on the trajectory $y = \phi(x)$ and in particular, they hold at (x_1, y_1). But (x_1, y_1) was an *arbitrary* point of R. Hence the equations hold in R and we get Theorem 6. The condition $r \neq 0$ follows from $F_y \neq 0$.

(h) A second-order equation

We now discuss the substitution $p = y'$, $y'' = p\, dp/dy$ that was used in Chapter 2 to simplify the equation

$$(13) \qquad\qquad y'' = g(y, y').$$

Confusion between the role of p as an independent variable and its role as $p = dy/dx$ can be avoided by use of differentials. The property needed here is the fact that, if f is a differentiable function, $df(u)$ has the same form no matter whether u is an independent variable or is a differentiable function $\phi(x)$ of another variable x. Namely,

$$df(\phi(x)) = f'(\phi(x))\phi'(x)\, dx = f'(u)\, du.$$

Let $y = \phi(x)$ satisfy (13), so that

$$(14) \qquad\qquad \phi''(x) = g(\phi(x), \phi'(x)).$$

In differential form, the equation $p\, dp/dy = g(y, p)$ is $p\, dp = g(y, p)\, dy$. With $p = \phi'(x)$ and $dp = \phi''(x)\, dx$ this becomes

$$(15) \qquad\qquad \phi'(x)\phi''(x)\, dx = g(\phi(x), \phi'(x))\phi'(x)\, dx.$$

Since (14) and (15) differ only by the factor $\phi'(x)dx$ we are led to:

THEOREM 7.   *Let $y = \phi(x)$ and $p = \phi'(x)$. Then the equations*

$$y'' = g(y, y'), \qquad p\, dp = g(y, p)\, dy$$

are equivalent on any interval on which $\phi'(x) \neq 0$.

To say that the equations are equivalent means if either holds, both do.

In most applications the zeros of $\phi'(x)$ are isolated. Equation (14) can be deduced from (15) at all points except these isolated zeros, and if both sides of (14) are continuous, taking limits shows that (14) holds without exception. Note, however, that if $\phi(x) = c$, a constant, then (15) always holds, but (14) requires $g(c, 0) = 0$.

PROBLEMS——

Assume all functions continuous and sufficiently differentiable to justify the calculations. The letters a, b, c, sometimes with subscripts, denote constants.

1. (abc) Check that the following are exact and find F as in Example 4:

$$3x^2y^2\,dx + 2x^3y\,dy \quad (xy^2 + x)\,dx + (yx^2 + y)\,dy \quad e^{xy}(dx + xy\,dx + x^2\,dy)$$

(defgh) Assuming (x, y) suitably restricted, do the same here:

$$\frac{x\,dx + y\,dy}{\sqrt{x^2 + y^2}} \quad \frac{y\,dx - x\,dy}{y^2} \quad \frac{x^2\,dx + y^2\,dy}{x^3 + y^3} \quad \frac{x\,dy - y\,dx}{x\sqrt{x^2 - y^2}} \quad \frac{2xy\,dy - y^2\,dx}{x^2 + y^4}$$

2. (abcdefgh) Explore the use of each equation (10) in Problem 1. Take $(x_0, y_0) = (0, 0)$ when possible, otherwise $(1, 0)$, otherwise $(0, 1)$.

3. If $dy/dx = f(ax + by)$ with $b \neq 0$ show that the substitution $u = ax + by$ leads to a separable equation on any interval on which $a + bf(u) \neq 0$. Show also that if $a + bf(u_0) = 0$ a solution is given by $ax + by = u_0$.

4. (a) Verify that each of these functions is homogeneous and find its degree:

$$x^2 + y^2 \quad (x + y)(x - y)^{-1} \quad (xy)^{-1/3} \quad (x + y)^e x^\pi \quad x^2 e^{y/x} \quad \ln x - \ln y$$

(b) If $f(x, y)$ is homogeneous of degree m and has continuous partial derivatives, derive Euler's equation $xf_x + yf_y = mf$.

Hint: Differentiate the defining equation with respect to t and set $t = 1$.

5. (abcd) Assuming (x, y) suitably restricted, solve by Theorem 4:

$$xy' = x + y \quad (x^2 + y^2)y' + 2xy = 0 \quad (x - \sqrt{xy})y' = y \quad xy' - y = \sqrt{x^2 - y^2}$$

(efgh) Find the solution through $(1, 1)$:

$$(x + y)y' = x - y \quad (x^3 - y^3)y' = x^2y \quad x^2y' = xy - y^2 \quad x^3y' = 2xy^2 - y^3$$

6. Change of origin. Let $a_1x_0 + b_1y_0 + c_1 = 0$ and $a_2x_0 + b_2y_0 + c_2 = 0$. Show that the substitution $u = x - x_0$, $v = y - y_0$ reduces the following equation, in general, to a homogeneous equation $dv/du = F(v/u)$:

$$\frac{dy}{dx} = f\left(\frac{a_1x + b_1y + c_1}{a_2x + b_2y + c_2}\right)$$

7. Let P and Q be homogeneous functions of the same degree m.

(a) If $P\,dx + Q\,dy$ is exact and $m > -1$ deduce from (10c) that

$$P\,dx + Q\,dy = dF \quad \text{where} \quad F(x,y) = \frac{xP(x,y) + yQ(x,y)}{m+1}$$

(b) Deduce the result (a) for all $m \neq -1$ by use of Problem 4(b).

8. (a) When is $(a_1x^2 + b_1xy + c_1y^2)\,dx + (a_2x^2 + b_2xy + c_2y^2)\,dy$ exact, and what is F in that case?

(b) The coefficients in Part (a) are homogeneous polynomials of degree 2. With this as a clue, state and solve a similar problem in which the coefficients are homogeneous polynomials of degree 3. No answer is provided.

9. (a) Show that the homogeneous equation $dy/dx = f(y/x)$ is invariant, that is, unchanged, if (x,y) is replaced by (ax, ay) with $a \neq 0$.

(b) Show that the equation on the left below is also invariant and that $u = y/x$, $v = dy/dx$ in general satisfy the first-order equation on the right:

$$x\frac{d^2y}{dx^2} = F\left(\frac{y}{x}, \frac{dy}{dx}\right), \qquad \frac{dv}{du} = \frac{F(u,v)}{v-u}$$

Hint: $dv/dx = (dv/du)(du/dx)$. Assume denominators nonzero as needed.

(c) If m is a constant that remains fixed throughout the discussion, check that the equation on the left below is invariant under $(x,y) \to (ax, a^m y)$ for $a \neq 0$. Also show that $u = y/x^m$ has the same invariance property and that, in general, it satisfies the separated equation on the right:

$$\frac{dy}{dx} = \frac{y}{x}f\left(\frac{y}{x^m}\right) \qquad \frac{du}{uf(u) - mu} = \frac{dx}{x}$$

In a branch of analysis known as the theory of Lie groups, it is found that invariance under substitutions such as $(x,y) \to (ax, ay)$ give a profound insight into the reasons why certain equations can be solved by quadrature.

10. Theorem 6 presupposes knowledge of the solution and does not give a practical method of finding an integrating factor r. But r can often be found by use of the equation $(rP)_y = (rQ)_x$ that results from Theorem 5.

(a) Show that $P\,dx + Q\,dy$ has an integrating factor $r = x^a y^b$ if, and only if,

$$xy(P_y - Q_x) = ayQ - bxP$$

(b) Show that $r(x, y) = r(x)$ and $r(x, y) = r(y)$ lead respectively to

$$\frac{r'}{r} = \frac{P_y - Q_x}{Q}, \qquad \frac{r'}{r} = \frac{Q_x - P_y}{P}$$

If the right-hand side is a function of x alone in the first case, or of y alone in the second, integration gives $\ln r$ and exponentiation yields r.

(c) If the equation on the left holds, obtain r as on the right:

$$\frac{P_y - Q_x}{Qy - Px} = h(xy), \qquad H' = h, \qquad r = e^{H(xy)}$$

(d) The Bernoulli equation is $y' + f(x)y = g(x)y^n$ with $n \neq 1$. Try $r(x, y) = s(x)y^m$, show that $m = -n$ is the right choice, and thus get

$$(y^{1-n}s)' = (1 - n)sg, \qquad F' = f, \qquad s = e^{(1-n)F(x)}$$

11. The d'Alembert-Lagrange equation. The equation $y = xf(p) + g(p)$ with $p = y'$ can be solved as follows. Differentiation with respect to x yields

$$p = xf'(p)\frac{dp}{dx} + f(p) + g'(p)\frac{dp}{dx}$$

Assuming $p \neq f(p)$, solve for dx/dp to get a linear equation. If the solution is $x = h(p)$ then a solution to the original equation is given parametrically, with the slope p as parameter, by

$$x = h(p), \qquad y = h(p)f(p) + g(p)$$

12. Clairaut's equation. For the equation $y = xy' + g(y')$ show that the result of Problem 11 reduces to $(dp/dx)(x + g'(p)) = 0$. Taking $p = c$ or $g'(p) = -x$ get $y = xc + g(c)$ or the parametric solution

$$x = -g'(p), \qquad y = g(p) - pg'(p)$$

The first of these represents a family of straight lines. Show that its envelope, if there is one, leads to the second solution with parameter c instead of p.

13. Factorization. With $D = d/dx$, $p = p(x)$, $q = q(x)$ show that the equation
$$(D^2 + pD + q)y = (D + u)(D + v)y$$
holds when $u + v = p$, $uv + v' = q$. These can be solved if, and only if, v satisfies the Riccati equation $v' = q - pv + v^2$. In that case $(D + v)y = 0$ yields a solution of $y'' + py' + qy = 0$.

14. This pertains to the Riccati equation $y' = A(x) + B(x)y + C(x)y^2$.

(a) If u is a given solution show that the substitution $y = u + v$ leads to a Bernoulli equation for v with exponent $n = 2$.

(b) Using the substitution $v = w^{1-n} = 1/w$ in Part (a) deduce that

$$y' = A + By + Cy^2 \iff w' + (B + 2Cu)w + C = 0, \qquad w \neq 0$$

Since w can be found by quadrature, you get a one-parameter family of solutions y (often the complete solution) from the single solution u.

(c) Show that the substitution $Cy = -z'/z$ with $Cz \neq 0$ yields

$$y' = A + By + Cy^2 \iff Cz'' - (BC + C')z' + AC^2z = 0$$

Hence, if either equation can be solved by quadrature, so can the other.

(d) Let u and v satisfy the linear equation of Part (c) with the initial conditions $u(x_0) = 1$, $u'(x_0) = 0$, $v(x_0) = 0$, $v'(x_0) = -C(x_0)$. Show that the solution of the Riccati equation satisfying $y(x_0) = c$ is given, in general, by

$$y(x) = -\frac{1}{C(x)} \frac{u'(x) + cv'(x)}{u(x) + cv(x)}$$

15. Change of the independent variable. In this problem $y(x) = z(t)$ where $t = H(x)$ and $H' = h > 0$. Primes on y, p, q, h denote differentiation with respect to x, but $z' = dz/dt$. We assume that $H'' = h'$ is continuous. In an obvious notation the chain rule gives $D_x = h(x)D_t$. Using this twice, also using $dx/dt = 1/h(x)$, show that the following equations are equivalent:

$$y'' + p(x)y' + q(x)y = 0 \qquad h(x)^2z'' + (h'(x) + p(x)h(x))z' + q(x)z = 0$$

Assuming $q > 0$ take $h^2 = q$ and deduce that z satisfies an equation with constant coefficients $(1, a, 1)$ if $q' + 2pq = 2aq^{3/2}$. A similar result follows with $h^2 = -q$ if $q < 0$. In either case, z and hence y can be found by quadrature.

16. Exact second-order equations. If $Ty = Py'' + Qy' + Ry$ can be written in the form $Ty = (Py')' + (qy)'$ for all y in the domain of T, the equation $Ty = f$ is said to be exact.

(a) Explain how to solve an exact equation $Ty = f$ by quadrature.

(b) Show that q can be found if, and only if, $(Q - P')' = R$ and in that case $q = Q - P'$ does the job.

(c) Show that the condition of Part (b) holds for rT if, and only if, $T^*r = 0$ where, by definition, $T^*r = (Pr)'' - (Qr)' + Rr$.

17P. The Lagrange identity. As in Problem 16 let

$$Ty = Py'' + Qy' + Ry, \qquad T^*y = (Py)'' - (Qy)' + Ry$$

where P, Q, R are functions of x. Show that

$$vTu - uT^*v = F' \quad \text{where} \quad F = uQv - uPv' - uP'v + u'Pv$$

The function $B(u, v) = F$ is called the *bilinear concomitant* of u and v and T^* is called the *adjoint* of T. Self-adjoint operators, that is, those for which $T^* = T$, form the main topic of Chapter 18. The above result generalizes the Lagrange identity as stated in Chapter 18, Section 1.

18. Show that if you have a nonvanishing solution of $T^*v = 0$ then $Tu = f$ can be solved by quadrature. Similarly, if you have a nonvanishing solution of $Tu = 0$ then $T^*v = f$ can be solved by quadrature.

Hint: If $T^*v = 0$ the equation $vTu = vf$ reduces to $F' = vf$. Integration gives a first-order linear equation for u.

19. For every choice of R show that F is constant on the trajectories of

$$x' = R(x, y, t)F_y(x, y), \qquad y' = -R(x, y, t)F_x(x, y)$$

If a nontrivial R can be found for which these equations are solvable by quadrature, or are easy to handle numerically, they give a means of computing the level curves of $z = F(x, y)$.

20. This problem assumes familiarity with line integrals. It shows that the restriction to a rectangular (or at least a simply connected) region in Theorem 5 is necessary. It also shows that the case $m = -1$ in Problem 7 is truly exceptional. Let

$$P\,dx + Q\,dy = \frac{x\,dy - y\,dx}{x^2 + y^2}$$

(a) Check that P and Q are homogeneous of degree -1, and that the equation $P_y = Q_x$ of Theorem 5 holds in any region not containing the origin.

(b) If C is given in polar coordinates by $x = \cos\theta$, $y = \sin\theta$ for $0 \le \theta \le 2\pi$ show that the line integral of $P\,dx + Q\,dy$ around C is 2π. Since this is not zero, the integrand cannot be exact in any ring-shaped region that excludes the origin but contains the circle $r = 1$.

21. This is difficult. Under the hypothesis of Theorem 5 interpret the following equation, prove it, and then prove Theorem 5 by use of indefinite integrals:

$$\frac{\partial}{\partial y} \int P(\xi, y)\, d\xi = \int P_y(\xi, y)\, d\xi$$

Here the first integral denotes any function F such that $F_x(x, y) = P(x, y)$ and the second denotes any function G such that $G_x(x, y) = P_y(x, y)$. You will need Theorem C and also the fact that R is a rectangle.

22. Differential equations often lead to differential inequalities that can be solved by quadrature even when the original equation cannot. This technique is explored in Problems 22–24. The letters ϵ, δ, K denote constants with $\epsilon \geq 0$, $\delta \geq 0$. All functions are assumed to be continuous on an interval $0 \leq t < a$ and we agree that $0 \leq t < a$ in both hypothesis and conclusion. The function E is defined by

$$E(t) = \epsilon \frac{e^{Kt} - 1}{K} + \delta e^{Kt}, \qquad K \neq 0; \qquad E(t) = \epsilon t + \delta, \qquad K = 0$$

(a) Solve $w' = Kw + f$, $w(0) = w_0$ by quadrature and check that the result is $w = E$ when $f = \epsilon$ and $w_0 = \delta$. That is where E comes from.

(b) If $f \leq \epsilon$ and $w(0) \leq \delta$ show that your formula gives $w \leq E$. Hence you have almost established the following:

$$w' \leq Kw + \epsilon, \qquad w(0) \leq \delta \Rightarrow w \leq E$$

By use of Rolle's theorem instead of integration, it can be shown that this holds without any continuity assumption on f and that the differential inequality is needed only at points t where $w(t) > 0$. These refinements are used in Problems 23 and 24.

23. Let $Tu = u' - f(t, u)$ where, for all relevant x, y,

$$(x - y)(f(t, x) - f(t, y)) \leq K(x - y)^2, \qquad 0 \leq t < a$$

(a) Deduce from Problem 22 with $w = u - v$ that

$$Tu - Tv \leq \epsilon, \qquad u(0) - v(0) \leq \delta \quad \Rightarrow \quad u - v \leq E$$

(b) Using this as it is, and also with u and v interchanged, get the estimate

$$|Tu - Tv| \leq \epsilon, \qquad |u(0) - v(0)| \leq \delta \quad \Rightarrow \quad |u - v| \leq E$$

This shows that any given solution of $u' = f(t, u)$, $u(0) = u_0$ is stable; that is, a small error in the equation and in the initial value leads to small error in the solution. Aside from its bearing on the theoretical problem of stability, the specific estimate of Part (b) is useful in numerical analysis.

24. Here x, y, u, v, f are vector-valued as in Chapter 14, Section 4. The expression (x, y) denotes the inner product and we use the Schwarz inequality from Chapter D, Problem 11. Let $Tu = u' - f(t, u)$ and suppose

$$(x - y, f(t, x) - f(t, y)) \le K\|x - y\|^2, \qquad 0 \le t < a$$

for $x \in R^m$ and $y \in R^m$. Taking $w^2 = \|u - v\|^2$ in Problem 22, show that

$$\|Tu - Tv\| \le \epsilon, \quad \|u(0) - v(0)\| \le \delta \quad \Rightarrow \quad \|u - v\| \le E$$

The hypothesis on f holds if $\|f(t, x) - f(t, y)\| \le K\|x - y\|$, that is, if f admits the Lipschitz constant K. But the above one-sided condition is much more general and it allows $K < 0$. When it holds, the operator T is said to be *dissipative*. Dissipative operators play a prominent role in current research.

Hint: $ww' = (u - v, u' - v')$, which can be written as

$$(u - v, Tu + f(t, u) - Tv - f(t, v)) = (u - v, f(t, u) - f(t, v)) + (u - v, Tu - Tv)$$

This yields $ww' \le Kw^2 + \epsilon w$. At points where $w > 0$ you get the inequality of Problem 22 and the result follows.

---------------------ANSWERS---------------------

1. (abcd) $\quad x^3 y^2, \quad (x^2 + 1)(y^2 + 1)/2, \quad xe^{xy}, \quad \sqrt{x^2 + y^2}$

   (efgh) $\quad x/y, \quad (1/3)\ln(x^3 + y^3), \quad \sin^{-1}(y/x), \quad \tan^{-1}(y^2/x)$

2. Aside from an additive constant, the answers agree with those for Problem 1. In some cases, however, (10) may lead to a divergent integral.

4. (a) $\quad 2, \quad 0, \quad -2/3, \quad e + \pi, \quad 2, \quad 0$

5. (a) $y = x \ln cx$ (e) $y^2 + 2xy - x^2 = 2$

   (b) $3x^2 y + y^3 = c$ (f) $3y^3 \ln y + x^3 = y^3$

   (c) $y = 0, \; 4x = y(\ln cy)^2$ (g) $y + y \ln x = x$

   (d) $y = x, \; y = -x, \; y = x\sin(\ln cx)$ (h) $y = x$

8. (a) $b_1 = 2a_2, \; b_2 = 2c_1$. Problem 7 gives F by inspection.

The world of mathematics is inexhaustible.

—Constance Reid

Chapter B

POWER SERIES

A *power series* (about $x = 0$) is an expression of the form

(1)
$$\sum_{j=0}^{\infty} a_j x^j = a_0 + a_1 x + a_2 x^2 + \cdots + a_n x^n + \cdots$$

where the a_j are real or complex numbers. Unless the contrary is said we assume that x is real. The extension to complex values causes no trouble, however, because the theory of limits for complex sequences is virtually identical to that for real sequences. Indeed, if

$$s_n = u_n + iv_n, \qquad s = u + iv$$

then, by definition,

$$\lim_{n \to \infty} s_n = s \quad \text{means} \quad \lim_{n \to \infty} u_n = u \quad \text{and} \quad \lim_{n \to \infty} v_n = v.$$

The symbol $\sum$, read "sigma," is a Greek capital S and denotes "sum." The phrase *sigma notation* distinguishes the condensed style of writing on the left of (1) from the expanded form on the right. The term a_0 does not involve x and is called the constant term. In order to make the constant term on the right agree with $a_0 x^0$ on the left we set

$$a_0 0^0 = \lim_{x \to 0} a_0 x^0 = a_0.$$

Some such convention is needed because 0^0 is otherwise undefined. In this book we make the further convention that

$$0 \cdot 0^j = \lim_{x \to 0} 0 \cdot x^j = 0 \quad \text{for all } j.$$

The algebra of power series has two aspects: the formal theory, in which no numerical value is attached to the series, and the analytic theory, in which the series has a definite numerical value called its sum. Both aspects are needed when power series are used to solve differential equations. Since the formal theory is easy (it is much like the algebra of polynomials) it is presented first.

(a) The formal theory

Two power series are said to be equal if corresponding coefficients agree. That is,

$$\sum_{j=0}^{\infty} a_j x^j = \sum_{j=0}^{\infty} b_j x^j \quad \text{means} \quad a_j = b_j, \qquad j = 0, 1, 2, \cdots.$$

A power series is multiplied by a constant k according to the rule

$$k \sum_{j=0}^{\infty} a_j x^j = \sum_{j=0}^{\infty} k a_j x^j$$

and two power series are added by adding corresponding coefficients:

$$\sum_{j=0}^{\infty} a_j x^j + \sum_{j=0}^{\infty} b_j x^j = \sum_{j=0}^{\infty} (a_j + b_j) x^j.$$

The product is defined in the same way as for polynomials. Namely,

$$(a_0 + a_1 x + a_2 x^2 + a_3 x^3 + \cdots)(b_0 + b_1 x + b_2 x^2 + b_3 x^3 + \cdots)$$

$$= a_0 b_0 + (a_0 b_1 + a_1 b_0)x + (a_0 b_2 + a_1 b_1 + a_2 b_0)x^2 + \cdots$$

or, in sigma notation,

$$\left(\sum_{j=0}^{\infty} a_j x^j \right) \left(\sum_{j=0}^{\infty} b_j x^j \right) = \sum_{j=0}^{\infty} c_j x^j$$

where
$$c_j = a_0 b_j + a_1 b_{j-1} + a_2 b_{j-2} + \cdots + a_j b_0.$$

This is called the *Cauchy product* after the French mathematician Augustin Louis Cauchy. When formulas such as this are used, any missing coefficients in the power series are considered to have the value 0. For example in the product

$$(a_0 + a_1 x)(b_0 + b_1 x + b_2 x^2 + b_3 x^3 + \cdots)$$

$$= a_0 b_0 + (a_0 b_1 + a_1 b_0)x + (a_0 b_2 + a_1 b_1)x^2 + (a_0 b_3 + a_1 b_2)x^3 + \cdots$$

the first factor $a_0 + a_1 x$ is thought of as an abbreviation for

$$a_0 + a_1 x + 0 x^2 + 0 x^3 + \cdots + 0 x^n + \cdots.$$

The Cauchy formula then gives the correct result,

$$c_0 = a_0 b_0, \qquad c_n = a_0 b_n + a_1 b_{n-1}, \qquad n \geq 1.$$

The quotient A/B can be defined by long division if the constant term b_0 of the denominator is not 0. But another method using geometric series (see Problem 9) is often more efficient. We first divide A and B by b_0, thus reducing the problem to the case $b_0 = 1$. If we then set $Q = 1 - B$, the formal reciprocal of B is given by

(2) $$\frac{1}{B} = \frac{1}{1-Q} = 1 + Q + Q^2 + Q^3 + \cdots$$

and so

$$\frac{A}{B} = A(1 + Q + Q^2 + Q^3 + \cdots).$$

As an illustration, let $B = (1-x)^2 = 1 - 2x + x^2$. Then $Q = 1 - B = 2x - x^2$. Hence the formal reciprocal of B is

$$\frac{1}{1 - (2x - x^2)} = 1 + (2x - x^2) + (2x - x^2)^2 + (2x - x^2)^3 + \cdots$$

$$= 1 + 2x + x^2(-1 + 4) + x^3(-4 + 8) + \cdots = 1 + 2x + 3x^2 + 4x^3 + \cdots.$$

As far as it goes, this agrees with the expansion of $(1 - x)^{-2}$ as given by the binomial theorem.

The derived series, or the formal derivative, is defined by

$$(a_0 + a_1 x + a_2 x^2 + a_3 x^3 + \cdots)' = a_1 + 2a_2 x + 3a_3 x^2 + \cdots.$$

In sigma notation the equation can be written

$$\left(\sum_{j=0}^{\infty} a_j x^j \right)' = \sum_{j=0}^{\infty} j a_j x^{j-1}$$

since the term with $j = 0$ in the sum on the right is 0. Higher derivatives are obtained by repetition of the same procedure. A power series has formal derivatives of all orders, even though the function represented by the series may be defined only at $x = 0$, hence not even once differentiable anywhere. As suggested by Problem 11, the formal derivative behaves much like the ordinary derivative.

(b) Convergence

The formal theory does not require that a power series have a numerical value at any given value of x. But a series of interest in differential equations generally does have a numerical value, which is obtained by considering the sequence $\{s_n\}$ of partial sums

$$a_0, \; a_0 + a_1 x, \; a_0 + a_1 x + a_2 x^2, \; a_0 + a_1 x + a_2 x^2 + a_3 x^3, \; \cdots .$$

The nth partial sum is

$$\sum_{j=0}^{n} a_j x^j = a_0 + a_1 x + a_2 x^2 + \cdots + a_n x^n.$$

If the nth partial sum tends to a limit as $n \to \infty$ the series is said to converge and its sum is the value of the limit. Thus, by definition,

$$\sum_{j=0}^{\infty} a_j x^j = \lim_{n \to \infty} \sum_{j=0}^{n} a_j x^j$$

provided the limit exists. If the limit does not exist the series is said to diverge and no value is assigned to the sum. Convergence expresses numerical equality, in contrast to the formal equality discussed above.

One might expect the set of points x where a power series converges to be a complicated set on the real axis. The following surprising theorem shows, however, that the set contains only the single point $x = 0$, or else it is an interval centered at the origin:

RADIUS OF CONVERGENCE. For any power series, one of the following holds: The series converges for $x = 0$ only; or the series converges for all x; or there is a positive number r such that the series converges when $|x| < r$ and diverges when $|x| > r$.

The number r is called the *radius of convergence* and the interval $|x| < r$ is called the *interval of convergence*. This terminology is extended by the convention that $r = 0$ if the series converges only for $x = 0$, and $r = \infty$ if the series converges for all x. The proof is outlined in Chapter 12, Section 3 where it is seen that the series not only converges for $|x| < r$, but converges absolutely; that is,

$$\sum_{j=0}^{\infty} |a_j||x|^j \quad \text{converges for} \quad |x| < r.$$

(c) Properties of convergent series

The same symbol is used to denote both a power series and, when the series is convergent, the value of its sum. This ambiguity causes no trouble because, if x is in the interval of convergence, the formal theory and the analytic theory are in agreement. With $r > 0$, the facts are as follows:

(i) Equality. If two power series converge to the same sum for $|x| < r$ then corresponding coefficients agree.

(ii) Sum, difference, and product. If two power series converge to $A(x)$ and $B(x)$ respectively for $|x| < r$ then the formal sum, difference, and product of the series converge, respectively, to $A(x)+B(x)$, $A(x)-B(x)$, and $A(x)B(x)$ for $|x| < r$.

(iii) Quotient. If in addition to the above hypothesis (ii) we have $B(0) \neq 0$, the formal quotient converges to $A(x)/B(x)$ on some interval $|x| < r_1$. Here r_1 is positive but may be smaller than r.

(iv) Differentiation. A power series and its formal derivative always have the same radius of convergence. If the common radius of convergence is r then for $|x| < r$ the series converges to a differentiable function $A(x)$ and the derived series converges to $A'(x)$. A similar result holds for integration.

(v) Rearrangement. If a power series converges for $|x| < r$ then for $|x| < r$ its terms can be rearranged in any manner without changing its sum.

(vi) Composite functions. If $f(u)$ has a convergent power series expansion for $|u| < \infty$ and $g(x)$ has an expansion for $|x| < r$ then $f(g(x))$ has a convergent expansion at least for $|x| < r$. The coefficients can be obtained by substituting $u = g(x)$ in the series for $f(u)$ and collecting terms.

Item (i) follows from Taylor's formula as discussed in Part (f) below. The statement (ii) for the sum and difference follows from Problem 7, and for the product it can be established along lines suggested by Problem 10. Though details will not be given here, (iii) can be deduced from (2) and from absolute convergence. The main point is that the condition $B(0) = 1$ gives $Q(0) = 0$ in (2). Hence $|Q(x)| < 1$ for small $|x|$ and the geometric series in (2) with $Q = Q(x)$ converges. Item (iv) is established in Chapter F.

Although the proof will not be given here, the statement (v) follows from the fact, mentioned above, that the series converges absolutely for $|x| < r$. The essential condition is that each term in the original series must occur once, and once only, in the rearrangement. For example we could form a series by taking all terms $a_j x^j$ with even j, and another by taking all terms with odd j. Statement (v) says that the two power series so obtained both converge at least for $|x| < r$, and that

$$\sum_{j=0}^{\infty} a_j x^j = \sum_{j=0}^{\infty} a_{2j} x^{2j} + \sum_{j=0}^{\infty} a_{2j+1} x^{2j+1}, \qquad |x| < r.$$

Item (vi) pertaining to $f(g(x))$ is a special case of the rearrangement theorem. If $g(0) = 0$ it suffices to assume that f and g have power-series expansions with positive radius of convergence, and the conclusion is that $f(g(x))$ also has such an expansion.

The condition $g(0) = 0$ greatly simplifies the calculations. Suppose, for example, that we want to use the expansion

$$e^u = 1 + u + \frac{u^2}{2!} + \frac{u^3}{3!} + \cdots, \qquad |u| < \infty,$$

given in Problem 2 to obtain an expansion of e^{1+x^2}. Collecting terms in

$$e^{1+x^2} = 1 + (1+x^2) + \frac{(1+x^2)^2}{2!} + \frac{(1+x^2)^3}{3!} + \cdots$$

leads to an infinite series for each coefficient and is a nearly hopeless task. The trouble is that $u = 1 + x^2$ does not vanish at $x = 0$. Far better to write, by inspection,

$$e^{1+x^2} = e e^{x^2} = e\left(1 + x^2 + \frac{x^4}{2!} + \frac{x^6}{3!} + \cdots\right).$$

(d) The ratio test

The simplest and often the most effective way to determine the radius of convergence is by means of the following theorem, which applies whether the c_j are real or complex:

RATIO TEST. The series with general term c_j converges if

$$\lim_{n \to \infty} \left| \frac{c_{n+1}}{c_n} \right| < 1.$$

If the limit is greater than 1, the series diverges. If the limit is 1 or does not exist, the ratio test gives no information.

For proof, denote the limit by r and choose R so that $r < R < 1$. If $k \geq N$, where N is sufficiently large, the hypothesis gives $|c_{k+1}/c_k| \leq R$. For $n > N$ this yields

$$\left| \frac{c_n}{c_N} \right| = \left| \frac{c_{N+1}}{c_N} \right| \left| \frac{c_{N+2}}{c_{N+1}} \right| \cdots \left| \frac{c_n}{c_{n-1}} \right| \leq \frac{R^n}{R^N}.$$

Hence $|c_n| \le CR^n$ where C is constant, and the series converges by comparison with the geometric series; see Problem 9. The comparison test upon which this argument depends is established in Chapter 12, Section 3.

To determine the radius of convergence of a power series, the ratio test is applied with such choices as

$$c_n = a_n x^n, \qquad c_n = a_{2n} x^{2n}, \qquad c_n = a_{2n+1} x^{2n+1}$$

depending on the structure of the series. For example, if the series is

$$2x^3 + 8x^6 + \cdots + n2^n x^{3n} + \cdots$$

it would not do to take $c_n = a_n x^n$ because $a_n = 0$ for infinitely many values of n. Instead we set

$$c_n = n2^n x^{3n}, \qquad x \ne 0$$

and obtain

$$\left| \frac{c_{n+1}}{c_n} \right| = \left| \frac{(n+1)2^{n+1} x^{3(n+1)}}{n2^n x^{3n}} \right| = \frac{n+1}{n} 2|x|^3.$$

The limit as $n \to \infty$ is $2|x|^3$. Hence the series converges if $2|x|^3 < 1$, diverges if $2|x|^3 > 1$, and the radius of convergence is given by $2r^3 = 1$. Whether mentioned or not, the condition $x \ne 0$ is always understood when the ratio test is applied to power series.

(e) Translation

Power series are often written in the form

(3)
$$\sum_{j=0}^{\infty} a_j (z - c)^j$$

where c is a real constant. It is said that the expansion is *about the point* c. When $c = 0$, as has been the case up to now, the expansion is about the origin. The equation $x = z - c$ that converts the one case into the other represents a translation.

The general case is easily reduced to the case $c = 0$. Suppose, for example, that $r > 0$ and that, in the sense of convergence,

$$f(z) = \sum_{j=0}^{\infty} a_j (z - c)^j, \qquad |z - c| < r.$$

Setting $x = z - c$ we get a series about $x = 0$, namely

$$f(x + c) = \sum_{j=0}^{\infty} a_j x^j, \qquad |x| < r.$$

Differentiation yields

$$\frac{d}{dx} f(x+c) = f'(x+c) = \sum_{j=0}^{\infty} j a_j x^{j-1}$$

or, since $x + c = z$,

$$f'(z) = \sum_{j=0}^{\infty} j a_j (z-c)^{j-1}.$$

Hence the series (3) can be differentiated term by term just as it could be in the case $c = 0$. Other results extend to the case $c \neq 0$ with equal ease. The parameter c will be retained in the rest of this chapter, but we take $c = 0$ in Chapter 12.

(f) Taylor series

Let us suppose that

$$f(x) = a_0 + a_1(x-c) + \cdots + a_n(x-c)^n + \cdots$$

for $|x - c| < r$, where $r > 0$. Differentiating n times yields

$$f^{(n)}(x) = n! a_n + \cdots$$

where the terms not written all have $(x - c)$ as a factor. When $x = c$ these terms vanish and we get

$$f^{(n)}(c) = n! a_n, \quad \text{hence} \quad a_n = \frac{f^{(n)}(c)}{n!}.$$

This is *Taylor's formula*, named after the English mathematician Brook Taylor who stated it in 1715. When the coefficients are given by Taylor's formula the resulting series, whether convergent or not, is called the *Taylor series* for f. Thus, every function that is infinitely differentiable at a single point has a Taylor series about that point. The special case $c = 0$ is often called "MacLaurin's series" though Taylor's work preceded MacLaurin's.

The above analysis shows that if *some* power series in $x - c$ converges to $f(x)$ for $|x - c| < r$, where $r > 0$, then that series necessarily coincides with the Taylor series. The analysis does not give any clue, however, to the question whether such an expansion exists. In fact a function $f(x)$ can be infinitely differentiable for all x, and yet its Taylor series may converge only when $x = c$. Or a Taylor series can be such that it converges for all x, yet converges to $f(x)$ only when $x = c$.

In contrast to these negative results we have the following:

CONVERGENCE TEST FOR TAYLOR SERIES. Suppose there are positive constants M, K and r such that

(4) $|f^{(n)}(t)| \le MK^n$, $|t - c| < r$, $n = 0, 1, 2, \cdots$.

Then the Taylor series for f converges to $f(x)$ for $|x - c| < r$.

This is a consequence of Lagrange's formula, which asserts that the difference between $f(x)$ and the first n terms of its Taylor series satisfies

(5) $$f(x) - \sum_{j=0}^{n-1} a_j (x - c)^j = \frac{f^{(n)}(t)}{n!}$$

for some t between c and x. Lagrange's formula holds if

  (i) $f(s)$ is real and continuous on a closed interval containing c and x,

  (ii) $f^{(n)}(s)$ exists at every interior point of the interval, and

  (iii) $a_j = f^j(c)/j!$, $j = 0, 1, \cdots, n - 1$.

In view of Problem 8, the convergence test follows when (5) is applied to the real and imaginary parts of the series.

  We now prove (5). Assuming without loss of generality that $c = 0$ and $x > 0$, let C be constant and define

$$P(t) = \sum_{j=0}^{n-1} a_j t^j, \qquad F(t) = f(t) - P(t) - Ct^n.$$

We regard x as fixed and we choose C so that $F(x) = 0$. Thus,

(6) $f(x) - P(x) = Cx^n$.

We have $f^{(j)}(0) = P^{(j)}(0)$ for $j = 0, 1, \cdots, n - 1$ and hence $F^{(j)}(0) = 0$ for these values of j. Since $F(0) = F(x) = 0$, Rolle's theorem gives a value t_1 between 0 and x such that $F'(t_1) = 0$. Since $F'(0) = F'(t_1) = 0$ Rolle's theorem gives t_2 between 0 and t_1 such that $F''(t_2) = 0$. Continuing in this way we get t_n such that

$$0 < t_n < t_{n-1} < \cdots < t_1 < x, \qquad F^{(n)}(t_n) = 0.$$

With $t = t_n$ this gives $f^{(n)}(t) = n!C$, and (5) now follows from (6).

PROBLEMS

1. Check that, as a formal power series,

$$\sum_{j=0}^{\infty} (2^{j+1} - (j+1)^2)x^j = 1 + 0x - x^2 + 0x^3 + 7x^4 + \cdots$$

This example shows that there is a certain ambiguity when an infinite series is described by giving only the first few terms. That is, from knowledge of the first four terms in the displayed series one could hardly guess that the fifth term is $7x^4$. Nevertheless we shall often describe infinite series by giving just a few terms, because in cases when we do this there is no difficulty in surmising what general term is intended.

2. Here are some Taylor expansions about the point $c = 0$:

(a) $\sin x = x - \dfrac{x^3}{3!} + \dfrac{x^5}{5!} - \cdots$ (d) $\sinh x = x + \dfrac{x^3}{3!} + \dfrac{x^5}{5!} + \cdots$

(b) $e^x = 1 + x + \dfrac{x^2}{2!} + \dfrac{x^3}{3!} + \cdots$ (e) $\cos x = 1 - \dfrac{x^2}{2!} + \dfrac{x^4}{4!} - \cdots$

(c) $\cosh x = 1 + \dfrac{x^2}{2!} + \dfrac{x^4}{4!} + \cdots$ (f) $\ln(1 + x) = x - \dfrac{x^2}{2} + \dfrac{x^3}{3} - \cdots$

Write each series in sigma notation, taking the range of summation to be from $j = 0$ to ∞.

3. (abcdef) Determine the radius of convergence of the series in Problem 2.

4. (abcdef) Without considering convergence, derive the Taylor expansions in Problem 2.

5. (abcde) Show that the Taylor expansions in Problem 2 (abcde) converge to the functions on the left for every value of x.

6P. (abcdefg) It is important to be able to pick out the coefficient of x^n in expressions of the form $x^m y^{(k)}$ where

$$y = \sum_{j=0}^{\infty} a_j x^j$$

This is best done by looking only at the general term, not at the whole series. For example, to find the coefficient of x^n in the formal expansions of y, y', y'', and xy'' we make the following table:

y	y'	y''	xy''
$a_j x^j$	$j a_j x^{j-1}$	$j(j-1)a_j x^{j-2}$	$j(j-1)a_j x^{j-1}$
a_n	$(n+1)a_{n+1}$	$(n+2)(n+1)a_{n+2}$	$(n+1)n a_{n+1}$

The second line gives the general term in the expansion of the first line and the third line gives the coefficient of x^n. The procedure is to take $j = n$, $j - 1 = n$, $j - 2 = n$, $j - 1 = n$ in the four entries of the second line to get the desired exponent n on x. Find the coefficient of x^n:

$$y, \quad xy', \quad x^2 y'', \quad x^{-1}y, \quad xy' + x^2 y'', \quad y' + xy'', \quad xy'' - y' + y + x^{-1}y$$

7P. If you let $n \to \infty$ in the following identities, you get three theorems about convergent infinite series. What theorems? No answer is provided:

$$c_n = \sum_{j=0}^{n} c_j - \sum_{j=0}^{n-1} c_j, \qquad \sum_{j=0}^{n} k c_j = k \sum_{j=0}^{n} c_j, \qquad \sum_{j=0}^{n}(a_j + b_j) = \sum_{j=0}^{n} a_j + \sum_{j=0}^{n} b_j$$

8. Let M and K be positive constants. Show that the series with general term $M K^n / n!$ converges and thus deduce $\lim_{n \to \infty} M K^n / n! = 0$.

9P. Geometric series. By inspection the two equations

$$s_n = 1 + r + r^2 + \cdots + r^n, \qquad r s_n = r + r^2 + \cdots + r^n + r^{n+1}$$

yield $s_n - r s_n = 1 - r^{n+1}$. Solve for s_n and let $n \to \infty$ to get

$$1 + r + r^2 + r^3 + \cdots + r^n + \cdots = \frac{1}{1 - r}, \qquad |r| < 1.$$

The quantity r gives the ratio of two successive terms and is called the *ratio* associated with the geometric series. What does the series do if the ratio satisfies $|r| \geq 1$?

10. This pertains to $(a_0 + a_1 x + a_2 x^2)(b_0 + b_1 x + b_2 x^2 + \cdots + b_n x^n + \cdots)$.

(a) Check that the terms up to x^n in the product are given by

$$\sum_{j=1}^{n} c_j x^j = a_0 \sum_{j=0}^{n} b_j x^j + a_1 x \sum_{j=0}^{n-1} b_j x^j + a_2 x^2 \sum_{j=0}^{n-2} b_j x^j$$

(b) Now let $n \to \infty$ to get, in an obvious notation,

$$C(x) = a_0 B(x) + a_1 x B(x) + a_2 x^2 B(x) = A(x)B(x).$$

In this argument what do A, B, and C stand for, what is the hypothesis, and what is the conclusion? No answer is provided.

11. Does Leibniz's rule $(uv)' = uv' + u'v$ hold as a formal identity for power series, whether convergent or not? Prove your answer.

12. Set $x = it$ in three of the series of Problem 2 to conclude that

    (a) $e^{it} = \cos t + i \sin t$ (b) $\cos it = \cosh t$ (c) $\sin it = i \sinh t$

13P. Suppose two power series

$$u(x) = a_0 + a_1(x-c) + a_2(x-c)^2 + \cdots, \quad v(x) = b_0 + b_1(x-c) + b_2(x-c)^2 + \cdots$$

both converge for $|x - c| < r$ where $r > 0$. Show that the Wronskian W satisfies $W(c) = a_0 b_1 - a_1 b_0$. Hence if $a_0 b_1 \neq a_1 b_0$ it follows that u and v are linearly independent on the interval $|x - c| < r$. A similar result applies to three or more series.

14. The limit of a complex sequence $\{s_n\}$ is often defined as follows:

$$\lim_{n \to \infty} s_n = s \quad \text{means} \quad \lim_{n \to \infty} |s_n - s| = 0$$

Show that this definition is equivalent to that in the text.

Hint: $\max(|u_n - u|, |v_n - v|) \leq |s_n - s| \leq |u_n - u| + |v_n - v|$.

--------------------------------ANSWERS--------------------------------

2. General term: (a) $(-1)^j x^{2j+1}/(2j+1)!$ (d) $x^{2j+1}/(2j+1)!$
                  (b) $x^j/j!$ (e) $(-1)^j x^{2j}/(2j)!$
                  (c) $x^{2j}/(2j)!$ (f) $(-1)^j x^{j+1}/(j+1)$

3. $r = \infty$ except: (f) $r = 1$

6. a_n, $n a_n$, $n(n-1)a_n$, a_{n+1}, $n^2 a_n$, $(n+1)^2 a_{n+1}$, $n^2 a_{n+1} + a_n$.

9. It diverges because the general term does not tend to 0.

11. Yes, and formal power series provide one of the classical approaches to the theory of functions of one or several complex variables.

We have adroitly defined the infinite in arithmetic by a love-knot, in this manner ∞, but we possess not therefore the clearer notion of it.

                                            —Voltaire

Chapter C

MATRICES

An m-by-n matrix is a set of mn real or complex numbers arranged in a rectangular array of m rows and n columns and enclosed by parentheses, thus:

$$A = \begin{pmatrix} a_{11} & a_{12} & \cdots & a_{1n} \\ a_{21} & a_{22} & \cdots & a_{2n} \\ \vdots & \vdots & \ddots & \vdots \\ a_{m1} & a_{m2} & \cdots & a_{mn} \end{pmatrix}.$$

This matrix can be also be described as follows:

$$A = (a_{ij}), \qquad i = 1, 2, \cdots, m; \quad j = 1, 2, \cdots, n.$$

The numbers a_{ij} are the elements or entries of the matrix and the ordered pair (m, n) gives its size, m-by-n. The size is often understood from context, in which case one writes $A = (a_{ij})$ without specifying the range of i and j. When $m = n$ the matrix is square and n is its order. In a square matrix the elements a_{ii} are *diagonal elements* and the remaining a_{ij} are off-diagonal.

If $m = 1$ the matrix consists of the first row only and is a row vector, while if $n = 1$ the matrix consists of the first column only and is a column vector. The elements of a row or column vector are *coordinates* of the vector and the number of elements in a row or column vector is its *dimension*. For legibility the coordinates of a row vector are often separated by commas, and this was done in Chapter 14. However, the commas have no mathematical significance and a row vector such as $(a\ b\ c)$ has the same meaning as $(a,\ b,\ c)$.

Here are four examples:

$$\begin{pmatrix} 1 & 2 & 3 \\ 4 & 5 & 6 \end{pmatrix}, \quad \begin{pmatrix} 3 & 4 \\ 5 & 6 \end{pmatrix}, \quad (a\ b\ c\ d), \quad \begin{pmatrix} p \\ q \end{pmatrix}$$

The first matrix is of size 2-by-3, the second is a square matrix of order 2, the third is a row vector of dimension 4, and the fourth is a column vector of dimension 2.

Two matrices are equal if they have the same size and corresponding elements agree. Thus

$$(a_{ij}) = (b_{ij}) \quad \text{means} \quad a_{ij} = b_{ij}$$

where it is understood that the matrices have the same size and that the equation on the right holds for all relevant i and j. This definition indicates that a matrix is not just a set of numbers, but involves the arrangement of the numbers as well. For example, no two of the following matrices are equal:

$$(1 \quad 2), \quad (2 \quad 1), \quad (1 \quad 2 \quad 0), \quad (0 \quad 1 \quad 2).$$

If the rows and columns of a matrix A are interchanged the resulting matrix is called the transpose of A and is written A^T. When the elements of A^T are replaced by their complex conjugates, the resulting matrix is called the conjugate transpose or adjoint of A and is written A^*. (Hence for real matrices the transpose and the adjoint coincide.) Here are some examples:

$$\begin{pmatrix} a & b \\ c & d \end{pmatrix}^* = \begin{pmatrix} \bar{a} & \bar{c} \\ \bar{b} & \bar{d} \end{pmatrix}, \qquad (a \quad b)^T = \begin{pmatrix} a \\ b \end{pmatrix}, \qquad \begin{pmatrix} p \\ q \end{pmatrix}^T = (p \quad q).$$

The transpose of a row vector is a column vector of the same dimension, and the transpose of a column vector is a row vector of the same dimension. Obviously

$$(A^T)^T = A, \qquad (A^*)^* = A.$$

A matrix A is said to be symmetric if $A = A^T$, self-adjoint if $A = A^*$. Self-adjoint matrices are also called Hermitian, after the nineteenth century French mathematician Charles Hermite. For example, with $i = \sqrt{-1}$ as usual, the first of the following is symmetric and Hermitian, the second is neither symmetric nor Hermitian, the third is symmetric but not Hermitian, and the fourth is Hermitian but not symmetric:

$$\begin{pmatrix} 1 & 3 \\ 3 & 4 \end{pmatrix}, \quad \begin{pmatrix} 1 & 2 \\ 3 & 4 \end{pmatrix}, \quad \begin{pmatrix} i & 1+i \\ 1+i & 2 \end{pmatrix}, \quad \begin{pmatrix} 1 & 1-i \\ 1+i & 2 \end{pmatrix}.$$

In a context involving matrices a real or complex number λ is often called a scalar. The chief operations on matrices are addition, multiplication by scalars, and matrix multiplication, symbolized respectively by

$$A + B, \qquad \lambda A, \qquad AB.$$

These operations are explained next.

(a) Addition and multiplication by scalars

Matrices of the same size are added by adding corresponding elements, and multiplication by a real or complex scalar λ is also done elementwise:

$$(a_{ij}) + (b_{ij}) = (a_{ij} + b_{ij}) \qquad \lambda(a_{ij}) = (\lambda a_{ij}).$$

We define $A\lambda = \lambda A$, $-A = (-1)A$, and $B - A = B + (-A)$. Here are two examples:

$$\begin{pmatrix} 1 & 2 \\ 3 & 4 \end{pmatrix} + \begin{pmatrix} 1 & 4 \\ 2 & 3 \end{pmatrix} = \begin{pmatrix} 2 & 6 \\ 5 & 7 \end{pmatrix}, \qquad 2\begin{pmatrix} 1 & 2 \\ 3 & 4 \end{pmatrix} = \begin{pmatrix} 2 & 4 \\ 6 & 8 \end{pmatrix}.$$

A matrix with all elements zero is denoted by 0, the same notation being used regardless of the size. The main property of 0 is that

$$A + 0 = A, \qquad 0 + A = A, \qquad A - A = 0$$

provided 0 has the same size as the matrix A in the equations. Only a little experience with matrices is needed to realize that *matrices of different sizes cannot be added,* and that is why the size of 0 need not be built into the notation.

For matrices A, B, C of the same size, you can easily convince yourself of the following:

$$A + B = B + A, \quad (A + B) + C = A + (B + C)$$
$$\lambda(A + B) = \lambda A + \lambda B, \quad (\lambda + \mu)A = \lambda A + \mu A, \quad (\lambda\mu)A = \lambda(\mu A)$$
$$(A + B)^T = A^T + B^T, \quad (A + B)^* = A^* + B^*, \quad (\lambda A)^T = \lambda A^T, \quad (\lambda A)^* = \bar{\lambda}A^*.$$

(b) The matrix product

If $R = (r_i)$ is a row vector and $C = (c_j)$ is a column vector of the same dimension, the product RC is a 1-by-1 matrix defined as follows:

$$RC = (r_1 c_1 + r_2 c_2 + \cdots + r_n c_n).$$

For example,

$$(a \quad b) \begin{pmatrix} p \\ q \end{pmatrix} = (ap + bq), \qquad (a \quad b \quad c) \begin{pmatrix} p \\ q \end{pmatrix} \quad \text{is undefined.}$$

Suppose next that A consists of row vectors R_i, and B consists of column vectors C_j. If the common dimension of the R_i is different from that of the C_j the product AB is undefined. But if the dimension of each R_i is the same as the dimension of each C_j the element p_{ij} in $P = AB$ is the product of the ith row of A with the jth column of B. The following example sheds light on this definition of the matrix product:

$$(1) \qquad \begin{pmatrix} a & b \\ c & d \end{pmatrix} \begin{pmatrix} p & 0 & 1 & 1 \\ q & 1 & 0 & 1 \end{pmatrix} = \begin{pmatrix} ap + bq & b & a & a + b \\ cp + dq & d & c & c + d \end{pmatrix}.$$

Taking this example as point of departure, we make some comments.

Each product $R_i C_j$ is a matrix. However, when a 1-by-1 matrix (a) is an element of another matrix, we agree not to distinguish (a) from a. Thus,

$$\begin{pmatrix} (a) & (b) \\ (c) & (d) \end{pmatrix} = \begin{pmatrix} a & b \\ c & d \end{pmatrix}.$$

That is why the elements of the matrix on the right of (1) need not be enclosed in parentheses. See also Problem 18.

If the rows of A have the same dimension as the columns of B we can form the product AB and it is said that A, B, in that order, are compatible. Thus the two matrices on the left of (1) are compatible in the order given, but if the order were reversed, they would not be compatible.

As suggested by (1), quite generally the product AB of two compatible matrices A, B can be obtained by applying A to each column of B. In symbols,

$$(2) \qquad A(C_1, C_2, \cdots, C_n) = (AC_1, AC_2, \cdots, AC_n).$$

Since each AC_j is a matrix, here too we need the above convention regarding omission of parentheses. With that convention, any matrix can be thought to be an ordered set of row vectors R_j or of column vectors C_j.

A counterpart of (2), also suggested by (1), is

$$(3) \qquad \begin{pmatrix} R_1 \\ R_2 \\ \vdots \\ R_m \end{pmatrix} B = \begin{pmatrix} R_1 B \\ R_2 B \\ \vdots \\ R_m B \end{pmatrix}.$$

This says that the kth row in the product AB is obtained when B is multiplied on the left by the kth row of A.

Equations such as (2) and (3) give a good deal of insight into matrix algebra. For example, if the kth row of A is multiplied by a constant, the kth row of AB is multiplied by the same constant. If two rows of A are interchanged, this has the effect of interchanging the same two rows in AB. These results follow from (3) and similar results for columns follow from (2).

(c) Algebraic properties

The algebra of matrices differs from the algebra of real or complex numbers in two very important respects:

(i) In general $AB \neq BA$ even if both products are defined.

(ii) It is possible to have $AB = 0$ even if neither A nor B is 0.

Both statements are illustrated by

$$\begin{pmatrix} 1 & 0 \\ 0 & 0 \end{pmatrix} \begin{pmatrix} 0 & 1 \\ 0 & 0 \end{pmatrix} = \begin{pmatrix} 0 & 1 \\ 0 & 0 \end{pmatrix}, \qquad \begin{pmatrix} 0 & 1 \\ 0 & 0 \end{pmatrix} \begin{pmatrix} 1 & 0 \\ 0 & 0 \end{pmatrix} = \begin{pmatrix} 0 & 0 \\ 0 & 0 \end{pmatrix}.$$

Otherwise the algebra of matrices is much like the algebra of numbers. This is seen by the equations

$$A(B+C) = AB + AC, \qquad (B+C)A = BA + CA$$
$$(\lambda A)B = \lambda(AB), \qquad A(\lambda B) = \lambda(AB), \qquad A(BC) = (AB)C$$

which hold if the sizes of A, B, C are such that the various expressions are meaningful. The first two equations are distributive laws, the next three are associative laws.

For proof, the formula for $R_i C_j$ shows that the product of compatible matrices A, B can be written in the form

$$(a_{ij})(b_{ij}) = (c_{ij}) \quad \text{where} \quad c_{ij} = \sum_{k=1}^{n} a_{ik} b_{kj}.$$

Here n is the common dimension of the row vectors of A and the column vectors of B. If B and C have the same size, and A, B are compatible, the distributive law $A(B + C) = AB + AC$ is obtained when the identity

$$a_{ik}(b_{kj} + c_{kj}) = a_{ik} b_{kj} + a_{ik} c_{kj}$$

is summed on k. If A, B and B, C are compatible, the associative law

$$A(BC) = (AB)C$$

can be deduced from

$$a_{ip}(b_{pq} c_{qj}) = (a_{ip} b_{pq}) c_{qj}$$

as follows: Sum the left-hand member first on q and then on p to get $A(BC)$. Sum the right-hand member first on p and then on q to get $(AB)C$. Since the sum is independent of the order in which the terms are added, the two sides after the summation are equal. Proof of the other equations above is similar and is left to you. It is also left for you to establish the following, which hold whenever AB is meaningful:

$$(AB)^T = B^T A^T, \qquad (AB)^* = B^* A^*.$$

A square matrix in which all the diagonal elements are 1 and all the other elements are 0 is called the *identity matrix* and is denoted by I, the same notation being used regardless of the size. The main property of I is that

$$AI = IA = A$$

whenever A is a square matrix of the same size as I. The proof follows from (2) and (3). Here is an illustration:

$$\begin{pmatrix} a & b & c \\ p & q & r \\ u & v & w \end{pmatrix} \begin{pmatrix} 1 & 0 & 0 \\ 0 & 1 & 0 \\ 0 & 0 & 1 \end{pmatrix} = \begin{pmatrix} a & b & c \\ p & q & r \\ u & v & w \end{pmatrix} = \begin{pmatrix} 1 & 0 & 0 \\ 0 & 1 & 0 \\ 0 & 0 & 1 \end{pmatrix} \begin{pmatrix} a & b & c \\ p & q & r \\ u & v & w \end{pmatrix}.$$

Only a little experience with matrices is needed to realize that *square matrices of different sizes cannot be multiplied* and that is why the order of I need not be built into the notation.

(d) The inverse

Throughout Part (d) we use A, B, M to denote square matrices of the same order. If there exists a matrix M such that $AM = MA = I$ then A is said to be invertible, M is called the inverse of A, and we write $M = A^{-1}$. Thus

$$A^{-1}A = AA^{-1} = I, \qquad A \quad \text{invertible.}$$

If A is invertible and $AM = I$, then $M = A^{-1}$. In other words A can have at most one inverse. To see why, let the equation $AM = I$ be multiplied *on the left* by A^{-1}. The result is

$$A^{-1}(AM) = A^{-1}I, \qquad (A^{-1}A)M = A^{-1}, \qquad IM = A^{-1}, \qquad M = A^{-1}.$$

We have included some unneeded parentheses to emphasize the role of the associative law. Since the matter has now been brought to your attention, unnecessary parentheses are omitted from now on. For example, $(AB)C$ and $A(BC)$ are denoted by ABC when defined. In the same spirit, repeated multiplication of the square matrix A is denoted by A^m, with $A^0 = I$ by definition. For example, $A^3 = AAA$.

One of the most important theorems about inverses is the following:

THEOREM 1. *If $AB = I$ then $BA = I$, so each matrix A, B is the inverse of the other. Hence to test whether a given matrix M is the inverse of A, it suffices to check one of the equations $AM = I$ or $MA = I$ and the other is an automatic consequence.*

Theorem 1 is somewhat deeper than anything we have stated up to now and its proof is postponed to Chapter E. It will be used, however, in both this chapter and the next.

If we try to construct A^{-1} by solving $AM = I$ for M we are led to a set of linear equations, one for each column of M. For example, the equation

$$\begin{pmatrix} a & b \\ c & d \end{pmatrix} \begin{pmatrix} p & r \\ q & s \end{pmatrix} = \begin{pmatrix} 1 & 0 \\ 0 & 1 \end{pmatrix}$$

is equivalent to the two linear systems

$$\begin{aligned} ap + bq &= 1 & ar + bs &= 0 \\ cp + dq &= 0 & cr + ds &= 1 \end{aligned}.$$

It is left for you to solve each system and thus to get the formula

$$\begin{pmatrix} a & b \\ c & d \end{pmatrix}^{-1} = \frac{1}{ad - bc} \begin{pmatrix} d & -b \\ -c & a \end{pmatrix}.$$

This assumes $ad \neq bc$; if $ad = bc$ the solution does not exist. By Theorem 1 the equation $MA = I$ holds automatically, so $M = A^{-1}$. The result is summarized as follows:

RULE FOR TWO-BY-TWO INVERSES. To get the inverse of an invertible 2-by-2 matrix, interchange the diagonal elements, change the sign of the off-diagonal elements, and divide by the determinant.

Obviously, the rule cannot be followed if the determinant is 0; but in that case the inverse does not exist.

This rule provides by far the most efficient way to solve systems of two equations in two unknowns. We illustrate in an example.

EXAMPLE 1. To solve the linear system $u + 2v = p$, $3u + 4v = q$ write the coefficient matrix, the inverse, and the solution:

$$\begin{pmatrix} 1 & 2 \\ 3 & 4 \end{pmatrix}, \qquad -\frac{1}{2}\begin{pmatrix} 4 & -2 \\ -3 & 1 \end{pmatrix}, \qquad \begin{pmatrix} u \\ v \end{pmatrix} = \begin{pmatrix} -2 & 1 \\ 3/2 & -1/2 \end{pmatrix}\begin{pmatrix} p \\ q \end{pmatrix}.$$

(e) Calculations with matrices

The following examples are intended chiefly to illustrate the technique of matrix calculation. However, the results have independent interest and will used whenever convenient. We use A, B, M to denote square matrices of order n.

EXAMPLE 2. Let A be invertible and let Y be any matrix with the same number of rows as A. (For example, Y could be a square matrix of order n or a column vector of dimension n.) Show that the equation $AX = Y$ has a unique solution, and find it.

If the equation $AX = Y$ is multiplied *on the left* by A^{-1} the result is

$$A^{-1}AX = A^{-1}Y, \qquad IX = A^{-1}Y, \qquad X = A^{-1}Y.$$

This shows that the only possible solution is $X = A^{-1}Y$. The following calculation shows that $X = A^{-1}Y$ is in fact a solution:

$$AX = AA^{-1}Y = IY = Y.$$

EXAMPLE 3. Suppose the equation $Ax = y$ has at least one solution x for each n-dimensional column vector y. Show that A is invertible (and hence the conclusion of Example 2 holds).

By (2), the equation $AM = I$ gives an equation of the form $Ax = y$ for the first column of M, where y is the first column of I. This can be solved by hypothesis. In a like manner $AM = I$ gives an equation of the form $Ax = y$ for the second column of M, and so on. This technique was already illustrated for 2-by-2 inverses in Part (d) and will be developed systematically in Chapter D.

EXAMPLE 4.   Quotient law. Suppose $Ax = Bx$ for all n-dimensional column vectors x. Show that $A = B$.

If we let x be the first column of I, then the second column of I, and so on, the hypothesis gives $AI = BI$ and hence $A = B$.

EXAMPLE 5. If both A and B are invertible show that AB is also invertible and find its inverse.

Assuming that AB has an inverse M we write $(AB)M = I$ and solve for M as follows:

$$ABM = I, \qquad BM = A^{-1}, \qquad M = B^{-1}A^{-1}.$$

We multiplied *on the left* first by A^{-1} and then by B^{-1}. Thus the only possibility is

$$(AB)^{-1} = B^{-1}A^{-1}.$$

That this is in fact the inverse follows from

$$ABB^{-1}A^{-1} = AIA^{-1} = AA^{-1} = I.$$

EXAMPLE 6.   If A is invertible show that A^{-1}, A^T, and A^* are also invertible and

$$(A^{-1})^{-1} = A, \qquad (A^T)^{-1} = (A^{-1})^T, \qquad (A^*)^{-1} = (A^{-1})^*.$$

Let $B = A^{-1}$. Then $AB = I$, $B^T A^T = (AB)^T = I$, $B^* A^* = (AB)^* = I$. In view of Theorem 1, all three statements follow from this.

PROBLEMS

1. Find AB_1, AB_2, AB if:

$$A = \begin{pmatrix} 1 & 2 & 3 \\ 0 & 1 & 2 \\ 0 & 0 & 1 \end{pmatrix} \quad B_1 = \begin{pmatrix} 0 \\ -4 \\ 3 \end{pmatrix} \quad B_2 = \begin{pmatrix} 0 \\ -3 \\ 2 \end{pmatrix} \quad B = \begin{pmatrix} 0 & 0 & 1 \\ -4 & -3 & -2 \\ 3 & 2 & 1 \end{pmatrix}$$

2. Find B^*, $(A - B)^2$, AC, AD, $D^T A$ and in two ways find $D^T AD$, if

$$A = \begin{pmatrix} 1 & 1 \\ 0 & 1 \end{pmatrix} \quad B = \begin{pmatrix} 1 & -1 \\ -1 & 1 \end{pmatrix} \quad C = \begin{pmatrix} 1 & 2 \\ 0 & 0 \end{pmatrix} \quad D = \begin{pmatrix} u \\ v \end{pmatrix}$$

3. Find A^*, AB, BA and in two ways find $C^T AC$ if

$$A = \begin{pmatrix} 0 & 1 & 2 \\ -1 & 0 & 3 \\ -2 & -3 & 0 \end{pmatrix} \quad B = \begin{pmatrix} 9 & -6 & 3 \\ -6 & 4 & -2 \\ 3 & -2 & 1 \end{pmatrix} \quad C = \begin{pmatrix} u \\ v \\ w \end{pmatrix}$$

4. Let

$$A = \begin{pmatrix} 1 & 0 & 1 \\ 1 & -1 & 0 \end{pmatrix} \qquad B = \begin{pmatrix} 1 & 0 \\ 1 & 1 \end{pmatrix} \qquad C = (1 \quad 0 \quad -1)$$

Decide which of the following products are meaningful and compute those that are meaningful in two ways, using the associative law:

$$ABC \qquad ABC^* \qquad BAC^* \qquad C^*BA \qquad BCA^* \qquad CA^*B$$

5. Compute the products

$$\begin{pmatrix} c & d \\ a & b \end{pmatrix} \begin{pmatrix} 1 & 0 & 1 & p \\ 1 & 1 & 0 & q \end{pmatrix} \qquad \begin{pmatrix} p & 0 & 1 & 1 \\ q & 1 & 0 & 1 \end{pmatrix}^T \begin{pmatrix} a & c \\ b & d \end{pmatrix}$$

6. If $i^2 = -1$ classify as Hermitian, symmetric, or neither:

$$\begin{pmatrix} i & 2 & 3 \\ 2 & 8 & i \\ 3 & i & 9 \end{pmatrix} \qquad \begin{pmatrix} 1 & 1+i & 7 \\ 1-i & 3 & 8 \\ 7 & 8 & 5 \end{pmatrix} \qquad \begin{pmatrix} 1 & 2 & 1 \\ 1 & 2 & 1 \\ 1 & 2 & 1 \end{pmatrix}$$

7. Solve as in Example 1 and check by substitution:

(a) $\begin{cases} u + 2v & = 1 \\ 3u + 5v & = 1 \end{cases}$ (b) $\begin{cases} 2u + 3v & = p \\ 4u + 5v & = q \end{cases}$ (c) $\begin{cases} u\cos\theta + v\sin\theta & = p \\ -u\sin\theta + v\cos\theta & = q \end{cases}$

8P. A square matrix (a_{ij}) in which all the off-diagonal elements are zero is called a diagonal matrix and is written $\operatorname{diag}(a_{ii})$. Prove that

$$\operatorname{diag}(\lambda_i)\operatorname{diag}(\mu_i) = \operatorname{diag}(\lambda_i\mu_i)$$

9P. If $\Lambda = \operatorname{diag}(\lambda_i)$ deduce from Problem 8 that $\Lambda^m = \operatorname{diag}(\lambda_i^m)$ for nonnegative integers m.

10P. Let $A = E\Lambda E^{-1}$ where E and Λ are square matrices of the same size with E invertible. Verify the equation $A^m = E\Lambda^m E^{-1}$ for $m = 3$, and show that it holds for all positive integers m.

11. (abcd) If A, B, C, X are square matrices of the same size, and if the needed inverses exist, solve for X:

$$AXB = C \qquad A(X - B) = C \qquad BX^*A^* = (CB^*)^* \qquad XA = XB + C$$

12P. A real matrix A is said to be orthogonal if $A^T A = I$, and a real or complex matrix A is said to be unitary if $A^*A = I$. Show that the following are orthogonal. Are they unitary?

$$\begin{pmatrix} 0 & 0 & 1 \\ 1 & 0 & 0 \\ 0 & 1 & 0 \end{pmatrix} \qquad \frac{1}{\sqrt{6}} \begin{pmatrix} 1 & -2 & 1 \\ \sqrt{2} & \sqrt{2} & \sqrt{2} \\ \sqrt{3} & 0 & -\sqrt{3} \end{pmatrix} \qquad \begin{pmatrix} 1 & 0 & 0 \\ 0 & \cos\theta & \sin\theta \\ 0 & -\sin\theta & \cos\theta \end{pmatrix}$$

13. In this problem A, B, E are square matrices of order n and u, v are column vectors of dimension n.

(a) Show that AB is orthogonal if A and B are each orthogonal, and AB is unitary if A and B are each unitary.

(b) If E is unitary show that $E^* A E = B \iff A = E B E^*$.

(c) If A is Hermitian show that $(u^* A v)^* = v^* A u$.

14. Block multiplication. The rule for multiplication of matrices can be extended to allow the elements a_{ij} themselves to be matrices. For example, if A_i and B_i are matrices such that the right-hand side is meaningful,

$$\begin{pmatrix} A_1 & A_2 \\ A_3 & A_4 \end{pmatrix} \begin{pmatrix} B_1 & B_2 \\ B_3 & B_4 \end{pmatrix} = \begin{pmatrix} A_1 B_1 + A_2 B_3 & A_1 B_2 + A_2 B_4 \\ A_3 B_1 + A_4 B_3 & A_3 B_2 + A_4 B_4 \end{pmatrix}$$

(a) Let $A_1 = (3)$, $A_2 = (-1 \quad 1)$, $B_1 = (2)$, $B_2 = (2 \quad 0)$, and

$$A_3 = \begin{pmatrix} p \\ q \end{pmatrix}, \qquad A_4 = \begin{pmatrix} 1 & 2 \\ 2 & 3 \end{pmatrix}, \qquad B_3 = \begin{pmatrix} u \\ v \end{pmatrix} \qquad B_4 = \begin{pmatrix} -1 & 0 \\ 3 & 1 \end{pmatrix}$$

Obtain 3-by-3 matrices by inserting these values for A_i and B_i on the left-hand side of the above equation and dropping parentheses. Then show that the product of these 3-by-3 matrices agrees with the right-hand side.

(b) Make a similar check when A_i and B_i are 2-by-2 matrices of your own choosing; the left-hand side is now a product of 4-by-4 matrices. Your example should lead to easy calculations, yet should not have so many zero elements that it becomes trivial.

—————————————————————ANSWERS———————————————————————

1. First column in A^T, second column in A^T, and A^T itself.

2. B, $2I$, C, $-$, $-$, $(u^2 + uv + v^2)$ 3. $-A$, 0, 0, 0

4. Partial answer: Two are meaningful, so you have to compute eight matrix products. The computation is self-checking.

5. (a) Interchange the two rows, and the first and last column, in the matrix on the right of (1). (b) Transpose of (1).   6. SHN

11. (a) $A^{-1} C B^{-1}$ (b) $B + A^{-1} C$ (c) $A^{-1} C$ (d) $C(A - B)^{-1}$

12. Yes, because $A^T = A^*$ for real matrices A.

———

The essence of mathematics is its freedom.

—Georg Cantor

Chapter D
LINEAR EQUATIONS

Throughout this chapter $A = (a_{ij})$ is a square matrix of order n, and x and y are column vectors of dimension n. The linear equations

$$\sum_{j=1}^{n} a_{ij}x_j = y_i, \qquad i = 1,\, 2,\, \cdots,\, n$$

can be written in the compact form $Ax = y$. This follows from the definition of matrix multiplication. For example, the system

$$a_{11}x_1 + a_{12}x_2 + a_{13}x_3 = y_1$$
$$a_{21}x_1 + a_{22}x_2 + a_{23}x_3 = y_2$$
$$a_{31}x_1 + a_{32}x_2 + a_{33}x_3 = y_3$$

is equivalent to

$$\begin{pmatrix} a_{11} & a_{12} & a_{13} \\ a_{21} & a_{22} & a_{23} \\ a_{31} & a_{32} & a_{33} \end{pmatrix} \begin{pmatrix} x_1 \\ x_2 \\ x_3 \end{pmatrix} = \begin{pmatrix} y_1 \\ y_2 \\ y_3 \end{pmatrix}.$$

Matrix notation gives a very efficient proof of the following:

THEOREM 1. *Suppose the equation $Ax = y$ has two distinct solutions. Then it has infinitely many.*

In other words, a linear system always has no solution, or exactly one solution, or infinitely many. By contrast, a quadratic equation in one unknown generally has exactly two solutions.

For proof let u and v be distinct solutions, so that $Au = y$, $Bv = y$, and $u \neq v$. If λ is any constant we have

$$A(u + \lambda(u - v)) = Au + \lambda(Au - Av) = y + \lambda(y - y) = y.$$

This gives infinitely many solutions $x = u + \lambda(u - v)$ and completes the proof. Theorem 1 holds even if A is not square and even if y stands for a matrix rather than a vector. The proof is unchanged.

Linear systems can be solved by a procedure that is called the method of pivot elements, the method of detached coefficients, or Gauss-Jordan elimination. (The latter is named after Gauss and the German geodesist Wilhelm Jordan.) Instead of using letters to represent the unknowns, one specifies the system by giving the augmented matrix (A, y). This is the coefficient matrix A with an extra column y at the right. As an illustration, the augmented matrix of the 3-by-3 system above is

$$\begin{pmatrix} a_{11} & a_{12} & a_{13} & y_1 \\ a_{21} & a_{22} & a_{23} & y_2 \\ a_{31} & a_{32} & a_{33} & y_3 \end{pmatrix}.$$

The following example illustrates the main idea.

EXAMPLE 1.   Solve the system on the left below, with the augmented matrix at the right:

$$\begin{array}{ll} u + 3v = 5 & \begin{pmatrix} 1^* & 3 & 5 \\ 2 & 4 & 6 \end{pmatrix} \\ 2u + 4v = 6 \end{array}$$

We subtract twice the first equation from the second, which means we subtract twice the first row of the augmented matrix from the second row. The element 1^* in the augmented matrix plays a special role in this calculation and is called a pivot; a formal definition is given later. The result of this first step is the left-hand matrix below:

$$\begin{pmatrix} 1 & 3 & 5 \\ 0 & -2 & -4 \end{pmatrix} \quad \begin{pmatrix} 1 & 3 & 5 \\ 0 & 1^* & 2 \end{pmatrix} \quad \begin{pmatrix} 1 & 0 & -1 \\ 0 & 1 & 2 \end{pmatrix}$$

The next matrix is obtained when the second equation is divided by -2. The last matrix is obtained when -3 times the second equation is added to the first equation. The linear system corresponding to the last matrix yields $u = -1$, $v = 2$, which gives the solution of the original system.

(a) Elementary row operations

Here are the main algebraic operations used in calculations such as those above:

(i) Multiply an equation by a nonzero constant.

(ii) Add a multiple of one equation to another equation.

(iii) Interchange two equations.

These correspond to operations on the augmented matrix that are called *elementary row operations* and are defined as follows:

(i) Multiply a row by a nonzero constant.

(ii) Add a multiple of one row to another row.

(iii) Interchange two rows.

The objective is to get an equivalent system that can be solved by inspection, as in the following example.

EXAMPLE 2. Solve by elementary row operations:

$$x_1 - x_2 + x_3 = 3$$
$$x_1 + 2x_2 + 4x_3 = 9$$
$$2x_1 + 3x_2 + 4x_3 = 7$$

The augmented matrix is on the left below, and a simplification is on the right:

$$\begin{pmatrix} 1^* & -1 & 1 & 3 \\ 1 & 2 & 4 & 9 \\ 2 & 3 & 4 & 7 \end{pmatrix} \begin{matrix} \\ R_2 - R_1 \\ R_3 - 2R_1 \end{matrix} \begin{pmatrix} 1 & -1 & 1 & 3 \\ 0 & 3 & 3 & 6 \\ 0 & 5 & 2 & 1 \end{pmatrix} R_2/3 \begin{pmatrix} 1 & -1 & 1 & 3 \\ 0 & 1^* & 1 & 2 \\ 0 & 5 & 2 & 1 \end{pmatrix}.$$

The element 1^* was used as a pivot to eliminate all other elements from the first column. The symbol $R_2 - R_1$ between the first two matrices indicates that the first row was subtracted from the second. It is placed opposite the second row because the second row was changed by the operation. Interpretation of $R_3 - 2R_1$ and $R_2/3$ is similar.

To simplify the second column we add the second row to the first and we subtract 5 times the second row from the third. The result is the first matrix on the left below:

$$\begin{pmatrix} 1 & 0 & 2 & 5 \\ 0 & 1 & 1 & 2 \\ 0 & 0 & -3 & -9 \end{pmatrix} \begin{matrix} \\ \\ -R_3/3 \end{matrix} \begin{pmatrix} 1 & 0 & 2 & 5 \\ 0 & 1 & 1 & 2 \\ 0 & 0 & 1 & 3 \end{pmatrix} \begin{matrix} R_1 - 2R_3 \\ R_2 - R_3 \\ \end{matrix} \begin{pmatrix} 1 & 0 & 0 & -1 \\ 0 & 1 & 0 & -1 \\ 0 & 0 & 1 & 3 \end{pmatrix}.$$

Continuing the process we get the matrix on the right, which corresponds to the system $x_1 = -1$, $x_2 = -1$, $x_3 = 3$. Merely writing this system gives the solution to the original equations.

(b) Pivots

In solving a linear system by Gauss-Jordan reduction, we always simplify from left to right, and we require that subsequent simplifications shall not disrupt columns already simplified. This suggests the following definition:

DEFINITION. A pivot is any element of the coefficient matrix that is the first nonzero element of its row.

The value of a pivot is usually reduced to 1 by division, but this requirement is not built into the definition.

For example in the row

$$0 \quad 0 \quad 0 \quad 4^* \quad 0 \quad 0 \quad 1 \quad 3 \quad 5 \quad 7$$

the element 4 could be used as a pivot but the following 1 could not. If 1 is used, the 4 on the left would disrupt the column in which 4 lies.

A column that has a pivot and at least one other nonzero element is unreduced. All other columns are reduced. We can now state the following:

RULE OF REDUCTION FOR LINEAR SYSTEMS. Find the leftmost unreduced column, reduce it by use of any pivot in it, and repeat until every column is reduced.

In the next example we have chosen the numbers so that no row has to be divided by a constant. Such division complicates the arithmetic but does not illustrate any new concepts.

EXAMPLE 3. Here is a possible choice of pivots:

$$\begin{pmatrix} -1 & 0 & -4 & 2 & 7 \\ -2 & 0 & -8 & 3 & 9 \\ 1^* & 1 & 6 & 0 & 4 \\ 1 & 2 & 8 & 1 & 10 \end{pmatrix} \begin{matrix} R_1 + R_3 \\ R_2 + 2R_3 \\ \\ R_4 - R_3 \end{matrix} \begin{pmatrix} 0 & 1 & 2 & 2 & 11 \\ 0 & 2 & 4 & 3 & 17 \\ 1 & 1 & 6 & 0 & 4 \\ 0 & 1^* & 2 & 1 & 6 \end{pmatrix} \begin{matrix} R_1 - R_4 \\ R_2 - 2R_4 \\ R_3 - R_4 \\ \\ \end{matrix}$$

Subtracting the fourth row from the others, as indicated, we get

$$\begin{pmatrix} 0 & 0 & 0 & 1 & 5 \\ 0 & 0 & 0 & 1^* & 5 \\ 1 & 0 & 4 & -1 & -2 \\ 0 & 1 & 2 & 1 & 6 \end{pmatrix} \begin{matrix} R_1 - R_2 \\ \\ R_3 + R_2 \\ R_4 - R_2 \end{matrix} \begin{pmatrix} 0 & 0 & 0 & 0 & 0 \\ 0 & 0 & 0 & 1 & 5 \\ 1 & 0 & 4 & 0 & 3 \\ 0 & 1 & 2 & 0 & 1 \end{pmatrix}.$$

The final system is $x_4 = 5$, $x_1 + 4x_3 = 3$, $x_2 + 2x_3 = 1$. Hence we can set $x_3 = c$ where c is any constant and express the solution as follows:

$$x_1 = 3 - 4c, \qquad x_2 = 1 - 2c, \qquad x_3 = c, \qquad x_4 = 5.$$

There are infinitely many solutions, and the above formulas give them all. If we had set $x_1 = c$ or $x_2 = c$ the resulting formulas would look different, but would represent the same set of solutions.

(c) Row-echelon form

The rule of reduction does not require the third elementary row operation, interchange of two rows. However, for many purposes it is desirable to arrange the reduced matrix in row-echelon form. This is a form in which the leftmost pivot is in the top row only, the next leftmost pivot is in the second row only, and so on. For example, the row-echelon form of the final matrix obtained in Example 3 is

$$\begin{array}{c} R_3 \\ R_4 \\ R_2 \\ R_1 \end{array} \begin{pmatrix} 1 & 0 & 4 & 0 & 3 \\ 0 & 1 & 2 & 0 & 1 \\ 0 & 0 & 0 & 1 & 5 \\ 0 & 0 & 0 & 0 & 0 \end{pmatrix}.$$

Here the R_3 opposite the first row means that the top row is the third row of the original matrix, and similarly in other cases. Going from a reduced matrix to a matrix in row-echelon form requires only one more step, no matter how large the matrix may be.

As defined above, a row-echelon form is not unique. A matrix in row-echelon form is in *reduced row-echelon form* if each pivot element is equal to 1 and is the only nonzero element in its column. Informally, this means that the process of simplification has been carried as far as possible.

It can be shown that the reduced row-echelon form is unique. In particular, if each column has a pivotal element, the reduced row-echelon form of the coefficient matrix is I, the reduced augmented matrix has the form (I, z), and the solution of the corresponding linear system is $x = z$. For example the augmented matrix

$$\begin{pmatrix} 1 & 0 & 0 & z_1 \\ 0 & 1 & 0 & z_2 \\ 0 & 0 & 1 & z_3 \end{pmatrix}$$

corresponds to the system $x_1 = z_1$, $x_2 = z_2$, $x_3 = z_3$, which gives $x = z$.

(d) Computation of the inverse

The following two features of the Gauss-Jordan method are helpful when linear systems are used to compute the inverse of a given matrix A:

(i) As soon as the coefficient matrix is reduced to I, you can read the solution by looking at the right-hand column of the augmented matrix.

(ii) The procedure by which the reduction (i) is effected is independent of the right-hand column of the augmented matrix.

Because of these features the n linear systems needed to get the columns of M from $AM = I$ can be solved all at once, as in the following example.

EXAMPLE 4.   We shall compute the inverse of the 3-by-3 matrix that forms the left half of the first matrix below. The initial steps are:

$$\begin{pmatrix} 1^* & 0 & -1 & 1 & 0 & 0 \\ 2 & 1 & 1 & 0 & 1 & 0 \\ 1 & 1 & 3 & 0 & 0 & 1 \end{pmatrix} \begin{array}{c} \\ R_2 - 2R_1 \\ R_3 - R_1 \end{array} \begin{pmatrix} 1 & 0 & -1 & 1 & 0 & 0 \\ 0 & 1^* & 3 & -2 & 1 & 0 \\ 0 & 1 & 4 & -1 & 0 & 1 \end{pmatrix}$$

The next step is to subtract the second row from the third. Thus we get the matrix on the left below, and the process is completed to give the matrix on the right:

$$\begin{pmatrix} 1 & 0 & -1 & 1 & 0 & 0 \\ 0 & 1 & 3 & -2 & 1 & 0 \\ 0 & 0 & 1^* & 1 & -1 & 1 \end{pmatrix} \begin{matrix} R_1 + R_3 \\ R_2 - 3R_3 \end{matrix} \begin{pmatrix} 1 & 0 & 0 & 2 & -1 & 1 \\ 0 & 1 & 0 & -5 & 4 & -3 \\ 0 & 0 & 1 & 1 & -1 & 1 \end{pmatrix}.$$

The right half of the last matrix is the desired inverse. As a check we compute

$$\begin{pmatrix} 1 & 0 & -1 \\ 2 & 1 & 1 \\ 1 & 1 & 3 \end{pmatrix} \begin{pmatrix} 2 & -1 & 1 \\ -5 & 4 & -3 \\ 1 & -1 & 1 \end{pmatrix} = \begin{pmatrix} 1 & 0 & 0 \\ 0 & 1 & 0 \\ 0 & 0 & 1 \end{pmatrix}.$$

The general principle illustrated here is that if you start with the augmented matrix $(A\ I)$ and reduce the coefficient matrix A to I by row operations, in the final augmented matrix $(I\ M)$ the right-hand half M gives A^{-1}. Though surprising, this result does not require a new theory. It is just an application of the procedure developed in Part (c) to several systems at once.

Since the process produces A^{-1}, it is bound to fail if A is not invertible; that is, A cannot be reduced to I by row operations in that case. In Chapter E you are invited to show, conversely, that if A is invertible then A can always be reduced to I by row operations.

PROBLEMS

1P. Suppose $Ax = 0$ has a solution $x = u \neq 0$. Show that $Ax = y$ either has no solution or infinitely many; it cannot have exactly one solution. Hint: Consider $x + \lambda u$.

2P. Suppose $Ax = y$ has two distinct solutions $x = u$ and $x = v$. Show $Ax = 0$ has infinitely many solutions. Hint: $x = \lambda(u - v)$.

3. (abc) The following are augmented matrices (A, y) for a linear system $Ax = y$ of three equations in three unknowns. Solve by row operations. You need not reduce to echelon form:

$$\begin{pmatrix} 1^* & 2 & 3 & 0 \\ 2 & 3 & 1 & 4 \\ 3 & 4 & 2 & 5 \end{pmatrix} \quad \begin{pmatrix} 4 & 7 & -1 & 3 \\ 3 & 2 & 3 & 6 \\ 2^* & 4 & 6 & 8 \end{pmatrix} \quad \begin{pmatrix} 3 & 2 & 8 & 4 \\ 4^* & 8 & 0 & 0 \\ 2 & 4 & 7 & 7 \end{pmatrix}$$

4. (abc) If possible, proceed as in Problem 3:

$$\begin{pmatrix} 1 & -1 & 2 & 1 \\ 3 & 4 & 2 & 13 \\ 5 & 1 & 4 & 11 \end{pmatrix} \quad \begin{pmatrix} 1 & 2 & 3 & 1 \\ 3 & 7 & 10 & 4 \\ 2 & 4 & 6 & 5 \end{pmatrix} \quad \begin{pmatrix} 1 & 2 & 3 & 10 \\ 2 & 4 & 5 & 19 \\ 1 & 3 & 6 & 15 \end{pmatrix}$$

5P. (abcde) For a homogeneous system $Ax = 0$ the rightmost column of the augmented matrix remains 0 during the reduction and need not be written down. Here are coefficient matrices for homogeneous systems with $n = 3$. After reduction, obtain a nontrivial solution with $x_3 = 1$ if possible. If not, try $x_2 = 1$. If that fails too, try $x_1 = 1$:

$$\begin{pmatrix} 1 & 0 & 0 \\ 0 & 0 & 0 \\ 0 & 1 & 4 \end{pmatrix} \begin{pmatrix} 0 & 0 & 1 \\ 1 & 2 & 0 \\ 0 & 0 & 0 \end{pmatrix} \begin{pmatrix} 0 & 0 & 0 \\ 1 & 0 & 1 \\ 0 & 1 & 1 \end{pmatrix} \begin{pmatrix} 0 & 1 & 2 \\ 0 & 1 & 1 \\ 0 & 1 & 3 \end{pmatrix} \begin{pmatrix} 1 & -1 & -6 \\ 0 & 0 & 0 \\ 1 & 2 & 3 \end{pmatrix}$$

6P. To get the inverse of the 3-by-3 matrix at the left, you can start as follows:

$$\begin{pmatrix} 0 & 3 & 3 & 1 & 0 & 0 \\ 0 & 1 & 2 & 0 & 1 & 0 \\ 3 & 6 & 9 & 0 & 0 & 1 \end{pmatrix} R_3/3, R_{13} \begin{pmatrix} 1 & 2 & 3 & 0 & 0 & 1/3 \\ 0 & 1 & 2 & 0 & 1 & 0 \\ 0 & 3 & 3 & 1 & 0 & 0 \end{pmatrix}$$

The notation $R_3/3$, R_{13} means that the third row is divided by 3 and that the first and third rows are then interchanged. The next step would be indicated by writing $R_1 - 2R_2$ after the first row and $R_3 - 3R_2$ after the third. Complete the calculation and check by computing AA^{-1} or $A^{-1}A$. Since you are asked to check, no answer is provided.

7. (abcde) As in Problem 6, find the inverse and check:

$$\begin{pmatrix} 1 & 2 & 3 \\ 0 & 1 & 4 \\ 0 & 0 & 1 \end{pmatrix} \begin{pmatrix} 1 & 0 & 0 \\ 3 & 1 & 0 \\ 7 & 4 & 1 \end{pmatrix} \begin{pmatrix} 1 & 0 & 1 \\ 2 & 3 & 1 \\ 1 & 1 & 1 \end{pmatrix} \begin{pmatrix} 2 & 2 & 1 \\ 3 & 2 & 1 \\ 4 & 3 & 2 \end{pmatrix} \begin{pmatrix} 2 & 1 & 1 \\ 3 & 3 & 2 \\ 4 & 5 & 5 \end{pmatrix}$$

8. (abcd) Compute the inverse by reducing (A, I) and check:

$$\begin{pmatrix} 1 & 2 & 3 & 4 \\ 0 & -1 & 2 & 3 \\ 0 & 0 & 1 & 5 \\ 0 & 0 & 0 & -1 \end{pmatrix} \begin{pmatrix} 1 & 0 & 0 & 0 \\ 3 & 1 & 0 & 0 \\ 4 & 3 & 1 & 0 \\ 5 & 1 & 1 & 1 \end{pmatrix} \begin{pmatrix} 1 & 1 & 0 & 0 \\ 1 & 0 & 1 & 0 \\ 1 & 1 & 0 & 1 \\ 1 & 0 & 0 & 0 \end{pmatrix} \begin{pmatrix} 1 & 1 & 0 & 0 \\ 1 & 0 & 1 & 0 \\ 1 & 1 & 0 & 1 \\ 1 & 1 & 1 & 1 \end{pmatrix}$$

9P. The inner product. Two vectors x, y of dimension n are said to be (complex) orthogonal if $(x, y) = 0$ where, by definition,

$$(x, y) = x_1\bar{y}_1 + x_2\bar{y}_2 + \cdots + x_n\bar{y}_n$$

(a) If $\alpha = (x, y)$ show that $y^*x = (\alpha)$ for column vectors and $xy^* = (\alpha)$ for row vectors. The expression (x, y) is called the *inner product* of x and y.

(b) Check that $\|x + y\|^2 = \|x\|^2 + \|y\|^2$ holds for real vectors x, y if, and only if, they are orthogonal.

10. The length of x is $\|x\| = (x, x)^{1/2}$. If E is unitary, that is, if $E^*E = I$, show that the columns of E are orthogonal and have length 1. Is it necessarily true that the rows of E are orthogonal and have length 1?

11. The Cauchy-Buniakowsky-Schwarz inequality. The inequality

$$|(x, y)| \le \|x\| \|y\|$$

was established first by Cauchy, extended later to integrals by Buniakowsky, and again still later by Schwarz. Hence the elaborate name. Since it is trivial when $y = 0$ we take $\|y\| > 0$. Fill in the details of the following proofs of the CBS inequality for real vectors (a) and complex vectors (b):

(a) We have $(x, x) + 2\lambda(x, y) + \lambda^2(y, y) = \|x + \lambda y\|^2 \ge 0$. Hence the minimum with respect to λ is also ≥ 0.

(b) Apply the real case (a) to $X = (|x_i|)$ and $Y = (|y_i|)$.

12. Let A be a matrix of order n and x, y column vectors of dimension n.

(a) Show that $(Ax, y) = (x, A^*y)$. This equation occurs in a branch of mathematics, known as the theory of operator algebras, which generalizes the theory of matrices. The operator A^* is the *adjoint* of A and that is why the conjugate transpose A^* is called the adjoint.

Hint: The desired equation is equivalent to $y^*Ax = (A^*y)^*x$.

(b) If $Ax = y$ has a solution x, show that y must be orthogonal to every vector z satisfying $A^*z = 0$. The converse is also true; that is, if y has the stated property then $Ax = y$ has a solution. This gives a profound insight into the deeper properties of linear systems.

Hint: Part (a) gives $(y, z) = (Ax, z) = (x, A^*z)$.

---------------------------**ANSWERS**---------------------------

3. (a) 1, 1, -1 (b) 1, 0, 1   (c) -2, 1, 1
4. (a) 1, 2, 1   (b) no solution   (c) 3, 2, 1
5. $x^T = (0 \ -4 \ 1)$ $(-2 \ 1 \ 0)$ $(-1 \ -1 \ 1)$ $(1 \ 0 \ 0)$ $(3 \ -3 \ 1)$
10. Yes, because Theorem 1 gives $EE^* = I$.

Rigor is no foe of simplicity.

—Otto Blumenthal

Chapter E
DETERMINANTS

Let $A = (a_{ij})$ be a square matrix of order at least 2 and let the parentheses ( ) in the displayed form of A be replaced by vertical bars | |. The result is a numerical-valued function of A, which is written $\det A$ or $|A|$ and is called the determinant. Here are two examples:

$$(1) \qquad \begin{vmatrix} a_1 & a_2 \\ b_1 & b_2 \end{vmatrix} = a_1 b_2 - a_2 b_1$$

$$(2) \qquad \begin{vmatrix} a_1 & a_2 & a_3 \\ b_1 & b_2 & b_3 \\ c_1 & c_2 & c_3 \end{vmatrix} = a_1 \begin{vmatrix} b_2 & b_3 \\ c_2 & c_3 \end{vmatrix} - a_2 \begin{vmatrix} b_1 & b_3 \\ c_1 & c_3 \end{vmatrix} + a_3 \begin{vmatrix} b_1 & b_2 \\ c_1 & c_2 \end{vmatrix}.$$

Note that the determinants on the right of (2) can be evaluated by (1). Note also that the vertical bars | | replace the parentheses ordinarily associated with a matrix; we do not use | | in addition to ( ).

Although the value of a determinant is a number, not a matrix, terminology is extended in a natural way from A to $\det A$. For example, the determinant on the left of (2) is of order 3, the first row of the determinant (1) is $(a_1\ a_2)$, the elements of a determinant are the elements of the corresponding matrix, and so on. A determinant of order 1 has only a single element and is defined by $\det(a) = a$. The notation $|a|$ is not used in this case because it could be confused with the absolute value.

By analogy to (1) and (2), a determinant of order $n \geq 2$ consists of a square array

$$(3) \qquad \det A = |A| = \begin{vmatrix} a_{11} & a_{12} & \cdots & a_{1n} \\ a_{21} & a_{22} & \cdots & a_{2n} \\ \vdots & \vdots & \ddots & \vdots \\ a_{n1} & a_{n2} & \cdots & a_{nn} \end{vmatrix}$$

to which a numerical value is attached as follows: Let M_{1i} be the determinant obtained when the first row and the ith column of the original determinant are deleted. Then

$$(4) \quad |A| = a_{11} M_{11} - a_{12} M_{12} + a_{13} M_{13} - a_{14} M_{14} + \cdots + (-1)^{n+1} a_{1n} A_{1n}.$$

This definition is inductive; a determinant of order n is defined in terms of those of order $n-1$. You can easily convince yourself that the definition gives (1) and (2) when $n = 2$ and $n = 3$, respectively. It should be emphasized that *only a square matrix can have a determinant* and, with that in mind, we use $A = (a_{ij})$ throughout this chapter to denote a square matrix of order n. We also use x and y to denote column vectors of dimension n.

The following example shows how determinants are reduced to determinants of lower order by successive expansion on the first row.

EXAMPLE 1. No matter what numbers are represented by the stars,

$$\begin{vmatrix} a & 0 & 0 & 0 \\ * & b & 0 & 0 \\ * & * & c & 0 \\ * & * & * & d \end{vmatrix} = a \begin{vmatrix} b & 0 & 0 \\ * & c & 0 \\ * & * & d \end{vmatrix} = ab \begin{vmatrix} c & 0 \\ * & d \end{vmatrix} = abcd.$$

(a) Laplace expansion

The determinant M_{ij} obtained when the ith row and jth column of (3) are deleted is the *minor* of the element a_{ij} and the signed minor

$$A_{ij} = (-1)^{i+j} M_{ij}$$

is the *cofactor* of a_{ij}. As an illustration, the 2-by-2 determinants on the right of (2) are the minors of the elements a_1, a_2, a_3 respectively, and when provided with alternating signs $+$, $-$, $+$ they become the corresponding cofactors. Hence (2) can be written

$$|A| = a_1 A_{11} + a_2 A_{12} + a_3 A_{13}.$$

In a like manner, the definition (4) is

$$(5) \qquad\qquad \det A = a_{11} A_{11} + a_{12} A_{12} + \cdots + a_{1n} A_{1n}.$$

Equation (5) is a *Laplace expansion* on the first row, named after the French mathematician Pierre Simon de Laplace. More generally we can make a Laplace expansion on the ith row:

$$|A| = a_{i1} A_{i1} + a_{i2} A_{i2} + \cdots + a_{in} A_{in},$$

or also on the jth column:

$$|A| = a_{1j} A_{1j} + a_{2j} A_{2j} + \cdots + a_{nj} A_{nj}.$$

The fact that all such expansions give the same value is the content of the following theorem, which provides a very convenient approach to the theory of determinants:

EXPANSION THEOREM. *A determinant can be evaluated by Laplace expansion on any row or on any column.*

The main ideas needed for the proof are outlined in the problems, though a complete proof will not be given in this book. The following examples show how the theorem is used.

EXAMPLE 2. Evaluate the determinant D below:

$$D = \begin{vmatrix} 1 & 0 & -1 & 2 \\ 6 & 0 & 4 & 3 \\ 4 & 7 & 0 & 2 \\ 2 & 0 & 2 & 3 \end{vmatrix} \qquad S = \begin{vmatrix} + & - & + & - \\ - & + & - & + \\ + & - & + & - \\ - & + & - & + \end{vmatrix}$$

The sign matrix S at the right interprets the $(-1)^{i+j}$ rule for going from minors to cofactors. Expanding D on the second column, and using signs as given by the second column of S, we get

$$D = 0(-M_{12}) + 0(+M_{22}) + 7(-M_{32}) + 0(+M_{42}).$$

The only surviving minor M_{32} is obtained by crossing out the row and column containing the element 7 in D. This gives the first equality below. The second is obtained when M_{32} is expanded on its first row:

$$D = (-7) \begin{vmatrix} 1 & -1 & 2 \\ 6 & 4 & 3 \\ 2 & 2 & 3 \end{vmatrix} = (-7)(6 + 12 + 8) = -182.$$

EXAMPLE 3. No matter what numbers are represented by the stars,

$$\begin{vmatrix} 1 & 2 & 0 & * \\ * & * & 4 & * \\ 2 & 1 & 0 & * \\ 0 & 0 & 0 & 3 \end{vmatrix} = -4 \begin{vmatrix} 1 & 2 & * \\ 2 & 1 & * \\ 0 & 0 & 3 \end{vmatrix} = -12 \begin{vmatrix} 1 & 2 \\ 2 & 1 \end{vmatrix} = 36.$$

Here the 4-by-4 determinant was expanded on the third column and the 3-by-3 determinant was then expanded on the third row.

(b) Elementary properties

The following properties of determinants are of fundamental importance:

(i) If each element of a row or column is 0, the determinant is 0.

(ii) If each element of a row or column is multiplied by λ the determinant is multiplied by λ.

(iii) If each element of a row or column is a sum of two terms the determinant is a sum of two corresponding determinants; for example,

$$(6) \qquad \begin{vmatrix} a_1 & a_2 & a_3 \\ b_1 + d_1 & b_2 + d_2 & b_3 + d_3 \\ c_1 & c_2 & c_3 \end{vmatrix} = \begin{vmatrix} a_1 & a_2 & a_3 \\ b_1 & b_2 & b_3 \\ c_1 & c_2 & c_3 \end{vmatrix} + \begin{vmatrix} a_1 & a_2 & a_3 \\ d_1 & d_2 & d_3 \\ c_1 & c_2 & c_3 \end{vmatrix}.$$

These follow by making a Laplace expansion on the row or column in question. For example the determinant on the left of (6) is expanded on the second row as follows:

$$(b_1 + d_1)A_{21} + (b_2 + d_2)A_{22} + (b_3 + d_3)A_{23}.$$

The terms with b_j give a Laplace expansion of the first determinant on the right and those with d_j give a Laplace expansion of the second. This illustrates the proof of (iii). Obviously if we expand on a row or column that contains only 0, we will get 0. This gives (i). It is left for you to discuss (ii) along the same lines. Properties (ii) and (iii) are summarized by saying that *a determinant is a linear function of any row or column in it.*

(iv) If two rows or two columns are interchanged the determinant changes sign.

(v) If rows and columns are interchanged the determinant is unaltered.

In the language of matrices, Property (v) means $\det A = \det A^T$.

These statements are easily checked for 2-by-2 determinants and are proved for $n > 2$ by mathematical induction. For example, if two rows are interchanged, expand on any row not involved in the interchange. The cofactors used in this expansion are of order $n - 1$ and each of them has two of its rows interchanged. By the induction hypothesis each of them changes sign and this gives property (iv). Property (v) says that $|A^T| = |A|$. To prove it, expand the original determinant $|A|$ on its first row and the transposed determinant $|A^T|$ on its first column. Again the conclusion follows by induction.

(vi) If two rows are equal, or two columns are equal, the determinant is 0.

(vii) If one row is a multiple of another, or one column is a multiple of another, the determinant is 0.

(viii) If a multiple of one row is added to another row, or a multiple of one column is added to another column, the determinant is unchanged.

(ix) If A and B are square matrices of the same size, $|AB| = |A||B|$.

These follow from results already established. If two equal columns are interchanged the determinant is unchanged, and yet it goes to its negative by (iv). Hence $|A| = -|A|$ and this yields (vi). Property (vii) is obtained by combining (vi) with (ii), and (viii) by combining (vii) with (iii). When $n = 3$ the reasoning is illustrated by

$$\begin{vmatrix} a_1 + \lambda a_3 & a_2 & a_3 \\ b_1 + \lambda b_3 & b_2 & b_3 \\ c_1 + \lambda c_3 & c_2 & c_3 \end{vmatrix} = \begin{vmatrix} a_1 & a_2 & a_3 \\ b_1 & b_2 & b_3 \\ c_1 & c_2 & c_3 \end{vmatrix} + \begin{vmatrix} \lambda a_3 & a_2 & a_3 \\ \lambda b_3 & b_2 & b_3 \\ \lambda c_3 & c_2 & c_3 \end{vmatrix}.$$

The second determinant on the right is 0 by (vii).

To prove (ix) let $A = (a_{ij})$ and $B = (b_{ij})$, so that

$$|AB| = |a_{i1}b_{1j} + a_{i2}b_{2j} + \cdots + a_{in}b_{nj}|.$$

If we think of j as fixed and of i as variable, the sum gives the jth column in AB. Writing A_j for the jth column of A we get

$$|AB| = |A_1 b_{1j} + A_2 b_{2j} + \cdots + A_n b_{nj}|.$$

When $j = 1$ the sum gives the first column in $|AB|$, when $j = 2$ it gives the second, and so on.

It is seen presently that $|AB|$ has the form

$$|AB| = P(b)|A|$$

where $P(b)$ denotes a polynomial in the elements of B. Once we know this, the polynomial $P(b)$ is found by setting $A = I$. The result is $P(b) = |B|$, hence $|AB| = |B||A| = |A||B|$ as desired.

Thus everything depends on the equation $|AB| = P(b)|A|$. To make the general case more understandable, we first discuss the case $n = 2$. Here

$$|AB| = |A_1 b_{11} + A_2 b_{21} \quad A_1 b_{12} + A_2 b_{22}|.$$

Repeated use of Property (iii) gives a sum of four determinants:

$$|A_1 b_{11} \ A_1 b_{12}| + |A_1 b_{11} \ A_2 b_{22}| + |A_2 b_{21} \ A_1 b_{12}| + |A_2 b_{21} \ A_2 b_{22}|.$$

The first and fourth of these are 0 by Property (vii). When the b_{ij} are factored out of their columns, the second and third are seen to be

$$b_{11}b_{22}|A| \quad \text{and} \quad b_{21}b_{12}(-|A|).$$

Obviously, the sum of all four determinants has the form $P(b)|A|$ mentioned above. A similar calculation for $n = 3$ is outlined in Problem 6.

In the general case, repeated use of Property (iii) gives $|AB|$ as a sum of n^n determinants. Properties (ii), (iv), and (vii) show that each term in that sum has the form

$$(\text{product of } n \text{ factors } b_{ij}) \text{ times } (|A|, \ 0, \ \text{or} \ -|A|).$$

Thus the sum has the form $|AB| = P(b)|A|$ and this completes the proof.

These properties show that a determinant can be simplified by elementary operations on the matrix, and we can operate on columns as well as on rows. Here is an example.

EXAMPLE 4. Explain this reduction:

$$\begin{vmatrix} 1 & 2 & 3 & 4 \\ 3 & 6 & 6 & 9 \\ 4 & 6 & 8 & 10 \\ 3 & 4 & 5 & 6 \end{vmatrix} = 6 \begin{vmatrix} 1 & 2 & 3 & 4 \\ 1^* & 2 & 2 & 3 \\ 2 & 3 & 4 & 5 \\ 3 & 4 & 5 & 6 \end{vmatrix} = 6 \begin{vmatrix} 0 & 0 & 1 & 1 \\ 1 & 2 & 2 & 3 \\ 0 & -1 & 0 & -1 \\ 0 & -2 & -1 & -3 \end{vmatrix}.$$

In the first step 3 was factored out of the second row and 2 out of the third row. In the next step the element 1^* was used as a pivot to reduce the first column by row operations; this does not change the value of the determinant. We now expand the determinant on elements of the first column. It is left for you to complete the calculation and show that the determinant is 0.

(c) Determinants and linear systems

The following theorem is of fundamental importance both in linear algebra and in the theory of differential equations:

THEOREM 1. *If $|A| = 0$ the equation $Ax = 0$ has a solution other than the trivial solution $x = 0$. Conversely, existence of a nontrivial solution implies $|A| = 0$.*

When $n = 1$ the equation is $0x_1 = 0$, which obviously has a nontrivial solution. Taking $n \geq 2$, we assume that the first statement in Theorem 1 is known for matrices of order $n - 1$ and we use mathematical induction. There are two cases.

Case 1. If the first column of A is 0 we can choose $x_1 = 1$ and all the other unknowns to be 0. This gives a nontrivial solution.

Case 2. If at least one element $a_{i1} \neq 0$ we assume, by interchanging rows, that $a_{11} \neq 0$. Dividing by a_{11} makes $a_{11} = 1$. Let $a_{11} = 1^*$ be used as a pivotal element to reduce the first column of the augmented matrix. The right-hand column is still 0, and the coefficient matrix now looks like this:

$$C = \begin{pmatrix} 1 & a_{12} & \cdots & a_{1n} \\ 0 & b_{22} & \cdots & b_{2n} \\ \vdots & \vdots & \ddots & \vdots \\ 0 & b_{n2} & \cdots & b_{nn} \end{pmatrix}.$$

The hypothesis $|A| = 0$ yields $|C| = 0$ and, expanding on the first column, we see that $|C| = |B|$ where $B = (b_{ij})$. Hence $|B| = 0$. By the induction hypothesis we can choose $x_2, x_3, \cdots, x_n$, not all 0, satisfying the system associated with B. If x_1 is now chosen so as to satisfy the first equation, we get a nontrivial solution for the system associated with C and hence for the original system. This completes the proof of the first statement. The second statement is taken up after the proof of Theorem 2.

A counterpart to Theorem 1 is:

THEOREM 2. *If $|A| \neq 0$ the equation $Ax = y$ has a unique solution x for each y. Conversely, if $Ax = y$ has a unique solution for at least one y, then $|A| \neq 0$.*

Let $|A| \neq 0$. When $n = 1$ the equation is $ax_1 = y_1$ with $a \neq 0$. This has the unique solution $x_1 = y_1/a$. Taking $n \geq 2$, we assume that the first statement in Theorem 2 is known for matrices of order $n - 1$ and we use mathematical induction. If the first column of A is 0 then $|A| = 0$, which is contrary to the hypothesis. Hence Case 1 above is ruled out automatically. The rest of the argument is similar to that in Case 2 and is left to you.

So far, we have proved (a) that there is a nontrivial solution u of $Au = 0$ if $|A| = 0$, and (b) that there is a unique solution x of $Ax = y$ if $|A| \neq 0$. These proofs were similar but independent.

Result (b) shows that $x = 0$ is the sole solution of $Ax = 0$ when $|A| \neq 0$, and this gives the second statement in Theorem 1. When $|A| = 0$, Result (a) shows that the equation $Ax = y$ has infinitely many solutions if it has any; namely, it has solutions $x + cu$ for any constant c. This gives the second statement in Theorem 2 and completes the proof of both theorems.

When $|A| \neq 0$, Part (b) of the following gives a formula for the unique solution in Theorem 2:

CRAMER'S RULE. Let D_i be the determinant obtained when the ith column of $|A|$ is replaced by y.

(a) *If $Ax = y$ then $|A|x_i = D_i$ for $i = 1, 2, \cdots, n$.*

(b) *If $|A| \neq 0$, the vector x in Part (a) satisfies $Ax = y$.*

The first statement is valid if $|A| = 0$, and it is also valid in many algebraic systems in which division is not possible. For example, it applies when the elements of A are polynomials, with constant coefficients, in the differentiation symbol $D = d/dt$. See Chapter 15, Section 4 and Part (f) below.

All essential features of the proof are contained in the case $n = 3$. In view of this, we consider the simultaneous equations

$$
\begin{aligned}
a_1x_1 + a_2x_2 + a_3x_3 &= y_1 \\
b_1x_1 + b_2x_2 + b_3x_3 &= y_2 \, . \\
c_1x_1 + c_2x_2 + c_3x_3 &= y_3
\end{aligned}
$$

(7)

By Rules (ii) and (viii) of Part (a)

$$
x_1 \begin{vmatrix} a_1 & a_2 & a_3 \\ b_1 & b_2 & b_3 \\ c_1 & c_2 & c_3 \end{vmatrix} = \begin{vmatrix} x_1a_1 & a_2 & a_3 \\ x_1b_1 & b_2 & b_3 \\ x_1c_1 & c_2 & c_3 \end{vmatrix} = \begin{vmatrix} x_1a_1 + x_2a_2 + x_3a_3 & a_2 & a_3 \\ x_1b_1 + x_2b_2 + x_3b_3 & b_2 & b_3 \\ x_1c_1 + x_2c_2 + x_3c_3 & c_2 & c_3 \end{vmatrix} .
$$

Hence, if (7) holds,

$$x_1 \begin{vmatrix} a_1 & a_2 & a_3 \\ b_1 & b_2 & b_3 \\ c_1 & c_2 & c_3 \end{vmatrix} = \begin{vmatrix} y_1 & a_2 & a_3 \\ y_2 & b_2 & b_3 \\ y_3 & c_2 & c_3 \end{vmatrix}.$$

In the notation of Cramer's rule, the above result and the corresponding results for x_2 and x_3 are

$$Dx_1 = D_1, \qquad Dx_2 = D_2, \qquad Dx_3 = D_3.$$

This gives Part (a) and Theorem 2 then gives Part (b).

EXAMPLE 5. In the system on the left below, the value of x_2 is given by the formula on the right:

$$\begin{aligned} x_1 + 2x_3 &= 4 \\ x_1 + x_2 &= 5 \\ 2x_2 + 4x_3 &= 6 \end{aligned} \qquad \begin{vmatrix} 1 & 0 & 2 \\ 1 & 1 & 0 \\ 0 & 2 & 4 \end{vmatrix} x_2 = \begin{vmatrix} 1 & 4 & 2 \\ 1 & 5 & 0 \\ 0 & 6 & 4 \end{vmatrix}.$$

The determinants are easily evaluated when the first row is subtracted from the second. The result is $8x_2 = 16$ or $x_2 = 2$.

In much of the theory of differential equations, Theorem 1, Theorem 2, and Cramer's rule constitute virtually all one has to know about linear algebra. For example, Theorems 1 and 2 yield the Fredholm alternative, which provides the mathematical basis for Chapter 11. That is the reason why a separate discussion, with few prerequisites, was given here in Part (c). Another proof of Theorem 2 and of Cramer's rule will be presented in Part (d) below.

(d) The inverse revisited

Let us recall that when the elements of a row or column are combined with the corresponding cofactors, as in the expansion theorem, the result is the determinant. For example

$$(8) \qquad \begin{vmatrix} a_{11} & a_{12} & a_{13} \\ a_{21} & a_{22} & a_{23} \\ a_{31} & a_{32} & a_{33} \end{vmatrix} = a_{12}A_{12} + a_{22}A_{22} + a_{32}A_{32}.$$

It is of considerable interest that if the elements of one row or column are combined with the cofactors of another row or column the result is 0.

The meaning of this statement and the reason for its truth are both explained by the calculation

(9)
$$\begin{vmatrix} a_{11} & a_{13} & a_{13} \\ a_{21} & a_{23} & a_{23} \\ a_{31} & a_{33} & a_{33} \end{vmatrix} = a_{13}A_{12} + a_{23}A_{22} + a_{33}A_{32}.$$

The sum on the right of (9) is the Laplace expansion of the determinant on the second column and is 0 because the determinant has two equal columns. This same sum would be obtained by combining elements of the third column of the determinant (8) with cofactors of the second column.

A similar result, with similar proof, is valid in general. In general if we combine elements of the ith column with cofactors of the jth column, where $i \neq j$, the result is an expansion of a determinant with two equal columns and is therefore 0. The same applies to rows.

These facts are conveniently summarized by use of the Kronecker delta, which was introduced in Chapter 11. Namely,

$$\delta_{ij} = 1 \quad \text{if } i = j, \qquad \delta_{ij} = 0 \quad \text{if } i \neq j.$$

In terms of this symbol we have

(10)
$$\sum_{k=1}^{n} a_{ik}A_{jk} = \delta_{ij}|A| = \sum_{k=1}^{n} a_{ki}A_{kj}.$$

When $i = j$ the result of the summation is $|A|$ by the expansion theorem, and when $i \neq j$ the result is 0 by the remarks above.

The left-hand sum in (10) would agree with the formula for a matrix product if the summand were $a_{ik}A_{kj}$ instead of $a_{ik}A_{jk}$. Similarly for the sum on the right. Since $(A_{jk})^T = (A_{kj})$ we introduce the *cofactor transpose*, which is denoted by adj A and is defined as follows:

$$\text{adj } A = (A_{ij})^T.$$

The abbreviation adj A is in deference to earlier terminology, in which the cofactor transpose was called the adjoint. (The term "adjoint" is now used mainly for the matrix A^* introduced in Chapter C.) When the sums (10) are interpreted by matrix products the result is

(11)
$$A(\text{adj } A) = (\text{adj } A)A = |A|I.$$

For example,

$$\begin{pmatrix} a_{11} & a_{12} & a_{13} \\ a_{21} & a_{22} & a_{23} \\ a_{31} & a_{32} & a_{33} \end{pmatrix} \begin{pmatrix} A_{11} & A_{21} & A_{31} \\ A_{12} & A_{22} & A_{32} \\ A_{13} & A_{23} & A_{33} \end{pmatrix} = \begin{pmatrix} |A| & 0 & 0 \\ 0 & |A| & 0 \\ 0 & 0 & |A| \end{pmatrix}.$$

Theorem 1 shows that if $|A| = 0$ then A has no inverse; otherwise $Ax = 0$ would have the unique solution $x = A^{-1}0 = 0$. Hence, if A is invertible, we can divide (11) by $|A|$ to get the following generalization of the rule for 2-by-2 inverses given in Chapter C:

THEOREM 3.   *To obtain the inverse of an invertible matrix A, write the matrix of cofactors (A_{ij}), take its transpose, and divide by the determinant.*

It is important that the matrix $A^{-1} = |A|^{-1}\text{adj } A$ given by Theorem 3 satisfies *both* equations

$$A^{-1}A = I, \qquad AA^{-1} = I.$$

This follows from (11). We illustrate Theorem 3 in three examples.

EXAMPLE 6.   For the matrix A below, the matrix of minors is M and the matrix of cofactors is C:

$$A = \begin{pmatrix} 1 & 2 & 3 \\ 0 & 1 & 1 \\ 2 & 0 & 4 \end{pmatrix}, \quad M = \begin{pmatrix} 4 & -2 & -2 \\ 8 & -2 & -4 \\ -1 & 1 & 1 \end{pmatrix}, \quad C = \begin{pmatrix} 4 & 2 & -2 \\ -8 & -2 & 4 \\ -1 & -1 & 1 \end{pmatrix}.$$

When the cofactor transpose adj $A = C^T$ is multiplied by A we get

$$(\text{adj } A)A = \begin{pmatrix} 4 & -8 & -1 \\ 2 & -2 & -1 \\ -2 & 4 & 1 \end{pmatrix} \begin{pmatrix} 1 & 2 & 3 \\ 0 & 1 & 1 \\ 2 & 0 & 4 \end{pmatrix} = \begin{pmatrix} 2 & 0 & 0 \\ 0 & 2 & 0 \\ 0 & 0 & 2 \end{pmatrix}.$$

The equation shows that $|A| = 2$. Dividing C^T by 2 yields the inverse.

EXAMPLE 7.   Cramer's rule. If $|A| \neq 0$, the equation $Ax = y$ has the unique solution $x = A^{-1}y$ where A^{-1} is given by Theorem 3. This yields

$$|A|x = (\text{adj } A)y$$

which is equivalent to

$$|A|x_i = y_1 A_{1i} + y_2 A_{2i} + \cdots + y_n A_{ni}, \qquad i = 1, 2, \cdots, n.$$

The right-hand side is the expansion of the determinant obtained when the ith column of A is replaced by y. Thus we have a new proof of Part (b) of Cramer's rule. It is left for you to deduce Part (a) from (11).

EXAMPLE 8.   Let A and B be square matrices such that $AB = I$. Since $|A||B| = |AB| = |I| = 1$ neither $|A|$ nor $|B|$ is 0, and Theorem 3 shows that A and B are both invertible. Multiplying $AB = I$ on the left by A^{-1}, we get $B = A^{-1}$. Hence $BA = I$, which yields Chapter C, Theorem 1.

(e) The expanded form

A *permutation* of the sequence 1, 2, $\cdots$, n is a rearrangement of its terms. A permutation can be regarded as a function p whose domain and range both consist of the set $\{1, 2, \cdots, n\}$. For example, the permutation taking 1, 2, 3, 4 into 4, 2, 3, 1 is the function p defined as follows:

$$(12) \qquad p(1) = 4, \qquad p(2) = 2, \qquad p(3) = 3, \qquad p(4) = 1.$$

In general, $p(1)$ can be any of the n numbers 1, 2, $\cdots$, n. After $p(1)$ is chosen there remain $n - 1$ choices for $p(2)$, and so on. Hence there are $n!$ different permutations p.

In a sequence such as 4, 2, 3, 1 an inversion is a pair of numbers that are not in their natural order; that is, the first number of the pair is larger than the second. This particular sequence has five inversions,

$$(4, 2), \ (4, 3) \ (4, 1) \ (2, 1) \ (3, 1).$$

The sign of a permutation p is plus if the number of inversions in it is even, minus if it is odd. For example the permutation (12) has sign $p = -$ since there are 5 inversions.

In this terminology the expanded form of an nth-order determinant is

$$(13) \qquad\qquad |A| = \sum \text{sign } p \ a_{1\,p(1)} a_{2\,p(2)} \cdots a_{n\,p(n)}$$

where the sum is over all $n!$ permutations p. Thus a determinant is a sum of $n!$ products of n elements a_{ij}, as follows:

(i) Each product has exactly one factor a_{ij} from each row and exactly one from each column.

(ii) Each product is preceded by a sign, plus or minus, according to a definite rule.

For the case $n = 3$ see Problem 1. The proof will not be given here, though some aspects of the proof are considered in the problems.

With no claim to precision, we make a few remarks about computational complexity. All we want to do is to give some idea of the effectiveness of the pivotal method as compared with direct expansion. Multiplication is much more time-consuming than addition and we consider multiplication only. Although each term in the expanded form (13) has n factors, there is a good deal of duplication, and we take $n!$ as a crude estimate of the complexity of direct calculation.

For the pivotal method, reduction of the first column requires about n^2 multiplications. Since the same must be repeated on each column, and since we need only reduce the matrix to triangular form, a crude estimate for the total complexity is given by the sum

$$n^2 + (n-1)^2 + \cdots + 3^2 + 2^2 + 1 = \frac{n(n+1)(2n+1)}{6}.$$

For large n this is about $n^3/3$.

To fix ideas suppose our computer can do 10,000,000 multiplications a second, each to an accuracy of sixteen digits. Taking this as a measure of computing speed we get the following table. The second row gives the time for direct computation and the third for the pivot method:

$n = 15$	$n = 20$	$n = 25$	$n = 500$
36 hours	77 centuries	4.9×10^8 centuries	too long
.0001 seconds	.0003 seconds	.0005 seconds	4 seconds

(f) Fields, integral domains, and rings

Chapters CDE are intended to be a minicourse in linear algebra. The following remarks indicate how this brief introduction fits into a broader context. They also shed light on the algebraic principles underlying the analysis of Chapters 6, 8, 9, 15, 17.

The real numbers are said to form a field because they satisfy the familiar laws of elementary algebra, including those connected with division. The complex numbers also form a field. By contrast, although the integers satisfy most of the laws of algebra, they do not form a field, because division by nonzero elements is not always possible. For example, $2x = 1$ has no solution in integers.

The set of integers is an example of an integral domain. If D is the differentiation operator $D = d/dt$, the polynomials $P(D)$ with constant real or complex coefficients also form an integral domain. Such polynomials satisfy most of the laws of algebra, but division by an expression containing D is not allowed. If A is a given square matrix with real or complex elements, the polynomials $P(A)$ with real or complex coefficients, do not, as a rule, form either a field or an integral domain. However, they form a more general algebraic system called a commutative ring.

When we speak of "the familiar laws of algebra" we mean: the commutative and associative laws of addition and multiplication, the left and right distributive laws, the identity laws $a + 0 = a$, $a1 = a$, and the laws pertaining to existence of additive and multiplicative inverses:

$$(14) \qquad a + (-a) = 0, \qquad b(b^{-1}) = 1 \quad \text{for} \quad b \neq 0.$$

A *field* satisfies all of these. An *integral domain* satisfies all of these except possibly the last, which is replaced by the following weaker condition:

(15)   if $ab = 0$ then $a = 0$ or $b = 0$.

A *commutative ring* satisfies the commutative, associative, and distributive laws, but (15) is not required, and there may be no unit element 1 such that $a1 = a$.

The matrices considered in Chapters CDE have real or complex elements, and these elements are subject to no restriction; for instance, they are not required to be integers. It is said, briefly, that the matrices are "over the real field" or "over the complex field" as the case may be. Matrices can also be considered over an integral domain or over a commutative ring. For example, we can require the elements to be integers, or to be polynomials in the differentiation symbol D. Many of the foregoing results involve addition, subtraction, and multiplication only, without division. Such results often remain valid for matrices over a commutative ring, or at least over an integral domain. Other results make essential use of division and cannot be so extended. The question whether the results do or do not extend is answered by looking at the proofs.

In Chapter C we said that the square matrix A is invertible if it has an inverse A^{-1}. Here we introduce the further definition that A is *nonsingular* if $|A| \neq 0$. Over the domain of integers it is possible to have $|A| \neq 0$ and yet have no inverse A^{-1}. (The trouble is that A^{-1} might involve fractions.) The same sort of thing happens whenever matrices are considered over a commutative ring that is not a field.

These remarks may serve to put the following theorem into perspective:

THEOREM 4.   *Over the real or complex field, a matrix is invertible if, and only if, it is nonsingular.*

Theorem 4 is an immediate consequence of Theorems 1 and 3.

In the following problems the matrices are over the real or complex field, and results of Chapters CDE should be used without hesitation.

PROBLEMS————————————————————————————

1P. By Equation (2) the determinant of $A = (a_{ij})$ is

$$a_{11}a_{22}a_{33} + a_{12}a_{23}a_{31} + a_{13}a_{21}a_{32} - a_{11}a_{23}a_{32} - a_{13}a_{22}a_{31} - a_{12}a_{21}a_{33}$$

when $n = 3$. Open the book to a random page and note the last digit of the page number. Then expand the determinant (2) according to the following scheme:

$(0, 1) = 2R$ $(2, 3) = 3R$ $(4, 5) = 1C$ $(6, 7) = 2C$ $(8, 9) = 3C$

For example, if the last digit of the page number is 6 or 7, expand the determinant on its second column. If your class is large, this problem will probably lead to a proof of the expansion theorem for $n = 3$.

2. Evaluate by using rules (ii) and (viii) to get a row with three zeros:

(a) $\begin{vmatrix} 1 & 2 & 3 & 4 \\ 3 & 6 & 6 & 9 \\ 4 & 6 & 8 & 10 \\ 3 & 4 & 6 & 5 \end{vmatrix}$
(b) $\begin{vmatrix} 1 & 1 & 1 & 1 \\ 2 & 3 & 4 & 1 \\ 3 & 4 & 1 & 2 \\ 4 & 1 & 2 & 3 \end{vmatrix}$
(c) $\begin{vmatrix} 3 & 5 & 7 & 9 \\ 7 & 9 & 3 & 5 \\ 5 & 9 & 3 & 7 \\ 9 & 7 & 5 & 3 \end{vmatrix}$

3. (abc) Evaluate the determinants above by using rules (ii) and (viii) to get a column with three zeros.

4. Write equations corresponding to the following augmented matrices and solve by Cramer's rule:

(a) $\begin{pmatrix} 3 & 2 & 1 & 6 \\ 1 & 1 & 1 & 2 \\ 1 & 2 & 4 & 1 \end{pmatrix}$
(b) $\begin{pmatrix} 2 & 2 & 1 & 8 \\ 2 & 8 & 1 & 8 \\ 2 & 8 & 4 & 5 \end{pmatrix}$
(c) $\begin{pmatrix} 3 & 1 & 4 & 1 \\ 5 & 9 & 2 & 6 \\ 5 & 3 & 5 & 8 \end{pmatrix}$

5P. (abcde) Find the inverse as in Example 6. Since the method incorporates a check, no answer is provided:

$\begin{pmatrix} 1 & 0 & 1 \\ 1 & 1 & 0 \\ 0 & 1 & 1 \end{pmatrix}$
$\begin{pmatrix} 1 & 0 & -1 \\ 2 & 1 & 0 \\ 0 & 1 & 1 \end{pmatrix}$
$\begin{pmatrix} 1 & 2 & 3 \\ 3 & 2 & 1 \\ 2 & 1 & 3 \end{pmatrix}$
$\begin{pmatrix} 1 & 1 & 1 \\ 1 & 2 & 1 \\ 1 & 1 & 3 \end{pmatrix}$
$\begin{pmatrix} 1 & 2 & 3 \\ 4 & 5 & 6 \\ 7 & 8 & 9 \end{pmatrix}$

6. In the proof of Property (ix), if $n = 3$ check that $|AB|$ is

$$|A_1 b_{11} + A_2 b_{21} + A_3 b_{31} \quad A_1 b_{12} + A_2 b_{22} + A_3 b_{32} \quad A_1 b_{13} + A_2 b_{23} + A_3 b_{33}|$$

By Property (iii) this is a sum of 27 determinants, a typical one of which is

$$|A_2 b_{21} \quad A_1 b_{12} \quad A_3 b_{33}| = b_{21} b_{12} b_{33} |A_2 \quad A_1 \quad A_3|$$

Using these remarks as a guide, explain why the sum of all 27 determinants has the form $P(b)|A|$ described in the text.

7. Let A, B, C be square matrices of the same size and suppose ABC is nonsingular. Show that A, B, C are each nonsingular.

Hint: $|ABC| = |A||B||C|$.

8P. If the elements a_{ij} of a matrix $A = (a_{ij})$ are functions of a real variable t, the operations of differentiation and integration are performed elementwise; for example $A' = (a'_{ij})$ where the prime denotes differentiation.

(a) Let A, B be compatible matrices such that A' and B' exist. What result do you get by summing the following identity on k?

$$\frac{d}{dt} a_{ik} b_{kj} = a'_{ik} b_{kj} + a_{ik} b'_{kj}$$

(b) Investigate $(A + B)' = A' + B'$ and $(\lambda A)' = \lambda A' + \lambda' A$.

(c) If the elements of the determinant $|A|$ are differentiable functions, show that its derivative $|A|'$ can be obtained by differentiating each row separately and adding the results, or also by differentiating each column separately and adding the results.

Hint: Differentiate the defining relation (5) and use mathematical induction, or proceed as in Chapter 11, Section 3, Problem 5. The fact that you can differentiate by columns follows from $|A^T| = |A|$.

9. If $|A| \neq 0$ and if the elements of A are differentiable functions of t, Theorem 3 shows that the inverse $B = A^{-1}$ is differentiable. Differentiate the equation $AB = I$ to get the formula $B' = -BA'B$.

10. Show that the following product in the expansion of a determinant of order 9 should be prefixed by a plus sign:

$$a_{31}a_{44}a_{79}a_{53}a_{16}a_{88}a_{92}a_{67}a_{25}$$

Suggestion: Rearrange the factors so that the first indices i on a_{ij} are in the normal order and look at the sequence that is then formed by the second indices j.

11. Here are two determinants of order 5:

$$\begin{vmatrix} 0 & 0 & 0 & a & 0 \\ 0 & b & 0 & 0 & 0 \\ * & * & * & * & * \\ * & * & * & * & * \\ * & * & * & * & * \end{vmatrix} \qquad \begin{vmatrix} 0 & 0 & a & 0 & 0 \\ 0 & 0 & 0 & 0 & b \\ * & * & * & * & * \\ * & * & * & * & * \\ * & * & * & * & * \end{vmatrix}$$

No matter what numbers are put in place of the stars, show in each case that expansion on the first row agrees with expansion on the second row. Show also that expansion on the first row agrees with expansion on the column containing a.

Suggestion: Let D denote the determinant of order 3 obtained when you cross out the two rows and the two columns containing a and b. This problem requires very little writing.

12. (a) Show that a nonsingular matrix A remains nonsingular after an elementary row operation is performed on its matrix.

(b) Deduce that if A is nonsingular, then A can be reduced to I by elementary row operations.

13. Prove by mathematical induction that the expansion of a determinant has $n!$ terms, each of which is a product like that in (13) preceded by a plus or minus sign. In other words, get (13) except for correctness of the factor sign p. The same proof works for expansion on any row or column.

14. Challenge problem. By Problem 13 the expansion theorem will follow if you show that the sign of any given term is the same, no matter whether that term is obtained by expansion on the ith row or jth column. Since the sign of a given term is unchanged if all the factors a_{ij} not occurring in it are replaced by 0, it suffices to consider the case of a determinant in which each column and each row has only one nonzero element. By mathematical induction reduce this case to a case like one of those in Problem 11 and prove the expansion theorem.

————————————————————**ANSWERS**————————————————————

2. 12, 16, 0

4. (a) $(1, 2, -1)$ (b) $(9/2, 0, -1)$ (c) $(191, -67, -122)/18$

8. (a) $(AB)' = A'B + AB'$. Since matrix multiplication is not commutative, the order of the factors on the right is important.

(b) The equations hold if A, B are the same size and A, B, λ are all differentiable.

> As the sun eclipses the stars by his brilliancy, so the man of knowledge will eclipse the fame of others in assemblies of the people if he proposes algebraic problems, and still more if he solves them.
>
> —Brahmagupta

Chapter F

UNIFORM CONVERGENCE

If finitely many continuous functions are added, the sum is continuous and the integral of the sum can be obtained by adding the individual integrals. The question arises whether this property remains valid for infinite series. In a like manner, one may want to get the derivative of an infinite series by differentiating the series term by term. As it happens, such operations are not always valid and can lead to incorrect results. The study of conditions under which they are valid leads to a special type of convergence called uniform convergence.

To illustrate the main idea, let us consider the sequence of functions

$$s_1(x), \; s_2(x), \; \cdots, \; s_n(x), \; \cdots$$

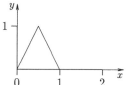

shown graphically in Figure 1. Each function $s_n(x)$ is defined if $0 \leq x \leq 2$ and its graph consists of three straight line segments joining the points $(0,0)$, $(1/(2n),1)$, $(1/n,0)$, and $(2,0)$. The graph coincides with the x axis except when it forms two sides of a triangle centered on the interval $(0,1/n)$. The maximum of each function is taken on at the apex of the triangle and has the value 1.

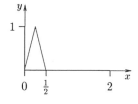

If x is fixed and $0 < x \leq 2$, the triangle lies wholly to the left of x when $1/n < x$, that is, when $n > 1/x$. For such n we have $s_n(x) = 0$ and hence the limit as $n \to \infty$ is 0. The same holds for $x = 0$ since $s_n(0) = 0$. Thus the sequence has the remarkable property that

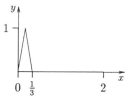

$$\lim_{n \to \infty} s_n(x) = 0, \qquad 0 \leq x \leq 2$$

even though the maximum in each case is

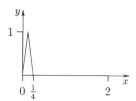

$$\max_{0 \leq x \leq 2} s_n(x) = 1.$$

The sequence $\{s_n(x)\}$ converges to 0 for $0 \leq x \leq 2$ but not uniformly. This terminology is suggested by the fact that $|s_n(x)|$

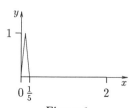

Figure 1

can be made small at each given x by choosing n large enough, but it cannot be made uniformly small for all x at once.

The integral of $s_n(x)$ from 0 to 2 is the area of the triangle, which is $1/(2n)$. This tends to 0 as $n \to \infty$ and hence

$$\lim_{n \to \infty} \int_0^2 s_n(x)\, dx = \int_0^2 \lim_{n \to \infty} s_n(x)\, dx$$

both values being 0. But if we put the apex of the triangle at $(1/(2n), n)$ instead of at $(1/(2n), 1)$ the integral of $s_n(x)$ is $1/2$ for each n and the above equation fails. If we put the apex at $(1/(2n), n^2)$ the integral is $n^2/2$ which tends to infinity. Nevertheless the limit of $s_n(x)$ is still 0 for $0 \le x \le 2$.

The remedy is suggested by the example. Let $\{s_n(x)\}$ be a sequence of functions defined on an interval I. The sequence converges uniformly to $s(x)$ on I if the maximum error

$$\max_{x \in I} |s_n(x) - s(x)|$$

tends to 0 as $n \to \infty$. Denoting the maximum by ϵ_n we get the two properties

(1) $\qquad |s_n(x) - s(x)| \le \epsilon_n, \qquad x \in I; \qquad \lim_{n \to \infty} \epsilon_n = 0.$

This formulation is meaningful even for functions of the kind that do not attain their maximum. It may serve to motivate the following definition:

DEFINITION OF UNIFORM CONVERGENCE. Let $\{s_n(x)\}$ be a sequence of real- or complex-valued functions defined on an interval I. The sequence converges uniformly to $s(x)$ on I, and we write

$$\lim_{n \to \infty} s_n(x) = s(x) \quad \text{uniformly on} \quad I,$$

if there exists a sequence of constants $\{\epsilon_n\}$ satisfying (1).

The statement that the ϵ_n are constant means that they do not depend on x, though in general they do depend on n. If we had $\epsilon_n(x)$ instead of ϵ_n conditions (1) would define, not uniform convergence, but ordinary convergence. As an illustration, you are invited to convince yourself that

$$\lim_{n \to \infty} \frac{nx}{1 + nx} = 1$$

uniformly for $x \ge \delta > 0$, where δ is fixed, but the convergence is not uniform for all $x > 0$.

(a) Integration

One of the most important theorems about uniform convergence pertains to integration and is as follows:

THEOREM 1.   *Let $s_n(x)$ be continuous functions on an interval I and suppose $\{s_n(x)\}$ converges uniformly to $s(x)$ on I. Then $s(x)$ is continuous and for a, b on I*

$$\lim_{n \to \infty} \int_a^b s_n(x)\, dx = \int_a^b s(x)\, dx.$$

To establish continuity let c be a given point of I and let a positive number ϵ be given. We have to find a positive constant δ such that the two conditions

$$x \in I, \qquad |x - c| < \delta$$

together imply $|s(x) - s(c)| < \epsilon$. If we do this, we have shown that $s(x)$ is continuous at $x = c$. Since c is arbitrary, this gives continuity on I.

The result follows from the identity

$$s(x) - s(c) = (s(x) - s_n(x)) - (s(c) - s_n(c)) + (s_n(x) - s_n(c)).$$

By the definition of uniform convergence we have

$$|s(x) - s_n(x)| \leq \epsilon_n, \qquad |s(c) - s_n(c)| \leq \epsilon_n$$

where $\epsilon_n \to 0$. Choose n so that $\epsilon_n < \epsilon/3$, then choose δ so that

$$x \in I \quad \text{and} \quad |x - c| < \delta \quad \text{implies} \quad |s_n(x) - s_n(c)| < \frac{\epsilon}{3}.$$

This can be done because $s_n(x)$ is continuous. The conditions $x \in I$ and $|x - c| < \delta$ in the above identity now give the desired inequality,

$$|s(x) - s(c)| < \frac{\epsilon}{3} + \frac{\epsilon}{3} + \frac{\epsilon}{3} = \epsilon.$$

Assuming without loss of generality that $b \geq a$, we establish the part of Theorem 1 pertaining to integration. Since $s(x)$ is continuous it is integrable and a familiar property of integrals gives

$$\left| \int_a^b s(x)\, dx - \int_a^b s_n(x)\, dx \right| \leq \int_a^b |s_n(x) - s(x)|\, dx.$$

By uniform convergence the integrand on the right is at most ϵ_n where ϵ_n tends to 0. Hence the integral is at most $\epsilon_n(b - a)$. Since this tends to 0, Theorem 1 follows.

(b) Differentiation

A counterpart of Theorem 1 for differentiation is:

THEOREM 2. *Let $\{s_n(x)\}$ be a sequence of differentiable functions that converges to $s(x)$ on an interval I. If the sequence $\{s_n'(x)\}$ converges uniformly on I then it converges to $s'(x)$.*

In this book Theorem 2 is needed only under the additional hypothesis that the derivatives $s_n'(x)$ are continuous, and we give the proof for that case. By Theorem 1 the limit of $s_n'(x)$ is a continuous function, which we denote by $S(x)$. If a and x are on I we have

$$(2) \qquad \lim_{n \to \infty} \int_a^x s_n'(t)\, dt = \int_a^x S(t)\, dt$$

This follows from Theorem 1 applied to $\{s_n'(t)\}$, the role of b in Theorem 1 being taken by x. The fundamental theorem of calculus gives

$$(3) \qquad \lim_{n \to \infty} (s_n(x) - s_n(a)) = \int_a^x S(t)\, dt.$$

By hypothesis $s_n(t)$ converges to $s(t)$ for $t \in I$. Applying this with $t = x$ and with $t = a$ we see that (3) can be written

$$(4) \qquad s(x) - s(a) = \int_a^x S(t)\, dt.$$

Equation (4) gives $s'(x) = S(x)$, which completes the proof.

(c) The Weierstrass test

The above results for sequences are easily translated into similar results for infinite series. If $a_j(x)$ are functions defined on an interval I the equation

$$(5) \qquad \sum_{j=0}^{\infty} a_j(x) = s(x)$$

means $\lim\limits_{n \to \infty} s_n(x) = s(x)$ where $\{s_n(x)\}$ is the sequence of partial sums,

$$s_n(x) = a_0(x) + a_1(x) + \cdots + a_n(x).$$

It is said that the series converges uniformly to $s(x)$ for $x \in I$ if the convergence of $s_n(x)$ to $s(x)$ is uniform.

One of the most important tests for uniform convergence is the Weierstrass M test, which reads as follows:

M TEST FOR SERIES. Let $a_j(x)$ be real- or complex-valued functions defined on an interval I. Let $|a_j(x)| \le M_j$ for $x \in I$ where M_j are constants such that

(6)
$$\sum_{j=0}^{\infty} M_j \quad \text{converges.}$$

Then the series (5) converges to a well-defined sum $s(x)$ for $x \in I$ and the convergence is uniform.

The proof depends on the theorems of Chapter 12, Section 3, and familiarity with those theorems is assumed. The comparison test shows that the series is absolutely convergent and hence the sum $s(x)$ is well defined. Uniform convergence follows from

$$|s(x) - s_n(x)| = \left| \sum_{j=n+1}^{\infty} a_j(x) \right| \le \sum_{j=n+1}^{\infty} |a_j(x)| \le \sum_{j=n+1}^{\infty} M_j.$$

If we denote the latter sum by ϵ_n it is easily checked that $\epsilon_n \to 0$ and hence the criterion for uniform convergence is fulfilled. This completes the proof.

A somewhat different form of the Weierstrass M test applies to sequences and reads as follows:

M TEST FOR SEQUENCES. Let $\{s_n(x)\}$ be a sequence of real- or complex-valued functions defined on an interval I and suppose

$$|s_n(x) - s_{n-1}(x)| \le M_n, \qquad x \in I$$

where $\{M_n\}$ satisfies (6). Then $\lim_{n \to \infty} s_n(x) = s(x)$ exists for $x \in I$, the convergence is uniform, and

(7)
$$|s_n(x) - s(x)| \le \sum_{j=n+1}^{\infty} M_j, \qquad x \in I.$$

For proof, use the following identity together with the M test for series:

$$s_n - s_0 = (s_1 - s_0) + (s_2 - s_1) + (s_3 - s_2) + \cdots + (s_n - s_{n-1}).$$

As an illustration of the M test, suppose a power series

(8)
$$\sum_{j=0}^{\infty} a_j x^j$$

has a positive radius of convergence r, and let $0 < r_0 < r$. Then the power series and its formal derivative both converge uniformly for $|x| \leq r_0$. To see why, choose R satisfying $r_0 < R < r$. By Chapter 12, Section 3, Theorem 1 there is a constant M such that

$$|a_n| \leq \frac{M}{R^n}.$$

The condition $|x| \leq r_0$ yields

$$|a_n x^n| \leq M \left(\frac{r_0}{R} \right)^n, \qquad |n a_n x^{n-1}| \leq \frac{Mn}{R} \left(\frac{r_0}{R} \right)^{n-1}.$$

Uniform convergence of both the original and the derived series now follows from the M test.

If the sum and partial sums of the above series are $s(x)$ and $s_n(x)$ respectively, Theorem 2 gives $\lim_{n \to \infty} s'_n(x) = s'(x)$. In other words

$$\sum_{n=0}^{\infty} n a_n x^{n-1} = \frac{d}{dx} \sum_{n=0}^{\infty} a_n x^n.$$

This is the reason why a power series can be differentiated term by term within its interval of convergence. Since the derived series has the same radius of convergence as the original series, the differentiation can be repeated as often as we like.

PROBLEMS

1. Suppose the apex of the triangle in the function $s_n(x)$ of Figure 1 is at $(1/n, a_n)$ instead of $(1/n, 1)$. What conditions on the sequence $\{a_n\}$ are necessary and sufficient for the following?

(a) $\lim_{n \to \infty} s_n(x) = 0$, $0 \leq x \leq 2$;   (b) The convergence in (a) is uniform;

$$\text{(c)} \lim_{n \to \infty} \int_0^2 s_n(x) \, dx = \int_0^2 \lim_{n \to \infty} s_n(x) \, dx$$

Answers: (a) no condition   (b) $\lim_{n \to \infty} a_n = 0$ (c) $\lim_{n \to \infty} a_n/n = 0$

A Hair perhaps divides the False and True.

—Omar Khayyam

Chapter G

GENERALIZED SOLUTIONS

As in Chapter 8 let T be defined by

$$Ty = y^{(n)} + p_{n-1}y^{(n-1)} + \cdots + p_1 y' + p_0 y$$

where the p_j are real or complex constants. The equation $Ty = f$ with a discontinuous input f is often encountered in applications; square-wave and sawtooth-wave inputs are only two of many examples. It was seen in Chapter 22 that the Laplace transform provides a very effective method of dealing with such problems. Sooner or later, however, we must come to grips with the following surprising theorem:

THEOREM 1. *Let $n \geq 1$, let I be an open interval, and let f have a simple discontinuity at some point of I. Then the equation $Ty = f$ has no classical solution on I.*

At a *simple discontinuity* the right-hand and left-hand limits exist but are unequal. A *classical solution* is a function $y = \phi(t)$ that satisfies the differential equation at every point of I. The condition $n \geq 1$ ensures that T actually involves differentiation; for $n = 0$ the theorem is obviously false.
    The proof is outlined in the problems. Here we extend the concept of "solution" to allow discontinuous inputs, and we develop the theory of the Laplace transform within the context of that extension.

DEFINITION 1. *A function $y = \phi(t)$ is a generalized solution of $Ty = f$ on an interval I if:*

 (i) *$y, y', y'', \cdots, y^{(n-1)}$ are continuous on I and*

 (ii) *$Ty = f(t)$ holds at all points $t \in I$ where $f(t)$ is continuous.*

At points where f is discontinuous we allow the differential equation to fail, and we do not even assume that $y^{(n)}$ exists. The condition (i) means that all of the bad behavior of f is absorbed by the highest derivative $y^{(n)}$.
    Appropriate conditions on f are given by:

DEFINITION 2. *A real or complex-valued function f is admissible on a finite interval I if:*

 (i) *f is defined and continuous except at finitely many points of I, and*

 (ii) *f is bounded on the set of points in I where it is defined.*

By definition, the function f is admissible on an infinite interval if it is admissible on every finite subinterval.

(a) Analytic properties

Several familiar theorems of calculus extend, with minor modifications, to the class of admissible functions. Here are three examples:

THEOREM 2. *On an interval I let $F(t)$ be continuous and suppose $F'(t) = 0$ except at finitely many points t_j, where $F'(t)$ need not be defined. Then $F(t)$ is constant on I.*

THEOREM 3. *Let f be admissible on an interval I and let $a \in I$. Then*

$$F(t) = \int_a^t f(s)\,ds, \qquad t \in I$$

is continuous and $F'(t) = f(t)$ at all points $t \in I$ where $f(t)$ is continuous.

THEOREM 4. *Suppose f, g, f', and g' are admissible and fg is continuous on an interval I. Then for $a \in I$ and $b \in I$*

$$\int_a^b f(s)g'(s)\,ds = f(s)g(s)\Big|_a^b - \int_a^b f'(s)g(s)\,ds.$$

To establish Theorem 2, note that the exceptional points t_j divide the interval I into intervals I_j on each of which $F'(t) = 0$. Hence $F(t) = c_j$, a constant, on I_j. The constants c_j must be equal, since otherwise $F(t)$ would have a discontinuity at some point t_j. This completes the proof.

Theorem 3 depends on an extension of the concept of integration. The Riemann integral of $f(t)$ on an interval $[a, b]$ is defined by use of sums

$$\sum_{j=1}^n f(\bar{t}_j)\Delta t_j$$

where the points $a = t_0 < t_1 < t_2 < \dots < t_n = b$ form a partition of $[a, b]$ and

$$\Delta t_j = t_j - t_{j-1}, \qquad t_{j-1} \le \bar{t}_j \le t_j.$$

If $f(t)$ is merely admissible, one of the intermediate points $\bar{t}_j$ might be a point at which $f(t)$ is undefined and the above sum would have no meaning.

To deal with this problem, avoid the points where $f(t)$ is undefined in choosing the t_j, and proceed as in the usual definition of the integral. Theorem 3 then follows from the usual proof.

To establish Theorem 4, replace b by a variable t and let $F(t)$ be the difference between the left- and right-hand sides of the stated equality, thus:

$$F(t) = \int_a^t f(s)g'(s)\,ds - f(t)g(t) + f(a)g(a) + \int_a^t f'(s)g(s)\,ds.$$

At any point where f, g, f', and g' are all continuous, Theorem 3 and the rule for differentiating a product give

$$F'(t) = f(t)g'(t) - f(t)g'(t) - f'(t)g(t) + f'(t)g(t) = 0.$$

This holds except in a finite set, and Theorem 3 also shows that $F(t)$ is continuous. Hence $F(t)$ is constant by Theorem 2. Since $F(a) = 0$ the constant value is 0, and setting $t = b$, we get Theorem 4.

The existence-uniqueness theorem of Chapter 8, Section 4 depends on an integral formulation to which the foregoing results apply. Hence that theorem remains valid for generalized solutions when f is admissible, a fact used below. Although details will not be given here, it can be said that the principal results of Chapter 14 also depend on integral formulas, and they too extend to generalized solutions.

(b) Functions of exponential type

A real- or complex-valued function $f(t)$ is said to be of *exponential type*, and we write $f \in E$, if there are positive constants A and B, depending on f, such that:

(i) $f(t)$ is admissible on the interval $[0, \infty)$

(ii) $|f(t)| \le Ae^{Bt}$ at all points $t \in [0, \infty)$ where $f(t)$ is defined.

In the theory of the Laplace transform as developed in Chapters 21–23, the following was stated without proof:

THEOREM 5.   *Let y be a generalized solution of $Ty = f$ on $[0, \infty)$. If $f \in E$ then y and its derivatives up to $y^{(n)}$ are also in E.*

Theorem 5 is applicable not only to this book but to a comprehensive literature of circuit analysis, systems science, feedback, and control.

Let us begin by showing that the class E is closed under the formation of sums and products and also under the operation of integration. If f and g belong to E the sum, product, and integral satisfy the condition (i) of having only finitely many discontinuities on a finite interval. (In fact, the integral has no discontinuities at all.) To check the inequality (ii) suppose

$$|f(t)| \le Ae^{Bt}, \qquad |g(t)| \le Ae^{Bt}$$

for $t \geq 0$ where A, B are positive constants; use of the same constants for f and g involves no loss of generality. Then

$$|f(t) + g(t)| \leq 2Ae^{Bt}, \qquad |f(t)g(t)| \leq A^2 e^{2Bt}, \qquad \left| \int_0^t f(s) \, ds \right| \leq \frac{A}{B} e^{Bt}$$

and hence the sum, product, and integral belong to E.

Next let us note that the operator $T(a, c)$ defined by

$$T(a, c)f = e^{at} \int_0^t e^{-as} f(s) \, ds + ce^{at}$$

is formed from f and from e^{at} by addition, multiplication, and integration. Hence

$$f \in \mathrm{E} \Rightarrow T(a, c)f \in \mathrm{E}$$

for all choices of a and c. In Chapter 8, Section 4 it was seen that we can get from f to y by a succession of such operations. Since each operation preserves the class E, they all do and hence

$$f \in \mathrm{E} \Rightarrow y \in \mathrm{E}.$$

This is the main assertion in Theorem 5.

To show that the derivatives belong to E we use mathematical induction. For $n = 1$ the equation says $y' = f - p_0 y$. Since f and y are in E, so is y'. Assume now that the conclusion holds for equations of order $n - 1$. If we set $u = y'$ in the equation $Ty = f$ the result is an equation of order $n - 1$ in u with right-hand member $f - p_0 y$. Since $f - p_0 y$ is in E, the induction hypothesis indicates that u together with its first $n - 1$ derivatives is in E. Hence y together with its first n derivatives is in E, and this completes the proof.

(c) A uniqueness theorem

The following theorem was stated without proof in Chapter 21. It ensures existence of the inverse Laplace transform, L^{-1}, by which $f = L^{-1}F$ is recovered from $F = Lf$:

THEOREM 6. *If f and g are functions of class* E *and if $F(s) = G(s)$ for all large s, then $f(t) = g(t)$ wherever both functions are continuous.*

Since Lf is unchanged when f is altered at an isolated point, the ambiguity at points of discontinuity is inherent in the problem.

The proof depends on another theorem that is of independent interest:

THEOREM 7.   *Let q be continuous for $0 \leq x \leq 1$. Then the following condition implies $q(x) = 0$ for $0 \leq x \leq 1$:*

$$\int_0^1 x^n q(x)\, dx = 0, \qquad n = 0,\ 1,\ 2,\ \cdots.$$

We deduce Theorem 6 from Theorem 7 and then we prove Theorem 7.

Given f and g as in Theorem 6, let u, v, w with transforms U, V, W be defined by

$$u(t) = f(t) - g(t), \qquad v(t) = \int_0^t u(\tau)\, d\tau, \qquad w(t) = e^{-ct} v(t).$$

The hypothesis of Theorem 6 gives $U(s) = 0$ for all large s and the conclusion is equivalent to the statement that $u(t) = 0$ at all points of continuity.

Entry k of the table gives $V(s) = 0$ for large s, and if we show that $v = 0$, the fundamental theorem of calculus yields $u(t) = 0$ at all points where u is continuous. Hence, we can work with the continuous function v.

If the constant c is sufficiently large, entry H of the table shows that the transform of w is 0 for all $s > 0$ (and not only for large s). The condition $v \in E$ holds because $u \in E$ and, for large c, it yields

$$(1) \qquad\qquad\qquad \lim_{t \to \infty} w(t) = 0.$$

By the change of variable $x = e^{-t}$ we get

$$W(s) = \int_0^\infty e^{-st} w(t)\, dt = \int_0^1 x^{s-1} w(-\ln x)\, dx.$$

If we define $q(x) = w(-\ln x)$ for $0 < x \leq 1$ and $q(0) = 0$, Equation (1) shows that q is continuous for $0 \leq x \leq 1$. Since $W(s) = 0$ for all $s \geq 0$ we certainly have $W(s) = 0$ for $s = 1, 2, 3, \cdots$. The conclusion $w = 0$ now follows from Theorem 7.

To establish Theorem 7, suppose $q(x_0) \neq 0$ at some point of $(0, 1)$. Replacing q by $-q$, if necessary, we assume $q(x_0) > 0$. Continuity then gives positive constants ϵ, δ such that

$$q(x) \geq \epsilon \quad \text{for} \quad |x - x_0| \leq 2\delta.$$

Let m be an arbitrary positive integer and define

$$p(x) = 1 + 4\delta^2 - (x - x_0)^2, \qquad I_m = \int_0^1 p(x)^m q(x)\, dx.$$

By the binomial theorem p^m is a polynomial in x, hence $I_m = 0$ under the hypothesis of Theorem 7.

On the other hand $I_m = J_1 + J_2 + J_3$ where J_1, J_2, J_3 are integrals over the part of $[0, 1]$ in which

$$|x - x_0| < \delta, \qquad \delta \leq |x - x_0| \leq 2\delta, \qquad |x - x_0| > 2\delta$$

respectively. On these three intervals we have

$$p(x) \geq 1 + 3\delta^2, \qquad p(x) \geq 1, \qquad |p(x)| \leq 1$$

and also, if M is a sufficiently large constant,

$$q(x) \geq \epsilon, \qquad q(x) \geq \epsilon, \qquad |q(x)| \leq M;$$

see Chapter 10, Section 3, Theorem A. It is easily checked that

$$J_1 \geq (1 + 3\delta^2)^m \delta\epsilon, \qquad J_2 \geq 0, \qquad J_3 \geq -M.$$

The first expression tends to ∞ as $m \to \infty$, hence $I_m \to \infty$, and this contradicts the fact that $I_m = 0$.

PROBLEMS

1. A theorem of Darboux. Let G be real and differentiable on an open interval I and suppose $g(t) = G'(t)$ for $t \in I$. In other words g is a derivative; this is not the same as saying that g has a derivative. Show that, if $g(a) > c > g(b)$ for two points a, $b \in I$, then $g(t) = c$ for some t between a and b. This is the *intermediate-value property*. Since a continuous function is the derivative of its integral, the intermediate-value property for continuous functions follows as a special case.

Hint: Deduce from Chapter 10, Section 3, Theorem A that $G(t) - ct$ assumes a maximum or minimum at some point between a and b.

2. Using Problem 1 with $G = y^{(n-1)}$ and $g = y^{(n)}$, prove Theorem 1. The same proof works if the p_j are continuous functions of t.

Our idea of what constitutes a valid proof is culturally conditioned, just like our idea of what constitutes virtue.

—Gordon Raisbeck

INDEX

An informal reminder

If introduced without explanation, c, C, c_j, C_j denote constants

Complete solution: Family containing all solutions

General solution: Enough solutions to satisfy initial conditions

Implicit solution: Solution defined by $F(x, y) = c$

Solution by quadrature: Solution expressed by integrals

Linear operator: $T(u + v) = Tu + Tv, \quad T(cu) = cTu$

Linear equation: $Ty = f$ where T is a linear operator

Linear homogeneous equation: $Ty = 0$ with T as above

Change of independent variable: If $y' = p$ then $y'' = p\, dp/dy$

Exponential shift theorem: $P(D)e^{kt}u = e^{kt}P(D + k)u$

Euler-de Moivre formula: $e^{ix} = \cos x + i \sin x$

Integrals 102,104   Complex numbers 132   Taylor series 669

Bounded: $|f(p)| \leq M$ for all relevant p

Lipschitzian: $|f(x, y) - f(x, z)| \leq K|y - z|$ for all relevant x, y, z

Existence: The problem has at least one solution

Uniqueness: The problem has at most one solution

Fundamental matrix: Solution of $U' = AU$ with $\det U \neq 0$

Eigenvalue: Scalar λ such that $Ac = \lambda c$ has a solution $c \neq 0$

Eigenvector: Nonzero vector c as above

Admissible: Finitely many discontinuities and bounded

Class E: Admissible on $[0, a]$ for each a and of exponential growth

12 inches = 1 foot   3 feet = 1 yard   5280 feet = 1 mile

1 meter is about 1.1 yards   1 kilometer is about 0.6 miles

One kilogram at the earth's surface weighs about 2.2 pounds